Engineering Mathematics with MATLAB Applications: Vol. 1, Fundamentals

Hasan Kurtaran

Engineering Mathematics with MATLAB Applications: Vol. 1, Fundamentals

Springer

Hasan Kurtaran
Adana Alparslan Türkeş Science and Technology University
Adana, Türkiye

ISBN 978-3-031-81868-4 ISBN 978-3-031-81869-1 (eBook)
https://doi.org/10.1007/978-3-031-81869-1

This Springer imprint is published by the registered company Springer Nature Switzerland AG
The registered company address is: Gewerbestrasse 11, 6330 Cham, Switzerland

This book is dedicated to my wife Sema *and my kids* Zeynep, Kerem, *and* Elif *as well as to all the people who have contributed to my life.*

Preface

Engineers design structures or systems by conducting mechanical, thermal, flow, electrical, or other types of analyses through theoretical or computer methods. Mathematical methods and their ingredients in such analyses are referred to as engineering mathematics. Engineering mathematics should be learned very well in order to take its maximum advantage in real applications. In the solution of engineering problems, the problem is first expressed with a mathematical equation and then solved with an appropriate method. Mathematical expression of the engineering problem is also called the mathematical model of the problem. Mathematical models are often expressed in algebraic, differential, or optimization form. Engineering problems are transformed into such forms using intermediate methods such as derivative, integral, interpolation, and curve fitting. Problems expressed in such forms are then solved using appropriate mathematical (Analytical, Approximate/Semi-analytical, or Numerical) methods. Solution method is determined based on the size, complexity (difficulty) of the engineering problem, available time, and the desired precision (accuracy) in the solution.

In this book, classical and modern mathematical methods within the context of engineering mathematics are introduced in detail. Content of the book is tailored specifically for the needs of engineers and enriched with worked out examples from the engineering applications. The book is organized in 16 chapters in two volumes. Volume 1 covers basic topics while Volume 2 focuses on advanced topics. Volumes 1 and 2 together constitute a complete and comprehensive reference source in engineering mathematics.

Volume 1 (this book) covers basic topics in ten chapters. Computer tools such as MATLAB software are increasingly used in the solution of engineering problems. Therefore, in Chap. 1, MATLAB software is introduced. In this chapter, readers learn how to use MATLAB commands and write their own programs to solve engineering problems. In Chap. 2, basic calculus and mathematical information are summarized for the convenience of readers. Various types of error such as modeling, cutting, and rounding error occur in the expression and solution stages of engineering problems. Chapter 3 explains error source and types in computer solutions.

Engineering problems are expressed in final form as linear or nonlinear algebraic equations. Therefore, Chap. 4 describes solution of linear algebraic equations using direct methods of Gauss elimination, Decomposition, Cramer and matrix inverse, and iterative methods of Jacobi and Gauss-Seidel. In Chap. 5, root finding and solution of nonlinear algebraic equations are explained. In this chapter, root finding with bracketing methods of Bisection and False position and point-based methods of Simple iteration, Newton-Raphson, and Secant methods are covered. In Chap. 6, interpolation with single and multivariate functions is explained using global interpolation methods of Newton-divided difference, Lagrange, and Hermite and local interpolation methods of Linear, Parabolic, and Cubic splines. Chapter 7 explains curve fitting with single and multivariate functions. In this chapter, Response Surface Methodology, Artificial Neural Networks (ANNs), and Kriging methods are covered in detail. Chapter 8 introduces numerical derivative methods. In this chapter, expressions of ordinary and partial derivatives using classical Finite Difference and relatively new Differential Quadrature methods are explained. In Chap. 9, numerical integral methods are covered. In this chapter, numerical calculation of single and multiple integrals on regions with and without regular boundaries using Newton-Cotes methods of Trapezoidal and Simpson and quadrature methods of Gauss, Gauss-Lobatto, and Gauss-Legendre is explained. Chapter 10 explains numerical solution of eigenvalue problems, which are a special type of algebraic equations. Polynomial and iterative methods are covered in this chapter.

Who is this book for?

This book (Volume 1) mainly targets *Undergraduate* students in engineering fields. It can be used as a *textbook* or *supplementary book* for the following courses in the Undergraduate education:

- Introduction to MATLAB and Programming
- Basic Calculus (or Calculus I)
- Linear Algebra
- Numerical Methods (or Numerical Analysis)

Introduction to MATLAB and Programming course: Chap. 1 can be used.

Basic Calculus (or Calculus I) course: Chap. 2 can be used.

Linear Algebra course: Chaps. 2, 4, and 10 (some parts) can be used.

Numerical Methods (or Numerical Analysis) course: Chaps. 1–4 and 5–9 (only the sections related to single-variable functions) can be selected.

This book is based on teaching and research experience of the author over very long years. The book includes both classical and modern methods in engineering mathematics. It has unique topics not covered in the similar books. For example, Differential Quadrature method is uniquely included in numerical derivative chapter. Simple and clear language is used in explaining the topics. Instead of expressing mathematical concepts directly, they are derived in an understandable and systematic way to teach the concepts and improve learning. Example problems are solved in each chapter with both hand calculations and MATLAB commands to deepen the concepts and enhance learning. Additional problems are provided at the end of each chapter to allow more practice of readers. In addition to MATLAB built-in commands, MATLAB programs developed by the author are also provided in appendices to implement the methods and show the examples of program development for engineering applications. *Undergraduate* students can significantly benefit from this book. Starting from Mechanical, Aerospace, Automotive, and Civil engineering areas, this book addresses all engineering fields.

The author has spent every effort to make the book free of errors. But, some corrections or improvements might be still needed. Therefore, any comment or suggestion about the book will be appreciated. Messages can be sent to the email addresses of hasan@atu.edu.tr and kurtaranhasan@gmail.com.

Adana, Turkey Hasan Kurtaran
September 2025

Acknowledgments

I would like to thank my family for their patience and support during the writing process of this book. Also, I would like to thank Prof. Dr. Mehmet Ali Arslan at Gebze Technical University and Dr. Murat Büyük for their careful reviews and valuable suggestions in improving the quality of the manuscript. Dr. Murat Büyük was also very active in editing the book spending long hours. Therefore, I would like to send him my special thanks for his hard work.

September 2025 Hasan Kurtaran

Contents

1 **Introduction to MATLAB** 1

2 **Basic Mathematics** 49

3 **Sources of Error in Engineering Calculations** 79

4 **Solution of Linear Equations** 95

5 **Root Finding and Solution of Nonlinear Equations** 139

6 **Interpolation** 183

7 **Curve Fitting (Regression)** 251

8 **Numerical Derivative** 425

9 **Numerical Integral** 481

10 **Solution of Eigenvalue Problems** 577

About the Author

Hasan Kurtaran obtained his B.Sc. and M.Sc. degrees in Aerospace Engineering from Istanbul Technical University in Turkey. He earned a scholarship for graduate study abroad after a nationwide organized examination among engineering students. Based on this scholarship, he came to the USA to pursue his Ph.D. degree. He obtained his Ph.D. degree from Mechanical Engineering department at George Washington University. During his Ph.D. education, he also worked as a researcher at the National Crash Analysis Center (NCAC) at George Washington University. Upon completion of Ph.D. program, Prof. Kurtaran returned to Gebze Technical University in Turkey. After working 16 years at Gebze Technical University, Prof. Kurtaran joined at Adana Alparslan Turkes Science and Technology University (ATU). He served as Vice Rector and Department Chair of Mechanical Engineering at ATU. Prof. Kurtaran specialized in structural mechanics, design optimization, and computational methods. Prof. Kurtaran has supervised over 30 graduate theses. Prof. Kurtaran also served in technical committees of Scientific and Technological Research Council of Turkey (TUBITAK). He gave lectures on Finite Element Analysis for design engineers in industry. Prof. Kurtaran has been in the list of "World's top 2 percent scientists" prepared by Elsevier and Stanford University.

Chapter 1
Introduction to MATLAB

1.1 Introduction

MATLAB software is increasingly used as a computational tool and a programming language in solving mathematical and engineering problems. Because of the common use, MATLAB software is introduced briefly in this section. More comprehensive introduction of MATLAB can be found in the books and notes given in the references of [1–10].

When MATLAB program is started, a window as shown in Fig. 1.1 opens up. The window has several sections in it. The most functional section in this window is the *Workspace* section. Built-in commands or the name of a previously prepared script file is typed after >> symbol to be executed. Commands should be separated by semicolon (;). Commands can be entered on separate lines or on the same line separated by semicolon (;).

Workspace can be used for two purposes in MATLAB:

1. Executing built-in commands:

   Built-in commands can be executed in Workspace by typing their names and then pressing the enter key. This approach is preferred when the number of commands to be executed is limited or when the commands are executed only once. If the commands are typed with semicolon at the end, they are executed but their responses are not displayed on the screen as shown in Table 1.1.

   However, if the commands are typed *without* semicolon at the end, their responses are immediately displayed on the screen upon execution as shown in Table 1.2.
2. Executing a file:

   Contents of a file can be executed by entering the file name in Workspace. For this, commands to perform a task should be saved in a separate file such as *FileName.m*. Then *FileName* is written in the Workspace to be executed as below:

H. Kurtaran, *Engineering Mathematics with MATLAB Applications: Vol. 1, Fundamentals*, https://doi.org/10.1007/978-3-031-81869-1_1

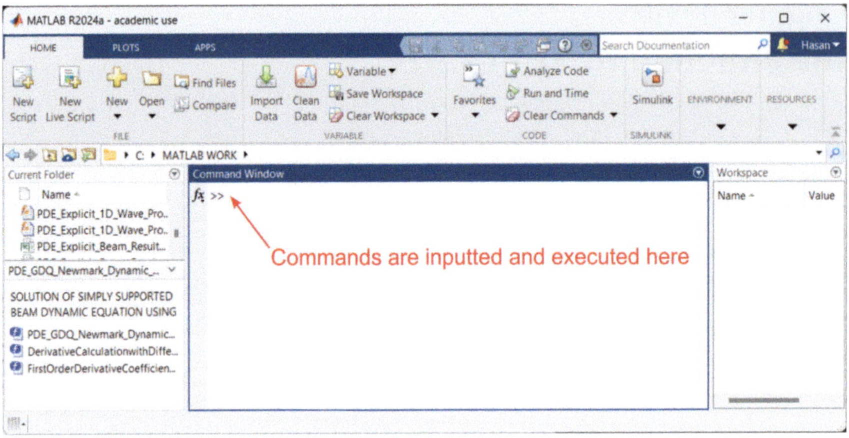

Fig. 1.1 MATLAB window upon starting

Table 1.1 Entering built-in commands in Workspace with semicolon

Workspace
`>> Command 1;`
`Response 1`
`>> Command 2;`
`Response 2`
…
`>> Command n;`
`Response n`

Table 1.2 Entering built-in commands in Workspace without semicolon

Workspace
`>> Command 1`
`Response 1`
`>> Command 2`
`Response 2`
…
`>> Command n`
`Response n`

```
>> FileName
```

Script files in MATLAB are divided into two groups:

(a) General script files: These files do not need any external input to run. An example of such files is shown in Table 1.3.
(b) Function script files: These files need external input to run and supply output upon running as shown in Table 1.4. Unlike general script files, function script files have the word "function" in the first line.

Table 1.3 General script file (FileName.m)

File = FileName.m

`Command 1;`
`Command 2;`
`Command 3;`
`...`
`Command n;`

Table 1.4 Function script file (FileName.m)

File=FileName.m

`function [Output]=FileName(Input)`
 	
`Command 1=Input;`
`Command 2;`
`Command 3;`
...

Note 1: In MATLAB, expressions or commands in any line after % sign are not executed. They are just considered as comments or information.

Note 2: In MATLAB, expressions or commands written after >> symbol without semicolon (`;`) at the end are both executed and printed on the screen.

1.2 Variables

In MATLAB, variables are divided into two groups as string (character) and numeric variables. For example, *name='Adam'* defines a character or string variable as *name* assigning *Adam* to it. Similarly, $x = 30$ defines a numerical variable of x assigning 30 to it.

The following rules should be respected when defining variables in MATLAB:

1. Variable name must start with a letter character.
2. Mathematical operators (such as + − * / ^) and punctuation marks (such as . , ; :) should not be included in the variable.
3. The number of characters in a variable should not exceed 63. Using more than 63 characters are ignored by MATLAB.
4. Non-English characters such as ç, ş ö, ü should not be used in variables.
5. Underscore (_) can be used in variables such as x_1, x_2, name_1, name_2.
6. Lower and upper case letters are considered differently. For example, Radius = 2 and radius = 2 are considered as two different variables in MATLAB.

7. Reserved words in MATLAB such as *for*, *end*, *if*, *then*, *else*, *switch*, and *case* cannot be used alone as a variable.

In MATLAB, variable generation and value assignment can be done simultaneously. For example, when $x = 30$ is written, a variable x is created and a value of 30 is assigned to it.

Semicolon (;) is used to separate executable expressions in MATLAB. When semicolon (;) is not used after a variable or expression, the result is immediately printed to the screen after the expression is executed. For example, when $x = 30$ is typed in a command line without semicolon (;), 30 is assigned to x and printed to the screen.

Example 1.1: Which of the following variables obey MATLAB rules?

(a) Student_name, StudentNo
(b) Student_1and2, Student_1&2
(c) Student_1, 1stStudent
(d) Month_1, Year-2022
(e) sum_Num1 + Num2, sum_Num1_Num2
(f) for, switch
(g) forx, switchval

Variables in *(a)* and *(g)* options comply with MATLAB rules. However, *b*, *c*, *d*, *e*, *f* options are not compatible with MATLAB rules. Because *(b)* option (second variable) includes a logical operator (&), *(c)* option (second variable) starts with a number (1), *(d)* option (second variable) includes the mathematical operator of (−) sign, *(e)* option (first variable) includes the mathematical operator of (+) sign and *(f)* option only consists of reserved words of *for* and *switch*.

1.3 Data Input and Output

1.3.1 Variable Assignment and Initialization

Assignment and initialization of a variable can be done simultaneously by typing *variable* = *value* such as $x = 30$ or using the *input* command. If the *input* command is used, the message to be displayed on the screen is written inside this command. When the *input* command is executed, message is displayed on the screen and the inputted value is assigned to the variable.

Example 1.2: Assign the value of 10 to x variable using the *input* command.

Value of 10 can be assigned to an x variable by typing the following command in the workspace.

```
>> x=input('enter x value:');
enter x value: 10
```

1.3.2 Printing Output to Screen

Message or variable value can be printed to the screen with *disp* or *fprintf* command.

disp command:

With *disp* command, desired message or variable value can be printed to the screen as *disp('message')* or *disp(variable)*. Message and variable value can be printed simultaneously with *disp("message" + variable)* command as shown in Table 1.5.

fprintf command:

fprintf command can be used to print variables in the formatted form such as formatted variable value or message+variable value to the screen or to a file. In formatted form, *%f* format is used to print fractional numbers, *%i* and *%d* formats are used to print integer values, and *%e* is used to print variables in 10-based format. *%f* format can be specified in more detail as *%m.nf* such as 5.2f. The *m* value here indicates the total number of digits reserved for the variable and *n* value indicates the number of digits for the decimal part. Use of *fprintf* command is given in Table 1.6.

1.4 Functions

1.4.1 Function Expressions

In MATLAB, functions such as *f*(*x*) function can be expressed in several ways to calculate their values. If the function expression is simple, it can be expressed in the workspace simply by typing *f* = *'function(x)'*, *f* = *@(x)function(x),* or *f* = *inline*

Table 1.5 Printing message and variable values to the screen

Message	Variable	Message and variable

`>>disp('x');`
`x`
	`>>x=2;`
	`>>disp(x);`
	`2`
	`>>name='Adam';`
	`>>disp(name);`
	`Adam`
		`>>x=2;`
		`>>disp("x="+x);`
		`x=2`

Table 1.6 Printing variables in formatted form

With *%f* format	With *%i* format	With *%e* format

`>>x=2;`
`>>fprintf('x= %f',x);`
`x= 2.000000`
`>>y=2.13298;`
`>>fprintf('y= %5.2f',`
`y);`
`y= 2.13`
	`>>x=2;`
	`>>fprintf('x=%i',`
	`x);`
	`x= 2`
		`>>x=0.0000055;`
		`>>fprintf('x=%e',`
		`x);`
		`x= 5.500000e-06`

('function'). Value of this function at an x_0 point can be calculated using *feval* command as *fvalue = feval (f,x0)*.

Example 1.3: Calculate the value of $f(x) = x^3 - e^x + 1$ function at $x = 1$ using *feval* command in MATLAB.

Value of *f(x)* function at $x = 1$ can be calculated in several ways as following:

Alternative 1:

```
>> f=@(x) x^3-exp(x)+1; fval=feval(f,1);
>> fval
fval =
  -0.7183
```

Symbol @(*x*) above indicates that *f* function depends on *x* variable.

Alternative 2:

```
>> f=inline('x^3-exp(x)+1'); fval=feval(f,1);
>> fval
fval =
  -0.7183
```

1.4.2 Function Plots

In MATLAB, *plot* or *fplot* command can be used to plot the graphic of a one-variable *f(x)* function, whereas *surf* or *mesh* commands can be used to plot the graphic of a two-variable *f(x,y)* function.

(a) Plot of f(x) function (one-variable function)
plot command:

To plot the graphic of an *f(x)* function, a set of *x* values and corresponding *f* values should be generated as $x = [x1, x2, \ldots, x_n]$ and $f = [f1, f2, \ldots, f_n]$. *x* values can be easily created in the following ways:

1. *x = initialvalue:increment:finalvalue* expression
2. *x = linspace(initial_value,final_value,number)* command

For example, *x* points with 0.5 increment in [0,3] interval can be created using *x = 0:0.5:3* expression as *x* = [0 0.5 1.0 1.5 2.0 2.5 3.0]. The same points can also be created using *x = linspace(0,3,7)* command, where 7 indicates the number of points in [0,3] interval.

After obtaining *x* and *f values*, graphic of *f(x)* function can be easily plotted by using *plot(x,f)* command. *plot(x,f)* command can be specified in more detail as *plot(x, f,'tsc')* where *t* indicates line style, *s* indicates line marker and *c* indicates line color.

Table 1.7 Style, marker, and color of lines in function plots

Styles		Markers		Colors	
Solid	–	Dot	.	Blue	b
Dotted	:	Circle	o	Green	g
Dash dot	–.	x-mark	x	Red	r
Dashed	–	Plus	+	Yellow	y
		Star	*	Black	k
		Square	s	Others	…
		Diamond	d		
		Others	…		

Style, marker, and color of lines in MATLAB can be expressed using the options in Table 1.7.

Axes can be named in plots if desired using *xlabel('AxisLabel')* and *ylabel ('AxisLabel')* commands. A title can also be defined in plots using *title ('GraphicLabel')* command.

fplot command:

Graphic of an *f(x)* function can also be plotted in an [a,b] interval using *fplot* command. This command is expressed as *fplot(@(x) f(x)_expression,[a,b])* for plotting.

Example 1.4: Plot the graphic of $f = x^2 + 1$ function in the interval/range of $-2 \leq x \leq 2$ using *plot* command and name axes as x and f.

First, x points to be used in plotting in the range of $[-2, 2]$ are determined. For example, 20 points in this range (including -2 and 2) can be easily created with *linspace* command by typing x = *linspace(−2,2,20)* in the Workspace. f values corresponding to x values can be obtained by typing *f* = *x.^2 + 1* (*Note:* dot (.) before ^ operation is required to create an f vector corresponding to x vector). Axes can bc named as x and f using *xlabel('x')* and *ylabel('f')* commands. Title of the plot can be expressed as '*f* = *x^2 + 1*' using *title*('*f* = *x^2 + 1*') command. Using this information, graphic of *f(x)* function can be plotted as in Fig. 1.2 by entering the following commands in the Workspace.

```
>> x=linspace(-2,2,20); f=x.^2+1;
>> plot(x,f,'--*r');
>> xlabel('x'); ylabel('f'); title('f=x^2+1');
```

Note: Graphic of $f = x^2 + 1$ function in $x = [-2, 2]$ interval can also be plotted by running *MATLAB Program 1.1: Plot of f* = *x^2 + 1 function* in Appendix.

Example 1.5: Plot the graphic of $f = x^2 + 1$ function in the range of $-2 \leq x \leq 2$ using *fplot* command and name axes as x and f.

Similar to Example 1.4, graphic of *f*(*x*) function can be plotted by entering the following commands in the Workspace.

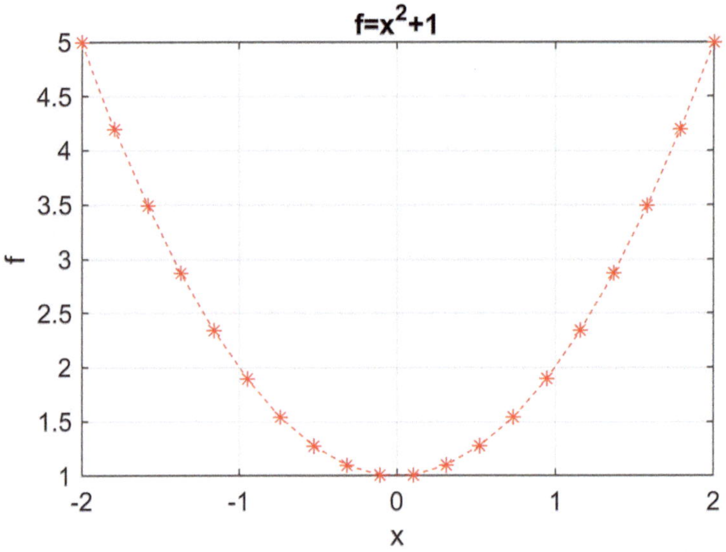

Fig. 1.2 Graphic of $f = x^2 + 1$ function

Fig. 1.3 Mesh points generated over a rectangular domain for plotting

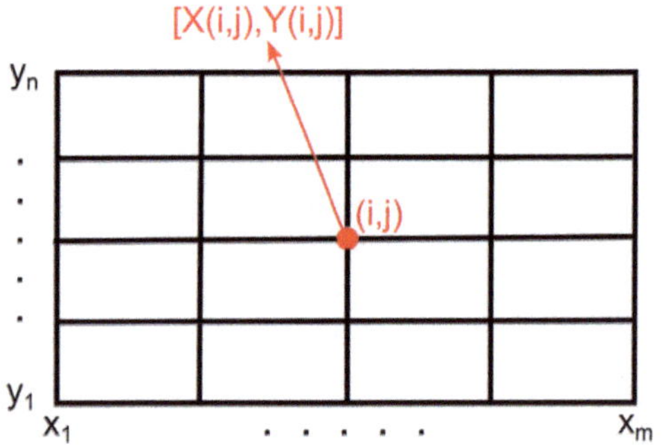


```
>> fplot(@(x) x.^2+1,[-2 2]);
>> xlabel('x'); ylabel('f'); title('f=x^2+1');
```

(b) Plot of f(x, y) function (two-variable function)

Two-variable (bivariate) functions such as $z = f(x, y)$ represent surfaces. Graphic of a surface function $z = f(x, y)$ can be plotted in ranges of $x_1 \leq x \leq x_m$ and $y_1 \leq y \leq y_n$ over a rectangular domain using *surf(z)* or *mesh(z)* commands. For plotting, first, a set of points called as mesh points or mesh is determined in the *x-y* plane as in Fig. 1.3. Set of points can be easily obtained using *[X, Y] = meshgrid(x, y)* command. Then, function values on *Z-axis* are calculated at mesh points. Finally, using *surf(X, Y, Z)* or *mesh(X, Y, Z)* command, graphic of *Z* versus *X* and *Y* values are plotted.

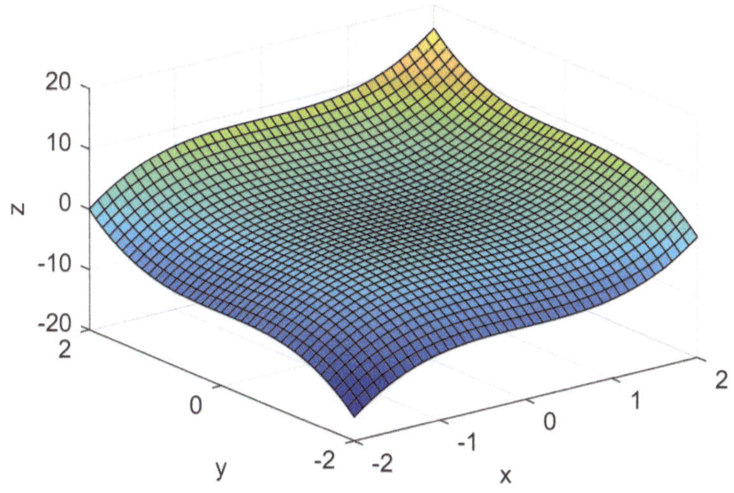

Fig. 1.4 Graphic of $z = x^3 + y^3$ function using mesh points with 0.1 distance from each other in x and y directions

Example 1.6: Plot the graphic of $z = x^3 + y^3$ function in intervals of $-2 \leq x \leq 2$ and $-2 \leq y \leq 2$ using the *surf* command in MATLAB and name the axes as x, y, and z.

To plot z (x, y) function, mesh points with 0.1 distance from each other in x and y directions can be created. Using these points, graphic of z (x, y) function can be plotted as in Fig. 1.4 by entering the following commands in the Workspace of MATLAB.

```
>> x=-2:0.1:2; y=-2:0.1:2; [X,Y]=meshgrid(x,y); Z=X.^3+Y.^3;
>> surf(X,Y,Z);
>> xlabel('x'); ylabel('y'); zlabcl('z');
```

Note: Graphic of $z = x^3 + y^3$ function in $x = [-2, 2]$ and $y = [-2, 2]$ intervals can also be plotted by running *MATLAB Program 1.2: Plot of z = x^3 + y^3 function* in Appendix.

1.5 Mathematical Operations

Mathematical operations can be carried out in MATLAB using the symbols in Table 1.8.

Table 1.8 Symbols for basic mathematical operations in MATLAB

Mathematical operations	Symbols	Examples
Addition	+	$2 + 3 = 5$
Subtraction	−	$3 - 2 = 1$
Multiplication	*	$2*3 = 6$
Division	/	$4/2 = 2$
Exponent	^	$3\text{^}2 = 9$

1.5.1 Vector and Matrix Operations in MATLAB

Vectors and matrices are expressed with square brackets [] in MATLAB. For example, a vector A can be expressed with its elements separated by spaces as $A = [1\ 2\ 3]$ or separated by commas (,) as A = [1, 2, 3]. Unlike vectors, matrices consist of both rows and columns. Rows in matrices are separated by semicolons (;). Each row in matrices can be thought of as a vector. Therefore, rows are expressed in vector form in matrices.

Example 1.7: Express vector $A = [3\ 2\ 1]$ and matrix $B = \begin{bmatrix} 3 & 2 & 1 \\ 6 & 5 & 4 \\ 9 & 8 & 7 \end{bmatrix}$ in MATLAB.

Vector A and matrix B can be expressed in MATLAB as following:

```
A=[3 2 1]; or A=[3, 2, 1];
B=[3 2 1; 6 5 4; 9 8 7]; or B=[3, 2, 1; 6, 5, 4; 9, 8, 7];
```

Member or element of a vector is expressed by specifying its row or column number. For example, $A(2)$ represents second member of vector A. In general form, i-th member of a vector A can be expressed as $A(i)$. Unlike vectors, member of a matrix is represented by both row and column numbers. For example, $B(3,2)$ represents the member of B matrix in third row and second column. In general, member of a B matrix in the i-th row and j-th column can be expressed as $B(i,j)$.

Special Matrices

Some matrices may have special properties. For example, all member of a matrix can be zero or one. In other case, only diagonal member of a matrix may be non-zero and equal to one. Such matrices are called as special matrices. Special matrices are defined in MATLAB as following (m: row number, n: column number):

All members of A matrix are *zero* → `A = zeros(m,n);`

All members of A matrix are *one* → `A = ones(m,n);`

Unit matrix (diagonal members are one, other members are zero) → `A = eye(m, n);`

Some other matrices can be generated from special matrices. For example, $A = \begin{bmatrix} 3 & 3 \\ 3 & 3 \end{bmatrix}$ matrix can be created from *ones* matrix as: `A = 3*ones(2,2).`

Extracting a Submatrix from a Main Matrix

Sometimes it may be desirable to obtain a submatrix from a main matrix. This can be done in MATLAB by either writing a script or using index expression of matrices. For example, a submatrix between *i-j* rows and *k-l* columns of *A* matrix can be obtained by using the index expression of *A*(*i*:*j*,*k*:*l*).

Example 1.8: *A* vector and *B* matrix are given as $A = [1\ 2\ 3\ 4]$ and $B = \begin{bmatrix} 1 & 2 & 3 & 4 \\ 5 & 6 & 7 & 8 \\ 9 & 10 & 11 & 12 \\ 13 & 14 & 15 & 16 \end{bmatrix}$. Obtain the sub-vector between 2 and 3 columns of *A* vector and the submatrix between 2 and 3 rows and 3 and 4 columns of *B* matrix.

If sub-vector is named as *C* vector and submatrix as *D* matrix, they can be obtained using the following expressions in MATLAB Workspace.

```
>> A=[1,2,3,4]; B=[1,2,3,4; 5,6,7,8; 9,10,11,12; 13,14,15,16];
>> C=A(2:3)
C =
   2   3
>> D=B(2:3,3:4)
D =
   7   8
  11  12
```

Example 1.9: *A* and *B* matrices are given as $A = \begin{bmatrix} 1 & 2 \\ 3 & 4 \end{bmatrix}$ and $B = \begin{bmatrix} 9 & 10 \\ 13 & 14 \end{bmatrix}$. Express the $C = \begin{bmatrix} 1 & 2 & 0 & 0 \\ 3 & 4 & 0 & 0 \\ 9 & 10 & 1 & 0 \\ 13 & 14 & 0 & 1 \end{bmatrix}$ matrix using *A*, *B* and special matrices (*Zero* and *Unit* matrices).

It is seen that upper right part of *C* matrix consists of a zero matrix of 2x2 size and lower right part consists of a unit matrix of 2x2 size. Zero and Unit matrices can be expressed with *Z* = *zeros*(2,2) and *I* = *eye*(2,2) expressions (other names can also be used for *Z* and *I* matrices). In this case, *C* matrix can be generated from the following submatrices in MATLAB.

```
>> A=[1 2; 3 4]; B=[9 10; 13 14];
>> Z=zeros(2,2); I=eye(2,2); C=[A Z; B I];
>> C
C =
   1   2   0   0
   3   4   0   0
   9  10   1   0
  13  14   0   1
```

Mathematical Operations on Matrices

Addition, subtraction, and multiplication operations are frequently applied on vectors and matrices. Such operations can be performed in MATLAB as below:

1. *Addition and Subtraction*:

   Addition and subtraction of two matrices (A and B matrices) can be simply expressed as $C = A + B$ and $C = A\text{-}B$ in MATLAB. Two matrices must have the same size (same column and row number) for addition or subtraction.

Example 1.10: A and B vectors are given as $A = [1 \quad 2 \quad 3]$ and $B = [4 \quad 5 \quad 6]$. Find sum and difference of these vectors in MATLAB.

Sum (C) and difference (D) of A and B vectors can be expressed in MATLAB as following:

```
>> A=[1 2 3]; B=[4 5 6]; C=A+B; D=A-B;
>> C
C =
   5   7   9
>> D
D =
  -3  -3  -3
```

2. *Multiplication*:

   Multiplication of two matrices such as A and B matrices can be expressed in MATLAB as $C = A*B$. For multiplication, internal dimensions of A and B matrices must be the same (column number of A and row number of B must be the same).

Example 1.11: A and B vectors are given as $A = [1 \quad 2 \quad 3]$ and $B = \begin{bmatrix} 4 \\ 5 \\ 6 \end{bmatrix}$. Find the product (multiplication) of these vectors in MATLAB.

Product (C) of A and B vectors can be expressed in MATLAB as following:

```
>> A=[1 2 3]; B=[4; 5; 6]; C=A*B;
>> C
C =
  32
```

3. *Operations on individual members*:

   Operations on individual members of a vector or matrix can be carried out using dot (.) symbol in front of the main operator.

Table 1.9 Extraction of special values from vectors and matrices

abs(A)	To calculate absolute value of vector/matrix *A*
min(A)	To calculate minimum value of vector/matrix *A*
max(A)	To calculate maximum value of vector/matrix *A*
sum(A)	To calculate sum of members of vector/matrix *A*
sort(A)	To sort members of a vector/matrix *A* from smallest to largest
inv(A)	To calculate inverse of matrix *A*
det(A)	To calculate determinant of matrix *A*
transpose(A)	To calculate transpose of matrix *A*

Example 1.12: *A* vector and *B* matrix are given as $A = [1 \quad 2 \quad 3]$ and $B = \begin{bmatrix} 3 & 2 & 1 \\ 6 & 5 & 4 \\ 9 & 8 & 7 \end{bmatrix}$. Find the square of individual members of *A* and *B* matrices in MATLAB.

Square of each member of *A* and *B* matrices can be expressed in MATLAB as below:

```
>> A=[1 2 3]; B=[3 2 1; 6 5 4; 9 8 7];
>> C=A.^2; D=B.^2;
>> C
C =
   1   4   9
>> D
D =
   9   4   1
  36   25  16
  81  64  49
```

4. *Other operations*:

   In some cases, extraction of a special value such as absolute value, minimum value, maximum value, and sum from a vector or matrix can be desired. Such values can be obtained in MATLAB using special commands as in Table 1.9.

1.6 Solution of Linear Equations

Linear equations can be expressed in algebraic and matrix forms as below:

$$\left.\begin{array}{l} a_{11}x_1 + a_{12}x_2 + \ldots + a_{1n}x_n = b_1 \\ a_{21}x_1 + a_{22}x_2 + \ldots + a_{2n}x_n = b_2 \\ \ldots \\ a_{n1}x_1 + a_{n2}x_2 + \ldots + a_{nn}x_n = b_n \end{array}\right\} \Rightarrow \underbrace{\begin{bmatrix} a_{11} & a_{12} & \cdots & a_{1n} \\ a_{21} & a_{22} & \cdots & a_{2n} \\ \cdots & \cdots & \ddots & \cdots \\ a_{n1} & a_{n2} & \cdots & a_{nn} \end{bmatrix}}_{A} \underbrace{\begin{Bmatrix} x_1 \\ x_2 \\ \cdots \\ x_n \end{Bmatrix}}_{X} = \underbrace{\begin{Bmatrix} b_1 \\ b_2 \\ \cdots \\ b_n \end{Bmatrix}}_{B}$$

$$\rightarrow AX = B$$

After forming A matrix and B vector, solution (X) of linear system can be obtained in MATLAB using inverse (*inv*), division (\), or *linsolve* commands as following:

1. $X = inv(A) * B$
2. $X = A\backslash B$
3. $X = linsolve(A, B)$

Example 1.13: Solve the following linear equation system in MATLAB:

$$\begin{aligned} &2x_1 + 3x_2 - 4x_3 = 6 \\ &-x_1 - 2x_2 + 2x_3 = 2 \\ &4x_1 + x_2 - x_3 = 12 \end{aligned}$$

For the solution in MATLAB, linear system should be expressed in matrix form as below:

$$\begin{array}{l} 2x_1 + 3x_2 - 4x_3 = 6 \\ -x_1 - 2x_2 + 2x_3 = 2 \\ 4x_1 + x_2 - x_3 = 12 \end{array} \rightarrow \underbrace{\begin{bmatrix} 2 & 3 & -4 \\ -1 & -2 & 2 \\ 4 & 1 & -1 \end{bmatrix}}_{A} \underbrace{\begin{Bmatrix} x_1 \\ x_2 \\ x_3 \end{Bmatrix}}_{X} = \underbrace{\begin{Bmatrix} 6 \\ 2 \\ 12 \end{Bmatrix}}_{B} \rightarrow AX = B$$

After forming A matrix and B vector, solution (X) of linear system can be obtained in MATLAB as following:

Alternative 1: $X = inv(A)*B$

```
>> A=[2 3 -4; -1 -2 2; 4 1 -1]; B=[6; 2; 12]; X=inv(A)*B;
>> X
X =
   3.7143
 -10.0000
  -7.1429
```

Alternative 2: $X = A\backslash B$

```
>> A=[2 3 -4; -1 -2 2; 4 1 -1]; B=[6; 2; 12]; X=A\B;
>> X
X =
   3.7143
 -10.0000
  -7.1429
```

Alternative 3: $X = linsolve(A,B)$

```
>> A=[2 3 -4; -1 -2 2; 4 1 -1]; B=[6; 2; 12]; X=linsolve(A,B);
>> X
X =
   3.7143
 -10.0000
  -7.1429
```

1.7 Root Finding

For an $f(x)$ function, finding the x values that make $f(x) = 0$ is called as root finding. Roots of a polynomial function expressed in $a_n x^n + \ldots + a_2 x^2 + a_1 x + a_0 = 0$ form (order is decreasing) can be obtained using *roots* command in MATLAB. For this, coefficients of the polynomial are expressed in square bracket []. Using these coefficients, roots are obtained with $x =$ *roots*([Coefficients]) command.

Example 1.14: Calculate roots of $x^2 + 3x + 2 = 0$ function in MATLAB.

Coefficients of the polynomial (highest order first, lowest order last) are expressed as [1 3 2]. Using these coefficients, roots (X) are obtained using *roots* command as following:

```
>> X=roots([1 3 2])
X =
   -2
   -1
```

1.8 Solution of Nonlinear Equations

In MATLAB, a single nonlinear equation can be solved using *fzero* command and a nonlinear equation system can be solved using *fsolve* command.

1.8.1 Solving Single Nonlinear Equation

To solve a single nonlinear equation in MATLAB, it should be written in zero ($f(x) = 0$) form. Then, solution can be obtained using *fzero(f,x0)* command. Here, f represents the nonlinear equation in zero form and x_0 indicates the initial value to start the solution. With *fzero(f,x0)* command, solution value which is nearest to the initial value x_0 is obtained.

Example 1.15: Solve the nonlinear equation $x^3 - e^x + 1 = 0$ with MATLAB using the initial value of $x_0 = 1$.

Solution of the nonlinear equation can be obtained in MATLAB by defining f as $f = x^3 - e^x + 1$ as following:

Alternative 1: f = 'x^3-exp(x) + 1'

```
>> f='x^3-exp(x)+1'; z=fzero(f,1);
>> z
z =
  1.5450
```

Alternative 2: f = @(x) x^3-exp(x) + 1

```
>> f=@(x) x^3-exp(x)+1; z=fzero(f,1);
>> z
z =
  1.5450
```

Alternative 3: f = inline('x^3-exp(x) + 1')

```
>> f=inline('x^3-exp(x)+1'); z=fzero(f,1);
>> z
z =
  1.5450
```

1.8.2 Solution of Nonlinear Equation System

Nonlinear equation system can be solved using *fsolve* command in MATLAB. For solution, nonlinear equations should be written in zero (=0) form and saved in a separate function file such as File.m. Then, using this file and initial values of $x_0 = [x_{10}; x_{20};\ldots; x_{n0}]$ (n is the number of variables), solution can be obtained with *fsolve('Equations_File_Name',x0)* command.

Table 1.10 Function file for equations

File=Equations.m

`function f=Equations(x)`
 	
`f(1)=x(1)^2+x(2)^2-9 ;`
`f(2)=x(1)+x(2)-1 ;`

Example 1.16: Find the solution of $x_1^2+x_2^2=9$ and $x_1 + x_2 = 1$ equations with MATLAB using initial values of $(x_{10}, x_{20}) = (1, 1)$.

Nonlinear equations in zero form can be named as $f(1)=x_1^2+x_2^2-9$ and $f(2) = x_1 + x_2 - 1$. These functions can be saved in a function file such as Equations.m as in Table 1.10 (*Note*: file name must be the same name after the *function* word in the file).

Using this file and initial values of x_1 and x_2 variables as $x_0 = [1\ 1]$, solution can be obtained with *fsolve('Equations_File_Name',x0)* command as following:

```
>> x0=[1 1]; x=fsolve('Equations',x0);
>> x
x =
   2.5616  -1.5616
```

Note: *fsolve* command can be used to solve a single nonlinear equation as well.

1.9 Interpolation

Estimation of a function value at an intermediate point in a data set is called as interpolation. In interpolation, a function is passed through the data points and then using this function intermediate value is estimated. *Interp1* command can be used for interpolation in MATLAB. *Interp1* command is expressed in detailed form as *interp1(x,f,x0,'method')*. Here, *x* indicates the set of points to be used in interpolation, *f* denotes corresponding function values, x_0 indicates the intermediate point and *method* specifies the type of interpolation polynomial or spline such as linear polynomial or cubic spline.

Example 1.17: For an *f(x)* function, *x* and *f* values are given as $x = [0\ 1\ 2\ 4\ 5\ 6]$ and $f = [0\ 2\ 3\ 15\ 27\ 35]$. Estimate the *f* value at $x_0 = 3$ point using MATLAB.

With *interp1(x,f,x0,'method')* command in MATLAB, *f* value at $x_0 = 3$ point can be predicted using different interpolation methods as in Table 1.11.

Table 1.11 MATLAB commands for interpolation

```
>> x=[0 1 2 4 5 6] ; f=[0 2 3 15 27 35] ;
>> f0=interp1(x,f,3,'linear')
f0 =
          9
>> f0=interp1(x,f,3,'spline')
f0 =
          6.7812
```

Table 1.12 MATLAB commands for curve fitting

```
>> x=[0 1 2 3 4 5]; f=[0 2 3 10 15 27] ;
>> L=polyfit(x,f,1) % To fit linear polynomial
L =
          5.1714      -3.4286
>> P=polyfit(x,f,2) % To fit parabolic polynomial
P =
          1.1786      -0.7214          0.5000
```

1.10 Curve Fitting

Passing a pre-assumed function or curve through a data set in a balanced way is called as curve fitting. Function value at an intermediate point can be estimated using the fitted curve. Linear and parabolic polynomials are commonly used as curve fitting functions. In MATLAB, the *polyfit* command can be used to fit a polynomial function to a data set. The *polyfit* command is expressed in detailed form as *polyfit(x, f,n)*. Here, x indicates the set of points to be used in curve fitting, f denotes corresponding function values and n indicates the order of fitting polynomial. Upon execution, *polyfit* command yields the coefficients of the fitting polynomial.

Example 1.18: For an $f(x)$ function, x and f values are given as $x = [0\ 1\ 2\ 4\ 5\ 6]$ and $f = [0\ 2\ 3\ 15\ 27\ 35]$. Fit linear and parabolic polynomials to f values with MATLAB.

To fit linear and parabolic polynomials to f values, L = *polyfit(x,f,1)* and P = *polyfit(x,f,2)* commands can be used in MATLAB Workspace as illustrated in Table 1.12.

Above, L shows the coefficients of linear polynomial and P shows the coefficients of parabolic polynomial in descending order. Thus, after executing *polyfit* commands, fitted linear polynomial will be as $f = 5.1714x - 3.4286$ and parabolic polynomial will be as $f = 1.1786x^2 - 0.7214x + 0.5$.

1.11 Derivative Calculation

In MATLAB, derivative of an $f(x)$ function can be obtained with *diff(f,x,n)* command where x denotes independent variable and n denotes derivative order. For symbolic derivative calculation, first, independent variable is specified after *syms* expression.

Then, derivative of the function is obtained using *diff(f,x,n)* command. Variables specified after *syms* expression are considered as symbolic variables in MATLAB.

Example 1.19: Obtain first and second derivatives of $f(x) = x^2 + 2x + 1$ function with respect to x variable with MATLAB.

Using *diff* command in MATLAB, first (denoted by *df1*) and second (denoted by *df2*) derivatives of *f(x)* function can be obtained as below:

```
>> syms x; f=x^2+2*x+1; df1=diff(f,x,1); df2= diff(f,x,2);
>> df1
df1 =
2*x + 2
>> df2
df2 =
2
```

1.12 Integral Calculation

In MATLAB, integral of an *f(x)* function in $[x_1, x_2]$ interval can be calculated analytically or numerically as following.

1.12.1 Analytical Integral Calculation

The *int* command can be used for analytical (closed form) integral calculations in MATLAB. For this, first, symbolic variable is specified after the *syms* expression. Then, integral of the function is obtained using *int* command. When integral limits are not specified, integral of *f(x)* function is obtained using *int(f,x)* command. When integral limits are specified with [a, b] interval, integral is calculated with *int(f,x,a,b)* command.

Example 1.20: Find the integral of $f(x) = x^2 + 2x + 1$ function using MATLAB.

Integral of $f(x) = x^2 + 2x + 1$ function can be obtained with MATLAB using the *int* command as following (*intval*: integral result):

```
>> syms x; f=x^2+2*x+1; intval=int(f,x);
>> intval
intval =
(x*(x^2 + 3*x + 3))/3
```

Example 1.21: Find the integral of $f(x) = x^2 + 2x + 1$ function in [0,1] interval using MATLAB.

Integral of $f(x) = x^2 + 2x + 1$ function in [0,1] interval can be calculated with MATLAB using the *int* command as following (*intval*: integral result):

```
>> syms x; f=x^2+2*x+1; intval=int(f,x,0,1);
>> intval
intval =
7/3
```

1.12.2 Numerical Integral Calculation

Unfortunately, not all integrals can be calculated analytically (in closed form). When analytical integral is not possible, numerical integral methods can be used. In MATLAB, *trapz* or *quad* commands can be used for numerical integration. The *trapz* command uses trapezoidal method while *quad* command uses Simpson's method in numerical integral calculations.

(a) Integral with ***trapz*** command:

To calculate the integral of an *f(x)* function in [a,b] interval using *trapz* command, a set of *x* points in [a,b] interval is created using *x = a:h:b* (*h*: step size between points) expression or *x = linspace(a,b,n)* (*n*: number of points) command. Then, *f* values corresponding to *x* points are calculated. *x* points and *f* values can be represented in vector form as *x* = [x1 x2 … xn] and *f* = [f1 f2 … fn]. Integral value is then calculated using *intval = trapz(x,f)* command where *intval* indicates the integral result.

(b) Integral with ***quad*** command:

Integral of an *f(x)* function in [a,b] interval can be calculated using *quad* command as *intval = quad(f,a,b)*. *Reminding*: *f(x)* function should be expressed to produce *f* values in vector form. To achieve this, function expression should include (.^) operator instead of just (^) operator for exponents. Similarly, function expression should include (./) operator instead of just (/) operator for division. For example, *f = x.^2 + 2*x* expression should be used instead of *f = x^2 + 2*x*.

Example 1.22: Calculate the integral of $f(x) = x^3 + 2x + 1$ function in [0,1] interval using a) *trapz* and b) *quad* commands in MATLAB.

To calculate the integral of $f(x) = x^3 + 2x + 1$ function in [0,1] interval, a set of *x* points in [0,1] interval with 0.1 distance can be created using *x* = 0:0.1:1 expression. *f* values corresponding to *x* values can be obtained in vector form using *f = x.^3 + 2*x + 1* expression. *Note*: To obtain *f* values in vector form (.^) operator is used instead of (^) operator in *f* expression.

After this preparation, integral of *f(x)* function can be calculated numerically using *trapz* and *quad* commands as following:

(a) Using *trapz* command:

```
>> x=0:0.01:1; f=x.^3+2*x+1; intval=trapz(x,f);
>> intval
intval =
  2.2500
```

(b) Using *quad* command:

```
>> f='x.^3+2*x+1'; intval=quad(f,0,1);
>> intval
intval =
  2.2500
```

1.13 Solution of Eigenvalue Problems

Eigenvalue problems are expressed as $\underbrace{\begin{bmatrix} a_{11} & a_{12} & \cdots & a_{1n} \\ a_{21} & a_{22} & \cdots & a_{2n} \\ \cdots & \cdots & \ddots & \cdots \\ a_{n1} & a_{n2} & \cdots & a_{nn} \end{bmatrix}}_{A} \underbrace{\begin{Bmatrix} x_1 \\ x_2 \\ \cdots \\ x_n \end{Bmatrix}}_{X} = \lambda \underbrace{\begin{Bmatrix} x_1 \\ x_2 \\ \cdots \\ x_n \end{Bmatrix}}_{X}$ or shortly as $AX = \lambda X$. Solution of eigenvalue problems can be obtained using *eig* command in MATLAB. Upon solution, eigenvalues are obtained as $\lambda = \lambda_1, \lambda_2, \ldots, \lambda_n$, number of which is equal to the size of matrix A. Eigenvalues and eigenvectors can be obtained using *eig* command in *[Xmat,Lam] = eig(A)* form, where *Xmat* gives eigenvectors in column form and *Lam* gives eigenvalues.

Example 1.23: For the eigenvalue problem of $AX = \lambda X$, A matrix is given as $A = \begin{bmatrix} 1 & 2 & 3 \\ 4 & 5 & 6 \\ 7 & 8 & 9 \end{bmatrix}$. Find eigenvalues λ ($\lambda = \lambda_1, \lambda_2, \lambda_3$) and their corresponding eigenvectors X ($X = \begin{Bmatrix} x_1 \\ x_2 \\ x_3 \end{Bmatrix}$) with MATLAB.

Eigenvalues λ ($\lambda = \lambda_1, \lambda_2, \lambda_3$) and their corresponding eigenvectors can be obtained using *eig* command in MATLAB as below:

```
>> A=[1 2 3 ; 4 5 6; 7 8 9]; [Xmat,Lam]=eig(A);
>> Xmat
Xmat =
  -0.2320  -0.7858   0.4082
  -0.5253  -0.0868  -0.8165
  -0.8187   0.6123   0.4082
>> Lam
Lam =
  16.1168        0        0
        0  -1.1168        0
        0        0  -0.0000
```

In above matrices, diagonal members of *Lam* matrix give the eigenvalues and columns of *Xmat* give the eigenvectors. In this case, eigenvalues are obtained as following:

$\lambda_1 = 16.1168, \lambda_2 = -1.1168, \lambda_3 = -0.0000.$

Eigenvectors are obtained from *Xmat* matrix as below:

For $\lambda_1 = 16.1168$, eigenvector is obtained as: $X = \begin{Bmatrix} -0.2320 \\ -0.5253 \\ -0.8187 \end{Bmatrix}.$

For $\lambda_2 = -1.1168$, eigenvector is obtained as: $X = \begin{Bmatrix} -0.7858 \\ -0.0868 \\ 0.6123 \end{Bmatrix}.$

For $\lambda_3 = -0.0000$, eigenvector is obtained as: $X = \begin{Bmatrix} 0.4082 \\ -0.8165 \\ 0.4082 \end{Bmatrix}.$

1.14 Solution of Ordinary Differential Equations

An equation consisting of function derivatives is called as differential equation. *D* symbol is used to express derivative in differential equations in MATLAB. The number after *D* symbol indicates the order of derivative and the variable after it indicates the function to derive. For example, second derivative $\frac{d^2y(x)}{dx^2}$ is expressed by *D*2y term in MATLAB. In the solution of differential equations, analytical and numerical methods can be used. Analytical methods are used for the linear equations that can be easily solved. In contrast, numerical methods are used for the equations that are difficult or cannot be solved analytically.

1.14.1 Analytical Solution of Differential Equations

In MATLAB, linear ordinary differential equations can be solved analytically using *dsolve* command. *dsolve* command is used in *dsolve ('DifferentialEquation','Variable')* form for a single differential equation. When initial or boundary conditions (shortly conditions) are specified, *dsolve* command can be expressed as *dsolve('DifferentialEquation','Conditions','Variable')*.

If there are multiple differential equations (i.e., differential equation system) to be solved, then *dsolve* command can be expressed as below:

```
dsolve('Equation 1','Equation 2',
...,'Equationn','Conditions','Variable')
```

Note: Since higher order differential equation systems are difficult to solve analytically, MATLAB may not always provide analytical solutions. In that case, numerical methods can be used to solve such equations.

Example 1.24: Find the solution of differential equation $\frac{d^2y(x)}{dx^2} + y = 1$ with MATLAB.

Analytical solution of $\frac{d^2y(x)}{dx^2} + y = 1$ equation can be obtained by using the *dsolve* command in MATLAB as following:

```
>> y=dsolve('D2y+y=1','x')
y =
C1*cos(x) + C2*sin(x) + 1
```

Alternatively, the above differential equation can be solved by representing the differential equation with a string variable (e.g., *Eqn* = *'D2y* + *y* = *1'*) and then calling this variable with *dsolve* command as below:

```
>> Eqn='D2y+y=1';
>> y=dsolve(Eqn,'x')
y =
C1*cos(x) + C2*sin(x) + 1
```

Example 1.25: Find the solution of the differential equation $\frac{d^2y(x)}{dx^2} + y = 1$ with MATLAB using the initial conditions of $y(0) = 2$, $\frac{dy(0)}{dx} = 1$.

Solution of the differential equation can be obtained using *dsolve* command in MATLAB as below:

```
>> y=dsolve('D2y+y=1','y(0)=2,Dy(0)=1','x')
y =
cos(x) + sin(x) + 1
```

Alternatively, differential equation can be solved by representing the differential equation and initial conditions with string variables of *Eqn* and *Cond*, respectively, and then calling these variables with *dsolve* command as below:

```
>> Eqn='D2y+y=1';
>> Cond='y(0)=2,Dy(0)=1';
>> y=dsolve(Eqn,Cond,'x')
y=
cos(x)+sin(x)+1
```

Example 1.26: Find the solutions of the differential equation system of $\frac{dx_1}{dt} = x_2$ and $\frac{dx_2}{dt} = 10\sin(2\pi t) - 0.1x_2 - x_1$ with MATLAB using the initial conditions of $x_1(0) = 0$, $x_2(0) = 1$ and plot the graphic of the solution functions within [0 5] interval.

In this example, a differential equation system consisting of two equations is solved. Solution can be obtained by representing the differential equations and initial conditions with string variables of *Eqn1*, *Eqn2*, *Cond* and then calling them with *dsolve* command as following:

```
>> Eqn1='Dx1=x2'; Eqn2='Dx2=10*sin(2*pi*t)-0.1*x2-1*x1';
>> Cond='x1(0)=0,x2(0)=1';
>> [x1,x2]=dsolve(Eqn1,Eqn2,Cond,'t');
>> x1
x1=
((exp(-t/20)*sin((399^(1/2)*t)/20))/20 + (399^(1/2)*exp(-t/20)*cos
((399^(1/2)*t)/20))/20)*(((100*pi^3*exp(t/20)*(7980*cos(2*pi*t)
*sin((399^(1/2)*t)/20) - 20*399^(1/2)*cos(2*pi*t)*cos((399^(1/2)
*t)/20)))/399 - (100*pi^2*exp(t/20)*(399*sin(2*pi*t)*sin((399^(1/2)
*t)/20) + 199*399^(1/2)*cos((399^(1/2)*t)/20)*sin(2*pi*t)))/399 -
(100*pi*exp(t/20)*(1995*cos(2*pi*t)*sin((399^(1/2)*t)/20) + 5*399^
(1/2)*cos(2*pi*t)*cos((399^(1/2)*t)/20)))/399 + (5000*399^(1/2)*exp
(t/20)*cos((399^(1/2)*t)/20)*sin(2*pi*t))/399)/(400*pi^4 - 199*pi^2
+ 25) + (399^(1/2)*(400*pi^4 + 2000*pi^3 - 199*pi^2 + 500*pi + 25))/(399*
(400*pi^4 - 199*pi^2 + 25))) - ((exp(-t/20)*cos((399^(1/2)*t)/20))/20
- (399^(1/2)*exp(-t/20)*sin((399^(1/2)*t)/20))/20)*((400*pi^4 +
2000*pi^3 - 199*pi^2 - 500*pi + 25)/(400*pi^4 - 199*pi^2 + 25) +
((100*pi^2*exp(t/20)*(399*cos((399^(1/2)*t)/20)*sin(2*pi*t) -
199*399^(1/2)*sin(2*pi*t)*sin((399^(1/2)*t)/20)))/399 -
(100*pi^3*exp(t/20)*(7980*cos(2*pi*t)*cos((399^(1/2)*t)/20) +
20*399^(1/2)*cos(2*pi*t)*sin((399^(1/2)*t)/20)))/399 + (100*pi*exp
(t/20)*(1995*cos(2*pi*t)*cos((399^(1/2)*t)/20) - 5*399^(1/2)*cos
(2*pi*t)*sin((399^(1/2)*t)/20)))/399 + (5000*399^(1/2)*exp(t/20)
*sin(2*pi*t)*sin((399^(1/2)*t)/20))/399)/(400*pi^4 - 199*pi^2 + 25))
```

```
>> x2
x2 =

exp(-t/20)*cos((399^(1/2)*t)/20)*((400*pi^4 + 2000*pi^3 - 199*pi^2 -
500*pi + 25)/(400*pi^4 - 199*pi^2 + 25) + ((100*pi^2*exp(t/20)*(399*cos
((399^(1/2)*t)/20)*sin(2*pi*t) - 199*399^(1/2)*sin(2*pi*t)*sin
((399^(1/2)*t)/20)))/399 - (100*pi^3*exp(t/20)*(7980*cos(2*pi*t)
*cos((399^(1/2)*t)/20) + 20*399^(1/2)*cos(2*pi*t)*sin((399^(1/2)
*t)/20)))/399 + (100*pi*exp(t/20)*(1995*cos(2*pi*t)*cos((399^(1/2)
*t)/20) - 5*399^(1/2)*cos(2*pi*t)*sin((399^(1/2)*t)/20)))/399 +
(5000*399^(1/2)*exp(t/20)*sin(2*pi*t)*sin((399^(1/2)*t)/20))/
399)/(400*pi^4 - 199*pi^2 + 25)) - exp(-t/20)*sin((399^(1/2)*t)/20)*
(((100*pi^3*exp(t/20)*(7980*cos(2*pi*t)*sin((399^(1/2)*t)/20) -
20*399^(1/2)*cos(2*pi*t)*cos((399^(1/2)*t)/20)))/399 -
(100*pi^2*exp(t/20)*(399*sin(2*pi*t)*sin((399^(1/2)*t)/20) +
199*399^(1/2)*cos((399^(1/2)*t)/20)*sin(2*pi*t)))/399 - (100*pi*exp
(t/20)*(1995*cos(2*pi*t)*sin((399^(1/2)*t)/20) + 5*399^(1/2)*cos
(2*pi*t)*cos((399^(1/2)*t)/20)))/399 + (5000*399^(1/2)*exp(t/20)
*cos((399^(1/2)*t)/20)*sin(2*pi*t))/399)/(400*pi^4 - 199*pi^2 + 25) +
(399^(1/2)*(400*pi^4 + 2000*pi^3 - 199*pi^2 + 500*pi + 25))/(399*
(400*pi^4 - 199*pi^2 + 25)))
```

Solution functions can be plotted within [0 5] interval as in Fig. 1.5 using the following MATLAB commands:

```
>> fplot(x1,[0 5]); hold on ;
>> fplot(x2,[0 5]) ;
>> xlabel('t') ; ylabel('x1, x2') ;
>> legend('x1','x2') ;
```

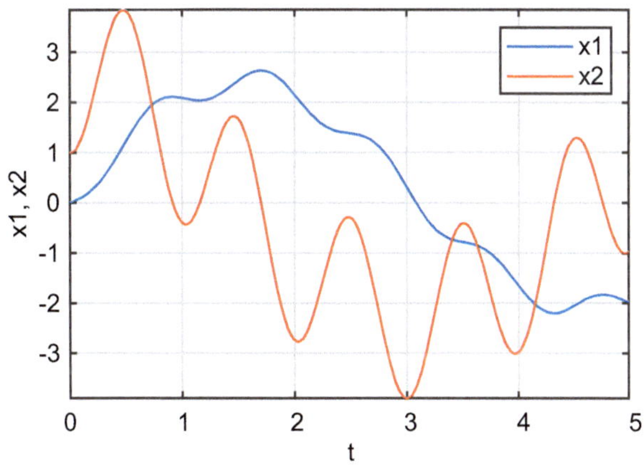

Fig. 1.5 Graphics of $x_1(t)$ and $x_2(t)$ functions

Note: *hold on* and *legend* commands above can be explained as below:

- *hold on* command ensures that first graphic is not deleted when the second graphic is created in the same figure.
- *legend('Label1','Label2')* command labels the plotted $x_1(t)$ and $x_2(t)$ functions in the same figure with names of *Label*1 and *Label*2, respectively.

1.14.2 Numerical Solution of Differential Equations

Not every differential equation can be solved analytically. For example, nonlinear differential equations usually cannot be solved analytically. Differential equations that are difficult or impossible to solve analytically are solved with numerical methods. Differential equations are divided into two groups as initial value and boundary value problems. In initial value problems, solution information is known at a single point while in boundary value problems solution information is known at more points.

Runge-Kutta methods are widely used in the numerical solution of first-order initial value type ordinary differential equations. There are several types of Runge-Kutta methods differing in accuracy and computational cost. Runge-Kutta solutions can be obtained with *ode23* (for low accuracy) and *ode45* (for high accuracy) commands in MATLAB. Higher order initial value problems can be converted into a set of first-order initial value problems which then can be solved with *ode23* or *ode45* command.

Steps of the numerical solution of differential equations can be summarized as following: First, differential equations are rearranged so that derivatives are kept in one side and non-derivative terms are kept in other side. Then, non-derivative terms are given names such as *Equation(1), Equation(2),…, Equation(n)* (any other name can be used). Named equations are then saved in a function file such as *File.m*. Function file is finally called by *ode23* or *ode45* commands for the solution.

Example 1.27: Find the numerical solution of the differential equation system of $\frac{dx_1}{dt} - x_2 = 0$ and $\frac{dx_2}{dt} - 10\sin(2\pi t) + 0.1x_2 + x_1 = 0$ for the initial conditions of $x_1(0) = 0$, $x_2(0) = 1$ using *ode45* command in MATLAB and plot the solutions within [0 5] interval.

First, differential equations are rearranged to isolate derivative terms in one side as below:

$$\frac{dx_1}{dt} = x_2, \frac{dx_2}{dt} = 10\sin(2\pi t) - 0.1x_2 - x_1$$

Table 1.13 MATLAB file for the solution of differential equations

File=DiffEqs.m

`function dxdt=DiffEqs(t,x)`
 	
`dxdt=zeros(2,1);`
`dxdt(1)=x(2);`
`dxdt(2)=10*sin(2*pi*t)-0.1*x(2)-1*x(1);`

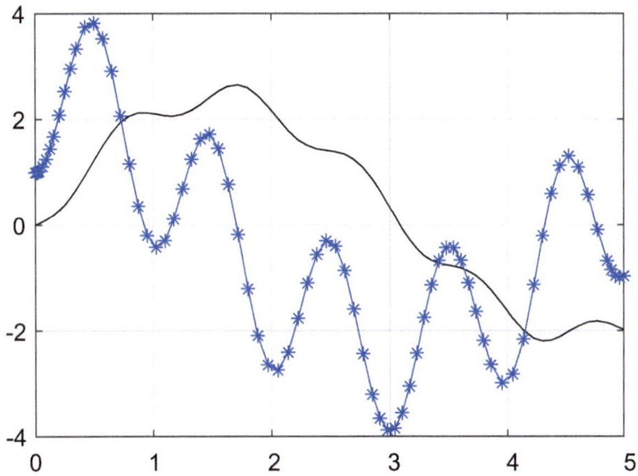

Fig. 1.6 Graphics of $x_1(t)$ and $x_2(t)$ solutions

After rearrangement of differential equations, non-derivative terms are given names such as *Equation(1)* = *expression*, *Equation(2)* = *expression* and saved in a file such as *DiffEqs.m* (any other name can be used) as in Table 1.13.

DiffEqs.m file can be called by *ode45* command (to use fourth-order Runge-Kutta method) for the numerical solution of differential equations as below:

```
>> tmin=0; tmax=5; x0=[0; 1];
>> [t,x]=ode45('DiffEqs',[tmin tmax],x0);
```

x0 vector above indicates the initial conditions, *tmin* and *tmax* indicate the lower and upper limits of the solution. Upon solution, graphics of the solution functions can be plotted as in Fig. 1.6 using the following MATLAB commands:

```
>> plot(t,x(:,1),'-k',t,x(:,2),'-b*');
```

1.15 Optimization

Finding minimum or maximum value of a function is called as optimization. The function to be minimized or maximized is called as objective function. Sometimes it is desirable to meet some conditions while finding the minimum/maximum value of a function. These conditions are called as constraints. Variables that objective and constraint functions depend on are called as optimization variables or shortly variables. Objective and constraint functions are expressed in terms of optimization variables.

Optimization problems can be expressed in general form by grouping the constraints as linear and nonlinear as following:

$$\begin{aligned}
&f(X=x_1,x_2,\ldots,x_n)\rightarrow \min\\
&\left.\begin{aligned}&AX\leq B\\&A_{eq}X=Beq\end{aligned}\right\}\text{LinearConstraints}\\
&\left.\begin{aligned}&g(X)\leq 0\\&h(X)=0\end{aligned}\right\}\text{NonlinearConstraints}\\
&X_L\leq X\leq X_U\}\ \text{SideConstraints}=\text{Constraints}\quad\text{on}\quad\text{Variables}
\end{aligned}\tag{1.1}$$

In Eq. (1.1), $f(X)$ specifies the objective function, $g(X)$ inequality-type constraint, $h(X)$ equality-type constraint, X optimization variables, X_L and X_U lower and upper limits of variables.

If the objective and constraint functions are linear, optimization problem is called linear. If one of these functions is nonlinear, optimization problem is said to be nonlinear. Examples of linear and nonlinear optimization problems are given in Table 1.14.

While linear optimization problems are mostly encountered in economics and industrial applications, nonlinear optimization problems are mostly encountered in engineering applications. Different methods are used for linear and nonlinear optimization problems to obtain their solutions efficiently.

Table 1.14 Examples of linear and nonlinear optimization problems

Linear optimization problem	Nonlinear optimization problem
$f=2x_1+3x_2\rightarrow \min$ $\left.\begin{aligned}&2x_1+3x_2\leq 20\\&x_1-x_2\leq 0\\&3x_1+2x_2=10\end{aligned}\right\}\text{Linear}\quad\text{Constraints}$ $\left.\begin{aligned}&x_1\geq 0\\&x_2\geq 0\end{aligned}\right\}\text{Side}\quad\text{Constraints}$	$f=(x_1-2)^2+(x_2-2)^2+2\rightarrow \min$ $\left.\begin{aligned}&{x_1}^2+{x_2}^2\leq 0\\&-x_1-x_2\leq 0\end{aligned}\right\}\text{Nonlinear}\quad\text{Constraints}$ $\left.\begin{aligned}&-4\leq x_1\leq 4\\&-4\leq x_2\leq 4\end{aligned}\right\}\text{Side}\quad\text{Constraints}$

1.15.1 Solution of Linear Optimization Problems

Linear optimization problems can be expressed in general form as following:

$$
\begin{aligned}
&f(X) \rightarrow \min \\
&AX \leq B \\
&A_{eq}X = B_{eq} \\
&X_L \leq X \leq X_U
\end{aligned}
\tag{1.2}
$$

In MATLAB, *linprog* command is used for the solution of linear optimization problems. *linprog* command can be used in variety of forms according to the information available for the solution as following:

$X =$ *linprog(f,A,b,Aeq,beq,*X_L*,*X_U*,*X_O*)*;
$X =$ *linprog(f,A,b,Aeq,beq,*X_L*,*X_U*)*; (If there is no X_O initial value)
$X =$ *linprog(f,A,b,Aeq,beq)*; (If there is no X_L, X_U and X_O value)

Example 1.28: Solve the linear optimization problem of $f = 2x_1 + 3x_2 \rightarrow \min$, $2x_1 + 3x_2 \leq 20$, $x_1 - x_2 \leq 0$, $3x_1 + 2x_2 = 10$, $0 \leq x_1, x_2 \leq 10$ with MATLAB.

Expression of the linear optimization problem and its solution with *linprog* command can be given as in Table 1.15.

1.15.2 Solution of Nonlinear Optimization Problems

Nonlinear optimization problems can be expressed in general form as following:

Table 1.15 Expression of linear optimization problem and its solution with MATLAB

Expression of linear optimization problem	Solution of linear optimization problem with MATLAB

$f = 2x_1 + 3x_2 \rightarrow \min \Rightarrow f = \underbrace{[2 \;\; 3]}_{F}\underbrace{\begin{Bmatrix} x_1 \\ x_2 \end{Bmatrix}}_{X} \rightarrow \min$	`>> F=[2 3] ;`
	`>> A=[2    3; 1 -1];     B=`
	`[20 ;0];`

$\left.\begin{matrix} 2x_1 + 3x_2 \leq 20 \\ x_1 - x_2 \leq 0 \end{matrix}\right\} \Rightarrow \underbrace{\begin{bmatrix} 2 & 3 \\ 1 & -1 \end{bmatrix}}_{A}\underbrace{\begin{Bmatrix} x_1 \\ x_2 \end{Bmatrix}}_{X} \leq \underbrace{\begin{Bmatrix} 20 \\ 0 \end{Bmatrix}}_{B}$	`>> Aeq=[3 2]; Beq=[10];`
	`>> XL=[0; 0]; XU=[10; 10];`
	`>> X=linprog(F,A,B,Aeq,Beq,`
	`XL,XU);`

$3x_1 + 2x_2 = 10 \Rightarrow \underbrace{[3 \;\; 2]}_{A_{eq}}\underbrace{\begin{Bmatrix} x_1 \\ x_2 \end{Bmatrix}}_{X} = \underbrace{\{10\}}_{B_{eq}}$	`>> X`
	`X =`
	`     2`
	`     2`

$\left.\begin{matrix} 0 \leq x_1 \leq 10 \\ 0 \leq x_2 \leq 10 \end{matrix}\right\} \Rightarrow \underbrace{\begin{Bmatrix} 0 \\ 0 \end{Bmatrix}}_{X_L} \leq \underbrace{\begin{Bmatrix} x_1 \\ x_2 \end{Bmatrix}}_{X} \leq \underbrace{\begin{Bmatrix} 10 \\ 10 \end{Bmatrix}}_{X_U}$	

$$\begin{aligned} & f(X = x_1, x_2, \ldots, x_n) \to \min \\ & \left. \begin{aligned} g(X) &\leq 0 \\ h(X) &= 0 \end{aligned} \right\} \text{NonlinearConstraints} \\ & X_L \leq X \leq X_U \} \text{ SideConstraints} \end{aligned} \tag{1.3}$$

Nonlinear optimization problems can be further divided into two groups as unconstrained (if there are no constraints) and constrained optimization problems depending on the existence of constraints.

(a) Unconstrained Optimization

In MATLAB, *fminbnd*, *fminunc*, and *ga* (Genetic Algorithm) commands can be used to solve unconstrained nonlinear optimization problems. Unlike other commands, *fminbnd* command is used to find the optimum of single variable functions such as *f (x)* function. In the solution of unconstrained optimization problems, objective function is saved in a function file. Then, this file is called by *fminbnd*, *fminunc*, or *ga* command for the solution. If the objective function has simple expression, it can be directly written to the workspace as *f* = *@(x)function_expression* or *f* = *inline ('function_expression')* without saving it in a function file. Determination of optimum values with *fminbnd*, *fminunc*, and *ga* commands can be explained below.

1. Optimum with ***fminbnd*** command:

   fminbnd command is used to find the minimum value of an *f(x)* function in an [a,b] interval. This command is used in *Xopt* = *fminbnd(f,a,b)* form. Here, *Xopt* indicates the optimum variable value calculated.

Example 1.29: Find the minimum value of $f = x^2 + 4x + 6$ function in [−2.2] interval using the *fminbnd* command in MATLAB.

Minimum value (e.g., using *fmin* command) can be obtained by typing the following expressions in the workspace of MATLAB.

```
>> f=@(x)x^2-4*x+6; xopt=fminbnd(f,-2,2); fmin=feval(f,xopt);
>> xopt
xopt =
  1.9999
>> fmin
fmin =
  2.0000
```

2. Optimum with ***fminunc*** command:

   In MATLAB, *fminunc* command can be used to find the minimum of a multivariate objective function. *fminunc* command can be used in variety of forms. The simplest form is expressed as *[Xopt,fopt]* = *fminunc(@FUN,X0)*. Here *FUN* is the expression or the file name of the objective function, X_0 is the initial value to start the optimization. *Xopt* and *fopt* correspond to optimum variable values calculated and corresponding objective value, respectively. If the objective function has simple expression, it can be written directly to the workspace as f = *@(x)function_expression* or f = *inline('function_expression')* without saving it in a separate function file.

3. Optimum with ***ga*** command:

   In MATLAB, *ga* command (it implements Genetic Algorithm) can also be used to find the optimum of a multivariate objective function, and it is expressed as *[Xopt,fopt]* = *ga(@FUN,NVARS)* in its simplest form. Here, *FUN* correspond to the objective function expression or the file name and *NVARS* correspond to the number of variables. *Xopt* and *fopt* correspond to optimum variable values calculated and corresponding objective value, respectively.

Example 1.30: Find minimum value of $f = (x_1 - 2)^2 + (x_2 - 2)^2 + 2$ function using a) *fminunc* and b) *ga* commands in MATLAB.

(a) Optimum with *fminunc* command: Since the function has very simple expression, it can be directly written to the MATLAB workspace without saving it in a separate function file. In this case, optimum value with *fminunc* command can be obtained by typing the following expressions in the workspace:

```
>> f=@(x)(x(1)-2)^2+(x(2)-2)^2+2;
>> x0=[0; 0]; % initial values
>> [x,f]=fminunc(f,x0);
>> x
x =
   2
   2
>> f
f =
   2
```

(b) Optimum with *ga* command: Optimum value with *ga* command can be obtained by typing the following expressions in the workspace:

```
>> f=@(x)(x(1)-2)^2+(x(2)-2)^2+2;
>> [x,f]=ga(f,2);
Optimization terminated: average change in the fitness value less
than options.FunctionTolerance.
>> x
x =
   2.0882   2.0112
>> f
f =
   2.0079
```

Note: After this, additional information between a command and its response such as "`Optimization terminated: average change in the fitness value less than options.FunctionTolerance.`" will not be shown to save space.

(b) Constrained Optimization

In the solution of constrained optimization problems, objective and constraint functions are saved in separate function files before the solution. Then, these files are called by commands such as *fmincon* or *ga* for the solution.

1. Optimum with ***fmincon*** command:

   In MATLAB, *fmincon* command can be expressed in the simplest form as below:

```
[Xopt,fopt]=fmincon(@FUN,X0,A,B,Aeq,Beq,LB,UB,@NONLCON)
```

   Here, *FUN* represents the file name of the objective function, *NONLCON* file name of constraint functions, X_0 initial values to start the optimization. *A* and *B* indicate the coefficients of inequality type linear constraints. *Aeq* and *Beq* denote coefficients of equality type linear constraints. *LB* and *UB* indicate lower and upper limits of variables. If there are no linear constraints in the optimization problem, null coefficients can be specified as *A* = [], *B* = [] and *Aeq* = [], *Beq* = []. Upon execution, *fmincon* command yields the optimum variable values (*Xopt*) and their corresponding objective value (*fopt*).
2. Optimum with ***ga*** command:

   Solution of the constrained optimization problems with Genetic Algorithm can be obtained using *ga* command in MATLAB. *ga* command is expressed in the simplest from as following:

```
[Xopt,fopt]=ga(@FUN,NVARS,A,B,Aeq,Beq,LB,UB,@NONLCON)
```

   Here, *NVARS* indicates the number of variables, other terms have the same meaning as in *fmincon* command. Upon execution, *ga* command yields the optimum variable values (*Xopt*) and their corresponding objective value (*fopt*).

Example 1.31: Obtain the solution of the following optimization problem using a) *fmincon* and b) *ga* commands in MATLAB.

$$f=(x_1-2)^2+(x_2-2)^2+2\rightarrow \min$$
$$-x_1-x_2\leq 0$$
$$0\leq x_1,x_2\leq 4$$

Before the solution of the constrained optimization problem, first, objective and constraint functions are saved in separate function files such as *Objective.m* and

Table 1.16 MATLAB files for the objective and constraint functions

File=Objective.m	File=Constraints.m

`function f=Objective(x)`
 	
`f=(x(1)-2)^2+(x(2)-2)^2+2 ;`
	`function [g,h]=Constraints(x)`
	
	`g(1)=-x(1)-x(2) ;`
	`h=[] ;`

Constraints.m as in Table 1.16. These files are then called by the optimization commands for the solution.

(a) Optimum with ***fmincon*** command:

Optimum solution with *fmincon* command can be obtained by typing the following expressions in the workspace of MATLAB:

```
>> x0=[0; 0]; % Initial values of variables
>> lb=[0; 0]; % Lower bounds of variables
>> ub=[4; 4]; % Upper bounds of variables
>> [x,f]=fmincon(@Objective,x0,[],[],[],[],lb,ub,
@Constraints);
>> x
x =
  2.0000
  2.0000
>> f
f =
  2
```

(b) Optimum with ***ga*** command:

Optimum solution with *ga* command can be obtained by typing the following expressions in the workspace of MATLAB:

```
>> x0=[0; 0]; % Initial values of variables
>> lb=[0; 0]; % Lower bounds of variables
>> ub=[4; 4]; % Upper bounds of variables
>> [x,f]=ga(@Objective,2,[],[],[],[],lb,ub,@Constraints);
>> x
x =
  2.0000  2.0000
>> f
f =
  2.0000
```

1.16 Programming with MATLAB

There may not be any built-in commands in MATLAB to perform some special operations or calculations. In such cases, it may be necessary to write a program to perform the specialized operations or calculations. For this purpose, MATLAB can be used as a programming language as well. In MATLAB programs, ProgramName is specified in the first line after *function* expression if it is a function file. Then, desired commands or expressions are written successively by separating them with semicolons (;). Upon filling the file with commands or expressions, it is saved using the same name after the *function* expression. Program is run by typing its name and then executing in the workspace. When the program is run, commands or expressions in the program file are executed one by one. In order to write a program in MATLAB or in any programming language, the following concepts should be known in advance.

1.16.1 Conditional Statements

Sometimes execution of a command or expression depends on some conditions to be satisfied. Such conditions are expressed with *if end* or *if else end* command block as shown in Table 1.17.

Conditions in MATLAB are expressed with logical operators such as && (and), || (or), == (equal), ~= (not equal), < (less than), <= (less than or equal), > (greater), >= (greater than or equal) as shown in Table 1.18.

Table 1.17 Expressing conditional statements with *if else end* command block in MATLAB

Single Action: *Command* is executed if *Condition* is satisfied	Double Action: if *Condition* is satisfied *Command1*, else *Command2* is executed
if (Condition) *Command* *end*	*if (Condition)* *Command1* *else* *Command2* *end*

Table 1.18 Examples for the expression of conditional operations in MATLAB

if (x~=2)
f=x^3 ;
end
	if (x>=0) && (x<=10)
	f=x^2 ;
	else
	f=0 ;
	end

1.16.2 Repetitive Operations (Loop-Based Operations)

In some calculations, commands or expressions are executed with a certain number of times. For example, operations on vectors and matrices require repetitive calculations with repetition number equal to vector or matrix size. Repetitive operations are expressed with loops in MATLAB as in all programming languages. There are two types of loops:

1. *Unconditional Loop*: In this loop, executions are repeated with a specified number of times using *for end* command block in MATLAB as below:

```
for i=m:q:n
   Command ;
end
```

   In *for end* loop, *i* is the counter index, *m* is the initial value of the counter, *q* is the increment, and *n* is the last value of the counter. *i*, *m*, *q*, *n* values must be integers.

2. *Conditional Loop*: In this loop, the number of repetition is not known in advance. Loop is executed until a specified condition is met. Conditional loops are expressed with *while end* command block in MATLAB as following:

```
while (Condition)
   Command ;
end
```

   Conditions in MATLAB are expressed with logical operators such as && (and), || (or), == (equal), ~= (not equal), < (less than), <= (less than or equal), > (greater), >= (greater than or equal).

Example 1.32: *A* vector is given as *A* = [1 2 3 4]. Write a MATLAB program that finds the sum of the elements in *A* vector.

Program can be named as *SumVectorA.m*. In this program, first the number of repetition (*n*) is determined from the length of vector *A*. Then, a variable such as *sumA* is defined to store the sum of members in vector *A*. An initial value of zero is assigned to *sumA* since members have not been added yet. Finally, using *for end* loop, all members of vector *A* are added to *sumA* one by one (sequentially) leading to the sum of *A* as in Table 1.19.

Program *SumVectorA.m* can be run by typing its name in the Workspace of MATLAB as following. Upon execution, output is printed on the screen.

```
>> SumVectorA
sumA =
  10
```

Table 1.19 MATLAB program for the summation of members of a vector

File=SumVectorA.m

```
A=[1 2 3 4] ;   % A vector
n=length(A) ;   % Length of A vector

sumA=0 ; % Variable to express sum of A

for i=1:n
   sumA=sumA+A(i) ; % Adding A(i) to the current value of sumA
end

sumA % Printing the sum value to the screen as sumA
```

Example 1.33: Numbers from *1* to *n* are added to make the sum <= 1000. Write a MATLAB program to determine the last number *n*.

Program can be named as *Findn.m.* In this program, first, a new variable such as *sumA* is defined to store the sum of numbers (i.e., total value). An initial value of zero is assigned to *sumA* since members have not been added yet. Then, using *while end* loop, numbers from 1 to *n* are added to *sumA* (total value) sequentially until sumA<=1000 condition is satisfied as shown in the MATLAB program in Table 1.20.

The program *Findn.m* can be run by typing its name in the Workspace of MATLAB as following. Upon execution, output is printed on the screen.

```
>> Findn
n =
  45
```

Example 1.34: Write a MATLAB program to find the maximum value of an *A* vector given as $A = [3\ -1\ 9\ 4\ 7]$.

Program can be named as *FindMax.m.* In this program, a new variable such as *maxA* is defined to store the maximum value of *A* vector. An initial value is assigned to *maxA*. First member of *A* vector (i.e., *A*(1)) is assigned to *maxA* as the initial value. Then, members of *A* vector ($A(i)$, $i = 2,3,4$) are compared with the current maximum sequentially. To scan the members of vector *A* sequentially, *for end* loop is used. For comparison of each member with the current maximum in the loop, *if end* condition is used. The MATLAB program that finds the maximum of an *A* vector is shown in Table 1.21.

Table 1.20 MATLAB program for the summation of numbers until 1000

File=Findn.m

```
sumA=0 ;  % Variable to express sum of A
i=0    ;  % initialization of counter variable i

while (sumA<=1000)
    i=i+1 ;          % Increasing counter by one
    sumA=sumA+i ; % Adding A(i) to the current value
of sumA
end

n=i  % Printing the last value of counter
```

Table 1.21 MATLAB program to determine the maximum value of an *A* vector

File=FindMax.m

```
A=[3 -1 9 4 7] ; % Vector A
n=length(A) ;    % Number of members in vector A
maxA=A(1) ;      % Initial value for the maximum

for i=2:n
    if A(i)>maxA
        maxA=A(i) ;
    end
end

maxA  % Printing the maximum value on the screen
```

The program *FindMax.m* can be run by typing its name in the Workspace of MATLAB as following. Upon execution, output is printed on the screen.

```
>> FindMax
maxA =
  9
```

Example 1.35: Write a MATLAB program to sum *A* and *B* vectors that are given as $A = [3\ -1\ 9\ 4\ 7]$ and $B = [2\ 4\ 3\ 0\ 6]$.

Program can be named as *FindSumAB.m*. In this program, a new variable (Vector) such as *C* is defined to store the summation of *A* and *B* vectors. Then, corresponding members of *A* and *B* are added to each other to obtain *C* vector. This operation is represented as $C(i) = A(i) + B(i)$ for the *i*-th member. Since there are

Table 1.22 MATLAB program to sum *A* and *B* vectors

File=FindSumAB.m

```
A=[3 -1 9 4 7] ; % Vector A
B=[2  4 3 0 6] ; % Vector B

n=length(A) ;     % Number of members in vector A

for i=1:n
    C(i)=A(i)+B(i) ;
end

C  % Printing the summation on the screen
```

four members in *A* and *B* vectors, summation operation is repeated four times through *for end* loop in MATLAB. The MATLAB program that sums *A* and *B* vectors is shown in Table 1.22.

The program *FindSumAB.m* can be run by typing its name in the Workspace of MATLAB as following. Upon execution, output is printed on the screen.

```
>> FindSumAB
C =
   5   3   12   4   13
```

Example 1.36: Write a MATLAB program to determine the number of members in an *A* vector that satisfy the condition of $3 \leq A(i) < 8$. *A* vector is given as $A = [3\ -1\ 9\ -4\ 7\ 6\ 12\ 0\ 5]$.

Program can be named as *Findn38.m*. In this program, a new variable such as *count* is defined to store the number of members satisfying $3 \leq A(i) < 8$ condition in *A* vector. An initial value of zero is assigned to *count*. Then, members of *A* vector ($A(i)$, $i = 1,\ldots,9$) are checked for the satisfaction of this condition sequentially through *for end* loop in MATLAB. If this condition is satisfied, *count* is increased by 1 in the loop. MATLAB program that finds the number of members in the *A* vector that satisfy the condition of $3 \leq A(i) < 8$ is shown in Table 1.23.

The program *Findn38.m* can be run by typing its name in the Workspace of MATLAB as following. Upon execution, output is printed on the screen.

```
>> Findn38
count =
   4
```

Table 1.23 MATLAB program to determine the number of members in an A vector that satisfy the condition of $3 \le A(i) < 8$

File=Findn38.m

```
A=[3 -1 9 -4 7 6 12 0 5] ; % A Vector

n=length(A) ;  % Number of members in A vector
count=0 ;  % Initialization of counter value

for i=1:n
    if (A(i)>=3) && (A(i)<8)   % Condition check
        count=count+1 ;
    end
end

count   % Printing the number of elements on the
screen
```

1.16.3 Multi-conditional Statements

In some calculations, only one of the statements among several ones is executed based on the specified condition. Such multi-conditional case is expressed with *switch end* command block in MATLAB as in Table 1.24.

In the multi-conditional case above, *Expression 1* is executed if $x = x1$ condition is satisfied, *Expression 2* is executed if $x = x2$ condition is satisfied,… ., *Expression n* is executed if $x = xn$ condition is satisfied and *Expression x* is executed if no condition is satisfied.

Example 1.37: Seasons are numbered as 1 for winter, 2 for spring, 3 for summer, and 4 for autumn. Write a program that prints the name of the season according to the number inputted on the screen.

Program can be named as *Season.m*. Using *switch end* command in this program, names of the seasons can be printed on the screen as in the MATLAB file in Table 1.25.

Program *Season.m* can be run by typing its name in the Workspace of MATLAB as following:

```
>>Season
Enter season number: 3
```

Table 1.24 Expression of multi-conditional case in MATLAB

```
switch (x)
        case x1
                Expression 1 ;
        case x2
                Expression 2 ;
        case x3
                Expression 3 ;
        ...
        case xn
                Expression n ;
        otherwise
                Expression x ;
end
```

Table 1.25 MATLAB file to the print season name based on the input number

File=Season.m

```
season=input(Enter season number:
') ;

switch season
    case 1
        disp('Winter') ;
    case 2
        disp('Spring') ;
    case 3
        disp('Summer') ;
    case 4
        disp('Fall') ;
    otherwise
        disp('No    season    like
this')
end
```

1.17 Summary

- MATLAB is a mathematical software with built-in commands and also a programming platform.
- Built-in commands can be typed and executed in Workspace after >> symbol.
- When there are multiple commands to be executed, they can be stored in a separate file and this file can be executed by typing its name in the Workspace after >> symbol.

- There are two types of files in MATLAB. Those that do not need any input to be executed. Those that need input from outside to be executed and produce output upon execution. These files are called as *function* files.
- For mathematical or other operations, variables are used in MATLAB. Predefined or calculated values can be assigned to variables in *variable = value* form.
- Variables are classified as string (character) such as *name = 'Adam'* and numeric variables such as *x = 10* in MATLAB.
- Variable should start with a letter and should not include any mathematical operators (such as + − * / ^) and punctuation symbols (such as . , ; :) in it. For example, *SumValue* is a valid variable. However, *SumValue+* or *SumValue** is not a valid variable name.
- Reserved words in MATLAB such as *for*, *end*, *if*, *then*, *else*, *switch*, *case* should not be used alone as variable.
- Assigning value to a variable can be done using *input* command.
- Calculated value or a message can be outputted to the screen using *disp* or *fprintf* command. *fprintf* command allows printing format of a variable.
- For mathematical operations, functions can be defined in MATLAB using *f = 'function(x)'*, *f = @(x)function(x)* or *f = inline('function')* command.
- Graphics of functions can be plotted on the screen using *plot* or *fplot* command for single variable functions such as *f(x)*, *surf* or *mesh* command for two-variable functions such as *f(x,y)*.
- Vectors and matrices are used to store the same type of data under a single variable name.
- Vectors and matrices are defined using the values in [] bracket such as *A* = [1 2 3].
- Mathematical operations of vectors and matrices can be easily performed by typing the operation in Workspace. For example, *A* = [1 2 3] and *B* = [4 5 6] vectors can be easily summed by typing *C* = *A* + *B* in the Workspace of MATLAB.
- Submatrix between *i-j* rows and *k-l* columns of *A* matrix can be obtained by using the index expression of *A(i:j,k:l)*.
- Linear equation system can be expressed in algebraic and matrix forms as below:

$$\left.\begin{array}{l} a_{11}x_1 + a_{12}x_2 + \ldots + a_{1n}x_n = b_1 \\ a_{21}x_1 + a_{22}x_2 + \ldots + a_{2n}x_n = b_2 \\ \ldots \\ a_{n1}x_1 + a_{n2}x_2 + \ldots + a_{nn}x_n = b_n \end{array}\right\} \Rightarrow \underbrace{\begin{bmatrix} a_{11} & a_{12} & \cdots & a_{1n} \\ a_{21} & a_{22} & \cdots & a_{2n} \\ \cdots & \cdots & \ddots & \cdots \\ a_{n1} & a_{n2} & \cdots & a_{nn} \end{bmatrix}}_{A} \underbrace{\begin{Bmatrix} x_1 \\ x_2 \\ \cdots \\ x_n \end{Bmatrix}}_{X} = \underbrace{\begin{Bmatrix} b_1 \\ b_2 \\ \cdots \\ b_n \end{Bmatrix}}_{B}$$

$$\rightarrow AX = B$$

  Linear equation system of $AX = B$ can be solved using *X = inv(A)*B*, *X = A\B* or *X = linsolve(A,B)* command.
- Roots of a polynomial function expressed in $a_nx^n + \ldots + a_2x^2 + a_1x + a_0 = 0$ form (order is decreasing) can be obtained using *roots* command in MATLAB. For

example, roots of $x^2 + 3x + 2 = 0$ equation can be obtained by using X = *roots*([1 3 2]) command.

- To solve a nonlinear equation in MATLAB, it should be written in zero (*f(x) = 0*) form. Then, solution can be obtained using *fzero(f,x0)* command.
- Nonlinear equation system can be solved using *fsolve* command in MATLAB. For this, nonlinear equations should be written in zero (=0) form and saved in a separate function file such as *FileName.m*. Then, using this file and initial values of *x0* = [*x10*; *x20*;…; *xn0*] (*n* is the number of variables), solution can be obtained with *fsolve('FileName',x0)* command.
- Intermediate value at an *x0* point in a data set can be predicted using interpolation methods in MATLAB. For interpolation, *interp1(x,f,x0,'method')* command can be used.
- Curve fitting to a data set can be carried out using *polyfit* command in MATLAB.
- Derivative of an *f(x)* function can be obtained with *diff(f,x,n)* command where *x* denotes independent variable and *n* denotes the derivative order.
- Integral of an *f(x)* function in an [*a*,*b*] interval can be calculated analytically (in closed form) using *int(f,x,a,b)* command.
- Integral of an *f(x)* function in an [*a*,*b*] interval can be calculated numerically using *trapz* and *quad* commands.
- Eigenvalue problems can be expressed as $\underbrace{\begin{bmatrix} a_{11} & a_{12} & \cdots & a_{1n} \\ a_{21} & a_{22} & \cdots & a_{2n} \\ \cdots & \cdots & \ddots & \cdots \\ a_{n1} & a_{n2} & \cdots & a_{nn} \end{bmatrix}}_{A} \underbrace{\begin{Bmatrix} x_1 \\ x_2 \\ \cdots \\ x_n \end{Bmatrix}}_{X} = \lambda \underbrace{\begin{Bmatrix} x_1 \\ x_2 \\ \cdots \\ x_n \end{Bmatrix}}_{X}$

  or shortly as $AX = \lambda X$. Eigenvalue problems can be solved using *eig* command in *[Xmat,Lam] = eig(A)* form where *Xmat* gives the eigenvectors in column form and *Lam* gives the eigenvalues.
- Differential equations can be solved analytically (in closed form) using *dsolve* command in MATLAB.
- Differential equations can be solved numerically using *ode23* and *ode45* commands in MATLAB.
- Finding the minimum or maximum value of a function is called as optimization. When there are conditions (i.e., constraints) to be satisfied, optimization problems can be expressed in general form as following:

$$
\begin{aligned}
& f(X = x_1, x_2, \ldots, x_n) \rightarrow \min \\
& \left.\begin{aligned} & AX \leq B \\ & A_{eq}X = Beq \end{aligned}\right\} \text{LinearConstraint} \\
& \left.\begin{aligned} & g(X) \leq 0 \\ & h(X) = 0 \end{aligned}\right\} \text{NonlinearConstraint} \\
& X_L \leq X \leq X_U \} \ \text{SideConstraint} = \text{Constraint} \quad \text{on} \quad \text{Variable}
\end{aligned}
$$

- When only linear constraints exist (linear optimization problem), optimization problem can be solved using *linprog* command.
- When there are no constraint (unconstrained optimization problem), optimization problem can be solved using *fminbnd*, *fminunc*, or *ga* (Genetic Algorithm) command.
- When there are constraints (constrained optimization problem, no limitation on constraints), optimization problem can be solved using *fmincon* command.
- MATLAB can also be used as a programming language. Written program can be saved in a separate file and executed by typing its name in the Workspace.
- Conditional expressions can be stated using *if end* or *if else end* command block.
- Multi-conditional expressions can be stated more efficiently using with *switch end* command block.
- Repetitive operations are expressed with loops in MATLAB. Unconditional loops are expressed using *for end* command block while conditional loops are expressed using *while end* command block.

Appendix: MATLAB Programs

MATLAB Program 1.1: Plot of f = x^2 + 1 function

File=Plot_x2plus1.m

```
function Plot_x2plus1
% NOTE: This program plots graphic of f=x^2+1 function

x0=-2 ; % initial value to start plotting
xf=2 ;  % final value to finish plotting
n=20 ;  % number of points in plotting

% GENERATION OF DATA FOR PLOTTING
dx=(xf-x0)/(n-1) ; % distance between points
for i=1:n
    x(i)=x0+(i-1)*dx ;  % coordinate of the i-th point
    f(i)=x(i)^2+1 ;  % function value at i-th point
end

% PLOT OF FUNCTION
plot(x,f,'--*r');
xlabel('x');
ylabel('f');
title('f=x^2+1');
```

MATLAB Program 1.2: Plot of z = x^3 + y^3 function

File=Plot_x3plusy3.m

```
function Plot_x3plusy3
% NOTE: This program plots graphic of z=x^3+y^3 function
x0=-2 ;   % initial value to start plotting in x direction
xf=2 ;    % final value to finish plotting in x direction
nx=20 ;   % number of points in plotting in x direction

y0=-2 ;   % initial value to start plotting in y direction
yf=2 ;    % final value to finish plotting in y direction
ny=20 ;   % number of points in plotting in y direction

% GENERATION OF DATA FOR PLOTTING
dx=(xf-x0)/(nx-1) ; % distance between points in x direction
dy=(yf-y0)/(ny-1) ; % distance between points in y direction
for i=1:nx
    for j=1:ny
        X(i)=x0+(i-1)*dx ;   % x coordinate of the (i,j)-th
point
        Y(j)=y0+(j-1)*dy ;   % y coordinate of the (i,j)-th
point
        Z(i,j)=X(i)^3+Y(j)^3 ; % function value at (i,j)-th
point
    end
end
% PLOT OF FUNCTION
surf(X,Y,Z);
xlabel('x');
ylabel('y');
zlabel('z');
```

Problems

Problem 1.1: Calculate the value of $f(x) = x^2 + 2 \sin x \cos x$ function at $x = 2$ using MATLAB command.

Problem 1.2: A and B vectors are given as $A = [3\ 4\ 0\ 1]$ and $B = [-2\ 5\ 3\ 6]$. Find the result of 2*A-3*B using MATLAB command.

Problem 1.3: A matrix is given as $A = \begin{bmatrix} 1 & 3 & 2 \\ 4 & 6 & 5 \\ 7 & 9 & 8 \end{bmatrix}$. Calculate determinant of A matrix using MATLAB command.

Problem 1.4: A matrix is given as $A = \begin{bmatrix} 1 & 3 & 2 \\ 4 & 6 & 5 \\ 7 & 9 & 8 \end{bmatrix}$. Calculate the inverse of A matrix using MATLAB command.

Problem 1.5: Solve the following linear equation system using MATLAB commands.

$$
\begin{aligned}
&3x_1 - 4x_2 + 5x_3 - 2x_4 = 1\\
&-x_1 + 2x_2 + x_3 - x_4 = 2\\
&5x_1 - x_2 + 3x_3 - 4x_4 = 5\\
&6x_1 + 3x_2 + 7x_3 - 5x_4 = 10
\end{aligned}
$$

Problem 1.6: Calculate roots of the polynomial function of $3x^3 - 4x^2 + 5x - 4 = 0$ using MATLAB command.

Problem 1.7: Solve the nonlinear equation of $2x^3 + \sqrt{x} - e^x = 4$ with MATLAB using the initial value of $x_0 = 1$.

Problem 1.8: Find the solution of $x_1^2 + x_2^2 = 4$ and $-2x_1 + 3x_2^2 = 5$ equations with MATLAB command using initial values of $(x_{10}, x_{20}) = (1, 1)$.

Problem 1.9: For an *f(x)* function, *x* and *f* values are given as $x = [0\ 1\ 2\ 3\ 5\ 6\ 7]$ and $f = [1\ 2\ 3\ 15\ 27\ 35\ 49]$. Estimate the *f* value at $x_0 = 4$ point using MATLAB command.

Problem 1.10: For an *f(x)* function, *x* and *f* values are given as $x = [0\ 1\ 2\ 3\ 5\ 6\ 7]$ and $f = [1\ 3\ 4\ 14\ 28\ 36\ 51]$. Fit linear and parabolic polynomials to *f* values with MATLAB.

Problem 1.11: Calculate the integral of $f(x) = x^3 + 2\sqrt{x} - e^x$ function numerically in [1, 3] interval using MATLAB command.

Problem 1.12: For the eigenvalue problem of $AX = \lambda X$, A matrix is given as $A = \begin{bmatrix} 1 & 3 & 2 \\ 4 & 6 & 5 \\ 7 & 9 & 8 \end{bmatrix}$. Find eigenvalues λ ($\lambda = \lambda_1, \lambda_2, \lambda_3$) and their corresponding eigenvectors X ($X = \begin{Bmatrix} x_1 \\ x_2 \\ x_3 \end{Bmatrix}$) with MATLAB command.

Problem 1.13: Find the solution of the differential equation $\frac{d^2y(x)}{dx^2} + 2\frac{dy(x)}{dx} + 3y = x$ with MATLAB command using the initial conditions of $y(0) = 2$, $\frac{dy(0)}{dx} = 1$.

Problem 1.14: Find the analytical solution of the differential equation system of $\frac{dx_1}{dt} = -x_1 + 2x_2$ and $\frac{dx_2}{dt} = 3x_1 + x_2$ with MATLAB using the initial conditions of $x_1(0) = 0$, $x_2(0) = 1$ and plot the graphic of the solution functions within [0 5] interval.

Problem 1.15: Find the numerical solution of the differential equation system of $\frac{dx_1}{dt} = -x_1 + 2x_2$ and $\frac{dx_2}{dt} = 3x_1 + x_2$ for the initial conditions of $x_1(0) = 0, x_2(0) = 1$ using *ode45* command in MATLAB and plot the solutions within [0 5] interval.

Problem 1.16: Solve the linear optimization problem of $f = x_1 + x_2 \rightarrow \min$, $2x_1 + 3x_2 \leq 4$, $x_1 - x_2 \leq 0$, $0 \leq x_1, x_2 \leq 10$ with MATLAB.

Problem 1.17: Find the minimum value of $f = x^2 - 6x + 12$ function in $[-4, 4]$ interval using the *fminbnd* command in MATLAB.

Problem 1.18: Find minimum value of $f = x_1^2 + 4x_2^2 - 4x_1 - 4x_2 + 8$ function using (a) *fminunc* and (b) *ga* commands in MATLAB.

Problem 1.19: Obtain the solution of the following optimization problem using (a) *fmincon* and (b) *ga* commands in MATLAB.

$$f = (3x_1 - 1)^2 + (x_2 - 2)^2 + 4 \rightarrow \min$$
$$-x_1 - x_2 \leq 0$$
$$0 \leq x_1, x_2 \leq 5$$

Problem 1.20: *A* vector is given as $A = [1\ 5\ 3\ 8\ 9\ -1\ 4\ -3\ 22\ 7\ -6]$. Write a MATLAB program that finds the number of members of A vector that are greater than 4.

Problem 1.21: *A* vector is given as $A = [9\ 1\ 8\ 2\ -3\ 7\ 4\ -6]$. Write a MATLAB program that finds the difference of maximum and minimum values of A vector (i.e., max(A)-min(A)).

References

1. Yang, W. Y., Cao, W., Chung, T.-S., & Morris, J. (2005). *Applied numerical methods using MATLAB*. Wiley.
2. Beers, K. J. (2007). *Numerical methods for chemical engineering: Applications in MATLAB*. Cambridge University Press.
3. Houcque, D. (2005). *Introduction to MATLAB for engineering students [lecture notes, PDF]*. Northwestern University. Retrieved from https://www.mccormick.northwestern.edu/documents/students/undergraduate/introduction-to-matlab.pdf.
4. Gilat, A. (2008). *MATLAB: An introduction with applications*. Wiley.
5. Kattan, P. I. (2010). *MATLAB for beginners: A gentle approach-revised edition*. CreateSpace Independent Publishing Platform.
6. Siemers, T. (2011). *An introduction to MATLAB and Mathcad*. APEX Calculus.
7. Bayen, A. M., & Siauw, T. (2015). *An introduction to Matlab programming and numerical methods for engineers*. Elsevier.

8. Nagar, S. (2017). *Introduction to MATLAB for engineers and scientists: Solutions for numerical computation and modeling*. Apress.
9. Turk, I. (2018). *MATLAB programming for beginners and professionals*. CreateSpace Independent Publishing Platform.
10. Kharab, A., & Guenther, R. B. (2018). *An introduction to numerical methods: A MATLAB approach*. CRC Press.

Chapter 2
Basic Mathematics

2.1 Introduction

This chapter provides basic mathematical information that will prepare the reader for the next chapters. Brief information about limit, derivative, integral, and matrix operations is given in this chapter. More comprehensive information can be found in the books given in the references of [1–4].

2.2 Limit

For some variable values, function values can be indeterminate. For example, for $x = 2$, value of $f(x) = \frac{x^2+x-6}{x-2}$ function is $f(2) = \frac{0}{0}$ that is indeterminate. However, this function will have a value when x is very close to or approaches 2. For example, function $f(x)$ will have $f(1.999999) = 4.999999$ and $f(2.000001) = 5.000001$ for the values of $x = 1.999999$ and $x = 2.000001$, respectively. From here, it is seen that $f(x)$ function approaches 5 when x approaches 2.

As in the above example, if an $f(x)$ function will have indeterminate value such as $f(x^*) = \frac{0}{0}$ or $f(x^*) = \frac{\infty}{\infty}$, removing uncertainty and obtaining a function value when x is very close to x^* is called limit operation or limit in mathematics and is represented as $\lim_{x \to x^*} f(x)$. In limit operation, there are various methods to remove the uncertainty such as factoring both the numerator and denominator and L'Hôpital's rule. These methods are described below with examples.

Example 2.1 Calculate the limit value of $f(x) = \frac{5+2x}{3-x}$ function for $x \to \infty$ (i.e., when x approaches ∞).

For $x = \infty$, function value becomes $f(\infty) = \frac{5+2\infty}{3-\infty} = -\frac{\infty}{\infty}$, which is indeterminate. To remove the uncertainty, function expression can be rearranged by creating a common factor x, both in the numerator and the denominator as

H. Kurtaran, *Engineering Mathematics with MATLAB Applications: Vol. 1, Fundamentals*, https://doi.org/10.1007/978-3-031-81869-1_2

$f(x)=\frac{5+2x}{3-x}=\frac{x\left(\frac{5}{x}+2\right)}{x\left(\frac{3}{x}-1\right)}=\frac{\left(\frac{5}{x}+2\right)}{\left(\frac{3}{x}-1\right)}$. After cancelling out x terms in the numerator and denominator and substituting $x=\infty$ value into the resulting expression, function value is obtained as $f(x)=\frac{\left(\frac{5}{x}+2\right)}{\left(\frac{3}{x}-1\right)}=\frac{\left(\underbrace{\frac{5}{\infty}}_{0}+2\right)}{\left(\underbrace{\frac{3}{\infty}}_{0}-1\right)}=\frac{(0+2)}{(0-1)}=-2$. From the result, it is seen that $f(x)$ value converges to -2 as x approaches to infinity.

Solution with MATLAB command:

Limit value of $f(x)$ function can be obtained by using the *limit*(*Function*, *Variable*, *Variable_value*) command in MATLAB as below:

```
>> syms x; f=(5+2*x)/(3-x); limit(f,x,inf)
ans =
-2
```

2.2.1 Hopital (L'Hôpital) Rule

If $f(x)$ value is indeterminate as $\frac{0}{0}$ or $\frac{\infty}{\infty}$ for an x value, another alternative to remove the uncertainty is to use Hopital (L'Hôpital) (Hopital's) rule. In the Hopital's rule, derivatives of numerator and denominator are taken and then x value is substituted into the resulting expression to obtain the limit value of $f(x)$ function.

Example 2.2: Calculate the limit value of $f(x)=\frac{x^2+x-6}{x-2}$ function for x → 2 (i.e., when x approaches 2) using the Hopital's rule.

For $x=2$, function value becomes $f(2)=\frac{0}{0}$, which is indeterminate. If derivatives of the numerator and denominator are taken and then $x=2$ is substituted into the resulting expression, limit value of $f(x)$ function is obtained as below:

$$\lim_{x\to 2} f(x)=\frac{x^2+x-6}{x-2}\to \lim_{x\to 2}\frac{Numerator\_derivative}{Denumerator\_derivative}=\frac{\frac{d\left(x^2+x-6\right)}{dx}}{\frac{d(x-2)}{dx}}$$

$$\to \lim_{x\to 2}\frac{2x+1}{1}=\frac{2(2)+1}{1}=5$$

Solution with MATLAB command:

Limit value of $f(x)$ function can be obtained using the *limit* command in MATLAB as following:

```
>> syms x; f=(x^2+x-6)/(x-2); limit(f,x,2)
ans =
5
```

Example 2.3: Calculate the limit value of $f(x) = \frac{x^3+2x^2-3x}{x^2-3x+2}$ function for x → 1 (i.e., when x approaches 1) using the Hopital's rule.

For $x = 1$, function value becomes $f(1) = \frac{0}{0}$, which is indeterminate. If derivatives of the numerator and denominator are taken and then $x = 1$ is substituted into the resulting expression, limit value of $f(x)$ function is obtained as below:

$$\lim_{x\to 1} f(x) = \frac{x^3 + 2x^2 - 3x}{x^2 - 3x + 2} \to \lim_{x\to 1} \frac{Numerator\_derivative}{Denumerator\_derivative}$$

$$= \frac{\frac{d(x^3+2x^2-3x)}{dx}}{\frac{d(x^2-3x+2)}{dx}} \to \lim_{x\to 1} \frac{3x^2+4x-3}{2x-3} = \frac{3(1)^2+4(1)-3}{2(1)-3} = -4$$

Solution with MATLAB command:

Limit value of $f(x)$ function can be obtained using the *limit* command in MATLAB as following:

```
>> syms x; f=(x^3+2*x^2-3*x)/(x^2-3*x+2); limit(f,x,1)
ans =
-4
```

2.3 Derivative

Derivatives express rate of changes in function values. For example, derivative of an $f(x)$ function expresses how $f(x)$ value changes with x and is expressed as $f'(x) = \frac{df(x)}{dx}$. First derivative of an $f(x)$ function expresses its slope as in Fig. 2.1.

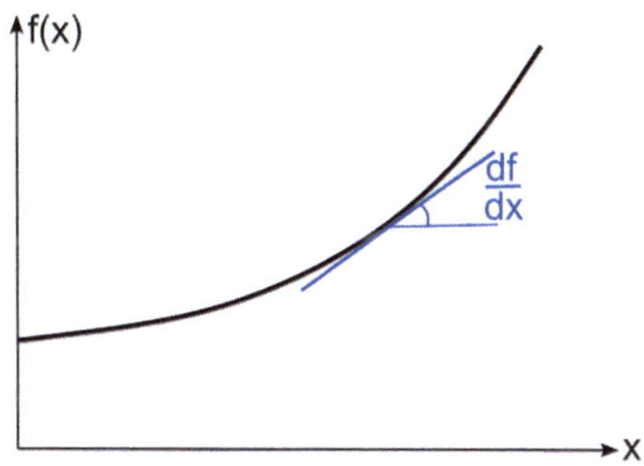

Fig. 2.1 Expressing slope of a function with its derivative

Derivatives of polynomial, trigonometric, and exponential functions are explained below.

2.3.1 Derivative of Polynomial Functions

Derivative of polynomial function $f(x) = x^n$ is expressed with the product of the exponent (n) and the modified function expression where the exponent is decreased by 1 as $f'(x) = nx^{n-1}$.

Example 2.4: Express the derivative of $f(x) = 3x^4 + \sqrt{x}$ function.

$f(x) = 3x^4 + \sqrt{x}$ function can also be written as $f(x) = 3x^4 + x^{1/2}$. Using new expression, derivative of the function can be obtained by applying the derivative rule to each term as below:

$$f'(x) = 12x^3 + \frac{1}{2}x^{(1/2)-1} \rightarrow f'(x) = 12x^3 + \frac{1}{2}x^{-(1/2)} \rightarrow f'(x) = 12x^3 + \frac{1}{2}\frac{1}{\sqrt{x}}$$

Solution with MATLAB command:

Derivative of an $f(x)$ function can be obtained using the *diff(Function, Variable)* command in MATLAB as following (representing the result of derivative with *dfdx*):

```
>> dfdx=diff(3*x^4+sqrt(x),x)
dfdx=1/(2*x^(1/2)) + 12*x^3
```

2.3.2 Derivative of Trigonometric Functions

sine and *cosine* functions and other functions derived from these functions are called as trigonometric functions. For example, *tangent* function can be obtained by dividing the *sine* and *cosine* functions as $\tan(x) = \frac{\sin(x)}{\cos(x)}$ and *cotangent* function can be obtained by dividing *cosine* and *sine* functions as $\cot(x) = \frac{\cos(x)}{\sin(x)}$.

Following relationships exist among trigonometric functions:

$$(\sin(x))^2 + (\cos(x))^2 = 1,\ \sin(2x) = 2\sin(x)\cos(x),\ \cos(2x) = 2(\cos(x))^2 - 1$$
$$= 1 - 2(\sin(x))^2$$

For brevity, *sine* and *cosine* functions can be shown with *sin* and *cos* terms. In this case, derivative of *sin*(x) and *cos*(x) functions can be expressed as below:

$$f(x) = \sin(x) \rightarrow f'(x) = \cos(x), f(x) = \cos(x) \rightarrow f'(x) = -\sin(x),$$

Trigonometric functions may also include internal functions such as $f(x) = \sin(u(x))$ and $f(x) = \cos(u(x))$, where $u(x)$ indicates the internal function. We can call such functions as general trigonometric functions.

Derivatives of general trigonometric functions can be expressed as the product of derivative of the internal function and derivative of the basic trigonometric function as following:

$$f(x) = \sin(u(x)) \rightarrow f'(x) = u'(x)\cos(u(x)),$$
$$f(x) = \cos(u(x)) \rightarrow f'(x) = -u'(x)\sin(u(x))$$

Example 2.5: Express the derivative of $f(x) = \sin(3x^4 + \sqrt{x})$ function.

If $u(x) = 3x^4 + \sqrt{x}$ is considered as the internal function, then derivative of $f(x)$ function can be expressed as below

$$f(x) = \sin(u(x)) \rightarrow f'(x) = u'(x)\cos(u(x)) \rightarrow f'(x) = \underbrace{\left(12x^3 + \frac{1}{2}\frac{1}{\sqrt{x}}\right)}_{u'(x)} \cos\left(\underbrace{3x^4 + \sqrt{x}}_{u(x)}\right)$$

Solution with MATLAB command:

Derivative of $f(x)$ function can be obtained using the *diff* command in MATLAB as below (representing the result of derivative with *dfdx*):

```
>> dfdx=diff(sin(3*x^4+sqrt(x)),x)
dfdx = cos(x^(1/2) + 3*x^4)*(1/(2*x^(1/2)) + 12*x^3)
```

2.3.3 Derivative of Exponential Functions

$e = 2.71\ldots$. value-based functions such as $f(x) = e^x$ are called exponential functions. Derivative of $f(x) = e^x$ function is expressed as $f'(x) = e^x$. An exponential function including an internal function of $u(x)$ such as $f(x) = e^{u(x)}$ is called as general exponential function. Derivative of general exponential function $e^{u(x)}$ can be expressed as the product of the internal function and the main exponential function as following:

$f(x) = e^{u(x)} \rightarrow f'(x) = u'(x)e^{u(x)}$

Example 2.6: Express the derivative of $f(x) = e^{\left(3x^4 + \sqrt{x}\right)}$ function.

If $u(x) = 3x^4 + \sqrt{x}$ is considered as the internal function, then derivative of $f(x)$ function can be expressed as below:

$$f(x)=e^{u(x)}\rightarrow f'(x)=u'(x)e^{u(x)}\rightarrow f'(x)=\underbrace{\left(12x^3+\frac{1}{2}\frac{1}{\sqrt{x}}\right)}_{u'(x)}e^{\overbrace{(3x^4+\sqrt{x})}^{u(x)}}$$

Solution with MATLAB command:

Derivative of *f*(*x*) function can be obtained using the *diff* command in MATLAB as below (representing the result of derivative with *dfdx*):

```
>> dfdx=diff(exp(3*x^4+sqrt(x)),x)
dfdx = exp(x^(1/2) + 3*x^4)*(1/(2*x^(1/2)) + 12*x^3)
```

2.3.4 Derivative of Logarithmic Functions

Logarithmic functions are used to express a value or a number in a different scale. In logarithmic functions, function is expressed with respect to the power of a base value. An *f*(*x*) function is expressed as *f*(*x*) = log (*x*) using logarithm with base 10 and as *f*(*x*) = ln (*x*) with base *e* = 2.71.... Derivative of the logarithmic function *f*(*x*) = ln (*x*) is expressed as $f'(x)=\frac{1}{x}$. Derivative of general logarithmic function *f*(*x*) = ln (*u*(*x*)) including the internal function of *u*(*x*) is expressed as $f'(x)=\frac{u'(x)}{u(x)}$.

Example 2.7: Express the derivative of $f(x)=\ln\left(e^{x^2}\right)$ function.

If $u(x)=e^{x^2}$ is considered as the internal function, then derivative of the internal function can be expressed as $u'(x)=2xe^{x^2}$. In this case, derivative of the logarithmic function $f(x)=\ln\left(e^{x^2}\right)$ is expressed as $f'(x)=\frac{u'(x)}{u(x)}=\frac{2xe^{x^2}}{e^{x^2}}=2x$.

Solution with MATLAB command:

Derivative of *f*(*x*) function can be obtained using the *diff* command in MATLAB as below (representing the result of derivative with *dfdx*):

```
>> dfdx=diff(log(exp(x^2)),x)
dfdx = 2*x
```

2.3.5 Derivative of Product Functions

(a) **Derivative of product of two functions**

Assume that an *h*(*x*) function is expressed as the product of *f*(*x*) and *g*(*x*) functions as *h*(*x*) = *f*(*x*)*g*(*x*). In this case, derivative of *h*(*x*) function can be expressed as the sum of the derivative of $f(x)$ multiplied by $g(x)$ and derivative of $g(x)$ multiplied by $f(x)$ as following:

$$h(x) = f(x)g(x) \rightarrow h'(x) = f'(x)g(x) + f(x)g'(x)$$

Example 2.8: Express the derivative of $h(x) = \sin(2x)e^{(-x^3)}$ function.

If $\sin(2x)$ is denoted as $f(x)$ and $e^{(-x^3)}$ is denoted as $g(x)$ function, then, derivative of the product function $h(x) = \sin(2x)e^{(-x^3)}$ can be expressed as below:

$$h'(x) = f'(x)g(x) + f(x)g'(x) \rightarrow h'(x) = \underbrace{[2\cos(2x)]}_{f'(x)}\underbrace{\left[e^{(-x^3)}\right]}_{g(x)} + \underbrace{[\sin(2x)]}_{f(x)}\underbrace{\left[(-3x^2)e^{(-x^3)}\right]}_{g'(x)}$$

Solution with MATLAB command:

Derivative of $h(x) = \sin(2x)e^{(-x^3)}$ function can be obtained using the *diff* command in MATLAB as below (representing the result of derivative with *dhdx*):

```
>> dhdx=diff(sin(2*x)*exp(-x^3),x)
dhdx =
2*cos(2*x)*exp(-x^3) - 3*x^2*sin(2*x)*exp(-x^3)
```

Example 2.9: Express the derivative of $h(x) = (x^3 + 3x^2 + 2x + 1) \cos x$ function.

If $x^3 + 3x^2 + 2x + 1$ is denoted as $f(x)$ and $\cos x$ is denoted as $g(x)$ function, then derivative of the product function $h(x) = f(x)g(x) = \underbrace{(x^3 + 3x^2 + 2x + 1)}_{f(x)}\underbrace{\cos x}_{g(x)}$ can be expressed as below:

$$h'(x) = f'(x)g(x) + f(x)g'(x) \rightarrow h'(x) = \underbrace{[3x^2 + 6x + 2]}_{f'(x)}\underbrace{[\cos x]}_{g(x)} + \underbrace{[x^3 + 3x^2 + 2x + 1]}_{f(x)}\underbrace{[-\sin x]}_{g'(x)}$$

Solution with MATLAB command:

Derivative of $h(x) = (x^3 + 3x^2 + 2x + 1) \cos x$ function can be obtained using the *diff* command in MATLAB as below (representing the result of derivative with *dhdx*):

```
>> dhdx=diff((x^3+3*x^2+2*x+1)*cos(x),x)
dhdx =
cos(x)*(3*x^2 + 6*x + 2) - sin(x)*(x^3 + 3*x^2 + 2*x + 1)
```

(b) **Derivative of division of two functions**

Assume that an $h(x)$ function is expressed as the division of $f(x)$ and $g(x)$ functions as $h(x) = \frac{f(x)}{g(x)}$. In this case, derivative of $h(x)$ function can be expressed as below:

$$h(x) = \frac{f(x)}{g(x)} \rightarrow h'(x) = \frac{f'(x)g(x) - f(x)g'(x)}{g(x)^2}$$

Example 2.10: Express the derivative of $h(x) = \tan x$ function.

Function $h(x) = \tan x$ can be expressed as the division of $f(x) = \sin x$ and $g(x) = \cos x$ functions. In this case, derivative of $h(x)$ function can be obtained as following:

$$h(x) = \frac{f(x) = \sin x}{g(x) = \cos x} \rightarrow h'(x) = \frac{f'(x)g(x) - f(x)g'(x)}{g(x)^2} \rightarrow h'(x)$$
$$= \frac{\cos x \cos x - (-\sin x)\sin x}{(\cos x)^2} = \frac{(\cos x)^2 + (\sin x)^2}{(\cos x)^2} = 1 + (\tan x)^2$$

Solution with MATLAB command:

Derivative of $h(x) = \tan x$ function can be obtained using the *diff* command in MATLAB as following (representing the result of derivative with *dhdx*):

```
>> dhdx=diff(tan(x),x)
dhdx = tan(x)^2 + 1
```

Example 2.11: Express the derivative of $h(x) = \cot x$ function.

Function $h(x) = \cot x$ can be expressed as the division of $f(x) = \cos x$ and $g(x) = \sin x$ functions. In this case, derivative of $h(x)$ function can be obtained as following:

$$h(x) = \frac{f(x) = \cos x}{g(x) = \sin x} \rightarrow h'(x) = \frac{f'(x)g(x) - f(x)g'(x)}{g(x)^2} \rightarrow h'(x)$$
$$= \frac{(-\sin x)\sin x - \cos x \cos x}{(\sin x)^2} = -\frac{(\sin x)^2 + (\cos x)^2}{(\sin x)^2} = -1 - (\cot x)^2$$

Solution with MATLAB command:

Derivative of the $h(x)$ function can be obtained using the *diff* command in MATLAB as following (representing the result of derivative with *dhdx*):

```
>> dhdx=diff(cot(x),x)
dhdx = - cot(x)^2 - 1
```

2.3.6 Derivatives of Nested Functions

Some functions can be expressed in nested form of other functions. For example, a $z(x)$ function can be expressed by nesting $h(x)$, $g(h)$ and $z(g)$ functions as $z(x) = f(g(h(x)))$. In this case, derivative of the nested function $z(x)$ can be expressed using the chain rule as below:

$$z(x) = f(g(h(x))) \rightarrow z'(x) = \frac{dz(x)}{dx} = \frac{dz(x)}{dg(x)} \frac{dg(x)}{dh(x)} \frac{dh(x)}{dx}$$

Example 2.12: Express the derivative of $z(x) = e^{\sin(x^2)}$ function using the chain rule.

When $h(x)$, $g(h)$ and $z(g)$ functions are represented with $h(x) = x^2$, $g(h) = \sin(h)$, and $z(g) = e^g$ functions, $z(x) = e^{\sin(x^2)}$ function can be expressed in nested form as $z(x) = f(g(h(x)))$. Derivative of $z(x)$ function can be expressed using the chain rule as following:

$$z(x) = f(g(h(x))) \rightarrow z'(x) = \frac{dz(x)}{dx} = \frac{dz(x)}{dg(x)} \frac{dg(x)}{dh(x)} \frac{dh(x)}{dx} = e^g \cos(h)(2x).$$

When $g(h)$ and $z(g)$ functions are expressed in terms of x as $g = \sin(h) = \sin(x^2)$ and $z(g) = e^g = e^{\sin(x^2)}$, then derivative of $z(x)$ function can be obtained in terms of x as below:

$$z'(x) = e^g \cos(h)(2x) = e^{\sin(x^2)} \cos(x^2)(2x)$$

Solution with MATLAB command:

Derivative of $z(x) = e^{\sin(x^2)}$ function can be obtained using the *diff* command in MATLAB as below (representing the result of derivative with *dzdx*):

```
>> dzdx=diff(exp(sin(x^2)),x)
dzdx = 2*x*cos(x^2)*exp(sin(x^2))
```

2.4 Integral

Integral of a function represents the area under it as seen in Fig. 2.2. Integral also corresponds to the inverse of derivative operation. Integral of an $f(x)$ function within an [a,b] interval is expressed as $I = \int_a^b f(x)dx = F(b) - F(a)$ where $F(a)$ and $F(b)$ denote function values at $x = a$ and $x = b$. If integral limits are not specified (this is

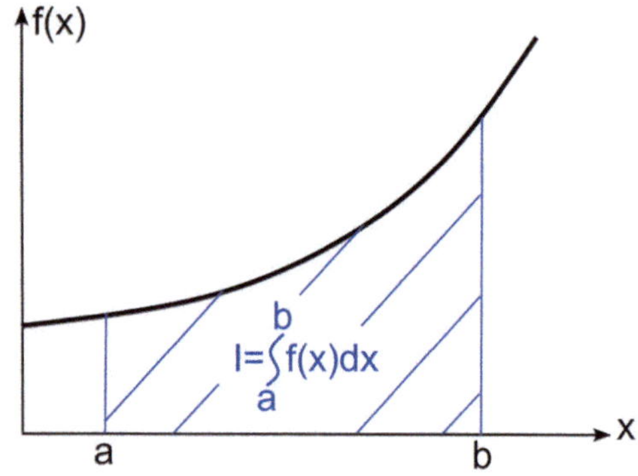

Fig. 2.2 Expressing the area under an $f(x)$ function using its integral

called as indefinite integral), then integral of $f(x)$ function is simply expressed as $I = \int f(x)dx$.

Integrals of trigonometric functions can be expressed in terms of each other. For example, integrals of *sine* (*sin*) and *cosine* (*cos*) functions can be expressed as below:

$$I = \int \sin(x)dx = -\cos(x), I = \int \cos(x)dx = \sin(x)$$

2.4.1 Integral of Polynomial Functions

Integral of the polynomial function $f(x) = x^n$ is obtained increasing the exponent value of the function by 1 and dividing the resulting expression with the new exponent value as $I = \int f(x)dx = \int x^n dx = \frac{x^{n+1}}{n+1}$. If derivative of the integral expression I is taken, again original function $f(x) = x^n$ is obtained as following:

$$I = \frac{x^{n+1}}{n+1} \rightarrow \frac{dI}{dx} = (n+1)\frac{x^{n+1-1}}{n+1} = x^n \rightarrow \frac{dI}{dx} = f(x)$$

From the above operations, it is seen that derivative and integral operations are inverse of each other.

Example 2.13: Calculate $I = \int (3x^4 + \sqrt{x})dx$ integral.

When $f(x)$ function is expressed as $f(x) = 3x^4 + \sqrt{x}$, then integral of this function can be obtained by applying the integral rule to each term as below:

$$I = \int (3x^4 + \sqrt{x})dx = \int \left(3x^4 + x^{1/2}\right)dx = 3\frac{x^{4+1}}{4+1} + \frac{x^{(1/2)+1}}{(1/2)+1} = 3\frac{x^5}{5} + \frac{x^{(3/2)}}{(3/2)}$$
$$= \frac{3x^5}{5} + \frac{2x^{(3/2)}}{3}$$

Solution with MATLAB command:

Integral *I* can be calculated using the *int* command in MATLAB as following:

```
>> I=int(3*x^4+sqrt(x))
I = (2*x^(3/2))/3 + (3*x^5)/5
```

Example 2.14: Calculate $I = \int_1^2 \frac{4x}{\sqrt{x^2-1}} dx$ integral.

When $x^2 - 1$ term in the denominator is denoted with a new *u* variable as $u = x^2 - 1$, *du* can be written as $du = 2xdx$. In this case, integral can be expressed in terms of *u* variable and calculated as following:

$$I = \int_1^2 \frac{4x}{\sqrt{x^2-1}} dx = \int \frac{2}{\sqrt{u}} du = \int 2u^{-1/2} du = 4\sqrt{u}$$

When *u* variable is replaced with $x^2 - 1$ again, integral result in $x = [1, 2]$ interval can be calculated as below:

$$I = 4\sqrt{u} = 4\sqrt{x^2-1}\Big|_1^2 = 4\sqrt{2^2-1} - 4\sqrt{1^2-1} = 4\sqrt{3}$$

Solution with MATLAB command:

Integral *I* can be calculated using the *int* command in MATLAB as following:

```
>> f=4*x/sqrt(x^2-1); I=int(f,x,1,2)
I =
4*3^(1/2)
```

2.4.2 Integral of Trigonometric Functions

Among trigonometric functions, integrals of *sine* and *cosine* functions are expressed as below:

$$I = \int \sin x dx = -\cos x, I = \int \cos x dx = \sin x$$

Transformation of variables is used in the integration of more complex trigonometric functions. This way integral expression is converted into a simpler form.

Example 2.15: Calculate $I = \int_0^{\pi/4} (\sin x)^3 \cos x dx$ integral.

If sin*x* is represented with a new *u*(*x*) variable as $u(x) = \sin x$, then *du* can be expressed as $du = \cos x dx$. In this case, using the new variable *u*(*x*), integral expression can be converted into the integral of polynomial function as below:

$$I = \int \Big(\underbrace{\sin x}_{u}\Big)^3 \underbrace{\cos x dx}_{du} = \int u^3 du = \frac{u^4}{4} = \frac{(\sin x)^4}{4}\Bigg|_0^{\pi/4} = \frac{(\sin(\pi/4))^4}{4} - \frac{(\sin 0)^4}{4} = \frac{1}{16}$$

Solution with MATLAB command:
Integral *I* can be calculated using the *int* command in MATLAB as following:

```
>> syms x
>> f=(sin(x))^3*cos(x) ;
>> I=int(f,x,0,pi/4)
I =
1/16
```

Example 2.16: Calculate $I = \int_0^{\pi/2} \frac{\cos x}{\sqrt{1+(\cot x)^2}} dx$ integral.

Integral expression can be simplified by expressing cot*x* as $\cot x = \frac{\cos x}{\sin x}$. Using $\cot x = \frac{\cos x}{\sin x}$ expression, integral can be expressed as following:

$$I = \int_0^{\pi/2} \frac{\cos x}{\sqrt{1+(\cot x)^2}} dx = \int_0^{\pi/2} \frac{\cos x}{\sqrt{1+\left(\frac{\cos x}{\sin x}\right)^2}} dx$$

$$= \int_0^{\pi/2} \frac{\cos x}{\sqrt{\frac{\overbrace{(\sin x)^2 + (\cos x)^2}^{1}}{(\sin x)^2}}} dx = \int_0^{\pi/2} \frac{\cos x}{\sqrt{\frac{1}{(\sin x)^2}}} dx \rightarrow$$

$$I = \int_0^{\pi/2} \underbrace{\sin x}_{u} \underbrace{\cos x dx}_{du} = \frac{u^2}{2} \rightarrow I = \frac{(\sin x)^2}{2}\Bigg|_0^{\pi/2} = \frac{(\sin(\pi/2))^2}{2} - \frac{(\sin 0)^2}{2} = \frac{1}{2}$$

Solution with MATLAB command:
Integral *I* can be calculated using the *int* command in MATLAB as following:

```
>> syms x
>> f=cos(x)/sqrt(1+(cot(x))^2);
>> I=int(f,x,0,pi/2)
I =
1/2
```

2.4.3 Integral of Exponential Functions

Integral of the exponential function $f(x) = e^x$ is expressed as following:

$$I = \int e^x dx = e^x$$

In the integral of more complex exponential functions, transformations of variables can be used. This way integral expression is converted into a simpler form.

Example 2.17: Calculate $I = \int e^{(x^2+2)} 2x dx$ integral.

If $x^2 + 2$ is represented with $u(x)$ function as $u(x) = x^2 + 2$, then integral I can be calculated using u and du expressions as below:

$$I = \int e^{\overbrace{(x^2+2)}^{u}} \underbrace{2xdx}_{du} = \int e^u du = e^u = e^{(x^2+2)}$$

Solution with MATLAB command:

Integral I can be calculated using the *int* command in MATLAB as following:

```
>> I=int(exp(x^2+2)*2*x)
I = exp(x^2)*exp(2)
```

2.5 Vector and Matrix Operations

Some variables contain multiple data of the same type. These variables are referred to as vectors or matrices. Vectors contain data only in row or column form while matrices contain data both in row and column form as below:

$$\text{Vector Example}: A = [2\ \ 3\ \ 5]\ \text{(in column form)}, A = \begin{Bmatrix} 2 \\ 3 \\ 5 \end{Bmatrix}\ \text{(in row form)}$$

$$\text{Matrix Example}: A = \begin{bmatrix} 3 & 2 & 4 \\ 8 & 5 & 6 \\ 7 & 9 & 1 \end{bmatrix}$$

Vectors can be considered as special forms of matrices since they consist of just rows or columns.

Member at *i*-th row or column of vector *A* can be indicated as *A*(*i*). Similarly member at *i*-th row and *j*-th column of matrix *A* can be indicated as *A*(*i*, *j*). An example of *A* matrix consisting of two rows and three columns is shown below:

$$A = \begin{bmatrix} \overbrace{A(1,1)}^{Column:1} & \overbrace{A(1,2)}^{Column:2} & \overbrace{A(1,3)}^{Column:3} \\ A(2,1) & A(2,2) & A(2,3) \end{bmatrix} \begin{matrix} \leftarrow Row:1 \\ \leftarrow Row:2 \end{matrix} = \begin{bmatrix} 1 & 2 & 3 \\ 4 & 5 & 6 \end{bmatrix}$$

In matrix *A*, *A*(2, 3) term represents the member at the intersection of second row and third column and its value is given as 6, (i.e., *A*(2, 3) = 6).

Matrices are also defined with their size in the form of *Row*x*Column*. For example, matrix *A* can be expressed as 2 × 3 matrix ($A_{2\ \times\ 3}$) since it consists of two rows and three columns.

2.5.1 *Norm of Vector*

Norm of a vector expresses its magnitude and is often denoted with a double line parenthesis as ||||. It is expressed as the square root of the sum of squares of its members. Norm of a vector *A* consisting of three elements can be expressed as following:

$$A = [A(1)\quad A(2)\quad A(3)] \rightarrow \text{Norm}: ||A|| = \sqrt{A(1)^2 + A(2)^2 + A(3)^2}$$

Example 2.18: Calculate the norm of $A = [3\quad 5\quad 7\quad 9]$ vector.

Norm of *A* vector can calculated as below:

$$\text{Norm}: ||A|| = \sqrt{A(1)^2 + A(2)^2 + A(3)^2 + A(4)^2} = \sqrt{3^2 + 5^2 + 7^2 + 9^2} = 12.8062$$

Solution with MATLAB command:

Norm of $A = \begin{bmatrix} 3 & 5 & 7 & 9 \end{bmatrix}$ vector can be calculated using the *norm* command in MATLAB as below:

```
>> A=[ 3 5 7 9];
>> norm(A)
ans =
  12.8062
```

Note: Norm of the vector can also be calculated using *MATLAB Program 2.1: Norm of Vector* in appendix.

2.5.2 Summation of Matrices

For summation (addition) of matrices, their dimensions must be compatible (i.e., their corresponding row and column numbers must be the same). For example, consider the following 2 × 3 matrices of *A* and *B*.

$$A = \begin{bmatrix} A(1,1) & A(1,2) & A(1,3) \\ A(2,1) & A(2,2) & A(2,3) \end{bmatrix}, B = \begin{bmatrix} B(1,1) & B(1,2) & B(1,3) \\ B(2,1) & B(2,2) & B(2,3) \end{bmatrix}$$

Sum of *A* and *B* matrices can be expressed with a new *C* matrix by adding the values in the same row and column as following:

$$C = A + B = \begin{bmatrix} A(1,1) & A(1,2) & A(1,3) \\ A(2,1) & A(2,2) & A(2,3) \end{bmatrix} + \begin{bmatrix} B(1,1) & B(1,2) & B(1,3) \\ B(2,1) & B(2,2) & B(2,3) \end{bmatrix}$$
$$= \begin{bmatrix} A(1,1)+B(1,1) & A(1,2)+B(1,2) & A(1,3)+B(1,3) \\ A(2,1)+B(2,1) & A(2,2)+B(2,2) & A(2,3)+B(2,3) \end{bmatrix}$$

Example 2.19: *A* and *B* matrices are given as $A = \begin{bmatrix} 1 & 2 & 3 \\ 4 & 5 & 6 \\ 7 & 8 & 9 \end{bmatrix}$, $B = \begin{bmatrix} 1 & 0 & 2 \\ 0 & 2 & 9 \\ 3 & 6 & 4 \end{bmatrix}$. Find the sum of *A* and *B* matrices.

Sum of *A* and *B* matrices can be expressed with a new matrix *C* as following:

$$C = A + B = \begin{bmatrix} 1 & 2 & 3 \\ 4 & 5 & 6 \\ 7 & 8 & 9 \end{bmatrix} + \begin{bmatrix} 1 & 0 & 2 \\ 0 & 2 & 9 \\ 3 & 6 & 4 \end{bmatrix} = \begin{bmatrix} 1+1 & 2+0 & 3+2 \\ 4+0 & 5+2 & 6+9 \\ 7+3 & 8+6 & 9+4 \end{bmatrix} = \begin{bmatrix} 2 & 2 & 5 \\ 4 & 7 & 15 \\ 10 & 14 & 13 \end{bmatrix}$$

Solution with MATLAB command:

```
>> A=[1 2 3; 4 5 6; 7 8 9]; B=[1 0 2; 3 2 9; 6 5 4] ;
>> C=A+B
C =
     2    2    5
     4    7   15
    10   14   13
```

Note: Summation (addition) of matrices can also be calculated by using *MATLAB Program 2.2: Addition of Matrices* in appendix.

2.5.3 Subtraction of Matrices

Subtraction of matrices is carried out similar to summation/addition. For subtraction of matrices, their dimensions must be compatible (i.e., their corresponding row and column numbers must be the same). For example, consider the following 2 × 3 matrices of *A* and *B*:

$$A=\begin{bmatrix} A(1,1) & A(1,2) & A(1,3) \\ A(2,1) & A(2,2) & A(2,3) \end{bmatrix}, B=\begin{bmatrix} B(1,1) & B(1,2) & B(1,3) \\ B(2,1) & B(2,2) & B(2,3) \end{bmatrix}$$

Subtraction of *B* matrix from *A* matrix can be expressed as a new *C* matrix by subtracting the values in the same row and column as below:

$$C=A-B=\begin{bmatrix} A(1,1) & A(1,2) & A(1,3) \\ A(2,1) & A(2,2) & A(2,3) \end{bmatrix}-\begin{bmatrix} B(1,1) & B(1,2) & B(1,3) \\ B(2,1) & B(2,2) & B(2,3) \end{bmatrix}$$
$$=\begin{bmatrix} A(1,1)-B(1,1) & A(1,2)-B(1,2) & A(1,3)-B(1,3) \\ A(2,1)-B(2,1) & A(2,2)-B(2,2) & A(2,3)-B(2,3) \end{bmatrix}$$

Example 2.20: *A* and *B* matrices are given as $A=\begin{bmatrix} 1 & 2 & 3 \\ 4 & 5 & 6 \\ 7 & 8 & 9 \end{bmatrix}$ and $B=\begin{bmatrix} 1 & 0 & 2 \\ 0 & 2 & 9 \\ 3 & 6 & 4 \end{bmatrix}$. Subtract *B* matrix from *A* matrix.

Subtraction of *B* matrix from *A* matrix can be expressed as a new matrix *C* as following:

$$C = A - B = \begin{bmatrix} 1 & 2 & 3 \\ 4 & 5 & 6 \\ 7 & 8 & 9 \end{bmatrix} - \begin{bmatrix} 1 & 0 & 2 \\ 0 & 2 & 9 \\ 3 & 6 & 4 \end{bmatrix} = \begin{bmatrix} 1-1 & 2-0 & 3-2 \\ 4-0 & 5-2 & 6-9 \\ 7-3 & 8-6 & 9-4 \end{bmatrix} = \begin{bmatrix} 0 & 2 & 1 \\ 4 & 3 & -3 \\ 4 & 2 & 5 \end{bmatrix}$$

Solution with MATLAB command:

```
>> A=[1 2 3; 4 5 6; 7 8 9]; B=[1 0 2; 3 2 9; 6 5 4] ;
>> C=A-B
C =
      0    2    1
      4    3   -3
      4    2    5
```

Note: Subtraction of matrices can also be calculated by using *MATLAB Program 2.3: Subtraction of Matrices* in appendix.

2.5.4 Multiplication (Product) of Matrices

In order to multiply two matrices (e.g., A and B matrices), internal dimensions must be the same. That is, column number of A matrix must be equal to the row number of B matrix. If the product of A and B matrices are expressed with a new $C = A \times B$ matrix, then $C(i, j)$ member can be obtained by multiplying i-th row of A with j-th column of B.

For example, assume A and B matrices are given as following:

$$A = \begin{bmatrix} A(1,1) & A(1,2) & A(1,3) \\ A(2,1) & A(2,2) & A(2,3) \end{bmatrix}, B = \begin{bmatrix} B(1,1) & B(1,2) \\ B(2,1) & B(2,2) \\ B(3,1) & B(3,2) \end{bmatrix}$$

In this case, multiplication (product) of these matrices can be expressed with a new C matrix as below:

$$C = \begin{bmatrix} C(1,1) & C(1,2) \\ C(2,1) & C(2,2) \end{bmatrix} = AxB = \begin{bmatrix} A(1,1) & A(1,2) & A(1,3) \\ A(2,1) & A(2,2) & A(2,3) \end{bmatrix} \begin{bmatrix} B(1,1) & B(1,2) \\ B(2,1) & B(2,2) \\ B(3,1) & B(3,2) \end{bmatrix}$$

$$C = \begin{bmatrix} A(1,1)B(1,1) + A(1,2)B(2,1) + A(1,3)B(3,1) & A(1,1)B(1,2) + A(1,2)B(2,2) + A(1,3)B(3,2) \\ A(2,1)B(1,1) + A(2,2)B(2,1) + A(2,3)B(3,1) & A(2,1)B(1,2) + A(2,2)B(2,2) + A(2,3)B(3,2) \end{bmatrix}$$

Order of the matrices is important in matrix multiplication. Results of AxB and BxA differ.

Example 2.21: *A* and *B* matrices are given as $A=\begin{bmatrix}2 & 3\\ 1 & 4\end{bmatrix}$ and $B=\begin{bmatrix}5 & 1 & 4\\ 3 & 6 & 2\end{bmatrix}$. Obtain the multiplication (product) of *A* and *B* matrices.

Multiplication (product) of *A* and *B* matrices can be expressed with a new *C* matrix as below:

$$C=AxB=\begin{bmatrix}2 & 3\\ 1 & 4\end{bmatrix}\begin{bmatrix}5 & 1 & 4\\ 3 & 6 & 2\end{bmatrix}=\begin{bmatrix}2x5+3x3 & 2x1+3x6 & 2x4+3x2\\ 1x5+4x3 & 1x1+4x6 & 1x4+4x2\end{bmatrix}=\begin{bmatrix}19 & 20 & 14\\ 17 & 25 & 12\end{bmatrix}$$

Solution with MATLAB command:

```
>> A=[2 3; 1 4]; B=[5 1 4; 3 6 2] ;
>> C=A*B
C=
  19   20   14
  17   25   12
```

Note: Multiplication (product) of matrices can also be calculated by using *MATLAB Program 2.4: Multiplication of Matrices* in appendix.

2.5.5 Determinant of Matrix

Determinant represents the characteristic value of a matrix. For determinant calculation, matrix must be a square matrix (row and column numbers must be the same). Determinant is indicated by a double line parenthesis ||. Determinant of an *A* matrix with 2 × 2 size can be expressed as following:

$$A=\begin{bmatrix}A(1,1) & A(1,2)\\ A(2,1) & A(2,2)\end{bmatrix}\rightarrow \text{Determinant : } |A|=A(1,1)A(2,2)-A(2,1)A(1,2)$$

Example 2.22: Calculate the determinant of $A=\begin{bmatrix}4 & 2\\ 7 & 3\end{bmatrix}$ matrix.

Determinant of *A* matrix can be calculated as following:

$$|A|=A(1,1)A(2,2)-A(2,1)A(1,2)=4x3-7x2=-2$$

Determinant of a matrix bigger than 2 × 2 size is obtained by using determinants of submatrices obtained from the main matrix. In the determinant calculation, first a row or column is determined as the pivot (reference) row or column. Then, a pivot member is selected in this pivot row or column. A submatrix is generated by excluding the row and column in the main matrix crossing the pivot member. This way, submatrices with equal number of row or column of main matrix are obtained. Multiplying determinants of the submatrices with pivot values and powers of -1, determinant of the main matrix can be calculated. To explain method, consider an A matrix of 3×3 size as below:

$$A = \begin{bmatrix} A(1,1) & A(1,2) & A(1,3) \\ A(2,1) & A(2,2) & A(2,3) \\ A(3,1) & A(3,2) & A(3,3) \end{bmatrix}$$

When first row is selected as the pivot row, submatrices can be generated by excluding rows and columns crossing the members of $A(1,1)$, $A(1,2)$, and $A(1,3)$, respectively. For example, *Submatrix 1* can be obtained by deleting the first row and first column of A matrix. For *Submatrix 1*, $A(1,1)$ will correspond to the pivot element or member. Similarly, *Submatrix 2* can be obtained by deleting the first row and second column of A matrix. For *Submatrix 2*, $A(1,2)$ will correspond to the pivot element or member. *Submatrix 3* can be obtained by deleting the first row and third column of A matrix. For *Submatrix 3*, $A(1,3)$ will correspond to the pivot element or member. In general case, *Submatrix j* can be obtained by deleting the first row and j-th column of A matrix and $A(1,j)$ corresponds to pivot element here. In this case, determinant of A matrix can be calculated by multiplying the determinants of submatrices with their pivots and power of -1 in the form of $|Submatrix\_j|(-1)^{1+j}A(1,j))$ and then summing them as below:

$$|A| = (-1)^{1+1}A(1,1)\underbrace{\begin{vmatrix} A(2,2) & A(2,3) \\ A(3,2) & A(3,3) \end{vmatrix}}_{Submatrix\bar{1}} + (-1)^{1+2}A(1,2)\underbrace{\begin{vmatrix} A(2,1) & A(2,3) \\ A(3,1) & A(3,3) \end{vmatrix}}_{Submatrix\bar{2}} +$$

$$(-1)^{1+3}A(1,3)\underbrace{\begin{vmatrix} A(2,1) & A(2,2) \\ A(3,1) & A(3,2) \end{vmatrix}}_{Submatrix\bar{3}}$$

Example 2.23: Calculate the determinant of $A = \begin{bmatrix} 3 & 1 & 2 \\ 5 & 6 & 7 \\ 9 & 2 & 8 \end{bmatrix}$ matrix.

Determinant of A matrix can be calculated as following:

$$|A| = (-1)^{1+1}3\underbrace{\begin{vmatrix} 6 & 7 \\ 2 & 8 \end{vmatrix}}_{Submatrix\bar{1}} + (-1)^{1+2}1\underbrace{\begin{vmatrix} 5 & 7 \\ 9 & 8 \end{vmatrix}}_{Submatrix\bar{2}} + (-1)^{1+3}2\underbrace{\begin{vmatrix} 5 & 6 \\ 9 & 2 \end{vmatrix}}_{Submatrix\bar{3}} \rightarrow$$

$$|A| = 3[6x8 - 2x7] - 1[5x8 - 9x7] + 2[5x2 - 9x6] = 37$$

Solution with MATLAB command:

Determinant of A matrix can be calculated using the *det* command in MATLAB as below:

```
>> A=[3 1 2; 5 6 7; 9 2 8];
>> det(A)
ans =
  37.0000
```

2.5.6 Inverse of Matrix

In order to calculate the inverse of a matrix, it must be a square matrix (rows and columns must be equal). To find the inverse of an A matrix with 2×2 size, first, a new matrix A' is generated by swapping the top-left ($A(1,1)$) and bottom-right ($A(2,2)$) members and by changing the signs of the bottom-left ($A(2,1)$) and top-right ($A(1,2)$) members of A matrix. Then, inverse of A matrix (i.e., A^{-1}) is obtained by dividing the new matrix A' with the determinant of A matrix as below:

$$A = \begin{bmatrix} A(1,1) & A(1,2) \\ A(2,1) & A(2,2) \end{bmatrix} \rightarrow A' = \begin{bmatrix} A(2,2) & -A(1,2) \\ -A(2,1) & A(1,1) \end{bmatrix}, |A| = A(1,1)A(2,2) - A(2,1)A(1,2)$$

$$A^{-1} = \frac{A'}{|A|} = \frac{\begin{bmatrix} A(2,2) & -A(1,2) \\ -A(2,1) & A(1,1) \end{bmatrix}}{|A|}$$

Example 2.24: Calculate the inverse (A^{-1}) of $A = \begin{bmatrix} 4 & 2 \\ 7 & 3 \end{bmatrix}$ matrix.

Inverse of A matrix (i.e., A^{-1}) can be calculated as following:

$$A' = \begin{bmatrix} A(2,2) & -A(1,2) \\ -A(2,1) & A(1,1) \end{bmatrix} = \begin{bmatrix} 3 & -2 \\ -7 & 4 \end{bmatrix}, \det[A] = 4x3 - 7x2 = -2,$$

$$A^{-1} = \frac{A'}{\det[A]} = \frac{\begin{bmatrix} 3 & -2 \\ -7 & 4 \end{bmatrix}}{-2} = \begin{bmatrix} \frac{3}{-2} & \frac{-2}{-2} \\ \frac{-7}{-2} & \frac{4}{-2} \end{bmatrix} = \begin{bmatrix} -\frac{3}{2} & 1 \\ \frac{7}{2} & -2 \end{bmatrix}$$

For matrices bigger than 2×2 size, other methods are used to calculate the inverse. Most well-known of these methods is the Adjoint method. In the Adjoint method, a Cofactor matrix *Cof*(A) is created first by using the determinants of the submatrices derived from A matrix. The (i, j) element of the Cofactor matrix *Cof*(A) is obtained from the determinant of the submatrix obtained by deleting i-th row and j-th column of A matrix. Then, Adjoint matrix is obtained by taking the transpose of Cofactor matrix. Finally, Adjoint matrix is divided by the determinant of A matrix leading to the inverse of A matrix.

To explain the inverse calculation of bigger matrices, consider an A matrix of 3×3 size as below:

$$A = \begin{bmatrix} A(1,1) & A(1,2) & A(1,3) \\ A(2,1) & A(2,2) & A(2,3) \\ A(3,1) & A(3,2) & A(3,3) \end{bmatrix}$$

Cofactor matrix $Cof(A)$ can be obtained using the submatrices of A matrix as following:

$$Cof(A) = \begin{bmatrix} (-1)^{1+1}\begin{vmatrix} A(2,2) & A(2,3) \\ A(3,2) & A(3,3) \end{vmatrix} & (-1)^{1+2}\begin{vmatrix} A(2,1) & A(2,3) \\ A(3,1) & A(3,3) \end{vmatrix} & (-1)^{1+3}\begin{vmatrix} A(2,1) & A(2,2) \\ A(3,1) & A(3,2) \end{vmatrix} \\ (-1)^{2+1}\begin{vmatrix} A(1,2) & A(1,3) \\ A(3,2) & A(3,3) \end{vmatrix} & (-1)^{2+2}\begin{vmatrix} A(1,1) & A(1,3) \\ A(3,1) & A(3,3) \end{vmatrix} & (-1)^{2+3}\begin{vmatrix} A(2,1) & A(2,1) \\ A(3,1) & A(3,2) \end{vmatrix} \\ (-1)^{3+1}\begin{vmatrix} A(1,2) & A(1,3) \\ A(2,2) & A(2,3) \end{vmatrix} & (-1)^{3+2}\begin{vmatrix} A(1,1) & A(1,3) \\ A(3,1) & A(3,3) \end{vmatrix} & (-1)^{3+3}\begin{vmatrix} A(1,1) & A(1,2) \\ A(2,1) & A(2,2) \end{vmatrix} \end{bmatrix}$$

$$= \begin{bmatrix} A(2,2)A(3,3) - A(3,2)A(2,3) & -(A(2,1)A(3,3) - A(3,1)A(2,3)) & A(2,1)A(3,2) - A(3,1)A(2,2) \\ -(A(1,2)A(3,3) - A(3,2)A(1,3)) & A(1,1)A(3,3) - A(3,1)A(1,3) & -(A(2,1)A(3,2) - A(3,1)A(2,1)) \\ A(1,2)A(2,3) - A(2,2)A(1,3) & -(A(1,1)A(3,3) - A(3,1)A(1,3)) & A(1,1)A(2,2) - A(2,1)A(1,2) \end{bmatrix}$$

In general form, coefficient in the Cofactor matrix corresponding to i-th row and j-th column can be expressed as $(-1)^{i+j}$. Using the transpose of the Cofactor matrix, Adjoint matrix $Adj(A)$ can be obtained as below:

$$Adj(A) = transpose(Cof(A))$$

$$= \begin{bmatrix} A(2,2)A(3,3) - A(3,2)A(2,3) & -(A(1,2)A(3,3) - A(3,2)A(1,3)) & A(1,2)A(2,3) - A(2,2)A(1,3) \\ -(A(2,1)A(3,3) - A(3,1)A(2,3)) & A(1,1)A(3,3) - A(3,1)A(1,3) & -(A(1,1)A(3,3) - A(3,1)A(1,3)) \\ A(2,1)A(3,2) - A(3,1)A(2,2) & -(A(2,1)A(3,2) - A(3,1)A(2,1)) & A(1,1)A(2,2) - A(2,1)A(1,2) \end{bmatrix}$$

Finally, inverse of A matrix is obtained by dividing the Adjoint matrix with the determinant of A matrix as following:

$$A^{-1} = \frac{Adj(A)}{|A|}$$

Example 2.25: Calculate the inverse of $A = \begin{bmatrix} 3 & 1 & 2 \\ 5 & 6 & 7 \\ 9 & 2 & 8 \end{bmatrix}$ matrix.

In calculating the inverse of A matrix, determinant of A and Adjoint matrix are needed. They can be calculated as below:

$$\text{Determinant of } A: |A| = (-1)^{1+1} 3 \underbrace{\begin{vmatrix} 6 & 7 \\ 2 & 8 \end{vmatrix}}_{Submatrix\bar{1}} + (-1)^{1+2} 1 \underbrace{\begin{vmatrix} 5 & 7 \\ 9 & 8 \end{vmatrix}}_{Submatrix\bar{2}} + (-1)^{1+3} 2 \underbrace{\begin{vmatrix} 5 & 6 \\ 9 & 2 \end{vmatrix}}_{Submatrix\bar{3}} \rightarrow$$

$$|A| = 3[6x8 - 2x7] - 1[5x8 - 9x7] + 2[5x2 - 9x6] = 37$$

Cofactor matrix *Cof(A)* is obtained using the submatrices of A as following:

$$Cof(A) = \begin{bmatrix} (-1)^{1+1}\begin{vmatrix} 6 & 7 \\ 2 & 8 \end{vmatrix} & (-1)^{1+2}\begin{vmatrix} 5 & 7 \\ 9 & 8 \end{vmatrix} & (-1)^{1+3}\begin{vmatrix} 5 & 6 \\ 9 & 2 \end{vmatrix} \\ (-1)^{2+1}\begin{vmatrix} 1 & 2 \\ 2 & 8 \end{vmatrix} & (-1)^{2+2}\begin{vmatrix} 3 & 2 \\ 9 & 8 \end{vmatrix} & (-1)^{2+3}\begin{vmatrix} 3 & 1 \\ 9 & 2 \end{vmatrix} \\ (-1)^{3+1}\begin{vmatrix} 1 & 2 \\ 6 & 7 \end{vmatrix} & (-1)^{3+2}\begin{vmatrix} 3 & 2 \\ 5 & 7 \end{vmatrix} & (-1)^{3+3}\begin{vmatrix} 3 & 1 \\ 5 & 6 \end{vmatrix} \end{bmatrix} = \begin{bmatrix} 34 & 23 & -44 \\ 4 & 6 & 3 \\ -5 & -11 & 13 \end{bmatrix}$$

Adjoint matrix is expressed as the transpose of the Cofactor matrix as below:

$$Adj(A) = Transpoze(Cof(A)) = \begin{bmatrix} 34 & 4 & -5 \\ 23 & 6 & -11 \\ -44 & 3 & 13 \end{bmatrix}$$

Finally, inverse of A matrix is obtained by dividing the Adjoint matrix with the determinant of A as following:

$$A^{-1} = \frac{Adj(A)}{|A|} = \frac{1}{37}\begin{bmatrix} 34 & 4 & -5 \\ 23 & 6 & -11 \\ -44 & 3 & 13 \end{bmatrix} = \begin{bmatrix} 0.9189 & -0.1081 & -0.1351 \\ 0.6216 & 0.1622 & -0.2973 \\ -1.1892 & 0.0811 & 0.3514 \end{bmatrix}$$

Solution with MATLAB command:

Inverse of A matrix can be obtained using the *inv* command in MATLAB as below:

```
>> A=[3 1 2; 5 6 7; 9 2 8];
>> inv(A)
ans =
   0.9189  -0.1081  -0.1351
   0.6216   0.1622  -0.2973
  -1.1892   0.0811   0.3514
```

2.6 Summary

- If an $f(x)$ function have indeterminate value such as $f(x^*) = \frac{0}{0}$ or $f(x^*) = \frac{\infty}{\infty}$ for $x = x^*$, removing uncertainty and obtaining function value when x is very close to x^* is called limit operation or limit in mathematics.
- Hopital's rule is widely used in limit calculation. In Hopital's rule, derivatives of numerator and denominator are taken and then x value is substituted into the resulting expression to obtain the limit value of $f(x)$ function.
- Derivatives express rate of changes in function values. First derivative of an $f(x)$ function expresses its slope.
- Derivative of a polynomial function $f(x) = x^n$ is expressed by writing the exponent as the premultiplier coefficient and decreasing the exponent by one as $f'(x) = nx^{n-1}$.
- Derivatives of trigonometric functions can be expressed in terms of each other. For example, derivative of *sine* function can be expressed as *cosine* function, and derivative of co*sine* function is expressed as $-$*sine* function.
- Derivative of exponential function $f(x) = e^x$ is expressed by itself as $f'(x) = e^x$.
- Derivative of logarithmic function $f(x) = \ln(x)$ is expressed as $f'(x) = \frac{1}{x}$. Derivative of general logarithmic function $f(x) = \ln(u(x))$ including the inner function $u(x)$ is expressed as $f'(x) = \frac{u'(x)}{u(x)}$.
- Derivative of the product function $h(x) = f(x)g(x)$ can be expressed as the sum of derivative of $f(x)$ multiplied by $g(x)$ and derivative of $g(x)$ multiplied by $f(x)$ as $h'(x) = f'(x)g(x) + f(x)g'(x)$.
- Derivative of division function $h(x) = \frac{f(x)}{g(x)}$ is expressed as $h'(x) = \frac{f'(x)g(x) - f(x)g'(x)}{g(x)^2}$.
- Derivative of nested function $z(x) = f(g(h(x)))$ can be expressed using chain rule as $z'(x) = \frac{dz(x)}{dx} = \frac{dz(x)}{dg(x)} \frac{dg(x)}{dh(x)} \frac{dh(x)}{dx}$.
- Integral of an $f(x)$ function represents the area under it.
- Integral of polynomial function $f(x) = x^n$ is expressed as $I = \int x^n dx = \frac{x^{n+1}}{n+1}$.
- Integrals of trigonometric functions can be expressed in terms of each other. For example, integrals of *sine*(*sin*) and *cosine* (*cos*) functions are expressed as below:

$$I = \int \sin(x)dx = -\cos(x), I = \int \cos(x)dx = \sin(x)$$

- Integral of exponential function $f(x) = e^x$ is expressed as $I = \int e^x dx = e^x$.
- Some variables contain more than one data of the same type and they are referred to as vectors or matrices. Vectors contain data only in row or column form while matrices contain data both in row and column form.
- Norm or magnitude of a vector such as $A = [A(1) \quad A(2) \quad A(3)]$ is expressed as square root of sum of squares of its members as following:

$$\text{Norm} : ||A|| = \sqrt{A(1)^2 + A(2)^2 + A(3)^2}$$

- For addition and subtraction of matrices, their dimensions must be compatible (their corresponding row and column numbers must be the same).
- Addition or subtraction of matrices is performed element by element as following:

$$C = A + B = \begin{bmatrix} A(1,1) & A(1,2) & A(1,3) \\ A(2,1) & A(2,2) & A(2,3) \end{bmatrix} + \begin{bmatrix} B(1,1) & B(1,2) & B(1,3) \\ B(2,1) & B(2,2) & B(2,3) \end{bmatrix}$$
$$= \begin{bmatrix} A(1,1)+B(1,1) & A(1,2)+B(1,2) & A(1,3)+B(1,3) \\ A(2,1)+B(2,1) & A(2,2)+B(2,2) & A(2,3)+B(2,3) \end{bmatrix}$$

$$C = A - B = \begin{bmatrix} A(1,1) & A(1,2) & A(1,3) \\ A(2,1) & A(2,2) & A(2,3) \end{bmatrix} - \begin{bmatrix} B(1,1) & B(1,2) & B(1,3) \\ B(2,1) & B(2,2) & B(2,3) \end{bmatrix}$$
$$= \begin{bmatrix} A(1,1)-B(1,1) & A(1,2)-B(1,2) & A(1,3)-B(1,3) \\ A(2,1)-B(2,1) & A(2,2)-B(2,2) & A(2,3)-B(2,3) \end{bmatrix}$$

- In order to multiply two matrices such as *A* and *B*, internal dimensions of these matrices must be the same. That is, column number of *A* matrix must be equal to the row number of *B* matrix.
- Matrix multiplication of *A***B* is obtained with multiplication of rows of *A* with columns of *B* as following:

$$C = \begin{bmatrix} C(1,1) & C(1,2) \\ C(2,1) & C(2,2) \end{bmatrix} = AxB = \begin{bmatrix} A(1,1) & A(1,2) & A(1,3) \\ A(2,1) & A(2,2) & A(2,3) \end{bmatrix} \begin{bmatrix} B(1,1) & B(1,2) \\ B(2,1) & B(2,2) \\ B(3,1) & B(3,2) \end{bmatrix}$$

$$C = \begin{bmatrix} A(1,1)B(1,1)+A(1,2)B(2,1)+A(1,3)B(3,1) & A(1,1)B(1,2)+A(1,2)B(2,2)+A(1,3)B(3,2) \\ A(2,1)B(1,1)+A(2,2)B(2,1)+A(2,3)B(3,1) & A(2,1)B(1,2)+A(2,2)B(2,2)+A(2,3)B(3,2) \end{bmatrix}$$

- Determinant represents characteristic value of a matrix. Determinants of a 2 × 2 matrix such as $A = \begin{bmatrix} A(1,1) & A(1,2) \\ A(2,1) & A(2,2) \end{bmatrix}$ can be calculated in a simple way as $|A| = A(1,1)A(2,2) - A(2,1)A(1,2)$. However, for determinant of bigger matrices, determinants of submatrices are used.
- Inverse of a 2 × 2 matrix such as $A = \begin{bmatrix} A(1,1) & A(1,2) \\ A(2,1) & A(2,2) \end{bmatrix}$ can be calculated by replacing diagonal terms and their signs and then dividing the determinant value as $A^{-1} = \frac{A'}{|A|} = \frac{\begin{bmatrix} A(2,2) & -A(1,2) \\ -A(2,1) & A(1,1) \end{bmatrix}}{|A|}$. However, for inverse of bigger matrices, more sophisticated methods based on determinants of submatrices are used.

Appendix: MATLAB Programs

MATLAB Program 2.1: Norm of Vector

```
function Norm_of_Vector

% THIS PROGRAM CALCULATES NORM OF A VECTOR

A=[3  5  7  9] ;

% Number of elements in Vector (Vector length): n
n=length(A) ;

% Norm Value: NormValue
Sum=0 ;
for i=1:n
    Sum=Sum+A(i)^2 ;
end
NormValue=sqrt(Sum) ;

fprintf('Norm Value= %f \n', NormValue) ;
```

MATLAB Program 2.2: Addition of Matrices

```
function MatrixAddition

% THIS PROGRAM CALCULATES ADDITION OF A AND B MATRICES (C=A+B)

A=[1  2  3 ;
   4  5  6 ;
   7  8  9 ] ;

B=[1  0  2 ;
   0  2  9 ;
   3  6  4 ] ;

% Matrix dimension (m: row number, n: column number)
[m,n]=size(A) ;

% C=A+B
for i=1:m
    for j=1:n
        C(i,j)=A(i,j)+B(i,j) ;
    end
end

fprintf('Addition Matrix: C \n') ;
C
```

MATLAB Program 2.3: Subtraction of Matrices

```
function MatrixSubtraction
% THIS PROGRAM CALCULATES SUBTRACTION OF B MATRIX FROM A MATRIX (C=A-B)

A=[1  2  3 ;
   4  5  6 ;
   7  8  9 ] ;

B=[1  0  2 ;
   0  2  9 ;
   3  6  4 ] ;

% Matrix dimension (m: row number, n: column number)
[m,n]=size(A) ;

% C=A-B
for i=1:m
    for j=1:n
        C(i,j)=A(i,j)-B(i,j) ;
    end
end

fprintf('Subtraction Matrix: C \n') ;
C
```

MATLAB Program 2.4: Multiplication of Matrices

```
function MatrixMultiplication
% THIS PROGRAM CALCULATES MULTIPLICATION OF A AND B MATRICES (C=A*B)

A=[2  3 ;
   1  4 ] ;

B=[5  1  4 ;
   3  6  2 ] ;

% Matrix dimensions (m: row number, n: column number)
[mA,nA]=size(A) ; % Dimension of A matrix
[mB,nB]=size(B) ; % Dimension of B matrix

% For matrix multiplication column number of A matrix must be equal
% to row number of B matrix
% Dimension check for matrix multiplication
if nA~=mB
    error('Column number of A is not equal to row number of B')
end

% C=A*B
for i=1:mA
    for j=1:nB
        C(i,j)=0 ;
        for k=1:nA
           C(i,j)=C(i,j)+A(i,k)*B(k,j) ;
        end
    end
end

fprintf('Multiplication Matrix: C \n') ;
C
```

Problems

Problem 2.1: Calculate the limit value of $f(x) = \frac{5x+6}{x-4}$ function for x $\to \infty$ (i.e., when x approaches ∞).

Problem 2.2: Calculate the limit value of $f(x) = \frac{x^2-7x+12}{x-3}$ function for x $\to$ 3 (i.e., when x approaches 3) using Hopital's rule.

Problem 2.3: Express the derivative of $f(x) = 3x^4 - x^2 + \frac{1}{\sqrt{x}}$ function.

Problem 2.4: Express the derivative of $f(x) = \sin(2x^3 + \sqrt{x}) + \cos\left(3x^2 - \frac{1}{2\sqrt{x}}\right)$ function.

Problem 2.5: Express the derivative of $f(x) = e^{\left(3x^4 - x^2 + \frac{1}{\sqrt{x}}\right)}$ function.

Problem 2.6: Express the derivative of $f(x) = \ln\left(e^{x^3 - 2x^2 + \frac{1}{\sqrt{x}}}\right)$ function.

Problem 2.7: Express the derivative of $h(x) = \sin(3x - 1)e^{\left(x^3 - 2x^2 + \frac{1}{\sqrt{x}}\right)} + \cos(2x + 4)e^{\left(-x^3\right)}$ function.

Problem 2.8: Express the derivative of $h(x) = \tan(x) \cot x$ function.

Problem 2.9: Express the derivative of $z(x) = e^{\cos\left(x^2 - 2x + 1\right)}$ function using the chain rule.

Problem 2.10: Calculate $I = \int (3x^4 + \sqrt{x})dx$ integral.

Problem 2.11: Calculate $I = \int (\sin(x))^3 \cos(x)dx$ integral.

Problem 2.12: Calculate $I = \int e^{(x^2+2)} 2xdx$ integral.

Problem 2.13: Calculate $I = \int_0^{\pi/2} \frac{\sin x}{\sqrt{1 + (\tan x)^2}} dx$ integral.

Problem 2.14: Calculate $I = \int_0^{\pi/2} \left(2(\sin x)^2 \sqrt{1 + (\cot x)^2}\right) dx$ integral.

Problem 2.15: Calculate $I = \int_0^1 3x^2 \tan x \cot x dx$ integral.

Problem 2.16: Calculate the norm of $A = \begin{bmatrix} 3 & 5 & 7 & 9 \end{bmatrix}$ vector.

Problem 2.17: A and B matrices are given as $A = \begin{bmatrix} 3 & 0 & 2 \\ 6 & 5 & 3 \\ 9 & 8 & 7 \end{bmatrix}$, $B = \begin{bmatrix} 5 & 6 & 4 \\ 3 & 2 & 9 \\ 7 & 6 & 8 \end{bmatrix}$.
Find the result of $2*A - 3*B$.

Problem 2.18: A and B matrices are given as $A = \begin{bmatrix} 1 & 6 & 4 \\ 4 & 2 & 9 \\ 5 & 6 & 8 \end{bmatrix}$, $B = \begin{bmatrix} 3 & 5 & 2 \\ 7 & 0 & 3 \\ 9 & 8 & 6 \end{bmatrix}$.
Obtain the multiplication (product) of A and B matrices.

Problem 2.19: Calculate the determinant of $A = \begin{bmatrix} 3 & 0 & 2 \\ 6 & 5 & 3 \\ 9 & 8 & 7 \end{bmatrix}$ matrix.

Problem 2.20: Calculate the inverse of $A = \begin{bmatrix} 5 & 6 & 4 \\ 3 & 2 & 9 \\ 7 & 6 & 8 \end{bmatrix}$ matrix.

References

1. Thomas, G. B. (2006). *Thomas calculus early transcendentals*. Pearson.
2. Morris, C. C., & Stark, R. M. (2015). *Fundamentals of calculus*. John Wiley & Sons.
3. Velleman, D. J. (2017). *Calculus: A rigorous first course*. Dover Publications.
4. McMullen, C. (2018). *Essential calculus skills*. Zishka Publishing.

Chapter 3
Sources of Error in Engineering Calculations

3.1 Introduction

Errors in the expression and solution of engineering problems can be grouped as (1) Modeling Error, (2) Numerical Error, and (3) Iteration Error (Method error) [1–8]. Therefore, the total error made in solving an engineering problem will be the sum of these errors. These errors can be controlled or reduced by following some rules. Errors in engineering calculations are explained in more detail below.

3.2 Modeling Error

This type of error occurs due to simplifying assumptions for the expression and modeling of engineering problems. Some simple examples of modeling errors are illustrated below.

3.2.1 Area of Circle

The area of a circle can be expressed with the addition of triangular areas as in Fig. 3.1. In this representation, the area of the circle will differ according to the number of triangles. Large number of triangles will represent the circle very closely yielding the accurate area of the circle. For example, among the three models in Fig. 3.1, four triangles (Model 1) will yield the most inaccurate area for the circle, while six triangles (Model 3) will yield the most accurate one. Therefore, increasing the number of triangles will reduce the modeling error.

H. Kurtaran, *Engineering Mathematics with MATLAB Applications: Vol. 1, Fundamentals*, https://doi.org/10.1007/978-3-031-81869-1_3

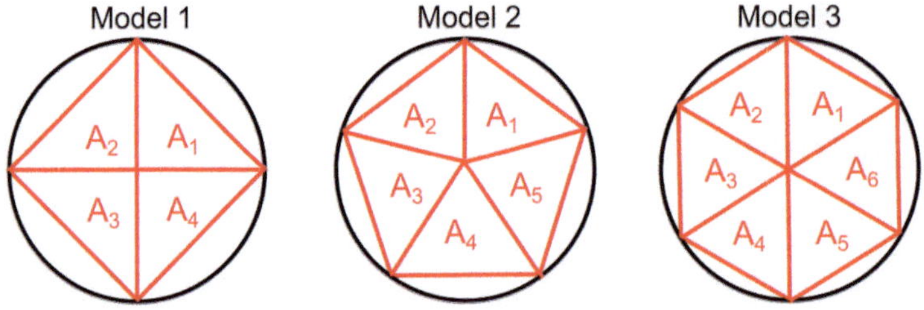

Fig. 3.1 Representation of the area of circle with addition of triangular areas

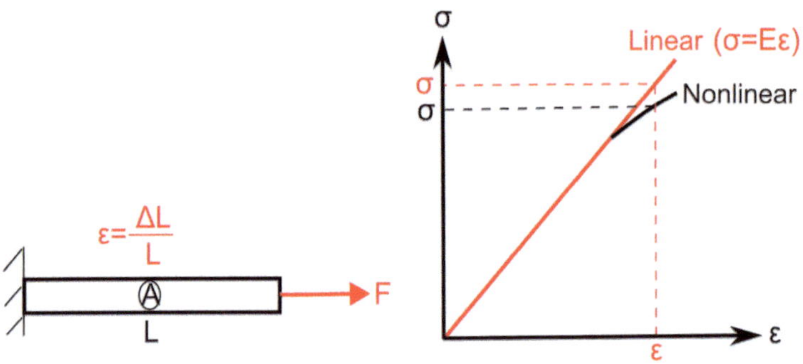

Fig. 3.2 Linear (approximate) and nonlinear (accurate) stress–strain curves for the bar loaded with force *F* at the free edge

3.2.2 *Stress in Bars*

Consider the bar loaded with force *F* at the free edge as shown in Fig. 3.2. When *F* is high, axial stress–strain ($\sigma - \varepsilon$) curve for this bar becomes nonlinear. However, the linear stress–strain curve known as *Hooke's Law* is generally used in engineering calculations. Therefore, the use of a linear stress–strain curve instead of a nonlinear one introduces modeling error in engineering applications when the applied force *F* is high.

3.2.3 *Mass-Spring-Damper Model of Vehicles*

In vehicle dynamics, mass-spring-damper models as shown in Fig. 3.3 are commonly used as simplified models to examine the driving behavior of vehicles. Two models are shown in Fig. 3.3 as ride models of vehicles. In Model 1, the entire vehicle is represented with a single mass-spring-damper model. In Model 2, vehicle is modeled in more detail using additional mass-spring-damper models for representing upper part (body) and lower parts (axle and wheels). Therefore,

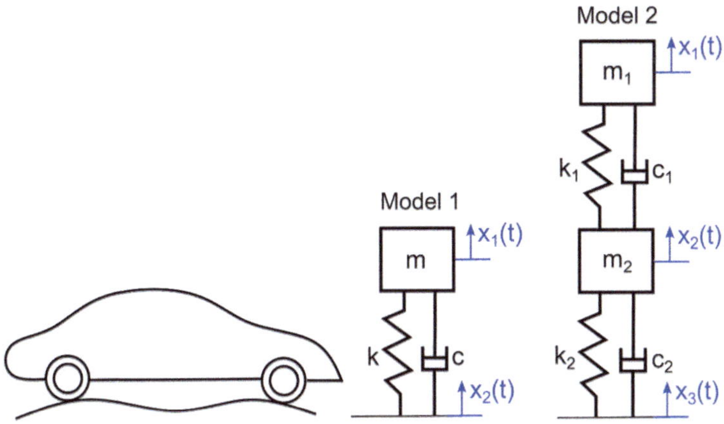

Fig. 3.3 Mass-spring-damper models to examine ride behavior of vehicles

Model 2 will contain less modeling error than Model 1 and will yield more accurate ride results.

Modeling errors as shown in the examples above can be reduced or avoided by including more details.

3.3 Numerical Errors

Numerical errors occur as a result of mathematical operations during the solution of engineering problems. These errors are divided into two groups: (1) Rounding errors and (2) Truncation errors.

3.3.1 Rounding Errors

In computer calculations, a certain number of digits are reserved for each variable in the computer memory. If the calculated value occupies a bigger space than allocated digits: (a) either the extra digits in the value are *trimmed/chopped* or (b) the value is rounded to fit the allocated digits in the memory. For example, $\sqrt{2}$ and π values will require an infinite number of digits for exact representation. However, due to limitation of computer memory, rounding or trimming must be applied to these values to store them in computer memory. Upon rounding and trimming, representation of π value in computer memory for varying digits is shown in Table 3.1.

Mathematical operations also may lead to new variables that do not fit to the allocated memory. For example, if x variable is given as $x = 0.0001$, squaring x will lead to the eight-digit value of $x^2 = 0.00000001$. If only four digits are available in

Table 3.1 Rounding and trimming of π value for varying memory digits

$\pi = 3.142857142857143\ldots$			
Rounded π (π_R)	Rounding Error ($\pi - \pi_R$)	Trimmed π (π_T)	Trimming Error ($\pi - \pi_T$)
3.14285714285714	−0.001264489267347	3.14285714285714	−0.001264489267347
3.1428571428571	−0.001264489267207	3.1428571428571	−0.001264489267307
3.142857142857	−0.001264489267307	3.142857142857	−0.001264489267207
3.14285714286	−0.001264489270207	3.14285714285	−0.001264489260207
3.1428571429	−0.001264489310207	3.1428571428	−0.001264489210207
3.142857143	−0.001264489410207	3.142857142	−0.001264488410207
3.14285714	−0.001264486410207	3.14285714	−0.001264486410207
3.1428571	−0.001264446410207	3.1428571	−0.001264446410207
3.142857	−0.001264346410207	3.142857	−0.001264346410207
3.14286	−0.001267346410207	3.14285	−0.001257346410207
3.1429	−0.001307346410207	3.1428	−0.001207346410207
3.143	−0.001407346410207	3.142	−0.000407346410207
3.14	0.001592653589793	3.14	0.001592653589793
3.1	0.041592653589793	3.1	0.041592653589793
3	0.141592653589793	3	0.141592653589793

the computer memory to store a variable, x^2 value will be stored as $x^2 =$

$0.0000\overbrace{\underbrace{0001}_{\text{Excessive digits}}}^{\text{Deleted}} \rightarrow x^2 = 0.0000$ upon rounding or trimming. This will cause rounding or trimming error (difference between the true and the rounded values) of 0.00000001.

Rounding error can be reduced or controlled by using double precision in mathematical operations or using high-precision computers such as workstations that have more allocated digits for variables.

Example 3.1: Assuming four digits can be allocated to a variable in computer memory, calculate $y = x^2 + \frac{x}{10^3} + 2000x$ value for $x = 0.0001$. Determine true and rounded values as well as rounding error for y.

By taking $x = 0.0001$, y is calculated as below:

$$\begin{aligned} y &= x^2 + \frac{x}{10^3} + 2000x = (0.0001)^2 + 0.0001/1000 + (2000x0.0001) \\ &= 0.00000001 + 0.0000001 + 0.2 \\ &= 0.20000011 \end{aligned}$$

If four digits can be allocated in computer memory to store variables, y variable is stored as $y = 0.2000$ upon rounding. In this case, rounding error is expressed as following:

$$\text{Rounding Error} = \text{True Value} - \text{Rounded Value} = 0.20000011 - 0.2000 \\ = 0.00000011$$

3.3.2 Storing Variables in Computers

Computers use binary system consisting of 0 and 1 numbers to store variables in the memory. Variables are represented with a certain number of digits (i.e., bits) called as words in the binary system. Digit and bit are interchangeably used in binary system to indicate members of a variable or word. These words are represented by powers of 2-bit such as 4-bit, 8-bit, 16-bit, 32-bit, and 64-bit. Mathematical operations on variables often lead to excessive digits in results. Computers with high bit capacity allow excessive digits to be stored without rounding or trimming. However, those with small bit capacity require excessive digits to be cut yielding rounding or trimming error.

Conversion of Binary Integers into Decimal Value

In binary system, integer variables are represented with words consisting of 0 and 1 bits such as 01101. A binary integer expressed with d_i coefficients can be converted into decimal value by multiplying bit numbers with varying powers of 2 as following:

$$x = \underbrace{d_{n-1}d_{n-2}\ldots d_1 d_0}_{\text{Bits}} \Rightarrow x = \underbrace{d_{n-1}2^{n-1} + d_{n-2}2^{n-2} + \ldots + d_1 2^1 + d_0 2^0}_{\text{Decimal Value}}$$

For example, a word x represented with 1011 bits can be converted into decimal value as $x = 1 \times 2^3 + 0 \times 2^2 + 1 \times 2^1 + 1 \times 2^0 = 11$.

In binary representation, the first digit in words is reserved for the sign of the variable. For example, when n digit is allocated for variables, first digit is used for the sign and remaining n-1 bits are used to indicate the value of the variable. Therefore, minimum decimal value of a binary integer consisting of n bits is expressed with $x = -2^{n-1}$ and maximum one is expressed with $x = 2^{n-1} - 1$. Table 3.2 shows limit values of binary integers.

Table 3.2 Minimum and maximum decimal values of binary integers

Bit (n)	Number range $(-2^{n-1}, 2^{n-1} - 1) = (\min, \max)$
4	$(-2^{4-1}, 2^{4-1} - 1) = (-8, 7)$
8	$(-2^{8-1}, 2^{8-1} - 1) = (-128, 127)$
16	$(-2^{16-1}, 2^{16} - 1) = (-32{,}768, 32{,}767)$
32	$(-2^{32-1}, 2^{32-1} - 1) = (-2{,}147{,}483{,}648, 2{,}147{,}483{,}647)$
64	$(-2^{64-1}, 2^{64-1} - 1) = (-9{,}223{,}372{,}036{,}854{,}775{,}808, 9{,}223{,}372{,}036{,}854{,}775{,}807)$
128	$(-2^{128-1}, 2^{128-1} - 1)$

If decimal value of an integer exceeds (min, max) range in Table 3.2, it cannot be represented exactly with binary system.

Conversion of Binary Fractionals into Decimal Values

Fractional numbers are more difficult to express than integers. Fractional numbers can be expressed in binary system using (a) fixed point and (b) floating point methods.

(a) Representation of fractional numbers with fixed point method:
In this method, fractional numbers are expressed in binary system as $x = \underbrace{d_{n-1}d_{n-2}\ldots d_1 d_0}_{\text{Integer part}}.\underbrace{d_{-1}d_{-2}\ldots d_{-m}}_{\text{Fractional part}}$, where $d_1, d_2, \ldots, d_n$ correspond to digits of 0 and 1. Decimal value of a binary fractional variable is calculated by expressing the fractional part with decreasing powers of 2 as following:

$$x = \underbrace{d_{n-1}2^{n-1} + d_{n-2}2^{n-2} + \ldots + d_1 2^1 + d_0 2^0}_{\text{Integer part}} + \underbrace{d_{-1}2^{-1} + d_{-2}2^{-2} + \ldots + d_{-m}2^{-m}}_{\text{Fractional part}} \tag{3.1}$$

Example 3.2: Calculate the decimal value of binary variable $x = 1011.101$ using fixed point method.

Decimal value of binary variable x expressed as $x = 1011.101 = \underbrace{d_3 d_2 d_1 d_0}_{\text{Integer part}}.\underbrace{d_{-1}d_{-2}d_{-3}}_{\text{Fractional part}}$ can be calculated using Eq. 3.1 as following:

$$\begin{aligned} x &= \underbrace{d_3 2^3 + d_2 2^2 + d_1 2^1 + d_0 2^0}_{\text{Integer part}} + \underbrace{d_{-1}2^{-1} + d_{-2}2^{-2} + d_{-3}2^{-3}}_{\text{Fractional part}} \\ &= \underbrace{1 \times 2^3 + 0 \times 2^2 + 1 \times 2^1 + 1 \times 2^0}_{\text{Integer part}} + \underbrace{1 \times 2^{-1} + 0 \times 2^{-2} + 1 \times 2^{-3}}_{\text{Fractional part}} \\ &= 8 + 0 + 2 + 1 + 0.5 + 0 + 0.125 = 11.625 \end{aligned}$$

Table 3.3 Floating point representation of fractional numbers in the binary system

		Integer part				Fractional part			
Sign	Exponent	2^n	...	2^1	2^0	2^{-1}	2^{-2}	...	2^{-m}
0 or 1	0 or 1	d_n	...	d_1	d_0	d_{-1}	d_{-2}	...	d_{-m}

(b) Representation of fractional numbers with floating point method:
Floating point notation has been developed to store very big and very small numbers (floating numbers) efficiently in computer memory. In this notation, a fractional variable is represented as $x =$ (Fractional part) $\times$ (Base)^{Exponent}. In floating point notation, a number is expressed to have a single digit before the fraction part. For example, number 25.34 can be represented with floating point system as 0.2534×10^2.

Binary fractional numbers are stored in computer memory as shown in Table 3.3 by assigning separate bits for the sign, exponent as well as integer and fractional parts. In binary system, first bit is used for the sign, next bit is used for the sign of the exponent and rest of the bits are used to represent the value of fractional number. In the binary system, positive and negative signs are represented with 0 and 1 bits, respectively.

For example, if an x variable is expressed with 11101010 digits (bits) in binary system, its floating point representation can be expressed as following:

$$x = \overbrace{\underbrace{1}_{\text{Variable}}\ \underbrace{1}_{\text{Exponent}}}^{\text{Sign}}\ \overbrace{\underbrace{1}_{d_1}\ \underbrace{0}_{d_0}}^{\text{Integer part}}\ \overbrace{\underbrace{1}_{d_{-1}}\ \underbrace{0}_{d_{-2}}\ \underbrace{1}_{d_{-3}}\ \underbrace{0}_{d_{-4}}}^{\text{Fractional part}} \rightarrow$$

$$x = \underbrace{-}_{\text{Variable Sign}} \underbrace{\left(1\times 2^{-1} + 0\times 2^{-2} + 1\times 2^{-3} + 0\times 2^{-4}\right)}_{\text{Fractional part}} \times 2^{\underbrace{-}_{\text{Exponent Sign}} \overbrace{\left(1\times 2^{1} + 0\times 2^{0}\right)}^{\text{Integer part}}} = -0.625\times 2^{-2}$$

In the above representation, sign of the fractional variable and its exponent are represented with digits of 1. This ensures that the sign and exponent of the fractional variable are negative.

Conversion of Integer Value to Binary System

Decimal value of an integer variable can be converted to binary form using remainders upon successive division of the variable by 2. For example, decimal value of 20 can be expressed in binary system using the remainders in Table 3.4 as following:

$$20 = 10100 \rightarrow 20 = 1 \times 2^4 + 0 \times 2^3 + 1 \times 2^2 + 0 \times 2^1 + 0 \times 2^0$$

3.3.3 *Truncation Error*

Truncation error is encountered in functions expressed with series. In engineering solutions, functions are often expressed with infinite series for convenience of mathematical calculations. Taylor, Maclaurin, and Fourier series are commonly used for this purpose. For example, an *f*(*x*) function can be expressed with Taylor series expansion using function and derivative values at an x_0 point as following:

$$\begin{aligned} f(x) &= f(x_0) + f'(x_0)(x - x_0) + \frac{f''(x_0)}{2!}(x - x_0)^2 + \ldots + \frac{f^{(n)}(x_0)}{n!}(x - x_0)^n + \ldots \\ &= \sum_{n=0} \frac{f^{(n)}(x_0)}{n!}(x - x_0)^n \end{aligned} \tag{3.2}$$

When $x_0 = 0$ is adopted, Taylor Series is called as Maclaurin Series and it can be expressed as below:

$$f(x) = f(0) + f'(0)x + \frac{f''(0)}{2!}x^2 + \frac{f''(0)}{3!}x^3 + \ldots + \frac{f^{(n)}(0)}{n!}x^n + \ldots = \sum_{n=0} \frac{f^{(n)}(0)}{n!}x^n \tag{3.3}$$

Since Taylor and Maclaurin Series have infinite terms, they have to be cut (truncated) after a certain number of terms in practical use. Truncated or neglected terms cause truncation errors. For example, when Maclaurin series is expressed with only three terms, truncation error can be expressed as following:

Maclaurin series with three terms: $f(x) = f(0) + f'(0)x + \frac{f''(0)}{2!}x^2$

Table 3.4 Expressing decimal value of 20 in binary form using remainders

Number/2	Quotient	Remainder
20/2	10	0
10/2	5	0
5/2	2	1
2/2	1	0
1/2	0.5 → 0 (taking integer part)	1

Truncation error: $\frac{f''(0)}{3!}x^3 + \ldots + \frac{f^{(n)}(0)}{n!}x^n + \ldots = \sum_{n=0} \frac{f^{(n)}(0)}{n!}x^n$

Truncation error can be reduced by taking more terms in the series expression. However, this will cause more mathematical operations and thereby more rounding or trimming errors due to an increasing number of terms. In short, increasing the number of terms in series decreases truncation errors but increases rounding errors.

Example 3.3: Express $f(x) = e^x$ function with Maclaurin series.

Derivatives of $f(x) = e^x$ function can be expressed as $f'(x) = f''(x) = f'''(x) = \ldots = f^{(n)}(x) = e^x$. Using these expressions, derivative values at $x_0 = 0$ point are obtained as $f(0) = f'(0) = f''(0) = f'''(0) = \ldots = f^{(n)}(0) = 1$. Thus, substituting the function and derivative values into Eq. 3.3 $f(x) = e^x$ function can be expressed with Maclaurin series as below:

$$e^x = 1 + x + \frac{x^2}{2!} + \frac{x^3}{3!} + \frac{x^4}{4!} + \ldots + \frac{x^n}{n!} = \sum_{n=0} \frac{x^n}{n!} = \sum_{n=1} \frac{x^{n-1}}{(n-1)!}$$

Example 3.4: Express $f(x) = \sin x$ function with Maclaurin series.

Derivatives of $f(x) = \sin x$ function can be expressed as below:

$f'(x) = \cos x, f''(x) = -\sin x, f'''(x) = -\cos x, f^{(4)}(x) = \sin x, \ldots,$

Using these expressions, derivative values at $x_0 = 0$ are obtained as $f(0) = \sin 0 = 0$, $f'(0) = \cos 0 = 1, f''(0) = -\sin 0 = 0, f'''(0) = -\cos 0 = -1, f^{(4)}(0) = \sin 0 = 0$, ...

Thus, substituting the function and derivative values into Eq. 3.3, $f(x) = \sin x$ function can be expressed with Maclaurin series as following:

$$\sin x = x - \frac{x^3}{3!} + \frac{x^5}{5!} - \frac{x^7}{7!} + \ldots = \sum_{n=0} \frac{(-1)^n}{(2n+1)!}x^{2n+1} = \sum_{n=1} \frac{(-1)^{n-1}}{(2n-1)!}x^{2n-1}$$

3.3.4 Calculation of Truncation Error

As shown in Examples 3.3 and 3.4, all functions can be expressed in terms of polynomial functions using Taylor or Maclaurin series. However, Taylor or Maclaurin series contain an infinite number of terms. Therefore, for practical use, they must be truncated after a predetermined number of terms. For example, Taylor series for an $f(x)$ function taking $n + 1$ terms can be written as following:

$$f(x) = f(x_0) + f'(x_0)(x - x_0) + \frac{f''(x_0)}{2!}(x - x_0)^2 + \ldots + \frac{f^{(n)}(x_0)}{n!}(x - x_0)^n \\ \underbrace{+\frac{f^{(n+1)}(x_0)}{(n+1)!}(x - x_0)^{n+1} + \ldots}_{R} \quad (3.4)$$

In this case, truncation error $R(x)$ can be expressed as below:

$$R(x) = \frac{f^{(n+1)}(x_0)}{(n+1)!}(x-x_0)^{n+1} + \ldots = \frac{f^{(n+1)}(\xi)}{(n+1)!}(x-x_0)^{n+1} \quad x_0 < \xi < x \tag{3.5}$$

In $R(x)$ expression, only controllable term is $(x - x_0)^{n+1}$. Therefore, truncation error can be reduced either increasing the number of terms in series or reducing $(x - x_0)^{n+1}$ value by reducing $(x - x_0)$.

Example 3.5: Express (a) e^2 and (b) $\sin\frac{\pi}{4}$ with Maclaurin series using different numbers of terms.

(a) Using Maclaurin series, e^x function can be expressed as following:

$$e^x = 1 + x + \frac{x^2}{2!} + \frac{x^3}{3!} + \frac{x^4}{4!} + \ldots + \frac{x^n}{n!}.$$

In this case, e^2 value can be expressed as $e^2 = 1 + 2 + \frac{2^2}{2!} + \frac{2^3}{3!} + \frac{2^4}{4!} + \ldots + \frac{2^n}{n!}$.

(b) Using Maclaurin series, sin$x$ can be expressed as $\sin x = x - \frac{x^3}{3!} + \frac{x^5}{5!} - \frac{x^7}{7!} + \ldots$.
In this case, $\sin\frac{\pi}{4}$ value can be expressed as following:

$$\sin(\pi/4) = (\pi/4) - \frac{(\pi/4)^3}{3!} + \frac{(\pi/4)^5}{5!} - \frac{(\pi/4)^7}{7!} + \ldots$$

Using different numbers of terms (N) in Maclaurin Series, e^2 and $\sin\frac{\pi}{4}$ values are obtained as in Table 3.5.

From Table 3.5 it is seen that using 9–10 terms for e^2 and 2–3 terms for $\sin\frac{\pi}{4}$ in Maclaurin series give very close results to the true values.

Note 1**:** e^2 value can be calculated using *MATLAB Program 3.1: Calculation of* e^x *Value Using Maclaurin Series* in Appendix.

Note 2**:** $\sin\frac{\pi}{4}$ value can be calculated using *MATLAB Program 3.2: Calculation of sin(x) Value Using Maclaurin Series* in Appendix.

3.4 Iteration Error (Method Error)

Linear, nonlinear and optimization problems are usually solved with numerical methods and iteratively (step-by-step way). Iterative methods try to achieve the true solution by using simple methods at each iteration. Iterative methods start from an estimated solution, called initial value, and update it until converging to the true solution. Error at each iteration can be called as iteration error. Iteration error gradually reduces in new iterations. Convergence to the true solution is ensured when iteration error cannot be decreased any further in new iterations. This can be equivalently ensured by checking the similarity of solution values in the last two iterations.

Table 3.5 Calculation of e^2 and $\sin\frac{\pi}{4}$ values using different numbers of terms (N) in Maclaurin Series

	$e^2 = 7.389056098930650$ (True value)		$\sin\frac{\pi}{4} = 0.707106781186547$ (True value)	
Number of terms (N)	e^2 value	Truncation Error (e^2)	$\sin\frac{\pi}{4}$ value	Truncation Error $\left(\sin\frac{\pi}{4}\right)$
1	1	6.389056098930650	0.785398163397448	0.785398163397448
2	3	4.389056098930650	0.704652651209168	0.002454129977380
3	5	2.389056098930650	0.707143045779360	−0.000036264592813
4	6.333333333333333	1.055722765597317		
5	7	0.389056098930650		
6	7.266666666666667	0.122389432263984		
7	7.355555555555555	0.033500543375095		
8	7.380952380952381	0.008103717978270		
9	7.387301587301587	0.001754511629064		
10	7.388712522045855	0.000343576884796		

Table 3.6 Iterative solution of equation $x^2 + 2x = 15$ using Simple Iteration (Substitution) and Newton-Raphson method

	Simple Iteration			Newton-Raphson		
Iteration (k)	x_{k-1}	x_k	$E=\left\lvert\frac{x_k-x_{k-1}}{x_{k-1}}\right\rvert$	x_{k-1}	x_k	$E=\left\lvert\frac{x_k-x_{k-1}}{x_{k-1}}\right\rvert$
1	2	3.7500	0.8750	2	3.1667	0.5833
2	3.7500	2.6087	0.3043	3.1667	3.0003	0.0516
3	2.6087	3.2547	0.2476	3.0003	3.0000	0.0011
4	3.2547	2.8546	0.1229	3.0000	3.0000	0.0000
5	2.8546	3.0899	0.0824			
6	3.0899	2.9470	0.0462			
7	2.9470	3.0321	0.0289			
8	3.0321	2.9809	0.0169			
9	2.9809	3.0115	0.0103			
10	3.0115	2.9931	0.0061			
11	2.9931	3.0004	0.0037			
12	3.0004	2.9975	0.0022			
13	2.9975	3.0015	0.0013			
14	3.0015	2.9991	0.0007			

If k represents the current iteration and k-1 represents the previous iteration, then iteration error at the k-th iteration can be expressed using function values as following:

1. Using the difference between last two iterations:

$$E = |f_k - f_{k-1}| \tag{3.6a}$$

2. Using the normalized difference between last two iterations:

$$E = \left| \frac{f_k - f_{k-1}}{f_{k-1}} \right| \tag{3.6b}$$

Equation 3.6b is used more commonly as the indicator of iteration error. If E value in Eq. 3.6b is less than a tolerance value ε ($E \leq \varepsilon$), then it is assumed that true solution has been reached (converged). $E \leq \varepsilon$ criterion is called as convergence criterion and is expressed in more detail as following:

$$\left| \frac{f_k - f_{k-1}}{f_{k-1}} \right| \leq \varepsilon \tag{3.7}$$

In Eq. 3.7, ε is never chosen as zero due to rounding, trimming, and truncation errors in computer calculations. Instead, ε is chosen a very small value such as $\varepsilon = 0.001$.

Depending on the iterative solution method, iteration error reduces at different speeds in new iterations. Iteration error can be reduced fast by choosing a good initial value (close to the true solution) to start the solution. Selection of a good initial value will decrease iteration errors and speed up the convergence to the true solution. However, if initial value is not chosen appropriately, some methods may even diverge from the true solution. Iteration error is usually larger than rounding, trimming, or truncation errors. Therefore, the most critical step in an iterative solution is to determine the most appropriate iterative method that will solve the problem accurately and efficiently.

Example 3.6: Obtain the solution of $x^2 + 2x = 15$ equation iteratively using Simple Iteration and Newton-Raphson methods and specify iteration errors.

Using the discriminant method $\left(x^2 + 2x = 15 \rightarrow \underbrace{1}_{a} x^2 + \underbrace{2}_{b} x \underbrace{- 15}_{c} = 0 \rightarrow x_{1,2} = \frac{-b \pm \sqrt{b^2 - 4ac}}{2a}\right)$, solution of this equation can be obtained as $x = -5$ and $x = 3$. These solutions can be obtained (only $x = 3$ solution is obtained here) using Simple Iteration and Newton-Raphson methods as well. These methods are described in detail in Chap. 5. Therefore, they will be only briefly explained here.

(a) Simple iteration (Substitution) method

In this method, an iterative equation is obtained by leaving x alone as below: $x^2 + 2x = 15 \rightarrow x(x + 2) = 15 \rightarrow x = \frac{15}{x+2}$

$x = \frac{15}{x+2}$ equation can be written in iterative form as $x_k = \frac{15}{x_{k-1}+2}$ (k: iteration number). If the initial value of $x_0 = 2$ is selected, iterative solution can be obtained as in Table 3.6.

(b) Newton-Raphson method

In this method, solution of a nonlinear equation is converted into the root finding form as $f(x) = 0$. Then, using linear Taylor series expansion for $f(x)$, iterative schema is obtained in the form of $x_k = x_{k-1} - \frac{f(x_{k-1})}{f'(x_{k-1})}$ for the solution of the nonlinear equation. For this example, iterative schema can be expressed as $x_k = x_{k-1} - \frac{x_{k-1}^2 + 2x_{k-1} - 15}{2x_{k-1}+2}$. Using initial value of $x_0 = 2$, iterative solution can be obtained as in Table 3.6.

From Table 3.6, it is seen that iteration errors in Newton-Raphson method are smaller than Simple iteration method. Therefore, Newton-Rapson method converges to the true solution faster than Simple Iteration method.

Similarly, $x = -5$ solution can also be obtained iteratively by selecting an appropriate (close) initial value.

3.5 Summary

- Different types of error occur in the solution of engineering problems.
- These errors can be grouped as (1) Modeling Error, (2) Numerical Error, and (3) Iteration Error (Method Error).
- Modeling error occurs due to crude model or inexact mathematical expression of engineering problem. It arises from the simplifying assumptions in the expression of the problem at hand.
- Numerical errors occur as the result of mathematical operations in the solution of engineering problems. These errors can be divided into two groups as (1) Rounding and (2) Truncation errors.
- Rounding or trimming error occurs in fitting values to the limited allocated digits in computer memory. For example, π value can be stored as 3.14285714285714 when large computer memory is available or it can be rounded as 3.14 when limited memory is available. In this case, due to rounding, error (difference between values) in storing π value will be as 0.001592653589793.
- Rounding or trimming errors can be controlled or reduced by using double precision arithmetic in mathematical operations.
- Computers use binary system consisting of 0 and 1 numbers to store variables in the memory.
- A binary integer expressed with d_i coefficients can be converted into decimal value by multiplying bit numbers with varying powers of 2 as following:

$$x = \underbrace{d_{n-1}d_{n-2}\ldots d_1 d_0}_{\text{Bits}} \Rightarrow x = \underbrace{d_{n-1}2^{n-1} + d_{n-2}2^{n-2} + \ldots + d_1 2^1 + d_0 2^0}_{\text{Decimal Value}}$$

- All functions can be expressed in polynomial form using Taylor or Maclaurin series.
- An $f(x)$ function can be expressed with Taylor series expansion using function and derivative values at an x_0 point as following:

$$f(x) = f(x_0) + f'(x_0)(x - x_0) + \frac{f''(x_0)}{2!}(x - x_0)^2 + \ldots + \frac{f^{(n)}(x_0)}{n!}(x - x_0)^n + \ldots$$
$$= \sum_{n=0} \frac{f^{(n)}(x_0)}{n!}(x - x_0)^n$$

- When $x_0 = 0$ is used in Taylor series, Maclaurin series is obtained as below:

$$f(x) = f(0) + f'(0)x + \frac{f''(0)}{2!}x^2 + \frac{f''(0)}{3!}x^3 + \ldots + \frac{f^{(n)}(0)}{n!}x^n + \ldots = \sum_{n=0} \frac{f^{(n)}(0)}{n!}x^n$$

- Truncation error occurs when a limited number of terms are taken in series expressions. Truncation error can be controlled or reduced by taking more terms in series expressions.
- Iteration error (method error) occurs when iterative methods are used in the solution of engineering problems. It can be reduced by choosing a good (close to true solution) initial value and appropriate iterative method.

Appendix: MATLAB Programs

MATLAB Program 3.1: Calculation of e^x Value Using Maclaurin Series

```
function ex_Calculation_with_Maclaurin_Series

% THIS PROGRAM CALCULATES e^x VALUE USING MACLAURIN SERIES

% e^x term is expressed with Maclaurin Series as following:
% e^x=1+x+x^2/2!+x^3/3!+x^4/4!+...+x^n/n!

% x value in e^x expression
x=2 ;

% Number of terms in Maclaurin Series: n
n=7 ;

% e^x value: Sum
Sum=0 ;
for i=1:n
    Sum=Sum+(x^(i-1))/factorial(i-1) ;
end

fprintf('e^x Value = %f \n',Sum) ;
```

MATLAB Program 3.2: Calculation of sin(x) Value Using Maclaurin Series

```
function sinx_Calculation_with_Maclaurin_Series

% THIS PROGRAM CALCULATES sin(x) VALUE USING MACLAURIN SERIES

% sin(x) term is expressed with Maclaurin Series as following:
% sin(x)=x-x^3/3!+x^5/5!-x^7/7!...+x^(2*n-1)/(2*n-1)!

% x value in sin(x) expression
x=pi/4 ;

% Number of terms in Maclaurin Series: n
n=3 ;

% sin(x) value: Sum
Sum=0 ;
for i=1:n
    Sum=Sum+(-1)^(i-1)*(x^(2*i-1))/factorial(2*i-1) ;
end

fprintf('e^x Value = %f \n',Sum) ;
```

Problems

Problem 3.1: Round $x = 0.97367259$ value to have four digits after the decimal point.

Problem 3.2: Round $x = 0.97367259$ value to have two digits after the decimal point.

Problem 3.3: Assuming four digits can be allocated to a variable in computer memory, calculate $y = x^2 - \frac{x}{10^2} + 1000x$ value for $x = 0.0002$. Determine true and rounded values as well as rounding error for y.

Problem 3.4: Calculate the decimal value of binary variable $x = 101101$.

Problem 3.5: Calculate the decimal value of binary variable $x = 1011.101$ using fixed point method.

Problem 3.6: Express $f(x) = e^{2x+1}$ function with Maclaurin series.

Problem 3.7: Express $f(x) = \cos x$ function with Maclaurin series.

Problem 3.8: Calculate $\cos \frac{\pi}{3}$ value with Maclaurin series using three terms.

Problem 3.9: Calculate $\sqrt{e}$ values with Maclaurin series using three terms.

References

1. Chapra, S. C., & Canale, R. P. (2020). *Numerical methods for engineers*. McGraw-Hill.
2. Bernatz, R. A. (2010). *Fourier series and numerical methods for partial differential equations*. John Wiley & Sons.
3. Everstine, G. C. (2010). *Numerical solution of partial differential equations* [Lecture notes, PDF]. George Washington University. https://www.academia.edu/31518755/Numerical_Solution_of_Partial_Differential_Equations
4. Epperson, J. F. (2013). *An introduction to numerical methods and analysis*. John Wiley & Sons.
5. Burden, R. L., & Faires, J. D. (2011). *Numerical analysis*. Cengage Learning.
6. Sert, C. (2006). *ME310 numerical methods* [Lecture notes, PDF]. Middle East Technical University. https://users.metu.edu.tr/csert/me310/me310_1_introduction.pdf
7. Grewal, B. S. (2019). *Numerical methods in engineering and science*. Mercury Learning and Information.
8. Sastry, S. S. (2012). *Introductory methods of numerical analysis*. PHI Learning.

Chapter 4
Solution of Linear Equations

4.1 Introduction

If an equation is in polynomial form and has terms with a maximum order of one, it is called as linear equation. Multiple linear equations are called as linear equation system. For example, a system of linear equations consisting of n unknown variables (x_1, x_2,…, x_n) and n equations can be written in general form as following:

$$
\begin{aligned}
&a_{11}x_1 + a_{12}x_2 + a_{13}x_3 + \ldots + a_{1n}x_n = b_1 \\
&a_{21}x_1 + a_{22}x_2 + a_{23}x_3 + \ldots + a_{2n}x_n = b_2 \\
&a_{31}x_1 + a_{32}x_2 + a_{33}x_3 + \ldots + a_{3n}x_n = b_3 \\
&\ldots \\
&a_{n1}x_1 + a_{n2}x_2 + a_{n3}x_3 + \ldots + a_{nn}x_n = b_n
\end{aligned}
\tag{4.1}
$$

The a_{ij} constants in Eq. 4.1 represent the coefficients of unknown variables. In order to solve a linear system, the number of *independent* equations must be equal to that of *unknown* variables.

In real applications, very few engineering problems (such as extension of spring system under loading, or static equilibrium of a truss system) are expressed directly by a system of linear equations. We often encounter linear equations in numerical solutions of engineering problems expressed with differential or integral equations. In numerical solutions, an engineering problem expressed with differential or integral equations is converted into a system of linear or nonlinear equations through methods such as Finite Difference (explained in Chap. 8) or Finite Element (explained in Chap. 6 of Vol. 2) as illustrated in Fig. 4.1.

Conversion of differential equations in mechanics and heat conduction problems into linear equations can be stated as below.

H. Kurtaran, *Engineering Mathematics with MATLAB Applications: Vol. 1, Fundamentals*, https://doi.org/10.1007/978-3-031-81869-1_4

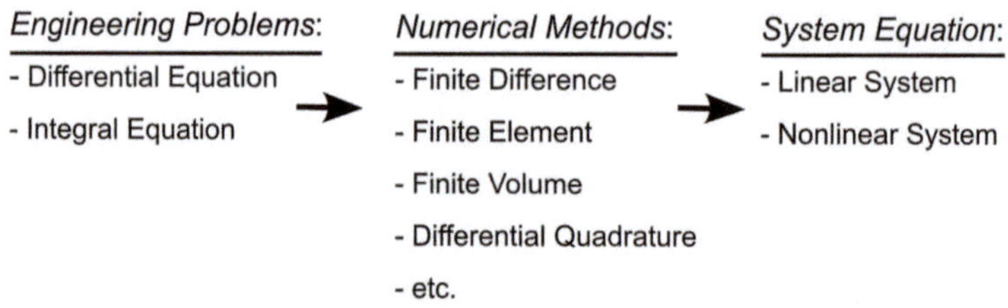

Fig. 4.1 Solution of differential and integral equations by converting them into linear or nonlinear equation systems

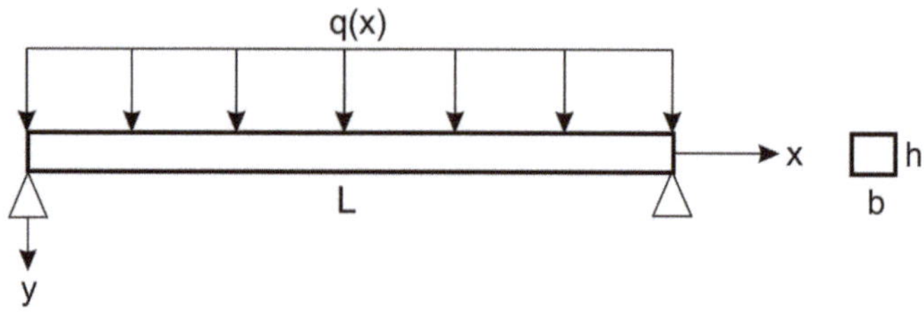

Fig. 4.2 Simply supported beam under distributed load

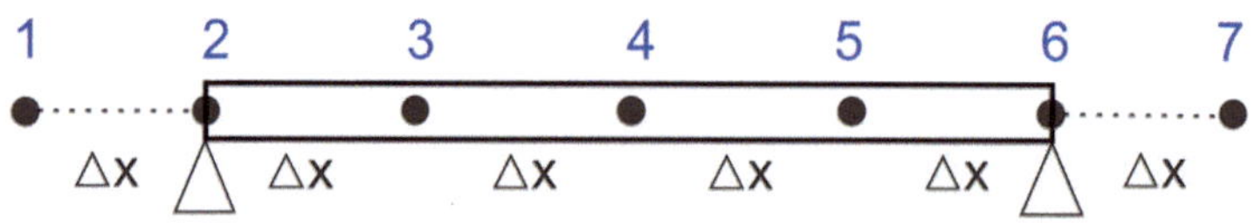

Fig. 4.3 Discretization of the beam for the solution of displacements with Finite Difference method

Beam Bending

Beam models are used in simplified bending analysis of *slender* structures in engineering. For example, wings of aircrafts, reinforcements, and bridges can be represented with these models for easy and efficient solution. An example of beam which is subjected to the distributed load of q is shown in Fig. 4.2. Differential equation and boundary conditions of this beam can be expressed as below (b: width, h: height, L: length, E: elastic modulus, q: distributed load).

Differential equation: $EI\frac{d^4v(x)}{dx^4}=q(x) \quad \left(I=\frac{bh^3}{12}\right)$

Boundary conditions (simply supported edges: displacement (v) and moment (M) are zero):

Left edge: $v(0)=0$, $M(0)=EI\frac{d^2v}{dx^2}(0)=0$

Right edge: $v(L)=0$, $M(L)=EI\frac{d^2v}{dx^2}(L)=0$.

For numerical solution, displacements on the beam can be discretized at points (i.e., nodes or grids) as in Fig. 4.3. Nodal displacement values can be obtained using a numerical method such as Finite Difference method. In Finite Difference method, derivative of an $f(x)$ function at a discretized point i is expressed in terms of function values at other points. For example, second- and fourth-order derivatives can be

Fig. 4.4 Bar subjected to the end temperatures of Ta and Tb

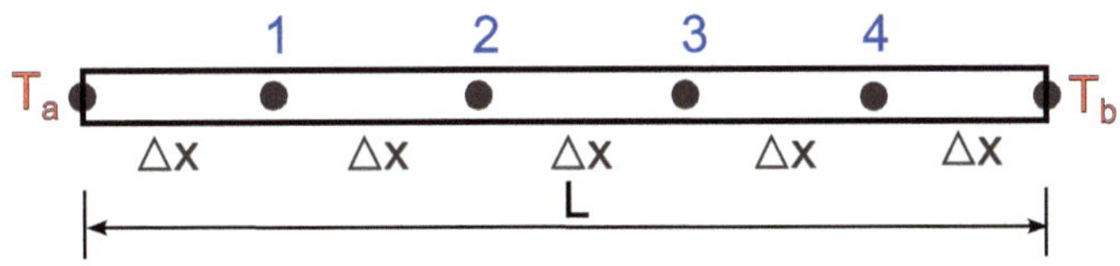

Fig. 4.5 Discretization of the bar for the solution of temperatures with Finite Difference method

expressed with $\frac{d^2f}{dx^2}(i) = \frac{f_{i-1} - 2f_i + f_{i+1}}{(\Delta x)^2}$ and $\frac{d^4f}{dx^4}(i) = \frac{f_{i-2} - 4f_{i-1} + 6f_i - 4f_{i+1} + f_{i+2}}{(\Delta x)^4}$ equations, respectively.

When Finite Difference method is used in expressing the differential equation at interior points and boundary conditions at boundary points, nodal displacements in Fig. 4.3 can be expressed in matrix form as a linear equation system as following:

$$\begin{bmatrix} 1 & -4 & 6 & -4 & 1 & 0 & 0 \\ 0 & 1 & -4 & 6 & -4 & 1 & 0 \\ 0 & 0 & 1 & -4 & 6 & -4 & 1 \\ 0 & 1 & 0 & 0 & 0 & 0 & 0 \\ 1 & -2 & 1 & 0 & 0 & 0 & 0 \\ 0 & 0 & 0 & 0 & 0 & 1 & 0 \\ 0 & 0 & 0 & 0 & 1 & -2 & 1 \end{bmatrix} \begin{Bmatrix} v_1 \\ v_2 \\ v_3 \\ v_4 \\ v_5 \\ v_6 \\ v_7 \end{Bmatrix} = \begin{Bmatrix} q_3(\Delta x)^4/EI \\ q_4(\Delta x)^4/EI \\ q_5(\Delta x)^4/EI \\ 0 \\ 0 \\ 0 \\ 0 \end{Bmatrix}$$

Heat Conduction

Heat conduction of slender structures such as bars is expressed with one-dimensional heat equation. For example, heat equation of the bar in Fig. 4.4 can be expressed as $-k\frac{d^2T}{dx^2} = q$ (q: heat flux applied to the bar).

For numerical solution, temperatures on the bar can be discretized at points (i.e., nodes or grids) as in Fig. 4.5. Nodal temperature values can then be determined using a numerical method such as Finite Difference method.

When Finite Difference method is used in expressing the differential equation at interior points, nodal temperatures in Fig. 4.5 can be expressed in matrix form as a linear equation system as following:

$$\begin{bmatrix} -2 & 1 & 0 & 0 \\ 1 & -2 & 1 & 0 \\ 0 & 1 & -2 & 1 \\ 0 & 0 & 1 & -2 \end{bmatrix} \begin{Bmatrix} T_1 \\ T_2 \\ T_3 \\ T_4 \end{Bmatrix} = \begin{Bmatrix} -T_a \\ 0 \\ 0 \\ -T_b \end{Bmatrix}$$

More examples from engineering can be given for the conversion of the differential equation into linear equation system as above. Since engineering problems are generally converted into linear equation systems for the numerical solution, solution efficiency highly depends on linear equation solvers. Therefore, several types of methods were developed to solve linear equation systems effectively [1–8]. Commonly used ones are introduced as following:

4.2 Existence of Solution in Linear Equation Systems

Solution of a linear equation system leads to a point corresponding to the intersection of all equations in the system. But, if all equations are not independent from each other, a unique solution may not be obtained. Therefore, before starting the solution, existence of a unique solution should be ensured. Two methods or approaches can be used for this purpose.

4.2.1 Graphical Method

This method is applied to linear equation systems with three or fewer equations, for which graphics of the equations can be drawn easily. In graphical method, graphics of all equations are plotted and intersection of these equations is determined as the solution point. In graphical method, graphics of linear equations must be drawn precisely for the correct solution. For example, graphical solution of linear equations of $2x_1 + 3x_2 = 4$ and $x_1 - x_2 = 1$ can be shown in Fig. 4.6.

From Fig. 4.6, intersection of these equations is obtained as $x_1 = \frac{7}{5} = 1.4$, $x_2 = \frac{2}{5} = 0.4$. Graphical method cannot be used for solving bigger linear systems since it is not possible to draw the graphics of more than three equations in the same plot.

4.2.2 Determinant Method

Since it is not possible to draw the graphics of more than three equations, a general strategy is needed to ensure the existence of a solution in big systems. Determinant method is an effective method for this purpose. In the determinant method, linear

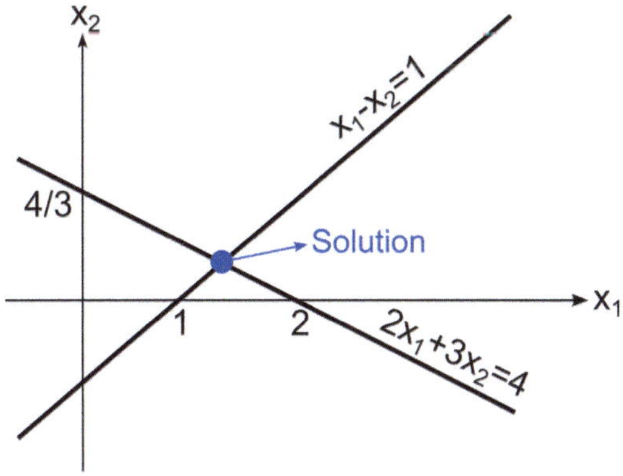

Fig. 4.6 Graphical solution of a linear equation system with two unknowns

equations are written in matrix form and determinant of the coefficients matrix is calculated. According to the determinant value, existence of a solution is ensured. To explain the determinant method, we can write general linear equation system in matrix form as below:

$$\left.\begin{array}{l} a_{11}x_1+a_{12}x_2+a_{13}x_3+\ldots+a_{1n}x_n=b_1 \\ a_{21}x_1+a_{22}x_2+a_{23}x_3+\ldots+a_{2n}x_n=b_2 \\ a_{31}x_1+a_{32}x_2+a_{33}x_3+\ldots+a_{3n}x_n=b_3 \\ \ldots \\ a_{n1}x_1+a_{n2}x_2+a_{n3}x_3+\ldots+a_{nn}x_n=b_n \end{array}\right\} \Rightarrow \underbrace{\begin{bmatrix} a_{11} & a_{12} & a_{13} & \ldots & a_{1n} \\ a_{21} & a_{22} & a_{23} & \ldots & a_{2n} \\ a_{31} & a_{32} & a_{33} & \ldots & a_{3n} \\ \ldots & \ldots & \ldots & \ldots & \ldots \\ a_{n1} & a_{n2} & a_{n3} & \ldots & a_{nn} \end{bmatrix}}_{A} \underbrace{\begin{Bmatrix} x_1 \\ x_2 \\ x_3 \\ \ldots \\ x_n \end{Bmatrix}}_{X} = \underbrace{\begin{Bmatrix} b_1 \\ b_2 \\ b_3 \\ \ldots \\ b_n \end{Bmatrix}}_{B} \tag{4.2}$$

Linear system in Eq. 4.2 can be shortly written in matrix form as $AX = B$. Here, A indicates coefficient matrix, X denotes unknown variables, and B indicates known constants.

Example: A linear equation system consisting of $2x_1 + x_2 = 12$ and $4x_1 - 3x_2 = 4$ equations can be written in matrix form as below:

$$\begin{array}{l} 2x_1 + x_2 = 12 \\ 4x_1 - 3x_2 = 4 \end{array} \rightarrow \underbrace{\begin{bmatrix} 2 & 1 \\ 4 & -3 \end{bmatrix}}_{A} \underbrace{\begin{Bmatrix} x_1 \\ x_2 \end{Bmatrix}}_{X} = \underbrace{\begin{Bmatrix} 12 \\ 4 \end{Bmatrix}}_{B}$$

There are three cases for the determinant of A (det[A] or $|A|$):

(a) det[A] $\neq 0$ (determinant is non-zero):

In this case, it can be said that all equations are independent from each other, and there is an intersection point (i.e., unique solution) for them.

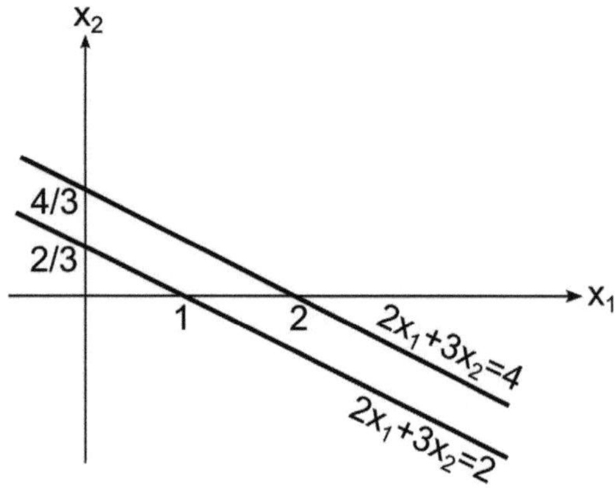

Fig. 4.7 A linear equation system where det[A] = 0 (parallel equations)

(b) det[A] = 0 (determinant is zero):

In this case, it can be said that not all equations are independent. Some equations are actually derived from others by multiplication, addition, or subtraction. In this case, determinant of the coefficient matrix will be zero. If det [A] = 0, A matrix is called as singular matrix. In graphical representation, some equations will be parallel to each other. Therefore, there is no intersection point (i.e., unique solution) for them.

Example: As can be seen from Fig. 4.7, $2x_1 + 3x_2 = 4$ and $2x_1 + 3x_2 = 2$ equations do not have a unique solution (intersection point). Determinant of the coefficient matrix for these equations can be shown to be equal to zero as following:

$$\begin{matrix} 2x_1 + 3x_2 = 4 \\ 2x_1 + 3x_2 = 2 \end{matrix} \rightarrow \underbrace{\begin{bmatrix} 2 & 3 \\ 2 & 3 \end{bmatrix}}_{A} \underbrace{\begin{Bmatrix} x_1 \\ x_2 \end{Bmatrix}}_{X} = \underbrace{\begin{Bmatrix} 4 \\ 2 \end{Bmatrix}}_{B} \rightarrow A = \begin{bmatrix} 2 & 3 \\ 2 & 3 \end{bmatrix} \rightarrow \det A = 0$$

(c) det[A] ≈ 0 (determinant is very close to zero):

When det[A] ≈ 0, the coefficient matrix A is called ill-conditioned. In this case, equations are assumed to intersect at infinity and existence of the solution depends on the sensitivity of the computer used for the solution. While an intersection point can be obtained with high-precision computers, it may not be possible with less-precision computers.

4.3 Solution of Linear Equations

Two types of methods called as Direct and Iterative (step-by-step correction) methods are available in the solution of linear equation systems. Direct methods are preferred in solving small systems. Direct methods guarantee the solution of a linear system if existence of a unique solution is proven. On the other hand, iterative methods are usually preferred in solving large systems. Iterative methods need initial values to start solution and they converge to a true solution in a step-by-step way, that is iteratively. Therefore, efficiency and effectiveness of the solution depend on the selection of good initial values. If initial values are chosen far from the true solution, iterative methods may diverge. Shortly, iterative methods do not guarantee the true solution. Direct and iterative methods are explained in detail below.

4.4 Direct Methods

4.4.1 Gaussian Elimination

Direct methods are generally based on transforming a linear system into independent single equations that can be solved one by one and successively. Gauss elimination method is also based on this strategy. In Gaussian elimination method, first, linear equation system is written in matrix form as $AX = B$. Then, applying (1) *Elimination* and (2) *Substitution* steps, this system is solved. In Elimination step, A matrix is converted into an upper triangular matrix where over diagonal terms are non-zero by applying multiplication/division as well as addition/subtraction operations in the rows of $AX = B$ system. At the end of Elimination step, $AX = B$ system is converted into a new form where equations can be solved independently. In Substitution step, equations are solved one by one from bottom to top by substituting the obtained solutions. Gauss elimination method can be simply explained with the following example.

Example: We can use a linear system of three equations to explain the method. Three linear equations and its matrix form can be expressed symbolically as below:

$$\begin{aligned} a_{11}x_1 + a_{12}x_2 + a_{13}x_3 &= b_1 \\ a_{21}x_1 + a_{22}x_2 + a_{23}x_3 &= b_2 \rightarrow \text{Matrix form}: \underbrace{\begin{bmatrix} a_{11} & a_{12} & a_{13} \\ a_{21} & a_{22} & a_{23} \\ a_{31} & a_{32} & a_{33} \end{bmatrix}}_{A} \underbrace{\begin{Bmatrix} x_1 \\ x_2 \\ x_3 \end{Bmatrix}}_{X} = \underbrace{\begin{Bmatrix} b_1 \\ b_2 \\ b_3 \end{Bmatrix}}_{B} \\ a_{31}x_1 + a_{32}x_2 + a_{33}x_3 &= b_3 \end{aligned}$$

Applying elimination and substitution steps to the above system, solution can be obtained as following.

(a) Elimination Step
In elimination step, all members below the diagonal terms in coefficient matrix A are made zero. Diagonal term is also called as *pivot*. In Gauss elimination method, zeroing under pivot terms are made systematically. First, members under the first diagonal term in column 1 are made zero. Then, members under the second diagonal

term in column 2 are made zero. This process is continued until all members below the diagonal terms are made zero.

For example, zeroing members under the first diagonal (first pivot) term can be done as following. a_{11} is the pivot term in the first column. To reset a_{21} to zero, first row is multiplied by $-\frac{a_{21}}{a_{11}}$ and added to the second row. Similarly, to reset a_{31} to zero, first row is multiplied by $-\frac{a_{31}}{a_{11}}$ and added to the third row. The zeroing procedure for the first column can be illustrated as below:

$$\begin{matrix} \\ a_{2j}-\left(\frac{a_{21}}{a_{11}}\right)a_{1j}\rightarrow \\ a_{3j}-\left(\frac{a_{31}}{a_{11}}\right)a_{1j}\rightarrow \end{matrix} \begin{bmatrix} \underbrace{a_{11}}_{\text{pivot}} & a_{12} & a_{13} \\ a_{21}-\left(\frac{a_{21}}{a_{11}}\right)a_{11} & a_{22}-\left(\frac{a_{21}}{a_{11}}\right)a_{12} & a_{23}-\left(\frac{a_{21}}{a_{11}}\right)a_{13} \\ a_{31}-\left(\frac{a_{31}}{a_{11}}\right)a_{11} & a_{32}-\left(\frac{a_{31}}{a_{11}}\right)a_{12} & a_{33}-\left(\frac{a_{31}}{a_{11}}\right)a_{13} \end{bmatrix} \begin{Bmatrix} x_1 \\ x_2 \\ x_3 \end{Bmatrix}$$

$$= \begin{Bmatrix} b_1 \\ b_2-\left(\frac{a_{21}}{a_{11}}\right)b_1 \\ b_3-\left(\frac{a_{31}}{a_{11}}\right)b_1 \end{Bmatrix}$$

The new matrix after zeroing members under the first diagonal term can be represented as below:

$$\rightarrow \begin{bmatrix} a_{11} & a_{12} & a_{13} \\ 0 & a'_{22} & a'_{23} \\ 0 & a'_{32} & a'_{33} \end{bmatrix} \begin{Bmatrix} x_1 \\ x_2 \\ x_3 \end{Bmatrix} = \begin{Bmatrix} b_1 \\ b'_2 \\ b'_3 \end{Bmatrix}$$

Elimination process can be repeated for the second column by taking a'_{22} as the new pivot. To reset a'_{32} to zero, the second row is multiplied by $-\frac{a'_{32}}{a'_{22}}$ and added to the third row leading to the following matrix:

$$a'_{3j}-\left(\frac{a'_{32}}{a'_{22}}\right)a'_{2j}\rightarrow \begin{bmatrix} a_{11} & a_{12} & a_{13} \\ 0 & \underbrace{a'_{22}}_{\text{pivot}} & a'_{23} \\ 0 & a'_{32}-\left(\frac{a'_{32}}{a'_{22}}\right)a'_{22} & a'_{33}-\left(\frac{a'_{32}}{a'_{22}}\right)a'_{23} \end{bmatrix} \begin{Bmatrix} x_1 \\ x_2 \\ x_3 \end{Bmatrix} = \begin{Bmatrix} b_1 \\ b'_2 \\ b'_3-\left(\frac{a'_{32}}{a'_{22}}\right)b'_2 \end{Bmatrix}$$

After the zeroing process (elimination step) is completed, an upper triangular matrix where below diagonal terms are zero is obtained as following:

$$\begin{bmatrix} a_{11} & a_{12} & a_{13} \\ 0 & a'_{22} & a'_{23} \\ 0 & 0 & a''_{33} \end{bmatrix} \begin{Bmatrix} x_1 \\ x_2 \\ x_3 \end{Bmatrix} = \begin{Bmatrix} b_1 \\ b'_2 \\ b''_3 \end{Bmatrix} \tag{4.3}$$

***Note*:** During elimination step, pivot terms may become zero even though initially they are not. If a pivot term becomes zero, division by zero occurs during the elimination step. In order to avoid this, row of the zero pivot is exchanged with row of the maximum coefficient. This process is also referred to as pivoting.

(b) Substitution Step

After elimination step, system equation turns to a new form where equations can be solved independently as below:

$$\begin{aligned} a_{11}x_1 + a_{12}x_2 + a_{13}x_3 &= b_1 \\ a'_{22}x_2 + a'_{23}x_3 &= b'_2 \qquad (\mathrm{x}_1 \text{ eliminated}) \\ a''_{33}x_3 &= b''_3 \qquad (\mathrm{x}_1 \text{ and } \mathrm{x}_2 \text{ eliminated}) \end{aligned}$$

Using the new form of the system equation, unknowns can be solved one by one from bottom to top as following:

$a'_{33}x_3 = b''_3$ (equation with one unknown) $\rightarrow x_3$.

$a'_{22}x_2 + a'_{23}x_3 = b'_2$ (equation with one unknown after substituting x_3) $\rightarrow x_2$.

$a_{11}x_1 + a_{12}x_2 + a_{13}x_3 = b_1$ (equation with one unknown after substituting x_2 and x_3) $\rightarrow x_1$.

Example 4.1: Obtain the solution of the following linear equation system using Gaussian Elimination method:

$$\begin{aligned} &18x_1 + 4x_2 - 3x_3 + 5x_4 = 91 \\ &-2x_1 - 24x_2 + 7x_3 - 3x_4 = -79 \\ &4x_1 - x_2 + 32x_3 + 5x_4 = 55 \\ &-9x_1 - 7x_2 - 3x_3 + 40x_4 = 20 \end{aligned}$$

Above linear equation system can be written in matrix form as $AX = B$ as below:

$$\underbrace{\begin{bmatrix} 18 & 4 & -3 & 5 \\ -2 & -24 & 7 & -3 \\ 4 & -1 & 32 & 5 \\ -9 & -7 & -3 & 40 \end{bmatrix}}_{A} \underbrace{\begin{Bmatrix} x_1 \\ x_2 \\ x_3 \\ x_4 \end{Bmatrix}}_{X} = \underbrace{\begin{Bmatrix} 91 \\ -79 \\ 55 \\ 20 \end{Bmatrix}}_{B}$$

Matrix system above can be solved by applying Elimination and Substitution steps as following.

Elimination:

In Elimination step, coefficient matrix A is converted into an upper triangular form (only above diagonal terms are non-zero) column by column. For this, first, all terms under the diagonal term (i.e., pivot) in the first column of A matrix are zeroed as below:

$$\begin{bmatrix} \underbrace{18}_{\text{pivot}} & 4 & -3 & 5 \\ -2 & -24 & 7 & -3 \\ 4 & -1 & 32 & 5 \\ -9 & -7 & -3 & 40 \end{bmatrix} \begin{Bmatrix} x_1 \\ x_2 \\ x_3 \\ x_4 \end{Bmatrix} = \begin{Bmatrix} 91 \\ -79 \\ 55 \\ 20 \end{Bmatrix} \rightarrow$$

$$\begin{bmatrix} 18 & 4 & -3 & 5 \\ -2-\left(\frac{-2}{18}\right)18 & -24-\left(\frac{-2}{18}\right)4 & 7-\left(\frac{-2}{18}\right)(-3) & -3-\left(\frac{-2}{18}\right)5 \\ 4-\left(\frac{4}{18}\right)18 & -1-\left(\frac{4}{18}\right)4 & 32-\left(\frac{4}{18}\right)(-3) & 5-\left(\frac{4}{18}\right)5 \\ -9-\left(\frac{-9}{18}\right)18 & -7-\left(\frac{-9}{18}\right)4 & -3-\left(\frac{-9}{18}\right)(-3) & 40-\left(\frac{-9}{18}\right)5 \end{bmatrix}$$

$$\times \begin{Bmatrix} x_1 \\ x_2 \\ x_3 \\ x_4 \end{Bmatrix} = \begin{Bmatrix} 91 \\ -79-\left(\frac{-2}{18}\right)91 \\ 55-\left(\frac{4}{18}\right)91 \\ 20-\left(\frac{-9}{18}\right)91 \end{Bmatrix}$$

Then, all terms under the pivot in the second column of A matrix are zeroed as below:

$$\begin{bmatrix} 18 & 4 & -3 & 5 \\ 0 & \underbrace{-23.5556}_{\text{pivot}} & 6.6667 & -2.4444 \\ 0 & -1.8889 & 32.6667 & 3.8889 \\ 0 & -5 & -4.5 & 42.5 \end{bmatrix} \begin{Bmatrix} x_1 \\ x_2 \\ x_3 \\ x_4 \end{Bmatrix} = \begin{Bmatrix} 91 \\ -68.8889 \\ 34.7778 \\ 65.5 \end{Bmatrix} \rightarrow$$

$$\begin{bmatrix} 18 & 4 & -3 & 5 \\ 0 & -23.5556 & 6.6667 & -2.4444 \\ 0 & -1.8889-\left(\frac{-1.8889}{-23.5556}\right)(-23.5556) & 32.6667-\left(\frac{-1.8889}{-23.5556}\right)6.6667 & 3.8889-\left(\frac{-1.8889}{-23.5556}\right)(-2.4444) \\ 0 & -5-\left(\frac{-5}{-23.5556}\right)(-23.5556) & -4.5-\left(\frac{-5}{-23.5556}\right)6.6667 & 42.5-\left(\frac{-5}{-23.5556}\right)(-2.4444) \end{bmatrix}.$$

$$\begin{Bmatrix} x_1 \\ x_2 \\ x_3 \\ x_4 \end{Bmatrix} = \begin{Bmatrix} 91 \\ -68.8889 \\ 34.7778-\left(\frac{-1.8889}{-23.5556}\right)(-68.8889) \\ 65.5-\left(\frac{-5}{-23.5556}\right)(-68.8889) \end{Bmatrix}$$

Then, all terms under the pivot in the third column of A matrix are zeroed as below:

$$\begin{bmatrix} 18 & 4 & -3 & 5 \\ 0 & -23.5556 & 6.6667 & -2.4444 \\ 0 & 0 & \underbrace{32.1321}_{\text{pivot}} & 4.0849 \\ 0 & 0 & -5.9151 & 43.0189 \end{bmatrix} \begin{Bmatrix} x_1 \\ x_2 \\ x_3 \\ x_4 \end{Bmatrix} = \begin{Bmatrix} 91 \\ -68.8889 \\ 40.3019 \\ 80.1226 \end{Bmatrix} \rightarrow$$

$$\begin{bmatrix} 18 & 4 & -3 & 5 \\ 0 & -23.5556 & 6.6667 & -2.4444 \\ 0 & 0 & 32.1321 & 4.0849 \\ 0 & 0 & -5.9151-\left(\frac{-5.9151}{32.1321}\right)32.1321 & 43.0189-\left(\frac{-5.9151}{32.1321}\right)4.0849 \end{bmatrix} \begin{Bmatrix} x_1 \\ x_2 \\ x_3 \\ x_4 \end{Bmatrix}$$

$$= \begin{Bmatrix} 91 \\ -68.8889 \\ 40.3019 \\ 80.1226-\left(\frac{-5.9151}{32.1321}\right)40.3019 \end{Bmatrix}$$

Thus, at the end of the elimination step, coefficient matrix A is converted into the upper triangular form as following:

$$\begin{bmatrix} 18 & 4 & -3 & 5 \\ 0 & -23.5556 & 6.6667 & -2.4444 \\ 0 & 0 & 32.1321 & 4.0849 \\ 0 & 0 & 0 & 43.7708 \end{bmatrix} \begin{Bmatrix} x_1 \\ x_2 \\ x_3 \\ x_4 \end{Bmatrix} = \begin{Bmatrix} 91 \\ -68.8889 \\ 40.3019 \\ 87.5417 \end{Bmatrix}$$

Substitution Step:
After elimination step, linear equations can be written in algebraic form as below:

$$18x_1 + 4x_2 - 3x_3 + 5x_4 = 91$$
$$-23.5556x_2 + 6.6667x_3 - 2.4444x_4 = -68.8889$$
$$32.1321x_3 + 4.0849x_4 = 40.3019$$
$$43.7708x_4 = 87.5417$$

Above equations can be solved one by one starting from the last equation to the first one as following:

x_4 can be solved immediately from Eq. 4.4 as below:

$$43.7708x_4 = 87.5417 \rightarrow x_4 = 2$$

After substituting $x_4 = 2$ into Eq. 4.3, x_3 can be solved as below:

$$32.1321x_3 + 4.0849x_4 = 40.3019 \rightarrow x_3 = 1$$

After substituting $x_3 = 1$ and $x_4 = 2$ in Eq. 4.2, x_2 can be solved as below:

$$-23.5556x_2 + 6.6667x_3 - 2.4444x_4 = -68.8889 \rightarrow x_2 = 3$$

Finally, after substituting $x_2 = 3$, $x_3 = 1$, and $x_4 = 2$ into Eq. 4.1, x_1 can be solved as below:

$$18x_1 + 4x_2 - 3x_3 + 5x_4 = 91 \rightarrow x_1 = 4.$$

Solution with MATLAB Commands:
Linear system can be solved using the division command in MATLAB in the form of $X = A\backslash B$ as below:

```
>> A=[18 4 -3 5; -2 -24 7 -3; 4 -1 32 5; -9 -7 -3 40]; B=[91; -79; 55; 20];
>> X=A\B;
>> X
X =
4.0000
3.0000
1.0000
2.0000
```

***Note*:** Solution of this problem with Gaussian Elimination Method can also be obtained using *MATLAB Program 4.1: Gaussian Elimination Method* in Appendix.

Gauss-Jordan Elimination:
Gauss-Jordan elimination is a special case of Gauss elimination method. In Gauss-Jordan elimination, Gauss elimination is applied to both below and above pivot term leading to a diagonal (unit) matrix for A as below:

$$\underbrace{\begin{bmatrix} a_{11} & a_{12} & a_{13} \\ a_{21} & a_{22} & a_{23} \\ a_{31} & a_{32} & a_{33} \end{bmatrix}}_{A} \underbrace{\begin{Bmatrix} x_1 \\ x_2 \\ x_3 \end{Bmatrix}}_{X} = \underbrace{\begin{Bmatrix} b_1 \\ b_2 \\ b_3 \end{Bmatrix}}_{B} \rightarrow$$

$$\text{After Gauss – Jordan elimination : } \begin{bmatrix} 1 & 0 & 0 \\ 0 & 1 & 0 \\ 0 & 0 & 1 \end{bmatrix} \begin{Bmatrix} x_1 \\ x_2 \\ x_3 \end{Bmatrix} = \begin{Bmatrix} b'_1 \\ b'_2 \\ b'_3 \end{Bmatrix}$$

After Gauss-Jordan elimination, solution of the linear equation system is immediately obtained as $x_1 = b'_1$, $x_2 = b'_2$, $x_3 = b'_3$ without using a substitution step.

4.4.2 Cramer Method

As stated previously a system of linear equations with three unknowns can be written in matrix form as following:

$$\underbrace{\begin{bmatrix} a_{11} & a_{12} & a_{13} \\ a_{21} & a_{22} & a_{23} \\ a_{31} & a_{32} & a_{33} \end{bmatrix}}_{A} \underbrace{\begin{Bmatrix} x_1 \\ x_2 \\ x_3 \end{Bmatrix}}_{X} = \underbrace{\begin{Bmatrix} b_1 \\ b_2 \\ b_3 \end{Bmatrix}}_{B}$$

From the matrix equation, A matrix and B vector can be rewritten as below:

$$A = \begin{bmatrix} a_{11} & a_{12} & a_{13} \\ a_{21} & a_{22} & a_{23} \\ a_{31} & a_{32} & a_{33} \end{bmatrix}, \quad B = \begin{Bmatrix} b_1 \\ b_2 \\ b_3 \end{Bmatrix}$$

In Cramer method, unknowns are calculated by using determinant ratios of a new A_i matrix and original A matrix. A_i matrix is obtained by modifying A matrix and its form varies according to the unknown to be solved. For example, if x_1 is to be solved, A_1 matrix is obtained by substituting B vector into the first column of A matrix. Solution of a linear system with three equations using Cramer method can be expressed in symbolic form as following:

$$x_1 = \frac{\overbrace{\begin{vmatrix} b_1 & a_{12} & a_{13} \\ b_2 & a_{22} & a_{23} \\ b_3 & a_{32} & a_{33} \end{vmatrix}}^{A_1}}{\underbrace{\begin{vmatrix} a_{11} & a_{12} & a_{13} \\ a_{21} & a_{22} & a_{23} \\ a_{31} & a_{32} & a_{33} \end{vmatrix}}_{A}} = \frac{|A_1|}{|A|}, \quad x_2 = \frac{\overbrace{\begin{vmatrix} a_{11} & b_1 & a_{13} \\ a_{21} & b_2 & a_{23} \\ a_{31} & b_3 & a_{33} \end{vmatrix}}^{A_2}}{\underbrace{\begin{vmatrix} a_{11} & a_{12} & a_{13} \\ a_{21} & a_{22} & a_{23} \\ a_{31} & a_{32} & a_{33} \end{vmatrix}}_{A}} = \frac{|A_2|}{|A|}, \quad x_3 = \frac{\overbrace{\begin{vmatrix} a_{11} & a_{12} & b_1 \\ a_{21} & a_{22} & b_2 \\ a_{31} & a_{32} & b_3 \end{vmatrix}}^{A_3}}{\underbrace{\begin{vmatrix} a_{11} & a_{12} & a_{13} \\ a_{21} & a_{22} & a_{23} \\ a_{31} & a_{32} & a_{33} \end{vmatrix}}_{A}} = \frac{|A_3|}{|A|} \tag{4.4}$$

Determinant calculation is usually time consuming for big systems. Therefore, Cramer method is mostly used for solving small linear systems. Although there is no rule, some engineering software consider systems to be small if the number of equations is less than 100,000 (one hundred thousand).

Example 4.2: Solve the following linear equation system using Cramer method

$$\begin{aligned} 9x_1 + 10x_2 + 7x_3 &= 5 \\ -x_1 + 5x_2 + 6x_3 &= 9 \\ 2x_2 + 3x_3 &= 12 \end{aligned}$$

For the solution with Cramer method, linear equations can be written in matrix form as following:

$$\underbrace{\begin{bmatrix} 9 & 10 & 7 \\ -1 & 5 & 6 \\ 0 & 2 & 3 \end{bmatrix}}_{A} \underbrace{\begin{Bmatrix} x_1 \\ x_2 \\ x_3 \end{Bmatrix}}_{X} = \underbrace{\begin{Bmatrix} 5 \\ 9 \\ 12 \end{Bmatrix}}_{B}$$

Using A matrix and B vector, unknowns can be solved using determinant ratios as below:

$$x_1 = \frac{\begin{vmatrix} 5 & 10 & 7 \\ 9 & 5 & 6 \\ 12 & 2 & 3 \end{vmatrix}}{\begin{vmatrix} 9 & 10 & 7 \\ -1 & 5 & 6 \\ 0 & 2 & 3 \end{vmatrix}} = 3.9767, \quad x_2 = \frac{\begin{vmatrix} 9 & 5 & 7 \\ -1 & 9 & 6 \\ 0 & 12 & 3 \end{vmatrix}}{\begin{vmatrix} 9 & 10 & 7 \\ -1 & 5 & 6 \\ 0 & 2 & 3 \end{vmatrix}} = -11.0233, \quad x_3 = \frac{\begin{vmatrix} 9 & 10 & 5 \\ -1 & 5 & 9 \\ 0 & 2 & 12 \end{vmatrix}}{\begin{vmatrix} 9 & 10 & 7 \\ -1 & 5 & 6 \\ 0 & 2 & 3 \end{vmatrix}} = 11.3488$$

Solution with MATLAB Commands:
Linear system can be solved using the determinant (*det*) command in MATLAB as following:

```
>> A=[9 10 7; -1 5 6; 0 2 3]; A1=[5 10 7; 9 5 6; 12 2 3];
>> A2=[9 5 7; -1 9 6; 0 12 3]; A3=[9 10 5; -1 5 9; 0 2 12];
>> x1=det(A1)/det(A); x2=det(A2)/det(A); x3=det(A3)/det(A);
>> x1
x1 =
3.9767
>> x2
x2 =
-11.0233
>> x3
x3 =
11.3488
```

4.4.3 Decomposition Method

Decomposition method like Gaussian elimination method is based on transforming the linear equation system into equations that can be solved independently. However, decomposition method performs the elimination step more efficiently compared to Gaussian Elimination method. In decomposition method, coefficient matrix A is decomposed into the product of two special matrices in the form of $A = LU$ where L indicates the lower triangular matrix and U indicates the upper triangular matrix as below:

$$L = \begin{bmatrix} L_{11} & 0 & 0 \\ L_{21} & L_{22} & 0 \\ L_{31} & L_{32} & L_{33} \end{bmatrix}, \quad U = \begin{bmatrix} 1 & U_{12} & U_{13} \\ 0 & 1 & U_{23} \\ 0 & 0 & 1 \end{bmatrix}$$

Decomposition of A matrix as $A = LU$ can be written using matrix multiplication as following:

$$\underbrace{\begin{bmatrix} a_{11} & a_{12} & a_{13} \\ a_{21} & a_{22} & a_{23} \\ a_{31} & a_{32} & a_{33} \end{bmatrix}}_{A} = \underbrace{\begin{bmatrix} L_{11} & 0 & 0 \\ L_{21} & L_{22} & 0 \\ L_{31} & L_{32} & L_{33} \end{bmatrix}}_{L} \underbrace{\begin{bmatrix} 1 & U_{12} & U_{13} \\ 0 & 1 & U_{23} \\ 0 & 0 & 1 \end{bmatrix}}_{U} \tag{4.5}$$

Members of L and U matrices can be obtained using the equality of left and right sides of Eq. 4.5 as below:

For row 1:

$a_{11} = L_{11} \rightarrow L_{11}$ is calculated.
$a_{12} = L_{11}U_{12} \rightarrow U_{12}$ is calculated by substituting L_{11}
$a_{13} = L_{11}U_{13} \rightarrow U_{13}$ is calculated by substituting L_{11}

For row 2:

$a_{21} = L_{21} \rightarrow L_{21}$ is calculated.
$a_{22} = L_{21}U_{12} + L_{22} \rightarrow L_{22}$ is calculated by substituting L_{21}, U_{12}
$a_{23} = L_{21}U_{13} + L_{22}U_{23} \rightarrow U_{23}$ is calculated by substituting L_{21}, L_{22}, U_{13}

For row 3:

$a_{31} = L_{31} \rightarrow L_{31}$ is calculated.
$a_{32} = L_{31}U_{12} + L_{32} \rightarrow L_{32}$ is calculated by substituting L_{31}, U_{12}
$a_{33} = L_{31}U_{13} + L_{32}U_{23} + L_{33} \rightarrow L_{33}$ is calculated by substituting L_{31}, L_{32}, U_{13}, U_{23}

$AX = B$ equation can alternatively be written as $LUX = B$ using $A = LU$ decomposition. Using new $LUX = B$ form, linear system can be converted into the solution of two linear subsystems expressed as $LY = B$ (Y temporary unknown) and $UX = Y$ (X main unknown). Both $LY = B$ and $UX = Y$ subsystems correspond to the linear equations that can be solved independently. From $LY = B$ equation, temporary unknowns Y are obtained first. Then, substituting Y values into $UX = Y$ equation, main unknowns X are solved. Steps of the decomposition method can be explained symbolically as following:

- *Step 1*: $LY=B$ solution
  From the solution of $LY=B$, $Y= y_1, y_2, y_3$ values are obtained as below:

$$\underbrace{\begin{bmatrix} L_{11} & 0 & 0 \\ L_{21} & L_{22} & 0 \\ L_{31} & L_{32} & L_{33} \end{bmatrix}}_{L} \underbrace{\begin{Bmatrix} y_1 \\ y_2 \\ y_3 \end{Bmatrix}}_{Y} = \underbrace{\begin{Bmatrix} b_1 \\ b_2 \\ b_3 \end{Bmatrix}}_{B} \rightarrow \begin{matrix} L_{11}y_1 = b_1 \Rightarrow y_1 \\ L_{21}y_1 + L_{22}y_2 = b_2 \Rightarrow y_2 \\ L_{31}y_1 + L_{32}y_2 + L_{33}y_3 = b_3 \Rightarrow y_3 \end{matrix}$$

- *Step 2*: $UX = Y$ solution
  Substituting the obtained Y values into $UX = Y$ equation, $X= x_1, x_2, x_3$ values are obtained as below:

$$\underbrace{\begin{bmatrix} 1 & U_{12} & U_{13} \\ 0 & 1 & U_{23} \\ 0 & 0 & 1 \end{bmatrix}}_{U} \underbrace{\begin{Bmatrix} x_1 \\ x_2 \\ x_3 \end{Bmatrix}}_{X} = \underbrace{\begin{Bmatrix} y_1 \\ y_2 \\ y_3 \end{Bmatrix}}_{Y} \rightarrow \begin{matrix} x_1 + U_{12}x_2 + U_{13}x_3 = y_1 \Rightarrow x_1 \\ x_2 + U_{23}x_3 = y_2 \Rightarrow x_2 \\ x_3 = y_3 \Rightarrow x_3 \end{matrix}$$

In the solution of engineering problems, sometimes A matrix has special forms such as banded and symmetric forms. Such special forms enhance the efficiency of the decomposition method as below.

A matrix has banded form:

$$A = \begin{bmatrix} a_{11} & a_{12} & 0 \\ a_{21} & a_{22} & a_{23} \\ 0 & a_{32} & a_{33} \end{bmatrix}$$

The banded form of A matrix is mostly encountered in the solution of engineering problems with numerical methods such as the Finite Difference method. In banded case, coefficient matrix A can be decomposed into the product of banded L (upper triangle) and U (lower triangle) matrices as $A = LU$ as below:

$$\underbrace{\begin{bmatrix} a_{11} & a_{12} & 0 \\ a_{21} & a_{22} & a_{23} \\ 0 & a_{32} & a_{33} \end{bmatrix}}_{A} = \underbrace{\begin{bmatrix} 1 & 0 & 0 \\ e_2 & 1 & 0 \\ 0 & e_3 & 1 \end{bmatrix}}_{L} \underbrace{\begin{bmatrix} f_1 & g_1 & 0 \\ 0 & f_2 & g_2 \\ 0 & 0 & f_3 \end{bmatrix}}_{U} \tag{4.6}$$

Again, using the equality of left and right sides in matrix multiplication, members of L and U matrices can be obtained as following:

Row 1: $a_{11} = f_1 \Rightarrow f_1,\ a_{12} = g_1 \Rightarrow g_1$
Row 2: $a_{21} = e_2 f_1 \Rightarrow e_2,\ a_{22} = e_2 g_1 + f_2 \Rightarrow f_2,\ a_{23} = g_2 \Rightarrow g_2$
Row 3: $a_{32} = e_3 f_2 \Rightarrow e_3,\ a_{33} = e_3 g_2 + f_3 \Rightarrow f_3$

After obtaining L and U matrices in banded forms, linear system of AX = B can be solved by converting it into the solution of two linear subsystems of LY = B and UX = Y. From the solution of linear subsystem $LY=B$, $Y= y_1, y_2, y_3$ values are obtained as below:

$$\begin{bmatrix} 1 & 0 & 0 \\ e_2 & 1 & 0 \\ 0 & e_3 & 1 \end{bmatrix} \begin{Bmatrix} y_1 \\ y_2 \\ y_3 \end{Bmatrix} = \begin{Bmatrix} b_1 \\ b_2 \\ b_3 \end{Bmatrix} \rightarrow \begin{matrix} y_1 = b_1 \Rightarrow y_1 \\ e_2 y_1 + y_2 = b_2 \Rightarrow y_2 \\ e_3 y_2 + y_3 = b_3 \Rightarrow y_3 \end{matrix}$$

Substituting the obtained Y values into the subsystem $UX = Y$ equation, $X= x_1, x_2, x_3$ values can be obtained as below:

$$\underbrace{\begin{bmatrix} f_1 & g_1 & 0 \\ 0 & f_2 & g_2 \\ 0 & 0 & f_3 \end{bmatrix}}_{U} \underbrace{\begin{Bmatrix} x_1 \\ x_2 \\ x_3 \end{Bmatrix}}_{X} = \underbrace{\begin{Bmatrix} y_1 \\ y_2 \\ y_3 \end{Bmatrix}}_{Y} \rightarrow \begin{matrix} f_1 x_1 + g_1 x_2 = y_1 \Rightarrow x_1 \\ f_2 x_2 + g_2 x_3 = y_2 \Rightarrow x_2 \\ f_3 x_3 = y_3 \Rightarrow x_3 \end{matrix}$$

A matrix has symmetry:

$$\underbrace{\begin{bmatrix} a_{11} & a_{12} & a_{13} & \cdots & a_{1n} \\ a_{12} & a_{22} & a_{23} & \cdots & a_{2n} \\ a_{13} & a_{23} & a_{33} & \cdots & a_{3n} \\ \cdots & \cdots & \cdots & \cdots & \cdots \\ a_{1n} & a_{2n} & a_{3n} & \cdots & a_{nn} \end{bmatrix}}_{A} \underbrace{\begin{Bmatrix} x_1 \\ x_2 \\ x_3 \\ \cdots \\ x_n \end{Bmatrix}}_{X} = \underbrace{\begin{Bmatrix} b_1 \\ b_2 \\ b_3 \\ \cdots \\ b_n \end{Bmatrix}}_{B} \rightarrow A = \begin{bmatrix} a_{11} & a_{12} & a_{13} & \cdots & a_{1n} \\ a_{12} & a_{22} & a_{23} & \cdots & a_{2n} \\ a_{13} & a_{23} & a_{33} & \cdots & a_{3n} \\ \cdots & \cdots & \cdots & \cdots & \cdots \\ a_{1n} & a_{2n} & a_{3n} & \cdots & a_{nn} \end{bmatrix}$$

The symmetrical form of A matrix is mostly encountered in the solution of engineering problems with numerical methods such as Finite Element method. In symmetric case, A matrix can be decomposed into the product of a lower triangular (L) matrix and its transpose as $[A] = [L][L]^{\mathrm{T}}$ as following:

$$\mathrm{A} = LL^T = \begin{bmatrix} L_{11} & 0 & 0 \\ L_{21} & L_{22} & 0 \\ L_{31} & L_{32} & L_{33} \end{bmatrix} \begin{bmatrix} L_{11} & L_{21} & L_{31} \\ 0 & L_{22} & L_{32} \\ 0 & 0 & L_{33} \end{bmatrix} \tag{4.7}$$

This decomposition method is also called as Cholesky Decomposition method. For Cholesky decomposition method to be applied, the coefficient matrix A must be positive definite. Positive definiteness of A matrix can be ensured in two ways:

1. $[A]$ matrix is positive definite if the absolute value of a diagonal in any row is greater than the sum of the absolute values of other members. This can be expressed mathematically as $|a_{ii}| > \sum_{\substack{j=1 \\ j \neq n}}^{n} |a_{ij}| \ (i = 1, \ldots, n)$
2. $[A]$ matrix is positive definite if all eigenvalues $\lambda = \lambda_1, \lambda_2, \lambda_3$ of A matrix are positive, i.e., $\lambda_i \geq 0$. In the case of a linear system with three equations, eigenvalues of $\lambda = \lambda_1, \lambda_2, \lambda_3$ can be calculated from the solution of $\det[A] = 0$ as below:

$$[A]\{x\} = \lambda I\{x\} \rightarrow [A - \lambda I]\{x\} = 0 \rightarrow \det[A - \lambda I] = 0 \rightarrow$$
$$\begin{vmatrix} a_{11} - \lambda & a_{12} & a_{13} \\ a_{21} & a_{22} - \lambda & a_{23} \\ a_{31} & a_{32} & a_{33} - \lambda \end{vmatrix} = 0 \rightarrow c_1 + c_2\lambda + c_3\lambda^2 + c_4\lambda^3 = 0 \Rightarrow \lambda_1, \lambda_2, \lambda_3$$

***Note*:** Decomposition method is particularly efficient in solving linear systems where A matrix does not change with change of B vector. In such problems, once A matrix is decomposed as $A = LU$, obtained L and U matrices can be used permanently even if B vector changes.

Example: Bending calculation of a cantilever beam as shown in Fig. 4.8 with elastic modulus E, length L, width b, height h under different F forces ($F1$, $F2$,

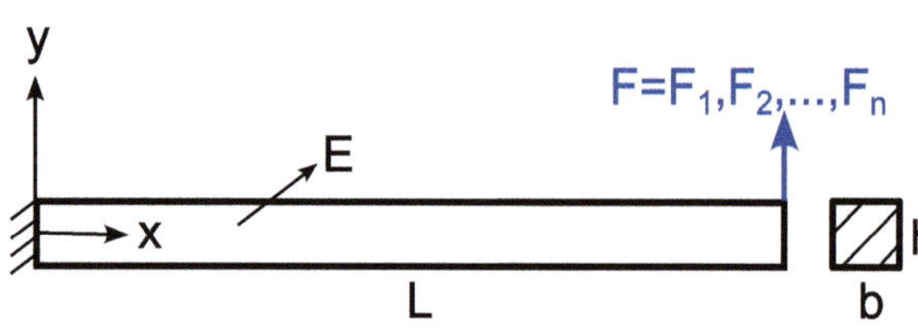

Fig. 4.8 Beam bending problem where only B changes in the linear system of $AX = B$

$F3, \ldots$) is an example of the above situation (A matrix is the same but B vector changes).

Example 4.3: Find the solution of the following linear equation system using the Cholesky Decomposition method

$$
\begin{aligned}
18x_1 + 4x_2 - 3x_3 + 5x_4 &= 91 \\
-2x_1 - 24x_2 + 7x_3 - 3x_4 &= -79 \\
4x_1 - x_2 + 32x_3 + 5x_4 &= 55 \\
-9x_1 - 7x_2 - 3x_3 + 40x_4 &= 20
\end{aligned}
$$

Linear equations above can be written in matrix form of $AX = B$ as below:

$$
\underbrace{\begin{bmatrix} 18 & 4 & -3 & 5 \\ -2 & -24 & 7 & -3 \\ 4 & -1 & 32 & 5 \\ -9 & -7 & -3 & 40 \end{bmatrix}}_{A}
\underbrace{\begin{Bmatrix} x_1 \\ x_2 \\ x_3 \\ x_4 \end{Bmatrix}}_{X}
=
\underbrace{\begin{Bmatrix} 91 \\ -79 \\ 55 \\ 20 \end{Bmatrix}}_{B}
$$

In Cholesky decomposition method, A matrix is decomposed into the product of lower triangular (L) and upper triangular (U) matrices as $A = LU$ as below:

$$
\begin{bmatrix} a_{11} & a_{12} & a_{13} & a_{14} \\ a_{21} & a_{22} & a_{23} & a_{24} \\ a_{31} & a_{32} & a_{33} & a_{34} \\ a_{41} & a_{42} & a_{43} & a_{44} \end{bmatrix}
=
\begin{bmatrix} L_{11} & 0 & 0 & 0 \\ L_{21} & L_{22} & 0 & 0 \\ L_{31} & L_{32} & L_{33} & 0 \\ L_{41} & L_{42} & L_{43} & L_{44} \end{bmatrix}
\begin{bmatrix} 1 & U_{12} & U_{13} & U_{14} \\ 0 & 1 & U_{23} & U_{24} \\ 0 & 0 & 1 & U_{34} \\ 0 & 0 & 0 & 1 \end{bmatrix}
$$

Members of L and U matrices can be obtained using the equality of left and right sides of the above decomposition as below:

Using first row of matrix A:

$$
\begin{aligned}
&a_{11} = L_{11} \rightarrow L_{11} = 18 \\
&a_{12} = L_{11}U_{12} \rightarrow 4 = 18U_{12} \rightarrow U_{12} = 0.2222 \\
&a_{13} = L_{11}U_{13} \rightarrow -3 = 18U_{13} \rightarrow U_{13} = -0.1667 \\
&a_{14} = L_{11}U_{14} \rightarrow 5 = 18U_{14} \rightarrow U_{14} = 0.2778
\end{aligned}
$$

Using second row of matrix A:

$$
\begin{aligned}
&a_{21} = L_{21} \rightarrow L_{21} = -2 \\
&a_{22} = L_{21}U_{12} + L_{22} \rightarrow -24 = -2(0.2222) + L_{22} \rightarrow L_{22} = -23.5556 \\
&a_{23} = L_{21}U_{13} + L_{22}U_{23} \rightarrow 7 = -2(-0.1667) + (-23.5556)U_{23} \rightarrow U_{23} = -0.2830 \\
&a_{24} = L_{21}U_{14} + L_{22}U_{24} \rightarrow -3 = -2(0.2778) + (-23.5556)U_{24} \rightarrow U_{24} = 0.1038
\end{aligned}
$$

Using third row of matrix A:

$a_{31} = L_{31} \rightarrow L_{31} = 4$
$a_{32} = L_{31}U_{12} + L_{32} \rightarrow -1 = 4(0.2222) + L_{32} \rightarrow L_{32} = -1.8889$
$a_{33} = L_{31}U_{13} + L_{32}U_{23} + L_{33} \rightarrow 32 = 4(-0.1667) + (-1.8889)(-0.2830) + L_{33} \rightarrow L_{33} = 32.1321$
$a_{34} = L_{31}U_{14} + L_{32}U_{24} + L_{33}U_{34} \rightarrow 5 = 4(0.2778) + (-1.8889)(0.1038) + 32.1321U_{34} \rightarrow U_{34} = 0.1271$

Using fourth row of matrix A:

$a_{41} = L_{41} \rightarrow L_{41} = -9$
$a_{42} = L_{41}U_{12} + L_{42} \rightarrow -7 = -9(0.2222) + L_{42} \rightarrow L_{42} = -5$
$a_{43} = L_{41}U_{13} + L_{42}U_{23} + L_{43} \rightarrow -3 = -9(-0.1667) + (-5.0002)(-0.2830) + L_{43} \rightarrow L_{43} = -5.9151$
$a_{44} = L_{41}U_{14} + L_{42}U_{24} + L_{43}U_{34} + L_{44} \rightarrow 40 = -9(0.2778) + (-5.0002)(0.1038) + (-5.9151)(0.1271) + L_{44} \rightarrow L_{44} = 43.7708$

Thus, L and U matrices are obtained as following:

$$L = \begin{bmatrix} 18 & 0 & 0 & 0 \\ -2 & -23.5556 & 0 & 0 \\ 4 & -1.8889 & 32.1321 & 0 \\ -9 & -5 & -5.9151 & 43.7708 \end{bmatrix}, \quad U = \begin{bmatrix} 1 & 0.2222 & -0.1667 & 0.2778 \\ 0 & 1 & -0.2830 & 0.1038 \\ 0 & 0 & 1 & 0.1271 \\ 0 & 0 & 0 & 1 \end{bmatrix}$$

From the solution of the subsystem $LY=B$, $Y= y_1, y_2, y_3, y_4$ values are obtained as below:

$$\begin{bmatrix} 18 & 0 & 0 & 0 \\ -2 & -23.5556 & 0 & 0 \\ 4 & -1.8889 & 32.1321 & 0 \\ -9 & -5 & -5.9151 & 43.7708 \end{bmatrix} \begin{Bmatrix} y_1 \\ y_2 \\ y_3 \\ y_4 \end{Bmatrix} = \begin{Bmatrix} 91 \\ -79 \\ 55 \\ 20 \end{Bmatrix} \rightarrow$$

$$18y_1 = 91 \rightarrow y_1 = 5.0556$$

$$-2y_1 + (-23.5556)y_2 = -79 \rightarrow y_2 = 2.9245$$

$$4y_1 + (-1.8889)y_2 + 32.1321y_3 = 55 \rightarrow y_3 = 1.2543$$

$$-9y_1 + (-5)y_2 - 5.9151y_3 + 43.7708y_4 = 20 \rightarrow y_4 = 2$$

Substituting the obtained Y values into the subsystem $UX = Y$, $X= x_1, x_2, x_3, x_4$ values are obtained as below:

$$\begin{bmatrix} 1 & 0.2222 & -0.1667 & 0.2778 \\ 0 & 1 & -0.2830 & 0.1038 \\ 0 & 0 & 1 & 0.1271 \\ 0 & 0 & 0 & 1 \end{bmatrix} \begin{Bmatrix} x_1 \\ x_2 \\ x_3 \\ x_4 \end{Bmatrix} = \begin{Bmatrix} 5.0556 \\ 2.9245 \\ 1.2543 \\ 2 \end{Bmatrix} \rightarrow$$

$$x_4 = 2$$

$$x_3 + (0.1271)x_4 = 1.2543 \rightarrow x_3 = 1$$

$$x_2 + (-0.2830)x_3 + (0.1038)x_4 = 2.9245 \rightarrow x_2 = 3$$

$$x_1 + 0.2222x_2 + (-0.1667)x_3 + (0.2778)x_4 = 5.0556 \rightarrow x_1 = 4$$

Solution with MATLAB Commands:
Linear system can be solved using the division command in MATLAB in the form of $X = A\backslash B$ as below:

```
>> A=[18 4 -3 5; -2 -24 7 -3; 4 -1 32 5; -9 -7 -3 40]; B=[91; -79; 55; 20];
>> X=A\B;
>> X
X =
4.0000
3.0000
1.0000
2.0000
```

***Note*:** Solution of this problem with the Cholesky decomposition method can also be obtained using *MATLAB Program 4.2: Cholesky Decomposition Method* in Appendix.

4.4.4 Inverse Matrix Method

In this method, solution of the linear system $AX = B$ is obtained by multiplying the inverse of A matrix with B vector as following:

$$AX = B \Rightarrow X = [A]^{-1}B \tag{4.8}$$

Although the method is simple, solution cost depends on the size of A matrix. Method can be computationally costly for big A matrices since obtaining inverse of A matrix ($[A]^{-1}$) can be difficult and time consuming. Adjoint matrix, Jacobi and Linear system solution methods are commonly used to calculate inverse of matrices.

Adjoint Matrix Method

Adjoint matrix method uses determinants of submatrices to calculate inverse of a matrix. Because of using determinants, method becomes computationally costly for big matrices. Therefore, it is preferred in the inverse calculation of small matrices. In the adjoint method, inverse of A matrix (A^{-1}) is obtained by dividing the adjoint matrix with the determinant of A matrix as $A^{-1} = \frac{adj(A)}{|A|} = \frac{C^T}{|A|}$. Adjoint matrix $adj(A)$ can be obtained by transposing the cofactor matrix C. Members of cofactor matrix C are obtained by using the determinants of the submatrices as $c_{ij} = (-1)^{i+j}|A_r|$ where A_r corresponds to the sub (reduced) matrix obtained by deleting i-th row and j-th column of A matrix.

Example 4.4: Find the inverse of the following matrix using the Adjoint method

$$A = \begin{bmatrix} 4 & 2 & 1 \\ 3 & 6 & 2 \\ 1 & 4 & 8 \end{bmatrix}$$

To calculate the inverse of A matrix, first Cofactor matrix C is obtained. Members of cofactor matrix C can be calculated by deleting specific row and columns in A matrix as below:

$$c_{11} = (-1)^{1+1}\begin{vmatrix} 6 & 2 \\ 4 & 8 \end{vmatrix} = 40, \quad c_{12} = (-1)^{1+2}\begin{vmatrix} 3 & 2 \\ 1 & 8 \end{vmatrix} = -22, \quad c_{13} = (-1)^{1+3}\begin{vmatrix} 3 & 6 \\ 1 & 4 \end{vmatrix} = 6,$$

$$c_{21} = (-1)^{2+1}\begin{vmatrix} 2 & 1 \\ 4 & 8 \end{vmatrix} = -12, \quad c_{22} = (-1)^{2+2}\begin{vmatrix} 4 & 1 \\ 1 & 8 \end{vmatrix} = 31, \quad c_{23} = (-1)^{2+3}\begin{vmatrix} 4 & 2 \\ 1 & 4 \end{vmatrix} = -14,$$

$$c_{31} = (-1)^{3+1}\begin{vmatrix} 2 & 1 \\ 6 & 2 \end{vmatrix} = -2, \quad c_{32} = (-1)^{3+2}\begin{vmatrix} 4 & 1 \\ 3 & 2 \end{vmatrix} = -5, \quad c_{33} = (-1)^{3+3}\begin{vmatrix} 4 & 2 \\ 3 & 6 \end{vmatrix} = 18,$$

Thus, Cofactor matrix C is obtained as following:

$$C = \begin{bmatrix} c_{11} & c_{12} & c_{13} \\ c_{21} & c_{22} & c_{23} \\ c_{31} & c_{32} & c_{33} \end{bmatrix} = \begin{bmatrix} 40 & -22 & 6 \\ -12 & 31 & -14 \\ -2 & -5 & 18 \end{bmatrix}$$

Adjoint matrix *adj(A)* can be obtained as the transpose of the Cofactor matrix C as below:

$$adj(A) = C^T = \begin{bmatrix} 40 & -22 & 6 \\ -12 & 31 & -14 \\ -2 & -5 & 18 \end{bmatrix}^T = \begin{bmatrix} 40 & -12 & -2 \\ -22 & 31 & -5 \\ 6 & -14 & 18 \end{bmatrix}$$

Thus, inverse of A matrix is obtained using $A^{-1} = \frac{adj(A)}{|A|}$ expression as following

$$A^{-1} = \frac{adj(A)}{|A|} = \frac{\begin{bmatrix} 40 & -12 & -2 \\ -22 & 31 & -5 \\ 6 & -14 & 18 \end{bmatrix}}{\begin{vmatrix} 4 & 2 & 1 \\ 3 & 6 & 2 \\ 1 & 4 & 8 \end{vmatrix}} = \frac{\begin{bmatrix} 40 & -12 & -2 \\ -22 & 31 & -5 \\ 6 & -14 & 18 \end{bmatrix}}{122}$$

$$= \begin{bmatrix} 0.3279 & -0.0984 & -0.0164 \\ -0.1803 & 0.2541 & -0.0410 \\ 0.0492 & -0.1148 & 0.1475 \end{bmatrix}$$

Gauss-Jordan Elimination Method

Gauss-Jordan elimination method uses Gaussian elimination method to calculate the inverse of a matrix. In this method, A matrix and a new Unit matrix I (I has the same size as A) are written next to each other. Gaussian elimination method is applied simultaneously to A and I matrices such that A matrix is transformed into Unit matrix and Unit matrix I is transformed into the inverse matrix A^{-1} as illustrated below:

$$\left[\underbrace{\begin{matrix} a_{11} & a_{12} & a_{13} \\ a_{21} & a_{22} & a_{23} \\ a_{31} & a_{32} & a_{33} \end{matrix}}_{A} \vdots \underbrace{\begin{matrix} 1 & 0 & 0 \\ 0 & 1 & 0 \\ 0 & 0 & 1 \end{matrix}}_{I}\right] \Rightarrow \left[\underbrace{\begin{matrix} 1 & 0 & 0 \\ 0 & 1 & 0 \\ 0 & 0 & 1 \end{matrix}}_{I} \vdots \underbrace{\begin{matrix} d_{11} & d_{12} & d_{13} \\ d_{21} & d_{22} & d_{23} \\ d_{31} & d_{32} & d_{33} \end{matrix}}_{A^{-1}}\right] \tag{4.9}$$

Example 4.5: Find the inverse of A matrix given below using Gauss-Jordan elimination method

$$A = \begin{bmatrix} 4 & 2 & 1 \\ 3 & 6 & 2 \\ 1 & 4 & 8 \end{bmatrix}$$

After writing A matrix and Unit matrix I next to each other and applying Gauss elimination simultaneously, inverse of A matrix is obtained as below:

$$\left[\underbrace{\begin{matrix} 4 & 2 & 1\vdots \\ 3 & 6 & 2\vdots \\ 1 & 4 & 8\vdots \end{matrix}}_{A} \quad \underbrace{\begin{matrix} 1 & 0 & 0 \\ 0 & 1 & 0 \\ 0 & 0 & 1 \end{matrix}}_{I}\right] \Rightarrow \left[\underbrace{\begin{matrix} 1 & 0 & 0\vdots \\ 0 & 1 & 0\vdots \\ 0 & 0 & 1\vdots \end{matrix}}_{I} \quad \underbrace{\begin{matrix} 0.3279 & -0.0984 & -0.0164 \\ -0.1803 & 0.2541 & -0.0410 \\ 0.0492 & -0.1148 & 0.1475 \end{matrix}}_{A^{-1}}\right]$$

***Note*:** Inverse matrix method is also effective like the decomposition method in solving linear problems $AX = B$ where only B vector changes. In such problems, even if B vector changes, inverse of A matrix is calculated only once.

Linear System Solution

In this method, inverse of an A matrix (A^{-1}) is obtained successively by solving systems of linear equations. Considering the product of an A matrix and its inverse (A^{-1}) will yield identity matrix (I) (i.e., $AA^{-1} = I$), $AA^{-1} = I$ expression can be transformed into the series of linear equation system each expressed in $AX = B$ form. Solution of each linear system in $AX = B$ form yields a column of the A^{-1} matrix. For example, when B is taken as the i-th column of I matrix, X solution corresponds to the i-th column of A^{-1} matrix.

To explain the method better, we can calculate inverse of an A matrix of 3x3 size using this method. For convenience, A matrix and its inverse (A^{-1}) can be represented in symbolic form as following:

$$A = \begin{bmatrix} a_{11} & a_{12} & a_{13} \\ a_{21} & a_{22} & a_{23} \\ a_{31} & a_{32} & a_{33} \end{bmatrix}, A^{-1} = \begin{bmatrix} d_{11} & d_{12} & d_{13} \\ d_{21} & d_{22} & d_{23} \\ d_{31} & d_{32} & d_{33} \end{bmatrix}$$

In this case, using A and A^{-1} matrices, $AA^{-1} = I$ equation can be written as below:

$$\underbrace{\begin{bmatrix} a_{11} & a_{12} & a_{13} \\ a_{21} & a_{22} & a_{23} \\ a_{31} & a_{32} & a_{33} \end{bmatrix}}_{A} \underbrace{\begin{bmatrix} d_{11} & d_{12} & d_{13} \\ d_{21} & d_{22} & d_{23} \\ d_{31} & d_{32} & d_{33} \end{bmatrix}}_{A^{-1}} = \underbrace{\begin{bmatrix} 1 & 0 & 0 \\ 0 & 1 & 0 \\ 0 & 0 & 1 \end{bmatrix}}_{I}$$

$AA^{-1} = I$ equation above can alternatively be expressed as three linear equation systems as below by considering the columns of I matrix one by one.

Linear system 1 (using 1-st column of I matrix):

$$\begin{bmatrix} a_{11} & a_{12} & a_{13} \\ a_{21} & a_{22} & a_{23} \\ a_{31} & a_{32} & a_{33} \end{bmatrix} \begin{Bmatrix} d_{11} \\ d_{21} \\ d_{31} \end{Bmatrix} = \begin{Bmatrix} 1 \\ 0 \\ 0 \end{Bmatrix} \rightarrow \text{Solution}: \begin{Bmatrix} d_{11} \\ d_{21} \\ d_{31} \end{Bmatrix} \ (\text{first column of } A^{-1} \text{ matrix})$$

Linear system 2 (using second column of I matrix):

$$\begin{bmatrix} a_{11} & a_{12} & a_{13} \\ a_{21} & a_{22} & a_{23} \\ a_{31} & a_{32} & a_{33} \end{bmatrix} \begin{Bmatrix} d_{12} \\ d_{22} \\ d_{32} \end{Bmatrix} = \begin{Bmatrix} 0 \\ 1 \\ 0 \end{Bmatrix} \rightarrow \text{Solution}: \begin{Bmatrix} d_{12} \\ d_{22} \\ d_{32} \end{Bmatrix} \ (\text{second column of } A^{-1} \text{ matrix})$$

Linear system 3 (using third column of I matrix):

$$\begin{bmatrix} a_{11} & a_{12} & a_{13} \\ a_{21} & a_{22} & a_{23} \\ a_{31} & a_{32} & a_{33} \end{bmatrix} \begin{Bmatrix} d_{13} \\ d_{23} \\ d_{33} \end{Bmatrix} = \begin{Bmatrix} 0 \\ 0 \\ 1 \end{Bmatrix} \rightarrow \text{Solution}: \begin{Bmatrix} d_{13} \\ d_{23} \\ d_{33} \end{Bmatrix} \ (\text{third column of } A^{-1} \text{ matrix})$$

Solutions of three linear systems yield columns of A^{-1} matrix. By combining these columns, inverse of A matrix is obtained.

***Note*:** From above expressions it is seen that inverse of an A matrix with nxn size requires solution of n independent linear systems. This explains why inverse of large matrices is computational costly.

Example 4.6: Find the solution of the following linear system using Inverse Matrix method:

$$\begin{aligned} &18x_1 + 4x_2 - 3x_3 + 5x_4 = 91 \\ &-2x_1 - 24x_2 + 7x_3 - 3x_4 = -79 \\ &4x_1 - x_2 + 32x_3 + 5x_4 = 55 \\ &-9x_1 - 7x_2 - 3x_3 + 40x_4 = 20 \end{aligned}$$

With inverse matrix method, solution of a linear system expressed with $AX = B$ equation can be obtained by multiplying both sides with the inverse of A matrix (A^{-1}) as $X = A^{-1} B$. Inverse of A matrix can be obtained using, for example, Cofactor matrix C as $A^{-1} = \frac{adj(A)}{|A|} = \frac{C^T}{|A|}$.

Cofactor matrix C can be expressed as $C=\begin{bmatrix} c_{11} & c_{12} & c_{13} & c_{14} \\ c_{21} & c_{22} & c_{23} & c_{24} \\ c_{31} & c_{32} & c_{33} & c_{34} \\ c_{41} & c_{42} & c_{43} & c_{44} \end{bmatrix}$ where its members are calculated as following:

$$c_{11}=(-1)^{1+1}\begin{vmatrix} -24 & 7 & -3 \\ -1 & 32 & 5 \\ -7 & -3 & 40 \end{vmatrix}=-31726,\quad c_{12}=(-1)^{1+2}\begin{vmatrix} -2 & 7 & -3 \\ 4 & 32 & 5 \\ -9 & -3 & 40 \end{vmatrix}=4853,$$

$$c_{13}=(-1)^{1+3}\begin{vmatrix} -2 & -24 & -3 \\ 4 & -1 & 5 \\ -9 & -7 & 40 \end{vmatrix}=5041,\quad c_{14}=(-1)^{1+4}\begin{vmatrix} -2 & -24 & 7 \\ 4 & -1 & 32 \\ -9 & -7 & -3 \end{vmatrix}=-5911,$$

$$c_{21}=(-1)^{2+1}\begin{vmatrix} 4 & -3 & 5 \\ -1 & 32 & 5 \\ -7 & -3 & 40 \end{vmatrix}=-6300,\quad c_{22}=(-1)^{2+2}\begin{vmatrix} 18 & -3 & 5 \\ -1 & 32 & 5 \\ -9 & -3 & 40 \end{vmatrix}=25305,$$

$$c_{23}=(-1)^{2+3}\begin{vmatrix} 18 & 4 & 5 \\ 4 & -1 & 5 \\ -9 & -7 & 40 \end{vmatrix}=1095,\quad c_{24}=(-1)^{2+4}\begin{vmatrix} 18 & 4 & -3 \\ 4 & -1 & 32 \\ -9 & -7 & -3 \end{vmatrix}=3093,$$

$$c_{31}=(-1)^{3+1}\begin{vmatrix} 4 & -3 & 5 \\ -24 & 7 & -3 \\ -7 & -3 & 40 \end{vmatrix}=-1254,\quad c_{32}=(-1)^{3+2}\begin{vmatrix} 18 & -3 & 5 \\ -2 & 7 & -3 \\ -9 & -3 & 40 \end{vmatrix}=-4902,$$

$$c_{33}=(-1)^{3+3}\begin{vmatrix} 18 & 4 & 5 \\ -2 & -24 & -3 \\ -9 & -7 & 40 \end{vmatrix}=-18240,\quad c_{34}=(-1)^{3+4}\begin{vmatrix} 18 & 4 & -3 \\ -2 & -24 & 7 \\ -9 & -7 & -3 \end{vmatrix}=-2508,$$

$$c_{41}=(-1)^{4+1}\begin{vmatrix} 4 & -3 & 5 \\ -24 & 7 & -3 \\ -1 & 32 & 5 \end{vmatrix}=3650,\quad c_{42}=(-1)^{4+2}\begin{vmatrix} 18 & -3 & 5 \\ -2 & 7 & -3 \\ 4 & 32 & 5 \end{vmatrix}=1904,$$

$$c_{43}=(-1)^{4+3}\begin{vmatrix} 18 & 4 & 5 \\ -2 & -24 & -3 \\ 4 & -1 & 5 \end{vmatrix}=1732,\quad c_{44}=(-1)^{4+4}\begin{vmatrix} 18 & 4 & -3 \\ -2 & -24 & 7 \\ 4 & -1 & 32 \end{vmatrix}=-13624,$$

$$\rightarrow C = \begin{bmatrix} -31726 & 4853 & 5041 & -5911 \\ -6300 & 25305 & 1095 & 3093 \\ -1254 & -4902 & -18240 & -2508 \\ 3650 & 1904 & 1732 & -13624 \end{bmatrix}$$

Thus, Adjoint matrix *adj (A)* is obtained as below:

$$adj(A) = C^T = \begin{bmatrix} -31726 & -6300 & -1254 & 3650 \\ 4853 & 25305 & -4902 & 1904 \\ 5041 & 1095 & -18240 & 1732 \\ -5911 & 3093 & -2508 & -13624 \end{bmatrix} \rightarrow$$

Finally, using the Adjoint matrix and determinant of A matrix, inverse matrix can be obtained as below:

$$A^{-1} = \frac{adj(A)}{|A|} \rightarrow A^{-1} = \begin{bmatrix} 0.0532 & 0.0106 & 0.0021 & -0.0061 \\ -0.0081 & -0.0424 & 0.0082 & -0.0032 \\ -0.0085 & -0.0018 & 0.0306 & -0.0029 \\ 0.0099 & -0.0052 & 0.0042 & 0.0228 \end{bmatrix}$$

In this case, using the inverse of A matrix, solution of the linear system is obtained as following:

$$X = A^{-1}B \rightarrow X = \begin{bmatrix} 0.0532 & 0.0106 & 0.0021 & -0.0061 \\ -0.0081 & -0.0424 & 0.0082 & -0.0032 \\ -0.0085 & -0.0018 & 0.0306 & -0.0029 \\ 0.0099 & -0.0052 & 0.0042 & 0.0228 \end{bmatrix} \begin{Bmatrix} 91 \\ -79 \\ 55 \\ 20 \end{Bmatrix} \rightarrow \begin{Bmatrix} x_1 \\ x_2 \\ x_3 \\ x_4 \end{Bmatrix} = \begin{Bmatrix} 4 \\ 3 \\ 1 \\ 2 \end{Bmatrix}$$

Solution with MATLAB Commands:

Linear system $AX = B$ can be solved using the inverse command (*inv*) in MATLAB as below:

```
>> A=[18 4 -3 5; -2 -24 7 -3; 4 -1 32 5; -9 -7 -3 40]; B=[91; -79; 55; 20];
>> X= inv (A)*B;
>> X
X =
4.0000
3.0000
1.0000
2.0000
```

4.5 Iterative Methods

In iterative methods, linear systems are solved iteratively using initial values. In iterative methods, single equation is created for each unknown in *unknown* = *terms* form. Initial values are substituted into the right sides and new values of unknowns are calculated in left sides. This process is repeated by substituting old values in right sides and calculating new values in left sides until left and right sides are sufficiently close to each other.

Iterative methods do not guarantee converging the true solution. Convergence depends on the type of iterative method, structure of the linear system (if diagonal values are dominant in rows), and initial values. Iterative methods are generally more efficient than direct methods. Therefore, they are preferred in solving big linear systems. Some common iterative methods are described in detail below using a linear system with three equations.

4.5.1 Jacobi Method

Linear system $AX = B$ with three equations can be expressed algebraically as below:

$$\begin{aligned} a_{11}x_1 + a_{12}x_2 + a_{13}x_3 &= b_1 \\ a_{21}x_1 + a_{22}x_2 + a_{23}x_3 &= b_2 \\ a_{31}x_1 + a_{32}x_2 + a_{33}x_3 &= b_3 \end{aligned}$$

Unknowns of x_1, x_2, x_3 can be left alone in these equations as below:

$$\begin{aligned} x_1 &= \frac{b_1 - (a_{12}x_2 + a_{13}x_3)}{a_{11}} \\ x_2 &= \frac{b_2 - (a_{21}x_1 + a_{23}x_3)}{a_{22}} \\ x_3 &= \frac{b_3 - (a_{31}x_1 + a_{32}x_2)}{a_{33}} \end{aligned} \tag{4.10}$$

From the above expressions, it can be easily seen that values in the denominators (i.e., diagonal members of A matrix) must be different from zero ($a_{11}, a_{22}, a_{33} \neq 0$) in order to have a solution.

In Jacobi method, initial values of x_1, x_2, x_3 are chosen first as $x_1^{(0)}, x_2^{(0)}, x_3^{(0)}$. Then, substituting these values into the right side of Eq. 4.10, corrected or improved values are obtained. This procedure is repeated until corrected values do not change. Iterations with Jacobi method can be symbolically expressed below:

- *Iteration 1*:

$$x_1^{(1)} = \frac{b_1 - \left(a_{12}x_2^{(0)} + a_{13}x_3^{(0)}\right)}{a_{11}}$$
$$x_2^{(1)} = \frac{b_2 - \left(a_{21}x_1^{(0)} + a_{23}x_3^{(0)}\right)}{a_{22}}$$
$$x_3^{(1)} = \frac{b_3 - \left(a_{31}x_1^{(0)} + a_{32}x_2^{(0)}\right)}{a_{33}}$$

- *Iteration 2*:

$$x_1^{(2)} = \frac{b_1 - \left(a_{12}x_2^{(1)} + a_{13}x_3^{(1)}\right)}{a_{11}}$$
$$x_2^{(2)} = \frac{b_2 - \left(a_{21}x_1^{(1)} + a_{23}x_3^{(1)}\right)}{a_{22}}$$
$$x_3^{(2)} = \frac{b_3 - \left(a_{31}x_1^{(1)} + a_{32}x_2^{(1)}\right)}{a_{33}}$$

- *Iteration k*:

$$x_1^{(k)} = \frac{b_1 - \left(a_{12}x_2^{(k-1)} + a_{13}x_3^{(k-1)}\right)}{a_{11}}$$
$$x_2^{(k)} = \frac{b_2 - \left(a_{21}x_1^{(k-1)} + a_{23}x_3^{(k-1)}\right)}{a_{22}}$$
$$x_3^{(k)} = \frac{b_3 - \left(a_{31}x_1^{(k-1)} + a_{32}x_2^{(k-1)}\right)}{a_{33}}$$

In iterative solutions, convergence condition or convergence criterion is used to stop iterations. Convergence criteria can be expressed as the normalized difference between x values of the last two iterations as below ($k-1$: previous iteration, k: current iteration).

Convergence Criterion:

$$\left|\frac{x_1^{(k)}-x_1^{(k-1)}}{x_1^{(k-1)}}\right|\leq\varepsilon,\quad\left|\frac{x_2^{(k)}-x_2^{(k-1)}}{x_2^{(k-1)}}\right|\leq\varepsilon,\quad\left|\frac{x_3^{(k)}-x_3^{(k-1)}}{x_3^{(k-1)}}\right|\leq\varepsilon \tag{4.11}$$

ε value in Eq. 4.11 indicates the tolerance value, and it is chosen very small such as $\varepsilon = 0.001$.

Jacobi method does not guarantee converging the true solution. For convergence, diagonal terms must be dominant in their rows as below:

$$\begin{aligned}|a_{11}|&>|a_{12}|+|a_{13}|\\|a_{22}|&>|a_{21}|+|a_{23}|\\|a_{33}|&>|a_{31}|+|a_{32}|\end{aligned} \tag{4.12}$$

Convergence condition can be expressed in general form as $|a_{ii}|>\sum_{\substack{j=1\\j\neq i}}^{n}|a_{ij}|$ is $i=1$, …, n. Fortunately, most engineering problems satisfy this condition. If diagonal terms are not dominant, changing rows of equations may help satisfy this condition. Changing the order of equations is often called as pivoting.

For example, consider the following linear system and its matrix form:

$$\begin{aligned}2x_2+3x_3&=12\\-x_1+5x_2+6x_3&=9\\9x_1+10x_2+7x_3&=5\end{aligned}\rightarrow\underbrace{\begin{bmatrix}0&2&3\\-1&5&6\\9&10&7\end{bmatrix}}_{A}\underbrace{\begin{Bmatrix}x_1\\x_2\\x_3\end{Bmatrix}}_{X}=\underbrace{\begin{Bmatrix}12\\9\\5\end{Bmatrix}}_{B}\rightarrow A=\begin{bmatrix}0&2&3\\-1&5&6\\9&10&7\end{bmatrix}$$

Above system cannot be solved directly with Jacobi method since the first diagonal term in A matrix is zero. However, replacing first and third equations will make first diagonal non-zero, therefore enabling the solution with Jacobi method.

$$\begin{aligned}9x_1+10x_2+7x_3&=5\\-x_1+5x_2+6x_3&=9\\2x_2+3x_3&=12\end{aligned}\rightarrow\underbrace{\begin{bmatrix}9&10&7\\-1&5&6\\0&2&3\end{bmatrix}}_{A}\underbrace{\begin{Bmatrix}x_1\\x_2\\x_3\end{Bmatrix}}_{X}=\underbrace{\begin{Bmatrix}5\\9\\12\end{Bmatrix}}_{B}\rightarrow A=\begin{bmatrix}9&10&7\\-1&5&6\\0&2&3\end{bmatrix}$$

4.5.2 Gauss-Siedel Method

Solution of linear equation systems with Gauss-Siedel method is very similar to Jacobi method. The only difference is the most recent values are used in iterations as below:

- *Iteration 1*:

$$x_1^{(1)} = \frac{b_1 - \left(a_{12}x_2^{(0)} + a_{13}x_3^{(0)}\right)}{a_{11}}$$

$$x_2^{(1)} = \frac{b_2 - \left(a_{21}x_1^{(1)} + a_{23}x_3^{(0)}\right)}{a_{22}}$$

$$x_3^{(1)} = \frac{b_3 - \left(a_{31}x_1^{(1)} + a_{32}x_2^{(1)}\right)}{a_{33}}$$

  From the above expressions, it is seen that in calculation of x_2, the most recent value of x_1 (i.e., $x_1^{(1)}$) is used. In calculation of x_3, the most recent values of x_1 and x_2 (i.e., $x_1^{(1)}$ and $x_2^{(1)}$) are used.
- *Iteration 2*:

$$x_1^{(2)} = \frac{b_1 - \left(a_{12}x_2^{(1)} + a_{13}x_3^{(1)}\right)}{a_{11}}$$

$$x_2^{(2)} = \frac{b_2 - \left(a_{21}x_1^{(2)} + a_{23}x_3^{(1)}\right)}{a_{22}}$$

$$x_3^{(2)} = \frac{b_3 - \left(a_{31}x_1^{(2)} + a_{32}x_2^{(2)}\right)}{a_{33}}$$

  From the above expressions, it is seen that in calculation of x_2, the most recent value of x_1 (i.e., $x_1^{(2)}$) is used. In calculation of x_3, the most recent values of x_1 and x_2 (i.e., $x_1^{(2)}$ and $x_2^{(2)}$) are used.
- *Iteration k*:

$$x_1^{(k)} = \frac{b_1 - \left(a_{12}x_2^{(k-1)} + a_{13}x_3^{(k-1)}\right)}{a_{11}}$$

$$x_2^{(k)} = \frac{b_2 - \left(a_{21}x_1^{(k)} + a_{23}x_3^{(k-1)}\right)}{a_{22}}$$

$$x_3^{(k)} = \frac{b_3 - \left(a_{31}x_1^{(k)} + a_{32}x_2^{(k)}\right)}{a_{33}}$$

From the above expressions, it is seen that in calculation of x_2, the most recent value of x_1 (i.e., $x_1^{(k)}$) is used. In calculation of x_3, the most recent values of x_1 and x_2 (i.e., $x_1^{(k)}$ and $x_2^{(k)}$) are used.

Stopping or convergence criterion explained in the Jacobi method can also be used in Gauss-Siedel method. Since Gauss-Siedel method uses the most recent values in each iteration, it converges faster than Jacobi method. Therefore, it is more efficient than Jacobi method in iterative solutions of linear systems. The convergence condition in Gauss-Siedel method is the same as with Jacobi method. For convergence, diagonal terms must be greater than the sum of other terms in that row as expressed with $|a_{ii}| > \sum_{\substack{j=1 \\ j \neq i}}^{n} |a_{ij}|$, $i = 1, \ldots, n$ equation.

Example 4.7: Find the solution of the following linear system using Gauss-Siedel method

$$
\begin{aligned}
&18x_1 + 4x_2 - 3x_3 + 5x_4 = 91 \\
&-2x_1 - 24x_2 + 7x_3 - 3x_4 = -79 \\
&4x_1 - x_2 + 32x_3 + 5x_4 = 55 \\
&-9x_1 - 7x_2 - 3x_3 + 40x_4 = 20
\end{aligned}
$$

Above linear system satisfies the convergence condition of $|a_{ii}| > \sum_{\substack{j=1 \\ j \neq i}}^{n} |a_{ij}|$ as shown below:

$$
\begin{aligned}
|a_{11}| > |a_{12}| + |a_{13}| + |a_{14}| &\rightarrow 18 > 4 + |-3| + 5, \\
|a_{22}| > |a_{21}| + |a_{23}| + |a_{24}| &\rightarrow |-24| > |-2| + 7 + |-3|, \\
|a_{33}| > |a_{31}| + |a_{32}| + |a_{34}| &\rightarrow 32 > 4 + |-1| + 5, \\
|a_{44}| > |a_{41}| + |a_{42}| + |a_{43}| &\rightarrow 40 > |-9| + |-7| + |-3|,
\end{aligned}
$$

Therefore, this system can be solved with Gauss-Siedel method. Iterative solution schema with Gauss-Siedel method can be expressed as below (k indicates iteration counter).

$$x_1^{(k)} = \frac{b_1 - \left(a_{12}x_2^{(k-1)} + a_{13}x_3^{(k-1)} + a_{14}x_4^{(k-1)}\right)}{a_{11}}$$

$$x_2^{(k)} = \frac{b_2 - \left(a_{21}x_1^{(k)} + a_{23}x_3^{(k-1)} + a_{24}x_4^{(k-1)}\right)}{a_{22}}$$

$$x_3^{(k)} = \frac{b_3 - \left(a_{31}x_1^{(k)} + a_{32}x_2^{(k)} + a_{34}x_3^{(k-1)}\right)}{a_{33}}$$

$$x_4^{(k)} = \frac{b_4 - \left(a_{41}x_1^{(k)} + a_{42}x_2^{(k)} + a_{43}x_3^{(k)}\right)}{a_{44}}$$

Selecting initial values as $x_1^0 = 0, x_2^0 = 0, x_3^0 = 0, x_4^0 = 0$, solution can be obtained iteratively as below:

- *Iteration 1*:

$$x_1^{(1)} = \frac{b_1 - \left(a_{12}x_2^{(0)} + a_{13}x_3^{(0)} + a_{14}x_4^{(0)}\right)}{a_{11}} = \frac{91}{18} = 5.0556$$

$$x_2^{(1)} = \frac{b_2 - \left(a_{21}x_1^{(1)} + a_{23}x_3^{(0)} + a_{24}x_4^{(0)}\right)}{a_{22}} = \frac{\text{-79-(-2)5.0556}}{\text{-24}} = 2.8704$$

$$x_3^{(1)} = \frac{b_3 - \left(a_{31}x_1^{(1)} + a_{32}x_2^{(1)} + a_{34}x_3^{(0)}\right)}{a_{33}} = \frac{\text{55-4(5.0556)} + \text{(-1)2.8704}}{32} = 1.1765$$

$$x_4^{(1)} = \frac{b_4 - \left(a_{41}x_1^{(1)} + a_{42}x_2^{(1)} + a_{43}x_3^{(1)}\right)}{a_{44}}$$
$$= \frac{\text{20-(-9)5.0556} + \text{(-7)2.8704} + \text{(-3)1.1765}}{40} = 2.2281$$

- *Iteration 2:*

$$x_1^{(2)} = \frac{b_1 - \left(a_{12}x_2^{(1)} + a_{13}x_3^{(1)} + a_{14}x_4^{(1)}\right)}{a_{11}}$$
$$= \frac{91 - 4(2.2281) + \text{(-3)1.1765} + 5(2.2281)}{18} = 3.9949$$

$$x_2^{(2)} = \frac{b_2 - \left(a_{21}x_1^{(2)} + a_{23}x_3^{(1)} + a_{24}x_4^{(1)}\right)}{a_{22}}$$
$$= \frac{\text{-79-(-2)3.9949} + 7(1.1765) + \text{(-3)2.2281}}{\text{-24}} = 3.0234$$

$$x_3^{(2)} = \frac{b_3 - \left(a_{31}x_1^{(2)} + a_{32}x_2^{(2)} + a_{34}x_3^{(1)}\right)}{a_{33}}$$
$$= \frac{55\text{-}4(3.9949) + (\text{-}1)3.0234 + 5(2.2281)}{32} = 0.9657$$

$$x_4^{(2)} = \frac{b_4 - \left(a_{41}x_1^{(2)} + a_{42}x_2^{(2)} + a_{43}x_3^{(2)}\right)}{a_{44}}$$
$$= \frac{20\text{-}(\text{-}9)3.9949 + (\text{-}7)3.0234 + (\text{-}3)0.9657}{40} = 2.0004$$

Similarly, other iterations can be conducted until convergence criteria is satisfied. Using convergence tolerance of $\varepsilon = 0.005$, convergence is achieved after four iterations as $x_1^{(4)} = 4.0033$, $x_2^{(4)} = 3.0005$, $x_3^{(4)} = 1.0002$,$x_4^{(4)} = 2.0009$.

Solution with MATLAB Commands:
Linear system $AX = B$ can be solved using the *linsolve* command in MATLAB as below:

```
>> A=[18 4 -3 5; -2 -24 7 -3; 4 -1 32 5; -9 -7 -3 40]; B=[91; -79; 55; 20];
>> X=linsolve(A,B);
>> X
X =
4.0000
3.0000
1.0000
2.0000
```

***Note*:** Solution of this problem with Gauss-Siedel method can also be obtained using *MATLAB Program 4.3: Gauss-Siedel Method* in Appendix.

4.5.3 Optimization Method

Optimization methods normally solve minimization or maximization problems using iterative approaches. However, if linear equations can be transformed into an optimization problem, they can be solved using optimization methods as well. To enable this, all equations in the linear system are written as equal to zero and expressions in the left side are considered as functions of $f_1, f_2, \ldots, f_n$ as following:

$$
\begin{array}{ll}
a_{11}x_1 + a_{12}x_2 + a_{13}x_3 - b_1 = 0 & f_1(x_1, x_2, x_3) = a_{11}x_1 + a_{12}x_2 + a_{13}x_3 - b_1 \\
a_{21}x_1 + a_{22}x_2 + a_{23}x_3 - b_2 = 0 \rightarrow & f_2(x_1, x_2, x_3) = a_{21}x_1 + a_{22}x_2 + a_{23}x_3 - b_2 \\
a_{31}x_1 + a_{32}x_2 + a_{33}x_3 - b_3 = 0 & f_3(x_1, x_2, x_3) = a_{31}x_1 + a_{32}x_2 + a_{33}x_3 - b_3
\end{array}
$$

Then a single objective function F is created by adding squares of $f_1, f_2,\ldots, f_n$ functions as $F = f_1^2 + f_2^2 + f_3^2$. Minimum of F function gives solution of the linear system as below:

$$
F = f_1^2 + f_2^2 + f_3^2 \Rightarrow \min \rightarrow x_1, x_2, \ldots, x_n \tag{4.13}
$$

From Eq. 4.13, it can be seen that minimum value of F can be zero ($F = 0$) since F is expressed by sum of squares of $f_1, f_2,\ldots, f_n$ functions. Minimum value of $F = 0$ is achieved only when $f_1 = 0, f_2 = 0,\ldots, f_n = 0$ are satisfied, leading to the original linear equation system. This proves that optimization problem and linear system are equivalent to each other. Preconditioned Conjugate Gradient (PCG) optimization method is commonly used to solve big linear systems. Optimization methods are explained in detail in Chap. 7 of Vol. 2.

Optimization methods can be divided into two main groups as below:

1. Local optimization methods that use gradient information
2. Global optimization methods that use only function values

Local optimization methods (methods that search the optimum near the initial values) are fast, but they are not successful in finding the global optimum (best of local optimums). Global optimization methods, on the other hand, are slower (requiring more computations), but they are likely to find a better optimum than local optimization methods.

Steepest-Descent method, which is the simplest of local optimization methods, solves an optimization problem iteratively (step by step), starting from an initial value. Iterative solution schema with the Steepest descent method can be written as following:

$$
x_1^{k+1} = x_1^k - \alpha\left(\frac{\partial F}{\partial x_1}\right)^k, \quad x_2^{k+1} = x_2^k - \alpha\left(\frac{\partial F}{\partial x_2}\right)^k, \quad x_3^{k+1} = x_3^k - \alpha\left(\frac{\partial F}{\partial x_3}\right)^k \tag{4.14}
$$

In Eq. 4.14, k shows the iteration counter and α the step length parameter. α parameter is usually determined from the solution of an auxiliary optimization problem. In a simple approach, α is given a pre-specified value and reduction of function value at new point x_i^{k+1} is checked using $f(x^{k+1}) < f(x^k)$ expression. If function value reduces, x_i^{k+1} point is accepted as the optimum point. If not, α is decreased (e.g., dividing it by half) successively until the function value at new point x_i^{k+1} is less than the function value at previous point x_i^k.

Example 4.8: Linear system is given with equations of $2x_1 + 3x_2 = 4$ and $-x_1 + x_2 = 9$. Express the solution of this system using optimization method.

After equations are written as equal to zero, non-zero sides can be given function names of $f_1(x_1, x_2)$ and $f_2(x_1, x_2)$ as below:

$$\begin{matrix} 2x_1 + 3x_2 - 4 = 0 \\ -x_1 + x_2 - 9 = 0 \end{matrix} \rightarrow \begin{matrix} f_1(x_1, x_2) = 2x_1 + 3x_2 - 4 \\ f_2(x_1, x_2) = -x_1 + x_2 - 9 \end{matrix}$$

A new objective function F is formed by adding squares of $f_1(x_1, x_2)$ and $f_2(x_1, x_2)$ functions. Minimization of F function will give the solution of linear system as following:

$$F = f_1(x_1, x_2)^2 + f_2(x_1, x_2)^2 = (2x_1 + 3x_2 - 4)^2 + (-x_1 + x_2 - 9)^2 \rightarrow \text{min}$$

4.6 Summary

- Numerical solutions of engineering problems are usually expressed with linear equations.
- Linear equations can be expressed in general form as below:

$$\left.\begin{matrix} a_{11}x_1 + a_{12}x_2 + \ldots + a_{1n}x_n = b_1 \\ a_{21}x_1 + a_{22}x_2 + \ldots + a_{2n}x_n = b_2 \\ \ldots \\ a_{n1}x_1 + a_{n2}x_2 + \ldots + a_{nn}x_n = b_n \end{matrix}\right\} \rightarrow AX = B$$

- Linear equations can be expressed in matrix form as $AX = B$.
- In order to solve linear equations, the number of independent equations must be equal to that of unknowns. This can be easily understood by checking if the determinant of A matrix is different from zero.
- Methods for solving linear equations are divided into two groups as Direct and Iterative methods.
- Direct methods (such as Gaussian Elimination, Decomposition, Cramer, Inverse Matrix) guarantee finding the solution.
- Since the number of operations is high in direct methods, they are generally preferred in the solution of small linear systems.
- In Gauss elimination method, linear equations are converted into upper triangular form which can be solved independently and easily as following:

$$\underbrace{\begin{bmatrix} a_{11} & a_{12} & a_{13} \\ a_{21} & a_{22} & a_{23} \\ a_{31} & a_{32} & a_{33} \end{bmatrix}}_{A} \underbrace{\begin{Bmatrix} x_1 \\ x_2 \\ x_3 \end{Bmatrix}}_{X} = \underbrace{\begin{Bmatrix} b_1 \\ b_2 \\ b_3 \end{Bmatrix}}_{B} \rightarrow \begin{bmatrix} a_{11} & a_{12} & a_{13} \\ 0 & a'_{22} & a'_{23} \\ 0 & 0 & a''_{33} \end{bmatrix} \begin{Bmatrix} x_1 \\ x_2 \\ x_3 \end{Bmatrix} = \begin{Bmatrix} b_1 \\ b'_2 \\ b''_3 \end{Bmatrix} \rightarrow \begin{matrix} \Rightarrow x_1 \\ \Rightarrow x_2 \\ \Rightarrow x_3 = b''_3/a''_3 \end{matrix}$$

- In Cramer method, unknowns are calculated by using determinant ratios of a new A_i matrix and original A matrix. A_i matrix is obtained by modifying A matrix and its form varies according to the unknown to be solved as below:

$$x_1 = \frac{\overbrace{\begin{vmatrix} b_1 & a_{12} & a_{13} \\ b_2 & a_{22} & a_{23} \\ b_3 & a_{32} & a_{33} \end{vmatrix}}^{A_1}}{\underbrace{\begin{vmatrix} a_{11} & a_{12} & a_{13} \\ a_{21} & a_{22} & a_{23} \\ a_{31} & a_{32} & a_{33} \end{vmatrix}}_{A}} = \frac{|A_1|}{|A|}, \quad x_2 = \frac{\overbrace{\begin{vmatrix} a_{11} & b_1 & a_{13} \\ a_{21} & b_2 & a_{23} \\ a_{31} & b_3 & a_{33} \end{vmatrix}}^{A_2}}{\underbrace{\begin{vmatrix} a_{11} & a_{12} & a_{13} \\ a_{21} & a_{22} & a_{23} \\ a_{31} & a_{32} & a_{33} \end{vmatrix}}_{A}} = \frac{|A_2|}{|A|}, \quad x_3 = \frac{\overbrace{\begin{vmatrix} a_{11} & a_{12} & b_1 \\ a_{21} & a_{22} & b_2 \\ a_{31} & a_{32} & b_3 \end{vmatrix}}^{A_3}}{\underbrace{\begin{vmatrix} a_{11} & a_{12} & a_{13} \\ a_{21} & a_{22} & a_{23} \\ a_{31} & a_{32} & a_{33} \end{vmatrix}}_{A}} = \frac{|A_3|}{|A|}$$

- In decomposition method, coefficient matrix A is decomposed into the product of two special matrices in the form of $A = LU$ where L indicates the lower triangular matrix and U indicates the upper triangular matrix as below:

$$\underbrace{\begin{bmatrix} a_{11} & a_{12} & a_{13} \\ a_{21} & a_{22} & a_{23} \\ a_{31} & a_{32} & a_{33} \end{bmatrix}}_{A} = \underbrace{\begin{bmatrix} L_{11} & 0 & 0 \\ L_{21} & L_{22} & 0 \\ L_{31} & L_{32} & L_{33} \end{bmatrix}}_{L} \underbrace{\begin{bmatrix} 1 & U_{12} & U_{13} \\ 0 & 1 & U_{23} \\ 0 & 0 & 1 \end{bmatrix}}_{U}$$

- In decomposition method, $AX = B$ equation can alternatively be written as $LUX = B$ using $A = LU$ definition. Using new $LUX = B$ form, linear system can be converted into the solution of two linear subsystems expressed as $LY = B$ (Y temporary unknown) and $UX = Y$ (X main unknown). Both $LY = B$ and $UX = Y$ subsystems correspond to the linear equations that can be solved independently. From $LY = B$ equation, temporary unknowns Y are obtained. Then, substituting Y values into $UX = Y$ equation, main unknowns X are solved.
- In inverse method, solution of the linear system $AX = B$ is obtained by multiplying the inverse of A matrix with B vector as $AX = B \Rightarrow X = [A]^{-1}B$.
- Inverse of A matrix can be calculated using Adjoint, Gauss-Jordan elimination, or decomposition method.
- Iterative methods (such as Jacobi, Gauss-Siedel, Optimization methods) solve linear systems iteratively starting from an initial value (solution estimation).

- Iterative methods do not guarantee finding the solution. Convergence condition must be satisfied to obtain a solution. For convergence, diagonal terms in A matrix must be dominant in their corresponding rows.
- Iterative methods can find solution faster than direct methods if a good initial value is chosen.
- Iterative methods are generally preferred in the solution of big linear systems.
- In Jacobi method, unknowns are left alone and iteratively solved (substituting *k-1* iteration values to calculate new *k* iteration values) as below:

$$\left.\begin{aligned} a_{11}x_1+a_{12}x_2+a_{13}x_3=b_1 \\ a_{21}x_1+a_{22}x_2+a_{23}x_3=b_2 \\ a_{31}x_1+a_{32}x_2+a_{33}x_3=b_3 \end{aligned}\right\} \Rightarrow \quad \begin{aligned} x_1^{(k)} &= \frac{b_1-\left(a_{12}x_2^{(k-1)}+a_{13}x_3^{(k-1)}\right)}{a_{11}} \\ x_2^{(k)} &= \frac{b_2-\left(a_{21}x_1^{(k-1)}+a_{23}x_3^{(k-1)}\right)}{a_{22}} \\ x_3^{(k)} &= \frac{b_3-\left(a_{31}x_1^{(k-1)}+a_{32}x_2^{(k-1)}\right)}{a_{33}} \end{aligned}$$

- Gauss-Siedel method is very similar to Jacobi method. The only difference is the most recent values are used in iterations as following:

$$\left.\begin{aligned} a_{11}x_1+a_{12}x_2+a_{13}x_3=b_1 \\ a_{21}x_1+a_{22}x_2+a_{23}x_3=b_2 \\ a_{31}x_1+a_{32}x_2+a_{33}x_3=b_3 \end{aligned}\right\} \Rightarrow \quad \begin{aligned} x_1^{(k)} &= \frac{b_1-\left(a_{12}x_2^{(k-1)}+a_{13}x_3^{(k-1)}\right)}{a_{11}} \\ x_2^{(k)} &= \frac{b_2-\left(a_{21}x_1^{(k)}+a_{23}x_3^{(k-1)}\right)}{a_{22}} \\ x_3^{(k)} &= \frac{b_3-\left(a_{31}x_1^{(k)}+a_{32}x_2^{(k)}\right)}{a_{33}} \end{aligned}$$

- Optimization methods originally solve minimization or maximization problems ($F \Rightarrow \min$) using iterative approaches. However, they can be used to solve linear equations by converting them into an optimization problem as below:

$$\begin{aligned} & a_{11}x_1+a_{12}x_2+a_{13}x_3-b_1=0 \quad f_1(x_1,x_2,x_3)=a_{11}x_1+a_{12}x_2+a_{13}x_3-b_1 \\ \text{Linear system:}\ & a_{21}x_1+a_{22}x_2+a_{23}x_3-b_2=0 \rightarrow f_2(x_1,x_2,x_3)=a_{21}x_1+a_{22}x_2+a_{23}x_3-b_2 \\ & a_{31}x_1+a_{32}x_2+a_{33}x_3-b_3=0 \quad f_3(x_1,x_2,x_3)=a_{31}x_1+a_{32}x_2+a_{33}x_3-b_3 \end{aligned}$$

$$\text{Optimization problem}: F=f_1^2+f_2^2+f_3^2 \Rightarrow \min \rightarrow x_1,x_2,\ldots,x_n.$$

Appendix: MATLAB Programs

MATLAB Program 4.1: Gaussian Elimination Method

```
function LinearEquation_Solution_GaussElimination

% SOLUTION OF LINEAR EQUATION SYSTEM WITH GAUSS ELIMINATION METHOD

A=[18     4     -3      5 ;
   -2   -24      7     -3 ;
    4    -1     32      5 ;
   -9    -7     -3     40 ]  ;

B=[91 ; -79 ;  55 ; 20 ] ;
% ======================================================================

n=length(B) ; % length of B vector

% 1- ELIMINATION STEP
for ii=1:n-1
    for i=ii+1:n
        PivotCoeff=(A(i,ii)/A(ii,ii)) ;

        for j=ii:n
            A(i,j)=A(i,j)-PivotCoeff*A(ii,j) ;
        end

        B(i)=B(i)-PivotCoeff*B(ii)   ;
    end
end

% 2- SUBSTITUTION STEP
X=zeros(n,1) ;
for i=n:-1:1
    sum=0 ;
    for j=1:n
        if j~=i
            sum=sum+A(i,j)*X(j) ;
        end
    end
    X(i)=(B(i)-sum)/A(i,i) ;
end

fprintf('Solution: X \n') ;
X
```

MATLAB Program 4.2: Cholesky Decomposition Method

```
function LinearEquation_Solution_LU_Decomposition

% SOLUTION OF LINEAR EQUATION SYSTEM WITH CHOLESKY L*U DECOMPOSITION METHOD

A=[18    4    -3     5 ;
   -2   -24    7    -3 ;
    4   -1     32    5 ;
   -9   -7    -3    40 ]  ;

B=[91 ; -79 ;  55 ; 20 ] ;
% =====================================================================

n=length(B) ; % length of B vector

L=zeros(n,n) ;
U=zeros(n,n) ;

for i=1:n
    U(i,i)=1 ;
end

% A=L*U Decomposition
for i=1:n
    for j=1:n
        if j<=i
            sum=0 ;
            for k=1:j-1
                sum=sum+L(i,k)*U(k,j) ;
            end

            L(i,j)=A(i,j)-sum ;
        else
            sum=0 ;
            for l=1:i-1
                sum=sum+L(i,l)*U(l,j) ;
            end

            U(i,j)=(A(i,j)-sum)/L(i,i) ;
        end
    end
end


% L*Y=B Solution
Y=zeros(n,1) ;
for i=1:n
    sum=0 ;
    for j=1:n
        if j~=i
            sum=sum+L(i,j)*Y(j) ;
        end
    end
    Y(i)=(B(i)-sum)/L(i,i) ;
end

% U*X=Y Solution
X=zeros(n,1) ;
for i=n:-1:1
    sum=0 ;
    for j=1:n
        if j~=i
            sum=sum+U(i,j)*X(j) ;
        end
    end
    X(i)=Y(i)-sum ;
end

fprintf('Solution: X \n') ;
X
```

MATLAB Program 4.3: Gaussian Siedel Method

```
function LinearEquation_Solution_Gauss_Siedel

% SOLUTION OF LINEAR EQUATION SYSTEM WITH GAUSS-SIEDEL ITERATION METHOD

A=[18     4     -3      5 ;
   -2   -24      7     -3 ;
    4    -1     32      5 ;
   -9    -7     -3     40 ]  ;

B=[91 ; -79 ;  55 ; 20 ] ;

n=length(B) ; % length of B vector

ConvTol=0.005 ; % Convergence tolerance
MaxIter=1000 ;  % Maximum iteration number
% =====================================================================

% Initial Values
X=zeros(n,1) ;
Conv=1e10 ;
it=0 ;

X0=X ;

while (Conv>=ConvTol) && (it<=MaxIter)
    it=it+1 ;

    for i=1:n
        sum=0 ;
        for j=1:n
            if j~=i
                sum=sum+A(i,j)*X(j) ;
            end
        end
        X(i)=(B(i)-sum)/A(i,i) ;
    end

    % Convergence Check
    for i=1:n
        if abs(X0(i))<=1e-10
            ConvVec(i)=abs(X(i)-X0(i)) ;
        else
            ConvVec(i)=abs((X(i)-X0(i))/X0(i)) ;
        end
    end

    Conv=max(ConvVec) ;
    X0=X ;
end

fprintf('Solution: X \n') ;
X
```

Problems

Problem 4.1: Write the following linear equation system in the matrix form of $A^*X = B$ and then check singularity of A matrix.

$$\begin{aligned} &3x_1 + x_2 - 3x_3 = 8 \\ &-x_1 - 2x_2 + 2x_3 = 6 \\ &4x_1 - 2x_2 - 2x_3 = 4 \end{aligned}$$

Problem 4.2: Solve the following linear equation system using Gaussian Elimination method.

$$\begin{aligned} &3x_1 + 4x_2 - 3x_3 = 7 \\ &-2x_1 - 4x_2 + 7x_3 = 2 \\ &4x_1 - x_2 + 2x_3 = 5 \end{aligned}$$

Problem 4.3: Solve the following linear equation system using Cramer method.

$$\begin{aligned} &2x_1 + 9x_2 + 7x_3 = 5 \\ &-3x_1 + 5x_2 + 6x_3 = 9 \\ &x_1 + 2x_2 + 3x_3 = 4 \end{aligned}$$

Problem 4.4: Decompose $A = \begin{bmatrix} 4 & 2 & 1 \\ 3 & 6 & 2 \\ 1 & 4 & 8 \end{bmatrix}$ matrix into L^*U form (L: lower triangular matrix, U upper triangular matrix).

Problem 4.5: Solve the following linear equation system using Cholesky (L^*U) decomposition method.

$$\begin{aligned} &3x_1 - 4x_2 + 2x_3 = 5 \\ &-2x_1 + 5x_2 + 3x_3 = 6 \\ &2x_1 - 2x_2 + x_3 = 4 \end{aligned}$$

Problem 4.6: Find the inverse of the following A matrix using Adjoint method.

$$A = \begin{bmatrix} 4 & 2 & 1 \\ 3 & 6 & 2 \\ 1 & 4 & 8 \end{bmatrix}$$

Problem 4.7: Find the inverse of the following A matrix using Gauss-Jordan elimination method.

$$A = \begin{bmatrix} 4 & 2 & 1 \\ 3 & 6 & 2 \\ 1 & 4 & 8 \end{bmatrix}$$

Problem 4.8: Find the solution of the following linear system using Inverse Matrix method.

$$\begin{aligned} &4x_1 - 4x_2 - 3x_3 = 4 \\ &-2x_1 + 6x_2 - 3x_3 = 5 \\ &4x_1 - 2x_2 + x_3 = 8 \end{aligned}$$

Problem 4.9: Can the following linear system be solved with iterative methods?

$$\begin{aligned} &4x_1 - 4x_2 - 3x_3 = 4 \\ &-2x_1 + 6x_2 - 3x_3 = 5 \\ &4x_1 - 2x_2 + x_3 = 8 \end{aligned}$$

Problem 4.10: Can the following linear system be solved with iterative Gauss-Siedel Method? If yes, solve it with Gauss-Siedel Method.

$$\begin{aligned} &20x_1 + 4x_2 - 3x_3 + 5x_4 = 11 \\ &-12x_1 - 34x_2 + 7x_3 - 3x_4 = -22 \\ &4x_1 - 5x_2 + 21x_3 + 3x_4 = 33 \\ &-8x_1 + 6x_2 - 3x_3 + 40x_4 = 40 \end{aligned}$$

References

1. Chapra, S. C., & Canale, R. P. (2020). *Numerical methods for engineers*. McGraw-Hill.
2. Bernatz, R. A. (2010). *Fourier series and numerical methods for partial differential equations*. John Wiley & Sons.
3. Everstine, G. C. (2010). *Numerical solution of partial differential equations* [Lecture notes, PDF]. George Washington University. https://www.academia.edu/31518755/Numerical_Solution_of_Partial_Differential_Equations
4. Epperson, J. F. (2013). *An introduction to numerical methods and analysis*. John Wiley & Sons.
5. Burden, R. L., & Faires, J. D. (2011). *Numerical analysis*. Cengage Learning.

6. Sert, C. (2006). *ME310 numerical methods* [Lecture notes, PDF]. Middle East Technical University. https://users.metu.edu.tr/csert/me310/me310_1_introduction.pdf
7. Grewal, B. S. (2019). *Numerical methods in engineering and science*. Mercury Learning and Information.
8. Sastry, S. S. (2012). *Introductory methods of numerical analysis*. PHI Learning.

Chapter 5
Root Finding and Solution of Nonlinear Equations

5.1 Introduction

Finding the x values that zero an $f(x)$ function (i.e., $f(x) = 0$) is called as root finding. In graphical representation, intersection of $f(x)$ function with x axis gives its roots as shown in Fig. 5.1.

Calculation of buckling load in buckling problems and calculation of natural frequencies in vibration problems all correspond to root finding problems in engineering. For example, critical buckling load P for a beam with simply supported boundary conditions as shown in Fig. 5.2 can be expressed as root finding problem in the form of $\sin\left(\sqrt{\frac{P}{EI}}L\right)=0$.

Similarly, calculation of natural frequency ω for various beams can be expressed as root finding problem as shown in Fig. 5.3.

Analytical (closed form) and numerical methods can be used for the solution of root finding problems [1–8]. Analytical methods are generally available for roots of low-order polynomials. For example, roots of a quadratic polynomial can be calculated analytically using *discriminant* method as following:

$$ax^2 + bx + c = 0 \rightarrow x_{1,2} = \frac{-b \mp \sqrt{b^2 - 4ac}}{2a} \tag{5.1}$$

However, analytical (closed-form) methods do not exist for higher order polynomial or non-polynomial functions. For this reason, numerical methods are used to find roots of such functions. Numerical methods are general and can be applied to any problem. Numerical methods can be divided into four groups: (1) Graphical method: In this method, roots are determined by drawing graphic of the function. (2) Screening method: In this method, starting from an initial point, root domain is scanned with very small steps until the root ($f(x) = 0$) can be detected. (3) Interval-based methods: In these methods, a predetermined interval or region is successively narrowed until the interval can be converted into root point. (4) Point-based

H. Kurtaran, *Engineering Mathematics with MATLAB Applications: Vol. 1, Fundamentals*, https://doi.org/10.1007/978-3-031-81869-1_5

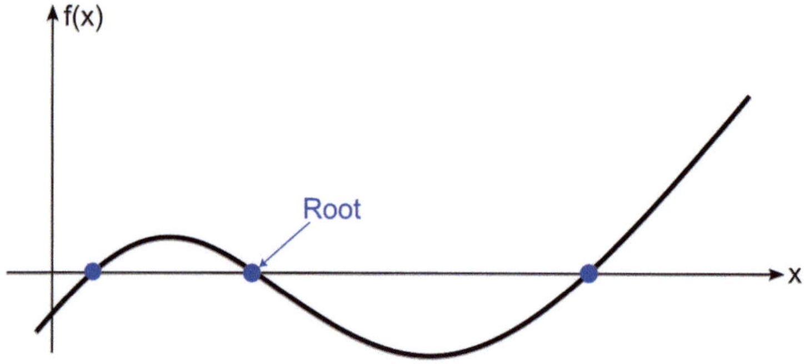

Fig. 5.1 Roots of an $f(x)$ function

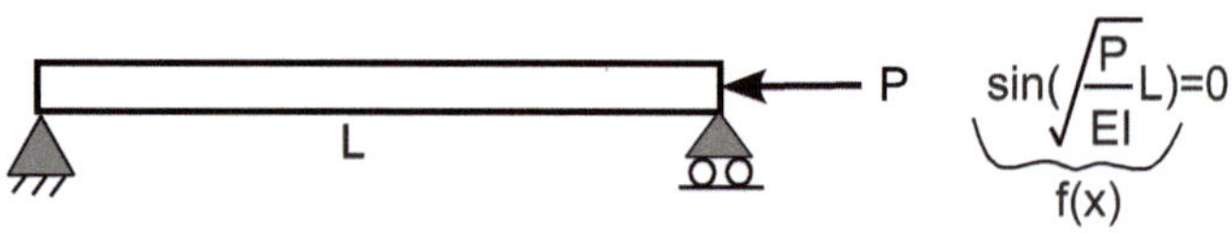

Fig. 5.2 Calculation of critical buckling load for a simply supported beam by solving root finding problem

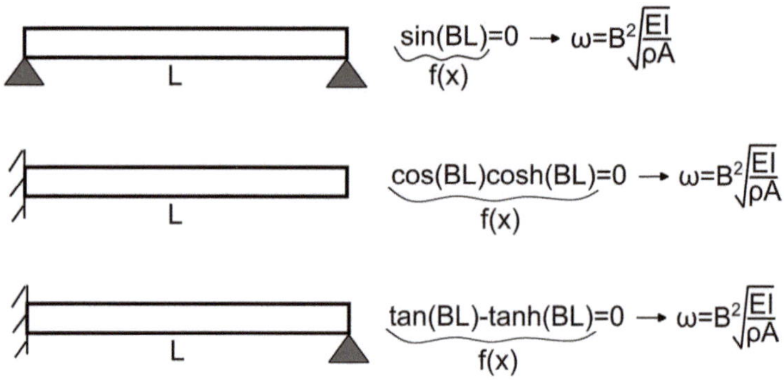

Fig. 5.3 Calculation of natural frequencies for beams with various boundary conditions by solving root finding problems

methods: In these methods, an initial estimate of root value (initial value) is successively (iteratively) corrected until the actual root is converged.

Interval-based methods guarantee finding a root value when it is bracketed. In contrast, convergence problem can be faced with point-based methods. If the initial value is not chosen sufficiently close to the actual root, root value may not be converged. In short, success of point-based methods highly depends on the feature of the function and the selection of an appropriate initial value. Root finding methods can also be extended into the solution of nonlinear equations. Root finding methods are explained in detail below.

5.2 Root Finding with Graphical Method

In this method, graphic of the function is drawn precisely as in Fig. 5.4, and the root is determined with an eye check on the graphic. The method is very simple but it requires a high-quality plot of the function.

5.3 Root Finding with Scanning Method

Starting from an initial point, function is scanned with small steps. Scanning is stopped when the function value is sufficiently close to zero. Method is very simple, but it requires very small steps to capture the root precisely. This can increase the computational cost of the root finding process. Therefore, scanning method is usually not preferred in the precise determination of an actual root value.

Implementation of scanning method is shown in Fig. 5.5. In implementation, an initial value x_0 is selected and function value $f(x_0)$ at this point is calculated. If $f(x_0)$ is not close to zero, a new function value $f(x_1 = x_0 + \Delta x)$ is calculated by increasing x_0 value with Δx step length. This process is repeated until the new function value $f(x_i = x_0 + i\Delta x)$ is sufficiently close to zero. When stopped, final x value of $x_i = x_0 + i\Delta x$ is accepted as the root value.

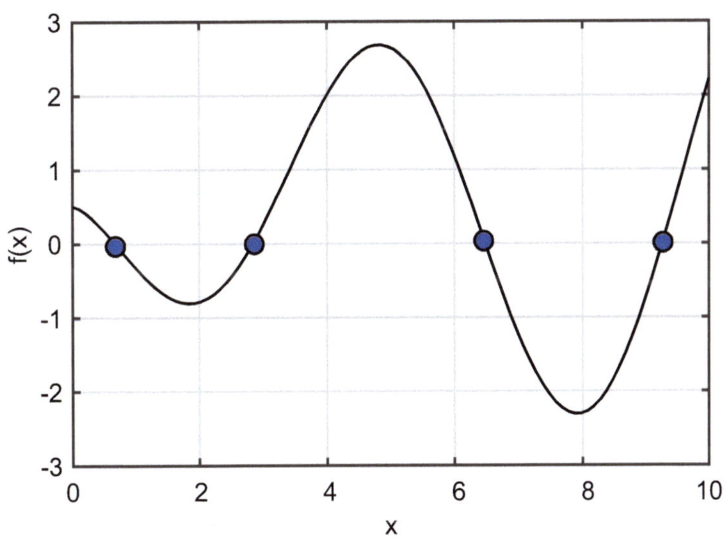

Fig. 5.4 Determining roots of a function by drawing its graphic

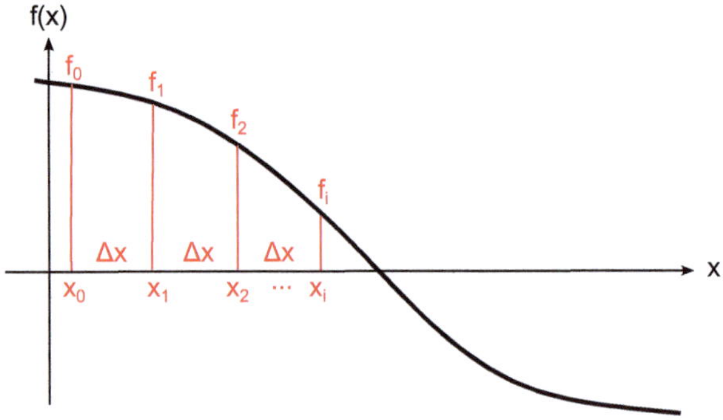

Fig. 5.5 Determining the root by scanning the function starting from an initial point

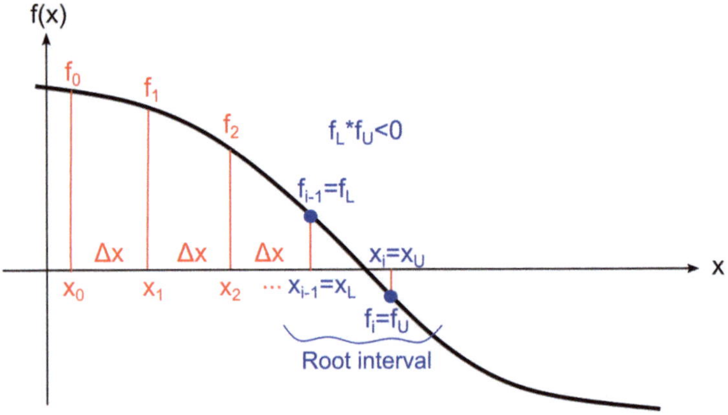

Fig. 5.6 Determination of the interval containing the root with scanning

5.4 Root Finding with Interval-Based Methods

Interval-based methods are more efficient than scanning method. In these methods, an interval that contains a root is determined first. For this purpose, scanning method can be used. As shown in Fig. 5.6, in determining the interval, two neighboring points (x_0 and $x_1 = x_0 + \Delta x$) are selected and their corresponding function values ($f(x_0)$ and $f(x_1) = f(x_0 + \Delta x)$) are calculated. If two function values have opposite signs (zero is between negative and positive values), then $[x_0, x_0 + \Delta x]$ interval is chosen as the interval of the root. Otherwise, new points are created until the last two points have function values with opposite signs. After proceeding like this, last two points $x_{i-1} = x_0 + (i - 1)\Delta x$ and $x_i = x_0 + i\Delta x$ will give the limits of the root interval. For convenience, lower and upper bounds of the root interval will be shown with $[x_L, x_U]$ instead of $[x_{i-1}, x_i]$ in this chapter.

Upon determination of the root interval, it is narrowed iteratively by excluding the part that does not contain the root. Narrowing continues until the updated interval becomes the root point. Interval-based methods guarantee root finding. However,

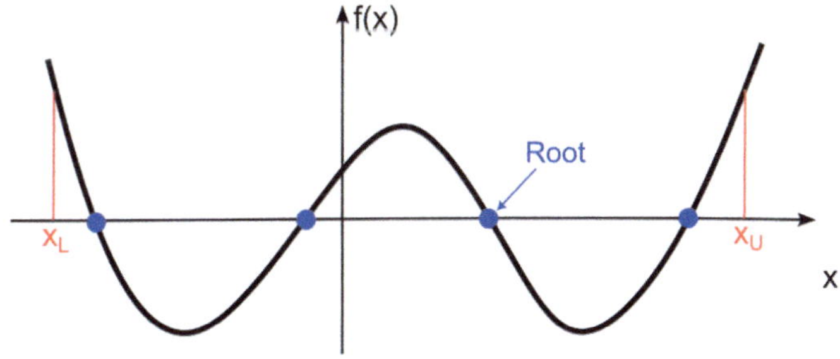

Fig. 5.7 Wide interval including multiple roots

their efficiency is lower than point-based methods. In these methods, if the root interval is chosen too wide as shown in Fig. 5.7, root finding process can be oscillatory since there may be multiple roots in it. Therefore, size of the root interval should be sufficiently small to include only a single root. Interval-based root finding methods are explained in detail below.

***Note*:** In order to determine the root interval, *MATLAB Program 5.1: Determining Root Interval* in Appendix can be used.

5.4.1 Interval Halving Method (Bisection Method)

Interval halving or Bisection method is an iterative root finding method. At each iteration, half of the interval that does not include the root is discarded. In Bisection method, root is assumed to be the middle of $[x_L, x_U]$ interval. Therefore, it is expressed as $x_M = \frac{x_L + x_U}{2}$. If the function value $f(x_M)$ in the middle of the interval is sufficiently close to zero, the midpoint becomes the root. Otherwise, half of the interval is discarded (narrowed) and root search continues within the remaining interval. In interval halving, sign of $f(x_M)$ is checked. If $f(x_M)f(x_L) > 0$ is satisfied, then root is considered to be in the right half of the interval defined by the range of $[x_M, x_U]$. Otherwise, root is considered to be in the left half of the interval defined by the range of $[x_L, x_M]$. Bisection iterations by narrowing intervals are repeated until the function value at new midpoint ($f(x_M)$) is close enough to zero. Figures 5.8 and 5.9 show first two iterations of the interval halving method.

In interval halving method, root search continues until the new interval becomes small enough or the function value in the middle is close to zero. In this method, one or both of the following convergence criteria can be used simultaneously to stop iterations (k: iteration number):

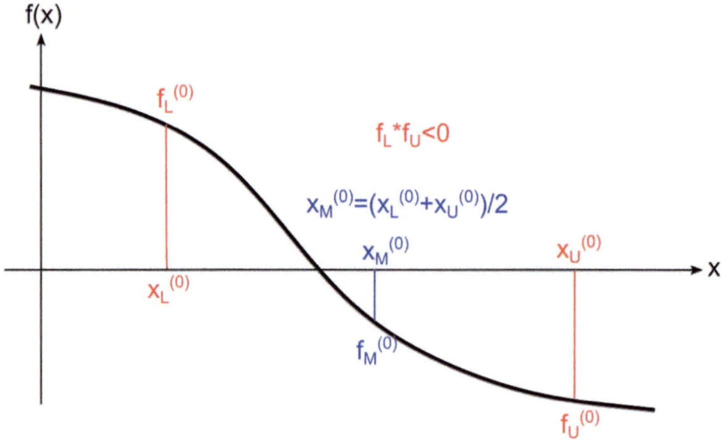

Fig. 5.8 First iteration: determination of the root value as the midpoint of the interval

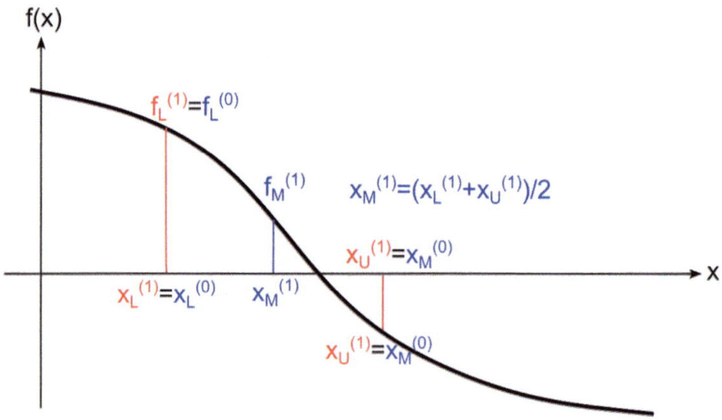

Fig. 5.9 Second iteration: narrowing the root interval and determining the new midpoint as the updated root estimate

1. Interval size (in normalized form):

$$\frac{\left|x_U^{(k)} - x_L^{(k)}\right|}{\left|x_U^{(0)} - x_L^{(0)}\right|} \leq \varepsilon \tag{5.2}$$

2. Function value at middle point:

$$|f(x_M)| \leq \varepsilon \tag{5.3}$$

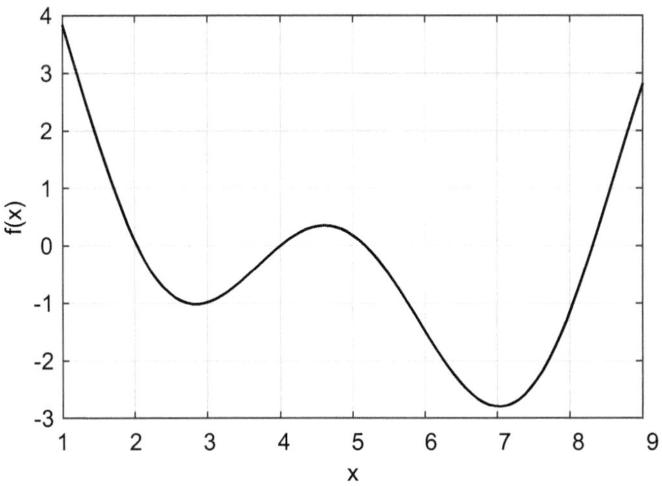

Fig. 5.10 Graphics of $f(x) = (2 - \sqrt{x})[4\cos x + 2\sin x]$ function

Example 5.1: Find the root of $f(x) = (2 - \sqrt{x})(4\cos x + 2\sin x)$ function using Bisection method.

Graphic of $f(x) = (2 - \sqrt{x})(4\cos x + 2\sin x)$ function can be potted as in Fig. 5.10 to identify the initial root interval.

Root search with Bisection method can be applied iteratively as following.

- *Iteration 1*:

  From Fig. 5.10, initial interval containing the root can be taken as $\left[x_L^{(1)}, x_U^{(1)}\right]$ = [0,3]. Function values for this interval are calculated as $\left[f\left(x_L^{(1)}\right), f\left(x_U^{(1)}\right)\right]$ = [8, −0.9854]. Midpoint of this interval and function value at this point can be calculated as $x_M^{(1)}$ = 1.5 and $f\left(x_M^{(1)}\right)$ = 1.7660. Since $f\left(x_M^{(1)}\right)$ value is not sufficiently close to zero, iteration continues. Interval is narrowed by half and new interval limits are determined as $\left[x_L^{(2)}, x_U^{(2)}\right]$ = [1.5,3] since $f\left(x_M^{(1)}\right) f\left(x_U^{(1)}\right) < 0$.
- *Iteration 2*:

  For the new interval $\left[x_L^{(2)}, x_U^{(2)}\right]$ = [1.5,3], function values are obtained as $\left[f\left(x_L^{(2)}\right), f\left(x_U^{(2)}\right)\right]$ = [1.7660, −0.9854]. Midpoint of this interval and function value at this point can be calculated as $x_M^{(2)} = 2.25$ and $f\left(x_M^{(2)}\right) = -0.4783$. Since

Table 5.1 Iteration results with Bisection method

Iteration	x_L	x_M	x_U	f_L	f_M	f_U
1	0	1.5000	3	8	1.7660	−0.9854
2	1.5000	2.2500	3	1.7660	−0.4783	−0.9854
3	1.5000	1.8750	2.2500	1.7660	0.4478	−0.4783
4	1.8750	2.0625	2.2500	0.4478	−0.0707	−0.4783
5	1.8750	1.9688	2.0625	0.4478	0.1752	−0.0707
6	1.9688	2.0156	2.0625	0.1752	0.0488	−0.0707
7	2.0156	2.0391	2.0625	0.0488	−0.0118	−0.0707
8	2.0156	2.0273	2.0391	0.0488	0.0183	−0.0118
9	2.0273	2.0332	2.0391	0.0183	0.0032	−0.0118
10	2.0332	2.0361	2.0391	0.0032	−0.0043	−0.0118

$f\left(x_M^{(2)}\right)$ value is not sufficiently close to zero, iteration continues. Interval is narrowed by half and new interval limits are determined as $\left[x_L^{(3)}, x_U^{(3)}\right]$ $= [1.5,2.25]$ since $f\left(x_L^{(2)}\right)f\left(x_M^{(2)}\right) < 0$. If iterations are continued until convergence, results shown in Table 5.1 are obtained.

***Note*:** For the calculation of root with Bisection Method, *MATLAB Program 5.2: Root Finding with Bisection Method* in Appendix can be used.

5.4.2 Linear Interpolation (False-Position) Method

In this method, linear polynomial approximation of the original function is constructed within the root interval of $[x_L, x_U]$. Root of the linear polynomial is accepted as the root value. Linear polynomial function passing through the boundaries of the interval $[x_L, x_U]$ can be generated using Lagrange interpolation method (method will be explained in Chap. 6) as below:

$$f(x) = \frac{x - x_U}{x_L - x_U} f_L + \frac{x - x_L}{x_U - x_L} f_U \tag{5.4}$$

Root of the linear $f(x)$ function can be calculated as x_R as following:

$$0 = \frac{x - x_U}{x_L - x_U} f_L + \frac{x - x_L}{x_U - x_L} f_U \rightarrow x = x_R = \frac{x_L f_U - x_U f_L}{f_U - f_L} \tag{5.5}$$

After calculating the root estimate x_R, function value $f_R = f(x_R)$ at this point is checked if it is close enough to zero ($f_R = f(x_R) \approx 0$). If not, interval is narrowed and root value is sought within the narrowed interval. In narrowing the interval, sign

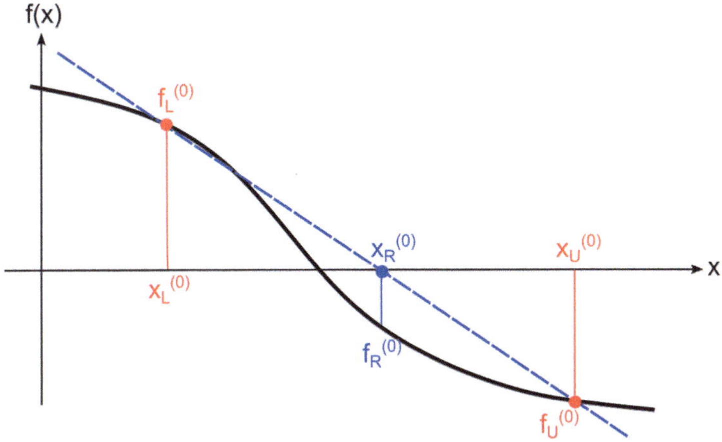

Fig. 5.11 First iteration: determination of the initial root interval and estimation of the root by linear interpolation

of f_R is compared with those of f_L and f_U. If f_R has the same sign with f_L, new interval containing the root is determined as $[x_R, x_U]$; otherwise, new interval containing the root is determined as $[x_L, x_R]$. Root finding and narrowing process are repeated until the function value at new estimated root point is close enough to zero or the convergence condition is satisfied. One or both of the following convergence criteria can be used to stop iterations (k: iteration number):

1. Interval size (in normalized form):

$$\frac{\left|x_U^{(k)} - x_L^{(k)}\right|}{\left|x_U^{(0)} - x_L^{(0)}\right|} \leq \varepsilon \tag{5.6}$$

2. Function value at root:

$$|f(x_R)| \leq \varepsilon \tag{5.7}$$

Figures 5.11 and 5.12 show first two iterations of root searching process with linear interpolation method.

Example 5.2: Find the root of $f(x) = (2 - \sqrt{x})(4\cos x + 2\sin x)$ function in [0,3] interval using Linear Interpolation (False-Position) method.

Root searching with linear interpolation (False-Position) method can be performed iteratively as following:

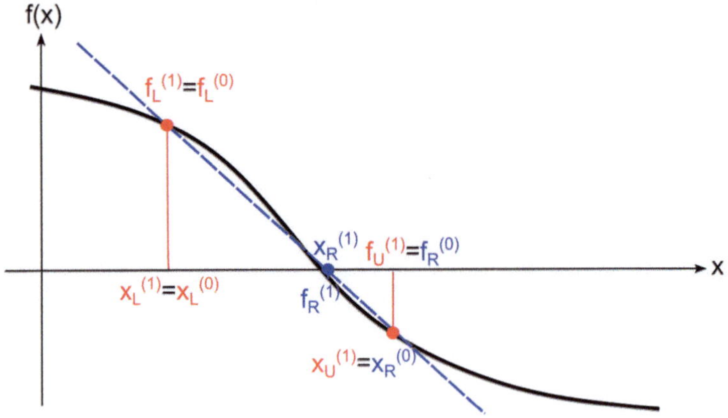

Fig. 5.12 Second iteration: root estimation by linear interpolation in the narrowed interval

- *Iteration 1*:

  For the initial root interval $\left[x_L^{(1)}, x_U^{(1)}\right] = [0,3]$, function values are calculated as $\left[f\left(x_L^{(1)}\right), f\left(x_U^{(1)}\right)\right] = [8, -0.9854]$. With linear interpolation method, the root point and corresponding function value are obtained as $x_R^{(1)} = 2.6710$ and $f\left(x_R^{(1)}\right) = -0.9721$. Since $f\left(x_R^{(1)}\right)$ value is not close enough to zero, iteration continues after narrowing the interval. Narrowed interval is obtained as $\left[x_L^{(2)}, x_U^{(2)}\right] = [0,2.6710]$ since $f\left(x_L^{(1)}\right)f\left(x_R^{(1)}\right) < 0$.

- *Iteration 2*:

  For the new interval $\left[x_L^{(2)}, x_U^{(2)}\right] = [0, 2.6710]$, function values are calculated as $\left[f\left(x_L^{(2)}\right), f\left(x_U^{(2)}\right)\right] = [8, -0.9721]$. With linear interpolation (False-Position) method, the root point and corresponding function value are obtained as $x_R^{(2)} = 2.3816$ and $f\left(x_R^{(2)}\right) = -0.6950$. Since $f\left(x_R^{(2)}\right)$ value is still not close enough to zero, iteration continues. Narrowed interval is obtained as $\left[x_L^{(3)}, x_U^{(3)}\right] = [0,2.3816]$ since $f\left(x_L^{(2)}\right)f\left(x_R^{(2)}\right) < 0$. If iterations are repeated until convergence, results shown in Table 5.2 are obtained.

Table 5.2 Iteration results with linear interpolation (False-position) method

Iteration	x_L	x_R	x_U	f_L	f_R	f_U
1	0	2.6710	3	8	−0.9721	−0.9854
2	0	2.3816	2.6710	8	−0.6950	−0.9721
3	0	2.1912	2.3816	8	−0.3629	−0.6950
4	0	2.0961	2.1912	8	−0.1523	−0.3629
5	0	2.0570	2.0961	8	−0.0570	−0.1523
6	0	2.0424	2.0570	8	−0.0204	−0.0570
7	0	2.0372	2.0424	8	−0.0072	−0.0204
8	0	2.0354	2.0372	8	−0.0025	−0.0072
9	0	2.0348	2.0354	8	−0.0009	−0.0025

***Note*:** For the calculation of root with Linear Interpolation (False-Position) Method, *MATLAB Program 5.3: Root Finding with Linear Interpolation (False-Position) Method* in Appendix can be used.

5.5 Root Finding with Point-Based Methods

Point-based methods require estimate of the root referred to as initial point. They start searching the root from the initial point and progress step by step (iteratively) towards the actual root value. Although they do not guarantee root finding, their convergence speed is generally higher than interval-based methods. By choosing a good initial point, convergence can be ensured and accelerated. Common point-based methods are explained below:

5.5.1 *Simple Iteration Method*

In simple iteration method, root equation $f(x) = 0$ is rewritten in $x = g(x)$ form. Then, initial value is selected for x, and this value is substituted in $g(x)$ leading to the corrected or improved x value. Substituting known or old x value in $g(x)$ function and obtaining the new or improved x value from $x = g(x)$ equation is repeated until the convergence or stopping condition is satisfied as following:

$$x^{(1)} = g\left(x^{(0)}\right) \rightarrow x^{(2)} = g\left(x^{(1)}\right) \rightarrow x^{(3)} = g\left(x^{(2)}\right) \rightarrow \ldots \rightarrow x^{(k)} = g\left(x^{(k-1)}\right) \tag{5.8}$$

Iterations are stopped when right and left sides of $x = g(x)$ equations are sufficiently close to each other or the following convergence criteria is satisfied.

Change in variable (in normalized form):

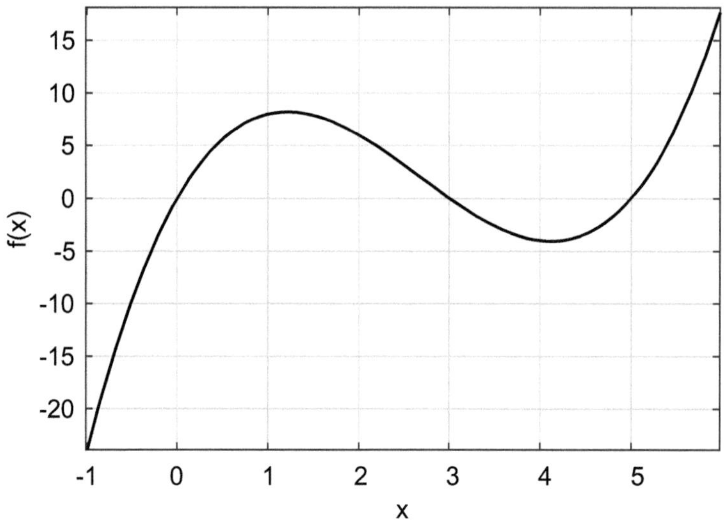

Fig. 5.13 Graphic of $f(x) = x^3 - 8x^2 + 15x$ function

$$\left|\frac{x^{(k)} - x^{(k-1)}}{x^{(k-1)}}\right| \leq \varepsilon \tag{5.9}$$

In Eq. 5.9, ε indicates the convergence tolerance. Simple iteration method although very simple may not always converge to the actual root. In simple iteration method, arrangement of $f(x) = 0$ equation in $x = g(x)$ form is very critical. If certain conditions are not satisfied for $g(x)$ function, iterations might even diverge from the actual root. In simple iteration method, convergence can be ensured by expressing $g(x)$ function such that its derivative is <1, i.e., $|g'(x)| < 1$.

Example 5.3: Find the root of $f(x) = x^3 - 8x^2 + 15x$ function with Simple Iteration method using the initial value of $x_0 = 4$.

Graphic of $f(x) = x^3 - 8x^2 + 15x$ function can be plotted as in Fig. 5.13 to identify the location of roots.

$f(x) = x^3 - 8x^2 + 15x$ function can be written in $x = g(x)$ form in several ways. One alternative can be expressed as below:

$$x^3 - 8x^2 + 15x = 0 \rightarrow x^3 + 15x = 8x^2 \rightarrow x = \underbrace{\sqrt{\frac{x^3 + 15x}{8}}}_{g(x)} \rightarrow g(x) = \sqrt{\frac{x^3 + 15x}{8}}$$

Using the above $g(x)$ function, iterative root finding schema for k-th iteration can be written as following:

Table 5.3 Iteration results with simple iteration method

Iteration	x	$g(x)$
1	4	3.9370
2	3.9370	3.8743
3	3.8743	3.8122
4	3.8122	3.7515
...	...	...
42	3.0090	3.0079
43	3.0079	3.0069

$$x^{(k)} = g\left(x^{(k-1)}\right) = \sqrt{\frac{\left(x^{(k-1)}\right)^3 + 15\left(x^{(k-1)}\right)}{8}}$$

Above iterative schema yields the following iteration results.

- *Iteration 1*:
  For Iteration 1, root finding schema is expressed as $x^{(1)} = g\left(x^{(0)}\right) = \sqrt{\frac{\left(x^{(0)}\right)^3 + 15\left(x^{(0)}\right)}{8}}$. Substituting the initial value of $x^{(0)} = 4$ in $g(x^{(0)})$ function, corrected root value is obtained as $x^{(1)} = 3.9370$.
- *Iteration 2*:
  For Iteration 2, root finding schema is expressed as $x^{(2)} = g\left(x^{(1)}\right) = \sqrt{\frac{\left(x^{(1)}\right)^3 + 15\left(x^{(1)}\right)}{8}}$. Substituting $x^{(1)} = 3.9370$ value in $g(x^{(1)})$ function, corrected root value is obtained as $x^{(2)} = 3.8743$. If iterations are repeated until convergence, results shown in Table 5.3 are obtained.

Solution with MATLAB commands:

(a) Root value close to the initial value of $x0 = 4$ can be obtained using *fzero* command in MATLAB as below:

```
>> f='x^3-8*x^2+15*x'; r=fzero(f,4);
>> r
r =
3.0000
```

(b) All roots (3 roots) can be obtained using *roots* command in MATLAB as following:

```
>> r= roots([1 -8 15 0]);
>> r
r =
0
5
3
```

***Note*:** For the calculation of root with Simple Iteration (Substitution) Method, *MATLAB Program 5.4: Root Finding with Simple Iteration (Substitution) Method* in Appendix can be used.

5.5.2 Newton–Raphson Method

In Newton–Raphson method, $f(x)$ function around the root is approximated with a *linear polynomial* tangent to the function as in Fig. 5.14. Root of the tangent function is calculated as the estimate of the actual root. In Newton–Raphson method, Taylor series expansion is used to obtain the linear approximation of $f(x)$ function.

Taylor series expansion forms the base of Newton–Raphson method. Taylor series expansion of an $f(x)$ function at point x_0 can be written using function and derivative values as following:

$$f(x) = f(x_0) + f'(x_0)(x - x_0) + \frac{f''(x_0)}{2!}(x - x_0)^2 + \frac{f^{(3)}(x_0)}{3!}(x - x_0)^3 + \ldots + \frac{f^{(n)}(x_0)}{n!}(x - x_0)^n \tag{5.10}$$

If first two terms (linear terms) are taken from Taylor series expansion in Eq. 5.10, following linear approximation of the function is obtained

$$f(x) \approx f(x_0) + f'(x_0)(x - x_0) \tag{5.11}$$

Above expression is also called as *linearization*. Root of the linearized function can be obtained as below:

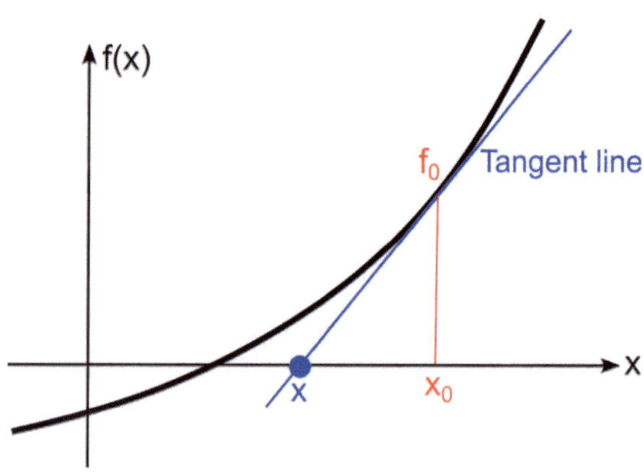

Fig. 5.14 Determination of the root by extending the tangent line in Newton–Raphson method

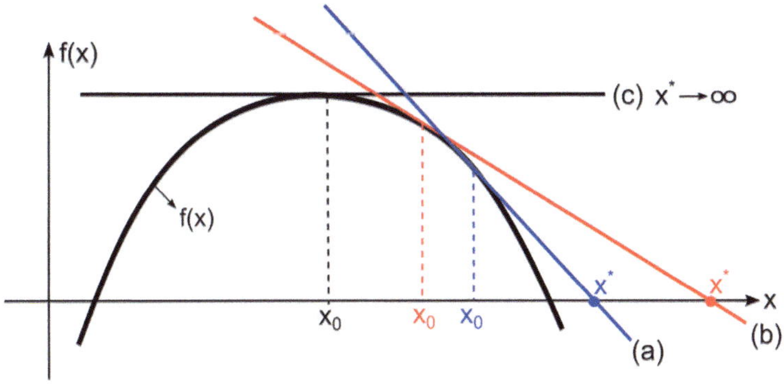

Fig. 5.15 Approximate root value (x^*) for different initial values (x_0) named as (a), (b), and (c) cases

$$f(x) \approx f(x_0) + f'(x_0)(x - x_0) = 0 \rightarrow x = x_0 - \frac{f(x_0)}{f'(x_0)} \tag{5.12}$$

Root value obtained with Eq. 5.12 corresponds to the intersection point of the tangent line with x axis. In Newton–Raphson method, since $f(x)$ function is approximated with a tangent line, that is with a linear function, root value obtained with Eq. 5.12 will be only the approximation of the true value. To converge to the true root value, tangent line approximation of $f(x)$ function should be repeated successively until convergence condition can be satisfied. For successive root search, Newton–Raphson method can be written in iterative form as following:

$$x^{(k)} = x^{(k-1)} - \frac{f\left(x^{(k-1)}\right)}{f'\left(x^{(k-1)}\right)} \quad (k : \text{iteration number}) \tag{5.13}$$

Stopping or convergence criterion for iterations can be expressed as following:

Change in variable in normalized form:

$$\left| \frac{x^{(k)} - x^{(k-1)}}{x^{(k-1)}} \right| \leq \varepsilon \tag{5.14}$$

Function value:

$$\left| f\left(x^{(k)}\right) \right| \leq \varepsilon \tag{5.15}$$

In Newton–Raphson method, Eq. 5.14 or Eq. 5.15 or both of them can be used simultaneously to stop iterations. In Newton–Raphson method to achieve convergence, initial value must be chosen carefully. Initial value should not make the denominator term zero in Eq. 5.13. Otherwise, approximate root value will be infinite. This case is illustrated in Fig. 5.15. In Fig. 5.15, approximate root value

(x^*) with different initial values named as (a), (b) and (c) cases are shown. From Fig. 5.15, it is seen that initial value corresponding to (c) case makes tangent zero. When tangent is zero as in (c) case, tangent line will never cross x axis leading to indefinite (infinite) or no root value.

Newton–Raphson method can be summarized as following:

- It needs an initial value to start the solution.
- Since first derivative of function is used, derivative expression should be easily obtained.
- Since iteration scheme includes first derivative in the denominator term, initial value should not make first derivative zero.
- It converges very quickly if the initial point is chosen close to the actual root.

Example 5.4: Find the root of $f(x) = (2 - \sqrt{x})(4\cos x + 2\sin x)$ function with Newton–Raphson method using the initial value of $x_0 = 1$.

Root of $f(x) = (2 - \sqrt{x})(4\cos x + 2\sin x)$ function can be expressed in iterative form using Newton–Raphson method as following:

$$x^{(k)} = x^{(k-1)} - \frac{f\left(x^{(k-1)}\right)}{f'\left(x^{(k-1)}\right)}$$

$$= x^{(k-1)} - \frac{\left[\left(2-\sqrt{x^{(k-1)}}\right)\left[4\cos x^{(k-1)} + 2\sin x^{(k-1)}\right]\right]}{\left[\left(-\frac{1}{2\sqrt{x^{(k-1)}}}\right)\left[4\cos x^{(k-1)} + 2\sin x^{(k-1)}\right] + \left(2-\sqrt{x^{(k-1)}}\right)\left[-4\sin x^{(k-1)} + 2\cos x^{(k-1)}\right]\right]}$$

Above iterative root expression can be applied as below:

- *Iteration 1*:

  Using initial value of $x^{(0)} = 1$, corrected root value and its corresponding function value are obtained as $x^{(1)} = x^{(0)} - \frac{f\left(x^{(0)}\right)}{f'\left(x^{(0)}\right)} = 1.9137$ and $f(x^{(1)}) = 0.3322$. Since new function value $f(x^{(1)})$ is not close enough to zero, iteration continues.
- *Iteration 2*:

  Using $x^{(1)} = 1.9137$, corrected root value and its corresponding function value are obtained as $x^{(2)} = x^{(1)} - \frac{f\left(x^{(1)}\right)}{f'\left(x^{(1)}\right)} = 2.0270$ and $f(x^{(2)}) = 0.0192$. Since new function value $f(x^{(2)})$ is not close enough to zero, iteration continues.
- *Iteration 3*:

  Using $x^{(2)} = 2.0270$, corrected root value and its corresponding function value are obtained as $x^{(3)} = x^{(2)} - \frac{f\left(x^{(2)}\right)}{f'\left(x^{(2)}\right)} = 2.0344$ and $f(x^{(3)}) = 0.0001$. Since new function value $f(x^{(3)})$ is close enough to zero, iteration is stopped and $x^{(3)} = 2.0344$ is accepted as the actual root value. Iteration results until convergence using Newton–Raphson method are summarized in Table 5.4.

Table 5.4 Iteration results with Newton–Raphson method

Iteration	x	$f(x)$	df/dx
1	1.9137	0.3322	−2.9324
2	2.0270	0.0192	−2.5889
3	2.0344	0.0001	−2.5656

Solution with MATLAB Commands:
Using initial value of $x0 = 1$, root of the function can be calculated using *fzero* command in MATLAB as below:

```
>> f='(2-sqrt(2))*(4*cos(x)+2*sin(x))'; r= fzero(f,1);
>> r
r =
2.0344
```

Note: For the calculation of root with Newton–Raphson Method, *MATLAB Program 5.5: Root Finding with Newton–Raphson Method* in Appendix can be used.

5.5.3 *Secant Method*

Newton–Raphson method has superior convergence property. However, it requires the expression of first derivative of the function which might be difficult to obtain for complicated functions. To alleviate this disadvantage, Secant method was developed. Secant method is obtained by replacing first derivative in Newton–Raphson method with Finite Difference expression using function values at two adjacent (x_0 and x_1) points as below:

$$f'(x_1) = \frac{f(x_1) - f(x_0)}{x_1 - x_0} \tag{5.16}$$

Using the derivative expression in Eq. 5.16, root of an $f(x)$ function with Secant method can be written as following:

$$x = x_1 - \frac{f(x_1)}{f'(x_1)} = x_1 - \frac{f(x_1)}{\frac{f(x_1) - f(x_0)}{x_1 - x_0}} \tag{5.17}$$

From Fig. 5.16, it is seen that $f(x)$ function in Secant method is approximated with a line passing through $f(x_0)$ and $f(x_1)$ points. Therefore, root value obtained with Eq. 5.17 corresponds to the intersection point of this line with x axis. Since $f(x)$ function is approximated with a line in Secant method, root value obtained with Eq. 5.17 will be only the approximation of the true value. To converge to the true

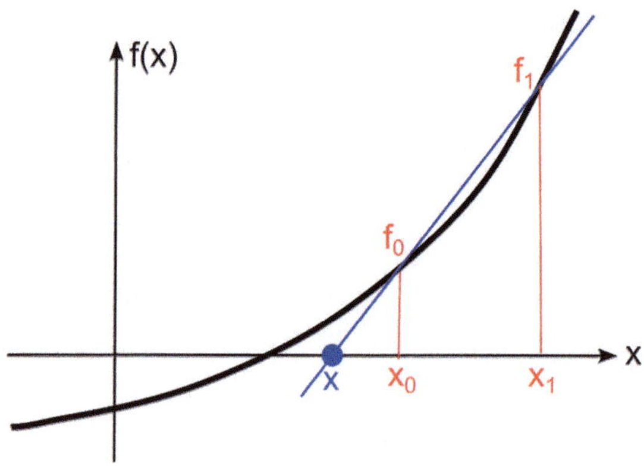

Fig. 5.16 Determining the root in Secant method by extending the line passing through $f(x_0)$ and $f(x_1)$ values

root value, line approximation of $f(x)$ function should be repeated successively until convergence condition is satisfied.

For successive root search, Secant method can be written in iterative form as following:

$$x^{(k)} = x^{(k-1)} - \frac{f\left(x^{(k-1)}\right)}{\frac{f(x^{(k-1)}) - f(x^{(k-2)})}{x^{(k-1)} - x^{(k-2)}}} \quad (k : \text{iteration number}) \tag{5.18}$$

Convergence criteria specified in Newton–Raphson method can also be used in Secant method to stop iterations. Features of secant method can be summarized as below:

- Two initial values (x_0 and x_1) are needed to start the solution.
- It is easier to implement than Newton–Raphson method since it does not directly use the derivative information.
- Convergence rate is slightly lower than Newton–Raphson method.

Example 5.5: Find the root of $f(x) = (2 - \sqrt{x})(4\cos x + 2\sin x)$ function with Secant method using the initial values of $x^{(-1)} = 0$ and $x^{(0)} = 1$.

Secant method can be written in iterative form for $f(x) = (2 - \sqrt{x}) \times (4\cos x + 2\sin x)$ function as below:

$$x^{(k)} = x^{(k-1)} - \frac{f\left(x^{(k-1)}\right)}{\frac{f\left(x^{(k-1)}\right) - f\left(x^{(k-2)}\right)}{x^{(k-1)} - x^{(k-2)}}}$$

$$= x^{(k-1)} - \frac{\left[\left(2-\sqrt{x^{(k-1)}}\right)\left[4\cos x^{(k-1)} + 2\sin x^{(k-1)}\right]\right]}{\frac{\left[\left(2-\sqrt{x^{(k-1)}}\right)\left[4\cos x^{(k-1)} + 2\sin x^{(k-1)}\right]\right] - \left[-\left(2-\sqrt{x^{(k-2)}}\right)\left[4\cos x^{(k-2)} + 2\sin x^{(k-2)}\right]\right]}{x^{(k-1)} - x^{(k-2)}}}$$

Above iterative root expression can be applied as below:

Table 5.5 Iteration results with Secant method

Iteration	$x^{(k-2)}$	$x^{(k-1)}$	$x^{(k)} = x$ (Root)	$f(x)$
1	0	1	1.9250	0.2992
2	1	1.9250	2.0031	0.0820
3	1.9250	2.0031	2.0326	0.0049
4	2.0031	2.0326	2.0344	0.0001

- *Iteration 1*:
  Using the initial values of $x^{(-1)} = 0$ and $x^{(0)} = 1$, corrected root value and its corresponding function value are obtained as $x^{(1)} = x^{(0)} - \frac{f(x^{(0)})}{\frac{f(x^{(0)}) - f(x^{(-1)})}{x^{(0)} - x^{(-1)}}} = 1.9250$ and $f(x^{(1)}) = 0.2992$. Since new function value $f(x^{(1)})$ is not close enough to zero, iteration continues.
- *Iteration 2*:
  Using new initial values of $x^{(0)} = 1$ and $x^{(1)} = 1.9250$, corrected root value and its corresponding function value are obtained as $x^{(2)} = x^{(1)} - \frac{f(x^{(1)})}{\frac{f(x^{(1)}) - f(x^{(0)})}{x^{(1)} - x^{(0)}}}$ $= 2.0031$ and $f(x^{(2)}) = 0.0820$. Since new function value $f(x^{(2)})$ is not close enough to zero, iteration continues.
- *Iteration 3*:
  Using new initial values of $x^{(1)} = 1.9250$ and $x^{(2)} = 2.0031$, corrected root value and its corresponding function value are obtained as $x^{(3)} = x^{(2)} - \frac{f(x^{(2)})}{\frac{f(x^{(2)}) - f(x^{(1)})}{x^{(2)} - x^{(1)}}} = 2.0326$ and $f(x^{(3)}) = 0.0049$. Since new function value $f(x^{(3)})$ is not close enough to zero, iteration continues.
- *Iteration 4*:
  Using new initial values of $x^{(2)} = 2.0031$ and $x^{(3)} = 2.0326$, corrected root value and its corresponding function value are obtained as $x^{(4)} = x^{(3)} - \frac{f(x^{(3)})}{\frac{f(x^{(3)}) - f(x^{(2)})}{x^{(3)} - x^{(2)}}} = 2.0344$ and $f(x^{(4)}) = 0.0001$. Since new function value $f(x^{(4)})$ is close enough to zero, iteration is stopped and $x^{(4)} = 2.0344$ is accepted as the actual root value. Iteration results until convergence using Secant method are summarized in Table 5.5.

Note: For the calculation of root with Secant Method, *MATLAB Program 5.6: Root Finding with Secant Method* in Appendix can be used.

5.6 Solution of Nonlinear Equations

When the order of a variable in an algebraic equation is different from 1 (one), it is called as nonlinear equation. When there are multiple equations like that, it is called as nonlinear algebraic equation system. Nonlinear equation system can be named according to the maximum power of unknowns as second order, third order, fourth order, *m*-th order, etc. For example, *m*-th order nonlinear equation system can be expressed in general form as below:

$$\begin{aligned} & a_{1,1}x_1^m + a_{1,2}x_2^m + \ldots + a_{1,n}x_n^m + a_{1,n+1}x_1x_1 + a_{1,n+2}x_1x_2 + \ldots + a_{1,2n}x_1x_n + a_{1,2n+1}x_1 + \ldots + a_{1,3n}x_n = b_1 \\ & \ldots \\ & a_{n,1}x_1^m + a_{n,2}x_2^m + \ldots + a_{n,n}x_n^m + a_{n,n+1}x_1x_1 + a_{n,n+2}x_1x_2 + \ldots + a_{n,2n}x_1x_n + a_{n,2n+1}x_1 + \ldots + a_{n,3n}x_n = b_n \end{aligned} \tag{5.19}$$

Unfortunately, many engineering problems are not originally in the algebraic form of Eq. 5.19. Rather, they are in differential equation form. Therefore, numerical methods are utilized to convert them into the algebraic form in Eq. 5.19 for easy solution. Examples of such conversions are given below:

Beam Bending (case of large displacement):
Consider the simply supported beam under the distributed load *q* in Fig. 5.17. Also, assume that beam is fixed (i.e., displacements are restrained) in *x* direction at support points. In this case, when the load is of a big value, large displacements occur on the beam leading to nonlinear differential equations as following [9, 10] (*b*: width, *h*: height, *L*: length, *E*: elastic modulus, *u*: displacement in *x* direction, *v*: displacement in *y* direction, *M*: moment around *z* axis)

$$\begin{aligned} & \frac{dN}{dx} = 0 \quad \left(N = EA\left(\frac{du}{dx} + \frac{1}{2}\left(\frac{dv}{dx}\right)^2 \right) \right) \\ & \frac{d^2M}{dx^2} + N\frac{d^2v}{dx^2} + q(x) = 0 \quad \left(M = -EI\frac{d^2v}{dx^2} \right) \end{aligned} \tag{5.20}$$

Equation 5.20 can be written in terms of displacements as following:

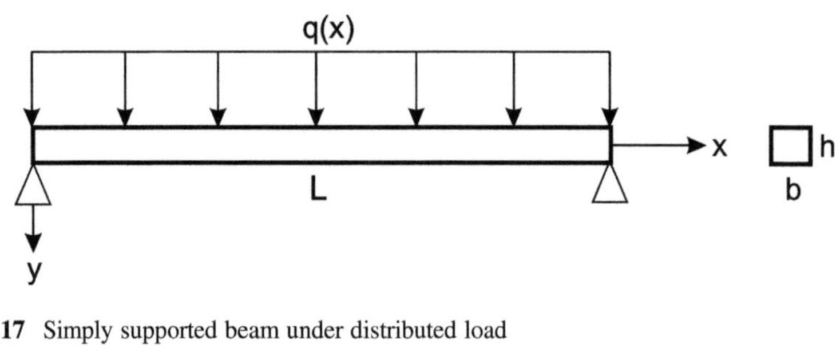

Fig. 5.17 Simply supported beam under distributed load

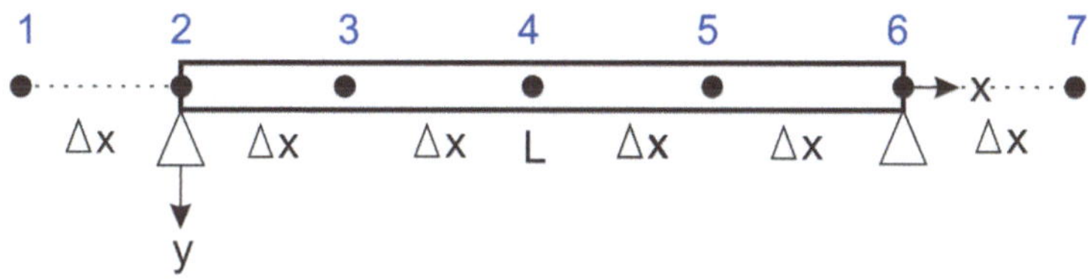

Fig. 5.18 Discretization of the beam for the solution of displacements with Finite Difference method

$$\begin{aligned} &\frac{d^2u}{dx^2}+\frac{dv}{dx}\frac{d^2v}{dx^2}=0 \\ &-EI\frac{d^4v}{dx^4}+EA\left(\frac{du}{dx}+\frac{1}{2}\left(\frac{dv}{dx}\right)^2\right)\frac{d^2v}{dx^2}+q(x)=0 \end{aligned} \tag{5.21}$$

For the simply supported beam, displacements and moment are zero at support points. In this case, boundary conditions can be expressed as following:

Left edge: $u(0) = 0$, $v(0) = 0$, $M(0)=EI\frac{d^2v}{dx^2}(0)=0$,

Right edge: $u(L) = 0$, $v(L) = 0$, $M(L)=EI\frac{d^2v}{dx^2}(L)=0$.

For a numerical solution, displacements on the beam can be discretized as in Fig. 5.18. Nodal displacement values can then be obtained using a numerical method such as Finite Difference method. In Finite Difference method (explained in Chap. 8), derivative of an $f(x)$ function at a discretized point i is expressed in terms of function values at other points (i.e., nodes). For example, first-, second-, and fourth-order derivatives can be expressed with $\frac{df}{dx}(i)=\frac{f_{i+1}-f_{i-1}}{2\Delta x}$, $\frac{d^2f}{dx^2}(i)=\frac{f_{i-1}-2f_i+f_{i+1}}{(\Delta x)^2}$ and $\frac{d^4f}{dx^4}(i)=\frac{f_{i-2}-4f_{i-1}+6f_i-4f_{i+1}+f_{i+2}}{(\Delta x)^4}$ equations, respectively.

In Fig. 5.18, only v values are unknowns at Node 1 and 7, whereas both u and v values are unknowns at other nodes. Using Finite Difference method, differential equations of the beam can be written at a general point i as below:

$$\frac{u_{i-1}-2u_i+u_{i+1}}{(\Delta x)^2}+\left(\frac{v_{i+1}-v_{i-1}}{2\Delta x}\right)\left(\frac{v_{i-1}-2v_i+v_{i+1}}{(\Delta x)^2}\right)=0 \tag{5.22a}$$

$$\begin{aligned} &-EI\left(\frac{v_{i-2}-4v_{i-1}+6v_i-4v_{i+1}+v_{i+2}}{(\Delta x)^4}\right) \\ &+EA\left(\frac{u_{i+1}-u_{i-1}}{2\Delta x}+\frac{1}{2}\left(\frac{v_{i+1}-v_{i-1}}{2\Delta x}\right)^2\right)\left(\frac{v_{i-1}-2v_i+v_{i+1}}{(\Delta x)^2}\right)+q_i \\ &=0 \end{aligned} \tag{5.22b}$$

When Finite Difference method is used in expressing the differential equation at interior points ($i = 3,4,5$) and boundary conditions at boundary points, following nonlinear algebraic equations are obtained.

$$i=3: \frac{u_2 - 2u_3 + u_4}{(\Delta x)^2} + \frac{1}{2}\left(\frac{v_4 - v_2}{2\Delta x}\right)\left(\frac{v_2 - 2v_3 + v_4}{(\Delta x)^2}\right) = 0$$

$$-EI\left(\frac{v_1 - 4v_2 + 6v_3 - 4v_4 + v_5}{(\Delta x)^4}\right) + EA\left(\frac{u_4 - u_2}{2\Delta x} + \frac{1}{2}\left(\frac{v_4 - v_2}{2\Delta x}\right)^2\right)\left(\frac{v_2 - 2v_3 + v_4}{(\Delta x)^2}\right)$$
$$+q_3 = 0 \tag{5.23a}$$

$$i=4: \frac{u_3 - 2u_4 + u_5}{(\Delta x)^2} + \frac{1}{2}\left(\frac{v_5 - v_3}{2\Delta x}\right)\left(\frac{v_3 - 2v_4 + v_5}{(\Delta x)^2}\right) = 0$$

$$-EI\left(\frac{v_2 - 4v_3 + 6v_4 - 4v_5 + v_6}{(\Delta x)^4}\right) + EA\left(\frac{u_5 - u_3}{2\Delta x} + \frac{1}{2}\left(\frac{v_5 - v_3}{2\Delta x}\right)^2\right)\left(\frac{v_3 - 2v_4 + v_5}{(\Delta x)^2}\right)$$
$$+q_4 = 0 \tag{5.23b}$$

$$i=5: \frac{u_4 - 2u_5 + u_6}{(\Delta x)^2} + \frac{1}{2}\left(\frac{v_6 - v_4}{2\Delta x}\right)\left(\frac{v_4 - 2v_5 + v_6}{(\Delta x)^2}\right) = 0$$

$$-EI\left(\frac{v_3 - 4v_4 + 6v_5 - 4v_6 + v_7}{(\Delta x)^4}\right) + EA\left(\frac{u_6 - u_4}{2\Delta x} + \frac{1}{2}\left(\frac{v_6 - v_4}{2\Delta x}\right)^2\right)\left(\frac{v_4 - 2v_5 + v_6}{(\Delta x)^2}\right)$$
$$+q_5 = 0 \tag{5.23c}$$

Boundary conditions with Finite Difference method:

Left edge: $u_2 = 0$, $v_2 = 0$,

$$M(0) = EI\frac{d^2v}{dx^2}(0) = \frac{v_1 - 2v_2 + v_3}{(\Delta x)^2} = 0 \tag{5.24a}$$

Right edge: $u_6 = 0$, $v_6 = 0$,

$$M(L) = EI\frac{d^2v}{dx^2}(L) = \frac{v_5 - 2v_6 + v_7}{(\Delta x)^2} = 0 \tag{5.24b}$$

When variables are renamed as $x_1 = v_1$, $x_2 = u_2$, $x_3 = v_2$,..., $x_{10} = u_6$, $x_{11} = v_6$, $x_{12} = v_7$, Eqs. 5.23a and 5.24a transform into a third-order nonlinear algebraic equation system similar to the form in Eq. 5.19.

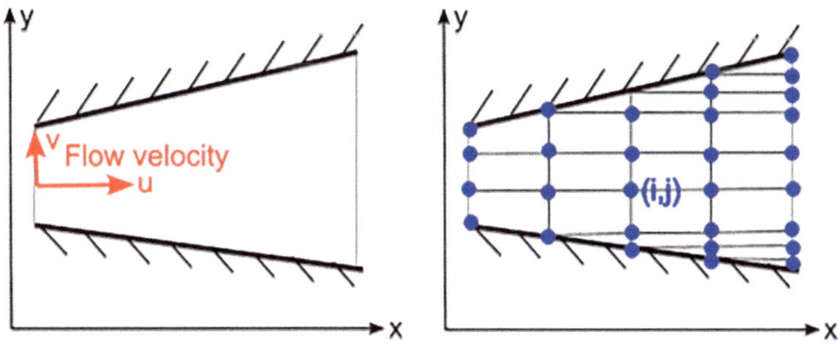

Fig. 5.19 (**a**, **b**) 2-D flow domain and its discretization

2-D Fluid flow:

Consider the steady-state 2-D flow in Fig. 5.19a where fluid is incompressible. In this case, continuity and momentum (Navier–Stokes) equations in x and y directions can be written as below:

$$\text{Continuity}: \frac{\partial u}{\partial x}+\frac{\partial v}{\partial y}=0 \tag{5.25a}$$

$$x\text{-momentum}: \rho\left(u\frac{\partial u}{\partial x}+v\frac{\partial u}{\partial y}\right)=-\frac{\partial P}{\partial x}+\mu\left(\frac{\partial^2 u}{\partial x^2}+\frac{\partial^2 u}{\partial y^2}\right) \tag{5.25b}$$

$$y\text{-momentum}: \rho\left(u\frac{\partial v}{\partial x}+v\frac{\partial v}{\partial y}\right)=-\frac{\partial P}{\partial y}+\mu\left(\frac{\partial^2 v}{\partial x^2}+\frac{\partial^2 v}{\partial y^2}\right) \tag{5.25c}$$

where μ, ρ, and P indicate viscosity coefficient, density, and pressure, whereas u and v denote flow velocities in x and y directions, respectively.

Equations 5.25a and 5.25b are nonlinear and they can be solved numerically. 2-D flow domain can be discretized as in Fig. 5.19b and then solved using appropriate numerical methods such as Finite Difference method. In Finite Difference method, derivatives in the differential equation are expressed with Finite Difference expressions. For example, first- and second-order derivatives of an $f(x, y)$ function at a general (i,j) point (i.e., node) can be expressed with Finite Difference method as $\frac{\partial f}{\partial x}(i,j)=\frac{f_{i+1,j}-f_{i-1,j}}{2\Delta x}$, $\frac{\partial f}{\partial y}(i,j)=\frac{f_{i,j+1}-f_{i,j-1}}{2\Delta y}$, $\frac{\partial^2 f}{\partial x^2}(i,j)=\frac{f_{i-1,j}-2f_{i,j}+f_{i+1,j}}{(\Delta x)^2}$ and $\frac{\partial^2 f}{\partial y^2}(i,j)=\frac{f_{i,j-1}-2f_{i,j}+f_{i,j+1}}{(\Delta y)^2}$, respectively, where Δx and Δy indicate the distance between nodes in x and y directions. Using Finite Difference expressions, Eq. 5.25a can be expressed at a general (i,j) point as following:

$$\frac{u_{i+1,j}-u_{i-1,j}}{2\Delta x}+\frac{v_{i,j+1}-v_{i,j-1}}{2\Delta y}=0 \tag{5.26a}$$

$$\rho\left(u_{i,j}\frac{u_{i+1,j}-u_{i-1,j}}{2\Delta x}+v_{i,j}\frac{u_{i,j+1}-u_{i,j-1}}{2\Delta y}\right)$$
$$=-\frac{P_{i+1,j}-P_{i-1,j}}{2\Delta x}+\mu\left(\frac{u_{i-1,j}-2u_{i,j}+u_{i+1,j}}{(\Delta x)^2}+\frac{u_{i,j-1}-2u_{i,j}+u_{i,j+1}}{(\Delta y)^2}\right) \tag{5.26b}$$

$$\rho\left(u_{i,j}\frac{v_{i+1,j}-v_{i-1,j}}{2\Delta x}+v_{i,j}\frac{v_{i,j+1}-v_{i,j-1}}{2\Delta y}\right)$$
$$=-\frac{P_{i,j+1}-P_{i,j-1}}{2\Delta y}+\mu\left(\frac{v_{i-1,j}-2v_{i,j}+v_{i+1,j}}{(\Delta x)^2}+\frac{v_{i,j-1}-2v_{i,j}+v_{i,j+1}}{(\Delta y)^2}\right) \tag{5.26c}$$

When the general schema in Eq. 5.26a is applied to the nodes in Fig. 5.19b, a second-order nonlinear algebraic equation system is obtained in terms of velocities (u, v) and pressure (P). More examples from engineering can be given for the conversion of the engineering problem into the nonlinear algebraic system as above.

To take the advantage of root finding methods in the solution of nonlinear equations, nonlinear equation system can be converted into root finding (i.e., =0) form as below:

$$\begin{aligned}
&a_{1,1}x_1^m+a_{1,2}x_2^m+\ldots+a_{1,n}x_n^m+a_{1,n+1}x_1x_1+a_{1,n+2}x_1x_2+\ldots+a_{1,2n}x_1x_n+a_{1,2n+1}x_1+\ldots+a_{1,3n}x_n-b_1=0\\
&\ldots\\
&a_{n,1}x_1^m+a_{n,2}x_2^m+\ldots+a_{n,n}x_n^m+a_{n,n+1}x_1x_1+a_{n,n+2}x_1x_2+\ldots+a_{n,2n}x_1x_n+a_{n,2n+1}x_1+\ldots+a_{n,3n}x_n-b_n=0
\end{aligned} \tag{5.27}$$

For convenience, left side of Eq. 5.27 can be given names such as $f(x)$, $g(x)$, and $h(x)$. This way nonlinear equations can be expressed as $f(x) = 0$, $g(x) = 0$, $h(x) = 0$, etc. which can be solved with Newton–Raphson method. When Newton–Raphson method is used in the solution of a nonlinear equation system, it is often called generalized Newton–Raphson method. Solution of a two-variable (consisting of two equations) and n-variable (consisting of n equations) nonlinear equation system using generalized Newton–Raphson method are explained in detail below:

5.6.1 Solution of Nonlinear System with Two Equations

When there are two nonlinear algebraic equations expressed in root finding form as $f(x,y) = 0$ and $g(x,y) = 0$, their solution can be devised as root finding problem. If values of $f(x,y)$ and $g(x,y)$ functions as well as their derivatives are known at a point (x^k, y^k), roots of these functions can be approximated using Newton–Raphson method. For this, $f(x, y)$ and $g(x, y)$ functions can be expanded into Taylor series at (x^k, y^k) point using the information in Fig. 5.20.

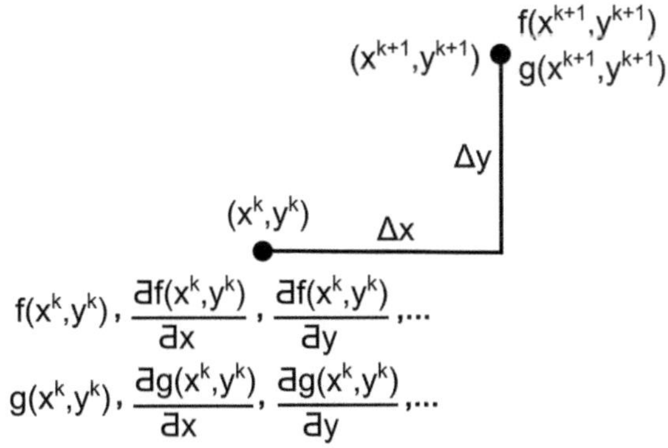

Fig. 5.20 Estimating bivariate function values at (x^{k+1}, y^{k+1}) point using function and derivative values at (x^k, y^k) point

Using Taylor series, bivariate function values at an (x^{k+1}, y^{k+1}) point can be obtained from function and derivative values at (x^k, y^k) point as following:

$$f(x^{k+1}, y^{k+1}) = f(x^k, y^k) + \frac{\partial f(x^k, y^k)}{\partial x}(x - x^k) + \frac{\partial f(x^k, y^k)}{\partial y}(y - y^k)$$
$$+ \frac{\partial^2 f(x^k, y^k)}{\partial x^2}(x - x^k)^2 + \frac{\partial^2 f(x^k, y^k)}{\partial y^2}(y - y^k)^2 + 2\frac{\partial^2 f(x^k, y^k)}{\partial x \partial y}(x - x^k)(y - y^k) + \ldots \tag{5.28a}$$

$$g(x^{k+1}, y^{k+1}) = g(x^k, y^k) + \frac{\partial g(x^k, y^k)}{\partial x}(x - x^k) + \frac{\partial g(x^k, y^k)}{\partial y}(y - y^k)$$
$$+ \frac{\partial^2 g(x^k, y^k)}{\partial x^2}(x - x^k)^2 + \frac{\partial^2 g(x^k, y^k)}{\partial y^2}(y - y^k)^2 + 2\frac{\partial^2 g(x^k, y^k)}{\partial x \partial y}(x - x^k)(y - y^k) + \ldots \tag{5.28b}$$

If only linear terms are taken from Taylor series expansions in Eq. 5.28a, $f(x, y)$ and $g(x, y)$ functions can be written as below:

$$f(x^{k+1}, y^{k+1}) \approx f(x^k, y^k) + \frac{\partial f(x^k, y^k)}{\partial x}(x - x^k) + \frac{\partial f(x^k, y^k)}{\partial y}(y - y^k)$$

$$g(x^{k+1}, y^{k+1}) \approx g(x^k, y^k) + \frac{\partial g(x^k, y^k)}{\partial x}(x - x^k) + \frac{\partial g(x^k, y^k)}{\partial y}(y - y^k)$$

Using distances of $\Delta x = x - x^k$ and $\Delta y = y - y^k$ between points, $f(x^{k+1}, y^{k+1})$ and $g(x^{k+1}, y^{k+1})$ can alternatively be expressed as following:

$$f(x^{k+1}, y^{k+1}) \approx f(x^k, y^k) + \frac{\partial f(x^k, y^k)}{\partial x}\Delta x + \frac{\partial f(x^k, y^k)}{\partial y}\Delta y$$

$$g\left(x^{k+1},y^{k+1}\right)\approx g\left(x^k,y^k\right)+\frac{\partial g\left(x^k,y^k\right)}{\partial x}\Delta x+\frac{\partial g\left(x^k,y^k\right)}{\partial y}\Delta y$$

For (x^{k+1}, y^{k+1}) point to be the actual root, $f(x^{k+1}, y^{k+1})$ and $g(x^{k+1}, y^{k+1})$ values must be equal to zero. In this case, root equations can be written as below:

$$0=f\left(x^k,y^k\right)+\frac{\partial f\left(x^k,y^k\right)}{\partial x}\Delta x+\frac{\partial f\left(x^k,y^k\right)}{\partial y}\Delta y$$

$$0=g\left(x^k,y^k\right)+\frac{\partial g\left(x^k,y^k\right)}{\partial x}\Delta x+\frac{\partial g\left(x^k,y^k\right)}{\partial y}\Delta y$$

By solving the above set of linear equations, distances of Δx and Δy can be obtained as following:

$$\begin{bmatrix}\frac{\partial f\left(x^k,y^k\right)}{\partial x} & \frac{\partial f\left(x^k,y^k\right)}{\partial y}\\ \frac{\partial g\left(x^k,y^k\right)}{\partial x} & \frac{\partial g\left(x^k,y^k\right)}{\partial y}\end{bmatrix}\begin{Bmatrix}\Delta x\\ \Delta y\end{Bmatrix}=-\begin{Bmatrix}f\left(x^k,y^k\right)\\ g\left(x^k,y^k\right)\end{Bmatrix}\rightarrow\Delta x,\ \ \Delta y \tag{5.29}$$

Equation 5.29 can be referred to as generalized Newton-Rapson method for bivariate functions. Using distances of Δx and Δy, corrected root value can be obtained as $x^{k+1} = x^k + \Delta x$ and $y^{k+1} = y^k + \Delta y$. Iterative schema in Eq. 5.29 is applied successively until $f(x^{k+1}, y^{k+1})$ and $g(x^{k+1}, y^{k+1})$ values are sufficiently close to zero (i.e., $f(x^{k+1}, y^{k+1}) \approx 0$, $g(x^{k+1}, y^{k+1}) \approx 0$). One or both of the following criteria can be used as convergence criterion

1. Increments:

$$|\Delta x|\leq\varepsilon,\quad |\Delta y|\leq\varepsilon \tag{5.30}$$

2. Function values:

$$\left|f\left(x_{i+1},y_{i+1}\right)\right|\leq\varepsilon,\quad \left|g\left(x_{i+1},y_{i+1}\right)\right|\leq\varepsilon \tag{5.31}$$

5.6.2 Solution of Nonlinear System with Multiple Equations

This time assume that there are n nonlinear algebraic equations to solve and they are given as $f^{(1)}(x_1,x_2,..,x_n)=0, f^{(2)}(x_1,x_2,..,x_n)=0,\ldots,f^{(n)}(x_1,x_2,..,x_n)=0$. Values of $f^{(1)}$, $f^{(2)}$,..., $f^{(n)}$ functions at an $\left(x_1^{(k+1)},x_2^{(k+1)},\ldots,x_n^{(k+1)}\right)$ point can be predicted using function and derivatives values at $\left(x_1^{(k)},x_2^{(k)},\ldots,x_n^{(k)}\right)$ point through Taylor series expansions. If only linear terms are used, Taylor series expansions of these functions can be expressed as following:

$$
\begin{aligned}
f^{(1)}\left(x_1^{(k+1)}, x_2^{(k+1)}, \ldots, x_n^{(k+1)}\right) \approx\ & f^{(1)}\left(x_1^{(k)}, x_2^{(k)}, \ldots, x_n^{(k)}\right) \\
& + \frac{\partial f^{(1)}\left(x_1^{(k)}, x_2^{(k)}, \ldots, x_n^{(k)}\right)}{\partial x_1} \underbrace{\left(x_1^{(k+1)} - x_1^{(k)}\right)}_{\Delta x_1} \\
& + \frac{\partial f^{(1)}\left(x_1^{(k)}, x_2^{(k)}, \ldots, x_n^{(k)}\right)}{\partial x_2} \underbrace{\left(x_2^{(k+1)} - x_2^{(k)}\right)}_{\Delta x_2} \\
& + \ldots + \frac{\partial f^{(1)}\left(x_1^{(k)}, x_2^{(k)}, \ldots, x_n^{(k)}\right)}{\partial x_n} \underbrace{\left(x_n^{(k+1)} - x_n^{(k)}\right)}_{\Delta x_n}
\end{aligned}
\tag{5.32a}
$$

$$
\begin{aligned}
f^{(2)}\left(x_1^{(k+1)}, x_2^{(k+1)}, \ldots, x_n^{(k+1)}\right) \approx\ & f^{(2)}\left(x_1^{(k)}, x_2^{(k)}, \ldots, x_n^{(k)}\right) \\
& + \frac{\partial f^{(2)}\left(x_1^{(k)}, x_2^{(k)}, \ldots, x_n^{(k)}\right)}{\partial x_1} \underbrace{\left(x_1^{(k+1)} - x_1^{(k)}\right)}_{\Delta x_1} \\
& + \frac{\partial f^{(2)}\left(x_1^{(k)}, x_2^{(k)}, \ldots, x_n^{(k)}\right)}{\partial x_2} \underbrace{\left(x_2^{(k+1)} - x_2^{(k)}\right)}_{\Delta x_2} \\
& + \ldots + \frac{\partial f^{(2)}\left(x_1^{(k)}, x_2^{(k)}, \ldots, x_n^{(k)}\right)}{\partial x_n} \underbrace{\left(x_n^{(k+1)} - x_n^{(k)}\right)}_{\Delta x_n}
\end{aligned}
\tag{5.32b}
$$

$$\ldots$$

$$f^{(n)}\left(x_1^{(k+1)},x_2^{(k+1)},\ldots,x_n^{(k+1)}\right) \approx f^{(n)}\left(x_1^{(k)},x_2^{(k)},\ldots,x_n^{(k)}\right) + \frac{\partial f^{(n)}\left(x_1^{(k)},x_2^{(k)},\ldots,x_n^{(k)}\right)}{\partial x_1}\underbrace{\left(x_1^{(k+1)}-x_1^{(k)}\right)}_{\Delta x_1} + \frac{\partial f^{(n)}\left(x_1^{(k)},x_2^{(k)},\ldots,x_n^{(k)}\right)}{\partial x_2}\underbrace{\left(x_2^{(k+1)}-x_2^{(k)}\right)}_{\Delta x_2} + \ldots + \frac{\partial f^{(n)}\left(x_1^{(k)},x_2^{(k)},\ldots,x_n^{(k)}\right)}{\partial x_n}\underbrace{\left(x_n^{(k+1)}-x_n^{(k)}\right)}_{\Delta x_n} \tag{5.32c}$$

For the new point $\left(x_1^{(k+1)},x_2^{(k+1)},\ldots,x_n^{(k+1)}\right)$ to be the root, function values at this point must be equal to zero. In this case, if Equations 5.32a are rearranged by specifying $f^{(1)}\left(x_1^{(k+1)},x_2^{(k+1)},\ldots,x_n^{(k+1)}\right)=0$, $f^{(2)}\left(x_1^{(k+1)},x_2^{(k+1)},\ldots,x_n^{(k+1)}\right)=0$, $f^{(n)}\left(x_1^{(k+1)},x_2^{(k+1)},\ldots,x_n^{(k+1)}\right)=0$, following iterative solution schema can be obtained.

$$\begin{bmatrix} \frac{\partial f^{(1)}\left(x_1^{(k)},x_2^{(k)},\ldots,x_n^{(k)}\right)}{\partial x_1} & \frac{\partial f^{(1)}\left(x_1^{(k)},x_2^{(k)},\ldots,x_n^{(k)}\right)}{\partial x_2} & \cdots & \frac{\partial f^{(1)}\left(x_1^{(k)},x_2^{(k)},\ldots,x_n^{(k)}\right)}{\partial x_n} \\ \frac{\partial f^{(2)}\left(x_1^{(k)},x_2^{(k)},\ldots,x_n^{(k)}\right)}{\partial x_1} & \frac{\partial f^{(2)}\left(x_1^{(k)},x_2^{(k)},\ldots,x_n^{(k)}\right)}{\partial x_2} & \cdots & \frac{\partial f^{(2)}\left(x_1^{(k)},x_2^{(k)},\ldots,x_n^{(k)}\right)}{\partial x_n} \\ \cdots & \cdots & \cdots & \cdots \\ \frac{\partial f^{(n)}\left(x_1^{(k)},x_2^{(k)},\ldots,x_n^{(k)}\right)}{\partial x_1} & \frac{\partial f^{(n)}\left(x_1^{(k)},x_2^{(k)},\ldots,x_n^{(k)}\right)}{\partial x2} & \cdots & \frac{\partial f^{(n)}\left(x_1^{(k)},x_2^{(k)},\ldots,x_n^{(k)}\right)}{\partial x_n} \end{bmatrix} \begin{Bmatrix} \Delta x_1 \\ \Delta x_2 \\ \cdots \\ \Delta x_n \end{Bmatrix} = -\begin{Bmatrix} f^{(1)}\left(x_1^{(k)},x_2^{(k)},\ldots,x_n^{(k)}\right) \\ f^{(2)}\left(x_1^{(k)},x_2^{(k)},\ldots,x_n^{(k)}\right) \\ \cdots \\ f^{(n)}\left(x_1^{(k)},x_2^{(k)},\ldots,x_n^{(k)}\right) \end{Bmatrix} \tag{5.33}$$

Equation 5.33 can be referred to as generalized Newton-Rapson method for multivariate functions. In Eq. 5.33, Δx_1, Δx_2,..., Δx_n indicate the distance or increment values between $\left(x_1^{(k+1)},x_2^{(k+1)},\ldots,x_n^{(k+1)}\right)$ and $\left(x_1^{(k)},x_2^{(k)},\ldots,x_n^{(k)}\right)$ points. Corrected solution or root values are obtained using increment values as following:

$$x_1^{(k+1)} = x_1^{(k)} + \Delta x_1, \quad x_2^{(k+1)} = x_2^{(k)} + \Delta x_2, \ldots, \quad x_n^{(k+1)} = x_n^{(k)} + \Delta x_n$$

Iterative schema in Eq. 5.33 is applied successively until $f^{(1)}\left(x_1^{(k+1)}, x_2^{(k+1)}, \ldots, x_n^{(k+1)}\right)$, $f^{(2)}\left(x_1^{(k+1)}, x_2^{(k+1)}, \ldots, x_n^{(k+1)}\right)$, ..., and $f^{(n)}\left(x_1^{(k+1)}, x_2^{(k+1)}, \ldots, x_n^{(k+1)}\right)$ values are sufficiently close to zero. One or both of the following criteria can be used as convergence criterion.

1. Increments:

$$|\Delta x_1| \le \varepsilon, \quad |\Delta x_2| \le \varepsilon, \ldots, \quad |\Delta x_n| \le \varepsilon \tag{5.34}$$

2. Function values:

$$\begin{aligned}
&\left|f^{(1)}\left(x_1^{(k+1)}, x_2^{(k+1)}, \ldots, x_n^{(k+1)}\right)\right| \le \varepsilon \\
&\left|f^{(2)}\left(x_1^{(k+1)}, x_2^{(k+1)}, \ldots, x_n^{(k+1)}\right)\right| \le \varepsilon \\
&\ldots \\
&\left|f^{(n)}\left(x_1^{(k+1)}, x_2^{(k+1)}, \ldots, x_n^{(k+1)}\right)\right| \le \varepsilon
\end{aligned} \tag{5.35}$$

Example 5.6: Solve the following set of nonlinear equations using generalized Newton–Raphson method.

$$\begin{aligned}
&18x_1^2 + 4x_1x_2 - 3x_3^3 + 5x_4^2 = 317 \\
&-2x_1x_2 - 24x_2^2 + 7x_3 - 3x_4 = -229 \\
&4x_1 - x_2^4 + 32x_3x_4 = -1 \\
&-9x_1^2 - 7x_2x_3 - 3x_3^3 + 40x_1x_4 = -50
\end{aligned}$$

By giving names to the equations, they can be written in root finding form as following:

$$f^{(1)}(x_1, x_2, x_3, x_4) = f(x_1, x_2, x_3, x_4) = 18x_1^2 + 4x_1x_2 - 3x_3^3 + 5x_4^2 - 317 = 0$$

$$f^{(2)}(x_1, x_2, x_3, x_4) = g(x_1, x_2, x_3, x_4) = -2x_1x_2 - 24x_2^2 + 7x_3 - 3x_4 + 229 = 0$$

$$f^{(3)}(x_1, x_2, x_3, x_4) = h(x_1, x_2, x_3, x_4) = 4x_1 - x_2^4 + 32x_3x_4 + 1 = 0$$

$$f^{(4)}(x_1, x_2, x_3, x_4) = p(x_1, x_2, x_3, x_4) = -9x_1^2 - 7x_2x_3 - 3x_3^3 + 40x_1x_4 + 50 = 0$$

Partial derivatives of functions can be written as below:

$$\frac{\partial f}{\partial x_1}=36x_1+4x_2,\quad \frac{\partial f}{\partial x_2}=4x_1,\quad \frac{\partial f}{\partial x_3}=-9x_3^2,\quad \frac{\partial f}{\partial x_4}=10x_4$$

$$\frac{\partial g}{\partial x_1}=-2x_2,\quad \frac{\partial g}{\partial x_2}=-2x_1-48x_2,\quad \frac{\partial g}{\partial x_3}=7,\quad \frac{\partial g}{\partial x_4}=-3$$

$$\frac{\partial h}{\partial x_1}=4,\quad \frac{\partial h}{\partial x_2}=-4x_2^3,\quad \frac{\partial h}{\partial x_3}=32x_4,\quad \frac{\partial h}{\partial x_4}=32x_3$$

$$\frac{\partial p}{\partial x_1}=-18x_1+40x_4,\quad \frac{\partial p}{\partial x_2}=-7x_3,\quad \frac{\partial p}{\partial x_3}=-7x_2-9x_3^2,\quad \frac{\partial p}{\partial x_4}=40x_1$$

Root of four functions around $\left(x_1^{(k)},x_2^{(k)},x_3^{(k)},x_4^{(k)}\right)$ point can be expressed in iterative form as following:

$$\begin{bmatrix} \frac{\partial f\left(x_1^{(k)},x_2^{(k)},x_3^{(k)},x_4^{(k)}\right)}{\partial x_1} & \frac{\partial f\left(x_1^{(k)},x_2^{(k)},x_3^{(k)},x_4^{(k)}\right)}{\partial x_2} & \frac{\partial f\left(x_1^{(k)},x_2^{(k)},x_3^{(k)},x_4^{(k)}\right)}{\partial x_3} & \frac{\partial f\left(x_1^{(k)},x_2^{(k)},x_3^{(k)},x_4^{(k)}\right)}{\partial x_4} \\ \frac{\partial g\left(x_1^{(k)},x_2^{(k)},x_3^{(k)},x_4^{(k)}\right)}{\partial x_1} & \frac{\partial g\left(x_1^{(k)},x_2^{(k)},x_3^{(k)},x_4^{(k)}\right)}{\partial x_2} & \frac{\partial g\left(x_1^{(k)},x_2^{(k)},x_3^{(k)},x_4^{(k)}\right)}{\partial x_3} & \frac{\partial g\left(x_1^{(k)},x_2^{(k)},x_3^{(k)},x_4^{(k)}\right)}{\partial x_4} \\ \frac{\partial h\left(x_1^{(k)},x_2^{(k)},x_3^{(k)},x_4^{(k)}\right)}{\partial x_1} & \frac{\partial h\left(x_1^{(k)},x_2^{(k)},x_3^{(k)},x_4^{(k)}\right)}{\partial x_2} & \frac{\partial h\left(x_1^{(k)},x_2^{(k)},x_3^{(k)},x_4^{(k)}\right)}{\partial x_3} & \frac{\partial h\left(x_1^{(k)},x_2^{(k)},x_3^{(k)},x_4^{(k)}\right)}{\partial x_4} \\ \frac{\partial p\left(x_1^{(k)},x_2^{(k)},x_3^{(k)},x_4^{(k)}\right)}{\partial x_1} & \frac{\partial p\left(x_1^{(k)},x_2^{(k)},x_3^{(k)},x_4^{(k)}\right)}{\partial x_2} & \frac{\partial p\left(x_1^{(k)},x_2^{(k)},x_3^{(k)},x_4^{(k)}\right)}{\partial x_3} & \frac{\partial p\left(x_1^{(k)},x_2^{(k)},x_3^{(k)},x_4^{(k)}\right)}{\partial x_4} \end{bmatrix}$$

$$\otimes\begin{Bmatrix}\Delta x_1\\ \Delta x_2\\ \Delta x_3\\ \Delta x_4\end{Bmatrix}=-\begin{Bmatrix} f\left(x_1^{(k)},x_2^{(k)},x_3^{(k)},x_4^{(k)}\right)\\ g\left(x_1^{(k)},x_2^{(k)},x_3^{(k)},x_4^{(k)}\right)\\ h\left(x_1^{(k)},x_2^{(k)},x_3^{(k)},x_4^{(k)}\right)\\ p\left(x_1^{(k)},x_2^{(k)},x_3^{(k)},x_4^{(k)}\right)\end{Bmatrix}$$

Solution of the above matrix equation leads to Δx_1, Δx_2, Δx_3, Δx_4 increments. Thus, corrected root value at $\left(x_1^{(k+1)},x_2^{(k+1)},x_3^{(k+1)},x_4^{(k+1)}\right)$ point are obtained using increment values as below:

$$x_1^{(k+1)}=x_1^{(k)}+\Delta x_1,\quad x_2^{(k+1)}=x_2^{(k)}+\Delta x_2,\quad x_3^{(k+1)}=x_3^{(k)}+\Delta x_3,\quad x_4^{(k+1)}=x_4^{(k)}+\Delta x_4.$$

If the above scheme is applied successively by taking the initial values of $x_1^{(0)}=2$, $x_2^{(0)}=2$, $x_3^{(0)}=2$, $x_4^{(0)}=2$, following iteration results are obtained.

- *Iteration 1*:

$$\begin{bmatrix} \frac{\partial f\left(x_1^{(0)},x_2^{(0)},x_3^{(0)},x_4^{(0)}\right)}{\partial x_1} & \frac{\partial f\left(x_1^{(0)},x_2^{(0)},x_3^{(0)},x_4^{(0)}\right)}{\partial x_2} & \frac{\partial f\left(x_1^{(0)},x_2^{(0)},x_3^{(0)},x_4^{(0)}\right)}{\partial x_3} & \frac{\partial f\left(x_1^{(0)},x_2^{(0)},x_3^{(0)},x_4^{(0)}\right)}{\partial x_4} \\ \frac{\partial g\left(x_1^{(0)},x_2^{(0)},x_3^{(0)},x_4^{(0)}\right)}{\partial x_1} & \frac{\partial g\left(x_1^{(0)},x_2^{(0)},x_3^{(0)},x_4^{(0)}\right)}{\partial x_2} & \frac{\partial g\left(x_1^{(0)},x_2^{(0)},x_3^{(0)},x_4^{(0)}\right)}{\partial x_3} & \frac{\partial g\left(x_1^{(0)},x_2^{(0)},x_3^{(0)},x_4^{(0)}\right)}{\partial x_4} \\ \frac{\partial h\left(x_1^{(0)},x_2^{(0)},x_3^{(0)},x_4^{(0)}\right)}{\partial x_1} & \frac{\partial h\left(x_1^{(0)},x_2^{(0)},x_3^{(0)},x_4^{(0)}\right)}{\partial x_2} & \frac{\partial h\left(x_1^{(0)},x_2^{(0)},x_3^{(0)},x_4^{(0)}\right)}{\partial x_3} & \frac{\partial h\left(x_1^{(0)},x_2^{(0)},x_3^{(0)},x_4^{(0)}\right)}{\partial x_4} \\ \frac{\partial p\left(x_1^{(0)},x_2^{(0)},x_3^{(0)},x_4^{(0)}\right)}{\partial x_1} & \frac{\partial p\left(x_1^{(0)},x_2^{(0)},x_3^{(0)},x_4^{(0)}\right)}{\partial x_2} & \frac{\partial p\left(x_1^{(0)},x_2^{(0)},x_3^{(0)},x_4^{(0)}\right)}{\partial x_3} & \frac{\partial p\left(x_1^{(0)},x_2^{(0)},x_3^{(0)},x_4^{(0)}\right)}{\partial x_4} \end{bmatrix}$$

$$\otimes \begin{Bmatrix} \Delta x_1 \\ \Delta x_2 \\ \Delta x_3 \\ \Delta x_4 \end{Bmatrix} = - \begin{Bmatrix} f\left(x_1^{(0)},x_2^{(0)},x_3^{(0)},x_4^{(0)}\right) \\ g\left(x_1^{(0)},x_2^{(0)},x_3^{(0)},x_4^{(0)}\right) \\ h\left(x_1^{(0)},x_2^{(0)},x_3^{(0)},x_4^{(0)}\right) \\ p\left(x_1^{(0)},x_2^{(0)},x_3^{(0)},x_4^{(0)}\right) \end{Bmatrix}$$

From the solution of the above set of equations, increment values are obtained as $\Delta x_1 = 4.0212$, $\Delta x_2 = 1.3382$, $\Delta x_3 = 1.2490$, $\Delta x_4 = -2.7218$. Then, using the increment values, corrected root point is obtained as $(x_1^{(1)}, x_2^{(1)}, x_3^{(1)}, x_4^{(1)}) = (6.0212, 3.3382, 3.2490, -0.7218)$. Function values at this point are obtained as $f = 315.7004$, $g = -53.7437$, $h = -174.1479$, $p = -628.9552$. Since function values are not sufficiently close to zero, iteration continues.

- *Iteration 2*:

$$\begin{bmatrix} \frac{\partial f\left(x_1^{(1)},x_2^{(1)},x_3^{(1)},x_4^{(1)}\right)}{\partial x_1} & \frac{\partial f\left(x_1^{(1)},x_2^{(1)},x_3^{(1)},x_4^{(1)}\right)}{\partial x_2} & \frac{\partial f\left(x_1^{(1)},x_2^{(1)},x_3^{(1)},x_4^{(1)}\right)}{\partial x_3} & \frac{\partial f\left(x_1^{(1)},x_2^{(1)},x_3^{(1)},x_4^{(1)}\right)}{\partial x_4} \\ \frac{\partial g\left(x_1^{(1)},x_2^{(1)},x_3^{(1)},x_4^{(1)}\right)}{\partial x_1} & \frac{\partial g\left(x_1^{(1)},x_2^{(1)},x_3^{(1)},x_4^{(1)}\right)}{\partial x_2} & \frac{\partial g\left(x_1^{(1)},x_2^{(1)},x_3^{(1)},x_4^{(1)}\right)}{\partial x_3} & \frac{\partial g\left(x_1^{(1)},x_2^{(1)},x_3^{(1)},x_4^{(1)}\right)}{\partial x_4} \\ \frac{\partial h\left(x_1^{(1)},x_2^{(1)},x_3^{(1)},x_4^{(1)}\right)}{\partial x_1} & \frac{\partial h\left(x_1^{(1)},x_2^{(1)},x_3^{(1)},x_4^{(1)}\right)}{\partial x_2} & \frac{\partial h\left(x_1^{(1)},x_2^{(1)},x_3^{(1)},x_4^{(1)}\right)}{\partial x_3} & \frac{\partial h\left(x_1^{(1)},x_2^{(1)},x_3^{(1)},x_4^{(1)}\right)}{\partial x_4} \\ \frac{\partial p\left(x_1^{(1)},x_2^{(1)},x_3^{(1)},x_4^{(1)}\right)}{\partial x_1} & \frac{\partial p\left(x_1^{(1)},x_2^{(1)},x_3^{(1)},x_4^{(1)}\right)}{\partial x_2} & \frac{\partial p\left(x_1^{(1)},x_2^{(1)},x_3^{(1)},x_4^{(1)}\right)}{\partial x_3} & \frac{\partial p\left(x_1^{(1)},x_2^{(1)},x_3^{(1)},x_4^{(1)}\right)}{\partial x_4} \end{bmatrix}$$

$$\otimes \begin{Bmatrix} \Delta x_1 \\ \Delta x_2 \\ \Delta x_3 \\ \Delta x_4 \end{Bmatrix} = - \begin{Bmatrix} f\left(x_1^{(1)},x_2^{(1)},x_3^{(1)},x_4^{(1)}\right) \\ g\left(x_1^{(1)},x_2^{(1)},x_3^{(1)},x_4^{(1)}\right) \\ h\left(x_1^{(1)},x_2^{(1)},x_3^{(1)},x_4^{(1)}\right) \\ p\left(x_1^{(1)},x_2^{(1)},x_3^{(1)},x_4^{(1)}\right) \end{Bmatrix}$$

From the solution of the above set of equations, increment values are obtained as $\Delta x_1 = -1.7432$, $\Delta x_2 = -0.3060$, $\Delta x_3 = -1.0582$, $\Delta x_4 = 1.0690$. Then, using the increment values, corrected root point is obtained as $(x_1^{(2)}, x_2^{(2)}, x_3^{(2)}, x_4^{(2)}) = (4.2779, 3.0322, 2.1908, 0.3472)$. Function values at this point are obtained as $f = 33.3588$, $g = -3.3144$, $h = -42.0864$, $p = -133.3468$. Since function values are not sufficiently close to zero, iteration continues. If iterations are repeated until convergence, iteration results in Table 5.6 are obtained.

Solution with MATLAB Commands:
For the solution, four equations are written in *equal to zero* form and saved in a function file such as *Equations.m* (*Note*: Name of the file should be the same after the *function* word in the file) as illustrated in Table 5.7.

Then, solution is obtained by taking the initial point as *x0* = [2 2 2 2] and using the *fsolve* command in the MATLAB Workspace as below:

```
>> x0=[2 2 2 2]; x=fsolve('Equations',x0);
>> x
x =
4.0000 3.0000 2.0000 1.0000
```

5.7 Summary

- Root finding or nonlinear equation solution is encountered in engineering problems directly or indirectly.
- The x value zeroing an $f(x)$ function (i.e., $f(x) = 0$) is called as root.
- Nonlinear equations can be expressed in root finding form (i.e., $f(x) = 0$). This way, root finding methods can be used in the solution of nonlinear equations.
- Analytical (closed-form) root finding methods are available for low-order polynomial functions such as second- and third-order polynomials.
- For higher order polynomials or other complicated functions, numerical methods are used for root finding.
- Numerical methods search roots iteratively (step by step).
- Numerical root finding methods are divided into two groups as interval based and point-based methods.
- In interval-based methods such as Bisection and Linear Interpolation (False-position) method, an interval containing a root is narrowed iteratively to find the root.
- Interval-based methods guarantee finding roots.
- In Bisection method, root is assumed as the middle of the interval $[x_L, x_U]$ as $x_M = \frac{x_L+x_U}{2}$. If $f(x_M)f(x_L) > 0$ is satisfied, then root is considered to be in the right half of the interval defined by the range of $[x_M, x_U]$. Otherwise, root is considered to be in the left half of the interval defined by the range of $[x_L, x_M]$. Half of the interval that does not include the root is discarded.
- In linear interpolation (False-position) method, linear polynomial approximation of the original function is used within the root interval of $[x_L, x_U]$. Root of the linear polynomial function $f(x) = \frac{x-x_U}{x_L-x_U} f_L + \frac{x-x_L}{x_U-x_L} f_U$ is calculated as $x_R = x = \frac{x_L f_U - x_U f_L}{f_U - f_L}$.

Table 5.6 Iteration results with the generalized Newton–Raphson method

Iteration	x_1	x_2	x_3	x_4	f	g	h	p
1	6.0212	3.3382	3.2490	−0.7218	315.7004	−53.7437	−174.1479	−628.9552
2	4.2779	3.0322	2.1908	0.3472	33.3588	−3.3144	−42.0864	−133.3468
3	4.0124	2.9998	1.9768	0.9450	2.2175	−0.0424	−4.1517	−7.9082
4	4.0003	3.0000	2.0010	1.0002	0.0073	0.0000	0.0428	−0.0385
5	4.0000	3.0000	2.0000	1.0000	0.0000	0.0000	0.0000	0.0000

Table 5.7 MATLAB file for nonlinear equations

File=Equations.m

```
function f=Equations(x)

f(1)=18*(x(1)^2)+4*x(1)*x(2)-3*x(3)^3+5*(x(4)^2)-317;
f(2)=-2*x(1)*x(2)-24*(x(2)^2)+7*x(3)-3*x(4)+229;
f(3)=4*x(1)-(x(2)^4)+32*x(3)*x(4)+1;
f(4)=-9*(x(1)^2)-7*x(2)*x(3)-3*(x(3)^3)+40*x(1)*x(4 )+50;
```

- In linear interpolation (False-position) method, sign of $f_R = f(x_R)$ is compared with those of f_L and f_U. If f_R has the same sign with f_L, root is assumed in $[x_R, x_U]$ interval (i.e., $[x_L, x_R]$ interval is discarded). Otherwise, root is assumed in $[x_L, x_R]$ interval (i.e., $[x_R, x_U]$ interval is discarded).
- In point-based methods, such as Simple iteration, Newton–Raphson, and Secant methods, one or more initial points are used to find roots iteratively.
- Point-based methods do not guarantee finding roots.
- For convergence of point-based methods, initial value should be chosen close to the actual root.
- Point-based methods generally converge faster than interval-based methods.
- In simple iteration method, root equation $f(x) = 0$ is rewritten in $x = g(x)$ form. Old x value is substituted into $g(x)$ function to obtain new x value as following ($k-1$: old value, k: new value)

$$x^{(k)} = g\left(x^{(k-1)}\right)$$

- In Newton–Raphson method, $f(x)$ function is successively approximated with *linear polynomials* which are tangent to the function. Root of linear function is calculated as $x^{(k)} = x^{(k-1)} - \frac{f\left(x^{(k-1)}\right)}{f'\left(x^{(k-1)}\right)}$ ($k-1$: old value, k: new value)
- Secant method is obtained by replacing first derivative in Newton–Raphson method with finite difference expression using function values at two adjacent points as below ($k-1$, $k-2$: old values, k: new value):

$$x^{(k)} = x^{(k-1)} - \frac{f\left(x^{(k-1)}\right)}{\frac{f\left(x^{(k-1)}\right) - f\left(x^{(k-2)}\right)}{x^{(k-1)} - x^{(k-2)}}}$$

- Nonlinear equations can also be solved with Newton–Raphson method (referred to as generalized Newton–Raphson method) by converting them into root finding problems. For example, two nonlinear equations expressed as $f(x,y) = 0$ and $g(x, y) = 0$ can be solved using the generalized Newton–Raphson method as following (k: old value, $k + 1$: new value):

$$\begin{bmatrix} \frac{\partial f(x^k,y^k)}{\partial x} & \frac{\partial f(x^k,y^k)}{\partial y} \\ \frac{\partial g(x^k,y^k)}{\partial x} & \frac{\partial g(x^k,y^k)}{\partial y} \end{bmatrix} \begin{Bmatrix} \Delta x \\ \Delta y \end{Bmatrix} = -\begin{Bmatrix} f(x^k,y^k) \\ g(x^k,y^k) \end{Bmatrix} \rightarrow \begin{matrix} \Delta x \\ \Delta y \end{matrix} \rightarrow \begin{matrix} x^{k+1} = x^k + \Delta x \\ y^{k+1} = y^k + \Delta y \end{matrix}$$

Appendix: MATLAB Programs

MATLAB Program 5.1: Determining Root Interval

```
function BracketingRoot

% THIS PROGRAM DETERMINES THE ROOT INTERVAL (INTERVAL THAT INCLUDES ROOT)
% WITH SCANNING STARTING FROM AN x0 VALUE AND MOVING WITH dx STEPS

x0=0 ;   % Initial value to start scanning
dx=0.5 ; % Increment value (step length) in scanning

% ---------------------------------------------------- -----------
x1=x0 ;
x2=x0+dx ;

% Function values
f1=FunctionValue(x1) ;
f2=FunctionValue(x2) ;

% Root interval: [xL,xU]
if f1*f2<0
    xL=x1 ;
    xU=x2 ;

    fL=f1 ;
    fU=f2 ;
else
    i=1 ;
    x0=x1 ;

    while (f1*f2>0)
        i=i+1 ;

        x1=x0+(i-1)*dx ;
        x2=x0+i*dx ;

        f1=FunctionValue(x1) ;
        f2=FunctionValue(x2) ;

        xL=x1 ;
        xU=x2 ;

        fL=f1 ;
        fU=f2 ;
    end
end

fprintf('Root Interval: [xL,xU]=[%f,%f] \n',xL,xU) ;

% -----------------------------------------------------------
% -------------    FUNCTION TO FIND ROOT    --------------
% -----------------------------------------------------------
function f=FunctionValue(x)
f=(2-sqrt(x))*(4*cos(x)+2*sin(x)) ;
```

MATLAB Program 5.2: Root Finding with Bisection Method

```
function Bisection_RootFinding

% THIS PROGRAM FINDS ROOT OF AN f(x) FUNCTION USING BISECTION METHOD

x0=0 ;  % Initial value to start braketing the root interval
dx=2 ;  % Incremental value in braketing the root interval

Tol=0.001 ;  %  Convergence tolerance

% ----------------------------------------------------------------------
% ----------------------------------------------------------------------
% PLOT OF FUNCTION
figure(1)
xinc=0.001 ;
xini=0 ;
for i=1:1000
    xvec(i)=xini+(i-1)*xinc ;
    fvec(i)=FunctionValue(xvec(i))   ;
end
plot(xvec,fvec) ;
xlabel('x') ;
ylabel('f(x)') ;
set(gcf, 'Color', 'White')


% Bracketing root interval using: [xL,xU] interval
[xL,xU,fL,fU]=BracketingRoot(x0,dx) ;

it=0 ;

L=xU-xL ;  % Interval length

xM=xL+L/2 ;
fM=FunctionValue(xM) ;
Conv=abs(fM) ;

while (Conv>=Tol)
    it=it+1 ;

    ResultTable(it,1)=it ;
    ResultTable(it,2)=xL ;
    ResultTable(it,3)=xM ;
    ResultTable(it,4)=xU ;
    ResultTable(it,5)=fL ;
    ResultTable(it,6)=fM ;
    ResultTable(it,7)=fU ;

    if fL*fM<0
        xL_new=xL ;
        xU_new=xM ;
        L_new=xU_new-xL_new ;
        xM_new=xL_new+L_new/2 ;

        fL_new=fL ;
        fU_new=fM ;
        fM_new=FunctionValue(xM_new) ;
    else
        xL_new=xM ;
        xU_new=xU ;
        L_new=xU_new-xL_new ;
        xM_new=xL_new+L_new/2 ;

        fL_new=fL ;
        fU_new=fM ;
        fM_new=FunctionValue(xM_new) ;
    end
```

```
    xL=xL_new ;
    xM=xM_new ;
    xU=xU_new ;

    fL=FunctionValue(xL) ;
    fM=FunctionValue(xM) ;
    fU=FunctionValue(xU) ;

    Conv=abs(fM_new) ;
end

fprintf('ROOT ITERATIONS:\n') ;
fprintf('    it          xL       xM        xU         fL         fM         fU
\n') ;
ResultTable

% ---------------------------------------------------------------------
% ---------------------------------------------------------------------
function [xL,xU,fL,fU]=BracketingRoot(x0,dx)

x1=x0 ;
x2=x0+dx ;

f1=FunctionValue(x1) ;
f2=FunctionValue(x2) ;

% Determining Root Interval: [xL,xU]
if f1*f2<0
    xL=x1 ;
    xU=x2 ;

    fL=f1 ;
    fU=f2 ;
else
    i=1 ;
    x0=x1 ;

    while (f1*f2>0)
        i=i+1 ;

        x1=x0+(i-1)*dx ;
        x2=x0+i*dx ;

        f1=FunctionValue(x1) ;
        f2=FunctionValue(x2) ;

        xL=x1 ;
        xU=x2 ;

        fL=f1 ;
        fU=f2 ;
    end
end
% -------------------------------------------------------
% -------------    FUNCTION TO FIND ROOT   --------------
% -------------------------------------------------------
function f=FunctionValue(x)
% f(x) function to find root
f=(2-sqrt(x))*(4*cos(x)+2*sin(x)) ;
```

MATLAB Program 5.3: Root Finding with Linear Interpolation (False-Position) Method

```
function FalsePosition_RootFinding

% THIS PROGRAM FINDS ROOT OF f(x) FUNCTION USING FALSE-POSITION
% (LINEAR INTERPOLATION) METHOD

xL=0 ;  % Lower limit of root interval
xU=3 ;  % Upper limit of root interval

Tol=0.001 ;  %  Convergence tolerance

% ----------------------------------------------------------------------------
% ----------------------------------------------------------------------------
Conv=1e10 ;
it=0 ;

fL=FunctionValue(xL) ;
fU=FunctionValue(xU) ;

while (Conv>=Tol)
    it=it+1 ;

    % Root value: xR
    xR=(xL*fU-xU*fL)/(fU-fL) ;
    fR=FunctionValue(xR) ;

    ResultTable(it,1)=it ;
    ResultTable(it,2)=xL ;
    ResultTable(it,3)=xR ;
    ResultTable(it,4)=xU ;
    ResultTable(it,5)=fL ;
    ResultTable(it,6)=fR ;
    ResultTable(it,7)=fU ;


    if fL*fR>0
        xL_new=xR ;
        xU_new=xU ;

        fL_new=fR ;
        fU_new=fU ;
    else
        xL_new=xL ;
        xU_new=xR ;

        fL_new=fL ;
        fU_new=fR ;
    end

    xL=xL_new ;
    xU=xU_new ;

    fL=FunctionValue(xL) ;
    fU=FunctionValue(xU) ;

    Conv=abs(fR) ;
end

fprintf('ROOT ITERATIONS:\n') ;
fprintf('     it          xL         xR         xU         fL          fR         fU
\n') ;
ResultTable

% -------------------------------------------------------
% -------------     FUNCTION TO FIND ROOT    -------------
% -------------------------------------------------------
function f=FunctionValue(x)
% f(x) function to find root
f=(2-sqrt(x))*(4*cos(x)+2*sin(x)) ;
```

MATLAB Program 5.4: Root Finding with Simple Iteration (Substitution) Method

```
function Simple_Iteration_RootFinding

% THIS PROGRAM FINDS ROOT OF f(x) FUNCTION USING SIMPLE ITERATION METHOD

x0=4 ;          % Initial value to start root finding
Tol=0.001 ;     % Convergence tolerance
maxit=200 ;     % Maximum iteration number

% -------------------------------------------------------------------------
% -------------------------------------------------------------------------
x(1)=x0 ;

Conv=1e10 ;
i=1 ;

while (Conv>=Tol) && (i<=maxit)
    x(i+1)=g_Function_Value(x(i))   ;

    ResultTable(i,1)=i ;
    ResultTable(i,2)=x(i) ;
    ResultTable(i,3)=x(i+1) ;

    Conv=abs(x(i+1)-x(i)) ;

    x(i)=x(i+1) ;

    i=i+1 ;
end

if (i-1)==maxit
    error('No convergence after  %i iterations',maxit) ;
end

fprintf('ROOT ITERATIONS:\n') ;
fprintf('     it          x       g       \n') ;
ResultTable

% ------------------------------------------------------
% -- FUNCTION TO FIND ROOT IS ARRANGED IN x=g(x) FORM --
% ------------------------------------------------------
function g=g_Function_Value(x)
% f(x)=0 equation is rewritten in x=g(x) form for root finding
g=(8*x^2-15*x)^(1/3) ;
```

MATLAB Program 5.5: Root Finding with Newton–Raphson Method

```
function NewtonRapson_RootFinding

% THIS PROGRAM FINDS ROOT OF f(x) FUNCTION USING NEWTON-RAPHSON METHOD

x0=1 ;        % Initial value to start root finding
Tol=0.001 ;  % Convergence tolerance
maxit=200 ;  % Maximum iteration number

% --------------------------------------------------------------------
% --------------------------------------------------------------------
x(1)=x0 ;
[f(1),df(1)]=Function_and_DerivativeValue(x(1)) ;

Conv=1e10 ;
i=1 ;

while (Conv>=Tol) && (i<=maxit)
    x(i+1)=x(i)-f(i)/df(i)  ;
    [f(i+1),df(i+1)]=Function_and_DerivativeValue(x(i+1)) ;

    ResultTable(i,1)=i ;
    ResultTable(i,2)=x(i+1) ;
    ResultTable(i,3)=f(i+1) ;
    ResultTable(i,4)=df(i+1) ;

    Conv=abs(f(i+1)) ;

    i=i+1 ;
end

if (i-1)==maxit
    error('No convergence after  %i iterations',maxit) ;
end

fprintf('ROOT ITERATIONS:\n') ;
fprintf('     it          x        f       df/dx\n') ;
ResultTable

% -------------------------------------------------------
% -----  FUNCTION TO FIND ROOT AND ITS DERIVATIVE  -----
% -------------------------------------------------------
function [f,df]=Function_and_DerivativeValue(x)
% f : function, df: derivative
f=(2-sqrt(x))*(4*cos(x)+2*sin(x)) ;
df=(-1/(2*sqrt(x)))*(4*cos(x)+2*sin(x))+(2-sqrt(x))*(-4*sin(x)+2*cos(x))
;
```

MATLAB Program 5.6: Root Finding with Secant Method

```
function SecantMethod_RootFinding

% THIS PROGRAM FINDS ROOT OF f(x) FUNCTION USING SECANT METHOD

x0=0 ;  % Initial value 1 to start root finding
x1=1 ;  % Initial value 2 to start root finding

Tol=0.001 ;  %  Convergence tolerance

% -----------------------------------------------------------------------
% -----------------------------------------------------------------------
x(1)=x0 ;
x(2)=x1 ;

f(1)=FunctionValue(x(1)) ;
f(2)=FunctionValue(x(2)) ;

Conv=1e10 ;
i=1 ;

while (Conv>=Tol)
    df=(f(i+1)-f(i))/(x(i+1)-x(i)) ;
    x(i+2)=x(i+1)-f(i+1)/df  ;
    f(i+2)=FunctionValue(x(i+2)) ;

    ResultTable(i,1)=i ;
    ResultTable(i,2)=x(i) ;
    ResultTable(i,3)=x(i+1) ;
    ResultTable(i,4)=x(i+2) ;
    ResultTable(i,5)=f(i+2) ;

    Conv=abs(f(i+2)) ;

    i=i+1 ;
end

fprintf('ROOT ITERATIONS:\n') ;
fprintf('     it         x0          x1          x         f   \n') ;
ResultTable

% -----------------------------------------------------
% -------------    FUNCTION TO FIND ROOT    -------------
% -----------------------------------------------------
function [f]=FunctionValue(x)
% f(x) function to find root
f=(2-sqrt(x))*(4*cos(x)+2*sin(x)) ;
```

Problems

Problem 5.1: Determine the root interval for $f(x) = x^3 - 5x^2 + 2x + 8$ function starting from $x = 0.5$ point and moving with step length of $\Delta x = 1$.

Problem 5.2: Find the root of $f(x) = x^3 - 5x^2 + 2x + 8$ function using Bisection method in [0,3] interval.

Problem 5.3: Find the root of $f(x) = x^3 - 8x^2 + 15x$ function in [1, 4] interval using Linear Interpolation (False-Position) method.

Problem 5.4: Find the root of $f(x) = x^3 - 9x^2 + 20x$ function with Simple Iteration method using the initial value of $x_0 = 3$.

Problem 5.5: Find the root of $f(x) = (2 - \sqrt{x})(\cos x - \sin x)$ function with Newton–Raphson method using the initial value of $x_0 = 0$.

Problem 5.6: Find the root of $f(x) = x^3 - 9x^2 + 20x$ function with Newton–Raphson method using the initial value of $x_0 = 2.5$.

Problem 5.7: Find the root of $f(x) = x^3 - 8x^2 + 15x$ function with Secant method using the initial values of $x^{(-1)} = 1$ and $x^{(0)} = 1.5$.

Problem 5.8: Find the root of $f(x) = (2 - \sqrt{x})(\cos x - \sin x)$ function with Secant method using the initial values of $x^{(-1)} = 0$ and $x^{(0)} = 0.25$.

Problem 5.9: Solve the following set of nonlinear equations using generalized Newton–Raphson method using the initial values of $x_1^{(0)} = 2$ and $x_2^{(0)} = 1$.

$$
\begin{aligned}
x_1^2 + 4x_2^2 &= 16 \\
x_1^2 + x_2^2 - 4x_1 &= 5
\end{aligned}
$$

Problem 5.10: Solve the following set of nonlinear equations using generalized Newton–Raphson method using the initial values of $x_1^{(0)} = 2.5$, $x_2^{(0)} = 1.5$, and $x_3^{(0)} = 0$.

$$
\begin{aligned}
x_1^2 - 2x_2^2 + x_3^2 &= 2 \\
2x_1^2 + x_2 - x_3^2 &= 19 \\
3x_1^2 + 3x_2^2 + x_3 &= 40
\end{aligned}
$$

Problem 5.11: Solve the following set of nonlinear equations using generalized Newton–Raphson method using the initial values of $x_1^{(0)} = 2.5$, $x_2^{(0)} = 1.5$, $x_3^{(0)} = 0$, and $x_4^{(0)} = 0.25$.

$$
\begin{aligned}
x_1^2 - 2x_2^2 + x_3^2 + x_4^2 &= 2 \\
2x_1^2 + x_2 - x_3^2 + 2x_4 &= 19 \\
3x_1^2 + 3x_2^2 + x_3 - x_4 &= 40 \\
x_1 + 2x_2^2 + x_3 - 2x_4^2 &= 12
\end{aligned}
$$

References

1. Chapra, S. C., & Canale, R. P. (2020). *Numerical methods for engineers*. McGraw-Hill.
2. Bernatz, R. A. (2010). *Fourier series and numerical methods for partial differential equations*. John Wiley & Sons.
3. Everstine, G. C. (2010). *Numerical solution of partial differential equations* [Lecture notes, PDF]. George Washington University. https://www.academia.edu/31518755/Numerical_Solution_of_Partial_Differential_Equations
4. Epperson, J. F. (2013). *An introduction to numerical methods and analysis*. John Wiley & Sons.
5. Burden, R. L., & Faires, J. D. (2011). *Numerical analysis*. Cengage Learning.
6. Sert, C. (2006). *ME310 numerical methods* [Lecture notes, PDF]. Middle East Technical University. https://users.metu.edu.tr/csert/me310/me310_1_introduction.pdf
7. Grewal, B. S. (2019). *Numerical methods in engineering and science*. Mercury Learning and Information.
8. Sastry, S. S. (2012). *Introductory methods of numerical analysis*. PHI Learning.
9. Reddy, J. N. (2004). *Mechanics of laminated composite plates and shells: Theory and analysis*. CRC Press.
10. Reddy, J. N. (2004). *An introduction to nonlinear finite element analysis*. Oxford University Press.

Chapter 6
Interpolation

6.1 Introduction

Prediction of intermediate values by passing a function through a set of data points is called as interpolation. The function used for this purpose is referred to as interpolation function. Interpolation as illustrated in Fig. 6.1 is commonly used to represent experimental results with a function. In interpolation, values in the data set are considered to be accurate (precise) and interpolation function is created using part or all of the data points. Interpolation is especially preferred in applications where generation of data is difficult, expensive, and time consuming.

Various methods were developed for interpolation [1–8]. *Polynomial* functions are widely used in these methods. Polynomial functions are used in two different forms in interpolation:

1. Global interpolation polynomial: A higher order polynomial (n-th order polynomial for $n + 1$ data points) is passed through the data points to obtain a global interpolation function.
2. Spline interpolation polynomial: Data points are divided into small groups and lower order polynomial functions referred to as spline interpolation polynomials are generated using each group of data. Linear, parabolic, and cubic polynomials are commonly used as spline functions.

6.2 Global Interpolation Polynomials

In this method, a higher order global interpolation polynomial function is passed through the data points to predict intermediate values. For example, n-th order polynomial function passing through (x_1,f_1), (x_2,f_2), (x_3,f_3),…, (x_{n+1},f_{n+1}) points can be expressed as following:

H. Kurtaran, *Engineering Mathematics with MATLAB Applications: Vol. 1, Fundamentals*, https://doi.org/10.1007/978-3-031-81869-1_6

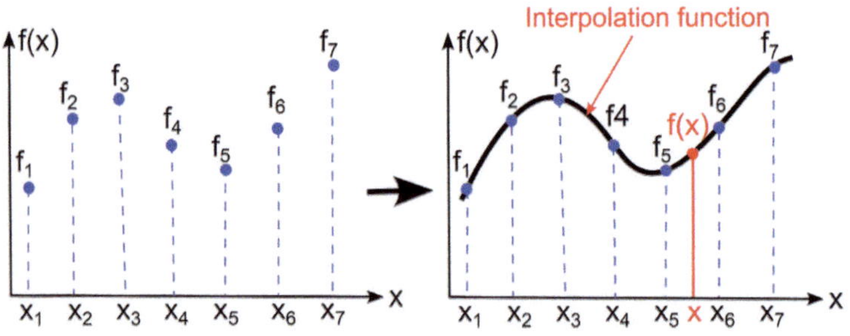

Fig. 6.1 Prediction of intermediate value by passing an interpolation function through a set of data points

$$f(x) = c_1 + c_2 x + c_3 x^2 + c_4 x^3 + \ldots c_{n+1} x^n \tag{6.1}$$

Constants c_i can be found by substituting the known function values in Eq. 6.1 as below:

$$f(x = x_1) = f_1 = c_1 + c_2 x_1 + c_3 x_1^2 + c_4 x_1^3 + \ldots c_{n+1} x_1^n$$

$$f(x = x_2) = f_2 = c_1 + c_2 x_2 + c_3 x_2^2 + c_4 x_2^3 + \ldots c_{n+1} x_2^n$$

$$f(x = x_3) = f_3 = c_1 + c_2 x_3 + c_3 x_3^2 + c_4 x_3^3 + \ldots c_{n+1} x_3^n$$

$$f(x = x_4) = f_4 = c_1 + c_2 x_4 + c_3 x_4^2 + c_4 x_4^3 + \ldots c_{n+1} x_4^n$$

$$\ldots$$

$$f(x = x_{n+1}) = f_{n+1} = c_1 + c_2 x_{n+1} + c_3 x_{n+1}^2 + c_4 x_{n+1}^3 + \ldots c_{n+1} x_{n+1}^n$$

The above equations can be rearranged in matrix form as following:

$$\underbrace{\begin{bmatrix} 1 & x_1 & x_1^2 & x_1^3 & \ldots & x_1^n \\ 1 & x_2 & x_2^2 & x_2^3 & \ldots & x_2^n \\ 1 & x_3 & x_3^2 & x_3^3 & \ldots & x_3^n \\ 1 & x_4 & x_4^2 & x_4^3 & \ldots & x_4^n \\ \ldots & \ldots & \ldots & \ldots & \ldots & \ldots \\ 1 & x_{n+1} & x_{n+1}^2 & x_{n+1}^3 & \ldots & x_{n+1}^n \end{bmatrix}}_{A} \underbrace{\begin{Bmatrix} c_1 \\ c_2 \\ c_3 \\ c_4 \\ \ldots \\ c_{n+1} \end{Bmatrix}}_{X} = \underbrace{\begin{Bmatrix} f_1 \\ f_2 \\ f_3 \\ f_4 \\ \ldots \\ f_{n+1} \end{Bmatrix}}_{B} \tag{6.2}$$

Equation 6.2 corresponds to a linear system in terms of unknown c_i coefficients which can be expressed in matrix form as $AX = B$. Solution of this system with one of the methods in Chap. 4 leads yields c_i values.

Global interpolation polynomial has the following drawbacks:

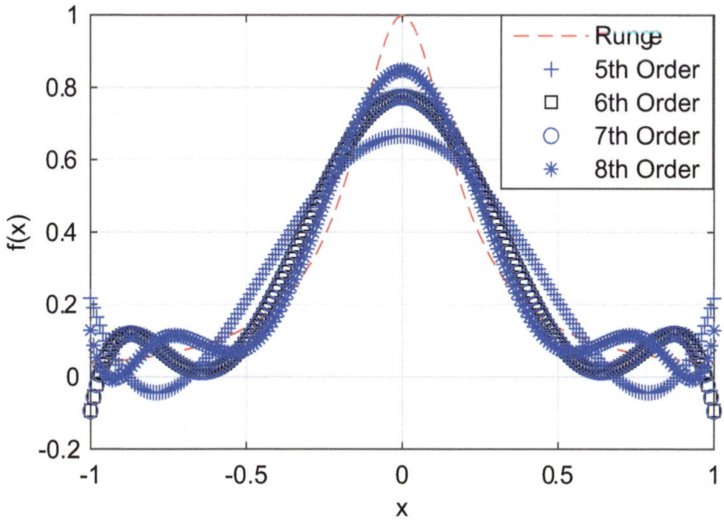

Fig. 6.2 Representation of the Runge function by global interpolation polynomials of different orders

1. *Physical difficulty*: Global interpolation polynomials generated using a large number of points will have highly oscillatory character. Highly oscillatory polynomials lead to incorrect estimation of intermediate values. Therefore, global interpolation polynomials up to third to fourth orders are usually used for accurate interpolation.

   Example: Runge function is given as $f(x) = \frac{1}{1+25x^2}$. This function can be approximated in the range [−1,1] using global interpolation functions with varying orders as in Fig. 6.2.

   From Fig. 6.2, it is seen that as the degree of the polynomial function increases, oscillation around the Runge function increases towards the ends.
2. *Mathematical difficulty*: In Eq. 6.1, coefficients matrix A is called as Vandermonde matrix. When the number of data points and thereby the degree of the interpolation polynomial is too high, this matrix becomes singular and its inverse cannot be calculated in the linear system solution. Therefore, for very high-order polynomials c_i constants cannot be calculated. To overcome this difficulty, alternative form of general $f(x) = c_1 + c_2x + c_3x^2 + c_4x^3 + \ldots c_{n+1}x^n$ polynomial such as Newton and Lagrange polynomials were developed.

6.2.1 Newton Interpolation Polynomial

Classical global interpolation polynomial $f(x) = c_1 + c_2x + c_3x^2 + c_4x^3 + \ldots c_{n+1}x^n$ function can be alternatively expressed with Newton method as below:

$$f(x) = a_1 + a_2(x-x_1) + a_3(x-x_1)(x-x_2) + a_4(x-x_1)(x-x_2)(x-x_3) + \ldots + a_{n+1}(x-x_1)(x-x_2)(x-x_3)\ldots(x-x_n) \tag{6.3}$$

Although classical global interpolation polynomial and Newton polynomial are equivalent (different expressions of the same polynomial), unknown constants can be found much easier in Newton form. Equivalency of two forms can be proven for linear and parabolic interpolation polynomials as below.

Linear Polynomial:
Assume that function values (x_1,f_1) and (x_2,f_2) are given. In this case, linear interpolation polynomial passing through these points in classical and Newton forms can be expressed as following:

Classical form: $f(x) = c_1 + c_2x$

Newton form: $f_N(x) = a_1 + a_2(x - x_1)$

We can write Newton form in more detail as below:

$$f_N(x) = a_1 + a_2(x-x_1) = \underbrace{a_1 - a_2x_1}_{\substack{c_1 = \text{Constant Term} \\ \text{Coefficient}}} + \underbrace{a_2}_{\substack{c_2 = \text{Linear Term} \\ \text{Coefficient}}} x = c_1 + c_2x$$

Since classical and Newton polynomials express the same linear polynomial, we can write the following relationships between the coefficients of two polynomials

$$c_1 = a_1 - a_2x_1, \quad c_2 = a_2$$

Parabolic Polynomial:
Assume that function values of (x_1,f_1), (x_2,f_2), and (x_3,f_3) are given. In this case, parabolic interpolation polynomial passing through these points in classical and Newton forms can be expressed as following:

Classical form: $f(x) = c_1 + c_2x + c_3x^2$

Newton form: $f_N(x) = a_1 + a_2(x - x_1) + a_3(x - x_1)(x - x_2)$

We can write Newton form in more detail as following:

$$\begin{aligned} f_N(x) &= a_1 + a_2(x-x_1) + a_3(x-x_1)(x-x_2) \\ &= a_1 + a_2x - a_2x_1 + a_3\left(x^2 - x_2x - x_1x + x_1x_2\right) = \\ &\underbrace{a_1 - a_2x_1 + a_3x_1x_2}_{\substack{c_1 = \text{Constant Term} \\ \text{Coefficient}}} + \underbrace{(a_2 - a_3x_1 - a_3x_2)}_{\substack{c_2 = \text{Linear Term} \\ \text{Coefficient}}}x + \underbrace{a_3}_{\substack{c_3 = \text{Parabolic Term} \\ \text{Coefficient}}} x^2 \\ &= c_1 + c_2x + c_3x^2 \end{aligned}$$

Since classical and Newton polynomials express the same parabolic polynomial, we can write the following relationships between the coefficients of two polynomials

$$c_1 = a_1 - a_2x_1 + a_3x_1x_2, \quad c_2 = a_2 - a_3x_1 - a_3x_2, \quad c_3 = a_3$$

After showing the equivalency of Newton and classical polynomials, we can explain general methodology for finding the coefficients of Linear, Parabolic, and Cubic Newton interpolation polynomials as following:

Linear Newton Polynomial:

$$f_N(x) = a_1 + a_2(x - x_1)$$

We can obtain coefficients of the linear Newton polynomial $f_N(x) = a_1 + a_2(x - x_1)$ passing through (x_1,f_1) and (x_2,f_2) points as below:

$$x = x_1 : f_1 = a_1 + a_2(x_1 - x_1) \rightarrow a_1 = f_1$$

$$x = x_2 : f_2 = a_1 + a_2(x_2 - x_1) \rightarrow a_2 = \left(\frac{f_2 - f_1}{x_2 - x_1}\right)$$

Parabolic Newton Polynomial:

$$f_N(x) = a_1 + a_2(x - x_1) + a_3(x - x_1)(x - x_2)$$

We can obtain coefficients of the parabolic Newtonian polynomial $f_N(x) = a_1 + a_2(x - x_1) + a_3(x - x_1)(x - x_2)$ passing through (x_1,f_1), (x_2,f_2), and (x_3,f_3) points as below:

$$x = x_1 : f_1 = a_1 + a_2(x_1 - x_1) + a_3(x_1 - x_1)(x - x_2) \rightarrow a_1 = f_1$$

$$x = x_2 : f_2 = a_1 + a_2(x_2 - x_1) + a_3(x_2 - x_1)(x_2 - x_2) \rightarrow a_2 = \left(\frac{f_2 - f_1}{x_2 - x_1}\right)$$

$$x = x_3 : f_3 = a_1 + a_2(x_3 - x_1) + a_3(x_3 - x_1)(x_3 - x_2) \rightarrow a_3 = \frac{\left(\frac{f_3 - f_2}{x_3 - x_2}\right) - \left(\frac{f_2 - f_1}{x_2 - x_1}\right)}{x_3 - x_1}$$

Cubic Newton Polynomial:

$$f_N(x) = a_1 + a_2(x - x_1) + a_3(x - x_1)(x - x_2) + a_4(x - x_1)(x - x_2)(x - x_3)$$

We can obtain coefficients of the cubic Newton polynomial passing through (x_1,f_1), (x_2,f_2), (x_3,f_3), and (x_4,f_4) points as below:

$$x=x_1: f_1=a_1+a_2(x_1-x_1)+a_3(x_1-x_1)(x-x_2)+a_4(x_1-x_1)(x_1-x_2)\times(x_1-x_3)\rightarrow a_1=f_1$$

$$\begin{aligned} x &= x_2: f_2 \\ &= a_1+a_2(x_2-x_1)+a_3(x_2-x_1)(x_2-x_2)+a_4(x_2-x_1)(x_2-x_2)(x_2-x_3) \\ &\longrightarrow \end{aligned}$$

$$a_2=\left(\frac{f_2-f_1}{x_2-x_1}\right)$$

$$\begin{aligned} x &= x_3: f_3 \\ &= a_1+a_2(x_3-x_1)+a_3(x_3-x_1)(x_3-x_2)+a_4(x_3-x_1)(x_3-x_2)(x_3-x_3) \\ &\longrightarrow \end{aligned}$$

$$a_3=\frac{\left(\frac{f_3-f_2}{x_3-x_2}\right)-\left(\frac{f_2-f_1}{x_2-x_1}\right)}{x_3-x_1}$$

$$x=x_4: f_4=a_1+a_2(x_4-x_1)+a_3(x_4-x_1)(x_4-x_2)+a_4(x_4-x_1)(x_4-x_2)(x_4-x_3)$$

$$a_3=\frac{\frac{\left(\frac{f_4-f_3}{x_4-x_3}\right)-\left(\frac{f_3-f_2}{x_3-x_2}\right)}{x_4-x_2}-\frac{\left(\frac{f_3-f_2}{x_3-x_2}\right)-\left(\frac{f_2-f_1}{x_2-x_1}\right)}{x_3-x_1}}{x_4-x_1}$$

From the above expressions, we can see that coefficients of Newton polynomials can be obtained using divided differences of function and variable values. For this reason, Newton interpolation method is also called as Newton Divided Difference method. Coefficients of Newton polynomials can be expressed using divided difference notations as below:

$$a_1=f_1,\quad a_2=f[x_2,x_1],\quad a_3=f[x_3,x_2,x_1]=\frac{f[x_3,x_2]-f[x_2,x_1]}{x_3-x_1},$$

$$a_4=f[x_4,x_3,x_2,x_1]=\frac{f[x_4,x_3,x_2]-f[x_3,x_2,x_1]}{x_4-x_1},\ldots,$$

$$a_n=f[x_n,\ldots,x_2,x_1]=\frac{f[x_n,\ldots,x_2]-f[x_{n-1},\ldots,x_2,x_1]}{x_n-x_1}$$

Using above information, general Newton polynomial of n-th order can be written in terms of divided difference terms as following:

Table 6.1 Newton Divided Difference table

x	f	$f[,]$	$f[,,]$	$f[,,,]$
x_1	f_1	$f[x_2,x_1]=\frac{f_2-f_1}{x_2-x_1}$	$f[x_3,x_2,x_1]$ $=\frac{f[x_3,x_2]-f[x_2,x_1]}{x_3-x_1}$	$f[x_4,x_3,x_2,x_1]$ $=\frac{f[x_4,x_3,x_2]-f[x_3,x_2,x_1]}{x_4-x_1}$
x_2	f_2	$f[x_3,x_2]=\frac{f_3-f_2}{x_3-x_2}$	$f[x_4,x_3,x_2]$ $=\frac{f[x_4,x_3]-f[x_3,x_2]}{x_4-x_2}$	
x_3	f_3	$f[x_4,x_3]=\frac{f_4-f_3}{x_4-x_3}$		
x_4	f_4			

$$\begin{aligned}f_N(x)=&a_1+a_2(x-x_1)+a_3(x-x_1)(x-x_2)+a_4(x-x_1)(x-x_2)(x-x_3)+\ldots\\&+a_n(x-x_1)(x-x_2)(x-x_3)\ldots(x-x_{n-1})\\&=f_1+f[x_2,x_1](x-x_1)+f[x_3,x_2,x_1](x-x_1)(x-x_2)+\\&f[x_4,x_3,x_2,x_1](x-x_1)(x-x_2)(x-x_3)+\ldots+f[x_n,\ldots,x_2,x_1](x-x_1)(x-x_2)(x-x_3)\ldots(x-x_{n-1})\end{aligned} \tag{6.4}$$

From Eq. 6.4, it is seen that Newton polynomial has similar expressions with Taylor series expansion. Derivatives in Taylor series are expressed with divided differences in Newton polynomials.

Advantages of Newton Polynomials:

- Coefficients of polynomials can be easily calculated using divided differences without linear system solution.
- Degree of polynomial can be easily increased by adding new points. Coefficients do not need to be recalculated from the beginning with addition of new points.

Since coefficients in Newton method are calculated with divided differences, preparation of a divided difference table in advance simplifies the calculation of coefficients. For example, Newton Divided Difference table using information for (x_1,f_1), (x_2,f_2), (x_3,f_3), and (x_4,f_4) points can be prepared as in Table 6.1.

Example 6.1: Function values (x_i,f_i) at different points are given as $(0,1)$, $(1,2)$, $(3,6)$, $(5,12)$. Predict function values at $x=2$ and $x=4$ using Newton Divided Difference method.

For given function values, Newton Divided Difference table can be created as in Table 6.2.

Cubic Newton interpolation polynomial for four points can be created taking the first point ($x=0$) as the reference point as below:

Table 6.2 Newton Divided Difference table for given data

x	f	$f[,]$	$f[,,]$	$f[,,,]$
0	1	$f[x_2,x_1]=\frac{2-1}{1-0}=1$	$f[x_3,x_2,x_1]=\frac{2-1}{3-0}=\frac{1}{3}$	$f[x_4,x_3,x_2,x_1]=\frac{\frac{1}{4}-\frac{1}{3}}{5-0}=-\frac{1}{60}$
1	2	$f[x_3,x_2]=\frac{6-2}{3-1}=2$	$f[x_4,x_3,x_2]=\frac{3-2}{5-1}=\frac{1}{4}$	
3	6	$f[x_4,x_3]=\frac{12-6}{5-3}=3$		
5	12			

$$\begin{aligned} f(x) &= f_1 + f[x_2,x_1](x-x_1) + f[x_3,x_2,x_1](x-x_1)(x-x_2) \\ &\quad + f[x_4,x_3,x_2,x_1](x-x_1)(x-x_2)(x-x_3) \\ f(x) &= 1 + 1(x-0) + \frac{1}{3}(x-0)(x-1) - \frac{1}{60}(x-0)(x-1)(x-3) \\ &= 1 + x + \frac{1}{3}x(x-1) - \frac{1}{60}x(x-1)(x-3) \end{aligned}$$

Using this polynomial, function values at $x = 2$ and $x = 4$ can be predicted as following:

$$f(2) = 1 + 2 + \frac{1}{3}2(2-1) - \frac{1}{60}2(2-1)(2-3) = 3.7$$

$$f(4) = 1 + 4 + \frac{1}{3}4(4-1) - \frac{1}{60}4(4-1)(4-3) = 8.8$$

***Note*:** Interpolation values can also be calculated using *MATLAB Program 6.1: Interpolation with Newton's Divided Forward Difference Method* in Appendix.

Special Case of Newton Polynomial:
When intervals between points are equal such as (x_1,f_1), $(x_2 = x_1 + h, f_2)$, $(x_3 = x_1 + 2h, f_3), \ldots, (x_{n+1} = x_1 + nh, f_{n+1})$, in this special case, divided differences can be written in simple form as below:

$$f[x_2,x_1] = \frac{f_2 - f_1}{x_2 - x_1} = \frac{\Delta f_1}{h} \quad (\Delta : \text{forward difference operator}),$$

$$\begin{aligned} f[x_3,x_2,x_1] &= \frac{f[x_3,x_2] - f[x_2,x_1]}{x_3 - x_1} = \frac{\left(\frac{f_3 - f_2}{x_3 - x_2}\right) - \left(\frac{f_2 - f_1}{x_2 - x_1}\right)}{x_3 - x_1} = \frac{\left(\frac{f_3 - f_2}{h}\right) - \left(\frac{f_2 - f_1}{h}\right)}{2h}, \\ &= \frac{\Delta f_2 - \Delta f_1}{2h^2} = \frac{\Delta^2 f_1}{2!h^2} \end{aligned}$$

$$\ldots$$

$$f[x_r, x_{r-1}, \ldots, x_2, x_1] = \frac{\Delta^r f_1}{r!h^r}$$

Table 6.3 Forward Finite Difference table for given data

x	f	Δf	$\Delta^2 f$	$\Delta^3 f$	$\Delta^4 f$
0	1	$\Delta f_1 = 3 - 1 = 2$	$\Delta^2 f_1 = 3 - 2 = 1$	$\Delta^3 f_1 = 1 - 1 = 0$	$\Delta^4 f_1 = 1 - 0 = 1$
2	3	$\Delta f_2 = 6 - 3 = 3$	$\Delta^2 f_2 = 4 - 3 = 1$	$\Delta^3 f_2 = 2 - 1 = 1$	
4	6	$\Delta f_3 = 10 - 6 = 4$	$\Delta^2 f_3 = 6 - 4 = 2$		
6	10	$\Delta f_4 = 16 - 10 = 6$			
8	16				

In this case, n-th order Newton polynomial can be written in terms of non-dimensional variable $\alpha = \frac{(x - x_1)}{h}$ as following:

$$\begin{aligned} f(x) &= f_1 + \Delta f_1 \alpha + \frac{\Delta^2 f_1}{2!} \alpha(\alpha - 1) + \frac{\Delta^3 f_1}{3!} \alpha(\alpha - 1)(\alpha - 2) + \ldots \\ &+ \frac{\Delta^{n+1} f_1}{(n+1)! h^{n+1}} \alpha(\alpha - 1)(\alpha - 2) \ldots (\alpha - n) \end{aligned} \tag{6.5}$$

Above expression is also called as Forward Newton-Gregory polynomial.

Example 6.2: Function values (x_i, f_i) at different points are given as $(0, 1)$, $(2, 3)$, $(4, 6)$, $(6, 10)$, $(8, 16)$. Predict function values at $x = 3$ and $x = 7$ using Newton Divided Difference method.

For given function values, Newton divided difference table can be created as in Table 6.3.

Newton interpolation polynomial can be written for the equidistant case as below $\left(\alpha = \frac{(x - x_1)}{h}\right)$

$$f(x) = f_1 + \Delta f_1 \alpha + \frac{\Delta^2 f_1}{2!} \alpha(\alpha - 1) + \frac{\Delta^3 f_1}{3!} \alpha(\alpha - 1)(\alpha - 2) + \frac{\Delta^4 f_1}{4!} \alpha(\alpha - 1)(\alpha - 2)(\alpha - 3)$$

Using this polynomial, function values at $x = 3$ and $x = 7$ can be predicted as following:

For $x = 3$:

$$\alpha = \frac{x - x_1}{h} = \frac{3 - 0}{2} = \frac{3}{2} \rightarrow$$

$$\begin{aligned} f(3) = 1 + 2\frac{3}{2} + \frac{1}{2!}\frac{3}{2}\left(\frac{3}{2} - 1\right) + \frac{0}{3!}\frac{3}{2}\left(\frac{3}{2} - 1\right)\left(\frac{3}{2} - 2\right) + \frac{1}{4!}\frac{3}{2}\left(\frac{3}{2} - 1\right)\left(\frac{3}{2} - 2\right) \\ \times \left(\frac{3}{2} - 3\right) = 4.3984 \end{aligned}$$

For $x = 7$:

$$\alpha = \frac{x - x_1}{h} = \frac{7-0}{2} = \frac{7}{2} \rightarrow$$

$$f(7) = 1 + 2\frac{7}{2} + \frac{1}{2!}\frac{7}{2}\left(\frac{7}{2}-1\right) + \frac{0}{3!}\frac{7}{2}\left(\frac{7}{2}-1\right)\left(\frac{7}{2}-2\right) + \frac{1}{4!}\frac{7}{2}\left(\frac{7}{2}-1\right)\left(\frac{7}{2}-2\right)$$
$$\times\left(\frac{7}{2}-3\right) = 12.6484$$

6.2.2 *Lagrange Interpolation Polynomial*

Like Newton polynomial, Lagrange interpolation polynomial is also a different form of classical polynomial expression. Polynomial coefficients in Lagrange method can also be determined without solving a linear system as in Newton method.

Linear and parabolic Lagrange polynomials can be obtained as following:

Linear Lagrange Polynomial:
Linear Lagrange polynomial passing through (x_1, f_1) and (x_2, f_2) points can be written as following:

$$f(x) = a_1(x - x_2) + a_2(x - x_1)$$

where a_1 and a_2 coefficients can be obtained by substituting the known function values into the polynomial expression as below:

$$f_1 = a_1(x_1 - x_2) + a_2(x_1 - x_1) \rightarrow a_1 = \frac{f_1}{x_1 - x_2}$$

$$f_2 = a_1(x_2 - x_2) + a_2(x_2 - x_1) \rightarrow a_2 = \frac{f_2}{x_2 - x_1}$$

After substituting a_1 and a_2 values, linear Lagrange polynomial is obtained as following:

$$f(x) = \underbrace{\frac{(x - x_2)}{(x_1 - x_2)}}_{L_1(x)} f_1 + \underbrace{\frac{(x - x_1)}{(x_2 - x_1)}}_{L_2(x)} f_2 = L_1(x)f_1 + L_2(x)f_2$$

$L_1(x)$, $L_2(x)$ functions are also called Lagrange functions, shape functions or weight functions. Lagrange functions take value of 1 or 0 at interpolation points as below:

$$L_1(x) = \begin{bmatrix} 1 & x = x_1 \\ 0 & x \neq x_1 \end{bmatrix}, \quad L_2(x) = \begin{bmatrix} 1 & x = x_2 \\ 0 & x \neq x_2 \end{bmatrix}$$

Linear Lagrange functions of $L_1(x)$ and $L_2(x)$ can be visualized as in Fig. 6.3.

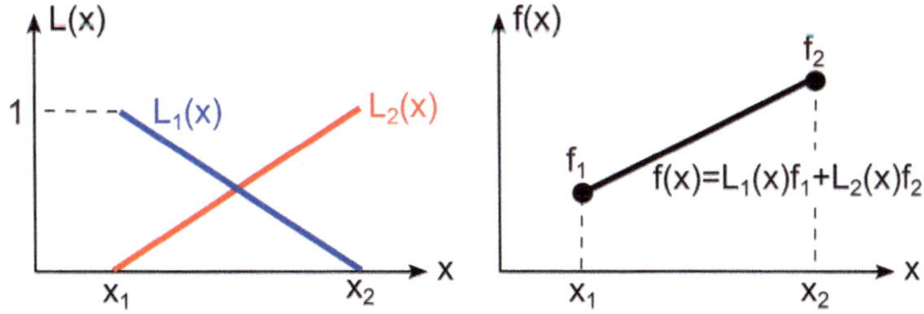

Fig. 6.3 Linear Lagrange functions and linear Lagrange interpolation polynomial

Parabolic Lagrange Polynomial:

Parabolic Lagrange polynomial passing through (x_1,f_1), (x_2,f_2), and (x_3,f_3) points can be written as following:

$$f(x) = a_1(x-x_2)(x-x_3) + a_2(x-x_1)(x-x_3) + a_3(x-x_1)(x-x_2)$$

where a_1, a_2, a_3 coefficients can be obtained by substituting the known function values into the polynomial expression as below:

$$f_1 = a_1(x_1-x_2)(x_1-x_3) + a_2(x_1-x_1)(x_1-x_3) + a_3(x_1-x_1)(x_1-x_2) \to a_1 = \frac{f_1}{(x_1-x_2)(x_1-x_3)}$$

$$f_2 = a_1(x_2-x_2)(x_2-x_3) + a_2(x_2-x_1)(x_2-x_3) + a_3(x_2-x_1)(x_2-x_2) \to a_2 = \frac{f_2}{(x_2-x_1)(x_2-x_3)}$$

$$f_3 = a_1(x_3-x_2)(x_3-x_3) + a_2(x_3-x_1)(x_3-x_3) + a_3(x_3-x_1)(x_3-x_2) \to a_3 = \frac{f_3}{(x_3-x_1)(x_3-x_2)}$$

After substituting a_1, a_2, a_3 values, parabolic Lagrange polynomial is obtained as following:

$$f(x) = \underbrace{\frac{(x-x_2)(x-x_3)}{(x_1-x_2)(x_1-x_3)}}_{L_1(x)} f_1 + \underbrace{\frac{(x-x_1)(x-x_3)}{(x_2-x_1)(x_2-x_3)}}_{L_2(x)} f_2 + \underbrace{\frac{(x-x_1)(x-x_2)}{(x_3-x_1)(x_3-x_2)}}_{L_3(x)} f_2$$
$$= L_1(x)f_1 + L_2(x)f_2 + L_3(x)f_3$$

$L_1(x)$, $L_2(x)$, $L_3(x)$ correspond to Lagrange functions which take the value of 1 or 0 at interpolation points as below:

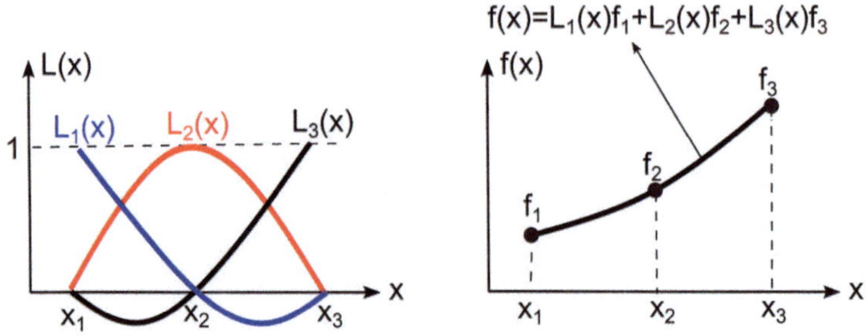

Fig. 6.4 Parabolic Lagrange functions and parabolic Lagrange interpolation polynomial

$$L_1(x)=\begin{bmatrix}1 & x=x_1\\ 0 & x\neq x_1\end{bmatrix},\quad L_2(x)=\begin{bmatrix}1 & x=x_2\\ 0 & x\neq x_2\end{bmatrix},\quad L_3(x)=\begin{bmatrix}1 & x=x_3\\ 0 & x\neq x_3\end{bmatrix}$$

Parabolic Lagrange functions $L_1(x)$, $L_2(x)$, and $L_3(x)$ can be visualized as in Fig. 6.4.

Similar to linear and parabolic Lagrange polynomials, a general n-th order Lagrange polynomial passing through (x_1,f_1), (x_2,f_2),..., (x_{n+1},f_{n+1}) points can be expressed as below:

$$f(x)=L_1(x)f_1+L_2(x)f_2+L_3(x)f_3+\ldots+L_{n+1}(x)f_{n+1} \tag{6.6}$$

where Lagrange functions $L_i(x)$ in Eq. 6.6 can be expressed as following:

$$L_i(x)=\frac{(x-x_1)(x-x_2)\ldots(x-x_{i-1})(x-x_{i+1})\ldots(x-x_n)}{(x_i-x_1)(x_i-x_2)\ldots(x_i-x_{i-1})(x_i-x_{i+1})\ldots(x_i-x_n)}\quad (i=1,2,\ldots,n+1) \tag{6.7}$$

$L_i(x)$ function takes the value of 1 or 0 at the interpolation points as below:

$$L_i(x)=\begin{bmatrix}1 & x=x_i\\ 0 & x\neq x_i\end{bmatrix}, \tag{6.8}$$

Note**:** Lagrange interpolation method presents the advantage that interpolation coefficients can be calculated directly without solving a set of equations.

Example 6.3 Function values at different points are given as $(x_1,f_1)=(0,1)$, $(x_2 f_2)=(3,5)$, $(x_3,f_3)=(6,18)$, $(x_4,f_4)=(9,36)$. Predict function values at $x=4$ and $x=7$ using Lagrange interpolation method.

Lagrange interpolation polynomial for given four points can be written symbolically as below:

$$\begin{aligned} f(x) &= \frac{(x-x_2)(x-x_3)(x-x_4)}{(x_1-x_2)(x_1-x_3)(x_1-x_4)}f_1 + \frac{(x-x_1)(x-x_3)(x-x_4)}{(x_2-x_1)(x_2-x_3)(x_2-x_4)}f_2 \\ &+ \frac{(x-x_1)(x-x_2)(x-x_4)}{(x_3-x_1)(x_3-x_2)(x_3-x_4)}f_3 + \frac{(x-x_1)(x-x_2)(x-x_3)}{(x_4-x_1)(x_4-x_2)(x_4-x_3)}f_4 \end{aligned}$$

By substituting numerical values, function values at $x = 4$ and $x = 7$ can be obtained as following:

$$\begin{aligned} f(4) &= \frac{(4-3)(4-6)(4-9)}{(0-3)(0-6)(0-9)}1 + \frac{(4-0)(4-6)(4-9)}{(3-0)(3-6)(3-9)}5 \\ &+ \frac{(4-0)(4-3)(4-9)}{(6-0)(6-3)(6-9)}18 + \frac{(4-0)(4-3)(4-6)}{(9-0)(9-3)(9-6)}36 \\ &= 8.5309 \end{aligned}$$

$$\begin{aligned} f(7) &= \frac{(7-3)(7-6)(7-9)}{(0-3)(0-6)(0-9)}1 + \frac{(7-0)(7-6)(7-9)}{(3-0)(3-6)(3-9)}5 \\ &+ \frac{(7-0)(7-3)(7-9)}{(6-0)(6-3)(6-9)}18 + \frac{(7-0)(7-3)(7-6)}{(9-0)(9-3)(9-6)}36 \\ &= 18.8942 \end{aligned}$$

Note: Interpolation values can also be calculated using *MATLAB Program 6.2: Lagrange Interpolation Method* in Appendix.

6.3 Spline Interpolation Polynomials

Previously, it was mentioned that higher order global polynomials passing through a large number of points have oscillatory character leading to inaccurate interpolation values. To reduce the drawback of global interpolation polynomials, spline (piecewise) interpolation polynomials were developed. In spline interpolation, data points are divided into small groups and lower order polynomial functions with smooth character such as linear, parabolic, or cubic polynomials are generated in each group of data points so as to satisfy continuity conditions at shared points. Spline interpolation functions are explained in detail below:

6.3.1 Linear Spline Interpolation Polynomial

In this method, adjacent points in a data set are connected with linear polynomial functions as seen in Fig. 6.5. In the case of n data points, linear spline functions between adjacent points can be expressed with Lagrange interpolation method as following:

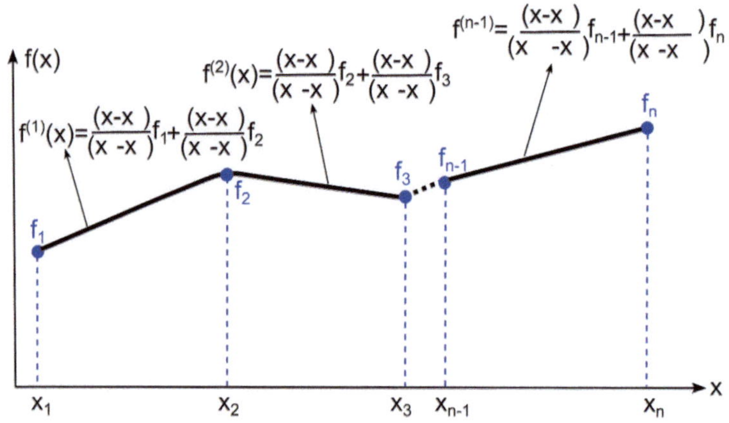

Fig. 6.5 Interpolation with linear piecewise functions

$$
\begin{aligned}
&f^{(1)}(x)=\left(\frac{x-x_2}{x_1-x_2}\right)f_1+\left(\frac{x-x_1}{x_2-x_1}\right)f_2 \quad (x_1\leq x\leq x_2)\\
&f^{(2)}(x)=\left(\frac{x-x_3}{x_2-x_3}\right)f_2+\left(\frac{x-x_2}{x_3-x_2}\right)f_3 \quad (x_2\leq x\leq x_3)\\
&\ldots\\
&f^{(n-1)}(x)=\left(\frac{x-x_n}{x_{n-1}-x_n}\right)f_{n-1}+\left(\frac{x-x_{n-1}}{x_n-x_{n-1}}\right)f_n \quad (x_{n-1}\leq x\leq x_n)
\end{aligned}
\tag{6.9}
$$

Linear spline polynomials do not have derivative (e.g., slope) continuities at shared points. In spite of this drawback, they are widely used in solving engineering problems due to their simplicity.

***Note*:** Interpolation values can also be calculated using *MATLAB Program 6.3: Interpolation with Linear Spline Functions* in Appendix.

6.3.2 Parabolic Spline Interpolation Polynomial

In this method, adjacent points in a data set are connected with parabolic polynomial functions called as parabolic splines as in Fig. 6.6. In the construction of parabolic splines, function and first-order derivative continuities at shared points are preserved.

In the case of n data points, parabolic spline functions between adjacent points can be expressed as following:

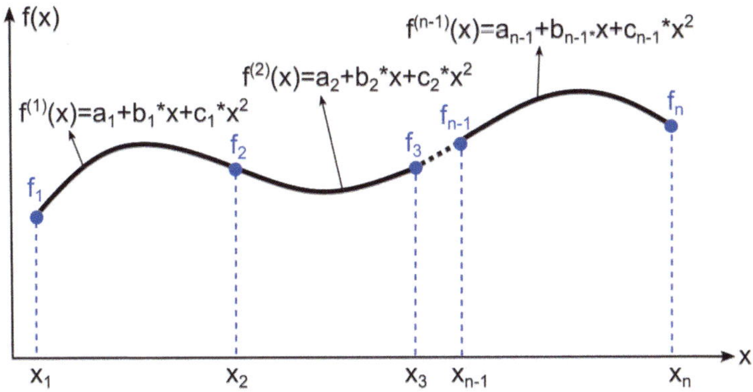

Fig. 6.6 Interpolation with parabolic spline functions

$$\begin{aligned} &f^{(1)}(x) = a_1 + b_1x + c_1x^2 \quad (x_1 \le x \le x_2) \\ &f^{(2)}(x) = a_2 + b_2x + c_2x^2 \quad (x_2 \le x \le x_3) \\ &\ldots \\ &f^{(n-1)}(x) = a_{n-1} + b_{n-1}x + c_{n-1}x^2 \quad (x_{n-1} \le x \le x_n) \end{aligned} \tag{6.10}$$

For n points, there will be $n-1$ parabolic spline functions yielding total $3(n-1) = 3n-3$ unknown coefficients to be determined. To determine these coefficients, $3n-3$ equations should be created.

Spline functions must satisfy function values at data points as following:

$$\begin{aligned} &f^{(i)}(x_i) = f_i, \quad i = 1, \ldots, n-1 \\ &f^{(i)}(x_{i+1}) = f_{i+1}, \quad i = 1, \ldots, n-1 \end{aligned} \tag{6.11}$$

Also, slope (first derivative) continuity can be enforced at shared (interior) points to make spline functions look as parts of a global smooth function as following:

$$\frac{df^{(i)}}{dx}(x_{i+1}) = \frac{df^{(i+1)}}{dx}(x_{i+1}) \quad i = 1, \ldots, n-2 \tag{6.12}$$

Using Eqs. 6.11 and 6.12, $3n-4$ equations can be created although there are $3n-3$ unknown constants. To solve unknown coefficients, the number of equations and unknown coefficients must be equal. Remaining one equation can be obtained from an end or edge condition. Edge condition can be expressed as zeroing second derivative at first point as below:

$$\frac{d^2f^{(1)}}{dx^2}(x_1) = 0 \tag{6.13}$$

Equations 6.11, 6.12, and 6.13 can be written more explicitly for n data points as following:

$$\left.\begin{aligned} &f^{(1)}(x_1)=f_1=a_1+b_1x_1+c_1x_1^2\\ &f^{(1)}(x_2)=f_2=a_1+b_1x_2+c_1x_2^2\\ &\dots\\ &f^{(n-1)}(x_{n-1})=f_{n-1}=a_{n-1}+b_{n-1}x_{n-1}+c_{n-1}x_{n-1}^2\\ &f^{(n-1)}(x_n)=f_n=a_{n-1}+b_{n-1}x_n+c_{n-1}x_n^2 \end{aligned}\right] \Rightarrow \text{Function Values} \tag{6.14}$$

$$\left.\begin{aligned} &\frac{df^{(1)}}{dx}(x_2)=\frac{df^{(2)}}{dx}(x_2)\Rightarrow b_1+2c_1x_2=b_2+2c_2x_2\\ &\frac{df^{(2)}}{dx}(x_3)=\frac{df^{(3)}}{dx}(x_3)\Rightarrow b_2+2c_2x_3=b_3+2c_3x_3\\ &\dots\\ &\frac{df^{(n-2)}}{dx}(x_{n-1})=\frac{df^{(n-1)}}{dx}(x_{n-1})\Rightarrow b_{n-2}+2c_{n-2}x_{n-1}=b_{n-1}+2c_{n-1}x_{n-1} \end{aligned}\right] \Rightarrow \begin{matrix}\text{1st Derivative}\\ \text{Continuities}\end{matrix} \tag{6.15}$$

$$\frac{d^2f^{(1)}}{dx^2}(x_1)=0\Rightarrow 2c_1=0 \text{ (Edge condition)} \tag{6.16}$$

Equations 6.14, 6.15, and 6.16 can be written in matrix form as below:

$$\begin{bmatrix} 1 & x_1 & x_1^2 & 0 & 0 & 0 & \cdots & 0 & 0 & 0\\ 1 & x_2 & x_2^2 & 0 & 0 & 0 & \cdots & 0 & 0 & 0\\ \cdots & \cdots & \cdots & \cdots & \cdots & \cdots & \cdots & \cdots & \cdots & \cdots\\ 0 & 0 & 0 & 0 & 0 & 0 & \cdots & 1 & x_{n-1} & x_{n-1}^2\\ 0 & 0 & 0 & 0 & 0 & 0 & \cdots & 1 & x_n & x_n^2\\ 0 & 1 & 2x_2 & 0 & -1 & -2x_2 & \cdots & 0 & 0 & 0\\ 0 & 0 & 0 & 0 & 1 & 2x_3 & \cdots & 0 & 0 & 0\\ \cdots & \cdots & \cdots & \cdots & \cdots & \cdots & \cdots & \cdots & \cdots & \cdots\\ 0 & 0 & 0 & 0 & 0 & 0 & \cdots & 0 & -1 & -2x_n\\ 0 & 0 & 2 & 0 & 0 & 0 & \cdots & 0 & 0 & 0 \end{bmatrix} \begin{Bmatrix} a_1\\ b_1\\ c_1\\ a_2\\ b_2\\ c_2\\ \cdots\\ a_{n-1}\\ b_{n-1}\\ c_{n-1} \end{Bmatrix} = \begin{Bmatrix} f_1\\ f_2\\ \cdots\\ f_{n-1}\\ f_n\\ 0\\ 0\\ \cdots\\ 0\\ 0 \end{Bmatrix} \tag{6.17}$$

From the solution of the linear system in Eq. 6.17, a, b, and c coefficients of parabolic splines can be easily obtained.

Example 6.4 Function values at different points are given as $(x_1,f_1) = (0,1)$, $(x_2,f_2) = (2,3)$, $(x_3,f_3) = (4,6)$, $(x_4,f_4) = (6,10)$. Predict function values at $x = 3$ and $x = 5$ using parabolic spline method.

Parabolic splines passing through the given points can be expressed as below:

$$\begin{aligned} f^{(1)}(x) - a_1 + b_1 x + c_1 x^2 \quad & (x_1 \le x \le x_2) \\ f^{(2)}(x) = a_2 + b_2 x + c_2 x^2 \quad & (x_2 \le x \le x_3) \\ f^{(3)}(x) = a_3 + b_3 x + c_3 x^2 \quad & (x_3 \le x \le x_4) \end{aligned}$$

Using above expressions, function values, derivative continuities, and edge condition can be written as following:

$$\left.\begin{aligned} f^{(1)}(x_1) &= f_1 = a_1 + b_1 x_1 + c_1 x_1^2 \\ f^{(1)}(x_2) &= f_2 = a_1 + b_1 x_2 + c_1 x_2^2 \\ f^{(2)}(x_2) &= f_2 = a_2 + b_2 x_2 + c_2 x_2^2 \\ f^{(2)}(x_3) &= f_3 = a_2 + b_2 x_3 + c_2 x_3^2 \\ f^{(3)}(x_3) &= f_3 = a_3 + b_3 x_3 + c_3 x_3^2 \\ f^{(3)}(x_4) &= f_4 = a_3 + b_3 x_4 + c_3 x_4^2 \end{aligned}\right] \Rightarrow \text{Function Values}$$

$$\left.\begin{aligned} \frac{df^{(1)}}{dx}(x_2) = \frac{df^{(2)}}{dx}(x_2) \Rightarrow b_1 + 2c_1 x_2 = b_2 + 2c_2 x_2 \\ \frac{df^{(2)}}{dx}(x_3) = \frac{df^{(3)}}{dx}(x_3) \Rightarrow b_2 + 2c_2 x_3 = b_3 + 2c_3 x_3 \end{aligned}\right] \Rightarrow \text{1st Derivative Continuities}$$

$$\left.\frac{d^2 f^{(1)}}{dx^2}(x_1) = 0 \Rightarrow 2c_1 = 0\right] \Rightarrow \text{Edge Condition}$$

Function values, first derivative continuities, and edge condition can be written in matrix form as following:

$$\begin{bmatrix} 1 & x_1 & x_1^2 & 0 & 0 & 0 & 0 & 0 & 0 \\ 1 & x_2 & x_2^2 & 0 & 0 & 0 & 0 & 0 & 0 \\ 0 & 0 & 0 & 1 & x_2 & x_2^2 & 0 & 0 & 0 \\ 0 & 0 & 0 & 1 & x_3 & x_3^2 & 0 & 0 & 0 \\ 0 & 0 & 0 & 0 & 0 & 0 & 1 & x_3 & x_3^2 \\ 0 & 0 & 0 & 0 & 0 & 0 & 1 & x_4 & x_4^2 \\ 0 & 1 & 2x_2 & 0 & -1 & -2x_2 & 0 & 0 & 0 \\ 0 & 0 & 0 & 0 & 1 & 2x_3 & 0 & -1 & -2x_3 \\ 0 & 0 & 2 & 0 & 0 & 0 & 0 & 0 & 0 \end{bmatrix} \begin{Bmatrix} a_1 \\ b_1 \\ c_1 \\ a_2 \\ b_2 \\ c_2 \\ a_3 \\ b_3 \\ c_3 \end{Bmatrix} = \begin{Bmatrix} f_1 \\ f_2 \\ f_2 \\ f_3 \\ f_3 \\ f_4 \\ 0 \\ 0 \\ 0 \end{Bmatrix}$$

Substituting numerical values into the matrix equation, polynomial coefficients can be determined as below:

a1 = 1.0000, b1 = 1.0000, c1 = 0, a2 = 2.0000, b2 = 0, c2 = 0.2500, a3 = −2.0000, b3 = 2.0000, c3 = 0.

Thus, parabolic spline polynomials are obtained as following:

$$f^{(1)}(x) = 1 + x \quad (0 \le x \le 2)$$
$$f^{(2)}(x) = 2 + 0.25x^2 \quad (2 \le x \le 4) f^{(3)}(x) = -2 + 2x \quad (4 \le x \le 6)$$

Using parabolic spline functions, function values *f*(3) and *f*(5) are obtained as below:

$$f^{(2)}(3) = 2 + 0.25(3)^2 = 4.25$$
$$f^{(3)}(5) = -2 + 2(5) = 8$$

Note: Interpolation values can also be calculated using *MATLAB Program 6.4: Interpolation with Parabolic Spline Functions* in Appendix.

6.3.3 Cubic Spline Interpolation

Cubic spline interpolation polynomial is one of the most widely used interpolation functions. In this method, adjacent points are connected with cubic polynomial functions as in Fig. 6.7. In the construction of cubic splines, function, first-order and second-order derivative continuities at shared (interior) points are used.

In the case of n data points, cubic spline functions between adjacent points can be expressed as following:

$$\begin{aligned} &f^{(1)}(x) = a_1 + b_1x + c_1x^2 + d_1x^3 \quad (x_1 \le x \le x_2) \\ &f^{(2)}(x) = a_2 + b_2x + c_2x^2 + d_2x^3 \quad (x_2 \le x \le x_3) \\ &\ldots \\ &f^{(n-1)}(x) = a_{n-1} + b_{n-1}x + c_{n-1}x^2 + d_{n-1}x^3 \quad (x_{n-1} \le x \le x_n) \end{aligned} \tag{6.18}$$

For n points, there will be $n-1$ cubic spline functions yielding total $4(n-1) = 4n-4$ unknown coefficients to be determined. Spline functions must satisfy function values at data points as following:

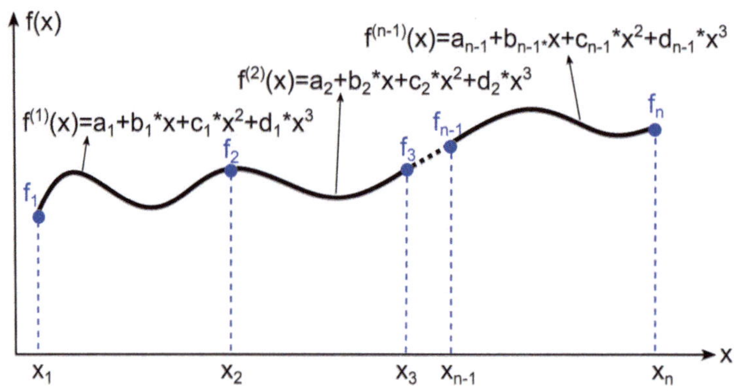

Fig. 6.7 Interpolation with cubic spline functions

$$\begin{aligned} f^{(i)}(x_i) &= f_i, \quad i=1,\ldots,n-1 \\ f^{(i)}(x_{i+1}) &= f_{i+1}, \quad i=1,\ldots,n-1 \end{aligned} \tag{6.19}$$

Also, first and second derivative continuities can be enforced at shared (interior) points to make spline functions look as parts of a global smooth function as following:

$$\begin{aligned} \frac{df^{(i)}}{dx}(x_{i+1}) &= \frac{df^{(i+1)}}{dx}(x_{i+1}) \quad i=1,\ldots,n-2 \\ \frac{d^2f^{(i)}}{dx^2}(x_{i+1}) &= \frac{d^2f^{(i+1)}}{dx^2}(x_{i+1}) \quad i=1,\ldots,n-2 \end{aligned} \tag{6.20}$$

Using Eqs. 6.19 and 6.20 total $4n-6$ equations are obtained although there are $4n-4$ unknown constants. To solve unknown coefficients, the number of equations and unknown coefficients must be equal. Remaining two equations can be obtained from edge conditions. Edge conditions can be expressed as zeroing second derivatives at end (first and last) points as below:

$$\left.\begin{aligned} \frac{d^2f^{(1)}}{dx^2}(x_1) &= 0 \\ \frac{d^2f^{(n-1)}}{dx^2}(x_n) &= 0 \end{aligned}\right. \quad \text{(Edge conditions)} \tag{6.21}$$

Equations 6.19, 6.20, and 6.21 can be written more explicitly for n data points as following:

$$\left.\begin{aligned} & f^{(1)}(x_1) - f_1 = a_1 + b_1x_1 + c_1x_1^2 + d_1x_1^3 \\ & f^{(1)}(x_2) = f_2 = a_1 + b_1x_2 + c_1x_2^2 + d_1x_2^3 \\ & \ldots \\ & f^{(n-1)}(x_{n-1}) = f_{n-1} = a_{n-1} + b_{n-1}x_{n-1} + c_{n-1}x_{n-1}^2 + d_{n-1}x_{n-1}^3 \\ & f^{(n-1)}(x_n) = f_n = a_{n-1} + b_{n-1}x_n + c_{n-1}x_n^2 + d_{n-1}x_n^3 \end{aligned}\right] \Rightarrow \text{Function Values} \tag{6.22}$$

$$\left.\begin{aligned} & \frac{df^{(1)}}{dx}(x_2) = \frac{df^{(2)}}{dx}(x_2) \Rightarrow b_1 + 2c_1x_2 + 3d_1x_2^2 = b_2 + 2c_2x_2 + 3d_2x_2^2 \\ & \ldots \\ & \frac{df^{(n-2)}}{dx}(x_{n-1}) = \frac{df^{(n-1)}}{dx}(x_{n-1}) \Rightarrow b_{n-2} + 2c_{n-2}x_{n-1} + 3d_{n-2}x_{n-1}^2 = b_{n-1} + 2c_{n-1}x_{n-1} + 3d_{n-1}x_{n-1}^2 \end{aligned}\right] \Rightarrow \begin{array}{c}\text{1st Derivative}\\ \text{Continuities}\end{array} \tag{6.23}$$

$$\left.\begin{array}{l}\dfrac{d^2f^{(1)}}{dx^2}(x_2)=\dfrac{d^2f^{(2)}}{dx^2}(x_2)\Rightarrow 2c_1+6d_1x_2=2c_2+6d_2x_2\\ \cdots\\ \dfrac{d^2f^{(n-2)}}{dx^2}(x_{n-1})=\dfrac{d^2f^{(n-1)}}{dx^2}(x_{n-1})\Rightarrow 2c_{n-2}+6d_{n-2}x_{n-1}=2c_{n-1}+6d_{n-1}x_{n-1}\end{array}\right]$$
$$\Rightarrow \begin{array}{c}\text{2nd Derivative}\\ \text{Continuities}\end{array} \tag{6.24}$$

$$\left.\begin{array}{l}\dfrac{d^2f^{(1)}}{dx^2}(x_1)=0\Rightarrow 2c_1+6d_1x_1=0\\ \dfrac{d^2f^{(n-1)}}{dx^2}(x_n)=0\Rightarrow 2c_{n-1}+6d_{n-1}x_n=0\end{array}\right]\Rightarrow \text{Edge Conditions} \tag{6.25}$$

Equations 6.22, 6.23, 6.24, and 6.25 can be written in matrix form as below:

$$\begin{bmatrix}
1 & x_1 & x_1^2 & x_1^3 & 0 & 0 & 0 & 0 & 0 & 0 & 0 & 0 & 0\\
1 & x_2 & x_2^2 & x_2^3 & 0 & 0 & 0 & 0 & 0 & 0 & 0 & 0 & 0\\
\cdots & \cdots & \cdots & \cdots & \cdots & \cdots & \cdots & \cdots & \cdots & \cdots & \cdots & \cdots & \cdots\\
0 & 0 & 0 & 0 & 0 & 0 & 0 & 0 & 0 & 1 & x_{n-1} & x_{n-1}^2 & x_{n-1}^3\\
0 & 0 & 0 & 0 & 0 & 0 & 0 & 0 & 0 & 1 & x_n & x_n^2 & x_{n-1}^3\\
0 & 1 & 2x_2 & 3x_2^2 & 0 & -1 & -2x_2 & -3x_2^2 & 0 & 0 & 0 & 0 & 0\\
\cdots & \cdots & \cdots & \cdots & \cdots & \cdots & \cdots & \cdots & \cdots & \cdots & \cdots & \cdots & \cdots\\
0 & 0 & 0 & 0 & 0 & 0 & 0 & 0 & 0 & 0 & -1 & -2x_{n-1} & -3x_{n-1}^2\\
0 & 0 & 2 & 6x_2 & 0 & 0 & -2 & -6x_2 & 0 & 0 & 0 & 0 & 0\\
\cdots & \cdots & \cdots & \cdots & \cdots & \cdots & \cdots & \cdots & \cdots & \cdots & \cdots & \cdots & \cdots\\
0 & 0 & 0 & 0 & 0 & 0 & 0 & 0 & 0 & 0 & 0 & -2 & -6x_{n-1}\\
0 & 0 & 2 & 6x_1 & 0 & 0 & 0 & 0 & 0 & 0 & 0 & 0 & 0\\
0 & 0 & 0 & 0 & 0 & 0 & & 0 & 0 & 0 & 0 & 2 & 6x_n
\end{bmatrix}
\begin{Bmatrix} a_1\\ b_1\\ c_1\\ d_1\\ a_2\\ b_2\\ c_2\\ d_2\\ \cdots\\ a_{n-1}\\ b_{n-1}\\ c_{n-1}\\ d_{n-1}\end{Bmatrix}
=\begin{Bmatrix} f_1\\ f_2\\ \cdots\\ f_{n-1}\\ f_n\\ 0\\ 0\\ \cdots\\ 0\\ 0\end{Bmatrix} \tag{6.26}$$

From the solution of the linear system in Eq. 6.26, *a*, *b*, *c*, and *d* coefficients of cubic splines can be easily obtained.

Example 6.5: Function values at different points are given as $(x_1,f_1) = (0,1)$, $(x_2 f_2) = (2,3)$, $(x_3,f_3) = (4,6)$, $(x_4,f_4) = (6,10)$. Predict function values at $x = 3$ and $x = 5$ using cubic spline method.

Cubic splines passing through the given points can be expressed as below:

$$
\begin{aligned}
f^{(1)}(x) &= a_1 + b_1 x + c_1 x^2 + d_1 x^3 \quad (x_1 \le x \le x_2) \\
f^{(2)}(x) &= a_2 + b_2 x + c_2 x^2 + d_2 x^3 \quad (x_2 \le x \le x_3) \\
f^{(3)}(x) &= a_3 + b_3 x + c_3 x^2 + d_3 x^3 \quad (x_3 \le x \le x_4)
\end{aligned}
$$

Using above expressions, function values, first and second derivative continuities as well as edge conditions can be written as following:

$$
\left.
\begin{aligned}
f^{(1)}(x_1) &= f_1 = a_1 + b_1 x_1 + c_1 x_1^2 + d_1 x_1^3 \\
f^{(1)}(x_2) &= f_2 = a_1 + b_1 x_2 + c_1 x_2^2 + d_1 x_2^3 \\
f^{(2)}(x_2) &= f_2 = a_2 + b_2 x_2 + c_2 x_2^2 + d_2 x_2^3 \\
f^{(2)}(x_3) &= f_3 = a_2 + b_2 x_3 + c_2 x_3^2 + d_2 x_3^3 \\
f^{(3)}(x_3) &= f_3 = a_3 + b_3 x_3 + c_3 x_3^2 + d_3 x_3^3 \\
f^{(3)}(x_4) &= f_3 = a_3 + b_3 x_4 + c_3 x_4^2 + d_3 x_4^3
\end{aligned}
\right] \Rightarrow \text{Function Values}
$$

$$
\left.
\begin{aligned}
\frac{df^{(1)}}{dx}(x_2) &= \frac{df^{(2)}}{dx}(x_2) \Rightarrow b_1 + 2c_1 x_2 + 3d_1 x_2^2 = b_2 + 2c_2 x_2 + 3d_2 x_2^2 \\
\frac{df^{(2)}}{dx}(x_3) &= \frac{df^{(3)}}{dx}(x_3) \Rightarrow b_2 + 2c_2 x_3 + 3d_2 x_3^2 = b_3 + 2c_3 x_3 + 3d_3 x_3^2
\end{aligned}
\right]
$$

$\Rightarrow$ 1st Derivative Continuities

$$
\left.
\begin{aligned}
\frac{d^2 f^{(1)}}{dx^2}(x_2) &= \frac{d^2 f^{(2)}}{dx^2}(x_2) \Rightarrow 2c_1 + 6d_1 x_2 = 2c_2 + 6d_2 x_2 \\
\frac{d^2 f^{(2)}}{dx^2}(x_3) &= \frac{d^2 f^{(3)}}{dx^2}(x_3) \Rightarrow 2c_2 + 6d_2 x_3 = 2c_3 + 6d_3 x_3
\end{aligned}
\right]
$$

$\Rightarrow$ 2nd Derivative Continuities

$$
\left.
\begin{aligned}
\frac{d^2 f^{(1)}}{dx^2}(x_1) &= 0 \Rightarrow 2c_1 + 6d_1 x_1 = 0 \\
\frac{d^2 f^{(3)}}{dx^2}(x_4) &= 0 \Rightarrow 2c_3 + 6d_3 x_4 = 0
\end{aligned}
\right] \Rightarrow \text{Edge Conditions}
$$

Function values, derivative continuities, and edge conditions can be written in matrix form as following:

$$\begin{bmatrix} 1 & x_1 & x_1^2 & x_1^3 & 0 & 0 & 0 & 0 & 0 & 0 & 0 & 0 \\ 1 & x_2 & x_2^2 & x_2^3 & 0 & 0 & 0 & 0 & 0 & 0 & 0 & 0 \\ 0 & 0 & 0 & 0 & 1 & x_2 & x_2^2 & x_2^3 & 0 & 0 & 0 & 0 \\ 0 & 0 & 0 & 0 & 1 & x_3 & x_3^2 & x_3^3 & 0 & 0 & 0 & 0 \\ 0 & 0 & 0 & 0 & 0 & 0 & 0 & 0 & 1 & x_3 & x_3^2 & x_3^3 \\ 0 & 0 & 0 & 0 & 0 & 0 & 0 & 0 & 1 & x_4 & x_4^2 & x_4^3 \\ 0 & 1 & 2x_2 & 3x_2^2 & 0 & -1 & -2x_2 & -3x_2^2 & 0 & 0 & 0 & 0 \\ 0 & 0 & 0 & 0 & 0 & 1 & 2x_3 & 3x_3^2 & 0 & -1 & -2x_3 & -3x_3^2 \\ 0 & 0 & 2 & 6x_2 & 0 & 0 & -2 & -6x_2 & 0 & 0 & 0 & 0 \\ 0 & 0 & 0 & 0 & 0 & 0 & 2 & 6x_3 & 0 & 0 & -2 & -6x_3 \\ 0 & 0 & 2 & 6x_1 & 0 & 0 & 0 & 0 & 0 & 0 & 0 & 0 \\ 0 & 0 & 0 & 0 & 0 & 0 & 0 & 0 & 0 & 0 & 2 & 6x_4 \end{bmatrix} \begin{Bmatrix} a_1 \\ b_1 \\ c_1 \\ d_1 \\ a_2 \\ b_2 \\ c_2 \\ d_2 \\ a_3 \\ b_3 \\ c_3 \\ d_3 \end{Bmatrix} = \begin{Bmatrix} f_1 \\ f_2 \\ f_2 \\ f_3 \\ f_3 \\ f_4 \\ 0 \\ 0 \\ 0 \\ 0 \\ 0 \\ 0 \end{Bmatrix}$$

Substituting numerical values in matrix equation, polynomial coefficients are obtained as below:

$$\begin{array}{llll} \text{a1} = 1.0000, & \text{b1} = 0.9000, & \text{c1} = 0, & \text{d1} = 0.0250, \\ \text{a2} = 1.2000, & \text{b2} = 0.6000, & \text{c2} = 0.1500, & \text{d2} = 0, \\ \text{a3} = 2.8000, & \text{b3} = \text{-}0.6000, & \text{c3} = 0.4500, & \text{d3} = \text{-}0.0250 \end{array}$$

Thus, cubic spline polynomials are obtained as following:

$$\begin{aligned} f^{(1)}(x) &= 1 + 0.9x + 0.0250x^3 && (0 \le x \le 2) \\ f^{(2)}(x) &= 1.2 + 0.6x + 0.15x^2 && (2 \le x \le 4) \\ f^{(3)}(x) &= 2.8 - 0.6x + 0.45x^2 - 0.025x^3 && (4 \le x \le 6) \end{aligned}$$

Using cubic spline functions, function values $f(3)$ and $f(5)$ are obtained as below:

$$\begin{aligned} f^{(2)}(3) &= 1.2 + 0.6(3) + 0.15(3)^2 = 4.35 \\ f^{(3)}(5) &= 2.8 - 0.6(5) + 0.45(5)^2 - 0.025(5)^3 = 7.925 \end{aligned}$$

Solution with MATLAB Commands:

For the data set of $x = [0\ 2\ 4\ 6]$ and $y = [1\ 3\ 6\ 10]$, interpolation values at $x0 = 3$ and $x0 = 5$ can be calculated using *interp1(x,y,x0,'method')* command in MATLAB as following:

```
>> x=[0 2 4 6]; y=[1 3 6 10];
>> y3=interp1(x,y,3,'spline');
>> y5=interp1(x,y,5,'spline');
>> y3
```

```
y3 =
4.3750
>> y5
y5 =
7.8750
```

Note: Interpolation values can also be calculated using *MATLAB Program 6.5: Interpolation with Cubic Spline Functions* in Appendix.

6.3.4 Hermite Interpolation Polynomial

If both function and first derivative values are known at data points, Hermite interpolation polynomials can be created between adjacent points as in Fig. 6.8. Among Hermite polynomials, cubic Hermite polynomial is the most common interpolation method. Cubic Hermite polynomial is commonly used as an interpolation function for thin (Euler) Beam elements in Finite Element solution of structural mechanics problems. Beam element is described in detail in Finite Element Method in Chap. 6 of Vol. 2.

Assume that function and first derivative values at $n + 1$ points are known as in Fig. 6.8. In this case, there will be total $2(n + 1)$ values. Using these values, an interpolation polynomial of $2n+1$ order can be created. This polynomial and its first derivative can be written in general form as following:

$$\begin{aligned} f(x) &= c_1 + c_2x + c_3x^2 + \ldots + c_{2(n+1)}x^{2n+1} \\ \frac{df(x)}{dx} &= c_2 + 2c_3x + \ldots + (2n+1)c_{2(n+1)}x^{2n} \end{aligned} \tag{6.27}$$

Function and its first derivative values can be expressed at data points as below:

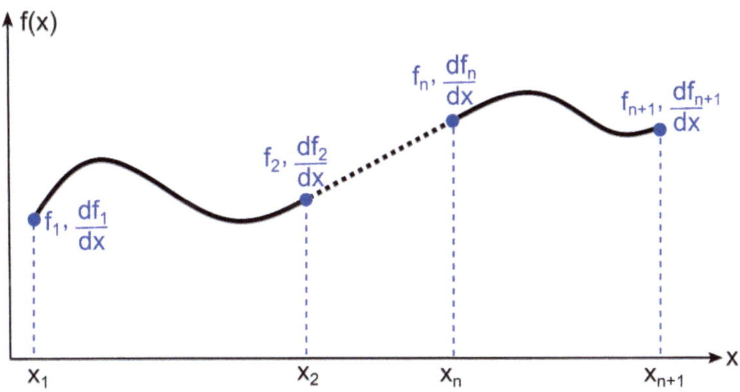

Fig. 6.8 Hermite interpolation polynomial

$$
\begin{aligned}
&f(x_1)=f_1=c_1+c_2x+c_3x^2+\ldots+c_{2(n+1)}x^{2n+1}\\
&\frac{df(x_1)}{dx}=\frac{df_1}{dx}=c_2+2c_3x+\ldots+(2n+1)c_{2(n+1)}x^{2n}\\
&f(x_2)=f_2=c_1+c_2x+c_3x^2+\ldots+c_{2(n+1)}x^{2n+1}\\
&\frac{df(x_2)}{dx}=\frac{df_2}{dx}=c_2+2c_3x+\ldots+(2n+1)c_{2(n+1)}x^{2n}\\
&\ldots\\
&f(x_{n+1})=f_{n+1}=c_1+c_2x+c_3x^2+\ldots+c_{2(n+1)}x^{2n+1}\\
&\frac{df(x_{n+1})}{dx}=\frac{df_{n+1}}{dx}=c_2+2c_3x+\ldots+(2n+1)c_{2(n+1)}x^{2n}
\end{aligned}
$$

Above equations can be arranged in matrix form as following:

$$
\begin{bmatrix}
1 & x_1 & x_1^2 & \ldots & x_1^{2n+1}\\
0 & 1 & 2x_2 & \ldots & (2n+1)x_2^{2n}\\
1 & x_2 & x_2^2 & \ldots & x_2^{2n+1}\\
0 & 1 & 2x_2 & \ldots & (2n+1)x_2^{2n}\\
\ldots & \ldots & \ldots & \ldots & \ldots\\
1 & x_{n+1} & x_{n+1}^2 & \ldots & x_{n+1}^{2n+1}\\
0 & 1 & 2x_{n+1} & \ldots & (2n+1)x_{n+1}^{2n}
\end{bmatrix}
\begin{Bmatrix} c_1\\ c_2\\ c_3\\ \ldots\\ c_{2(n+1)} \end{Bmatrix}
=
\begin{Bmatrix} f_1\\ \frac{df_1}{dx}\\ f_2\\ \frac{df_2}{dx}\\ \ldots\\ f_{n+1}\\ \frac{df_{n+1}}{dx} \end{Bmatrix}
\tag{6.28}
$$

Solution of Eq. 6.28 leads to the unknown coefficients of $c_1,\ldots, c_{2(n+1)}$.

***Note*:** For a large number of points, coefficient matrix in Eq. 6.28 may become ill-conditioned preventing the solution of linear system. As a remedy, an alternative method was developed to easily construct the Hermite interpolation polynomial similar to the Lagrange method. This method is explained below.

Alternative Method for Hermite Interpolation Polynomial:
In alternative method, Hermite interpolation polynomial and its derivative can be expressed using Lagrange functions as following:

$$
\begin{aligned}
f(x)&=U_1(x)f_1+U_2(x)f_2+\ldots+U_{n+1}(x)f_{n+1}+V_1(x)\frac{df_1}{dx}+V_2(x)\frac{df_2}{dx}+\ldots\\
&\quad+V_{n+1}(x)\frac{df_{n+1}}{dx}\\
&=\sum_{i=1}^{n+1}U_i(x)f_i+\sum_{i=1}^{n+1}V_i(x)\frac{df_i}{dx}
\end{aligned}
\tag{6.29}
$$

$$
\begin{aligned}
\frac{df(x)}{dx} = \frac{dU_1(x)}{dx} f_1 + \frac{dU_2(x)}{dx} f_2 + \ldots + \frac{dU_{n+1}(x)}{dx} f_{n+1} + \frac{dV_1(x)}{dx}\frac{df_1}{dx} \\
+ \frac{dV_2(x)}{dx}\frac{df_2}{dx} + \ldots \\
+ \frac{dV_{n+1}(x)}{dx}\frac{df_{n+1}}{dx} = \sum_{i=1}^{n+1} \frac{dU_i(x)}{dx} f_i + \sum_{i=1}^{n+1} \frac{dV_i(x)}{dx}\frac{df_i}{dx}
\end{aligned}
\tag{6.30}
$$

In Eq. 6.29, $U_i(x)$ and $V_i(x)$ serve as weight or shape functions for function values and its derivatives. Equations 6.29 and 6.30 can be applied to all points as following:

At $x = x_1$ point:

$$
\begin{aligned}
& f(x_1) = f_1 \longrightarrow \\
& U_1(x_1) = 1, \quad U_2(x_1) = 0, \ldots, \quad U_{n+1}(x_1) = 0, \quad V_1(x_1) = 0, \quad V_2(x_1) = 0, \ldots, \\
& V_{n+1}(x_1) = 0,
\end{aligned}
$$

$$
\begin{aligned}
& \frac{df(x_1)}{dx} = \frac{df_1}{dx} \longrightarrow \\
& \frac{dU_1(x_1)}{dx} = 0, \frac{dU_2(x_1)}{dx} = 0, \ldots, \frac{dU_{n+1}(x_1)}{dx} = 0, \frac{dV_1(x_1)}{dx} = 1, \ldots, \\
& \frac{dV_2(x_1)}{dx} = 0, \frac{dV_{n+1}(x_1)}{dx} = 0,
\end{aligned}
$$

At $x = x_2$ point:

$$
\begin{aligned}
& f(x_2) = f_2 \longrightarrow \\
& U_1(x_2) = 0, \quad U_2(x_2) = 1, \ldots, \quad U_{n+1}(x_2) = 0, \quad V_1(x_2) = 0, \quad V_2(x_2) = 0, \ldots, \\
& V_{n+1}(x_2) = 0,
\end{aligned}
$$

$$
\begin{aligned}
& \frac{df(x_2)}{dx} = \frac{df_2}{dx} \longrightarrow \\
& \frac{dU_1(x_2)}{dx} = 0, \frac{dU_2(x_2)}{dx} = 0, \ldots, \frac{dU_{n+1}(x_2)}{dx} = 0, \frac{dV_1(x_1)}{dx} = 0, \ldots, \\
& \frac{dV_2(x_2)}{dx} = 1, \frac{dV_{n+1}(x_1)}{dx} = 0
\end{aligned}
$$

$$\ldots$$

At $x = x_{n+1}$ point:

$$f(x_{n+1}) = f_{n+1} \longrightarrow$$
$$U_1(x_{n+1}) = 0, \quad U_2(x_{n+1}) = 0, \ldots, \quad U_{n+1}(x_{n+1}) = 1, \quad V_1(x_{n+1}) = 0,$$
$$V_2(x_{n+1}) = 0, \ldots, \quad V_{n+1}(x_{n+1}) = 0,$$

$$\frac{df(x_{n+1})}{dx} = \frac{df_{n+1}}{dx} \longrightarrow$$
$$\frac{dU_1(x_{n+1})}{dx} = 0, \frac{dU_2(x_{n+1})}{dx} = 0, \ldots, \frac{dU_{n+1}(x_{n+1})}{dx} = 0, \frac{dV_1(x_{n+1})}{dx} = 1, \ldots,$$
$$\frac{dV_2(x_{n+1})}{dx} = 0,$$

Above conditions can be written briefly in index notation as following:

$$U_i(x_j) = \left\{ \begin{array}{cc} 1 & (i=j) \\ 0 & (i \neq j) \end{array} \right\}, \quad V_i(x_j) = 0 \tag{6.31}$$

$$\frac{dU_i(x_j)}{dx} = 0, \quad \frac{dV_i(x_j)}{dx} = \left\{ \begin{array}{cc} 1 & (i=j) \\ 0 & (i \neq j) \end{array} \right\} \tag{6.32}$$

$U_i(x)$ and $V_i(x)$ functions can be written in terms of Lagrange functions of $L_i(x)$ as following:

$$U_i(x) = (a_i + b_i x)[L_i(x)]^2, \quad V_i(x) = (c_i + d_i x)[L_i(x)]^2 \tag{6.33}$$

In this case, substituting Eq. 6.33 into Eq. 6.29, Hermite polynomial function can be written as below:

$$f(x) = \sum_{i=1}^{n+1} U_i(x) f_i + \sum_{i=1}^{n+1} V_i(x) \frac{df_i}{dx} = \sum_{i=1}^{n+1} (a_i + b_i x)[L_i(x)]^2 f_i + \sum_{i=1}^{n+1}$$
$$\times (c_i + d_i x)[L_i(x)]^2 \frac{df_i}{dx} \tag{6.34}$$

Equation 6.34 can be written at x_i point as following (using $U_i(x_i) = 1$ and $V_i(x_i) = 0$ from Eqs. 6.31 and 6.32 as well as the property of Lagrange function as $L_i(x_i) = 1$),

$$\sum_{i=1}^{n+1} \underbrace{U_i(x_i)}_{1} f_i + \sum_{i=1}^{n+1} \underbrace{V_i(x_i)}_{0} \frac{df_i}{dx} = \sum_{i=1}^{n+1} (a_i + b_i x_i) \left[\underbrace{L_i(x_i)}_{1} \right]^2 f_i + \sum_{i=1}^{n+1} (c_i + d_i x_i) \left[\underbrace{L_i(x_i)}_{1} \right]^2 \frac{df_i}{dx} \tag{6.35}$$

Equivalency of the left and right sides of Eq. 6.35 leads to the following equations for a_i, b_i, c_i,d_i coefficients

$$\begin{aligned} 1 &= a_i + b_i x_i \\ 0 &= c_i + d_i x \end{aligned} \tag{6.36}$$

Substituting Eqs. 6.31 and 6.32 into Eq. 6.30, derivative of *f*(*x*) function at x_i point can be written as below:

$$\begin{aligned} \underbrace{\frac{df(x_i)}{dx}}_{\frac{df_i}{dx}} &= \sum_{i=1}^{n+1} \underbrace{\frac{dU_i(x_i)}{dx}}_{0} f_i + \sum_{i=1}^{n+1} \underbrace{\frac{dV_i(x_i)}{dx}}_{1} \frac{df_i}{dx} \\ &= \sum_{i=1}^{n+1} \left[b_i \Bigg(\underbrace{L_i(x_i)}_{=1} \Bigg)^2 + 2 \Bigg(\underbrace{a_i + b_i x_i}_{=1} \Bigg) \underbrace{L_i(x_i)}_{=1} \frac{dL_i(x_i)}{dx} \right] f_i \\ &+ \sum_{i=1}^{n+1} \left[d_i \Bigg(\underbrace{L_i(x_i)}_{=1} \Bigg)^2 + 2 \Bigg(\underbrace{c_i + d_i x_i}_{=0} \Bigg) \underbrace{L_i(x_i)}_{=1} \frac{dL_i(x_i)}{dx} \right] \frac{df_i}{dx} \end{aligned} \tag{6.37}$$

Equivalency of the left and right sides of Eq. 6.37 leads to the following equations for a_i, b_i, c_i,d_i coefficients

$$\begin{aligned} 0 &= b_i + 2\frac{dL_i(x_i)}{dx} \\ 1 &= d_i \end{aligned} \tag{6.38}$$

Using Eq. 6.36 and Eq. 6.38, coefficients a_i, b_i, c_i, d_i can be obtained as following:

$$\begin{aligned} a_i + b_i x_i &= 1 \\ c_i + d_i x_i &= 0 \\ b_i + 2\frac{dL_i(x_i)}{dx} &= 0 \\ d_i &= 1 \end{aligned} \quad \rightarrow \quad \begin{aligned} d_i &= 1 \\ b_i &= -2\frac{dL_i(x_i)}{dx} \\ a_i &= 1 + 2x_i \frac{dL_i(x_i)}{dx} \\ c_i &= -x_i \end{aligned} \tag{6.39}$$

Substituting a_i, b_i, c_i, d_i expressions of Eq. 6.39 into Eq. 6.34, Hermite polynomial functions can be obtained in terms of Lagrange functions as following:

$$f(x) = \sum_{i=1}^{n+1} \left[1 - 2(x - x_i)\frac{dL_i(x_i)}{dx} \right] [L_i(x)]^2 f_i + \sum_{i=1}^{n+1} (x - x_i)[L_i(x)]^2 \frac{df_i}{dx} \tag{6.40}$$

***Note*:** In the alternative form in Eq. 6.40, it is seen that Hermite interpolation polynomial can be directly obtained using Lagrange functions and their derivatives.

Example 6.6 Construct Hermite interpolation polynomial using the data in Table 6.4 and predict the function value at $x0 = 3$.

Cubic Hermite interpolation polynomial passing through x_1 and x_2 points can be expressed as below:

$$f(x)=\sum_{i=1}^{2}\left[1-2(x-x_i)\frac{dL_i(x_i)}{dx}\right][L_i(x)]^2 f_i+\sum_{i=1}^{n+1}(x-x_i)[L_i(x)]^2\frac{df_i}{dx}$$

$$=\left[1-2(x-x_1)\frac{dL_1(x_1)}{dx}\right][L_1(x)]^2 f_1+(x-x_1)[L_1(x)]^2\frac{df_1}{dx}+\left[1-2(x-x_2)\frac{dL_2(x_2)}{dx}\right][L_2(x)]^2 f_2$$

$$+(x-x_2)[L_2(x)]^2\frac{f_2}{dx}$$

Using the data in Table 6.4, Lagrange functions and their derivatives can be calculated as following:

$$L_1(x)=\frac{(x-x_2)}{(x_1-x_2)}=\frac{(x-4)}{(2-4)}=\frac{(4-x)}{2},\quad \frac{dL_1(x)}{dx}=\frac{(-1)}{2},\quad \frac{dL_1(x_1=2)}{dx}=\frac{-1}{2}$$

$$L_2(x)=\frac{(x-x_1)}{(x_2-x_1)}=\frac{(x-2)}{(4-2)}=\frac{(x-2)}{2},\quad \frac{dL_2(x)}{dx}=\frac{1}{2},\quad \frac{dL_2(x_2=4)}{dx}=\frac{1}{2}$$

In this case, using Lagrange functions and their derivatives, cubic Hermite interpolation polynomial can be obtained as following:

$$f(x)=\left[1-2(x-2)\frac{(-1)}{2}\right]\left[\frac{(4-x)}{2}\right]^2 4+(x-x_1)\left[\frac{(4-x)}{2}\right]^2(-8)$$

$$+\left[1-2(x-4)\frac{(1)}{2}\right]\left[\frac{(x-2)}{2}\right]^2 12+(x-x_2)\left[\frac{(x-2)}{2}\right]^2 4$$

Using the generated Hermite interpolation polynomial, function value at $x0 = 3$ is calculated as $f(3) = 5$.

Example 6.7 Construct Hermite interpolation polynomial using the data in Table 6.5 and predict the function value at $x0 = 6$.

Hermite interpolation polynomial passing through x_1, x_2, and x_3 points can be expressed as below:

Table 6.4 Data to construct Hermite interpolation polynomial

x	$f(x)$	$f'(x) = df(x)/dx$
2	4	−8
4	12	4

Table 6.5 Data to construct Hermite interpolation polynomial

x	$f(x)$	$f'(x) = df(x)/dx$
2	4	−8
4	12	4
8	24	−2

$$f(x)=\sum_{i=1}^{3}\left[1-2(x-x_i)\frac{dL_i(x_i)}{dx}\right][L_i(x)]^2 f_i+\sum_{i=1}^{n+1}(x-x_i)[L_i(x)]^2\frac{df_i}{dx}$$
$$=\left[1-2(x-x_1)\frac{dL_1(x_1)}{dx}\right][L_1(x)]^2 f_1+(x-x_1)[L_1(x)]^2\frac{df_1}{dx}+\left[1-2(x-x_2)\frac{dL_2(x_2)}{dx}\right][L_2(x)]^2 f_2$$
$$+(x-x_2)[L_2(x)]^2\frac{df_2}{dx}+\left[1-2(x-x_3)\frac{dL_3(x_3)}{dx}\right][L_3(x)]^2 f_3+(x-x_3)[L_3(x)]^2\frac{df_3}{dx}$$

Using the data in Table 6.5, Lagrange functions and their derivatives can be calculated as following:

$$L_1(x)=\frac{(x-x_2)(x-x_3)}{(x_1-x_2)(x_1-x_3)}=\frac{(x-4)(x-8)}{(2-4)(2-8)}=\frac{(x^2-12x+32)}{12},$$
$$\frac{dL_1(x)}{dx}=\frac{(2x-12)}{12}\rightarrow\frac{dL_1(x_1=2)}{dx}=\frac{-8}{12}$$
$$L_2(x)=\frac{(x-x_1)(x-x_3)}{(x_2-x_1)(x_2-x_3)}=\frac{(x-2)(x-8)}{(4-2)(4-8)}=-\frac{(x^2-10x+16)}{8},$$
$$\frac{dL_2(x)}{dx}=-\frac{(2x-10)}{8}\rightarrow\frac{dL_2(x_2=4)}{dx}=\frac{2}{8}$$
$$L_3(x)=\frac{(x-x_1)(x-x_2)}{(x_3-x_1)(x_3-x_2)}=\frac{(x-2)(x-4)}{(8-2)(8-4)}=\frac{(x^2-6x+8)}{24},$$
$$\frac{dL_3(x)}{dx}=\frac{(2x-6)}{24}\rightarrow\frac{dL_3(x_3=8)}{dx}=\frac{10}{24}$$

In this case, using Lagrange functions and their derivatives, Hermite interpolation polynomial can be obtained as following:

$$f(x)=\left[1-2(x-2)\frac{(-8)}{12}\right]\left[\frac{(x^2-12x+32)}{12}\right]^2 4+(x-2)\left[\frac{(x^2-12x+32)}{12}\right]^2(-8)$$
$$+\left[1-2(x-4)\frac{(2)}{8}\right]\left[-\frac{(x^2-10x+16)}{8}\right]^2 12+(x-4)\left[-\frac{(x^2-10x+16)}{8}\right]^2 4$$
$$+\left[1-2(x-8)\frac{(10)}{24}\right]\left[\frac{(x^2-6x+8)}{24}\right]^2 24+(x-8)\left[\frac{(x^2-6x+8)}{24}\right]^2(-2)$$

Using the generated Hermite interpolation polynomial, function value at $x0 = 6$ is calculated as $f(6) = 14.8148$.

6.4 Multi-Variable Global Interpolation Polynomials

Available data for an interpolation can sometimes be multivariate or multidimensional. If data points are distributed regularly in a domain such as in a rectangular area or cubic volume as shown in Fig. 6.9, then multidimensional interpolation functions can be easily created using products of one-dimensional interpolation functions. Global interpolation functions with two and three variables are described below:

6.4.1 Bivariate Classical Interpolation Polynomial

Two-dimensional (bivariate) classical interpolation polynomial function $f(x,y)$ can be formed by multiplying one dimensional interpolation polynomials in x and y direction as following:

$$f(x,y) = f(x)f(y) = \underbrace{\left(\sum_{i=0}^{m} c_i x^i\right)}_{f(x)} \underbrace{\left(\sum_{j=0}^{n} c_j y^j\right)}_{f(y)} = \sum_{i=0}^{m}\sum_{j=0}^{n} c_{ij} x^i y^j$$
$$= c_{00} + c_{10}x + c_{01}y + c_{11}xy + c_{20}x^2 + c_{02}y^2 + c_{21}x^2y + c_{12}xy^2 + c_{22}x^2y^2 + \ldots + c_{mn}x^m y^n \tag{6.41}$$

Example 6.8 Values of a bivariate (two-variable) function $f(x, y)$ are given at different points in a rectangular domain as in Fig. 6.10. Calculate $f(5, 3)$ value creating bivariate classical interpolation polynomial function.

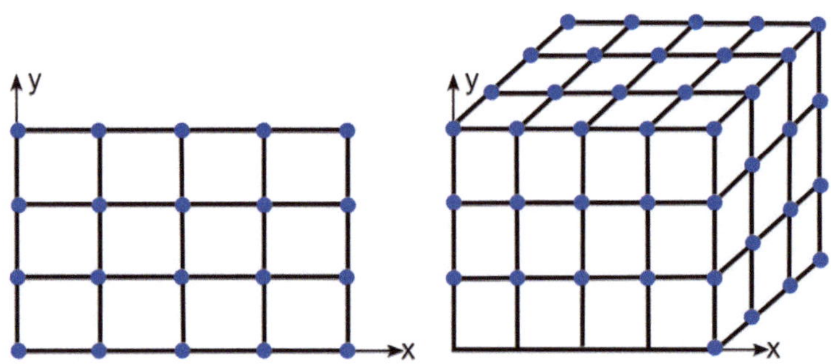

Fig. 6.9 Regularly distributed data points in rectangular (two-dimensional) and cubic (three-dimensional) domains

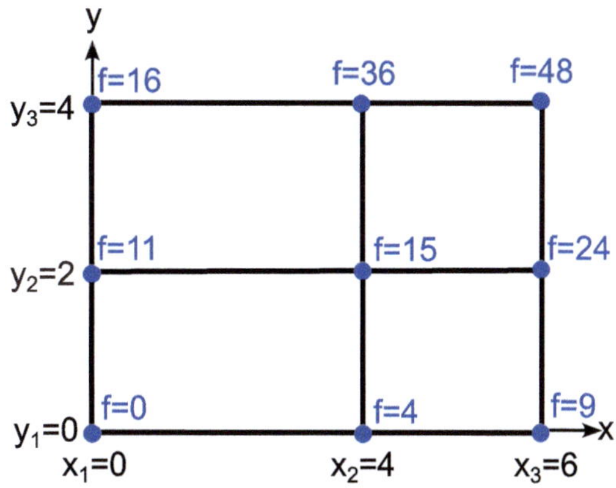

Fig. 6.10 Data points for two-dimensional interpolation

Bivariate classical parabolic function for given data points can be expressed as below:

$$f(x,y) = \sum_{i=0}^{2}\sum_{j=0}^{2} c_{ij}x^i y^j = c_{00} + c_{10}x + c_{01}y + c_{11}xy + c_{20}x^2 + c_{02}y^2$$
$$+ c_{21}x^2y + c_{12}xy^2 + c_{22}x^2y^2 \tag{6.42}$$

If the given values are substituted into Eq. 6.42, following equations are obtained:

$$f(x_1,y_1) = c_{00} + c_{10}x_1 + c_{01}y_1 + c_{11}x_1y_1 + c_{20}x_1^2 + c_{02}y_1^2 + c_{21}x_1^2y_1 + c_{12}x_1y_1^2 + c_{22}x_1^2y_1^2$$
$$f(x_1,y_2) = c_{00} + c_{10}x_1 + c_{01}y_2 + c_{11}x_1y_2 + c_{20}x_1^2 + c_{02}y_2^2 + c_{21}x_1^2y_2 + c_{12}x_1y_2^2 + c_{22}x_1^2y_2^2$$
$$f(x_1,y_3) = c_{00} + c_{10}x_1 + c_{01}y_3 + c_{11}x_1y_3 + c_{20}x_1^2 + c_{02}y_3^2 + c_{21}x_1^2y_3 + c_{12}x_1y_3^2 + c_{22}x_1^2y_3^2$$
$$f(x_2,y_1) = c_{00} + c_{10}x_2 + c_{01}y_1 + c_{11}x_2y_1 + c_{20}x_2^2 + c_{02}y_1^2 + c_{21}x_2^2y_1 + c_{12}x_2y_1^2 + c_{22}x_2^2y_1^2$$
$$f(x_2,y_2) = c_{00} + c_{10}x_2 + c_{01}y_2 + c_{11}x_2y_2 + c_{20}x_2^2 + c_{02}y_2^2 + c_{21}x_2^2y_2 + c_{12}x_2y_2^2 + c_{22}x_2^2y_2^2$$
$$f(x_2,y_3) = c_{00} + c_{10}x_2 + c_{01}y_3 + c_{11}x_2y_3 + c_{20}x_2^2 + c_{02}y_3^2 + c_{21}x_2^2y_3 + c_{12}x_2y_3^2 + c_{22}x_2^2y_3^2$$
$$f(x_3,y_1) = c_{00} + c_{10}x_3 + c_{01}y_1 + c_{11}x_3y_1 + c_{20}x_3^2 + c_{02}y_1^2 + c_{21}x_3^2y_1 + c_{12}x_3y_1^2 + c_{22}x_3^2y_1^2$$
$$f(x_3,y_2) = c_{00} + c_{10}x_3 + c_{01}y_2 + c_{11}x_3y_2 + c_{20}x_3^2 + c_{02}y_2^2 + c_{21}x_3^2y_2 + c_{12}x_3y_2^2 + c_{22}x_3^2y_2^2$$
$$f(x_3,y_3) = c_{00} + c_{10}x_3 + c_{01}y_3 + c_{11}x_3y_3 + c_{20}x_3^2 + c_{02}y_3^2 + c_{21}x_3^2y_3 + c_{12}x_3y_3^2 + c_{22}x_3^2y_3^2$$

Above equations can be rearranged in matrix form as following:

$$\begin{bmatrix} 1 & x_1 & y_1 & x_1y_1 & x_1^2 & y_1^2 & x_1^2y_1 & x_1y_1^2 & x_1^2y_1^2 \\ 1 & x_1 & y_2 & x_1y_2 & x_1^2 & y_2^2 & x_1^2y_2 & x_1y_2^2 & x_1^2y_2^2 \\ 1 & x_1 & y_3 & x_1y_3 & x_1^2 & y_3^2 & x_1^2y_3 & x_1y_3^2 & x_1^2y_3^2 \\ 1 & x_2 & y_1 & x_2y_1 & x_2^2 & y_1^2 & x_2^2y_1 & x_2y_1^2 & x_2^2y_1^2 \\ 1 & x_2 & y_2 & x_2y_2 & x_2^2 & y_2^2 & x_2^2y_2 & x_2y_2^2 & x_2^2y_2^2 \\ 1 & x_2 & y_3 & x_2y_3 & x_2^2 & y_3^2 & x_2^2y_3 & x_2y_3^2 & x_2^2y_3^2 \\ 1 & x_3 & y_1 & x_3y_1 & x_3^2 & y_1^2 & x_3^2y_1 & x_3y_1^2 & x_3^2y_1^2 \\ 1 & x_3 & y_2 & x_3y_2 & x_3^2 & y_2^2 & x_3^2y_2 & x_3y_2^2 & x_3^2y_2^2 \\ 1 & x_3 & y_3 & x_3y_3 & x_3^2 & y_3^2 & x_3^2y_3 & x_3y_3^2 & x_3^2y_3^2 \end{bmatrix} \begin{Bmatrix} c_{00} \\ c_{10} \\ c_{01} \\ c_{11} \\ c_{20} \\ c_{02} \\ c_{21} \\ c_{12} \\ c_{22} \end{Bmatrix} = \begin{Bmatrix} f(x_1,y_1) \\ f(x_1,y_2) \\ f(x_1,y_3) \\ f(x_2,y_1) \\ f(x_2,y_2) \\ f(x_2,y_3) \\ f(x_3,y_1) \\ f(x_3,y_2) \\ f(x_3,y_3) \end{Bmatrix} \tag{6.43}$$

After substitution of numerical values into Eq. 6.43, solution is obtained as following:

$c_{00} = 0$, $c_{10} = 0$, $c_{01} = 7$, $c_{11} = -2.4167$, $c_{20} = 0.25$, $c_{02} = -0.75$, $c_{21} = 0.3542$, $c_{12} = 0.875$, $c_{22} = -0.0938$. In this case, $f(5, 3)$ value is obtained as below:

$$f(5,3) = c_{00} + c_{10}.5 + c_{01}.3 + c_{11}.5.3 + c_{20}.5^2 + c_{02}.3^2 + c_{21}.5^2.3 + c_{12}.5.3^2 \\ + c_{22}.5^2.3^2 = 29.0937$$

6.4.2 Bivariate Lagrange Interpolation Polynomial

Bivariate Lagrange interpolation polynomial function $f(x,y)$ can be created using products of one-dimensional Lagrange interpolation polynomials in x and y directions as below:

$$\begin{aligned} f(x,y) &= \sum_{i=1}^{m}\sum_{j=1}^{n} L_{ix}(x)L_{jy}(y)f(x_i,y_j) \\ &= L_{1x}(x)\left[L_{1y}(y)+L_{2y}(y)+L_{3y}(y)+\ldots+L_{ny}(y)\right]+L_{2x}(x)\left[L_{1y}(y)+L_{2y}(y)+L_{3y}(y)+\ldots+L_{ny}(y)\right] \\ &+L_{3x}(x)\left[L_{1y}(y)+L_{2y}(y)+L_{3y}(y)+\ldots+L_{ny}(y)\right]+\ldots+L_{mx}(x)\left[L_{1y}(y)+L_{2y}(y)+L_{3y}(y)+\ldots+L_{ny}(y)\right] \end{aligned} \tag{6.44}$$

Above expression can be rearranged as following:

$$\begin{aligned} f(x,y) = {} & L_{1x}(x)L_{1y}(y)f(x_1,y_1) + L_{1x}(x)L_{2y}(y)f(x_1,y_2) + L_{1x}(x)L_{3y}(y)f(x_1,y_3) + \ldots + L_{1x}(x)L_{ny}(y)f(x_1,y_n) \\ & + L_{2x}(x)L_{1y}(y)f(x_2,y_1) + L_{2x}(x)L_{2y}(y)f(x_2,y_2) + L_{2x}(x)L_{3y}(y)f(x_2,y_3) + \ldots + L_{2x}(x)L_{ny}(y)f(x_2,y_n) \\ & + L_{3x}(x)L_{1y}(y)f(x_3,y_1) + L_{3x}(x)L_{2y}(y)f(x_3,y_2) + L_{3x}(x)L_{3y}(y)f(x_3,y_3) + \ldots + L_{3x}(x)L_{ny}(y)f(x_3,y_n) + \ldots \\ & + L_{mx}(x)L_{1y}(y)f(x_m,y_1) + L_{mx}(x)L_{2y}(y)f(x_m,y_2) + L_{mx}(x)L_{3y}(y)f(x_m,y_3) + \ldots + L_{mx}(x)L_{ny}(y)f(x_m,y_n) \end{aligned} \tag{6.45}$$

Example 6.9 Using the data points in Example 6.8, calculate $f(5, 3)$ value creating bivariate Lagrange interpolation polynomial function.

Bivariate parabolic Lagrange polynomial function for given data points can be expressed as below:

$$\begin{aligned} f(x,y) &= \sum_{i=1}^{3}\sum_{j=1}^{3} L_{ix}(x)L_{jy}(y)f(x_i,y_j) \\ &= L_{1x}(x)\left[L_{1y}(y)f(x_1,y_1) + L_{2y}(y)f(x_1,y_2) + L_{3y}(y)f(x_1,y_3)\right] \\ &\quad + L_{2x}(x)\left[L_{1y}(y)f(x_1,y_1) + L_{2y}(y)f(x_1,y_2) + L_{3y}(y)f(x_1,y_3)\right] \\ &\quad + L_3(x)\left[L_{1y}(y)f(x_1,y_1) + L_{2y}(y)f(x_1,y_2) + L_{3y}(y)f(x_1,y_3)\right] \end{aligned}$$

If the given values are substituted into the above function, Lagrange interpolation polynomial is obtained as following:

$$\begin{aligned} f(x,y) = {} & L_{1x}(x)L_{1y}(y)f(x_1,y_1) + L_{1x}(x)L_{2y}(y)f(x_1,y_2) + L_{1x}(x)L_{3y}(y)f(x_1,y_3) \\ & + L_{2x}(x)L_{1y}(y)f(x_2,y_1) + L_{2x}(x)L_{2y}(y)f(x_2,y_2) + L_{2x}(x)L_{3y}(y)f(x_2,y_3) \\ & + L_{3x}(x)L_{1y}(y)f(x_3,y_1) + L_{3x}(x)L_{2y}(y)f(x_3,y_2) + L_{3x}(x)L_{3y}(y)f(x_3,y_3) \end{aligned}$$

where $L_{ix}(x)$ and $L_{jy}(y)$ show univariate Lagrange functions in x and y directions and they are defined explicitly as below:

$$L_{1x}(x) = \frac{(x-x_2)(x-x_3)}{(x_1-x_2)(x_1-x_3)},\ L_{2x}(x) = \frac{(x-x_1)(x-x_3)}{(x_2-x_1)(x_2-x_3)},\ L_{3x}(x) = \frac{(x-x_1)(x-x_2)}{(x_3-x_1)(x_3-x_2)}$$

$$L_{1y}(y) = \frac{(y-y_2)(y-y_3)}{(y_1-y_2)(y_1-y_3)},\ L_{2y}(y) = \frac{(y-y_1)(y-y_3)}{(y_2-y_1)(y_2-y_3)},\ L_{3y}(y) = \frac{(y-y_1)(y-y_2)}{(y_3-y_1)(y_3-y_2)}$$

Thus, substituting numerical values into Lagrange functions and interpolation polynomial, $f(5, 3)$ value is calculated as following:

$$L_{1x}(5) = \frac{(5-4)(5-6)}{(0-4)(0-6)} = -0.0417,\quad L_{2x}(5) = \frac{(5-0)(5-6)}{(4-0)(4-6)} = 0.625,$$

$$L_{3x}(5) = \frac{(5-0)(5-4)}{(6-0)(6-4)} = 0.4167$$

$$L_{1y}(3) = \frac{(3-2)(3-4)}{(0-2)(0-4)} = -0.125,\quad L_{2y}(3) = \frac{(3-0)(3-4)}{(2-0)(2-4)} = 0.75,$$

$$L_{3y}(3) = \frac{(3-0)(3-2)}{(4-0)(4-2)} = 0.375$$

$$\begin{aligned} f(5,3) &= L_{1x}(5)L_{1y}(3)f(0,0) + L_{1x}(5)L_{2y}(3)f(0,2) + L_{1x}(5)L_{3y}(3)f(0,4) \\ &+ L_{2x}(5)L_{1y}(3)f(4,0) + L_{2x}(5)L_{2y}(3)f(4,2) + L_{2x}(5)L_{3y}(3)f(4,4) \\ &+ L_{3x}(5)L_{1y}(3)f(6,0) + L_{3x}(5)L_{2y}(3)f(6,2) + L_{3x}(5)L_{3y}(3)f(6,4) \\ &= 29.0938 \end{aligned}$$

6.4.3 Bivariate Newton Divided Difference Interpolation Polynomial

Interpolation polynomial with two variables can be written in Newton form as below (m: number of points in x direction, n: number of points in y direction)

$$\begin{aligned} f(x,y) = {} & c_{0,0} + c_{1,0}(x-x_1) + c_{0,1}(y-y_1) + c_{2,0}(x-x_1)(x-x_2) + c_{1,1}(x-x_1)(y-y_1) \\ & + c_{0,2}(y-y_1)(y-y_2) + \ldots + c_{n,0}(x-x_1)(x-x_2)\ldots(x-x_n) + c_{n-1,1}(x-x_1)(x-x_2)\ldots(x-x_{n-1})(y-y_1) \\ & + c_{n-2,2}(x-x_1)(x-x_2)\ldots(x-x_{n-2})(y-y_1)(y-y_2) + \ldots + c_{0,m,}(y-y_1)(y-y_2)\ldots(y-y_m) \end{aligned} \tag{6.46}$$

$c_{i,\,j}$ coefficients in Eq. 6.46 can be expressed using Newton Divided Difference terms as following:

$$c_{i-1,j-1} = f\left[x_1, x_2, \ldots, x_{i-1}; y_1, y_2, \ldots, y_{j-1}\right] \quad (i = 1,2,\ldots m, \quad j = 1,2,\ldots n) \tag{6.47}$$

Newton divided difference expressions in Eq. 6.47 are calculated as below:

$$\begin{aligned} & f\left[x_i; y_j\right] = f\left(x_i, y_j\right) \\ & f\left[x_1, x_2; y_j\right] = \frac{f\left[x_2; y_j\right] - f\left[x_1; y_j\right]}{x_2 - x_1} \\ & f\left[x_i; y_j, y_{j+1}\right] = \frac{f\left[x_i; y_{j+1}\right] - f\left[x_i; y_j\right]}{y_{j+1} - y_j} \\ & \ldots \\ & f\left[x_1, x_2, \ldots, x_i; y_1, y_2, \ldots, y_j\right] = \frac{f\left[x_1, x_2, \ldots, x_i; y_2, \ldots, y_j\right] - f\left[x_1, x_2, \ldots, x_i; y_1, y_2, \ldots, y_{j-1}\right]}{y_j - y_1} \\ & = \frac{f\left[x_2, x_2, \ldots, x_i; y_1, y_2, \ldots, y_j\right] - f\left[x_1, x_2, \ldots, x_{i-1}; y_1, y_2, \ldots, y_j\right]}{x_i - x_1} \end{aligned} \tag{6.48}$$

Example 6.10 Using the data points in Example 6.8, calculate $f(5, 3)$ value creating bivariate interpolation polynomial with Newton Divided Difference method.

Bivariate parabolic interpolation polynomial with Newton Divided Difference method for given data points can be expressed as below:

$$\begin{gathered}f(x,y)=c_{0,0}+c_{1,0}(x-x_1)+c_{0,1}(y-y_1)+c_{2,0}(x-x_1)(x-x_2)+c_{1,1}(x-x_1)(y-y_1)\\+c_{0,2}(y-y_1)(y-y_2)+c_{2,1}(x-x_1)(x-x_2)(y-y_1)+c_{1,2}(x-x_1)(y-y_1)(y-y_2)\\+c_{2,2}(x-x_1)(x-x_2)(y-y_1)(y-y_2)\end{gathered}$$

Divided Difference values to be used in $c_{i,\ j}$ coefficients can be calculated as following:

$$\begin{aligned}&f[x_1;y_1]=f(x_1,y_1)=0, \quad f[x_1;y_2]=f(x_1,y_2)=11, \quad f[x_1;y_3]=f(x_1,y_3)=16,\\&f[x_2;y_1]=f(x_2,y_1)=4, \quad f[x_2;y_2]=f(x_2,y_2)=15, \quad f[x_2;y_3]=f(x_2,y_3)=36,\\&f[x_3;y_1]=f(x_3,y_1)=9, \quad f[x_3;y_2]=f(x_3,y_2)=24, \quad f[x_3;y_3]=f(x_3,y_3)=48,\end{aligned}$$

$$f[x_1,x_2;y_1]=\frac{f[x_2;y_1]-f[x_1;y_1]}{x_2-x_1}=1, \quad f[x_1,x_2;y_2]=\frac{f[x_2;y_2]-f[x_1;y_2]}{x_2-x_1}=1,$$

$$f[x_1,x_2;y_3]=\frac{f[x_2;y_3]-f[x_1;y_3]}{x_2-x_1}=5,$$

$$f[x_2,x_3;y_1]=\frac{f[x_3;y_1]-f[x_2;y_1]}{x_3-x_2}=2.5, \quad f[x_2,x_3;y_2]=\frac{f[x_3;y_2]-f[x_2;y_2]}{x_3-x_2}=4.5,$$

$$f[x_2,x_3;y_3]=\frac{f[x_3;y_3]-f[x_2;y_3]}{x_3-x_2}=6,$$

$$f[x_1;y_1,y_2]=\frac{f[x_1;y_2]-f[x_1;y_1]}{y_2-y_1}=5.5, \quad f[x_2;y_1,y_2]=\frac{f[x_2;y_2]-f[x_2;y_1]}{y_2-y_1}=5.5,$$

$$f[x_3;y_1,y_2]=\frac{f[x_3;y_2]-f[x_3;y_1]}{y_2-y_1}=7.5,$$

$$f[x_1;y_2,y_3]=\frac{f[x_1;y_3]-f[x_1;y_2]}{y_3-y_2}=2.5, \quad f[x_2;y_2,y_3]=\frac{f[x_2;y_3]-f[x_2;y_2]}{y_3-y_2}=10.5,$$

$$f[x_3;y_2,y_3]=\frac{f[x_3;y_3]-f[x_3;y_2]}{y_3-y_2}=12,$$

$$f[x_1,x_2;y_2,y_3]=\frac{f[x_1,x_2;y_3]-f[x_1,x_2;y_2]}{y_3-y_2}=2,$$

$$f[x_2,x_3;y_2,y_3]=\frac{f[x_2,x_3;y_3]-f[x_2,x_3;y_2]}{y_3-y_2}=0.75,$$

$$f[x_2,x_3;y_1,y_2]=\frac{f[x_2,x_3;y_2]-f[x_2,x_3;y_1]}{y_2-y_1}=1,$$

$$f[x_1,x_2,x_3;y_1]=\frac{f[x_2,x_3;y_1]-f[x_1,x_2;y_1]}{x_3-x_1}=0.25,$$

$$f[x_1;y_1,y_2,y_3] = \frac{f[x_1;y_2,y_3] - f[x_1;y_1,y_2]}{y_3 - y_1} = -0.75,$$

$$f[x_1,x_2;y_1,y_2] = \frac{f[x_1,x_2;y_2] - f[x_1,x_2;y_1]}{y_2 - y_1} = \frac{f[x_2;y_1,y_2] - f[x_1;y_1,y_2]}{x_2 - x_1} = 0,$$

$$f[x_1,x_2,x_3;y_1,y_2] = \frac{f[x_2,x_3;y_1,y_2] - f[x_1,x_2;y_1,y_2]}{x_3 - x_1} = 0.1667,$$

$$f[x_1,x_2;y_1,y_2,y_3] = \frac{f[x_1,x_2;y_2,y_3] - f[x_1,x_2;y_1,y_2]}{y_3 - y_1} = 0.5,$$

$$f[x_2,x_3;y_1,y_2,y_3] = \frac{f[x_2,x_3;y_2,y_3] - f[x_2,x_3;y_1,y_2]}{y_3 - y_1} = -0.0625,$$

$$\begin{aligned} f[x_1,x_2,x_3;y_1,y_2,y_3] &= \frac{f[x_1,x_2,x_3;y_2,y_3] - f[x_1,x_2,x_3;y_1,y_2]}{y_3 - y_1} \\ &= \frac{f[x_2,x_3;y_1,y_2,y_3] - f[x_1,x_2;y_1,y_2,y_3]}{x_3 - x_1} \\ &= -0.0938 \end{aligned}$$

$c_{i,\,j}$ coefficients can be calculated using Divided Difference values as below:

$$c_{0,0} = f[x_1;y_1] = 0,$$

$$c_{1,0} = f[x_1,x_2;y_1] = 1, \quad c_{0,1} = f[x_1;y_1,y_2] = 5.5, \quad c_{1,1} = f[x_1,x_2;y_1,y_2] = 0,$$

$$c_{2,0} = f[x_1,x_2,x_3;y_1] = 0.25, \quad c_{0,2} = f[x_1;y_1,y_2,y_3] = -0.75, \quad c_{1,2} = f[x_1,x_2;y_1,y_2,y_3] = 0.5,$$

$$c_{2,1} = f[x_1,x_2,x_3;y_1,y_2] = 0.1667, \quad c_{2,2} = f[x_1,x_2,x_3;y_1,y_2,y_3] = -0.0938$$

Thus, substituting $c_{i,\,j}$ values into interpolation polynomial, $f(5, 3)$ value is calculated as following:

$$\begin{aligned} f(5,3) = {} & c_{0,0} + c_{1,0}(5-0) + c_{0,1}(3-0) + c_{2,0}(5-0)(5-4) + c_{1,1}(5-0)(3-0) + c_{0,2}(3-0)(3-2) \\ & + c_{2,1}(5-0)(5-4)(3-0) + c_{1,2}(5-0)(3-0)(3-2) + c_{2,2}(5-0)(5-4)(3-0)(3-2) \\ = {} & 29.0938 \end{aligned}$$

6.5 Multivariable Spline Interpolation Polynomials

In the previous section, it was stated that multidimensional interpolation functions can be created with products of one-dimensional interpolation functions if data points are spread regularly within a domain. However, if the domain does not

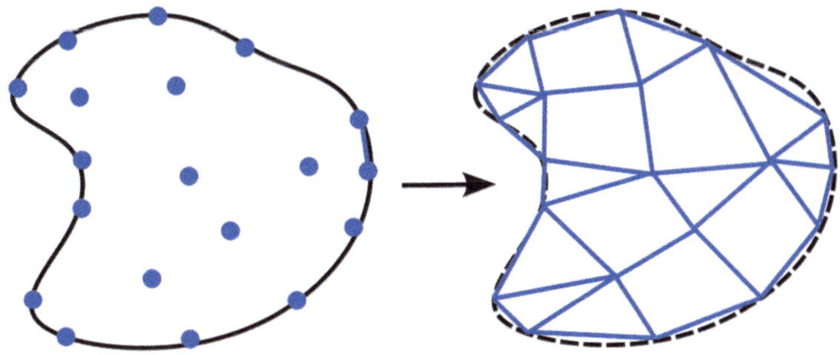

Fig. 6.11 Division of a general domain into simpler sub-domains with triangular and quadrilateral shapes

have a regular shape such as rectangular area or cubic volume, distribution of data points will not be regular either. In such cases, global multivariate (multivariable) interpolation function cannot be generated from the product of one-dimensional interpolation functions. As a remedy, domain can be divided into sub-domains with simpler shapes and spline interpolation functions can be generated over them as shown in Fig. 6.11. Usually, lower order interpolation polynomials are utilized over sub-domains. Interpolation polynomials over sub-domains are called as multivariate piecewise (spline) interpolation polynomials. Using multivariate piecewise polynomials, generation of global interpolation function over a general complex domain is facilitated. Moreover, accuracy of the global interpolation functions is protected from the oscillation effect.

Multivariate piecewise polynomials are widely used in creating two- and three-dimensional interpolation functions in Finite Element method (popular method in engineering) for the prediction of displacement, strain, stress, pressure, velocity, etc. in elements (sub-domains) in mechanical, fluid, and heat transfer problems. Multivariate piecewise polynomials can be obtained using Lagrange and Hermite interpolation methods as below:

6.6 Two-Dimensional Spline Interpolation Polynomials

When interpolation domain is two-dimensional and does not correspond to a regular shape, it can be divided into simpler triangular or quadrilateral sub-areas. These sub-areas are called as finite elements in Finite Element method. Over sub-areas, piecewise interpolation functions can be generated using Lagrange and Hermite methods as below.

6.6.1 Bivariate Lagrange Interpolation Functions

Lagrange Interpolation Function for Triangular Area

Two-dimensional piecewise interpolation function over a triangular area as shown in Fig. 6.12 can be expressed as following:

$$f(x,y) = c_1 + c_2x + c_3y = \begin{bmatrix} 1 & x & y \end{bmatrix} \begin{Bmatrix} c_1 \\ c_2 \\ c_3 \end{Bmatrix} \tag{6.49}$$

Function values at vertices of the triangle can be expressed using Eq. 6.49 as following:

$$\begin{aligned} f_1 &= c_1 + c_2x_1 + c_3y_1 \\ f_2 &= c_1 + c_2x_2 + c_3y_2 \\ f_3 &= c_1 + c_2x_3 + c_3y_3 \end{aligned}$$

If the above equations are written in matrix form and solved, unknown constants of c_1, c_2, c_3 can be obtained as below:

$$\begin{bmatrix} 1 & x_1 & y_1 \\ 1 & x_2 & y_2 \\ 1 & x_3 & y_3 \end{bmatrix} \begin{Bmatrix} c_1 \\ c_2 \\ c_3 \end{Bmatrix} = \begin{Bmatrix} f_1 \\ f_2 \\ f_3 \end{Bmatrix} \rightarrow \begin{Bmatrix} c_1 \\ c_2 \\ c_3 \end{Bmatrix} = \begin{bmatrix} 1 & x_1 & y_1 \\ 1 & x_2 & y_2 \\ 1 & x_3 & y_3 \end{bmatrix}^{-1} \begin{Bmatrix} f_1 \\ f_2 \\ f_3 \end{Bmatrix} \tag{6.50}$$

After substituting Eq. 6.50 into Eq. 6.49, interpolation function can be obtained in Lagrange form as following:

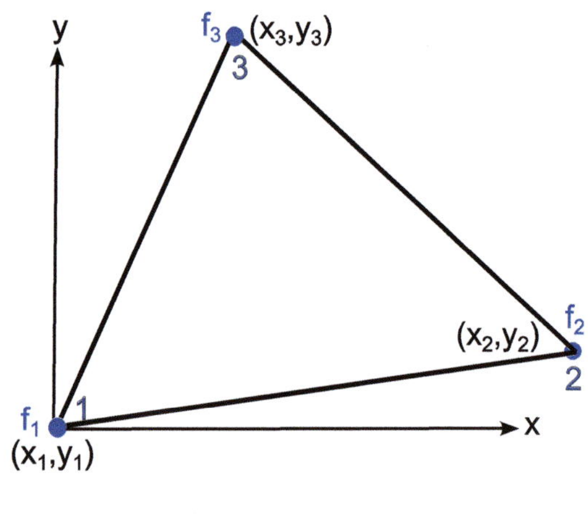

Fig. 6.12 Two-dimensional interpolation function over a triangular area

$$f(x,y)=\underbrace{\begin{bmatrix}1 & x & y\end{bmatrix}\begin{bmatrix}1 & x_1 & y_1\\ 1 & x_2 & y_2\\ 1 & x_3 & y_3\end{bmatrix}^{-1}}_{\begin{bmatrix}N_1 & N_2 & N_3\end{bmatrix}}\begin{Bmatrix}f_1\\ f_2\\ f_3\end{Bmatrix}=N_1(x,y)f_1+N_2(x,y)f_2+N_3(x,y)f_3 \quad (6.51)$$

$N_1(x,y)$, $N_2(x,y)$, $N_3(x,y)$ functions in Eq. 6.51 are often called as Lagrange or Shape functions and they are written explicitly as following:

$$\begin{aligned}N_1(x,y)&=\frac{1}{2A}[(x_2y_3-x_3y_2)+(y_2-y_3)x+(x_3-x_2)y]\\ N_2(x,y)&=\frac{1}{2A}[(x_3y_1-x_1y_3)+(y_3-y_1)x+(x_1-x_3)y]\\ N_3(x,y)&=\frac{1}{2A}[(x_1y_2-x_2y_1)+(y_1-y_2)x+(x_2-x_1)y]\end{aligned} \quad (6.52)$$

where A denotes the area of the triangle which can be calculated using the following determinant:

$$A=\frac{1}{2}\begin{vmatrix}1 & x_1 & y_1\\ 1 & x_2 & y_2\\ 1 & x_3 & y_3\end{vmatrix} \quad (6.53)$$

Reminding: One-dimensional Lagrange interpolation polynomial is expressed as $f(x) = \sum N_i(x)f_i$ using the product of Lagrange functions and function values. Two-dimensional Lagrange interpolation function is also expressed similarly as $f(x,y) = \sum N_i(x,y)f_i$.

Special Case: Right-Angled Triangle

In solving two-dimensional engineering problems (for example, with Finite Element method), interpolation on right-angled triangle (shortly right triangle) is frequently used. For the right triangle, interpolation function can be expressed in Lagrange form using special coordinates referred to as natural coordinates attached on the right triangle. Right triangle defined with natural coordinates of ξ and η is shown Fig. 6.13. Coordinates of the vertices are given in terms of natural coordinates in Table 6.6.

Interpolation function for the right triangle can be written in classical form as below:

$$f(\xi,\eta)=c_1+c_2\xi+c_3\eta=\begin{bmatrix}1 & \xi & \eta\end{bmatrix}\begin{Bmatrix}c_1\\ c_2\\ c_3\end{Bmatrix} \quad (6.54)$$

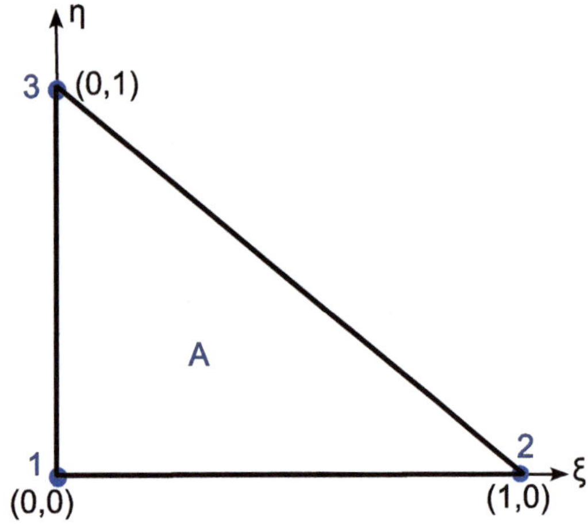

Fig. 6.13 Right triangle defined with natural coordinates of ξ and η

Table 6.6 Coordinates of the vertices of right triangle in terms of natural coordinates ξ and η

Node	ξ	η	$f(\xi,\eta)$
1	0	0	f_1
2	1	0	f_2
3	0	1	f_3

Applying Eq. 6.54 at vertices (nodes), unknown constants of c_1, c_2, c_3 can be obtained as below:

$$\begin{bmatrix} 1 & \xi_1 & \eta_1 \\ 1 & \xi_2 & \eta_2 \\ 1 & \xi_3 & \eta_3 \end{bmatrix} \begin{Bmatrix} c_1 \\ c_2 \\ c_3 \end{Bmatrix} = \begin{Bmatrix} f_1 \\ f_2 \\ f_3 \end{Bmatrix} \rightarrow \begin{Bmatrix} c_1 \\ c_2 \\ c_3 \end{Bmatrix} = \begin{bmatrix} 1 & \xi_1 & \eta_1 \\ 1 & \xi_2 & \eta_2 \\ 1 & \xi_3 & \eta_3 \end{bmatrix}^{-1} \begin{Bmatrix} f_1 \\ f_2 \\ f_3 \end{Bmatrix} \rightarrow$$
$$\begin{Bmatrix} c_1 \\ c_2 \\ c_3 \end{Bmatrix} = \begin{bmatrix} 1 & 0 & 0 \\ -1 & 1 & 0 \\ -1 & 0 & 1 \end{bmatrix} \begin{Bmatrix} f_1 \\ f_2 \\ f_3 \end{Bmatrix} \tag{6.55}$$

After substituting Eq. 6.55 into Eq. 6.54, interpolation function can be obtained in Lagrange form as following:

$$f(\xi,\eta) = \underbrace{\begin{bmatrix} 1 & \xi & \eta \end{bmatrix} \begin{bmatrix} 1 & 0 & 0 \\ -1 & 1 & 0 \\ -1 & 0 & 1 \end{bmatrix}}_{\begin{bmatrix} N_1 & N_2 & N_3 \end{bmatrix}} \begin{Bmatrix} f_1 \\ f_2 \\ f_3 \end{Bmatrix} \rightarrow$$

$$f(\xi,\eta)=\underbrace{(1-\xi-\eta)}_{N_1(\xi,\eta)}f_1+\underbrace{\xi}_{N_2(\xi,\eta)}f_2+\underbrace{\eta}_{N_3(\xi,\eta)}f_3=N_1(\xi,\eta)f_1+N_2(\xi,\eta)f_2+N_3(\xi,\eta)f_3 \quad (6.56)$$

$N_1(\xi,\eta)$, $N_2(\xi,\eta)$, $N_3(\xi,\eta)$ functions in Eq. 6.56 represent Lagrange or shape functions in terms of natural coordinates.

Lagrange Interpolation Function for Quadrilateral Area

Quadrilateral domain or area is shown in Fig. 6.14. It can be considered as the distorted form of the rectangular area. Similar to the triangular area, interpolation polynomial for the quadrilateral area can be expressed as following:

$$f(x,y)=c_1+c_2x+c_3y+c_4xy=\begin{bmatrix}1 & x & y & xy\end{bmatrix}\begin{Bmatrix}c_1\\c_2\\c_3\\c_4\end{Bmatrix} \quad (6.57)$$

Equation 6.57 can be applied at vertices (nodes) as following:

$$\begin{aligned}f_1&=c_1+c_2x_1+c_3y_1+c_4x_1y_1\\f_2&=c_1+c_2x_2+c_3y_2+c_4x_2y_2\\f_3&=c_1+c_2x_3+c_3y_3+c_4x_3y_3\\f_4&=c_1+c_2x_4+c_3y_4+c_4x_4y_4\end{aligned}$$

If the above equations are written in matrix form and solved, unknown constants of c_1, c_2, c_3, c_4 can be obtained as below:

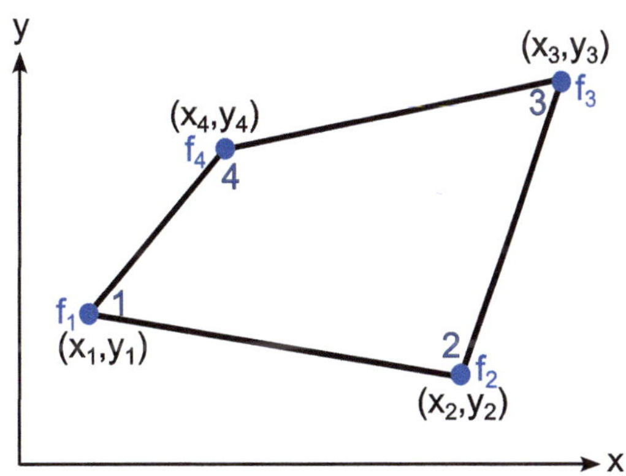

Fig. 6.14 Quadrilateral domain and its corners

$$\begin{bmatrix} 1 & x_1 & y_1 & x_1y_1 \\ 1 & x_2 & y_2 & x_2y_2 \\ 1 & x_3 & y_3 & x_3y_3 \\ 1 & x_4 & y_4 & x_4y_4 \end{bmatrix} \begin{Bmatrix} c_1 \\ c_2 \\ c_3 \\ c_4 \end{Bmatrix} = \begin{Bmatrix} f_1 \\ f_2 \\ f_3 \\ f_4 \end{Bmatrix} \rightarrow \begin{Bmatrix} c_1 \\ c_2 \\ c_3 \\ c_4 \end{Bmatrix} = \begin{bmatrix} 1 & x_1 & y_1 & x_1y_1 \\ 1 & x_2 & y_2 & x_2y_2 \\ 1 & x_3 & y_3 & x_3y_3 \\ 1 & x_4 & y_4 & x_4y_4 \end{bmatrix}^{-1} \begin{Bmatrix} f_1 \\ f_2 \\ f_3 \\ f_4 \end{Bmatrix} \tag{6.58}$$

After substituting Eq. 6.58 into Eq. 6.57, interpolation function can be obtained in Lagrange form as following:

$$\begin{aligned} f(x,y) &= \underbrace{\begin{bmatrix} 1 & x & y & xy \end{bmatrix} \begin{bmatrix} 1 & x_1 & y_1 & x_1y_1 \\ 1 & x_2 & y_2 & x_2y_2 \\ 1 & x_3 & y_3 & x_3y_3 \\ 1 & x_4 & y_4 & x_4y_4 \end{bmatrix}^{-1}}_{\begin{bmatrix} N_1 & N_2 & N_3 & N_4 \end{bmatrix}} \begin{Bmatrix} f_1 \\ f_2 \\ f_3 \\ f_4 \end{Bmatrix} \\ &= N_1(x,y)f_1 + N_2(x,y)f_2 + N_3(x,y)f_3 + N_4(x,y)f_4 \end{aligned} \tag{6.59}$$

$N_i(x, y)$ functions in Eq. 6.59 are called as Lagrange or Shape functions. When quadrilateral geometry corresponds to a *rectangle,* $N_i(x, y)$ functions can be written explicitly as following:

$$\begin{aligned} N_1(x,y) &= \frac{(x-x_2)(y-y_4)}{(x_1-x_2)(y_1-y_4)}, \quad & N_2(x,y) &= \frac{(x-x_1)(y-y_3)}{(x_2-x_1)(y_2-y_3)} \\ N_3(x,y) &= \frac{(x-x_4)(y-y_2)}{(x_3-x_4)(y_3-y_2)}, \quad & N_4(x,y) &= \frac{(x-x_3)(y-y_1)}{(x_4-x_3)(y_4-y_1)} \end{aligned} \tag{6.60}$$

Special Case: Square Area

In solving two-dimensional engineering problems more easily, sometimes quadrilateral area is converted into a special square area using mapping concept. Therefore, obtaining Lagrange interpolation polynomial for square geometry is important. Square area can be represented more easily with natural coordinate system using ξ and η variables as in Fig. 6.15. Natural coordinate system is located at the center of square geometry. Function values at vertices of square geometry can be given in terms of natural coordinates as in Table 6.7.

Interpolation function for the square area can be written in terms of natural coordinates in classical form as below:

$$f(\xi,\eta) = c_1 + c_2\xi + c_3\eta + c_4\xi\eta = \begin{bmatrix} 1 & \xi & \eta & \xi\eta \end{bmatrix} \begin{Bmatrix} c_1 \\ c_2 \\ c_3 \\ c_4 \end{Bmatrix} \tag{6.61}$$

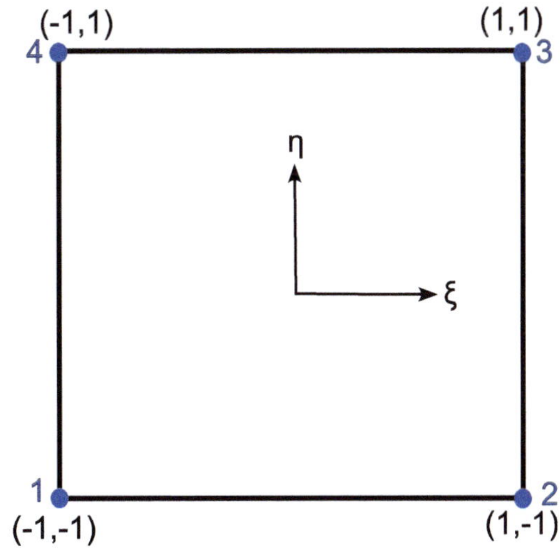

Fig. 6.15 Square geometry defined with natural coordinates of ξ and η

Table 6.7 Function values at vertices of square geometry in terms of natural coordinates ξ and η

Node	ξ	η	$f(\xi,\eta)$
1	-1	-1	f_1
2	1	-1	f_2
3	1	1	f_3
4	-1	1	f_4

Applying Eq. 6.61 at vertices (nodes), unknown constants of c_1, c_2, c_3, c_4 can be obtained as below:

$$\begin{bmatrix} 1 & \xi_1 & \eta_1 & \xi_1\eta_1 \\ 1 & \xi_2 & \eta_2 & \xi_2\eta_2 \\ 1 & \xi_3 & \eta_3 & \xi_3\eta_3 \\ 1 & \xi_4 & \eta_4 & \xi_4\eta_4 \end{bmatrix} \begin{Bmatrix} c_1 \\ c_2 \\ c_3 \\ c_4 \end{Bmatrix} = \begin{Bmatrix} f_1 \\ f_2 \\ f_3 \\ f_4 \end{Bmatrix} \rightarrow$$

$$\begin{Bmatrix} c_1 \\ c_2 \\ c_3 \\ c_4 \end{Bmatrix} = \begin{bmatrix} 1 & \xi_1 & \eta_1 & \xi_1\eta_1 \\ 1 & \xi_2 & \eta_2 & \xi_2\eta_2 \\ 1 & \xi_3 & \eta_3 & \xi_3\eta_3 \\ 1 & \xi_4 & \eta_4 & \xi_4\eta_4 \end{bmatrix}^{-1} \begin{Bmatrix} f_1 \\ f_2 \\ f_3 \\ f_4 \end{Bmatrix} \rightarrow \begin{Bmatrix} c_1 \\ c_2 \\ c_3 \\ c_4 \end{Bmatrix} = \begin{bmatrix} \frac{1}{4} & \frac{1}{4} & \frac{1}{4} & \frac{1}{4} \\ -\frac{1}{4} & \frac{1}{4} & \frac{1}{4} & -\frac{1}{4} \\ -\frac{1}{4} & -\frac{1}{4} & \frac{1}{4} & \frac{1}{4} \\ \frac{1}{4} & -\frac{1}{4} & \frac{1}{4} & -\frac{1}{4} \end{bmatrix} \begin{Bmatrix} f_1 \\ f_2 \\ f_3 \\ f_4 \end{Bmatrix} \quad (6.62)$$

After substituting Eq. 6.62 into Eq. 6.61, interpolation function can be obtained in Lagrange form as following:

$$f(\xi,\eta)=\underbrace{\begin{bmatrix}1 & \xi & \eta & \xi\eta\end{bmatrix}\begin{bmatrix}\frac{1}{4} & \frac{1}{4} & \frac{1}{4} & \frac{1}{4}\\ -\frac{1}{4} & \frac{1}{4} & \frac{1}{4} & -\frac{1}{4}\\ -\frac{1}{4} & -\frac{1}{4} & \frac{1}{4} & \frac{1}{4}\\ \frac{1}{4} & -\frac{1}{4} & \frac{1}{4} & -\frac{1}{4}\end{bmatrix}}_{\begin{bmatrix}N_1 & N_2 & N_3 & N_4\end{bmatrix}}\begin{Bmatrix}f_1\\ f_2\\ f_3\\ f_4\end{Bmatrix}$$

$$=\left(\frac{1}{4}-\frac{\xi}{4}-\frac{\eta}{4}+\frac{\xi\eta}{4}\right)f_1+\left(\frac{1}{4}+\frac{\xi}{4}-\frac{\eta}{4}-\frac{\xi\eta}{4}\right)f_2+\left(\frac{1}{4}+\frac{\xi}{4}+\frac{\eta}{4}+\frac{\xi\eta}{4}\right)f_3+\left(\frac{1}{4}-\frac{\xi}{4}+\frac{\eta}{4}-\frac{\xi\eta}{4}\right)f_4$$

After rearrangement, above expression can be written in terms of Lagrange or shape functions ($N_i(\xi,\eta)$) as following:

$$\begin{aligned}f(\xi,\eta) &= \underbrace{\frac{(1-\xi)(1-\eta)}{4}}_{N_1(\xi,\eta)}f_1+\underbrace{\frac{(1+\xi)(1-\eta)}{4}}_{N_2(\xi,\eta)}f_2+\underbrace{\frac{(1+\xi)(1+\eta)}{4}}_{N_3(\xi,\eta)}f_3+\underbrace{\frac{(1-\xi)(1+\eta)}{4}}_{N_4(\xi,\eta)}f_4\\ &= N_1(\xi,\eta)f_1+N_2(\xi,\eta)f_2+N_3(\xi,\eta)f_3+N_4(\xi,\eta)f_4\end{aligned} \tag{6.63}$$

6.6.2 Bivariate Hermite Interpolation Functions

As mentioned before, Hermite interpolation polynomials can be used when both function and derivative values are available simultaneously. Two-dimensional Hermite interpolation functions can be obtained by generalizing one-dimensional ones. Bivariate cubic Hermite interpolation polynomials are commonly used in the solution of thin plate (Kirchhoff plate) problems in structural mechanics based on Finite Element method. Therefore, cubic Hermite interpolation functions for triangular and quadrilateral domains/areas will be explained in this section.

Cubic Hermite Interpolation Function for Triangular Area

Triangular area with corner points is shown in Fig. 6.16. Function values f_i and derivative values $\partial f_i/\partial x$ and $\partial f_i/\partial y$ at corner points of this triangle are given in Table 6.8. In the solution of triangular plates in mechanics, f_i denotes lateral displacements and $\partial f_i/\partial x$ and $\partial f_i/\partial y$ denote rotations around x and y axes.

Bivariate cubic Hermite interpolation polynomial and its derivatives can be expressed for triangular geometry as following:

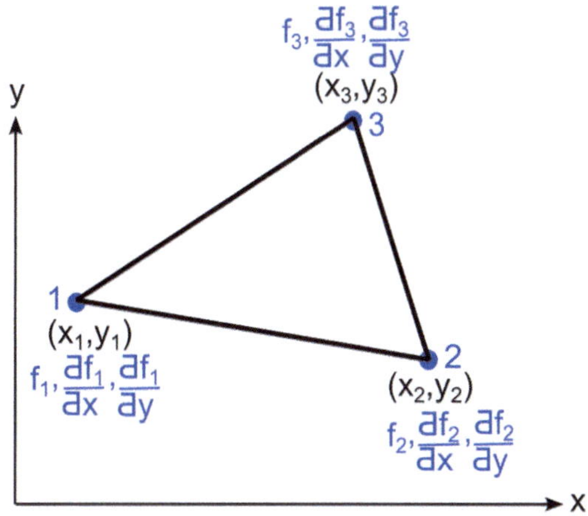

Fig. 6.16 Triangular geometry for Hermite interpolation

Table 6.8 Function values f_i and derivative values $\partial f_i/\partial x$ and $\partial f_i/\partial y$ at corner points of the triangular geometry

x	y	$f(x,y)$	$\partial f(x,y)/\partial x$	$\partial f(x,y)/\partial y$
x_1	y_1	f_1	$\frac{\partial f_1}{\partial x}$	$\frac{\partial f_1}{\partial y}$
x_2	y_2	f_2	$\frac{\partial f_2}{\partial x}$	$\frac{\partial f_2}{\partial y}$
x_3	y_3	f_3	$\frac{\partial f_3}{\partial x}$	$\frac{\partial f_3}{\partial y}$

$$\begin{aligned}
f(x,y) &= c_1 + c_2x + c_3y + c_4x^2 + c_5xy + c_6y^2 + c_7x^3 + c_8\left(x^2y + xy^2\right) + c_9y^3 \\
\frac{\partial f(x,y)}{\partial x} &= c_2 + 2c_4x + c_5y + 3c_7x^2 + c_8\left(2xy + y^2\right) \\
\frac{\partial f(x,y)}{\partial y} &= c_3 + c_5x + 2c_6y + c_8\left(x^2 + 2xy\right) + 3c_9y^2
\end{aligned} \tag{6.64}$$

Above equations can be written in matrix form as below:

$$\begin{Bmatrix} f(x,y) \\ \dfrac{\partial f(x,y)}{\partial x} \\ \dfrac{\partial f(x,y)}{\partial y} \end{Bmatrix} = \begin{bmatrix} 1 & x & y & x^2 & xy & y^2 & x^3 & x^2y + xy^2 & y^3 \\ 0 & 1 & 0 & 2x & y & 0 & 3x^2 & 2xy + y^2 & 0 \\ 0 & 0 & 1 & 0 & x & 2y & 0 & x^2 + 2xy & 3y^2 \end{bmatrix} \underbrace{\begin{Bmatrix} c_1 \\ c_2 \\ \cdots \\ c_9 \end{Bmatrix}}_{C} \tag{6.65}$$

Function and derivative values at vertices of the triangular domain can be expressed using Eq. 6.65 as following:

$$\underbrace{\begin{Bmatrix} f_1 \\ \frac{\partial f_1}{\partial x} \\ \frac{\partial f_1}{\partial y} \\ f_2 \\ \frac{\partial f_2}{\partial x} \\ \frac{\partial f_2}{\partial y} \\ f_3 \\ \frac{\partial f_3}{\partial x} \\ \frac{\partial f_3}{\partial y} \end{Bmatrix}}_{F} = \underbrace{\begin{bmatrix} 1 & x_1 & y_1 & x_1^2 & x_1y_1 & y_1^2 & x_1^3 & x_1^2y_1+x_1y_1^2 & y_1^3 \\ 0 & 1 & 0 & 2x_1 & y_1 & 0 & 3x_1^2 & 2x_1y_1+y_1^2 & 0 \\ 0 & 0 & 1 & 0 & x_1 & 2y_1 & 0 & x_1^2+2x_1y_1 & 3y_1^2 \\ 1 & x_2 & y_2 & x_2^2 & x_2y_2 & y_2^2 & x_2^3 & x_2^2y_2+x_2y_2^2 & y_2^3 \\ 0 & 1 & 0 & 2x_2 & y_2 & 0 & 3x_2^2 & 2x_2y_2+y_2^2 & 0 \\ 0 & 0 & 1 & 0 & x_2 & 2y_2 & 0 & x_2^2+2x_2y_2 & 3y_2^2 \\ 1 & x_3 & y_3 & x_3^2 & x_3y_3 & y_3^2 & x_3^3 & x_3^2y_3+x_3y_3^2 & y_3^3 \\ 0 & 1 & 0 & 2x_3 & y_3 & 0 & 3x_3^2 & 2x_3y_3+y_3^2 & 0 \\ 0 & 0 & 1 & 0 & x_3 & 2y_3 & 0 & x_3^2+2x_3y_3 & 3y_3^2 \end{bmatrix}}_{X} \underbrace{\begin{Bmatrix} c_1 \\ c_2 \\ c_3 \\ c_4 \\ c_5 \\ c_6 \\ c_7 \\ c_8 \\ c_9 \end{Bmatrix}}_{C} \quad (6.66)$$

Solution of unknown constants $C = c_1, c_2, \ldots, c_9$ in Eq. 6.66 can be expressed as $C = X^{-1}(F)$. Substituting $C = X^{-1}(F)$ expression into Eq. 6.65 yields the interpolation function and its derivatives in Lagrange form as below:

$$\begin{Bmatrix} f(x,y) \\ \frac{\partial f(x,y)}{\partial x} \\ \frac{\partial f(x,y)}{\partial y} \end{Bmatrix} = \underbrace{\begin{bmatrix} 1 & x & y & x^2 & xy & y^2 & x^3 & x^2y+xy^2 & y^3 \\ 0 & 1 & 0 & 2x & y & 0 & 3x^2 & 2xy+y^2 & 0 \\ 0 & 0 & 1 & 0 & x & 2y & 0 & x^2+2xy & 3y^2 \end{bmatrix}}_{\begin{bmatrix} N_1 & N_2 & \cdots & N_9 \end{bmatrix}} X^{-1}F = NF \quad (6.67)$$

In Eq. 6.67, $N = N_1, N_2, \ldots, N_9$ functions correspond to Lagrange or Shape functions. Equation 6.67 can be expressed more explicitly in terms of Lagrange functions as below:

$$\begin{aligned} f(x,y) &= N_1 f_1 + N_2\frac{\partial f_1}{\partial x} + N_3\frac{\partial f_1}{\partial y} + N_4 f_2 + N_5\frac{\partial f_2}{\partial x} + N_6\frac{\partial f_2}{\partial y} + N_7 f_3 \\ &\quad + N_8\frac{\partial f_3}{\partial x} + N_9\frac{\partial f_3}{\partial y} \\ \frac{\partial f(x,y)}{\partial x} &= \frac{\partial N_1}{\partial x} f_1 + \frac{\partial N_2}{\partial x}\frac{\partial f_1}{\partial x} + \frac{\partial N_3}{\partial x}\frac{\partial f_1}{\partial y} + \frac{\partial N_4}{\partial x} f_2 + \frac{\partial N_5}{\partial x}\frac{\partial f_2}{\partial x} \\ &\quad + \frac{\partial N_6}{\partial x}\frac{\partial f_2}{\partial y} + \frac{\partial N_7}{\partial x} f_3 + \frac{\partial N_8}{\partial x}\frac{\partial f_3}{\partial x} + \frac{\partial N_9}{\partial x}\frac{\partial f_3}{\partial y} \\ \frac{\partial f(x,y)}{\partial y} &= \frac{\partial N_1}{\partial y} f_1 + \frac{\partial N_2}{\partial y}\frac{\partial f_1}{\partial x} + \frac{\partial N_3}{\partial y}\frac{\partial f_1}{\partial y} + \frac{\partial N_4}{\partial y} f_2 + \frac{\partial N_5}{\partial y}\frac{\partial f_2}{\partial x} \\ &\quad + \frac{\partial N_6}{\partial y}\frac{\partial f_2}{\partial y} + \frac{\partial N_7}{\partial y} f_3 + \frac{\partial N_8}{\partial y}\frac{\partial f_3}{\partial x} + \frac{\partial N_9}{\partial y}\frac{\partial f_3}{\partial y} \end{aligned} \quad (6.68)$$

Cubic Hermite Interpolation Function for Quadrilateral Area

Quadrilateral area/domain with corner points is shown in Fig. 6.17. Function values f_i and derivative values $\partial f_i/\partial x$ and $\partial f_i/\partial y$ at corner points of the quadrilateral geometry are given in Table 6.9. In the solution of quadrilateral plates in mechanics, f_i denotes lateral displacements and $\partial f_i/\partial x$ and $\partial f_i/\partial y$ denote rotations around x and y axes.

Bivariate cubic Hermite interpolation polynomial and its derivatives can be expressed for quadrilateral geometry as following:

$$
\begin{aligned}
f(x,y) &= c_1 + c_2x + c_3y + c_4x^2 + c_5xy + c_6y^2 + c_7x^3 + c_8x^2y + c_9xy^2 \\
&\quad + c_{10}y^3 + c_{11}x^3y + c_{12}xy^3 \\
\frac{\partial f(x,y)}{\partial x} &= c_2 + 2c_4x + c_5y + 3c_7x^2 + 2c_8xy + c_9y^2 + 3c_{11}x^2y + c_{12}y \\
\frac{\partial f(x,y)}{\partial y} &= c_3 + c_5x + 2c_6y + c_8x^2 + 2c_9xy + 3c_{10}y^2 + c_{11}x^3 + 3c_{12}xy^2
\end{aligned}
\tag{6.69}
$$

Above equations can be written in matrix form as below:

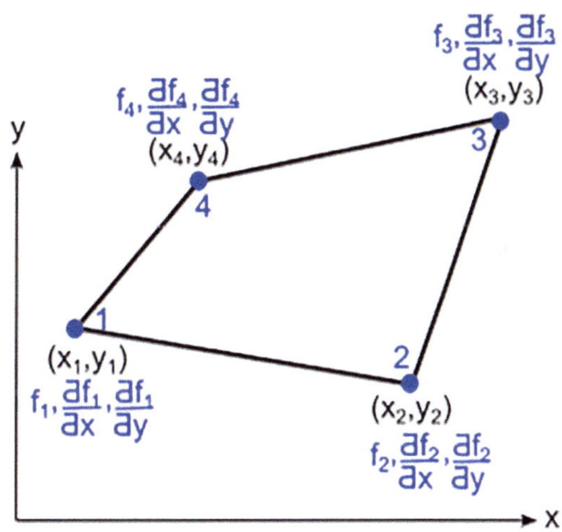

Fig. 6.17 Quadrilateral geometry for Hermite interpolation

Table 6.9 Function values f_i and derivative values $\partial f_i/\partial x$ and $\partial f_i/\partial y$ at corner points of the quadrilateral geometry

x	y	$f(x,y)$	$\partial f(x,y)/\partial x$	$\partial f(x,y)/\partial y$
x_1	y_1	f_1	$\frac{\partial f_1}{\partial x}$	$\frac{\partial f_1}{\partial y}$
x_2	y_2	f_2	$\frac{\partial f_2}{\partial x}$	$\frac{\partial f_2}{\partial y}$
x_3	y_3	f_3	$\frac{\partial f_3}{\partial x}$	$\frac{\partial f_3}{\partial y}$
x_4	y_4	f_4	$\frac{\partial f_4}{\partial x}$	$\frac{\partial f_4}{\partial y}$

$$\begin{Bmatrix} f(x,y) \\ \dfrac{\partial f(x,y)}{\partial x} \\ \dfrac{\partial f(x,y)}{\partial y} \end{Bmatrix} = \begin{bmatrix} 1 & x & y & x^2 & xy & y^2 & x^3 & x^2y & xy^2 & y^3 & x^3y & xy^3 \\ 0 & 1 & 0 & 2x & y & 0 & 3x^2 & 2xy & y^2 & 0 & 3x^2y & y^3 \\ 0 & 0 & 1 & 0 & x & 2y & 0 & x^2 & 2xy & 3y^2 & x^3 & 3xy^2 \end{bmatrix} \underbrace{\begin{Bmatrix} c_1 \\ c_2 \\ \cdots \\ c_{12} \end{Bmatrix}}_{C} \tag{6.70}$$

Function and derivative values at vertices of the quadrilateral domain can be expressed using Eq. 6.70 as following:

$$\underbrace{\begin{Bmatrix} f_1 \\ \dfrac{\partial f_1}{\partial x} \\ \dfrac{\partial f_1}{\partial y} \\ f_2 \\ \dfrac{\partial f_2}{\partial x} \\ \dfrac{\partial f_2}{\partial y} \\ f_3 \\ \dfrac{\partial f_3}{\partial x} \\ \dfrac{\partial f_3}{\partial y} \\ f_4 \\ \dfrac{\partial f_4}{dx} \\ \dfrac{\partial f_4}{\partial y} \end{Bmatrix}}_{F} = \underbrace{\begin{bmatrix} 1 & x_1 & y_1 & x_1^2 & x_1y_1 & y_1^2 & x_1^3 & x_1^2y_1 & x_1y_1^2 & y_1^3 & x_1^3y_1 & x_1y_1^3 \\ 0 & 1 & 0 & 2x_1 & y_1 & 0 & 3x_1^2 & 2x_1y_1 & y_1^2 & 0 & 3x_1^2y_1 & y_1 \\ 0 & 0 & 1 & 0 & x_1 & 2y_1 & 0 & x_1^2 & 2x_1y_1 & 3y_1^2 & x_1^3 & 3x_1y_1^2 \\ 1 & x_2 & y_2 & x_2^2 & x_2y_2 & y_2^2 & x_2^3 & x_2^2y_2 & x_2y_2^2 & y_2^3 & x_2^3y_2 & x_2y_2^3 \\ 0 & 1 & 0 & 2x_2 & y_2 & 0 & 3x_2^2 & 2x_2y_2 & y_2^2 & 0 & 3x_2^2y_2 & y_2 \\ 0 & 0 & 1 & 0 & x_2 & 2y_2 & 0 & x_2^2 & 2x_2y_2 & 3y_2^2 & x_2^3 & 3x_2y_2^2 \\ 1 & x_3 & y_3 & x_3^2 & x_3y_3 & y_3^2 & x_3^3 & x_3^2y_3 & x_3y_3^2 & y_3^3 & x_3^3y_3 & x_3y_3^3 \\ 0 & 1 & 0 & 2x_3 & y_3 & 0 & 3x_3^2 & 2x_3y_3 & y_3^2 & 0 & 3x_3^2y_3 & y_3 \\ 0 & 0 & 1 & 0 & x_3 & 2y_3 & 0 & x_3^2 & 2x_3y_3 & 3y_3^2 & x_3^3 & 3x_3y_3^2 \\ 1 & x_4 & y_4 & x_4^2 & x_4y_4 & y_4^2 & x_4^3 & x_4^2y_4 & x_4y_4^2 & y_4^3 & x_4^3y_4 & x_4y_4^3 \\ 0 & 1 & 0 & 2x_4 & y_4 & 0 & 3x_4^2 & 2x_4y_4 & y_4^2 & 0 & 3x_4^2y_4 & y_4 \\ 0 & 0 & 1 & 0 & x_4 & 2y_4 & 0 & x_4^2 & 2x_4y_4 & 3y_4^2 & x_4^3 & 3x_4y_4^2 \end{bmatrix}}_{X} \underbrace{\begin{Bmatrix} c_1 \\ c_2 \\ c_3 \\ c_4 \\ c_5 \\ c_6 \\ c_7 \\ c_8 \\ c_9 \\ c_{10} \\ c_{11} \\ c_{12} \end{Bmatrix}}_{C} \tag{6.71}$$

Solution of unknown constants $C = c_1,, \ldots, c_{12}$ in Eq. 6.71 can be expressed as $C = X^{-1}(F)$. Substituting $C = X^{-1}(F)$ expression into Eq. 6.70 yields the interpolation function and its derivatives in Lagrange form as below:

$$\begin{Bmatrix} f(x,y) \\ \dfrac{\partial f(x,y)}{\partial x} \\ \dfrac{\partial f(x,y)}{\partial y} \end{Bmatrix} = \underbrace{\begin{bmatrix} 1 & x & y & x^2 & xy & y^2 & x^3 & x^2y & xy^2 & y^3 & x^3y & xy^3 \\ 0 & 1 & 0 & 2x & y & 0 & 3x^2 & 2xy & y^2 & 0 & 3x^2y & y^3 \\ 0 & 0 & 1 & 0 & x & 2y & 0 & x^2 & 2xy & 3y^2 & x^3 & 3xy^2 \end{bmatrix} X^{-1}}_{N = \begin{bmatrix} N_1 & N_2 & \cdots & N_{12} \end{bmatrix}} F = NF \tag{6.72}$$

In Eq. 6.72, $N = N_1, N_2, \ldots, N_{12}$ functions correspond to Lagrange or Shape functions. Equation 6.72 can be expressed more explicitly in terms of Shape functions as below:

$$
\begin{aligned}
f(x,y) &= N_1 f_1 + N_2 \frac{\partial f_1}{\partial x} + N_3 \frac{\partial f_1}{\partial y} + N_4 f_2 + N_5 \frac{\partial f_2}{\partial x} + N_6 \frac{\partial f_2}{\partial y} + N_7 f_3 + N_8 \frac{\partial f_3}{\partial x} + N_9 \frac{\partial f_3}{\partial y} \\
&\quad + N_{10} f_4 + N_{11} \frac{\partial f_4}{\partial x} + N_{12} \frac{\partial f_4}{\partial y} \\
\frac{\partial f(x,y)}{\partial x} &= \frac{\partial N_1}{\partial x} f_1 + \frac{\partial N_2}{\partial x}\frac{\partial f_1}{\partial x} + \frac{\partial N_3}{\partial x}\frac{\partial f_1}{\partial y} + \frac{\partial N_4}{\partial x} f_2 + \frac{\partial N_5}{\partial x}\frac{\partial f_2}{\partial x} + \frac{\partial N_6}{\partial x}\frac{\partial f_2}{\partial y} + \frac{\partial N_7}{\partial x} f_3 \\
&\quad + \frac{\partial N_8}{\partial x}\frac{\partial f_3}{\partial x} + \frac{\partial N_9}{\partial x}\frac{\partial f_3}{\partial y} + \frac{\partial N_{10}}{\partial x} f_4 + \frac{\partial N_{11}}{\partial x}\frac{\partial f_4}{\partial x} + \frac{\partial N_{12}}{\partial x}\frac{\partial f_4}{\partial y} \\
\frac{\partial f(x,y)}{\partial y} &= \frac{\partial N_1}{\partial y} f_1 + \frac{\partial N_2}{\partial y}\frac{\partial f_1}{\partial x} + \frac{\partial N_3}{\partial y}\frac{\partial f_1}{\partial y} + \frac{\partial N_4}{\partial y} f_2 + \frac{\partial N_5}{\partial y}\frac{\partial f_2}{\partial x} + \frac{\partial N_6}{\partial y}\frac{\partial f_2}{\partial y} + \frac{\partial N_7}{\partial y} f_3 \\
&\quad + \frac{\partial N_8}{\partial y}\frac{\partial f_3}{\partial x} + \frac{\partial N_9}{\partial y}\frac{\partial f_3}{\partial y} + \frac{\partial N_{10}}{\partial y} f_4 + \frac{\partial N_{11}}{\partial y}\frac{\partial f_4}{\partial x} + \frac{\partial N_{12}}{\partial y}\frac{\partial f_4}{\partial y}
\end{aligned}
\tag{6.73}
$$

Special Case: Square Area

As mentioned before, quadrilateral area is sometimes converted into a special square area in the solution of engineering problems using mapping concept. Therefore, obtaining Hermite interpolation polynomial for square geometry is important. Square area can be represented more easily with natural coordinate system using ξ and η variables as in Fig. 6.18. Function and derivative values at vertices of square geometry can be given in terms of natural coordinates as in Table 6.10.

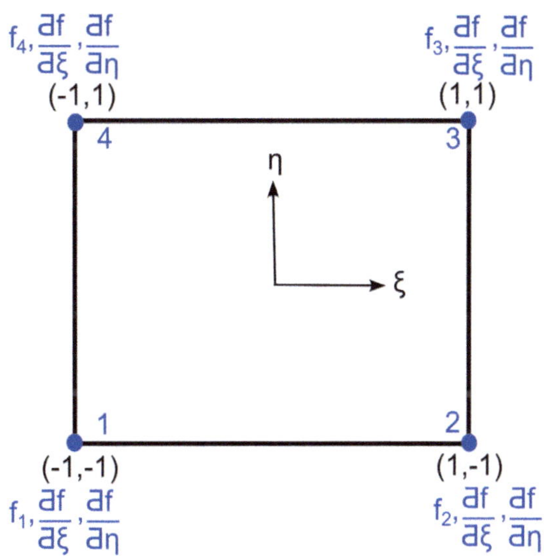

Fig. 6.18 Square geometry defined with natural coordinates ξ and η

Table 6.10 Function and derivative values at vertices of square geometry in terms of natural coordinates ξ and η

Node	ξ	η	$f(\xi,\eta)$	$\partial f(\xi,\eta)/\partial \xi$	$\partial f(\xi,\eta)/\partial \eta$
1	-1	-1	f_1	$\frac{\partial f_1}{\partial \xi}$	$\frac{\partial f_1}{\partial \eta}$
2	1	-1	f_2	$\frac{\partial f_2}{\partial \xi}$	$\frac{\partial f_2}{\partial \eta}$
3	1	1	f_3	$\frac{\partial f_3}{\partial \xi}$	$\frac{\partial f_3}{\partial \eta}$
4	-1	1	f_4	$\frac{\partial f_4}{\partial \xi}$	$\frac{\partial f_4}{\partial \eta}$

Bivariate cubic interpolation function for the square area can be written in terms of natural coordinates in classical form as below:

$$\begin{aligned} f(\xi,\eta) &= c_1 + c_2\xi + c_3\eta + c_4\xi^2 + c_5\xi\eta + c_6\eta^2 + c_7\xi^3 + c_8\xi^2\eta + c_9\xi\eta^2 + c_{10}\eta^3 + \\ &\quad c_{11}\xi^3\eta + c_{12}\xi\eta^3 \\ \frac{\partial f(\xi,\eta)}{\partial \xi} &= c_2 + 2c_4\xi + c_5\eta + 3c_7\xi^2 + 2c_8\xi\eta + c_9\eta^2 + 3c_{11}\xi^2\eta + c_{12}\eta \\ \frac{\partial f(\xi,\eta)}{\partial \eta} &= c_3 + c_5\xi + 2c_6\eta + c_8\xi^2 + 2c_9\xi\eta + 3c_{10}\eta^2 + c_{11}\xi^3 + 3c_{12}\xi\eta^2 \end{aligned} \tag{6.74}$$

Interpolation function and its derivatives in Eq. 6.74 can be written in matrix form as following:

$$\begin{Bmatrix} f(\xi,\eta) \\ \dfrac{\partial f(\xi,\eta)}{\partial \xi} \\ \dfrac{\partial f(\xi,\eta)}{\partial \eta} \end{Bmatrix} = \begin{bmatrix} 1 & \xi & \eta & \xi^2 & \xi\eta & \eta^2 & \xi^3 & \xi^2\eta & \xi\eta^2 & \eta^3 & \xi^3\eta & \xi\eta^3 \\ 0 & 1 & 0 & 2\xi & \eta & 0 & 3\xi^2 & 2\xi\eta & \eta^2 & 0 & 3\xi^2\eta & \eta^3 \\ 0 & 0 & 1 & 0 & \xi & 2\eta & 0 & \xi^2 & 2\xi\eta & 3\eta^2 & \xi^3 & 3\xi\eta^2 \end{bmatrix} \underbrace{\begin{Bmatrix} c_1 \\ c_2 \\ \cdots \\ c_{12} \end{Bmatrix}}_{C} \tag{6.75}$$

If the data values in Table 6.10 are substituted into Eq. 6.75, following equations are obtained

$$\underbrace{\begin{Bmatrix} f_1 \\ \frac{\partial f_1}{\partial \xi} \\ \frac{\partial f_1}{\partial \eta} \\ f_2 \\ \frac{\partial f_2}{\partial \xi} \\ \frac{\partial f_2}{\partial \eta} \\ f_3 \\ \frac{\partial f_3}{\partial \xi} \\ \frac{\partial f_3}{\partial \eta} \\ f_4 \\ \frac{\partial f_4}{\partial \xi} \\ \frac{\partial f_4}{\partial \eta} \end{Bmatrix}}_{F} = \underbrace{\begin{bmatrix} 1 & \xi_1 & \eta_1 & \xi_1^2 & \xi_1\eta_1 & \eta_1^2 & \xi_1^3 & \xi_1^2\eta_1 & \xi_1\eta_1^2 & \eta_1^3 & \xi_1^3\eta_1 & \xi_1\eta_1^3 \\ 0 & 1 & 0 & 2\xi_1 & \eta_1 & 0 & 3\xi_1^2 & 2\xi_1\nu_1 & \eta_1^2 & 0 & 3\xi_1^2\eta_1 & \eta_1 \\ 0 & 0 & 1 & 0 & \xi_1 & 2\eta_1 & 0 & \xi_1^2 & 2\xi_1\eta_1 & 3\eta_1^2 & \xi_1^3 & 3\xi_1\eta_1^2 \\ 1 & \xi_2 & \eta_2 & \xi_2^2 & \xi_2\eta_2 & \eta_2^2 & \xi_2^3 & \xi_2^2\eta_2 & \xi_2\eta_2^2 & \eta_2^3 & \xi_2^3\eta_2 & \xi_2\eta_2^3 \\ 0 & 1 & 0 & 2\xi_2 & \eta_2 & 0 & 3\xi_2^2 & 2\xi_2\eta_2 & \eta_2^2 & 0 & 3\xi_2^2\eta_2 & \eta_2 \\ 0 & 0 & 1 & 0 & \xi_2 & 2\eta_2 & 0 & \xi_2^2 & 2\xi_2\eta_2 & 3\eta_2^2 & \xi_2^3 & 3\xi_2\eta_2^2 \\ 1 & \xi_3 & \eta_3 & \xi_3^2 & \xi_3\eta_3 & \eta_3^2 & \xi_3^3 & \xi_3^2\eta_3 & \xi_3\eta_3^2 & \eta_3^3 & \xi_3^3\eta_3 & \xi_3\eta_3^3 \\ 0 & 1 & 0 & 2\xi_3 & \eta_3 & 0 & 3\xi_3^2 & 2\xi_3\eta_3 & \eta_3^2 & 0 & 3\xi_3^2\eta_3 & \eta_3 \\ 0 & 0 & 1 & 0 & \xi_3 & 2\eta_3 & 0 & \xi_3^2 & 2\xi_3\eta_3 & 3\eta_3^2 & \xi_3^3 & 3\xi_3\eta_3^2 \\ 1 & \xi_4 & \eta_4 & \xi_4^2 & \xi_4\eta_4 & \eta_4^2 & \xi_4^3 & \xi_4^2\eta_4 & \xi_4\eta_4^2 & \eta_4^3 & \xi_4^3\eta_4 & \xi_4\eta_4^3 \\ 0 & 1 & 0 & 2\xi_4 & \eta_4 & 0 & 3\xi_4^2 & 2\xi_4\eta_4 & \eta_4^2 & 0 & 3\xi_4^2\eta_4 & \eta_4 \\ 0 & 0 & 1 & 0 & \xi_4 & 2\eta_4 & 0 & \xi_4^2 & 2\xi_4\eta_4 & 3\eta_4^2 & \xi_4^3 & 3\xi_4\eta_4^2 \end{bmatrix}}_{X} \underbrace{\begin{Bmatrix} c_1 \\ c_2 \\ c_3 \\ c_4 \\ c_5 \\ c_6 \\ c_7 \\ c_8 \\ c_9 \\ c_{10} \\ c_{11} \\ c_{12} \end{Bmatrix}}_{C} \tag{6.76}$$

From Eq. 6.76, solution of unknown constants $C= c_1,\ldots, c_{12}$ can be expressed as $C=X^{-1}(F)$. Substituting $C=X^{-1}(F)$ expression into Eq. 6.75 yields bivariate cubic interpolation function and its derivatives in Lagrange form as following:

$$\begin{Bmatrix} f(\xi,\eta) \\ \frac{\partial f(\xi,\eta)}{\partial \xi} \\ \frac{\partial f(\xi,\eta)}{\partial \eta} \end{Bmatrix} = \underbrace{\begin{bmatrix} 1 & \xi & \eta & \xi^2 & \xi\eta & \eta^2 & \xi^3 & \xi^2\eta & \xi\eta^2 & \eta^3 & \xi^3\eta & \xi\eta^3 \\ 0 & 1 & 0 & 2\xi & \eta & 0 & 3\xi^2 & 2\xi\eta & \eta^2 & 0 & 3\xi^2\eta & \eta^3 \\ 0 & 0 & 1 & 0 & \xi & 2\eta & 0 & \xi^2 & 2\xi\eta & 3\eta^2 & \xi^3 & 3\xi\eta^2 \end{bmatrix} X^{-1}F}_{N=\begin{bmatrix} N_1 & N_2 & \cdots & N_{12} \end{bmatrix}} \\ = NF \tag{6.77}$$

In Eq. 6.77, $N= N_1, N_2,\ldots, N_{12}$ functions correspond to Lagrange or Shape functions. Equation 6.77 can be written more explicitly in terms of Shape functions as below:

$$\begin{aligned} f(\xi,\eta) =& N_1(\xi,\eta)f_1+N_2(\xi,\eta)\frac{\partial f_1}{\partial \xi}+N_3(\xi,\eta)\frac{\partial f_1}{\partial \eta}+N_4(\xi,\eta)f_2+N_5(\xi,\eta)\frac{\partial f_2}{\partial \xi}+N_6(\xi,\eta)\frac{\partial f_2}{\partial \eta} \\ &+N_7(\xi,\eta)f_3+N_8(\xi,\eta)\frac{\partial f_3}{\partial \xi}+N_9(\xi,\eta)\frac{\partial f_3}{\partial \eta}+N_{10}(\xi,\eta)f_4+N_{11}(\xi,\eta)\frac{\partial f_4}{\partial \xi}+N_{12}(\xi,\eta)\frac{\partial f_4}{\partial \eta} \end{aligned} \tag{6.78a}$$

$$\frac{\partial f(\xi,\eta)}{\partial \xi}=\frac{\partial N_1(\xi,\eta)}{\partial \xi}f_1+\frac{\partial N_2(\xi,\eta)}{\partial \xi}\frac{\partial f_1}{\partial \xi}+\frac{\partial N_3(\xi,\eta)}{\partial \xi}\frac{\partial f_1}{\partial \eta}+\frac{\partial N_4(\xi,\eta)}{\partial \xi}f_2+\frac{\partial N_5(\xi,\eta)}{\partial \xi}\frac{\partial f_2}{\partial \xi}+\frac{\partial N_6(\xi,\eta)}{\partial \xi}\frac{\partial f_2}{\partial \eta}$$
$$+\frac{\partial N_7(\xi,\eta)}{\partial \xi}f_3+\frac{\partial N_8(\xi,\eta)}{\partial \xi}\frac{\partial f_3}{\partial \xi}+\frac{\partial N_9(\xi,\eta)}{\partial \xi}\frac{\partial f_3}{\partial \eta}+\frac{\partial N_{10}(\xi,\eta)}{\partial \xi}f_4+\frac{\partial N_{11}(\xi,\eta)}{\partial \xi}\frac{\partial f_4}{\partial \xi}+\frac{\partial N_{12}(\xi,\eta)}{\partial \xi}\frac{\partial f_4}{\partial \eta} \tag{6.78b}$$

$$\frac{\partial f(\xi,\eta)}{\partial \eta}=\frac{\partial N_1(\xi,\eta)}{\partial \eta}f_1+\frac{\partial N_2(\xi,\eta)}{\partial \eta}\frac{\partial f_1}{\partial \xi}+\frac{\partial N_3(\xi,\eta)}{\partial \eta}\frac{\partial f_1}{\partial \eta}+\frac{\partial N_4(\xi,\eta)}{\partial \eta}f_2+\frac{\partial N_5(\xi,\eta)}{\partial \eta}\frac{\partial f_2}{\partial \xi}+\frac{\partial N_6(\xi,\eta)}{\partial \eta}\frac{\partial f_2}{\partial \eta}$$
$$+\frac{\partial N_7(\xi,\eta)}{\partial \eta}f_3+\frac{\partial N_8(\xi,\eta)}{\partial \eta}\frac{\partial f_3}{\partial \xi}+\frac{\partial N_9(\xi,\eta)}{\partial \eta}\frac{\partial f_3}{\partial \eta}+\frac{\partial N_{10}(\xi,\eta)}{\partial \eta}f_4+\frac{\partial N_{11}(\xi,\eta)}{\partial \eta}\frac{\partial f_4}{\partial \xi}+\frac{\partial N_{12}(\xi,\eta)}{\partial \eta}\frac{\partial f_4}{\partial \eta} \tag{6.78c}$$

where Lagrange or shape functions $N_i(\xi,\eta)$ in Eq. 6.78a can be expressed as following:

$$\begin{aligned}
&N_1(\xi,\eta)=-\frac{1}{8}(\xi-1)(\eta-1)\left(\xi^2+\xi+\eta^2+\eta-2\right), && N_2(\xi,\eta)=-\frac{1}{8}(\xi-1)(\eta-1)^2(\eta+1)\\
&N_3(\xi,\eta)=\frac{1}{8}(\xi-1)^2(\xi+1)(\eta-1), && N_4(\xi,\eta)=\frac{1}{8}(\xi+1)(\eta-1)\left(\xi^2-\xi+\eta^2+\eta-2\right)\\
&N_5(\xi,\eta)=\frac{1}{8}(\xi+1)(\eta-1)^2(\eta+1), && N_6(\xi,\eta)=\frac{1}{8}(\xi-1)(\xi+1)^2(\eta-1)\\
&N_7(\xi,\eta)=-\frac{1}{8}(\xi+1)(\eta+1)\left(\xi^2-\xi+\eta^2-\eta-2\right), && N_8(\xi,\eta)=\frac{1}{8}(\xi+1)(\eta-1)(\eta+1)^2\\
&N_9(\xi,\eta)=-\frac{1}{8}(\xi-1)(\xi+1)^2(\eta+1), && N_{10}(\xi,\eta)=\frac{1}{8}(\xi-1)(\eta+1)\left(\xi^2+\xi+\eta^2-\eta-2\right)\\
&N_{11}(\xi,\eta)=-\frac{1}{8}(\xi-1)(\eta-1)(\eta+1)^2, && N_{12}(\xi,\eta)=-\frac{1}{8}(\xi-1)^2(\xi+1)(\eta+1)
\end{aligned} \tag{6.79}$$

6.7 Three-Dimensional Spline Interpolation Polynomials

When interpolation domain is three-dimensional and does not correspond to a regular shape, it can be divided into simpler three-dimensional shapes such as tetrahedral and hexahedral shapes or geometries as shown in Fig. 6.19. Then, three-dimensional global interpolation function can be created by defining piecewise (spline) interpolation functions over these simple shapes so as to satisfy continuity. In this section, trivariate (three-variable) piecewise interpolation functions for tetrahedral and hexahedral geometries will be obtained.

6.7.1 Trivariate Interpolation Polynomial for Tetrahedral Geometry

Tetrahedral geometry can be shown as in Fig. 6.20. Three-variable interpolation polynomial for this geometry can be expressed in classical form as below:

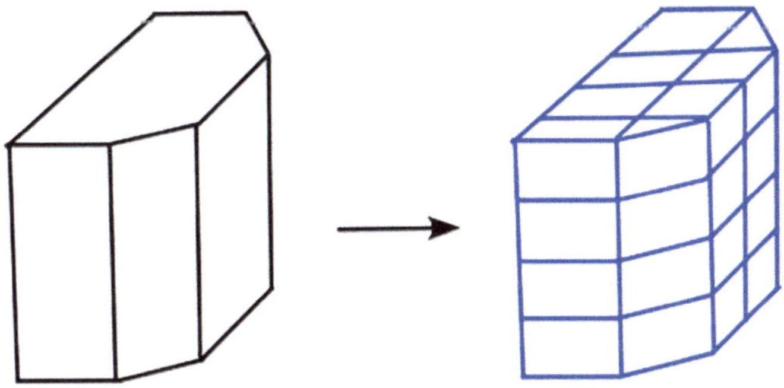

Fig. 6.19 Division of a complex three-dimensional domain into simpler three-dimensional basic shapes

Fig. 6.20 Tetrahedral geometry

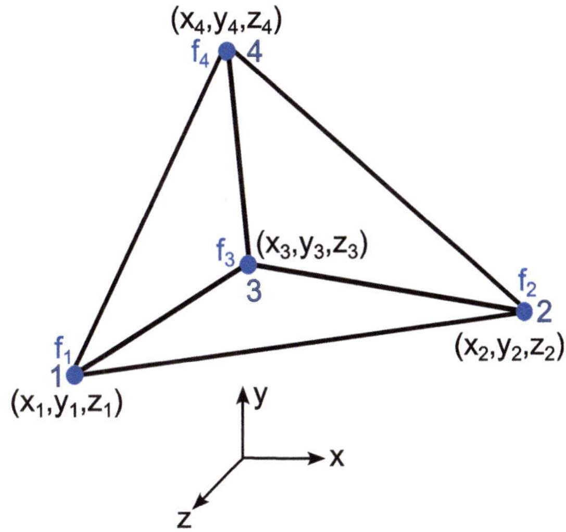

$$f(x,y,z) = c_1 + c_2x + c_3y + c_4z = \begin{bmatrix} 1 & x & y & z \end{bmatrix} \begin{Bmatrix} c_1 \\ c_2 \\ c_3 \\ c_4 \end{Bmatrix} \tag{6.80}$$

When Eq. 6.80 is applied to the vertices of the tetrahedral geometry, following equations are obtained.

$$\underbrace{\begin{Bmatrix} f_1 \\ f_2 \\ f_3 \\ f_4 \end{Bmatrix}}_{F} = \underbrace{\begin{bmatrix} 1 & x_1 & y_1 & z_1 \\ 1 & x_2 & y_2 & z_2 \\ 1 & x_3 & y_3 & z_3 \\ 1 & x_4 & y_4 & z_4 \end{bmatrix}}_{X} \underbrace{\begin{Bmatrix} c_1 \\ c_2 \\ c_3 \\ c_4 \end{Bmatrix}}_{C} \tag{6.81}$$

From Eq. 6.81, solution of unknown constants $C = c_1, c_2, c_3, c_4$ can be expressed as $C = X^{-1}(F)$. Substituting $C = X^{-1}(F)$ solution into Eq. 6.80 yields the interpolation function in Lagrange form as below:

$$f(x, y, z) = \underbrace{\begin{bmatrix} 1 & x & y & z \end{bmatrix} X^{-1}}_{N = \begin{bmatrix} N_1 & N_2 & N_3 & N_4 \end{bmatrix}} F \tag{6.82}$$

In Eq. 6.82, $N = N_1, N_2, N_3, N_4$ denote Lagrange or Shape functions. Interpolation function in Eq. 6.82 can be expressed more explicitly in terms of Lagrange functions as below:

$$f(x, y, z) = N_1(x, y, z) f_1 + N_2(x, y, z) f_2 + N_3(x, y, z) f_3 + N_4(x, y, z) f_4 \tag{6.83}$$

where N_1, N_2, N_3, N_4 are defined as following:

$$\begin{aligned} N_1(x, y, z) &= \frac{1}{6V}[p_1 + q_1 x + r_1 y + s_1 z], & N_2(x, y, z) &= \frac{1}{6V}[p_2 + q_2 x + r_2 y + s_2 z] \\ N_3(x, y, z) &= \frac{1}{6V}[p_3 + q_3 x + r_3 y + s_3 z], & N_4(x, y, z) &= \frac{1}{6V}[p_4 + q_4 x + r_4 y + s_4 z] \end{aligned} \tag{6.84}$$

$$p_1 = \begin{vmatrix} x_2 & y_2 & z_2 \\ x_3 & y_3 & z_3 \\ x_4 & y_4 & z_4 \end{vmatrix}, \; p_2 = \begin{vmatrix} x_3 & y_3 & z_3 \\ x_4 & y_4 & z_4 \\ x_1 & y_1 & z_1 \end{vmatrix}, \; p_3 = \begin{vmatrix} x_4 & y_4 & z_4 \\ x_1 & y_1 & z_1 \\ x_2 & y_2 & z_2 \end{vmatrix}, \; p_4 = \begin{vmatrix} x_1 & y_1 & z_1 \\ x_2 & y_2 & z_2 \\ x_3 & y_3 & z_3 \end{vmatrix} \tag{6.85a}$$

$$q_1 = -\begin{vmatrix} 1 & y_2 & z_2 \\ 1 & y_3 & z_3 \\ 1 & y_4 & z_4 \end{vmatrix}, \; q_2 = -\begin{vmatrix} 1 & y_3 & z_3 \\ 1 & y_4 & z_4 \\ 1 & y_1 & z_1 \end{vmatrix}, \; q_3 = -\begin{vmatrix} 1 & y_4 & z_4 \\ 1 & y_1 & z_1 \\ 1 & y_2 & z_2 \end{vmatrix}, \; q_4 = -\begin{vmatrix} 1 & y_1 & z_1 \\ 1 & y_2 & z_2 \\ 1 & y_3 & z_3 \end{vmatrix} \tag{6.85b}$$

$$r_1 = -\begin{vmatrix} y_2 & 1 & z_2 \\ y_3 & 1 & z_3 \\ y_4 & 1 & z_4 \end{vmatrix}, \; r_2 = -\begin{vmatrix} y_3 & 1 & z_3 \\ y_4 & 1 & z_4 \\ y_1 & 1 & z_1 \end{vmatrix}, \; r_3 = -\begin{vmatrix} y_4 & 1 & z_4 \\ y_1 & 1 & z_1 \\ y_2 & 1 & z_2 \end{vmatrix}, \; r_4 = -\begin{vmatrix} y_1 & 1 & z_1 \\ y_2 & 1 & z_2 \\ y_3 & 1 & z_3 \end{vmatrix} \tag{6.85c}$$

$$s_1 = -\begin{vmatrix} y_2 & z_2 & 1 \\ y_3 & z_3 & 1 \\ y_4 & z_4 & 1 \end{vmatrix}, \; s_2 = -\begin{vmatrix} y_3 & z_3 & 1 \\ y_4 & z_4 & 1 \\ y_1 & z_1 & 1 \end{vmatrix}, \; s_3 = -\begin{vmatrix} y_4 & z_4 & 1 \\ y_1 & z_1 & 1 \\ y_2 & z_2 & 1 \end{vmatrix}, \; s_4 = -\begin{vmatrix} y_1 & z_1 & 1 \\ y_2 & z_2 & 1 \\ y_3 & z_3 & 1 \end{vmatrix} \tag{6.85d}$$

In Eq. 6.84, V indicates the volume of the tetrahedral geometry which can be calculated using the following determinant

$$V = \frac{1}{6}\begin{vmatrix} 1 & x_1 & y_1 & z_1 \\ 1 & x_2 & y_2 & z_2 \\ 1 & x_3 & y_3 & z_3 \\ 1 & x_4 & y_4 & z_4 \end{vmatrix} \tag{6.86}$$

6.7.2 *Trivariate Interpolation Polynomial for Hexahedral Geometry*

A general hexahedral geometry can be shown as in Fig. 6.21. Coordinates of the vertices for this geometry can be defined as in Table 6.11.

Three-variable interpolation polynomial for the general hexahedral geometry in Fig. 6.21 can be written in classical form as below:

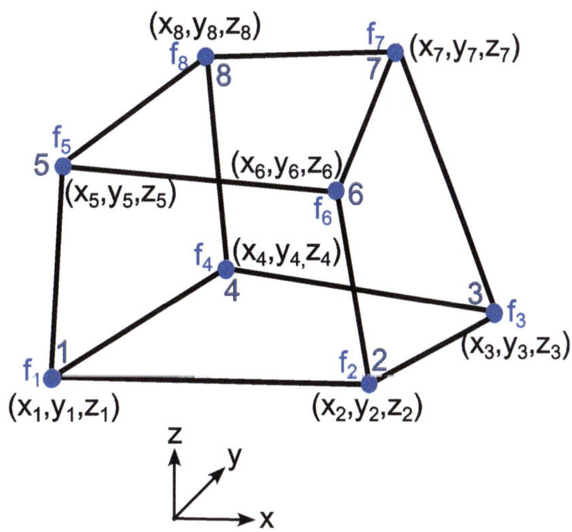

Fig. 6.21 General hexahedral geometry

Table 6.11 Coordinates of hexahedron at vertices

Node	x	y	z	$f(x,y,z)$
1	x_1	y_1	z_1	f_1
2	x_2	y_2	z_2	f_2
3	x_3	y_3	z_3	f_3
4	x_4	y_4	z_4	f_4
5	x_5	y_5	z_5	f_5
6	x_6	y_6	z_6	f_6
7	x_7	y_7	z_7	f_7
8	x_8	y_8	z_8	f_8

$$f(x,y,z) = c_1 + c_2x + c_3y + c_4z + c_5xy + c_6yz + c_7zx + c_8xyz$$
$$= \begin{bmatrix} 1 & x & y & z & xy & yz & zx & xyz \end{bmatrix} \begin{Bmatrix} c_1 \\ c_2 \\ \dots \\ c_8 \end{Bmatrix} \tag{6.87}$$

Function values at vertices of the hexahedral domain can be expressed using Eq. 6.87 as following:

$$\underbrace{\begin{Bmatrix} f_1 \\ f_2 \\ f_3 \\ f_4 \\ f_5 \\ f_6 \\ f_7 \\ f_8 \end{Bmatrix}}_{F} = \underbrace{\begin{bmatrix} 1 & x_1 & y_1 & z_1 & x_1y_1 & y_1z_1 & z_1x_1 & x_1y_1z_1 \\ 1 & x_2 & y_2 & z_2 & x_2y_2 & y_2z_2 & z_2x_2 & x_2y_2z_2 \\ 1 & x_3 & y_3 & z_3 & x_3y_3 & y_3z_3 & z_3x_3 & x_3y_3z_3 \\ 1 & x_4 & y_4 & z_4 & x_4y_4 & y_4z_4 & z_4x_4 & x_4y_4z_4 \\ 1 & x_5 & y_5 & z_5 & x_5y_5 & y_5z_5 & z_5x_5 & x_5y_5z_5 \\ 1 & x_6 & y_6 & z_6 & x_6y_6 & y_6z_6 & z_6x_6 & x_6y_6z_6 \\ 1 & x_7 & y_7 & z_7 & x_7y_7 & y_7z_7 & z_7x_7 & x_7y_7z_7 \\ 1 & x_8 & y_8 & z_8 & x_8y_8 & y_8z_8 & z_8x_8 & x_8y_8z_8 \end{bmatrix}}_{X} \underbrace{\begin{Bmatrix} c_1 \\ c_2 \\ c_3 \\ c_4 \\ c_5 \\ c_6 \\ c_7 \\ c_8 \end{Bmatrix}}_{C} \tag{6.88}$$

From Eq. 6.88, solution of unknown constants $C = c_1, c_2, \ldots, c_8$ can be expressed as $C = X^{-1}(F)$. Substituting $C = X^{-1}(F)$ solution into Eq. 6.87 yields the interpolation function in Lagrange form as below:

$$f(x,y,z) = \underbrace{[[1 \quad x \quad y \quad z \quad xy \quad yz \quad zx \quad xyz]]X^{-1}}_{N = \begin{bmatrix} N_1 & N_2 & \cdots & N_8 \end{bmatrix}} F \tag{6.89}$$

In Eq. 6.89, $N = N_1, N_2, \ldots, N_8$ correspond to Lagrange or Shape functions. Interpolation function in Eq. 6.89 can be expressed more explicitly in terms of Lagrange functions as below:

$$\begin{aligned} f(x,y,z) &= N_1(x,y,z)f_1 + N_2(x,y,z)f_2 + N_3(x,y,z)f_3 + N_4(x,y,z)f_4 \\ &+ N_5(x,y,z)f_5 + N_6(x,y,z)f_6 + N_7(x,y,z)f_7 + N_8(x,y,z)f_8 \end{aligned} \tag{6.90}$$

Special Case: Cube Geometry

Hexahedral geometry is sometimes converted into an ideal cube geometry in the solution of engineering problems using mapping concept. A special cube geometry that can be used for this purpose is shown in Fig. 6.22. This cube geometry can be easily defined with natural coordinate system using ξ and η variables assuming the coordinate system is placed at the center. Function values at the vertices of cube geometry can be given in terms of natural coordinates as in Table 6.12.

Interpolation polynomial for the special cube in Fig. 6.22 can be written in classical form as following:

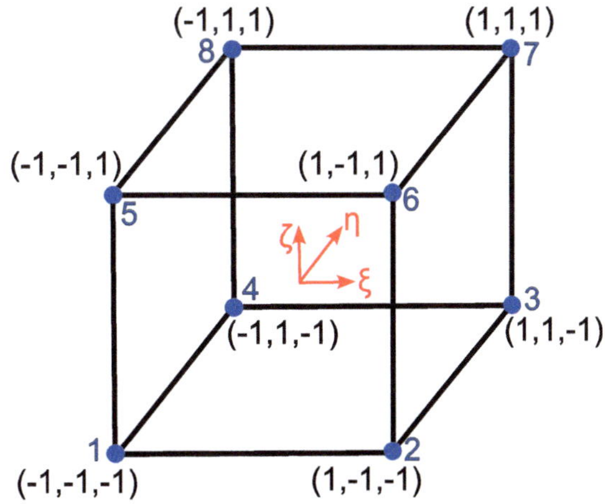

Fig. 6.22 Special cube geometry

Table 6.12 Function values at vertices of cube geometry in terms of natural coordinates ξ and η

Node	ξ	η	ζ	$f(\xi,\eta,\zeta)$
1	−1	−1	−1	f_1
2	1	−1	−1	f_2
3	1	1	−1	f_3
4	−1	1	−1	f_4
5	−1	−1	1	f_5
6	1	−1	1	f_6
7	1	1	1	f_7
8	−1	1	1	f_8

$$f(\xi,\eta,\zeta)=c_1+c_2\xi+c_3\eta+c_4\zeta+c_5\xi\eta+c_6\eta\zeta+c_7\zeta\xi+c_8\xi\eta\zeta \tag{6.91}$$

Substitution of $C= c_1,\ldots, c_8$ solutions in Eq. 6.91 yields the interpolation function in Lagrange form as below:

$$\begin{aligned} f(\xi,\eta,\zeta) &= N_1(\xi,\eta,\zeta)f_1+N_2(\xi,\eta,\zeta)f_2+N_3(\xi,\eta,\zeta)f_3+N_4(\xi,\eta,\zeta)f_4 \\ &+N_5(\xi,\eta,\zeta)f_5+N_6(\xi,\eta,\zeta)f_6+N_7(\xi,\eta,\zeta)f_7+N_8(\xi,\eta,\zeta)f_8 \end{aligned} \tag{6.92}$$

where $N_1,\ldots, N_8$ indicate Lagrange or shape functions and are expressed as following:

$$\begin{aligned} N_1(\xi,\eta,\zeta)&=\frac{(1-\xi)(1-\eta)(1-\zeta)}{8}, & N_2(\xi,\eta,\zeta)&=\frac{(1+\xi)(1-\eta)(1-\zeta)}{8} \\ N_3(\xi,\eta,\zeta)&=\frac{(1+\xi)(1+\eta)(1-\zeta)}{8}, & N_4(\xi,\eta,\zeta)&=\frac{(1-\xi)(1+\eta)(1-\zeta)}{8} \\ N_5(\xi,\eta,\zeta)&=\frac{(1-\xi)(1-\eta)(1+\zeta)}{8}, & N_6(\xi,\eta,\zeta)&=\frac{(1+\xi)(1-\eta)(1+\zeta)}{8} \\ N_7(\xi,\eta,\zeta)&=\frac{(1+\xi)(1+\eta)(1+\zeta)}{8}, & N_8(\xi,\eta,\zeta)&=\frac{(1-\xi)(1+\eta)(1+\zeta)}{8} \end{aligned} \tag{6.93}$$

6.8 Summary

- Prediction of an intermediate value in a data set is called as interpolation.
- Interpolation is especially desired in applications where generating data is difficult, expensive, or time consuming.
- In interpolation, an intermediate value is predicted by passing a polynomial function through the data points.
- Two types of polynomial function are used for interpolation:
  - Global interpolation polynomial: A higher order polynomial (n-th order polynomial for n+1 data points) in the form of $f(x) = c_1 + c_2x + c_3x^2 + c_4x^3 + \ldots c_{n+1}x^n$ is passed through the data points to obtain a global interpolation function.
  - Spline interpolation polynomial: Data points are divided into small groups and lower order polynomial functions referred to as spline interpolation polynomials are generated using each group of data. Linear, parabolic, and cubic polynomials are commonly used as spline interpolation functions.
- If higher order global interpolation polynomial $f(x) = c_1 + c_2x + c_3x^2 + c_4x^3 + \ldots c_{n+1}x^n$ is passed through the data points, data points should not have uniform distribution for the accuracy of the interpolation.
- Non-uniform data points can be generated using roots of the Chebyshev polynomials.
- Calculation of polynomial coefficients in global interpolation polynomial $f(x) = c_1 + c_2x + c_3x^2 + c_4x^3 + \ldots c_{n+1}x^n$ requires solution of a set of equations which can be difficult or erroneous for higher order interpolation polynomials.
- Global interpolation polynomials can be directly obtained with Newton Divided Difference and Lagrange methods as following:
  Newton Divided Difference:

$$f(x) = f_1 + \Delta f_1\alpha + \frac{\Delta^2 f_1}{2!}\alpha(\alpha-1) + \frac{\Delta^3 f_1}{3!}\alpha(\alpha-1)(\alpha-2) + \ldots$$
$$\ldots + \frac{\Delta^{n+1} f_1}{(n+1)!h^{n+1}}\alpha(\alpha-1)(\alpha-2)\ldots(\alpha-n) \qquad \alpha = \frac{(x-x_1)}{h}$$

  Lagrange method: $f(x) = L_1(x)f_1 + L_2(x)f_2 + L_3(x)f_3 + \ldots + L_{n+1}(x)f_{n+1}$ where Lagrange functions $L_i(x)$ is expressed as below:

$$L_i(x) = \frac{(x-x_1)(x-x_2)\ldots(x-x_{i-1})(x-x_{i+1})\ldots(x-x_n)}{(x_i-x_1)(x_i-x_2)\ldots(x_i-x_{i-1})(x_i-x_{i+1})\ldots(x_i-x_n)} (i = 1, 2, \ldots, n+1)$$

- When both function and derivative values are available simultaneously, global interpolation polynomial can be created using Hermite interpolation method as following ($L_i(x)$: Lagrange function).

$$f(x)=\sum_{i=1}^{n+1}\left[1-2(x-x_i)\frac{dL_i(x_i)}{dx}\right][L_i(x)]^2 f_i+\sum_{i=1}^{n+1}(x-x_i)[L_i(x)]^2\frac{df_i}{dx}$$

- When there are a large number of data points, data points can be divided into smaller groups and spline interpolation polynomials can be created in these groups. This approach prevents oscillations and increases accuracy of the interpolation function.
- Linear, parabolic, and cubic spline functions are commonly used for spline interpolation.
- In linear spline method, adjacent points are connected with linear polynomial functions as following:

$$f^{(1)}(x)=\left(\frac{x-x_2}{x_1-x_2}\right)f_1+\left(\frac{x-x_1}{x_2-x_1}\right)f_2 \quad (x_1\le x\le x_2)$$
$$\ldots$$
$$f^{(n-1)}(x)=\left(\frac{x-x_n}{x_{n-1}-x_n}\right)f_{n-1}+\left(\frac{x-x_{n-1}}{x_n-x_{n-1}}\right)f_n \quad (x_{n-1}\le x\le x_n)$$

- In parabolic spline method, adjacent points are connected with parabolic polynomial functions as following (*a*, *b*, *c* coefficients are determined using function values, continuity conditions and edge condition)

$$f^{(1)}(x)=a_1+b_1x+c_1x^2 \quad (x_1\le x\le x_2)$$
$$\ldots$$
$$f^{(n-1)}(x)=a_{n-1}+b_{n-1}x+c_{n-1}x^2 \quad (x_{n-1}\le x\le x_n)$$

- In cubic spline method, adjacent points are connected with cubic polynomial functions as following (*a*, *b*, *c*, *d* coefficients are determined using function values, continuity conditions and edge conditions)

$$f^{(1)}(x)=a_1+b_1x+c_1x^2+d_1x^3 \quad (x_1\le x\le x_2)$$
$$\ldots$$
$$f^{(n-1)}(x)=a_{n-1}+b_{n-1}x+c_{n-1}x^2+d_{n-1}x^3 \quad (x_{n-1}\le x\le x_n)$$

- In multivariable case, spline interpolation functions can be generated by dividing the interpolation domain into simpler sub-domains.
- Spline interpolation is widely used in Finite Element method in the prediction of displacement, strain, stress, pressure, velocity, etc. in elements (sub-domains).
- In two-variable case, sub-domain can be of triangular or quadrilateral shape.
- In three-variable case, sub-domain can be of tetrahedral or hexahedral shape.
- Lagrange method can be used to create spline interpolation functions in multi-variable cases.

- In two-variable case, spline interpolation polynomial can be expressed on a triangular sub-domain as $f(x, y) = N_1(x, y)f_1 + N_2(x, y)f_2 + N_3(x, y)f_3$.

  $N_1(x, y)$, $N_2(x, y)$, $N_3(x, y)$ functions above are given as below (A: area of triangular sub-domain)

$$N_1(x, y) = \frac{1}{2A}[(x_2y_3 - x_3y_2) + (y_2 - y_3)x + (x_3 - x_2)y]$$
$$N_2(x, y) = \frac{1}{2A}[(x_3y_1 - x_1y_3) + (y_3 - y_1)x + (x_1 - x_3)y]$$
$$N_3(x, y) = \frac{1}{2A}[(x_1y_2 - x_2y_1) + (y_1 - y_2)x + (x_2 - x_1)y]$$

- In two-variable case, spline interpolation polynomial can be expressed on a rectangular sub-domain as $N_1(x, y)f_1 + N_2(x, y)f_2 + N_3(x, y)f_3 + N_4(x, y)f_4$.

  $N_1(x, y)$, $N_2(x, y)$, $N_3(x, y)$, $N_4(x, y)$ functions above are given as below:

$$N_1(x, y) = \frac{(x - x_2)(y - y_4)}{(x_1 - x_2)(y_1 - y_4)}, \quad N_2(x, y) = \frac{(x - x_1)(y - y_3)}{(x_2 - x_1)(y_2 - y_3)}$$
$$N_3(x, y) = \frac{(x - x_4)(y - y_2)}{(x_3 - x_4)(y_3 - y_2)}, \quad N_4(x, y) = \frac{(x - x_3)(y - y_1)}{(x_4 - x_3)(y_4 - y_1)}$$

- In three-variable case, spline interpolation polynomial can be expressed on a tetrahedral sub-domain as following:

$$f(x, y, z) = N_1(x, y, z)f_1 + N_2(x, y, z)f_2 + N_3(x, y, z)f_3 + N_4(x, y, z)f_4.$$

  N_1, N_2, N_3, N_4 functions above are given as below (V: volume of tetrahedral sub-domain).

$$N_1(x, y, z) = \frac{1}{6V}[p_1 + q_1x + r_1y + s_1z], \quad N_2(x, y, z) = \frac{1}{6V}[p_2 + q_2x + r_2y + s_2z]$$
$$N_3(x, y, z) = \frac{1}{6V}[p_3 + q_3x + r_3y + s_3z], \quad N_4(x, y, z) = \frac{1}{6V}[p_4 + q_4x + r_4y + s_4z]$$

  p, q, r, s, and V values are obtained using Eqs. 6.85a and 6.86.

- In three-variable case, spline interpolation polynomial can be expressed on a hexahedral sub-domain (ideal cube) as below:

$$\begin{aligned} f(\xi, \eta, \zeta) = {} & N_1(\xi, \eta, \zeta)f_1 + N_2(\xi, \eta, \zeta)f_2 + N_3(\xi, \eta, \zeta)f_3 + N_4(\xi, \eta, \zeta)f_4 \\ & + N_5(\xi, \eta, \zeta)f_5 + N_6(\xi, \eta, \zeta)f_6 + N_7(\xi, \eta, \zeta)f_7 + N_8(\xi, \eta, \zeta)f_8. \end{aligned}$$

  $N_1, \ldots, N_8$ functions above are given as following:

$$N_1(\xi,\eta,\zeta)=\frac{(1-\xi)(1-\eta)(1-\zeta)}{8},\quad N_2(\xi,\eta,\zeta)=\frac{(1+\xi)(1-\eta)(1-\zeta)}{8}$$
$$N_3(\xi,\eta,\zeta)=\frac{(1+\xi)(1+\eta)(1-\zeta)}{8},\quad N_4(\xi,\eta,\zeta)=\frac{(1-\xi)(1+\eta)(1-\zeta)}{8}$$
$$N_5(\xi,\eta,\zeta)=\frac{(1-\xi)(1-\eta)(1+\zeta)}{8},\quad N_6(\xi,\eta,\zeta)=\frac{(1+\xi)(1-\eta)(1+\zeta)}{8}$$
$$N_7(\xi,\eta,\zeta)=\frac{(1+\xi)(1+\eta)(1+\zeta)}{8},\quad N_8(\xi,\eta,\zeta)=\frac{(1-\xi)(1+\eta)(1+\zeta)}{8}$$

Appendix: MATLAB Programs

MATLAB Program 6.1: Interpolation with Newton's Divided Forward Difference Method

```
function Interpolation_NewtonDividedDifference

% NOTE: THIS PROGRAM CALCULATES f(x) VALUE AT AN INTERMEDIATE POINT
% x0 USING INTERPOLATION METHOD BASED ON NEWTON DIVIDED DIFFERENCE

x=[0  1  3  5 ];  % x Points
f=[1  2  6  12];  % f(x) Values

x0=2 ; % Intermediate Point

% ==========================================================================
n=length(x) ; % Number of x points

% Generation of Newton Divided Difference Table: Df(j=point,i=order)
for j=1:n-1
    Df(j,1)=(f(j+1)-f(j))/(x(j+1)-x(j));
end

for i=1:n-1
    for j=1:n-(i+1)
        dx(j,i+1)=x(i+j+1)-x(j) ;
        Df(j,i+1)=(Df(j+1,i)-Df(j,i))/dx(j,i+1);
    end
end

% Function value at x0 point: f0
f0=f(1) ;
for i=1:n-1
    mult=1;
    for j=1:i
        mult=mult*(x0-x(j));
    end
    f0=f0+Df(1,i)*mult ;
end

fprintf('x0=%f, f0=%f \n',x0,f0) ;
```

MATLAB Program 6.2: Lagrange Interpolation Method

```
function Lagrange_Interpolation

% NOTE: THIS PROGRAM CALCULATES f(x) VALUE AT AN INTERMEDIATE POINT
% x0 USING LAGRANGE INTERPOLATION METHOD

x=[0   3   6   9 ];  % x Points
f=[1   5   18 36];   % f(x) Values

x0=4 ; % Intermediate Point

% ---------------------------------------------------------------------
n=length(x); % Number of x points

% Function value at x0 point: f0
f0=0 ;
for i=1:n
    mult=1 ;
    for j=1:n
        if(i~=j)
            mult=mult*(x0-x(j))/(x(i)-x(j));
        end
    end

    f0=f0+mult*f(i) ;
end

fprintf('x0=%f, f0=%f \n',x0,f0) ;
```

MATLAB Program 6.3: Interpolation with Linear Spline Functions

```
function LinearSpline

% NOTE: THIS PROGRAM CALCULATES f(x) VALUE AT AN INTERMEDIATE POINT
% x0 USING LINEAR SPLINE INTERPOLATION

x=[0   2   4   6] ;  % x Points
f=[1   3   6   10] ; % f(x) Values

x0=5 ; % Intermediate Point

% Determination of the interval including x0 point
i=1 ;
while x0>x(i)
    q=i ;
    i=i+1 ;
end

% Function value at x0 point: f0
f0=(x0-x(q+1))*f(q)/(x(q)-x(q+1))+(x0-x(q))*f(q+1)/(x(q+1)-x(q)) ;

fprintf('x0=%f, f0=%f \n',x0,f0) ;
```

MATLAB Program 6.4: Interpolation with Parabolic Spline Functions

```
function ParabolicSpline

% NOTE: THIS PROGRAM CALCULATES f(x) VALUE AT AN INTERMEDIATE POINT
% x0 USING PARABOLIC SPLINE INTERPOLATION

x=[0  2  4  6] ;  % x Points
f=[1  3  6  10] ; % f(x) Values

x0=3 ; % Intermediate Point
% ----------------------------------------
n=length(x) ; % Number of points

A=zeros(3*(n-1),3*(n-1)) ;
B=zeros(3*(n-1),1) ;

% Function values at x points: 2*(n-1) equations
for i=1:n-1
    A(2*i-1,3*i-2)=1  ;  A(2*i-1,3*i-1)=x(i) ;  A(2*i-1,3*i)=x(i)^2 ;
    A(2*i,3*i-2)=1    ;  A(2*i,3*i-1)=x(i+1) ;  A(2*i,3*i)=x(i+1)^2 ;
    B(2*i-1)=f(i) ;
    B(2*i)=f(i+1) ;
end

% First derivative continuities at shared points: n-2 equations
m=2*(n-1);
for i=1:n-2
    A(m+i,3*i-2)=0     ;   A(m+i,3*i-1)=1     ; A(m+i,3*i)=2*x(i+1) ;
    A(m+i,3*(i+1)-2)=0 ;  A(m+i,3*(i+1)-1)=-1;  A(m+i,3*(i+1))=-2*x(i+1)
;
    B(m+i)=0 ;
end

% End condition: Second derivative is zero at first point:  1 equation
A(3*(n-1),3)=2 ;
B(3*(n-1))=0 ;

% Solution of a,b,c coefficients
S=inv(A)*B ;
for i=1:n-1
    a(i)=S(3*i-2) ;
    b(i)=S(3*i-1) ;
    c(i)=S(3*i) ;
end

% Determination of the interval including x0 point
i=1 ;
while x0>=x(i)
    p=i ;
    i=i+1 ;
end

% Function value at x0 point: f0
f0=a(p)+b(p)*x0+c(p)*(x0^2) ;

fprintf('x0=%f, f0=%f \n',x0,f0) ;
```

MATLAB Program 6.5: Interpolation with Cubic Spline Functions

```
function CubicSpline

% NOTE: THIS PROGRAM CALCULATES f(x) VALUE AT AN INTERMEDIATE POINT
% x0 USING CUBIC SPLINE INTERPOLATION

x=[0  2  4  6] ;  % x Points
f=[1  3  6  10] ; % f(x) Values

x0=5 ; % Intermediate Point
% ----------------------------------------

n=length(x) ; % Number of points

A=zeros(4*(n-1),4*(n-1)) ;
B=zeros(4*(n-1),1) ;

% Function values at x points: 2*(n-1) equations
for i=1:n-1
    A(2*i-1,4*i-3)=1   ;  A(2*i-1,4*i-2)=x(i) ;  A(2*i-1,4*i-1)=x(i)^2 ;
A(2*i-1,4*i)=x(i)^3 ;
    A(2*i,4*i-3)=1     ;  A(2*i,4*i-2)=x(i+1) ;  A(2*i,4*i-1)=x(i+1)^2 ;
A(2*i,4*i)=x(i+1)^3 ;
    B(2*i-1)=f(i) ;
    B(2*i)=f(i+1) ;
end

% First derivative continuities at shared points: n-2 equations
m=2*(n-1);
for i=1:n-2
    A(m+i,4*i-3)=0 ;        A(m+i,4*i-2)=1;        A(m+i,4*i-1)=2*x(i+1)
;  A(m+i,4*i)=3*x(i+1)^2 ;
    A(m+i,4*(i+1)-3)=0 ;   A(m+i,4*(i+1)-2)=-1;  A(m+i,4*(i+1)-1)=-
2*x(i+1) ;  A(m+i,4*(i+1))=-3*x(i+1)^2 ;
    B(m+i)=0 ;
end

% Second derivative continuities at shared points: n-2 equations
m=2*(n-1)+n-2;
for i=1:n-2
    A(m+i,4*i-3)=0      ;  A(m+i,4*i-2)=0     ;  A(m+i,4*i-1)=2       ;
A(m+i,4*i)=6*x(i+1)       ;
    A(m+i,4*(i+1)-3)=0 ;  A(m+i,4*(i+1)-2)=0;  A(m+i,4*(i+1)-1)=-2 ;
A(m+i,4*(i+1))=-6*x(i+1) ;
    B(m+i)=0 ;
end

% End conditions: Second derivatives are zero at end points:  2 equations
A(4*n-5,1)=0     ;  A(4*n-5,2)=0     ;  A(4*n-5,3)=2     ;  A(4*n-
5,4)=6*x(1)      ;
A(4*n-4,4*n-7)=0 ;  A(4*n-4,4*n-6)=0 ;  A(4*n-4,4*n-5)=2 ;  A(4*n-4,4*n-
4)=6*x(n) ;
B(4*n-5)=0 ;
B(4*n-4)=0 ;

% Solution of a,b,c,d coefficients
S=inv(A)*B ;
for i=1:n-1
    a(i)=S(4*i-3) ;
    b(i)=S(4*i-2) ;
    c(i)=S(4*i-1) ;
    d(i)=S(4*i) ;
end

% Determination of the interval including x0 point
i=1 ;
while x0>=x(i)
    q=i ;
    i=i+1 ;
end

% Function value at x0 point: f0
f0=a(q)+b(q)*x0+c(q)*(x0^2)+d(q)*(x0^3) ;

fprintf('x0=%f, f0=%f \n',x0,f0) ;
```

Problems

Problem 6.1: Function values at different points are given as $(x_1,f_1) = (0,5)$, $(x_2 f_2) = (2,9)$, $(x_3,f_3) = (4,21)$. Predict the function value at $x = 3$ point using Newton Divided Difference method.

Problem 6.2: Function values at different points are given as $(x_1,f_1) = (0,5)$, $(x_2 f_2) = (2,9)$, $(x_3,f_3) = (4,21)$. Predict the function value at $x = 3$ point using Lagrange interpolation method.

Problem 6.3: Function values at different points are given as $(x_1,f_1) = (0,-1)$, $(x_2 f_2) = (2,3)$, $(x_3,f_3) = (4,15)$, $(x_4,f_4) = (6,35)$. Predict the function value at $x = 5$ point using Newton Divided Difference method.

Problem 6.4: Function values at different points are given as $(x_1,f_1) = (0,-1)$, $(x_2 f_2) = (2,3)$, $(x_3,f_3) = (4,15)$, $(x_4,f_4) = (6,35)$. Predict the function value at $x = 5$ point using Lagrange interpolation method.

Problem 6.5: Function values at different points are given as $(x_1,f_1) = (0,2)$, $(x_2 f_2) = (1,5)$, $(x_3,f_3) = (3,29)$, $(x_4,f_4) = (5,77)$. Predict the function value at $x = 4$ point using linear spline method.

Problem 6.6: Function values at different points are given as $(x_1,f_1) = (0,2)$, $(x_2 f_2) = (1,5)$, $(x_3,f_3) = (3,29)$, $(x_4,f_4) = (5,77)$. Predict the function value at $x = 4$ point using parabolic spline method.

Problem 6.7: Function values at different points are given as $(x_1,f_1) = (0,2)$, $(x_2 f_2) = (1,5)$, $(x_3,f_3) = (3,29)$, $(x_4,f_4) = (5,77)$. Predict the function value at $x = 4$ point using cubic spline method.

Problem 6.8: Construct Hermite interpolation polynomial using the data in Table 6.13 and predict the function value at $x0 = 3$.

Problem 6.9: Construct Hermite interpolation polynomial using the data in Table 6.14 and predict the function value at $x0 = 5$.

Table 6.13 Data to construct Hermite interpolation polynomial

x	$f(x)$	$f'(x) = df(x)/dx$
1	3	−2
4	8	2

Table 6.14 Data to construct Hermite interpolation polynomial

x	$f(x)$	$f'(x) = df(x)/dx$
2	3	−1
4	9	2
6	15	−2

Problem 6.10: Values of a bivariate (two-variable) $f(x, y)$ function are given at different points in a rectangular domain as in Fig. 6.23. Calculate $f(2, 1)$ value creating bivariate classical interpolation polynomial function.

Problem 6.11: Using the data points in Problem 6.10, calculate $f(2, 1)$ value creating bivariate Lagrange interpolation polynomial function.

Problem 6.12: Using the data points in Problem 6.10, calculate $f(2, 1)$ value creating bivariate interpolation polynomial with Newton Divided Difference method.

Problem 6.13: Values of a bivariate (two-variable) $f(x, y)$ function are given at different points in a rectangular domain as in Fig. 6.24. Calculate $f(4, 3)$ value creating bivariate classical interpolation polynomial function.

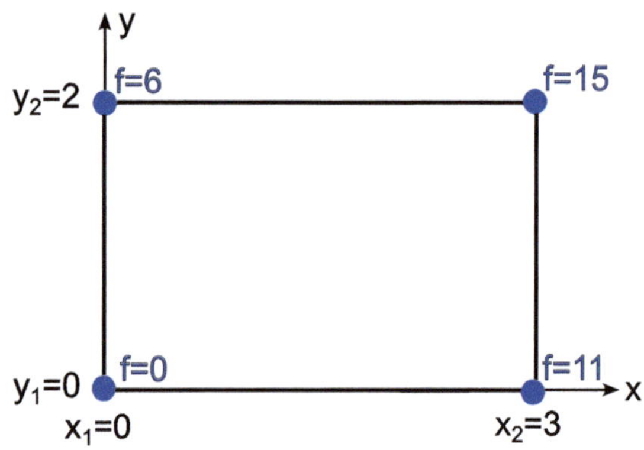

Fig. 6.23 Data points for two-dimensional interpolation

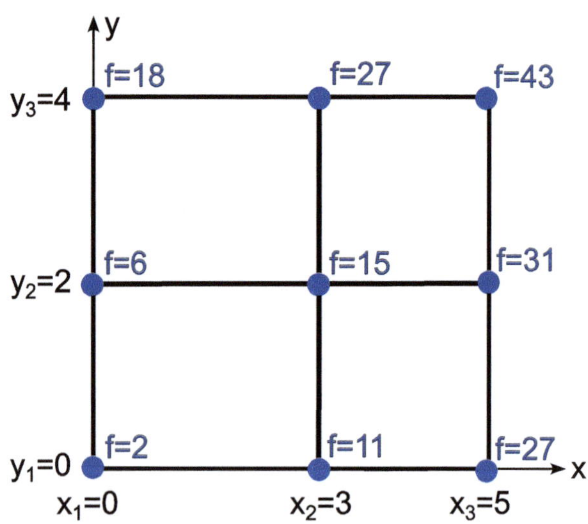

Fig. 6.24 Data points for two-dimensional interpolation

Problem 6.14: Using the data points in Problem 6.13, calculate $f(4,3)$ value creating bivariate Lagrange interpolation polynomial function.

Problem 6.15: Using the data points in Problem 6.13, calculate $f(4,3)$ value creating bivariate interpolation polynomial with Newton Divided Difference method.

References

1. Chapra, S. C., & Canale, R. P. (2020). *Numerical methods for engineers*. McGraw-Hill.
2. Bernatz, R. A. (2010). *Fourier series and numerical methods for partial differential equations*. John Wiley & Sons.
3. Everstine, G. C. (2010). *Numerical solution of partial differential equations* [Lecture notes, PDF]. George Washington University. https://www.academia.edu/31518755/Numerical_Solution_of_Partial_Differential_Equations
4. Epperson, J. F. (2013). *An introduction to numerical methods and analysis*. John Wiley & Sons.
5. Burden, R. L., & Faires, J. D. (2011). *Numerical analysis*. Cengage Learning.
6. Sert, C. (2006). *ME310 numerical methods* [Lecture notes, PDF]. Middle East Technical University. https://users.metu.edu.tr/csert/me310/me310_1_introduction.pdf
7. Grewal, B. S. (2019). *Numerical methods in engineering and science*. Mercury Learning and Information.
8. Sastry, S. S. (2012). *Introductory methods of numerical analysis*. PHI Learning.

Chapter 7
Curve Fitting (Regression)

7.1 Introduction

Regression is commonly known as curve fitting. More accurately, regression is called curve fitting in the univariate case and surface/hypersurface fitting in the multivariate case. However, in this section, curve fitting and surface fitting will be used interchangeably regardless of the number of variables. In regression, as in interpolation, a function that represents data points is obtained to estimate intermediate values [1–24]. Although regression and interpolation are similar, there are some differences between them regarding the reliability of the data. In interpolation, data is assumed to be accurate and free of errors such as measurement errors. Therefore, interpolation function is passed through the data points. However, in regression, data is implicitly assumed to include some uncontrollable errors such as measurement errors. Therefore, regression function is passed through the data points in a balanced way. Difference between interpolation and curve fitting (regression) is shown in Fig. 7.1 for different data distribution.

As seen from Figs. 7.1, 7.2, and 7.3 as the number of points increases, interpolation functions show oscillating character leading to the erroneous estimation of intermediate values. In curve fitting, fitting function has smooth (non-oscillatory) character and data points show a balanced distribution around this function. For example, data points are distributed around a line in Fig. 7.1b, around a parabola in Fig. 7.2b, and around a cubic polynomial in Fig. 7.3b.

In regression, the most critical thing is the selection of the function (i.e., mathematical model) to be fitted. This function should represent the trend of data very well. Low-order polynomial functions with some adjustable parameters or constants are often used in curve or surface fitting. Regression is divided into linear and nonlinear regression. In linear regression, regression funtion is linear in terms of model constants. However, in nonlinear regression, it is nonlinear in terms of model constants. Therefore, in linear regression, constants of the function are obtained with a solution of linear equation set, while in nonlinear regression, these constants are

H. Kurtaran, *Engineering Mathematics with MATLAB Applications: Vol. 1, Fundamentals*, https://doi.org/10.1007/978-3-031-81869-1_7

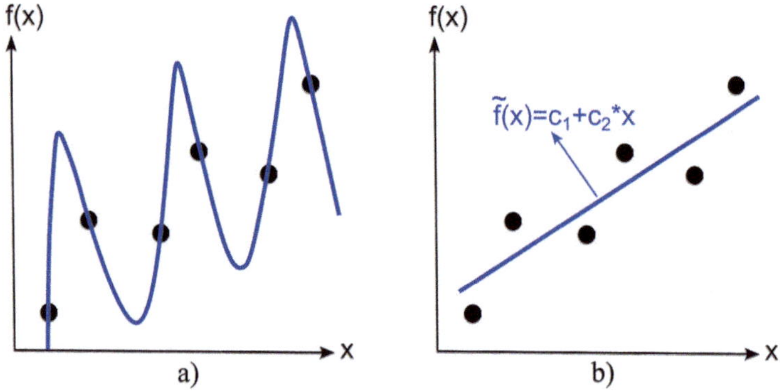

Fig. 7.1 For the same data points (**a**) interpolation, (**b**) curve fitting: Linear polynomial

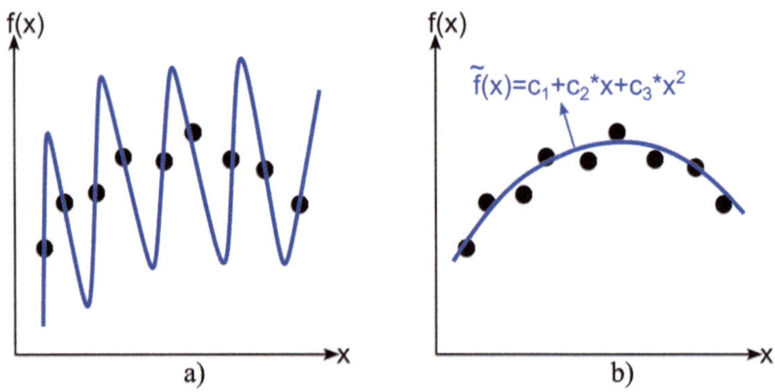

Fig. 7.2 For the same data points (**a**) interpolation, (**b**) curve fitting: Parabolic polynomial

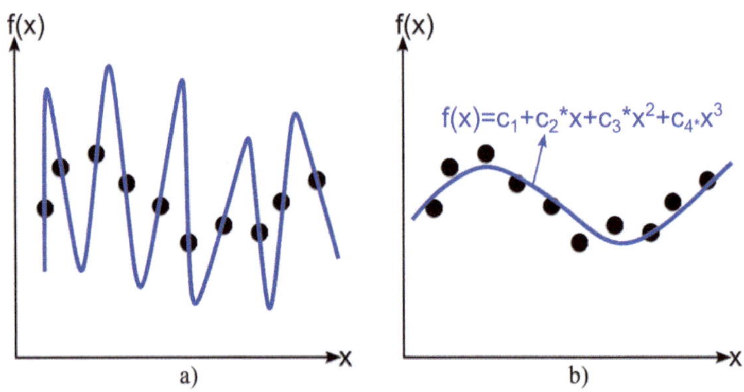

Fig. 7.3 For the same data points (**a**) interpolation, (**b**) curve fitting: Cubic polynomial

obtained with a solution of nonlinear equation set in iterative way. Nonlinear regression problems can sometimes be transformed into linear ones for easier calculation of model constants. In linear regression, simple models such as polynomial functions, Kriging function, and inverse distance approximation are used as regression function. In nonlinear regression, more complex and nonlinear models such as exponential functions and Artificial Neural Networks are used as regression function.

7.2 Linear Regression

In linear regression, low-order polynomials such as linear and parabolic functions are usually fitted to data points. Constants of these polynomials are determined with the solution of linear set of equations. These equations are obtained using Least Squares Method. Fitting of linear and parabolic polynomials can be explained as following:

7.2.1 Linear Polynomial Fitting

Assume that function values are known as $f_1, f_2, \ldots, f_n$ at $x_1, x_2, \ldots, x_n$ points as shown in Table 7.1. If, function values show a linear trend, linear polynomial function can be fitted to these values to estimate intermediate values. Linear polynomial function can be expressed as below:

$$\widetilde{f}(x) = c_1 + c_2 x \tag{7.1}$$

c_1 and c_2 constants in Eq. (7.1) serve as model parameters to be adjusted for best fit to the data using Least Squares method. In Least Squares method, sum of squares of error at data points are minimized to determine unknown model constants. Error value ε_i at a point is calculated as the difference between true and approximate function values. It is expressed as $\varepsilon_i = \widetilde{f}_i(x) - f_i(x)$, where $f_i(x)$ and $\widetilde{f}_i(x)$ denote true and approximate function value, respectively. When linear function is used for fitting, error at data points can be shown as in Table 7.1 and Fig. 7.4.

Table 7.1 Error at data points to fit linear polynomial model

Point	x	$f(x)$	$\widetilde{f}(x)$	$\varepsilon = \widetilde{f}(x) - f(x)$
1	x_1	f_1	$\widetilde{f}_1 = c_1 + c_2 x_1$	$\varepsilon_1 = c_1 + c_2 x_1 - f_1$
2	x_2	f_2	$\widetilde{f}_2 = c_1 + c_2 x_2$	$\varepsilon_2 = c_1 + c_2 x_2 - f_2$
3	x_3	f_3	$\widetilde{f}_3 = c_1 + c_2 x_3$	$\varepsilon_3 = c_1 + c_2 x_3 - f_3$
…	…	…	…	…
n	x_n	f_n	$\widetilde{f}_n = c_1 + c_2 x_n$	$\varepsilon_n = c_1 + c_2 x_n - f_n$

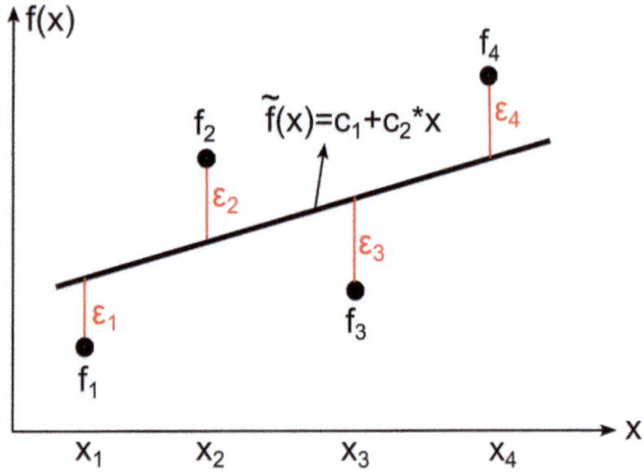

Fig. 7.4 Linear polynomial fitting

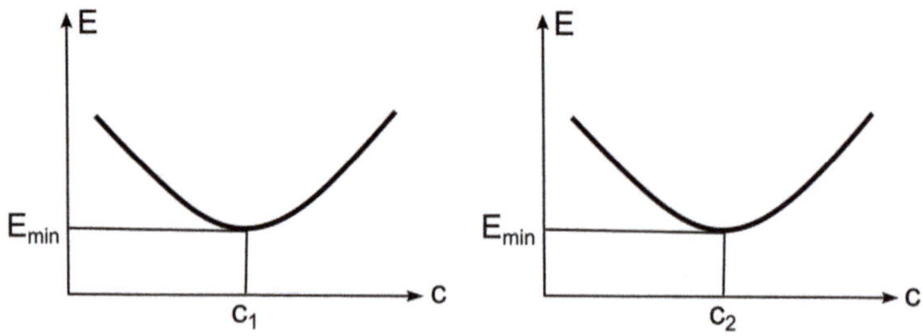

Fig. 7.5 Finding model constants by minimizing total error

Sum of squares of errors at points (shortly we can call it Total Error, E) can be written in terms of model constants as below:

$$\begin{aligned} E(c_1, c_2) &= \varepsilon_1^2 + \varepsilon_2^2 + \dots + \varepsilon_n^2 = \left(\tilde{f}_1 - f_1\right)^2 + \left(\tilde{f}_2 - f_2\right)^2 + \dots + \left(\tilde{f}_n - f_n\right)^2 \\ &= (c_1 + c_2 x_1 - f_1)^2 + (c_1 + c_2 x_2 - f_2)^2 + \dots + (c_1 + c_2 x_n - f_n)^2 \\ &= \sum_{i=1}^{n} (c_1 + c_2 x_i - f_i)^2 \end{aligned} \tag{7.2}$$

Minimizing total error E, model constants c_i can be determined. Minimization of total error can alternatively be expressed as zeroing of derivative of total error with respect to model constants as shown in Fig. 7.5 and calculated as following:

$$E(c_1, c_2) = \sum_{i=1}^{n} c_1 + c_2 x_i - f_i)^2 \rightarrow \min \rightarrow \frac{\partial E}{\partial c_1} = 0, \frac{\partial E}{\partial c_2} = 0 \tag{7.3}$$

Equation (7.3) can be implemented as below:

$$\frac{\partial E}{\partial c_1} = \sum_{i=1}^{n} 2(c_1 + c_2 x_i - f_i) = 0 \rightarrow \sum_{i=1}^{n} (c_1 + c_2 x_i - f_i) = 0$$

$$\frac{\partial E}{\partial c_2} = 2\sum_{i=1}^{n} (c_1 + c_2 x_i - f_i) x_i = 0 \rightarrow c_1 \sum_{i=1}^{n} x_i + c_2 \sum_{i=1}^{n} x_i^2 - \sum_{i=1}^{n} f_i x_i = 0$$

Above equations can be arranged in matrix form and solved using one of the linear methods in Chap. 4 as following:

$$\begin{bmatrix} n & \sum_{i=1}^{n} x_i \\ \sum_{i=1}^{n} x_i & \sum_{i=1}^{n} x_i^2 \end{bmatrix} \begin{Bmatrix} c_1 \\ c_2 \end{Bmatrix} = \begin{Bmatrix} \sum_{i=1}^{n} f_i \\ \sum_{i=1}^{n} f_i x_i \end{Bmatrix} \rightarrow c_1, c_2 \tag{7.4}$$

7.2.2 Parabolic Polynomial Fitting

Assume that function values are known as f_1, f_2, …, f_n at x_1, x_2, …, x_n points as shown in Table 7.2. If function values show a parabolic trend, parabolic polynomial function can be fitted to these values to estimate intermediate values. Parabolic polynomial function can be expressed as below:

$$\tilde{f}(x) = c_1 + c_2 x + c_3 x^2 \tag{7.5}$$

c_1, c_2, and c_3 constants in Eq. (7.5) serve as model parameters to be adjusted for best fit to the data points using Least Squares method. When parabolic function is used for fitting, error at points $\varepsilon_i = \tilde{f}_i(x) - f_i(x)$ can be shown as in Table 7.2 and Fig. 7.6.

Sum of squares of errors at points (shortly we can call it Total Error, E) can be written in terms of model constants as below:

Table 7.2 Error at data points to fit parabolic polynomial model

Point	x	$f(x)$	$\tilde{f}(x)$	$\varepsilon = \tilde{f}(x) - f(x)$
1	x_1	f_1	$\tilde{f}_1 = c_1 + c_2 x_1 + c_3 x_1^2$	$\varepsilon_1 = \tilde{f}_1 - f_1 = c_1 + c_2 x_1 + c_3 x_1^2 - f_1$
2	x_2	f_2	$\tilde{f}_2 = c_1 + c_2 x_2 + c_3 x_2^2$	$\varepsilon_2 = \tilde{f}_2 - f_2 = c_1 + c_2 x_2 + c_3 x_2^2 - f_2$
3	x_3	f_3	$\tilde{f}_3 = c_1 + c_2 x_3 + c_3 x_3^2$	$\varepsilon_3 = \tilde{f}_3 - f_3 = c_1 + c_2 x_3 + c_3 x_3^2 - f_3$
…	…	…	…	…
n	x_n	f_n	$\tilde{f}_n = c_1 + c_2 x_n + c_3 x_n^2$	$\varepsilon_n = \tilde{f}_n - f_n = c_1 + c_2 x_n + c_3 x_n^2 - f_n$

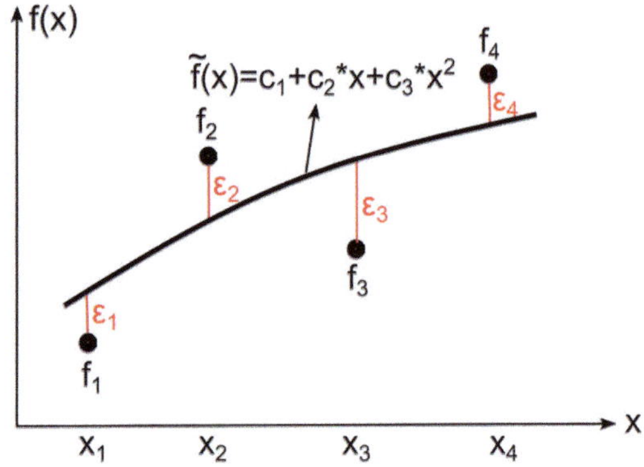

Fig. 7.6 Parabolic polynomial fitting

$$\begin{aligned} E(c_1, c_2, c_3) &= \left(c_1 + c_2x_1 + c_3x_1^2 - f_1\right)^2 + \left(c_1 + c_2x_2 + c_3x_3^2 - f_2\right)^2 \\ &+ \left(c_1 + c_2x_3 + c_3x_3^2 - f_3\right)^2 + \ldots \\ &+ \left(c_1 + c_2x_n + c_3x_n^2 - f_n\right)^2 = \sum_{i=1}^{n} \left(c_1 + c_2x_i + c_3x_i^2 - f_i\right)^2 \end{aligned} \quad (7.6)$$

Minimizing total error E, necessary equations to determine model constants c_i can be obtained as following:

$$\frac{\partial E}{\partial c_1} = 0 \rightarrow 2\sum_{i=1}^{n} \left(c_1 + c_2x_i + c_3x_i^2 - f_i\right) = 0 \rightarrow c_1 n + c_2\sum_{i=1}^{n} x_i + c_3\sum_{i=1}^{n} x_i^2 = \sum_{i=1}^{n} f_i$$

$$\frac{\partial E}{\partial c_2} = 0 \rightarrow 2\sum_{i=1}^{n} \left(c_1 + c_2x_i + c_3x_i^2 - f_i\right)x_i = 0 \rightarrow c_1\sum_{i=1}^{n} x_i + c_2\sum_{i=1}^{n} x_i^2 + c_3\sum_{i=1}^{n} x_i^3 = \sum_{i=1}^{n} f_i x_i$$

$$\frac{\partial E}{\partial c_3} = 0 \rightarrow 2\sum_{i=1}^{n} \left(c_1 + c_2x_i + c_3x_i^2 - f_i\right)x_i^2 = 0 \rightarrow c_1\sum_{i=1}^{n} x_i^2 + c_2\sum_{i=1}^{n} x_i^3 + c_3\sum_{i=1}^{n} x_i^4 = \sum_{i=1}^{n} f_i x_i^2$$

Above equations can be arranged in matrix form and solved using one of the linear methods in Chap. 4 as following:

$$\begin{bmatrix} n & \sum_{i=1}^{n} x_i & \sum_{i=1}^{n} x_i^2 \\ \sum_{i=1}^{n} x_i & \sum_{i=1}^{n} x_i^2 & \sum_{i=1}^{n} x_i^3 \\ \sum_{i=1}^{n} x_i^2 & \sum_{i=1}^{n} x_i^3 & \sum_{i=1}^{n} x_i^4 \end{bmatrix} \begin{Bmatrix} c_1 \\ c_2 \\ c_3 \end{Bmatrix} = \begin{bmatrix} \sum_{i=1}^{n} f_i \\ \sum_{i=1}^{n} f_i x_i \\ \sum_{i=1}^{n} f_i x_i^2 \end{bmatrix} \rightarrow c_1, c_2, c_3 \quad (7.7)$$

7.2.3 General Method for Finding Model Constants in Polynomial Function Fitting

As can be seen from fitting of linear and parabolic polynomials using Least Squares method, some steps are repeated. A general method can be devised to skip these steps and to facilitate the determination of model constants. To illustrate the development of general method, fitting of parabolic polynomial can be used again. Parabolic polynomial can alternatively be written as the product of variables and constants as following:

$$\tilde{f}(x) = c_1 + c_2 x + c_3 x^2 = \begin{bmatrix} 1 & x & x^2 \end{bmatrix} \begin{Bmatrix} c_1 \\ c_2 \\ c_3 \end{Bmatrix} \tag{7.8}$$

Using parabolic model, approximate function values ($\tilde{f}_i$) and resulting errors (ε_i) at design points of $x_1, x_2, \ldots, x_n$ can be expressed as following:

$$\underbrace{\begin{Bmatrix} \tilde{f}_1 \\ \tilde{f}_2 \\ \vdots \\ \tilde{f}_n \end{Bmatrix}}_{\tilde{F}} = \underbrace{\begin{bmatrix} 1 & x_1 & x_1^2 \\ 1 & x_2 & x_2^2 \\ \vdots & \vdots & \vdots \\ 1 & x_n & x_n^2 \end{bmatrix}}_{X} \underbrace{\begin{Bmatrix} c_1 \\ c_2 \\ c_3 \end{Bmatrix}}_{C} \rightarrow \text{Error}: \underbrace{\begin{Bmatrix} \varepsilon_1 \\ \varepsilon_2 \\ \vdots \\ \varepsilon_n \end{Bmatrix}}_{\varepsilon} = \overbrace{\underbrace{\begin{Bmatrix} \tilde{f}_1 \\ \tilde{f}_2 \\ \vdots \\ \tilde{f}_n \end{Bmatrix}}_{\tilde{F}}}^{Approximate} - \overbrace{\underbrace{\begin{Bmatrix} f_1 \\ f_2 \\ \vdots \\ f_n \end{Bmatrix}}_{F}}^{True} = \tilde{F} - F = XC - F \tag{7.9}$$

Note that X matrix in Eq. (7.9) indicates the values of model terms at design points. Using Eq. (7.9), sum of squares of the errors at all points (E) can be written in matrix form as following (Matrix rule: $A^2 = A^T A$):

$$E = \varepsilon^2 = \varepsilon^T \varepsilon = [XC - F]^T [XC - F] \tag{7.10}$$

Least squares method can be applied by taking the derivative of error with respect to the constants as below:

$$\frac{\partial E}{\partial C} = X^T [XC - F] + [XC - F]^T X = 0 \tag{7.11}$$

Using the information of $[XC - F]^T X = [X]^T ([XC - F]^T)^T = [X]^T [XC - F]$, Eq. (7.11) can be rearranged to yield model constant vector C as following:

$$2[X]^T[XC-F]=0_{\rightarrow}\ [X]^TXC=[X]^TF_{\rightarrow}\ C=\left[X^TX\right]^{-1}X^TF \tag{7.12}$$

Although Eq. (7.12) was obtained using parabolic model, it is valid for all kinds of polynomial models irrespective of their orders. Equation (7.12) can be conveniently used in determining model constants C for the selected model by modifying X matrix.

Example 7.1: For (x_1, y_1), (x_2, y_2), …, (x_n, y_n) points, function values are given as f_1, $f_2, \ldots, f_n$. Fit a two-variable (bivariate) parabolic function to these data using general fitting method.

Two-variable parabolic function can be expressed in general form as following:

$$\tilde{f}(x,y)=c_1+c_2x+c_3y+c_4x^2+c_5y^2+c_6xy$$

It can be expressed in matrix form as the product of constants and variables as below:

$$\tilde{f}(x,y)=\begin{bmatrix}1 & x & y & x^2 & y^2 & xy\end{bmatrix}\begin{Bmatrix}c_1\\c_2\\c_3\\c_4\\c_5\\c_6\end{Bmatrix}$$

Using two-variable parabolic model, approximate function values $(\tilde{f}_i)$ and resulting errors (ε_i) at (x_1, y_1), (x_2, y_2), …, (x_n, y_n) points can be expressed in matrix form as below:

$$\underbrace{\begin{Bmatrix}\tilde{f}_1\\\tilde{f}_2\\\tilde{f}_3\\\vdots\\\tilde{f}_n\end{Bmatrix}}_{\tilde{F}}=\underbrace{\begin{bmatrix}1 & x_1 & y_1 & x_1^2 & y_1^2 & x_1y_1\\1 & x_2 & y_2 & x_2^2 & y_2^2 & x_2y_2\\1 & x_3 & y_3 & x_3^2 & y_3^2 & x_3y_3\\\vdots & \vdots & \vdots & \vdots & \vdots & \vdots\\1 & x_n & y_n & x_n^2 & y_n^2 & x_ny_n\end{bmatrix}}_{X}\underbrace{\begin{Bmatrix}c_1\\c_2\\c_3\\\vdots\\c_6\end{Bmatrix}}_{C},\text{Error}:\underbrace{\begin{Bmatrix}\varepsilon_1\\\varepsilon_2\\\vdots\\\varepsilon_n\end{Bmatrix}}_{\varepsilon}=\underbrace{\overbrace{\begin{Bmatrix}\tilde{f}_1\\\tilde{f}_2\\\vdots\\\tilde{f}_n\end{Bmatrix}}^{Approximate}}_{\tilde{F}}-\underbrace{\overbrace{\begin{Bmatrix}f_1\\f_2\\\vdots\\f_n\end{Bmatrix}}^{True}}_{F}=\tilde{F}-F=XC-F$$

Using X matrix representing model terms and F vector representing true function values at data points, model constants C can be obtained using the general expression in Eq. (7.12) as $C=[X^TX]^{-1}X^TF$.

7.2.4 Determination of Fitting Quality

Fitting quality can be understood with eye check when fitting function can be easily plotted. However, when function cannot be plotted such as in tri-variable functions, a methodology is needed to quantify the fitting quality. R^2 method or criterion is one of these methods to determine the fitting quality. R^2 criterion calculates sum of squares of errors due to fitting and reference models and calculates their fractions. Reference model is selected as the simplest polynomial model to fit a data set. Constant model as expressed with $\widetilde{f}(x) = c_1$ function is used for this purpose.

Sum of squares of errors E_c due to constant model can be expressed as below:

$$E_c = (c_1 - f_1)^2 + (c_1 - f_2)^2 + (c_1 - f_3)^2 + \ldots + (c_1 - f_n)^2 = \sum_{i=1}^{n} (c_1 - f_i)^2$$

Unknown constant c_1 in the constant model can be determined using Least Squares method as the average of function values (f_{ave}) as following:

$$\text{Least Squares method}: E_c = \sum_{i=1}^{n} (c_1 - f_i)^2 \rightarrow \min \text{ or } \frac{\partial E_c}{\partial c_1} = 0$$

$$\frac{\partial E_c}{\partial c_1} = 0 \rightarrow 2\sum_{i=1}^{n} (c_1 - f_i) = 0 \rightarrow c_1 n = \sum_{i=1}^{n} f_i \rightarrow c_1 = \frac{\sum f_i}{n} = f_{ave} \tag{7.13}$$

Thus, for the constant model, sum of squares of errors E_c can be expressed as below:

$$\begin{aligned} E_c &= (c_1 - f_1)^2 + (c_1 - f_2)^2 + (c_1 - f_3)^2 + \ldots + (c_1 - f_n)^2 \\ &= (f_{ave} - f_1)^2 + (f_{ave} - f_2)^2 + (f_{ave} - f_3)^2 + \ldots + (f_{ave} - f_n)^2 = \sum_{i=1}^{n} (f_{ave} - f_i)^2 \end{aligned} \tag{7.14}$$

Indicating sum of squares of errors for the fitting model with E, R^2 value for the fitting model can be expressed in a dimensionless form as following:

$$R^2 = \frac{E_c - E}{E_c} \tag{7.15}$$

Meaning and calculation of R^2 criterion is illustrated in Fig. 7.7.

R^2 value varies between zero and one. When $R^2 = 1$, fitting function reflects the trend of data perfectly and it passes through all points. $R^2 = 1$ case is also called as perfect fitting. When $R^2 = 0$, it means data set do not show any trend. Simply data set is randomly distributed around the average value. $R^2 = 0$ case is also called as worst

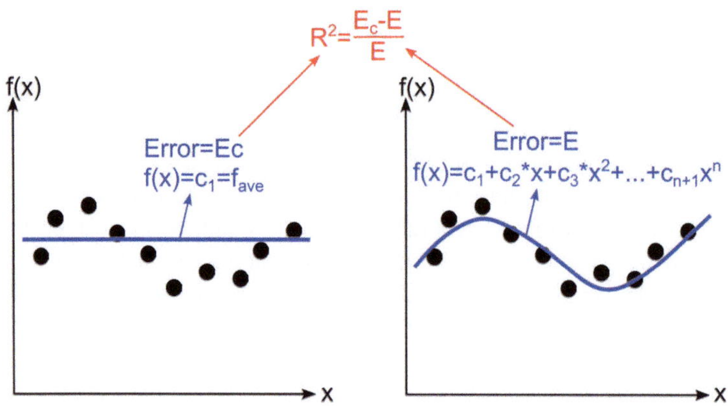

Fig. 7.7 Calculation of R^2 value

Table 7.3 Function values at different points

x	1	2	4	5
f	2	4	12	20

fitting. In fitting applications, R^2 value which is greater than 0.8 ($R^2 \geq 0.8$) is acceptable. It indicates that selected model fit to data set sufficiently well.

Note: Although the simplest form of R^2 criterion is mentioned above, more enhanced variants of this criterion are also available such as Adjusted R^2 criterion.

Example 7.2: Function values at different points are given in Table 7.3. Using function values: a) fit a linear polynomial function to the data set, b) fit a parabolic polynomial function to the data set.

(a) Fitting of linear polynomial function: $\tilde{f}(x) = c_1 + c_2 x$

Using least squares method, c_1 and c_2 constants in the model can be obtained as following:

The number of variables: $n = 4$, $\sum_{i=1}^{4} x_i = 1 + 2 + 4 + 5 = 12$, $\sum_{i=1}^{4} x_i^2 = 1^2 + 2^2 + 4^2 + 5^2 = 46$,

$$\sum_{i=1}^{4} f_i = 2 + 4 + 12 + 20 = 38, \sum_{i=1}^{4} f_i x_i = 2x1 + 4x2 + 12x4 + 20x5 = 158$$

$$\begin{bmatrix} n & \sum_{i=1}^{n} x_i \\ \sum_{i=1}^{n} x_i & \sum_{i=1}^{n} x_i^2 \end{bmatrix} \begin{Bmatrix} c_1 \\ c_2 \end{Bmatrix} = \begin{Bmatrix} \sum_{i=1}^{n} f_i \\ \sum_{i=1}^{n} f_i x_i \end{Bmatrix} \rightarrow \begin{bmatrix} 4 & \sum_{i=1}^{4} x_i \\ \sum_{i=1}^{4} x_i & \sum_{i=1}^{4} x_i^2 \end{bmatrix} \begin{Bmatrix} c_1 \\ c_2 \end{Bmatrix} = \begin{Bmatrix} \sum_{i=1}^{4} f_i \\ \sum_{i=1}^{4} f_i x_i \end{Bmatrix}$$

$$\begin{bmatrix} 4 & 12 \\ 12 & 46 \end{bmatrix} \begin{Bmatrix} c_1 \\ c_2 \end{Bmatrix} = \begin{Bmatrix} 38 \\ 158 \end{Bmatrix}_{\rightarrow} \quad c_1 = -3.7, c_2 = 4.4$$

Thus, linear fitting model is obtained as $f(x) = -3.7 + 4.4x$.

R^2 value for the linear model can be calculated as below:

$$E = \sum_{i=1}^{4} c_1 + c_2 x_i - f_i)^2 = 9.4, f_{ave} = \frac{1}{4} \sum_{i=1}^{4} f_i = 9.5_{\rightarrow} \ E_c$$

$$= \sum_{i=1}^{4} (f_{ave} - f_i)^2{}_{\rightarrow} \ R^2 = \frac{E_c - E}{E_c} = 0.9537$$

R^2 value was obtained close to 1. This shows that linear model fits to the data set quite well. That is, linear model represents the general trend of the data set good enough. Fitting of the linear model to the data set is shown in Fig. 7.8.

Solution with MATLAB commands:

Linear polynomial model can be fitted to the available data set using the *polyfit(x,f,n)* (n = polynomial degree) command in MATLAB. Using this command, coefficients ($L = c_2 \ c_1$) of the linear model can be obtained as following:

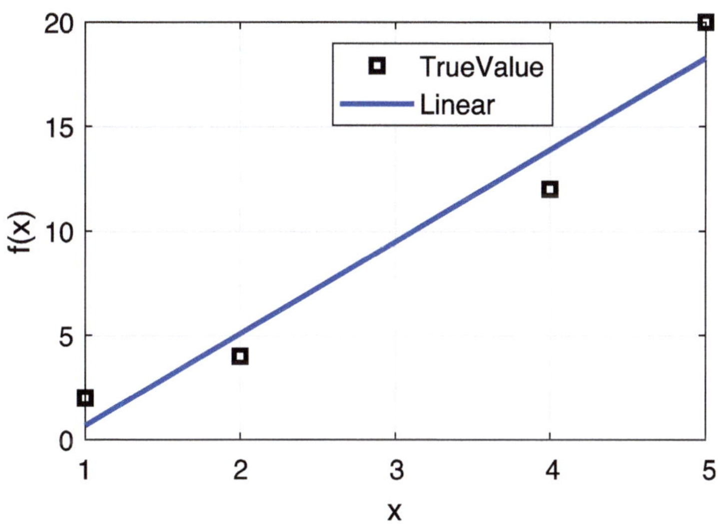

Fig. 7.8 Fitting of the linear model to the data set

```
>> x=[1 2 4 5]; f=[2 4 12 20] ;
>> L=polyfit(x,f,1)
L =
4.4000 -3.7000
```

L values show the coefficients of the linear polynomial model from high to low order. Thus, linear fitting model is obtained as $f(x) = 4.4x - 3.7$.

(b) Fitting of parabolic polynomial function: $\tilde{f}(x) = c_1 + c_2 x + c_3 x^2$

Using least squares method, c_1 and c_2 constants in the model can be obtained as following:

The number of variables: $n = 4$, $\sum_{i=1}^{4} x_i = 1+2+4+5 = 12$, $\sum_{i=1}^{4} x_i^2 = 1^2 + 2^2 + 4^2 + 5^2 = 46$,

$$\sum_{i=1}^{4} x_i^3 = 1^3 + 2^3 + 4^3 + 5^3 = 198, \ \sum_{i=1}^{4} x_i^4 = 1^4 + 2^4 + 4^4 + 5^4 = 898,$$

$$\sum_{i=1}^{4} f_i = 2 + 4 + 12 + 20 = 38, \ \sum_{i=1}^{4} f_i x_i = 2x1 + 4x2 + 12x4 + 20x5 = 158,$$

$$\sum_{i=1}^{4} f_i x_i^2 = 2x1^2 + 4x2^2 + 12x4^2 + 20x5^2 = 710$$

$$\begin{bmatrix} n & \sum_{i=1}^{n} x_i & \sum_{i=1}^{n} x_i^2 \\ \sum_{i=1}^{n} x_i & \sum_{i=1}^{n} x_i^2 & \sum_{i=1}^{n} x_i^3 \\ \sum_{i=1}^{n} x_i^2 & \sum_{i=1}^{n} x_i^3 & \sum_{i=1}^{n} x_i^4 \end{bmatrix} \begin{Bmatrix} c_1 \\ c_2 \\ c_3 \end{Bmatrix} = \begin{bmatrix} \sum_{i=1}^{n} f_i \\ \sum_{i=1}^{n} f_i x_i \\ \sum_{i=1}^{n} f_i x_i^2 \end{bmatrix} \rightarrow \begin{bmatrix} 4 & \sum_{i=1}^{4} x_i & \sum_{i=1}^{4} x_i^2 \\ \sum_{i=1}^{4} x_i & \sum_{i=1}^{4} x_i^2 & \sum_{i=1}^{4} x_i^3 \\ \sum_{i=1}^{4} x_i^2 & \sum_{i=1}^{4} x_i^3 & \sum_{i=1}^{4} x_i^4 \end{bmatrix}$$

$$\times \begin{Bmatrix} c_1 \\ c_2 \\ c_3 \end{Bmatrix} = \begin{bmatrix} \sum_{i=1}^{4} f_i \\ \sum_{i=1}^{4} f_i x_i \\ \sum_{i=1}^{4} f_i x_i^2 \end{bmatrix}$$

$$\begin{bmatrix} 4 & 12 & 46 \\ 12 & 46 & 198 \\ 46 & 198 & 898 \end{bmatrix} \begin{Bmatrix} c_1 \\ c_2 \\ c_3 \end{Bmatrix} = \begin{bmatrix} 38 \\ 158 \\ 710 \end{bmatrix}_{\rightarrow} \quad c_1 = 2.8, c_2 = -1.6, c_3 = 1$$

Thus, parabolic fitting model is obtained as $f(x) = 2.8 - 1.6x + x^2$.
R^2 value for the parabolic model can be calculated as below:

$$E = \sum_{i=1}^{4} \left(c_1 + c_2 x_i + c_3 x_i^2 - f_i\right)^2 = 0.4, f_{ort} = \frac{1}{4} \sum_{i=1}^{4} f_i = 9.5_{\rightarrow}\ E_c = \sum_{i=1}^{4} (f_{ort} - f_i)^2{}_{\rightarrow}\ R^2 = \frac{E_c - E}{E_c} = 0.998$$

R^2 value was obtained close to 1. This shows that parabolic model fits to the data set very well. Fitting of the parabolic model to the data set is shown in Fig. 7.9.

From R^2 values, it is seen that both linear and parabolic models fit the data set well. However, parabolic model fits to the data set better than the linear model since $R^2_{parabolic} = 0.998 > R^2_{linear} = 0.9537$.

Solution with MATLAB commands:

Parabolic polynomial model can be fitted to the available data set using the *polyfit(x,f,n)* (n = polynomial degree) command in MATLAB. Using this command, coefficients ($P = c_3\ c_2\ c_1$) of the parabolic model can be obtained as following:

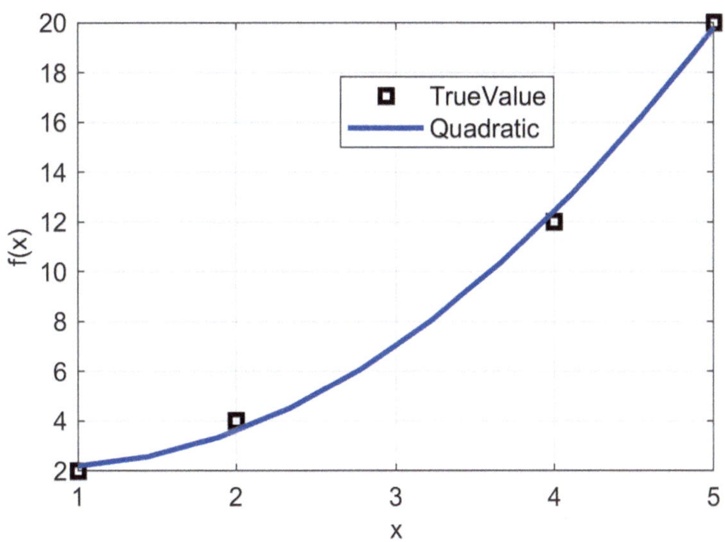

Fig. 7.9 Fitting of the parabolic model to the data set

Table 7.4 Function values for $f(x) = 10\sqrt{x}$ function in the range of [1, 10]

x	1	2	3	4	5	6	7	8	9	10
$f(x)$	10	14.1421	17.3205	20	22.3607	24.4949	26.4575	28.2843	30	31.6228

```
>> x=[1 2 4 5]; f=[2 4 12 20];
>> P=polyfit(x,f,2)
P =
1.0000 -1.6000 2.8000
```

P values show the coefficients of the parabolic polynomial model from high to low order. Thus, parabolic fitting model is obtained as $f(x) = x^2 - 1.6x + 2.8$.

Example 7.3: For $f(x) = 10\sqrt{x}$ function, function values at different points are given in Table 7.4. Fit (a) Linear: $\tilde{f}(x) = c\left(\frac{x}{a}\right)$, (b) Spherical: $\tilde{f}(x) = c\left[\frac{3}{2}\left(\frac{x}{a}\right) - \frac{1}{2}\left(\frac{x}{a}\right)^2\right]$, (c) Exponential: $\tilde{f}(x) = c\left[1 - \exp\left(\frac{-x}{a}\right)\right]$ models to the given function values and calculate fitting qualities using R^2 criterion (use $a = 10$).

Linear, spherical, and exponential models can be fitted to the given function values using Least Squares method. In Least Squares method, sum of squares of error (Total Error, E) is minimized to obtain model constants. To calculate minimum of error value, its derivative with respect to model constants are zeroed.

(a) Fitting Linear function: $\tilde{f}(x) = c\left(\frac{x}{a}\right)$

Constant of the linear model can be determined by minimizing sum of squares of error (Total Error, E) as below:

$$E = \sum_{i=1}^{10}\left[c\left(\frac{x_i}{a}\right) - f_i\right]^2 \rightarrow \min \rightarrow \frac{\partial E}{\partial c} = 0$$

$$\rightarrow 2\sum_{i=1}^{10}\left[c\left(\frac{x_i}{10}\right) - f_i\right]\left(\frac{x_i}{10}\right) = 0 \rightarrow c = \frac{\sum_{i=1}^{10} f_i\left(\frac{x_i}{10}\right)}{\sum_{i=1}^{10}\left(\frac{x_i}{10}\right)^2} = 37.0577$$

R^2 value for the linear model can be calculated as following:

$$E = \sum_{i=1}^{10}\left[37.0577\left(\frac{x_i}{a}\right) - f_i\right]^2 = 212.8862, \text{Average value}$$

$$: f_{ave} = \frac{1}{10}\sum_{i=1}^{10} f_i = 22.4683 \rightarrow E_c = \sum_{i=1}^{10}(f_{ave} - f_i)^2 = 451.7648 \rightarrow R^2 = \frac{E_c - E}{E_c}$$

$$= 0.5288$$

(b) Fitting Spherical function: $\tilde{f}(x) = c\left[\frac{3}{2}\left(\frac{x}{10}\right) - \frac{1}{2}\left(\frac{x}{10}\right)^2\right]$

Constant of the spherical model can be determined by minimizing sum of squares of error (Total Error, E) as below:

$$E = \sum_{i=1}^{n}\left[c\left(\frac{3}{2}\left(\frac{x_i}{10}\right) - \frac{1}{2}\left(\frac{x_i}{10}\right)^2\right) - f_i\right]^2 \rightarrow \min \rightarrow \frac{\partial E}{\partial c} = 0$$

$$\rightarrow 2\sum_{i=1}^{n}\left[c\left(\frac{3}{2}\left(\frac{x_i}{10}\right) - \frac{1}{2}\left(\frac{x_i}{10}\right)^2\right) - f_i\right]\left(\frac{3}{2}\left(\frac{x_i}{10}\right) - \frac{1}{2}\left(\frac{x_i}{10}\right)^2\right) = 0$$

$$\rightarrow c = \frac{\sum_{i=1}^{10} f_i\left(\frac{3}{2}\left(\frac{x_i}{10}\right) - \frac{1}{2}\left(\frac{x_i}{10}\right)^2\right)}{\sum_{i=1}^{10}\left(\frac{3}{2}\left(\frac{x_i}{10}\right) - \frac{1}{2}\left(\frac{x_i}{10}\right)^2\right)^2} = 33.7509$$

R^2 value for the spherical model can be calculated as following:

$$E = \sum_{i=1}^{n}\left[33.7509\left(\frac{3}{2}\left(\frac{x_i}{10}\right) - \frac{1}{2}\left(\frac{x_i}{10}\right)^2\right) - f_i\right]^2 = 79.6924, E_c = \sum_{i=1}^{10}(f_{ave} - f_i)^2$$

$$= 451.7648 \rightarrow R^2 = \frac{E_c - E}{E_c} = 0.8236$$

(c) Fitting Exponential function: $\tilde{f}(x) = c\left[1 - \exp\left(\frac{-x}{a}\right)\right]$

Constant of the exponential model can be determined by minimizing sum of squares of error (Total Error, E) as below:

$$E = \sum_{i=1}^{n}\left[c\left(1 - \exp\left(\frac{-x_i}{10}\right)\right) - f_i\right]^2 \rightarrow \min \rightarrow \frac{\partial E}{\partial c} = 0$$

$$\rightarrow 2\sum_{i=1}^{n}\left[c\left(1 - \exp\left(\frac{-x_i}{10}\right)\right) - f_i\right]\left(1 - \exp\left(\frac{-x_i}{10}\right)\right) = 0 \rightarrow c$$

$$= \frac{\sum_{i=1}^{10} f_i\left(1 - \exp\left(\frac{-x_i}{10}\right)\right)}{\sum_{i=1}^{10}\left(1 - \exp\left(\frac{-x_i}{10}\right)\right)^2} = 53.6647$$

R^2 value for the exponential model can be calculated as following:

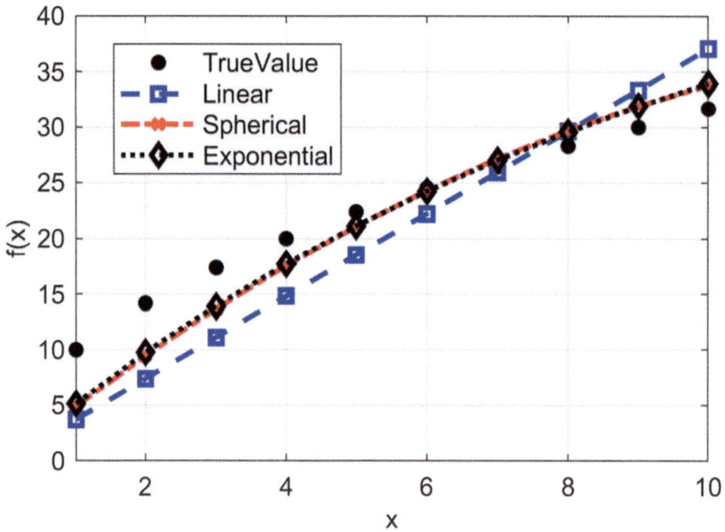

Fig. 7.10 Comparison of fitted models (Linear, Spherical, and Exponential) and true function

$$E = \sum_{i=1}^{n} \left[53.6647\left(1 - \exp\left(\frac{-x_i}{10}\right)\right) - f_i \right]^2 = 72.6399, E_c = \sum_{i=1}^{10} (f_{ave} - f_i)^2$$
$$= 451.7648 \rightarrow R^2 = \frac{E_c - E}{E_c} = 0.8392$$

True function $f(x) = 10\sqrt{x}$ and fitted linear, spherical, and exponential functions are shown in Fig. 7.10. R^2 values indicate that exponential and spherical model fits better than the linear model.

7.3 Response Surface Method

Surface fitting is the generalization of curve fitting. Curve fitting for multivariate functions are known as surface fitting. Surface fitting to the values of multivariate functions (multivariate regression) is frequently encountered in engineering and experimental studies. Surface fitting is also used like curve fitting to develop a simple mathematical model to predict intermediate values.

Response surface method is the more systematic form of surface fitting from the generation of data to the fitting. Response surface method uses special designs referred to as design of experiment methods to generate the data and fits simple polynomial models such as linear and parabolic models to the generated data [9–13]. Fitting model in response surface method is referred to as response surface model. Response surface model is especially desired when there is no real model or it is costly to produce an intermediate value. For example, quadratic (parabolic) response surface models are effectively used as fitting models in optimization applications by integrating an optimization method and finite element analysis.

In application of response surface method, initially it is assumed that no data is available. Data is generated using special methods to maximize the prediction quality of the selected model. Steps of the response surface method can be summarized as following:

1. *Model selection*: Linear or quadratic polynomial model is often selected for fitting.
2. *Selection of Design of Experiment Method*: Data points to fit the model are generated using design of experiment methods. Using design of experiment methods, location and distribution of data points are determined to maximize the prediction quality of the selected model.
3. *Surface Fitting with Least Squares Method*: After obtaining function values corresponding to data points, selected model is fitted to these values using Least Squares method.
4. *Determination of Fitting Quality*: Fitting quality of the selected model is determined using R^2 method or its variants.

Most of the steps of Response Surface method were already explained before. Remaining step regarding the Design of Experiment Methods is explained below:

7.4 Design of Experiment Methods

Design of experiment methods are also referred to as experimental design methods. Experiment here means data point or simply a point in the selected range. Design of experiment methods determine the location and the number of points so as to maximize the prediction quality of the fitting model. They are very influential on the accuracy and efficiency of the response surface model. They are specially designed to extract the most information from the least number of experiments. Within the context of design of experiment methods, Factorial, Composite, Box-Behnken and *D*-optimal experimental designs are widely used in the creation of linear and quadratic response surface models.

7.4.1 Factorial Experimental Design

Factorial Experimental Design is a method to create points in a structured way in the selected region defined by [lower, upper] bounds. Points are created in layers with regular intervals as shown in Fig. 7.11. Factorial Experimental Design method generates equally spaced and uniformly distributed points in the form of a grid

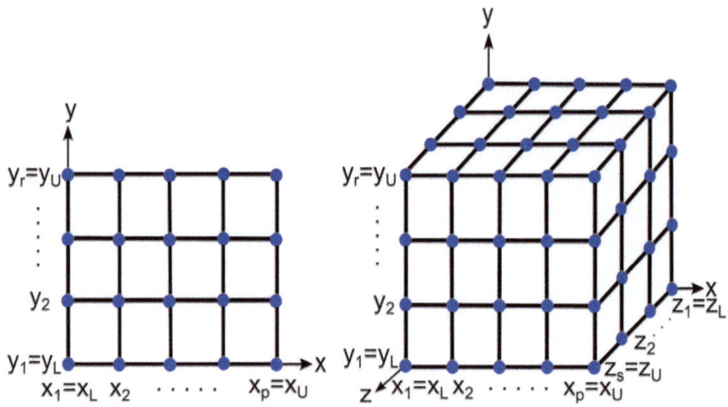

Fig. 7.11 Design points for two-variable *f*(*x*, *y*) and three-variable *f*(*x*, *y*, *z*) functions generated by Factorial Design methods

network. The number of data or design points within the range of [lower, upper] bounds in factorial design is called as *level*. Based on this definition, when x variable has p-level in the range of $[x_L, x_U]$, y variable has r-level in the range of $[y_L, y_U]$ and z variable has s-level in the range of $[z_L, z_U]$, design points can be expressed as following:

$$x = [x_1 = x_L, x_2, x_3, \ldots, x_p = x_U]_{\rightarrow} \; x : p \text{ level}$$

$$y = [y_1 = y_L, y_2, y_3, \ldots, y_r = y_U]_{\rightarrow} \; y : r \text{ level}$$

$$z = [z_1 = z_L, z_2, z_3, \ldots, z_s = z_U]_{\rightarrow} \; z : s \text{ level}$$

Combining levels in different directions, design points for multivariate functions can be generated with Factorial Design methods. Design points for a two-variable *f*(*x*,*y*) function obtained by combining levels in x and y directions and for a three-variable *f*(*x*, *y*, *z*) function by combining levels in x, y, and z directions are illustrated in Fig. 7.11.

In Factorial Design methods, the number of levels for a variable is determined by the highest order of this variable in the selected response surface model. For example, when x and y variables have the highest order terms of x^m and y^n in an *f*(*x*, *y*) model, numbers of levels are selected at least as p ($p \geq m + 1$) for x variable and r ($n \geq r + 1$) for y variable. Factorial experimental design methods to fit two-variable linear and quadratic response surface models are explained in detail below.

(a) **Factorial Design Methods for Two-Variable Response Surface Model**

Factorial experimental designs to fit two-variable linear and quadratic response surface models can be expressed as below:

Linear Response Surface Model:

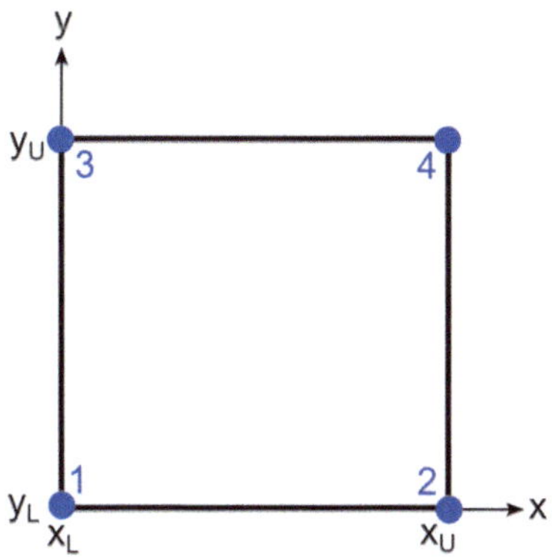

Point	x	y	f(x,y)
1	x_L	y_L	f_1
2	x_U	y_L	f_2
3	x_L	y_U	f_3
4	x_U	y_U	f_4

Fig. 7.12 Two-level factorial design to fit a two-variable linear response surface model

Two-variable linear response surface model can be expressed as $\widetilde{f}(x, y) = c_1 + c_2x + c_3y$. Since the function is linear in x and y variables, level numbers in x and y directions must be greater than or equal to 2. If we denote the limits of x variable with $[x_{\text{lower}}, x_{\text{upper}}] = [x_L, x_U]$ expression and limits of y variable with $[y_{\text{lower}}, y_{\text{upper}}] = [y_L, y_U]$ expression, the simplest Factorial design corresponding to only two levels in x and y directions can be shown as in Fig. 7.12. Factorial design in Fig. 7.12 can also be named as 2×2 Factorial design since there are two levels in each direction. In some sources, lower and upper levels of variables in two-level factorial design are represented with normalized values of -1 and $+1$. For example, in MATLAB software, lower and upper levels are represented with 1 and 2 values.

In two-variable linear response surface model of $\widetilde{f}(x, y) = c_1 + c_2x + c_3y$, there are just three constants (c_1, c_2, c_3) to be determined. If function values are known at three points, these constants can be determined easily. However, two-level factorial experimental design produces $2 \times 2 = 2^2 = 4$ points, which are more than needed (i.e., 3). In general case, the number of constants in a linear model will be $n + 1$ for n variable. Also, in general case, for n variables consisting of two levels each, number of design points generated by two-level factorial design will be 2^n. Since $2^n > n + 1$, factorial design always creates design points (2^n) more than needed ($n + 1$). 2^n points may correspond to an excessive number of points for big n values and producing so many experimental results or function evaluations may be impractical. This is the main disadvantage of factorial design method. To overcome this drawback, reduced factorial experimental design methods using fewer points can be employed to reduce the cost of model creation for big n values. Reduced factorial experimental designs can be obtained by using selection criteria such as *D-optimality* from a full factorial experimental design.

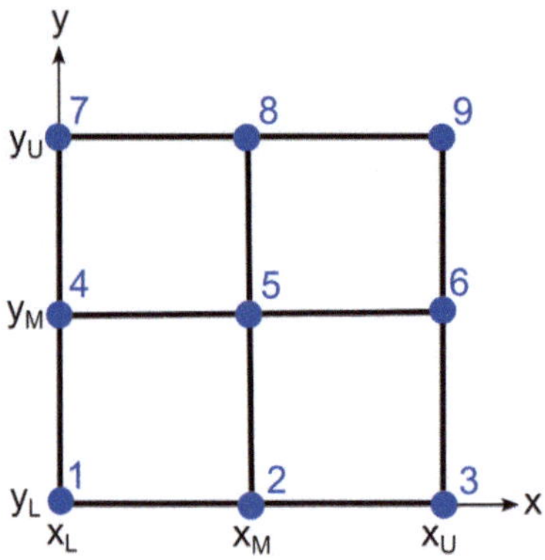

Point	x	y	f(x,y)
1	x_L	y_L	f_1
2	x_M	y_L	f_2
3	x_U	y_L	f_3
4	x_L	y_M	f_4
5	x_M	y_M	f_5
6	x_U	y_M	f_6
7	x_L	y_U	f_7
8	x_M	y_U	f_8
9	x_U	y_U	f_9

Fig. 7.13 Three-level factorial design to fit a two-variable quadratic response surface model

Parabolic (Quadratic) Response Surface Model:

Two-variable quadratic (parabolic) response surface model can be expressed as $\tilde{f}(x,y) = c_1 + c_2x + c_3y + c_4x^2 + c_5y^2 + c_6xy$. Since the function is quadratic (i.e., parabolic) in x and y variables, level numbers in x and y directions must be greater than or equal to 3. If we denote the levels of x variable with $[x_{\text{lower}}, x_{\text{middle}}, x_{\text{upper}}] = [x_L, x_M, x_U]$ expression and levels of y variable with $[y_{\text{lower}}, y_{\text{middle}}, y_{\text{upper}}] = [y_L, y_M, y_U]$ expression, the simplest factorial design corresponding to only three levels in x and y directions can be shown as in Fig. 7.13. Factorial design in Fig. 7.13 can also be named as 3 × 3 factorial design since there are three levels in each direction. In some sources, lower, middle, and upper levels of variables in three-level Factorial design are represented with normalized values of −1, 0, and +1. For example, in MATLAB software, lower, middle, and upper levels are represented with 1, 2, and 3 values.

In two-variable quadratic response surface model of $\tilde{f}(x,y) = c_1 + c_2x + c_3y + c_4x^2 + c_5y^2 + c_6xy$, there are six constants (c_1, c_2, …, c_6) to be determined. If function values are known at six points, these constants can be determined easily. However, three-level factorial experimental design produces $3 \times 3 = 3^2 = 9$ points, which is more than needed (i.e., 6). In general case, for n variables, the number of constants in a quadratic model will be $\frac{(n+1)(n+2)}{2}$. Also, in general case, for n variables consisting of three levels each, the number of design points generated by three-level factorial design will be 3^n. Since $3^n > \frac{(n+1)(n+2)}{2}$, factorial design always creates design points more than needed. 3^n points may correspond to an excessive number of points for big n values. For example, for ten variables with three levels, the number of constants in the model will be 66. However, factorial design will produce

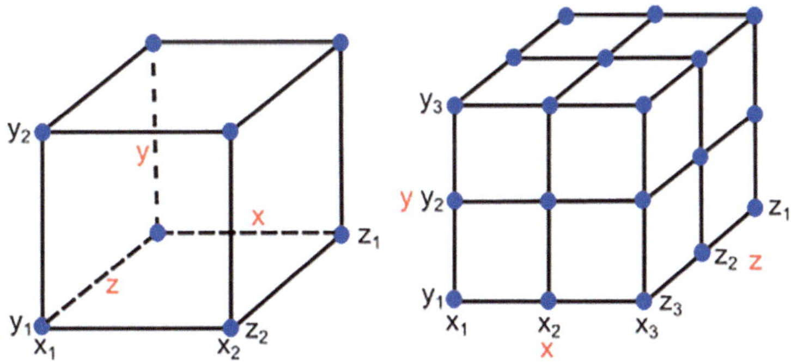

Fig. 7.14 Two- and three-level factorial experimental designs to fit three-variable linear and quadratic response surface models

54,049 points. Determining function values at so many points may be very costly or impractical. To overcome this drawback, reduced factorial experimental design methods using fewer points can be employed.

Note: In response surface method, usually the number of points in fitting is selected to be 50% higher than needed to provide smoothing effect on the model. For example, for ten variables with three levels, the number of constants in the model can be selected to be 50% higher than 66 which will be roughly 100.

(b) **Factorial Design Method for Three-Variable Response Surface Model**

Three-variable (trivariate) linear and quadratic response surface models can be expressed as following:

Linear response surface model:

$$\tilde{f}(x, y, z) = c_1 + c_2x + c_3y + c_4z$$

Parabolic (Quadratic) response surface model:

$$\tilde{f}(x, y, z) = c_1 + c_2x + c_3y + c_4z + c_5x^2 + c_6y^2 + c_7z^2 + c_8xy + c_9xz + c_{10}yz$$

Factorial experimental designs to fit linear and quadratic response models can be shown as in Fig. 7.14. In Fig. 7.14a, two-level factorial design consisting of $[x_1, x_2]$ values in x direction, $[y_1, y_2]$ values in y direction, and $[z_1, z_2]$ values in z direction to fit a linear model is illustrated. Whereas in Fig. 7.14b, three-level factorial design consisting of $[x_1, x_2, x_3]$ values in x direction, $[y_1, y_2, y_3]$ values in y direction, and $[z_1, z_2, z_3]$ values in z direction to fit a quadratic model is illustrated.

Numbers of model constants for linear and quadratic response surface models can be expressed in general form as in Table 7.5.

Note: Response Surface Program *MATLAB Program 7.1: Response Surface Model* in appendix can be used to fit linear and quadratic models. Example 7.4 is solved using this program.

Table 7.5 The number of constants in linear and quadratic response surface models

Number of variables	Model	Number of model constants	Number of points with factorial design
n	Linear	$n + 1$	2^n
n	Parabolic	$\frac{(n+1)(n+2)}{2}$	3^n

7.4.2 Central Composite Design

Central composite design is an efficient alternative of factorial design method in producing design points to fit quadratic response surface models. It yields fewer points than factorial experimental design. Due to its efficiency, central composite design is more commonly used in creating quadratic response surface models. Central composite design is a five-level experimental design method. It is obtained by adding a star (+) shaped experimental design to two-level factorial experimental design. Central composite design yields a total of $2^n + 2n + 1$ points for n variables. This number is less than three-level factorial experimental design leading to 3^n points in fitting a quadratic response surface model. In central composite experimental design, lower and upper points for n variables can be determined in normalized form as $-(2^n)^{1/4}$ and $(2^n)^{1/4}$.

Central Composite Design to Fit Two-variable Quadratic Response Surface Model:

Obtaining central composite design by adding a star (+)-shaped experimental design to two-level factorial experimental design to fit two-variable quadratic response surface models is illustrated in normalized form in Fig. 7.15. In Fig. 7.15, $-\alpha$ and α indicate coordinates of end points.

In Central Composite design, locations of end points vary according to the number of variables n as given in Table 7.6. For two variables ($n = 2$), design points in Central Composite design can be given as in Table 7.7.

In central composite experiment design, location of design points for two variables ($n = 2$) is obtained in normalized form in $[-\alpha, \alpha]$ interval as in Table 7.7.

Conversion of design points from x, y values in $[-\alpha, \alpha]$ interval in Table 7.7 to the actual values x_a, y_a in $[x_L, x_U]$ and $[y_L, y_U]$ intervals can be performed using Lagrange interpolation method as following:

$$x_a = \left(\frac{\alpha - x}{2\alpha}\right)x_L + \left(\frac{x + \alpha}{2\alpha}\right)x_U, \quad y_a = \left(\frac{\alpha - y}{2\alpha}\right)y_L + \left(\frac{y + \alpha}{2\alpha}\right)y_U \tag{7.16}$$

Central Composite Design to Fit Three-variable Quadratic Response Surface Model:

Obtaining central composite design by adding a star (+)-shaped experimental design to two-level factorial experimental design to fit three-variable quadratic response surface models is illustrated in normalized form in Fig. 7.16. In Fig. 7.16, $-\alpha$ and α indicate coordinates of end points.

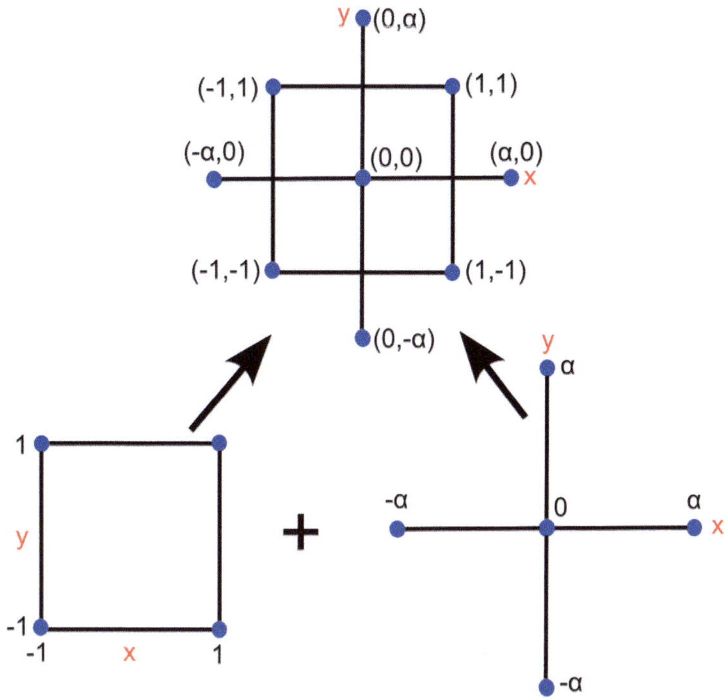

Fig. 7.15 Central composite experimental design for two variables

Table 7.6 Location of end points in normalized form according to the number of variables in central composite experimental design

n	α
2	$(2^2)^{1/4} = 1.414$
3	$(2^3)^{1/4} = 1.682$
4	$(2^4)^{1/4} = 2$

Conversion of design points from x, y, z values in $[-\alpha, \alpha]$ interval in Fig. 7.16 to the actual values x_a, y_a, z_a in $[x_L, x_U]$, $[y_L, y_U]$, and $[z_L, z_U]$ intervals can be performed using Lagrange interpolation method as following:

$$
\begin{aligned}
x_a &= \left(\frac{\alpha - x}{2\alpha}\right)x_L + \left(\frac{x + \alpha}{2\alpha}\right)x_U \\
y_a &= \left(\frac{\alpha - y}{2\alpha}\right)y_L + \left(\frac{y + \alpha}{2\alpha}\right)y_U \\
z_a &= \left(\frac{\alpha - z}{2\alpha}\right)z_L + \left(\frac{z + \alpha}{2\alpha}\right)z_U
\end{aligned}
\tag{7.17}
$$

Table 7.7 Central composite design in normalized form when $n = 2$

Experiment	x	y
1	−1	−1
2	−1	1
3	1	−1
4	1	1
5	$-\alpha$	0
6	α	0
7	0	$-\alpha$
8	0	α
9	0	0

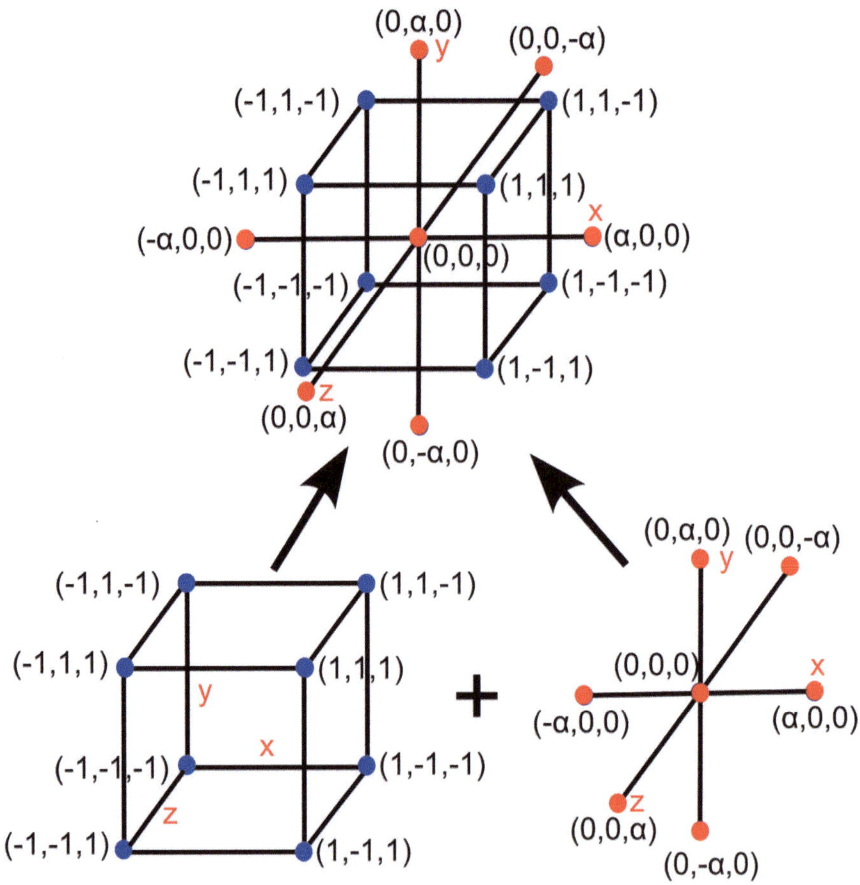

Fig. 7.16 Central Composite experimental design for three variables

7.4.3 Box-Behnken Experimental Design

Box-Behnken experimental design, like Central Composite design, is an efficient alternative of factorial design to create quadratic response surface models. In Box-Behnken experimental design, design points are located on a hypersphere around the center. Box-Behnken experimental design does not use extreme values at lower and upper limits of variables. Due to this feature, it is sometimes preferred over other experimental designs when it is difficult or expensive to conduct experiments at limits of variables.

In contrast to other experimental design methods, Box-Behnken experimental design is not available for two variables. Box-Behnken experimental design for three variables is shown in normalized form in the range of [−1, 1] in Fig. 7.17 and Table 7.8.

For high number of variables, Box-Behnken experimental design becomes complicated to create compared to other experimental designs. Box-Behnken design can be obtained by combining orthogonal block designs with a point at the center.

For example, Box-Behnken experimental design in Table 7.8 can be obtained using orthogonal three-block design [Block 1: 1–4 points, Block 2: 5–8 points, Block 3: 9–12 points]. In an orthogonal experimental design, change of rows or columns does not affect the prediction quality of the fitted response surface model. Because of this property, experimental designs are desired to have orthogonality. For example, in Table 7.8, order of the blocks can be changed in any order since Box-Behnken is an orthogonal design.

Conversion of design points from x, y, z values in [−1, 1] interval in Table 7.8 to the actual values x_a, y_a, z_a in $[x_L, x_U]$, $[y_L, y_U]$ and $[z_L, z_U]$ intervals can be performed using Lagrange interpolation method as following:

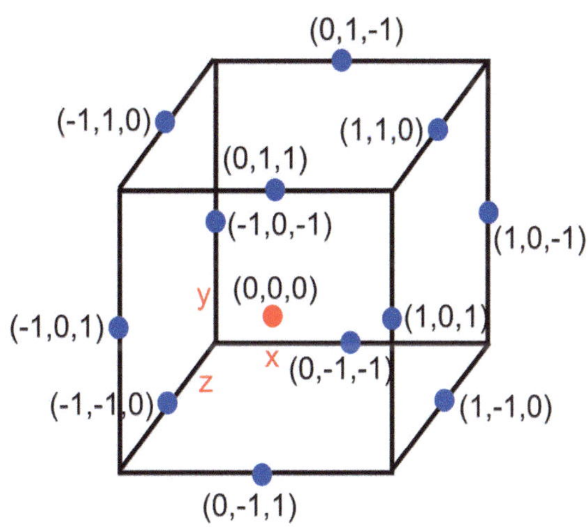

Fig. 7.17 Box-Behnken experimental design for three variables

Table 7.8 Box-Behnken experimental design for three variables

Experiment	x	y	z
1	−1	−1	0
2	1	−1	0
3	−1	1	0
4	1	1	0
5	0	−1	−1
6	0	−1	1
7	0	1	−1
8	0	1	1
9	−1	0	−1
10	−1	0	1
11	1	0	-1
12	1	0	1
13	0	0	0

$$x_a=\left(\frac{1-x}{2}\right)x_L+\left(\frac{x+1}{2}\right)x_U, y_a=\left(\frac{1-y}{2}\right)y_L+\left(\frac{y+1}{2}\right)y_U, z_a=\left(\frac{1-z}{2}\right)z_L+\left(\frac{z+1}{2}\right)z_U \quad (7.18)$$

7.4.4 D-*Optimal Experiment Design*

All of the experimental design methods described so far produce a large number of design points for high number of variables. This limits the use of these experimental design methods for a low number of variables. *D*-optimal experimental design is an efficient alternative to the aforementioned experimental design methods. *D*-optimal experimental design select a subset from a base experimental design using *D*-optimality criterion. Factorial and other experimental design methods can be used as base design. Using *D*-optimal experimental design, a desired number of points can be selected without any limitation to fit response surface model. Because of this feature, they are increasingly used in creating response surface models. Although *D*-optimal experimental designs are efficient experimental design methods, they do not have the orthogonality property. In *D*-optimal experimental design, design or test points are determined to maximize the determinant of $|X^TX|$ ($|X^TX| \Rightarrow$ max) where X matrix is obtained using values of response surface terms at design points.

Example 7.4: Using factorial experimental design method, fit a quadratic (parabolic) response surface model to the values of $f(x, y) = \sqrt{x} + \sqrt{y} + \sqrt{x}\sqrt{y}$ function within the ranges of $2 \le x \le 10$, $5 \le y \le 15$ and predict the function value at $(x, y) = (5, 9)$ point.

Quadratic response surface model can be created using the factorial experimental design in Fig. 7.18 which consists of three levels (lower, middle, upper) for x and y variables.

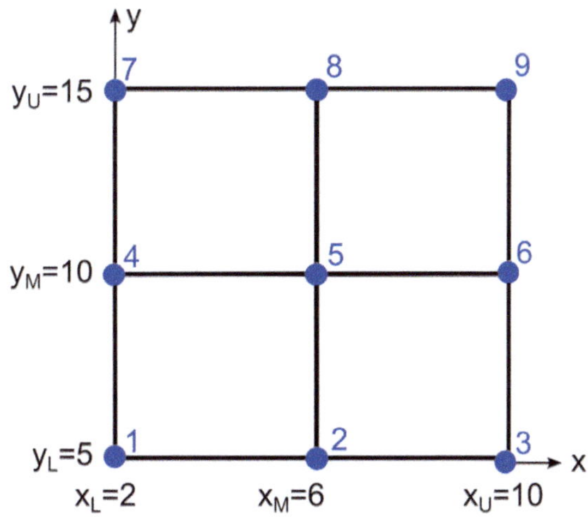

Fig. 7.18 Three-level factorial experimental design for two variables

Table 7.9 Location of points and corresponding function values in three-level factorial experimental design

Experiment (point)	x	y	$f=\sqrt{x}+\sqrt{y}+\sqrt{x}\sqrt{y}$
1	2	5	6.8126
2	6	5	10.1628
3	10	5	12.4694
4	2	10	9.0486
5	6	10	13.3577
6	10	10	16.3246
7	2	15	10.7644
8	6	15	15.8093
9	10	15	19.2827

Numerical values of design points and corresponding function values in the factorial experimental design can be given as in Table 7.9.

Quadratic response surface model to predict function values in Table 7.9 can be expressed as below:

$$\tilde{f}(x,y)=c_1+c_2x+c_3y+c_4x^2+c_5y^2+c_6xy=\begin{bmatrix}1 & x & y & x^2 & y^2 & xy\end{bmatrix}\begin{Bmatrix}c_1\\c_2\\c_3\\c_4\\c_5\\c_6\end{Bmatrix}$$

Using quadratic response surface model, approximate ($\tilde{f}_i$) and true (f_i) function values at design points in Table 7.9 can be expressed as following:

Approximate values:

$$\underbrace{\begin{Bmatrix} \tilde{f}_1 \\ \tilde{f}_2 \\ \tilde{f}_3 \\ \tilde{f}_4 \\ \tilde{f}_5 \\ \tilde{f}_6 \\ \tilde{f}_7 \\ \tilde{f}_8 \\ \tilde{f}_9 \end{Bmatrix}}_{\tilde{F}} = \underbrace{\begin{bmatrix} 1 & x_1 & y_1 & x_1^2 & y_1^2 & x_1y_1 \\ 1 & x_2 & y_2 & x_2^2 & y_2^2 & x_2y_2 \\ 1 & x_3 & y_3 & x_3^2 & y_3^2 & x_3y_3 \\ 1 & x_4 & y_4 & x_4^2 & y_4^2 & x_4y_4 \\ 1 & x_5 & y_5 & x_5^2 & y_5^2 & x_5y_5 \\ 1 & x_6 & y_6 & x_6^2 & y_6^2 & x_6y_6 \\ 1 & x_7 & y_7 & x_7^2 & y_7^2 & x_7y_7 \\ 1 & x_8 & y_8 & x_8^2 & y_8^2 & x_8y_8 \\ 1 & x_9 & y_9 & x_9^2 & y_9^2 & x_9y_9 \end{bmatrix}}_{X} \underbrace{\begin{Bmatrix} c_1 \\ c_2 \\ c_3 \\ c_4 \\ c_5 \\ c_6 \end{Bmatrix}}_{C} \rightarrow . \underbrace{\begin{Bmatrix} \tilde{f}_1 \\ \tilde{f}_2 \\ \tilde{f}_3 \\ \tilde{f}_4 \\ \tilde{f}_5 \\ \tilde{f}_6 \\ \tilde{f}_7 \\ \tilde{f}_8 \\ \tilde{f}_9 \end{Bmatrix}}_{\tilde{F}} = \underbrace{\begin{bmatrix} 1 & 2 & 5 & 4 & 25 & 10 \\ 1 & 6 & 5 & 36 & 25 & 30 \\ 1 & 10 & 5 & 100 & 25 & 50 \\ 1 & 2 & 10 & 4 & 100 & 20 \\ 1 & 6 & 10 & 36 & 100 & 60 \\ 1 & 10 & 10 & 100 & 100 & 100 \\ 1 & 2 & 15 & 4 & 225 & 30 \\ 1 & 6 & 15 & 36 & 225 & 90 \\ 1 & 10 & 15 & 100 & 225 & 150 \end{bmatrix}}_{X} \underbrace{\begin{Bmatrix} c_1 \\ c_2 \\ c_3 \\ c_4 \\ c_5 \\ c_6 \end{Bmatrix}}_{C}.$$

$$\text{True function values}: \underbrace{\begin{Bmatrix} f_1 \\ f_2 \\ f_3 \\ f_4 \\ f_5 \\ f_6 \\ f_7 \\ f_8 \\ f_9 \end{Bmatrix}}_{F} = \begin{Bmatrix} 6.8126 \\ 10.1628 \\ 12.4694 \\ 9.0486 \\ 13.3577 \\ 16.3246 \\ 10.7644 \\ 15.8093 \\ 19.2827 \end{Bmatrix}$$

In Eq. (7.12), it was shown that model constants C of any polynomial model can be determined using least squares method in the form of $C = [X^TX]^{-1}X^TF$ where X matrix and F vector represent model terms and true function values at data points. Therefore, using Eq. (7.12), model constants C for this example can be calculated as below:

$$\begin{aligned} C &= \left[X^TX\right]^{-1}X^TF \rightarrow C^T \\ &= \left[1.7383 \quad 1.0308 \quad 0.6205 \quad -0.0412 \quad -0.0144 \quad 0.0358\right] \rightarrow \end{aligned}$$

$$c_1 = 1.7383, c_2 = 1.0308, c_3 = 0.6205, c_4 = -0.0412, c_5 = -0.0144, c_6 = 0.0358$$

Thus, quadratic response surface model is obtained as following:

$$\tilde{f}(x,y) = 1.7383 + 1.0308x + 0.6205y - 0.0412x^2 - 0.0144y^2 + 0.0358xy$$

Fitting quality of the response surface model can be determined using R^2 criterion as following:

$$\text{Error of quadratic response surface model}: E = \sum_{i=1}^{9} (f_i - \tilde{f}_i)^2$$

Error of the simplest (constant) model:

$$\text{Average value}: f_{ave} = \frac{1}{9}\sum_{i=1}^{9} f_i \rightarrow E_c = \sum_{i=1}^{9} (f_{ave} - f_i)^2 \rightarrow R^2 = \frac{E_c - E}{E_c} = 0.9997$$

Since $R^2 = 0.9997$ is very close to 1, it is understood that quadratic response surface model fits function values very well. That is, quadratic response surface model is a very good approximation of $f(x, y) = \sqrt{x} + \sqrt{y} + \sqrt{x}\sqrt{y}$ function. Using the fitted quadratic response surface model, function value at $(x, y) = (5, 9)$ point is predicted as $f = 11.889303$.

Note: For the solution of this problem *MATLAB Program 7.1: Response Surface Model Fitting program* in appendix written in MATLAB language can also be used.

Example 7.5: Using Central Composite experimental design method, fit a quadratic (parabolic) response surface model to the values of $f(x, y) = \sqrt{x} + \sqrt{y} + \sqrt{x}\sqrt{y}$ function within the ranges of $2 \leq x \leq 10$, $5 \leq y \leq 15$.

To create a two-variable quadratic response surface model, Central Composite experimental design in normalized form as shown in Table 7.10 can be used.

Conversion of design points from x, y values in $[-1, 1]$ interval in Table 7.10 to the actual values x_a, y_a in $[x_L = 2, x_U = 10]$ and $[y_L = 5, y_U = 15]$ intervals can be performed using Lagrange interpolation method as following:

$$x_a = \left(\frac{1-x}{2}\right)x_L + \left(\frac{x+1}{2}\right)x_U, y_a = \left(\frac{1-y}{2}\right)y_L + \left(\frac{y+1}{2}\right)y_U$$

Using above conversion, actual location of points and corresponding function values in Central Composite experimental design are obtained as in Table 7.11.

Quadratic response surface model for two variables (x, y variables) can be expressed as below:

Table 7.10 Central Composite experimental design for two variables in normalized form

Experiment	x	y
1	−0.7071	−0.7071
2	−0.7071	0.7071
3	0.7071	−0.7071
4	0.7071	0.7071
5	−1	0
6	1	0
7	0	−1
8	0	1
9	0	0

Table 7.11 Location of points and corresponding function values in Central Composite experimental design

Experiment	x_a	y_a	f
1	3.1716	6.4645	8.8514
2	3.1716	13.5355	12.0120
3	8.8284	6.4645	13.0683
4	8.8284	13.5355	17.5818
5	2	10	9.0487
6	10	10	16.3245
7	6	5	10.1628
8	6	15	15.8093
9	6	10	13.3577

$$\widetilde{f}(x,y) = c_1 + c_2x + c_3y + c_4x^2 + c_5y^2 + c_6xy = \begin{bmatrix} 1 & x & y & x^2 & y^2 & xy \end{bmatrix} \begin{Bmatrix} c_1 \\ c_2 \\ c_3 \\ c_4 \\ c_5 \\ c_6 \end{Bmatrix}$$

Quadratic response surface model can be expressed at design points in Table 7.11 as following:

$$\underbrace{\begin{Bmatrix} \widetilde{f}_1 \\ \widetilde{f}_2 \\ \widetilde{f}_3 \\ \widetilde{f}_4 \\ \widetilde{f}_5 \\ \widetilde{f}_6 \\ \widetilde{f}_7 \\ \widetilde{f}_8 \\ \widetilde{f}_9 \end{Bmatrix}}_{\widetilde{F}} = \underbrace{\begin{bmatrix} 1 & x_1 & y_1 & x_1^2 & y_1^2 & x_1y_1 \\ 1 & x_2 & y_2 & x_2^2 & y_2^2 & x_2y_2 \\ 1 & x_3 & y_3 & x_3^2 & y_3^2 & x_3y_3 \\ 1 & x_4 & y_4 & x_4^2 & y_4^2 & x_4y_4 \\ 1 & x_5 & y_5 & x_5^2 & y_5^2 & x_5y_5 \\ 1 & x_6 & y_6 & x_6^2 & y_6^2 & x_6y_6 \\ 1 & x_7 & y_7 & x_7^2 & y_7^2 & x_7y_7 \\ 1 & x_8 & y_8 & x_8^2 & y_8^2 & x_8y_8 \\ 1 & x_9 & y_9 & x_9^2 & y_9^2 & x_9y_9 \end{bmatrix}}_{X} \underbrace{\begin{Bmatrix} c_1 \\ c_2 \\ c_3 \\ c_4 \\ c_5 \\ c_6 \end{Bmatrix}}_{C} \rightarrow$$

$$\underbrace{\begin{Bmatrix} \tilde{f}_1 \\ \tilde{f}_2 \\ \tilde{f}_3 \\ \tilde{f}_4 \\ \tilde{f}_5 \\ \tilde{f}_6 \\ \tilde{f}_7 \\ \tilde{f}_8 \\ \tilde{f}_9 \end{Bmatrix}}_{\tilde{F}} = \underbrace{\begin{bmatrix} 1 & 3.1716 & 6.4645 & (3.1716)^2 & (6.4645)^2 & (3.1716)(6.4645) \\ 1 & 3.1716 & 13.5355 & (3.1716)^2 & (13.5355)^2 & (3.1716)(13.5355) \\ 1 & 8.8284 & 6.4654 & (8.8284)^2 & (6.4654)^2 & (8.8284)(6.4654) \\ 1 & 8.8284 & 13.5355 & (8.8284)^2 & (13.5355)^2 & (8.8284)(13.5355) \\ 1 & 2 & 10 & 4 & 100 & 60 \\ 1 & 10 & 10 & 100 & 100 & 100 \\ 1 & 6 & 5 & 36 & 25 & 30 \\ 1 & 6 & 15 & 36 & 225 & 90 \\ 1 & 6 & 10 & 36 & 100 & 150 \end{bmatrix}}_{X} \underbrace{\begin{Bmatrix} c_1 \\ c_2 \\ c_3 \\ c_4 \\ c_5 \\ c_6 \end{Bmatrix}}_{C}$$

Using Table 7.11, true function values can be expressed as below:

$$\underbrace{\begin{Bmatrix} f_1 \\ f_2 \\ f_3 \\ f_4 \\ f_5 \\ f_6 \\ f_7 \\ f_8 \\ f_9 \end{Bmatrix}}_{F} = \begin{Bmatrix} 8.8514 \\ 12.0120 \\ 13.0683 \\ 17.5818 \\ 9.0487 \\ 16.3245 \\ 10.1628 \\ 15.8093 \\ 13.3577 \end{Bmatrix}$$

As expressed in Eq. (7.12), model constants C of any polynomial model can be determined using least squares method in the form of $C = [X^T X]^{-1} X^T F$ where X and F indicate model terms and true function values at data points. Therefore, substituting X and F into Eq. (7.12), model constants C for this example can be calculated as following:

$$\begin{aligned} C &= \left[X^T X\right]^{-1} X^T F \rightarrow C^T \\ &= \left[1.6617 \quad 1.0366 \quad 0.6312 \quad -0.0406 \quad -0.0140 \quad 0.0338\right] \rightarrow \end{aligned}$$

$$c_1 = 1.6617, c_2 = 1.0366, c_3 = 0.6312, c_4 = -0.0406, c_5 = -0.0140, c_6 = 0.0338$$

Thus, quadratic response surface model is obtained as following:

$$\tilde{f}(x, y) = 1.6617 + 1.0366x + 0.6312y - 0.0406x^2 - 0.0140y^2 + 0.0338xy$$

Fitting quality of the response surface model can be determined using R^2 criterion as following:

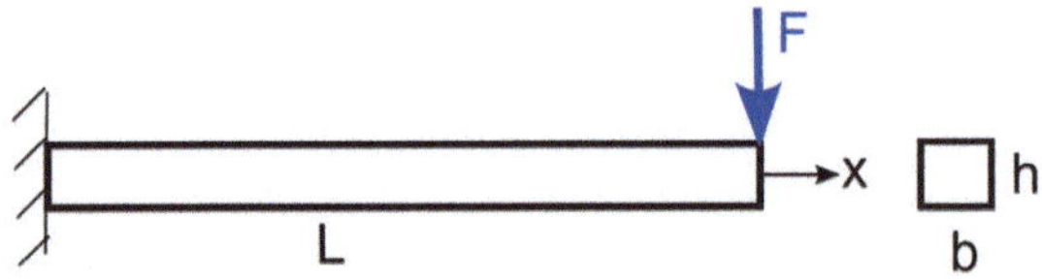

Fig. 7.19 Cantilever beam subjected to a tip force

$$\text{Error of quadratic response surface model}: E=\sum_{i=1}^{9}\left(f_i-\tilde{f}_i\right)^2$$

Error of the simplest (constant) model:

$$\text{Average value}: f_{ave}=\frac{1}{9}\sum_{i=1}^{9}f_i{}_{\rightarrow}\ E_c=\sum_{i=1}^{9}\left(f_{ave}-f_i\right)^2{}_{\rightarrow}\ R^2=\frac{E_c-E}{E_c}=0.9994$$

Since $R^2 = 0.9994$ is very close to 1, it can be concluded that quadratic response surface model fits function values very well.

Example 7.6: A cantilever beam is subjected to a tip force of $F = 1000$ N as shown in Fig. 7.19. Elastic modulus of beam is given as $E = 200$ GPa. Beam width, height, and length are represented with b, h, and L parameters, respectively. Using factorial experimental design method, create a quadratic (parabolic) response surface model for

(a) Tip deflection of the beam
(b) Maximum stress in the beam

within the ranges of 40 mm $\leq b \leq$ 120 mm, 50 mm $\leq h \leq$ 150 mm, 2000 mm $\leq L \leq$ 4000 mm.

Tip deflection (W) and maximum stress (S) for a cantilever beam subjected to a tip force can be expressed from Strength of Materials book as following:

$$\text{Tip deflection}: W=F\left(L^3\right)/(3EI),$$

$$\text{Maximum stress}: S=6FL/\left(bh^2\right), \text{where } I=bh^3/12.$$

To create quadratic response surface model, three-level factorial design can be used. Levels of b, h, and L variables in factorial design can be expressed as below:

$$b: [\text{lower, middle, upper}]=[b_1=40, b_2=80, b_3=120]$$

$$h: [\text{lower, middle, upper}]=[h_1=50, h_2=100, h_3=150]$$

$$L: [\text{lower, middle, upper}]=[L_1=2000, L_2=3000, L_3=4000]$$

Table 7.12 Factorial experimental design in symbolic form for b, h, and L variables

Point	b	h	L	W	S
1	b_1	h_1	L_1	W_1	S_1
2	b_2	h_1	L_1	W_2	S_2
3	b_3	h_1	L_1	W_3	S_3
4	b_1	h_2	L_1	W_4	S_4
5	b_2	h_2	L_1	W_5	S_5
6	b_3	h_2	L_1	W_6	S_6
7	b_1	h_3	L_1	W_7	S_7
8	b_2	h_3	L_1	W_8	S_8
9	b_3	h_3	L_1	W_9	S_9
10	b_1	h_1	L_2	W_{10}	S_{10}
11	b_2	h_1	L_2	W_{11}	S_{11}
12	b_3	h_1	L_2	W_{12}	S_{12}
13	b_1	h_2	L_2	W_{13}	S_{13}
14	b_2	h_2	L_2	W_{14}	S_{14}
15	b_3	h_2	L_2	W_{15}	S_{15}
16	b_1	h_3	L_2	W_{16}	S_{16}
17	b_2	h_3	L_2	W_{17}	S_{17}
18	b_3	h_3	L_2	W_{18}	S_{18}
19	b_1	h_1	L_3	W_{19}	S_{19}
20	b_2	h_1	L_3	W_{20}	S_{20}
21	b_3	h_1	L_3	W_{21}	S_{21}
22	b_1	h_2	L_3	W_{22}	S_{22}
23	b_2	h_2	L_3	W_{23}	S_{23}
24	b_3	h_2	L_3	W_{24}	S_{24}
25	b_1	h_3	L_3	W_{25}	S_{25}
26	b_2	h_3	L_3	W_{26}	S_{26}
27	b_3	h_3	L_3	W_{27}	S_{27}

Using levels of b, h, and L variables, Factorial experimental design for three variables can be expressed in symbolic form as in Table 7.12. In Table 7.12, last two columns indicate tip displacement (W) and maximum stress (S) values.

Numerical values of factorial design in Table 7.12 can be given as in Table 7.13.

(a) Quadratic response surface model for tip displacement:

Quadratic response surface model for tip displacement (W) can be expressed in terms of b, h, L variables as below:

Table 7.13 Numerical values of factorial design

Point	b (mm)	h (mm)	L (mm)	W (mm)	S (MPa)
1	40	50	2000	32.0000	12.0000
2	80	50	2000	16.0000	6.0000
3	120	50	2000	10.6667	4.0000
4	40	100	2000	4.0000	3.0000
5	80	100	2000	2.0000	1.5000
6	120	100	2000	1.3333	1.0000
7	40	150	2000	1.1852	1.3333
8	80	150	2000	0.5926	0.6667
9	120	150	2000	0.3951	0.4444
10	40	50	3000	108.0000	18.0000
11	80	50	3000	54.0000	9.0000
12	120	50	3000	36.0000	6.0000
13	40	100	3000	13.5000	4.5000
14	80	100	3000	6.7500	2.2500
15	120	100	3000	4.5000	1.5000
16	40	150	3000	4.0000	2.0000
17	80	150	3000	2.0000	1.0000
18	120	150	3000	1.3333	0.6667
19	40	50	4000	256.0000	24.0000
20	80	50	4000	128.0000	12.0000
21	120	50	4000	85.3333	8.0000
22	40	100	4000	32.0000	6.0000
23	80	100	4000	16.0000	3.0000
24	120	100	4000	10.6667	2.0000
25	40	150	4000	9.4815	2.6667
26	80	150	4000	4.7407	1.3333
27	120	150	4000	3.1605	0.8889

$$\tilde{W}(b,h,L) = c_1 + c_2 b + c_3 h + c_4 L + c_5 b^2 + c_6 h^2 + c_7 L^2 + c_8 bh + c_9 bL + c_{10} hL$$

$$= \begin{bmatrix} 1 & b & h & L & b^2 & h^2 & L^2 & bh & bL & hL \end{bmatrix} \begin{Bmatrix} c_1 \\ c_2 \\ c_3 \\ c_4 \\ c_5 \\ c_6 \\ c_7 \\ c_8 \\ c_9 \\ c_{10} \end{Bmatrix}$$

Using Table 7.13, quadratic response surface model for tip displacement can be expressed at design points as following:

$$\underbrace{\begin{Bmatrix} \tilde{W}_1 \\ \tilde{W}_2 \\ \vdots \\ \tilde{W}_{27} \end{Bmatrix}}_{<F} = \underbrace{\begin{bmatrix} 1 & b_1 & h_1 & L_1 & b_1^2 & h_1^2 & L_1^2 & b_1h_1 & b_1L_1 & h_1L_1 \\ 1 & b_2 & h_2 & L_2 & b_2^2 & h_2^2 & L_2^2 & b_2h_2 & b_2L_2 & h_2L_2 \\ \vdots & \vdots & \vdots & \vdots & \vdots & \vdots & \vdots & \vdots & \vdots & \vdots \\ 1 & b_{27} & h_{27} & L_{27} & b_{27}^2 & h_{27}^2 & L_{27}^2 & b_{27}h_{27} & b_{27}L_{27} & h_{27}L_{27} \end{bmatrix}}_{X} \underbrace{\begin{Bmatrix} c_1 \\ c_2 \\ \vdots \\ c_{10} \end{Bmatrix}}_{C}$$

Using W column in Table 7.13, true values for tip displacements at design points can be expressed as below:

$$\underbrace{\begin{Bmatrix} W_1 \\ W_2 \\ \vdots \\ W_{27} \end{Bmatrix}}_{F} = \begin{Bmatrix} 32.0000 \\ 16.0000 \\ \vdots \\ 3.1605 \end{Bmatrix}$$

After substituting the numerical values from Table 7.13 into X matrix, model constants C can be calculated using the general expression in Eq. (7.12) as below:

$$\begin{aligned} C &= \left[X^T X\right]^{-1} X^T F \rightarrow C^T \\ &= \left[68.95 \quad -1.25 \quad -2.19 \quad 0.07 \quad 0.01 \quad 0.01 \quad 0 \quad 0.01 \quad 0 \quad 0\right] \end{aligned}$$

$c_1 = 68.95, c_2 = -1.25, c_3 = -2.19, c_4 = 0.07, c_5 = 0.01, c_6 = 0.01, c_7 = 0, c_8 = 0.01, c_9 = 0, c_{10} = 0.$

Thus, quadratic response surface model for tip displacement is obtained as following:

$$\tilde{W} = 68.95 - 1.25b - 2.19h + 0.07L + 0.01b^2 + 0.01h^2 + 0.01bh$$

Fitting quality of the response surface model can be determined using R^2 criterion as following:

$$\text{Error of quadratic response surface model} : E = \sum_{i=1}^{27} \left(W_i - \tilde{W}_i\right)^2$$

Error of the simplest (constant) model:

$$\text{Average value}: W_{ave} = \frac{1}{27}\sum_{i=1}^{27} W_{i \rightarrow} \; E_c = \sum_{i=1}^{27} (W_{ort} - W_i)^2 {}_{\rightarrow} \; R^2 = \frac{E_c - E}{E_c}$$

$$= 0.889613$$

Since $R^2 = 0.8896$ is greater than 0.8, it can be concluded that quadratic response surface model fits function values sufficiently well within the given limits of variables. That is, response surface model can be used instead of analytical model to predict the tip displacement on the beam.

(b) Quadratic response surface model for maximum stress:

Quadratic response surface model for maximum stress (S) can be expressed in terms of b, h, L variables as below:

$$\tilde{S}(b,h,L) = c_1 + c_2 b + c_3 h + c_4 L + c_5 b^2 + c_6 h^2 + c_7 L^2 + c_8 bh + c_9 bL + c_{10} hL$$

$$= \begin{bmatrix} 1 & b & h & L & b^2 & h^2 & L^2 & bh & bL & hL \end{bmatrix} \begin{Bmatrix} c_1 \\ c_2 \\ c_3 \\ c_4 \\ c_5 \\ c_6 \\ c_7 \\ c_8 \\ c_9 \\ c_{10} \end{Bmatrix}$$

Using Table 7.13, quadratic response surface model for maximum stress can be expressed at design points as following

$$\underbrace{\begin{Bmatrix} \tilde{S}_1 \\ \tilde{S}_2 \\ \vdots \\ \tilde{S}_{27} \end{Bmatrix}}_{<F} = \underbrace{\begin{bmatrix} 1 & b_1 & h_1 & L_1 & b_1^2 & h_1^2 & L_1^2 & b_1 h_1 & b_1 L_1 & h_1 L_1 \\ 1 & b_2 & h_2 & L_2 & b_2^2 & h_2^2 & L_2^2 & b_2 h_2 & b_2 L_2 & h_2 L_2 \\ \vdots & \vdots & \vdots & \vdots & \vdots & \vdots & \vdots & \vdots & \vdots & \vdots \\ 1 & b_{27} & h_{27} & L_{27} & b_{27}^2 & h_{27}^2 & L_{27}^2 & b_{27} h_{27} & b_{27} L_{27} & h_{27} L_{27} \end{bmatrix}}_{X} \underbrace{\begin{Bmatrix} c_1 \\ c_2 \\ \vdots \\ c_{10} \end{Bmatrix}}_{C}$$

Using S column in Table 7.13, true values for maximum stress at design points can be expressed as below:

$$\underbrace{\begin{Bmatrix} S_1 \\ S_2 \\ \vdots \\ S_{27} \end{Bmatrix}}_{F} = \begin{Bmatrix} 12.0000 \\ 6.0000 \\ \vdots \\ 0.8889 \end{Bmatrix}$$

After substituting the numerical values from Table 7.13 into X matrix, model constants C can be calculated using the general expression in Eq. (7.12) as below:

$$\begin{aligned} C &= \left[X^T X\right]^{-1} X^T F \rightarrow C^T \\ &= \left[264.07 \quad -2.69 \quad -3.76 \quad 0.07 \quad 0.01 \quad 0.01 \quad 0 \quad 0.01 \quad 0 \quad 0\right] \end{aligned}$$

$c_1 = 264.07, c_2 = -2.69, c_3 = -3.76, c_4 = 0.07, c_5 = 0.01, c_6 = 0.01, c_7 = 0, c_8 = 0.01, c_9 = 0, c_{10} = 0.$

Thus, quadratic response surface model for maximum stress is obtained as following:

$$S(b, h, L) = 264.07 - 2.69b - 3.76h + 0.07L + 0.01b^2 + 0.01h^2 + 0.01bh$$

Fitting quality of the response surface model can be determined using R^2 criterion as following:

$$\text{Error of quadratic response surface model} : E = \sum_{i=1}^{27} \left(S_i - \tilde{S}_i\right)^2$$

Error of the simplest (constant) model:

$$\text{Average value} : S_{ave} = \frac{1}{27} \sum_{i=1}^{27} S_i \rightarrow E_c = \sum_{i=1}^{27} \left(S_{ort} - S_i\right)^2 \rightarrow R^2 = \frac{E_c - E}{E_c} = 0.9585$$

Since $R^2 = 0.9585$ is very close to 1, it can be concluded that quadratic response surface model fits function values very well within the given limits of variables. That is, response surface model can be used instead of analytical model to predict maximum stress values on the beam.

7.5 Analysis of Variance (ANOVA)

Analysis of Variance, shortly known as ANOVA, is used to interpret experimental design results. ANOVA study quantitatively reveals how function is affected by its variables. ANOVA uses statistical concepts for this purpose. To explain ANOVA, we can use the values of a function $f(x)$ as shown in Fig. 7.20. If $f(x)$ values vary with change of x values, we can say that x variable has an effect or influential on $f(x)$ function. ANOVA method uses the mean value concept of statistics to facilitate this study. In statistics, if function values at different points of a variable vary around the mean function value, the variable is considered to have an effect on the function.

Mathematically, effect of a variable on a function can be expressed with Sum of Squares of deviations from the mean value. For example, if function values are known as $f_1, f_2,\ldots, f_n$ at $x_1,x_2,\ldots, x_n$ points of x variable, Sum of Squares for the function can be expressed using average value as following:

$$\text{Average value}: f_m = \frac{f_1 + f_2 + \ldots + f_n}{n} = \frac{\sum_{i=1}^{n} f_i}{n} \tag{7.19}$$

$$\text{Sum of Squares}: S_T = (f_1 - f_m)^2 + (f_2 - f_m)^2 + \ldots + (f_n - f_m)^2 = \sum_{i=1}^{n} (f_i - f_m)^2 \tag{7.20}$$

Average and Sum of Squares concepts are extensively used in ANOVA as in regression analysis. ANOVA and regression analysis are compared in Fig. 7.21. From Fig. 7.21, ANOVA and regression are seen to examine the same data with different methods. Therefore, regression and ANOVA can sometimes be expressed as different forms of the relationship between variable and function.

In regression, a mathematical model is obtained between variable and function. Function values are assumed to have a normal distribution around the fitted model as shown in Fig. 7.21a. In ANOVA, effect of variables on the function are determined quantitatively using mean and sum of squares values. Function values are assumed

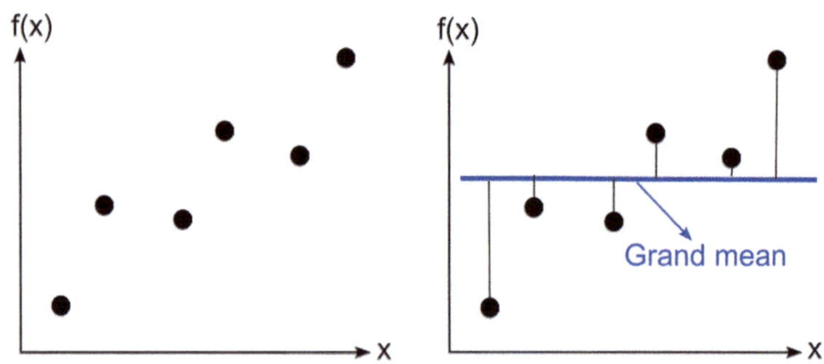

Fig. 7.20 Determining effect of a variable on a function through the deviation of function values from the mean value

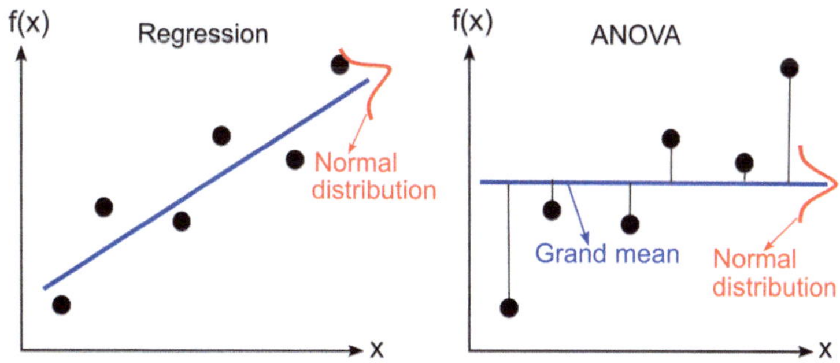

Fig. 7.21 Regression and ANOVA

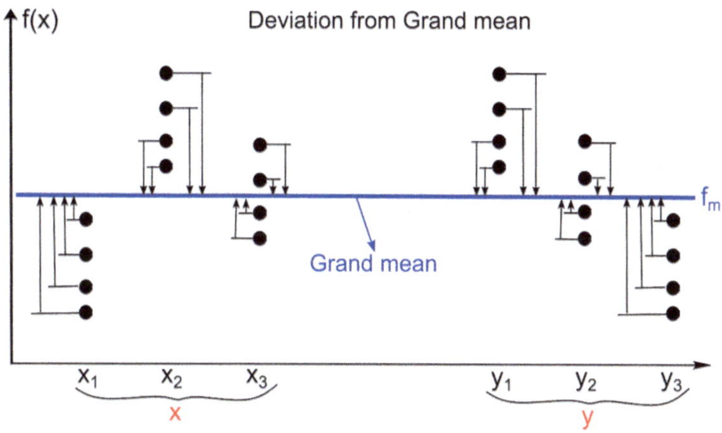

Fig. 7.22 Multiple function values at x_1, x_2, x_3 levels of x variable and y_1, y_2, y_3 levels of y variable

to have a normal distribution around the general mean as shown in Fig. 7.21b. Unlike regression, ANOVA method can work with non-numerical (string) variables such as gender, color, and smell. ANOVA analysis is also more advantageous than regression in examining the interaction between variable and function when multiple function values are available at the same point. Multiple values at the same point are frequently encountered in experimental studies.

In ANOVA study, levels of variables as in experimental design methods are used. To investigate the effect of a variable on a function, average function values at all levels (i.e., level means) as well as the average of all values (i.e., grand mean) as shown in Fig. 7.22 are used.

In ANOVA method, mean of all data is called as Grand average or Grand mean and mean at each level is called as Level Mean. Deviation from a Level Mean is called Level Error. Figure 7.23 shows level means of a two-variable function and deviations of level means from the grand mean. Figure 7.24 shows deviations of function values from level means.

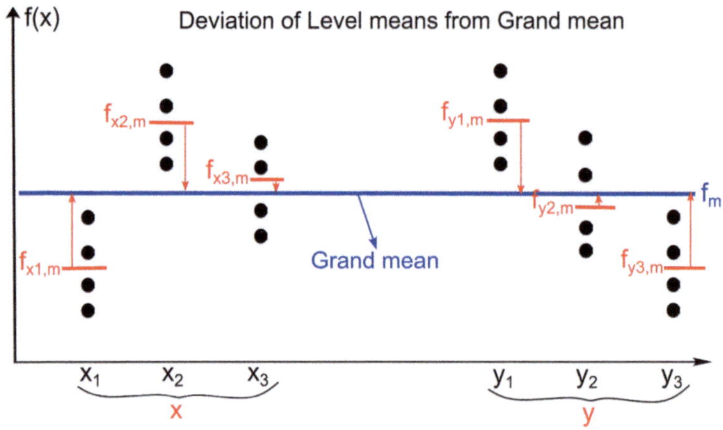

Fig. 7.23 Average values of variable levels. Level means of a two-variable function and deviations of level means from the grand mean

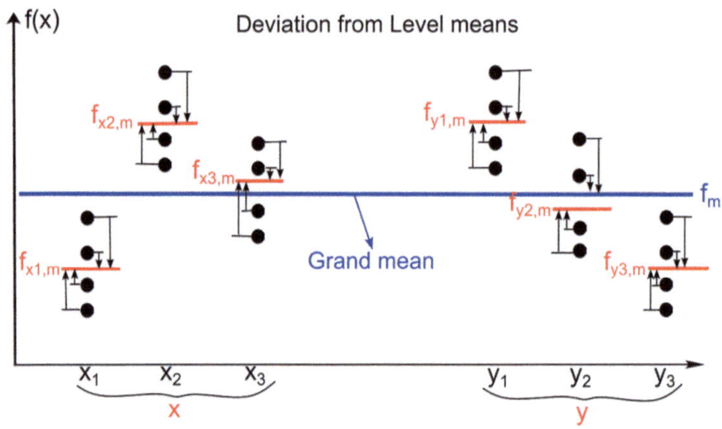

Fig. 7.24 Deviations of function values from level means (Error)

In ANOVA, deviation of function values from the grand mean is calculated by superposition (i.e., summation) of: (1) deviation of level means from the grand mean, and (2) deviation of function values from level means (i.e., error) as illustrated in Fig. 7.25. Deviation of level means from the grand mean corresponds to the effect of variables on the result, while deviation of function values from level means corresponds to the effect of measurement errors.

After introducing the basic concepts of ANOVA, we can explain its mathematical expressions. Assume that multiple values of a two-variable $f(x,y)$ function are known at $x_1, x_2, \ldots, x_k$ levels of x variable and $y_1, y_2, \ldots, y_l$ levels of y variable as following (function values are denoted with f):

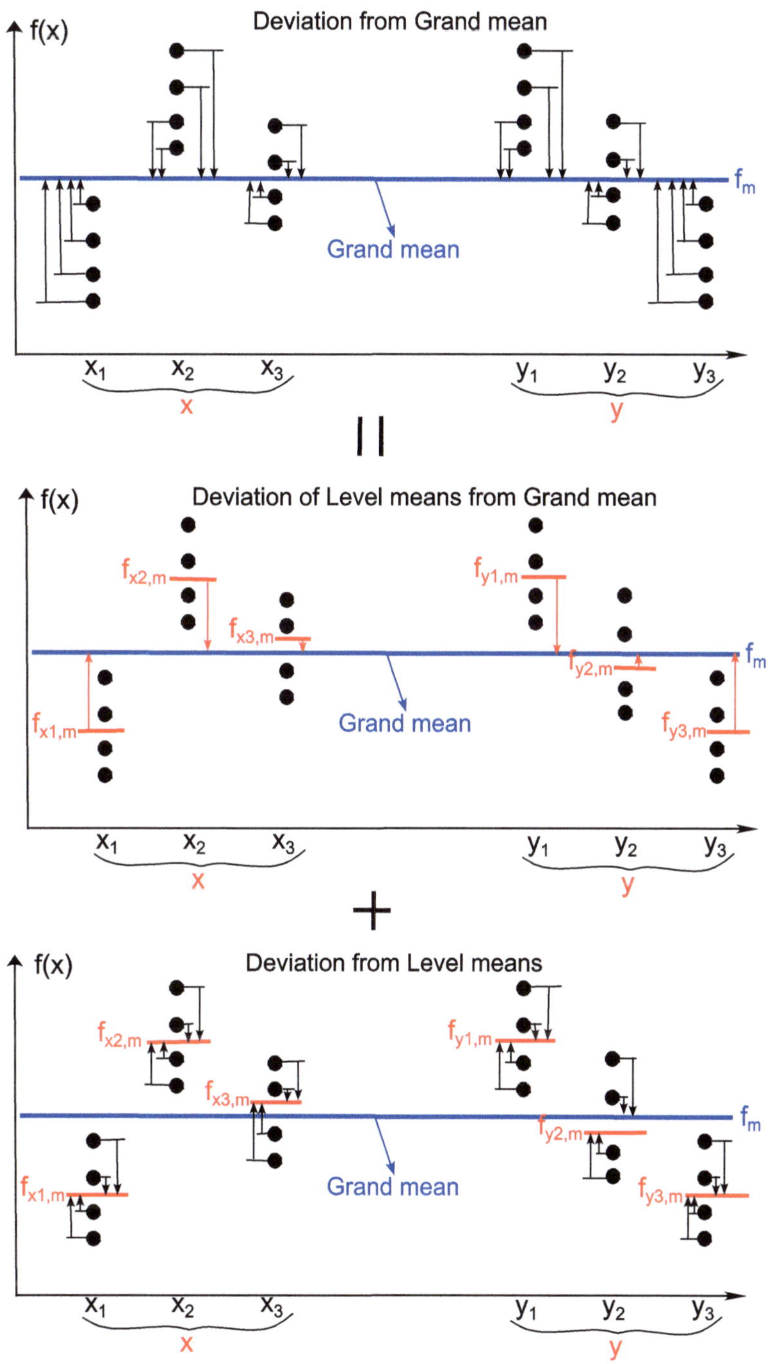

Fig. 7.25 Expressing deviation from the grand mean as the deviation of level mean from the grand mean and deviation of error from level mean

For x variable:

$$\begin{aligned} x_1 \text{ level} &: f_{x1,1}, f_{x1,2}, \ldots, f_{x1,p} \text{ (}p\text{ pieces of data)} x_2 \text{ level} \\ &: f_{x2,1}, f_{x2,2}, \ldots, f_{x2,q} \text{ (}q\text{ pieces of data)} \ldots x_k \text{ level} \\ &: f_{xk,1}, f_{xk,2}, \ldots, f_{xk,r} \text{ (}r\text{ pieces of data)} \end{aligned}$$

For y variable:

$$\begin{aligned} y_1 \text{ level} &: f_{y1,1}, f_{y1,2}, \ldots, f_{y1,s} \text{ (}s\text{ pieces of data)} y_2 \text{ level} \\ &: f_{y2,1}, f_{y2,2}, \ldots, f_{y2,t} \text{ (}t\text{ pieces of data)} \ldots y_l \text{ level} \\ &: f_{yl,1}, f_{yl,2}, \ldots, f_{yl,z} \text{ (}z\text{ pieces of data)} \end{aligned}$$

In this case, grand mean of function values can be expressed as below:

$$\text{Sum of function values} : f_T$$

$$\begin{aligned} f_T = &\underbrace{f_{x1,1} + f_{x1,2} + \ldots + f_{x1,p}}_{x1 \quad level} + \underbrace{f_{x2,1} + f_{x2,2} + \ldots + f_{x2,q}}_{x2 \quad level} + \ldots \\ &+ \underbrace{f_{xk,1} + f_{xk,2} + \ldots + f_{xk,r}}_{xk \quad level} + \underbrace{f_{y1,1} + f_{y1,2} + \ldots + f_{y1,s}}_{y1 \quad level} \\ &+ \underbrace{f_{y2,1} + f_{y2,2} + \ldots + f_{y2,t}}_{y2 \quad level} + \ldots + \underbrace{f_{yl,1} + f_{yl,2} + \ldots + f_{yl,z}}_{yl \quad level} \end{aligned} \tag{7.21}$$

$$\text{Grand mean} : f_m = \frac{f_T}{N = (p + q + \ldots + r) + (s + t + \ldots + z)} \tag{7.22}$$

Level means of variables can be expressed as following:
Level means of x variable:

$$\begin{aligned} x_1 \text{ level} &: f_{x1,m} = \frac{f_{x1,1} + f_{x1,2} + \ldots + f_{x1,p}}{p}, \quad x2 \text{ level} \\ &: f_{x2,m} = \frac{f_{x2,1} + f_{x2,2} + \ldots + f_{x2,q}}{q}, \ldots, x_k \text{ level} \\ &: f_{xk,m} = \frac{f_{xk,1} + f_{xk,2} + \ldots + f_{xk,r}}{r} \end{aligned}$$

Level means of y variable:

$$y_1 \text{ level} : f_{y1,m} = \frac{f_{y1,1} + f_{y1,2} + \ldots + f_{y1,s}}{s},$$
$$y2 \text{ level} : f_{y2,m} = \frac{f_{y2,1} + f_{y2,2} + \ldots + f_{y2,t}}{t}, \ldots,$$
$$y_l \text{ level} : f_{yl,m} = \frac{f_{yl,1} + f_{yl,2} + \ldots + f_{yl,z}}{z}$$

Using function values and grand mean, sum of squares of function values (S_T) can be calculated as below:

$$\begin{aligned} S_T = {} & \underbrace{(f_{x1,1} - f_m)^2 + (f_{x1,2} - f_m)^2 + \ldots + (f_{x1,p} - f_m)^2}_{x1 \quad level} \\ & + \underbrace{(f_{x2,1} - f_m)^2 + (f_{x2,2} - f_m)^2 + \ldots + (f_{x2,q} - f_m)^2}_{x2 \quad level} + \ldots \\ & + \underbrace{(f_{xk,1} - f_m)^2 + (f_{xk,2} - f_m)^2 + \ldots + (f_{xk,r} - f_m)^2}_{xk \quad level} \\ & + \underbrace{(f_{y1,1} - f_m)^2 + (f_{y1,2} - f_m)^2 + \ldots + (f_{y1,s} - f_m)^2}_{y1 \quad level} \\ & + \underbrace{(f_{y2,1} - f_m)^2 + (f_{y2,2} - f_m)^2 + \ldots + (f_{y2,t} - f_m)^2}_{y2 \quad level} + \ldots \\ & + \underbrace{(f_{yl,1} - f_m)^2 + (f_{yl,2} - f_m)^2 + \ldots + (f_{yl,z} - f_m)^2}_{yl \quad level} \end{aligned} \tag{7.23}$$

Sum of squares for x and y variables can be calculated as following:

$$\text{For } x \text{ variable} : S_x = p\underbrace{(f_{x1,.m} - f_m)^2}_{x1 \ level} + q\underbrace{(f_{x2,m} - f_m)^2}_{x2 \ level} + \ldots + r\underbrace{(f_{xk,m} - f_m)^2}_{xk \ level} \tag{7.24}$$

$$\text{For } y \text{ variable} : S_y = s\underbrace{(f_{y1,.m} - f_m)^2}_{y1 \quad level} + t\underbrace{(f_{y2,m} - f_m)^2}_{y2 \quad level} + \ldots + z\underbrace{(f_{yl,m} - f_m)^2}_{yl \quad level} \tag{7.25}$$

Sum of squares for error (S_e) can be determined using S_x, S_y, and S_T values as below:

$$S_T = S_x + S_y + S_e \rightarrow S_e = S_T - (S_x + S_y)$$

Using sum of squares values, Variance (Mean Square) of x, y variables and error can be obtained as following:

$$\text{For } x \text{ variable}: V_x = \frac{S_x}{d_x} = \frac{S_x}{k-1} \; (d_x: \text{degree of freedom for } x \text{ variable})$$

where d_x is calculated as d_x =(number of x levels) − 1.

$$\text{For } y \text{ variable}: V_y = \frac{S_y}{d_y} = \frac{S_y}{l-1} \; \left(d_y: \text{degree of freedom for } y \text{ variable}\right)$$

where d_y is calculated as d_y =(number of y levels) − 1.

$$\text{For error}: V_e = \frac{S_e}{d_e}$$
$$= \frac{S_e}{d_T - \left(d_x + d_y\right)} (d_T = N - 1: \text{total degrees of freedom}, N: \text{number of data})$$

Using Variance expressions, F-ratios for x and y variables can be expressed as below:

$$F-\text{ratio for } x \text{ variable}: F_x = \frac{V_x}{V_e}, \quad F-\text{ratio for } y \text{ variable}: F_y = \frac{V_y}{V_e}$$

Reliability of statistical results can be ensured with p-values which are obtained using cumulative distribution function based on F-ratios and degrees of freedoms. For example, p-value for x variable can be obtained with cumulative distribution function using F_x, d_x, d_e values. Similarly, p-value for y variable can be obtained with cumulative distribution function using F_y, d_y, d_e values. To obtain p values for x and y variables, MATLAB built-in command of *fcdf* can be used in the form of $p = fcdf(F_x, d_x, d_e$, 'upper') and $p = fcdf(F_y, d_y, d_e$, 'upper'). Calculated p values are then compared with critical values. Critical p value is often chosen as $p_{cr} = 0.05$ indicating $1 - 0.05 = 0.95$ (95%) confidence level. If p-value for a variable is smaller than the critical value ($p < p_{cr}$), it means that variable is effective on the function with $100(1 - p_{cr})$ % confidence level.

After ensuring confidence levels, effect of variables and error (i.e., effect of unconsidered variables or factors) on the function can be quantitatively determined using P (capital P) values as following:

$$\text{For } x \text{ variable}: P_x = \frac{S_x}{S_T} \tag{7.26}$$

$$\text{For } y \text{ variable}: P_y = \frac{S_y}{S_T} \tag{7.27}$$

$$\text{For error}: P_e = \frac{S_e}{S_T} \tag{7.28}$$

Effect of variables and error can be calculated more conservatively as below:

$$\text{For } x \text{ variable}: P_x = \frac{S_x - d_x V_e}{S_T} \tag{7.29}$$

$$\text{For } y \text{ variable}: P_y = \frac{S_y - d_y V_e}{S_T} \tag{7.30}$$

$$\text{For error}: P_e = \frac{S_e}{S_T} \tag{7.31}$$

P values in above equations indicate the effect of variables on the function as percentage. For reliability of P values, P_e value should be less than 0.5 (50%).

Example 7.7: Perform ANOVA analysis using Table 7.14 and determine the effect of x and y variables on f function quantitatively.

Using Table 7.14, grand mean can be calculated as below:

Sum of function values: $f_T = f_1 + f_2 + f_3 + \ldots + f_9 = 93$ ($N = 9$, number of experiments)

$$\text{Grand mean}: f_m = \frac{f_T}{N} = \frac{93}{9} = 10.3333$$

Indicating lower level with x_1, middle level with x_2, upper level with x_3 for x variable and lower level with y_1, middle level with y_2, upper level with y_3 for y variable, level means can be obtained as following:

Table 7.14 Factorial experimental design and corresponding function values

Point	x	y	$f(x, y)$
1	0	1	1
2	1	1	4
3	2	1	9
4	0	2	4
5	1	2	9
6	2	2	16
7	0	3	9
8	1	3	16
9	2	3	25

$$f_{x1,m}=\frac{f_1+f_4+f_7}{3}=4.6667, f_{x2,m}=\frac{f_2+f_5+f_8}{3}=9.6667, f_{x3,m}=\frac{f_3+f_6+f_9}{3}=16.6667$$

$$f_{y1,m}=\frac{f_1+f_2+f_3}{3}=4.6667, f_{y2,m}=\frac{f_4+f_5+f_6}{3}=9.6667, f_{y3,m}=\frac{f_7+f_8+f_9}{3}=16.6667$$

Using function values and grand mean, sum of squares of function values (S_T) can be calculated as below:

$$\begin{aligned} S_T &= (f_1-f_m)^2+(f_2-f_m)^2+(f_3-f_m)^2+(f_4-f_m)^2+(f_5-f_m)^2 \\ &+(f_6-f_m)^2+(f_7-f_m)^2+(f_8-f_m)^2+(f_9-f_m)^2=452 \end{aligned}$$

Using level means and grand mean, sum of squares for x and y levels can be calculated as following:

$$S_x=3\underbrace{(f_{x1,.m}-f_m)^2}_{x1\quad level}+3\underbrace{(f_{x2,m}-f_m)^2}_{x2\quad level}+3\underbrace{(f_{x3,m}-f_m)^2}_{x3\quad level}=218$$

$$S_y=3\underbrace{(f_{y1,.m}-f_m)^2}_{y1\quad level}+3\underbrace{(f_{y2,m}-f_m)^2}_{y2\quad level}+3\underbrace{(f_{y3,m}-f_m)^2}_{y3\quad level}=218$$

Sum of squares for error (S_e) can be determined using S_x, S_y, and S_T values as below:

$$S_T=S_x+S_y+S_e \rightarrow S_e=S_T-(S_x+S_y)=452-(218+218)=16$$

Degrees of freedom for variables and error can be calculated as following:

$$\text{Total degrees of freedom}: d_T=(\text{Number of Experiments})-1 \rightarrow d_T=N-1=9-1=8$$

$$\text{Degree of freedom for } x \text{ variable}: d_x=(\text{Number of } x \text{ levels})-1 \rightarrow d_x=3-1=2$$

$$\text{Degree of freedom for } y \text{ variable}: d_y=(\text{Number of } y \text{ levels})-1 \rightarrow d_y=3-1=2$$

$$\begin{aligned} \text{Degrees of freedom for error}: d_e &= (\text{Total degrees of freedom}) \\ &-(\text{Sum of degrees of freedom for all variables}) \\ &\rightarrow d_e=d_T-(d_x+d_y)=8-(2+2)=4 \end{aligned}$$

Using sum of squares values, Variances (Mean Squares) of x and y variables as well as Error can be obtained as following:

$$\text{For } x \text{ variable} : V_x = \frac{S_x}{d_x} = \frac{218}{2} = 109, \quad \text{For } y \text{ variable}$$

$$: V_y = \frac{S_y}{d_y} = \frac{218}{2} = 109, \quad \text{For error} : V_e = \frac{S_e}{d_e} = \frac{16}{4} = 4$$

Using Variance expressions, F-ratios for x and y variables can be expressed as below:

$$F - \text{ratio for } x \text{ variable} : F_x = \frac{V_x}{V_e} = \frac{109}{4} = 27.25, \quad F - \text{ratio for } y \text{ variable}$$

$$: F_y = \frac{V_y}{V_e} = \frac{109}{4} = 27.25$$

Confidence levels for x and y variables can be ensured with p-values which are obtained using cumulative distribution function $p(F_x, d_x, d_e)$ based on F-ratio and degrees of freedom. p values can be obtained using *fcdf* command in MATLAB as following:

For x variable:

```
p=fcdf(Fx, dx, de,'upper')→p=fcdf(27.25,2,4,'upper')=0.0047<0.05
```

For y variable:

```
p=fcdf(Fy, dy, de,'upper')→p=fcdf(27.25,2,4,'upper')=0.0047<0.05
```

Since p-values are smaller than the selected critical value of $p_{cr} = 0.05$ $(p < p_{cr})$, variables are assumed to be effective on the function with 95% confidence level.

After ensuring confidence levels, effect of variables and error (effect of unconsidered variables or factors) on the function can be quantitatively determined using P (capital P) values as following:

$$\text{For } x \text{ variable} : P_x = \frac{S_x}{S_T} = \frac{218}{452} = 0.4823 = \%48.23$$

$$\text{For } y \text{ variable} : P_y = \frac{S_y}{S_T} = \frac{218}{452} = 0.4823 = \%48.23$$

$$\text{For error} : P_e = \frac{S_e}{S_T} = \frac{218}{452} = 0.0354 = \%3.54$$

Effect of variables and error can be calculated more conservatively as below:

$$\text{For } x \text{ variable} : P_x = \frac{S_x - d_x V_e}{S_T} = \frac{218 - 2x4}{452} = 0.4646 = \%46.46$$

$$\text{For } y \text{ variable} : P_y = \frac{S_y - d_y V_e}{S_T} = \frac{218 - 2x4}{452} = 0.4646 = \%46.46$$

$$\text{For error} : P_e = \frac{S_e}{S_T} = \frac{218}{452} = 0.0354 = \%3.54$$

Regarding the reliability of P values since P_e value is less than 0.5 (50%), results can be trusted. From the results it is concluded that, x and y variables are equally effective on the $f(x,y)$ function. In contrast, Measurement error (error of unconsidered variables or factors) has a very small effect on the function.

7.6 Inverse Distance Method

Inverse distance method is one of the simplest methods to predict intermediate values. In this method, intermediate value is predicted with weighted sum of the function values in a limited region/domain centered around the intermediate point. Region is often assumed to have circular (in two-variable case) or spherical shape (in multivariable case) with radius of R. Weight coefficients are selected to be inversely proportional with distances from the intermediate point to the neighboring points in the region. Radius R indicates the domain of influence in inverse distance method.

Inverse distance method can be simply illustrated as in Fig. 7.26. In Fig. 7.26, function value at the intermediate point i can be estimated using function values at neighboring points in the circular domain with radius of R as following:

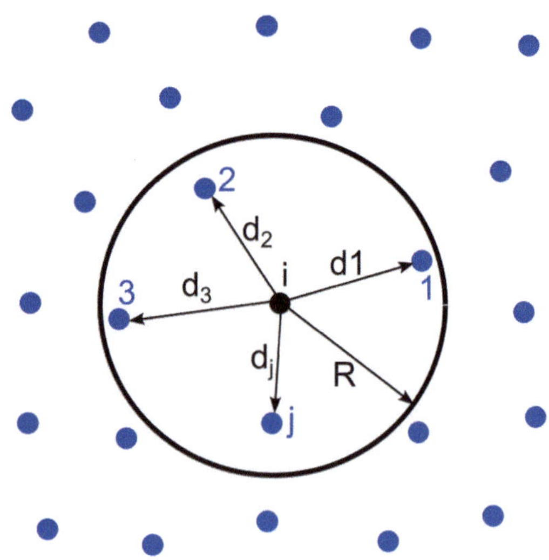

Fig. 7.26 Estimation of function value at a point i with inverse distance method

$$f_i = \frac{\frac{1}{d_1^p}f_1 + \frac{1}{d_2^p}f_2 + \frac{1}{d_3^p}f_3 + \frac{1}{d_4^p}f_4}{\frac{1}{d_1^p} + \frac{1}{d_2^p} + \frac{1}{d_3^p} + \frac{1}{d_4^p}} = \frac{\sum_{j=1}^{4} \frac{1}{d_j^p} f_j}{\sum_{j=1}^{4} \frac{1}{d_j^p}} \tag{7.32}$$

In Eq. (7.32), d_j denotes the distance between the intermediate point (i) and j-th neighboring point and $\frac{1}{d_j^p}$ shows the weight coefficient for the j-th point, which is inversely proportional with the distance. p is the coefficient indicating the influence of distance in the region. $p = 2$ is commonly used in applications. Distance between $i(x_i, y_i)$ and $j(x_j, y_j)$ points in a 2-D domain can be expressed in terms of coordinates as below:

$$d_{ij} = \sqrt{(x_j - x_i)^2 + (y_j - y_i)^2} \tag{7.33}$$

In inverse distance method, effect of neighboring points on the intermediate value can be increased or decreased by adjusting the size of radius R and p coefficient.

Example 7.8: Using function values in Fig. 7.27, estimate the function value at intermediate point s with Inverse Distance method.

Coordinates of the points in Fig. 7.27 and their distances from the intermediate point s can be calculated as in Table 7.15.

Taking the exponential coefficient as $p = 2$ and using the values in Table 7.15, function value at intermediate point s can be predicted as following:

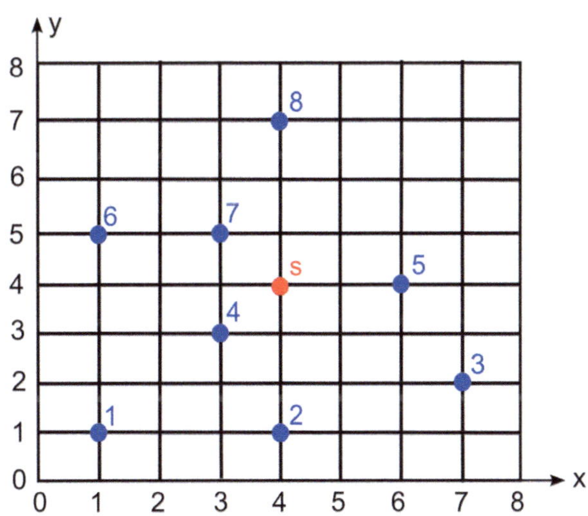

Fig. 7.27 Function values to be used in inverse distance method

Table 7.15 Coordinates of the points and their distance from the *s* point

Point	x	y	d (distance from s)	f
s	4	4		?
1	1	1	4.2426	40
2	4	1	3	60
3	7	2	3.6056	30
4	3	3	1.4142	20
5	6	4	2	24
6	1	5	3.1623	36
7	3	5	1.4142	28
8	4	7	3	42

$$f_s = f(4,4) = \frac{\frac{1}{d_1^2}f_1 + \frac{1}{d_2^2}f_2 + \frac{1}{d_3^2}f_3 + \frac{1}{d_4^2}f_4 + \frac{1}{d_5^2}f_5 + \frac{1}{d_6^2}f_6 + \frac{1}{d_7^2}f_7 + \frac{1}{d_8^2}f_8}{\frac{1}{d_1^2} + \frac{1}{d_2^2} + \frac{1}{d_3^2} + \frac{1}{d_4^2} + \frac{1}{d_5^2} + \frac{1}{d_6^2} + \frac{1}{d_7^2} + \frac{1}{d_8^2}} = 29.0158$$

7.7 Kriging Method

Kriging method is a method to predict intermediate values combining regression (curve fitting) and statistical techniques [14–18]. It is a linear regression method that has been increasingly used in recent years. Kriging method is a linear regression method developed by a South African Mining Engineer for modeling land forms and later expanded to other areas. In Kriging method, it is assumed that points with the same proximities will have similar values (with a statistical distribution around a certain value). In the Kriging method, covariance concept is used to express the relationship between position proximity and function value. Basic statistical concepts used in Kriging method can be introduced as following.

7.7.1 Basic Concepts of Statistics

Expected Value:

If a variable X has multiple values such as $X_1, X_2, X_3, \ldots, X_n$, expected value ($E$) for this variable can be expressed as the mean (μ) of function values as following:

$$E(X) = \mu = \frac{X_1 + X_2 + X_3 + \ldots + X_n}{n} \tag{7.34}$$

Variance:

Variance measures the difference between variable values. Variance can be most easily expressed as the deviation of variable values from the expected or mean value. For a random variable X with X_1, X_2, X_3, …, X_n values, sum of square of the difference (*SSD*) between variable value X_i and mean value μ can be expressed as below:

$$SSD = (X_1 - \mu)^2 + (X_2 - \mu)^2 + (X_3 - \mu)^2 + \ldots + (X_n - \mu)^2$$

Variance is expressed as the average of the sum of square of differences and often shown with σ^2 symbol as following:

$$\sigma^2 = \frac{SSD}{n} = \frac{(X_1 - \mu)^2 + (X_2 - \mu)^2 + (X_3 - \mu)^2 + \ldots + (X_n - \mu)^2}{n} \quad (7.35)$$

Definition of variance is one of the basic concepts in statistics. Variance can also be written in terms of expected value (*E*) as below:

$$\sigma^2 = E\left[(X - \mu)^2\right] = E\left[(X - E(X))^2\right] \quad (\mu = E(X)) \quad (7.36)$$

Variance can be simply interpreted as the measure of the scattering of variable values.

Covariance:

Similar to variance, covariance concept is also widely used in statistics. Covariance is used as the measure of relationship between two variables. Covariance is expressed using the joint variability of variables. For a random variable X with X_1, X_2, X_3, ..., X_n values and random variable Y with Y_1, Y_2, Y_3, ..., Y_n values, Covariance for these variables $Cov(X, Y)$ can be expressed as following:

$$Cov(X,Y) = \frac{(X_1 - \mu_X)(Y_1 - \mu_Y) + (X_2 - \mu_X)(Y_2 - \mu_Y) + \ldots + (X_n - \mu_X)(Y_n - \mu_Y)}{n} \quad (7.37)$$

Covariance can also be expressed in terms of expected values as below:

$$\begin{aligned} &Cov(X,Y) = E[(X - \mu_X)(Y - \mu_Y)] \\ &= E[XY - \mu_Y X - \mu_X Y + \mu_X \mu_X] \\ &= E(XY) - \mu_Y E(X) - \mu_X E(Y) + \mu_X \mu_Y \quad (\mu_X = E(X), \mu_Y = E(Y)) \\ &= E(XY) - \mu_Y \mu_X - \mu_X \mu_Y + \mu_X \mu_Y \\ &= E(XY) - \mu_X \mu_Y \end{aligned}$$

Using the above information, Covariance can be simply expressed as following:

$$Cov(X,Y) = E(XY) - \mu_X \mu_Y \quad (7.38)$$

where $E(XY)$ indicates the expected value of the product of X and Y variables. In obtaining Covariance expression above, summation and scalar multiplication properties of expected values such as $E(X + Y) = E(X) + E(Y)$, $E(cX) = cE(X)$ and $E(c) = c$ was used.

From Eq. (7.38), expected value of the product can be expressed in terms of Covariance as below:

$$E(XY) = Cov(X, Y) + \mu_X\mu_Y \tag{7.39}$$

When X and Y are assumed the same ($X = Y$ or $Y = X$), Covariance-Variance relationships can be expressed as $Cov(X, X) = Var(X)$ and $Cov(Y, Y) = Var(Y)$.

Using $X = Y$ or $Y = X$ in Eq. (7.39) as well as $Cov(X, X) = Var(X)$ and $Cov(Y, Y) = Var(Y)$ expressions, following equations can be obtained:

$$E\left(X^2\right) = Var(X) + (\mu_X)^2 \tag{7.40}$$

$$E\left(Y^2\right) = Var(Y) + (\mu_Y)^2 \tag{7.41}$$

Variance of the Difference of Two Variables:

Variance of the difference of two random variables $Var(X - Y)$ can be written in terms of the expected values as following:

$$Var(X - Y) = E\left[(X - Y)^2\right] = E\left[X^2 - 2XY + Y^2\right] = E\left(X^2\right) - 2E(XY) + E\left(Y^2\right) \tag{7.42}$$

Also, if $X = Y$ or $Y = X$ is considered, relationship between Variance and Covariance can be written as $Var(X) = Cov(X, X)$ and $Var(Y) = Cov(Y, Y)$. In this case, using the relationships of $Var(X) = Cov(X, X)$, $Var(Y) = Cov(Y, Y)$ and Eqs. (7.40)–(7.41), Eq. (7.42) can be rewritten as below:

$$\begin{aligned} Var(X - Y) &= E\left[(X - Y)^2\right] \\ &= E\left(X^2\right) - 2E(XY) + E\left(Y^2\right) \\ &= E\left(X^2\right) - 2(Cov(X, Y) + \mu_X\mu_X) + E\left(Y^2\right) \end{aligned} \tag{7.43}$$

In the literature, for convenience, semi-variance expression $\frac{1}{2}E\left[(X - Y)^2\right]$ instead of variance expression $E[(X - Y)^2]$ is used as following:

$$\gamma = \frac{1}{2}E\left[(X - Y)^2\right] = \frac{E\left(X^2\right) - 2(Cov(X, Y) + \mu_X\mu_X) + E\left(Y^2\right)}{2} \tag{7.44}$$

In Eq. (7.44), division by 2 is done to facilitate the mathematical operations in Kriging method.

Variogram:

Semi-variance expressed as function of the distance between points is called as Variogram function and the plotted graphic is shortly called as Variogram or Semi-Variogram. Variogram shows how the difference between two values changes width distance h. For this reason, Variogram is widely used in Kriging method as a measure of difference.

Consider $Z(s_j) = Z_j$ and $Z(s_i) = Z_i$ values of a Z variable at s_i and s_j points separated by h ($h = s_i - s_j$) distance. Taking $X = Z_i$ and $Y = Z_j$, Eq. (7.39) can be written as below:

$$E(Z_iZ_j) = C(s_i - s_j) + \mu^2 \tag{7.45}$$

Where C indicates Covariance expressed as $Cov(X, Y)$ in Eq. (7.39). Using $Z_i = Z_j$ and $Z_j = Z_i$ as well as $s_i = s_j$ in Eq. (7.45), following equations can be obtained:

$$E(Z_i^2) = C(0) + \mu^2 = \sigma^2 + \mu^2 \tag{7.46}$$

$$E\left(Z_j^2\right) = C(0) + \mu^2 = \sigma^2 + \mu^2 \tag{7.47}$$

Substituting Eqs. (7.46) and (7.47) into Eq. (7.44), Semi-Variance for Z_i and Z_j values separated by $h = s_i - s_j$ distance can be expressed as below:

$$\gamma(s_i - s_j) = \frac{C(0) + \mu^2 - 2(C(s_i - s_j) + \mu^2) + C(0) + \mu^2}{2} \tag{7.48}$$

Equation (7.48) can be rewritten as following:

$$\gamma(s_i - s_j) = C(0) - C(s_i - s_j) \tag{7.49}$$

$$C(s_i - s_j) = C(0) - \gamma(s_i - s_j) \tag{7.50}$$

Using $h = s_i - s_j$ expression, Eqs. (7.49)–(7.50) can be expressed as below:

$$\gamma(h) = C(0) - C(h) \tag{7.51}$$

$$C(h) = C(0) - \gamma(h) \tag{7.52}$$

When variogram function in Eq. (7.51) is potted, it is seen that variogram values increase with distance h. This trend is opposite for Covariogram function. Plots of idealized Variogram and Covariogram functions are shown in Fig. 7.28.

In practice, experimental Variogram values are usually not on a smooth curve like the idealized one in Fig. 7.28. They are often scattered around a theoretical curve. Therefore, to simplify calculations, a theoretical curve is fitted to the actual Variogram values to obtain a Variogram model as in Fig. 7.29.

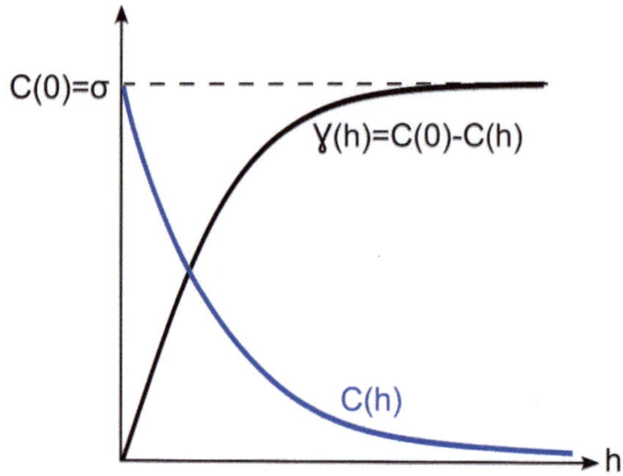

Fig. 7.28 Relationship between variogram and covariogram functions

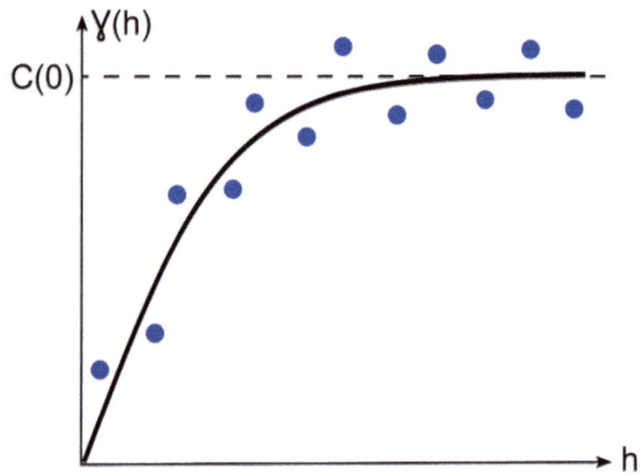

Fig. 7.29 Fitting a theoretical model to the experimental Variogram data

Linear, spherical, circular, exponential, and Gaussian functions can be used to fit Variogram values. These functions can be expressed as following:

$$\text{Linear}: \gamma(h) = \begin{cases} c\left(\dfrac{h}{a}\right), & h < a \\ c, & h \geq a \end{cases} \tag{7.53a}$$

$$\text{Spherical}: \gamma(h) = \begin{cases} c\left[\dfrac{3}{2}\left(\dfrac{h}{a}\right) - \dfrac{1}{2}\left(\dfrac{h}{a}\right)^2\right], & h < a \\ c, & h \geq a \end{cases} \tag{7.53b}$$

$$\text{Circular}: \gamma(h) = \begin{cases} c\left[1 - \dfrac{2}{\pi}\cos^{-1}\left(\dfrac{h}{a}\right) + \sqrt{1 - \left(\dfrac{h}{a}\right)^2}\right], & h < a \\ c, & h \geq a \end{cases} \tag{7.53c}$$

$$\text{Exponential}: \gamma(h) = \begin{cases} c\left[1 - \exp\left(\frac{-h}{a}\right)\right], & h < a \\ c, & h \geq a \end{cases} \tag{7.53d}$$

$$\text{Gauss}: \gamma(h) = \begin{cases} c\left[1 - \exp\left(\frac{-h^2}{a^2}\right)\right], & h < a \\ c, & h \geq a \end{cases} \tag{7.53e}$$

In the above expressions, a indicates the distance where maximum Variogram value is reached and c denotes the maximum Variogram value. Plots of Variogram models in Eqs. (7.53a)–(7.53e) are shown in Fig. 7.30.

In the case of noisy data, Variogram model does not start from zero. It is shifted with a c_0 value as shown in Fig. 7.31.

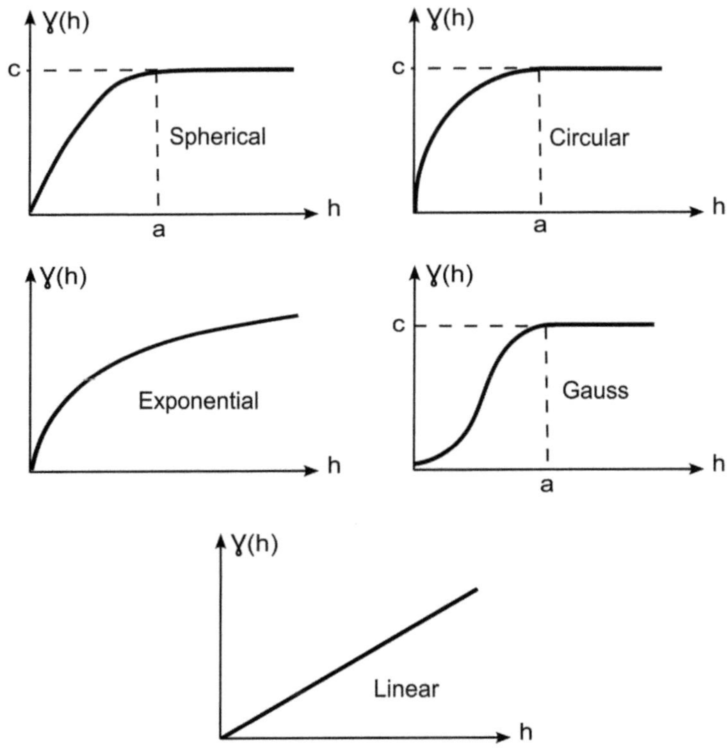

Fig. 7.30 Plots of linear, spherical, circular, exponential, and Gaussian Variogram models

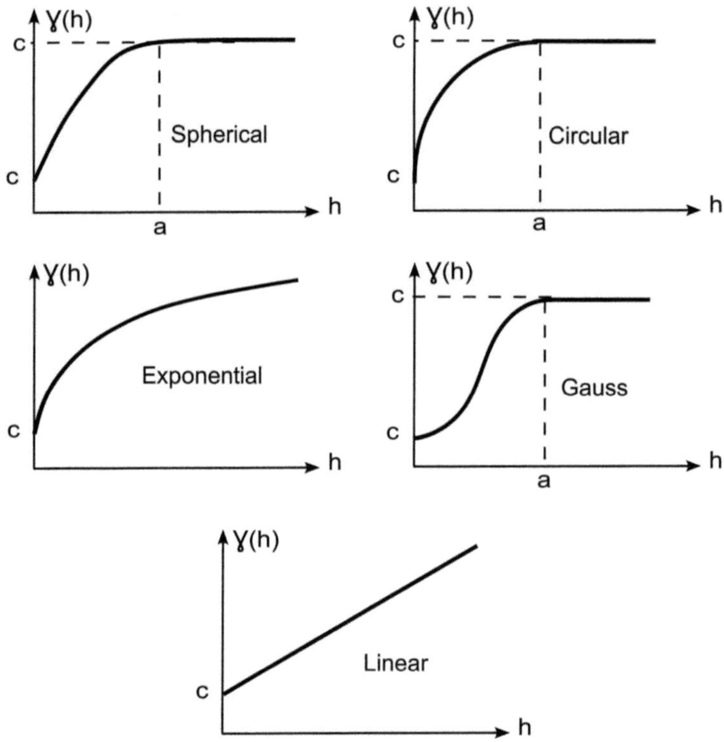

Fig. 7.31 Shifted Variogram models

7.7.2 Basic Assumptions in Kriging Methods

Several Kriging methods were developed such as Ordinary and Universal Kriging methods based on different simplifying assumptions. Common assumptions in Kriging methods can be stated as following:

1. Random variable shows normal distribution around a mean value at all points.
2. In Ordinary Kriging method, mean (expected value) of a random variable is assumed constant at all points. This assumption can be expressed for a Z variable at i-th point as $E(Z_i) = \mu =$ constant.
3. Variance of a random variable is assumed constant at all points. This assumption can be expressed for a Z variable at i-th point as $Var(Z_i) = \sigma^2 =$ constant.
4. Variance and Covariance can be expressed as function of h distance between points.
5. In Universal Kriging method, expected value at an s point can be expressed with a function such as $\mu(s) = \beta_0 f_0(s) + \beta_1 f_1(s) + \beta_2 f_2(s) + \ldots + \beta_p f_p(s)$. Linear and parabolic polynomials are widely used for this purpose.

Kriging methods use both local and global trend of data points to predict intermediate values. Ordinary and Universal Kriging methods are explained in detail below:

7.7.3 Ordinary Kriging Method

Ordinary Kriging method is one of the simplest Kriging methods. In this method, it is assumed that values of a $Z(s)$ variable at different s_i points are distributed normally around a constant mean value of μ as seen in Fig. 7.32. This assumption for a Z variable at an i-th point can be expressed as $E(Z_i) = \mu =$ constant and $Var(Z_i) = \sigma^2 =$ constant. In addition to this assumption, it is also accepted that the covariance of any two adjacent points (s_i, s_j) depends only on the distance between points (h), and this is stated as $Cov(Z_i, Z_j) = C(s_i - s_j) = C(h)$. In Ordinary Kriging method, constant mean value μ serves as the unknown to be determined.

If values of a Z variable are known at different points such as $Z(s_1) = Z_1$, $Z(s_2) = Z_2$, $Z(s_3) = Z_3$, …, $Z(s_n) = Z_n$, Z value at an intermediate point s_0 can be predicted using Ordinary Kriging method as illustrated in Fig. 7.33. True and approximate values at s_0 point can be represented with Z_0 and $\tilde{Z}_0$, respectively. Approximate value $\tilde{Z}_0$ can be obtained as the weighted sum of the variable values as following:

$$\tilde{Z}_0 = \tilde{Z}(s_0) = w_1 Z_1 + w_2 Z_2 + w_3 Z_3 + \ldots + w_n Z_n = \sum_{i=1}^{n} w_i Z_i \tag{7.54}$$

where w_i in Eq. (7.54) show unknown weight coefficients.

When Z variable shows normal distribution at all s_i points and mean values at these points are constant, it can be written as $E[Z_1] = E[Z_2] = E[Z_3] = . \ldots = E[Z_n] = E[Z_0] = \mu =$ constant. In this case, expected values of true (Z_0) and

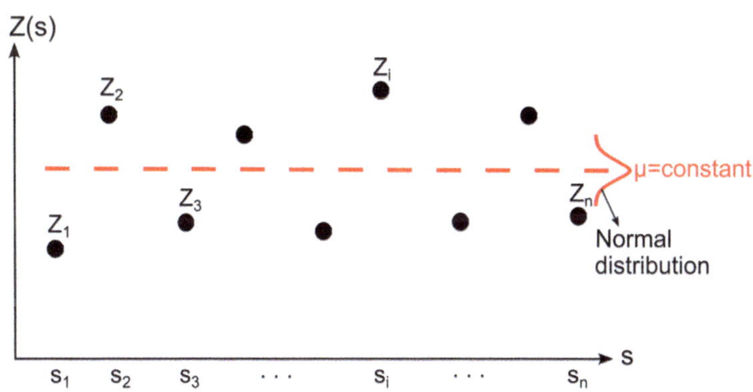

Fig. 7.32 Ordinary Kriging method that assumes normal distribution of data around a constant mean value

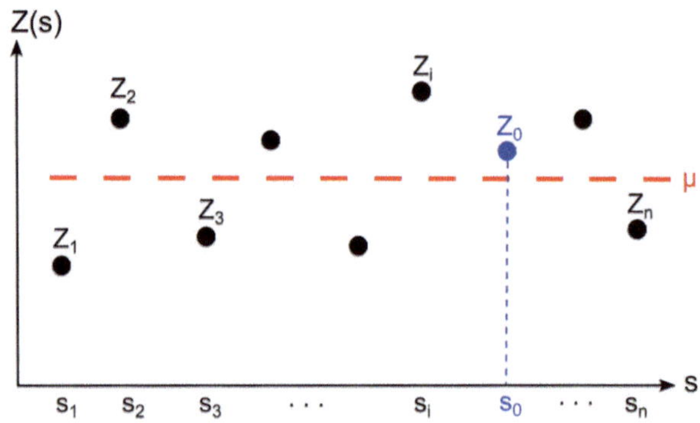

Fig. 7.33 Prediction of *Z* variable at an intermediate point using Ordinary Kriging method

approximate ($\tilde{Z}_0$) values of *Z* variable at s_0 point will be the same. This assumption can be expressed as following:

$$E\left[\tilde{Z}_0\right] = E[Z_0] \rightarrow E\left[\sum_{i=1}^{n} w_i Z_i\right] = E[Z_0] \rightarrow \sum_{i=1}^{n} w_i E[Z_i] = E[Z_0] \tag{7.55}$$

Since expected or mean values are assumed constant at all points including s_0 point in Ordinary Kriging method, then it can be written as $E[Z_i] = E[Z_0] = \mu$. Using this relationship, Eq. (7.55) can be rewritten as below:

$$\sum_{i=1}^{n} w_i \mu = \mu \rightarrow \mu \sum_{i=1}^{n} w_i = \mu \rightarrow \sum_{i=1}^{n} w_i = 1 \tag{7.56}$$

From Eq. (7.56), sum of the weight coefficients to be equal to one is obtained.

Closeness of true (Z_0) and approximate ($\tilde{Z}_0$) values of *Z* variable at s_0 point can be expressed as an optimization problem. In optimization form, this closeness can be statistically expressed as the minimization of the variance or expected value of the square of the difference between these values (shortly Mean Square Value = *MSE*) as following:

$$MSE = Var\left[Z_0 - \ddot{Z}_0\right] = E\left[\left(Z_0 - \ddot{Z}_0\right)^2\right] \rightarrow \min$$

$E\left[\left(Z_0 - \ddot{Z}_0\right)^2\right]$ term in the above equation can be written in expanded form as below:

$$\begin{aligned} E\left[\left(Z_0 - \ddot{Z}_0\right)^2\right] &= E\left[\left(Z_0 - \sum_{i=1}^{n} w_i Z_i\right)^2\right] = E\left[Z_0^2 + \left(\sum_{i=1}^{n}\sum_{j=1}^{n} w_i w_j Z_i Z_j - 2\sum_{i=1}^{n} w_i Z_i Z_0\right)^2\right] \\ &= E\left[Z_0^2\right] + \sum_{i=1}^{n}\sum_{j=1}^{n} w_i w_j E\left[Z_i Z_j\right] - 2\sum_{i=1}^{n} w_i E[Z_i Z_0] \end{aligned} \tag{7.57}$$

Using the expressions in Eqs. (7.45)–(7.47), $E\left[Z_0^2\right]$, $E[Z_iZ_j]$, and $E[Z_iZ_0]$ terms in Eq. (7.57) can be written in terms of Covariance as following:

$$E\left(Z_0^2\right) = C(0) + \mu^2$$
$$E\left(Z_iZ_j\right) = C\left(s_i - s_j\right) + \mu^2$$
$$E(Z_iZ_0) = C(s_i - s_0) + \mu^2$$

Using the above expressions, *MSE* can be rewritten in terms of Covariance as below:

$$\begin{aligned}
MSE &= E\left[\left(Z_0 - \ddot{Z}_0\right)^2\right] \\
&= C(0) + \mu^2 + \sum_{i=1}^{n}\sum_{j=1}^{n} w_iw_j\left[C\left(s_i - s_j\right) + \mu^2\right] - 2\sum_{i=1}^{n} w_i\left[C(s_i - s_0) + \mu^2\right] \\
&= C(0) + \mu^2 + \sum_{i=1}^{n}\sum_{j=1}^{n} w_iw_jC\left(s_i - s_j\right) + \sum_{i=1}^{n}\sum_{j=1}^{n} w_iw_j\mu^2 \\
&\quad - 2\sum_{i=1}^{n} w_iC(s_i - s_0) - 2\sum_{i=1}^{n} w_i\mu^2 \\
&= C(0) + \mu^2 + \sum_{i=1}^{n}\sum_{j=1}^{n} w_iw_jC\left(s_i - s_j\right) + \mu^2\sum_{i=1}^{n}\sum_{j=1}^{n} w_iw_j \\
&\quad - 2\sum_{i=1}^{n} w_iC(s_i - s_0) - 2\mu^2\sum_{i=1}^{n} w_i \\
&= C(0) + \mu^2 + \sum_{i=1}^{n}\sum_{j=1}^{n} w_iw_jC\left(s_i - s_j\right) + \mu^2\sum_{i=1}^{n} w_i\underbrace{\sum_{j=1}^{n} w_j}_{=1} \\
&\quad - 2\sum_{i=1}^{n} w_iC(s_i - s_0) - 2\mu^2\underbrace{\sum_{i=1}^{n} w_i}_{=1} \\
&= C(0) + \mu^2 + \sum_{i=1}^{n}\sum_{j=1}^{n} w_iw_jC\left(s_i - s_j\right) + \mu^2\underbrace{\sum_{i=1}^{n} w_i}_{=1} - 2\sum_{i=1}^{n} w_iC(s_i - s_0) - 2\mu^2 \\
&= C(0) + \sum_{i=1}^{n}\sum_{j=1}^{n} w_iw_jC\left(s_i - s_j\right) - 2\sum_{i=1}^{n} w_iC(s_i - s_0)
\end{aligned} \tag{7.58}$$

Using the general Variance-Covariance relationship of $C(s_i - s_j) = C(0) - \gamma(s_i - s_j)$, MSE expression can be written in terms of Variogram function $\gamma(s_i - s_j)$ as following:

$$
\begin{aligned}
MSE &= C(0) + \sum_{i=1}^{n}\sum_{j=1}^{n} w_i w_j C(s_i - s_j) - 2\sum_{i=1}^{n} w_i C(s_i - s_0) \\
&= C(0) + \sum_{i=1}^{n}\sum_{j=1}^{n} w_i w_j [C(0) - \gamma(s_i - s_j)] - 2\sum_{i=1}^{n} w_i [C(0) - \gamma(s_i - s_0)] \\
&= C(0) + C(0)\sum_{i=1}^{n}\sum_{j=1}^{n} w_i w_j - \sum_{i=1}^{n}\sum_{j=1}^{n} w_i w_j \gamma(s_i - s_j) - 2C(0)\sum_{i=1}^{n} w_i \\
&\quad + 2\sum_{i=1}^{n} w_i \gamma(s_i - s_0) \\
&= C(0) + C(0)\sum_{i=1}^{n} w_i \underbrace{\sum_{j=1}^{n} w_j}_{=1} - \sum_{i=1}^{n}\sum_{j=1}^{n} w_i w_j \gamma(s_i - s_j) - 2C(0)\underbrace{\sum_{i=1}^{n} w_i}_{=1} \\
&\quad + 2\sum_{i=1}^{n} w_i \gamma(s_i - s_0) \\
&= C(0) + C(0)\sum_{i=1}^{n} w_i \underbrace{\sum_{j=1}^{n} w_j}_{=1} - \sum_{i=1}^{n}\sum_{j=1}^{n} w_i w_j \gamma(s_i - s_j) - 2C(0)\underbrace{\sum_{i=1}^{n} w_i}_{=1} \\
&\quad + 2\sum_{i=1}^{n} w_i \gamma(s_i - s_0) \\
&= C(0) + C(0)\underbrace{\sum_{i=1}^{n} w_i}_{=1} - \sum_{i=1}^{n}\sum_{j=1}^{n} w_i w_j \gamma(s_i - s_j) - 2C(0) + 2\sum_{i=1}^{n} w_i \gamma(s_i - s_0) \\
&= -\sum_{i=1}^{n}\sum_{j=1}^{n} w_i w_j \gamma(s_i - s_j) + 2\sum_{i=1}^{n} w_i \gamma(s_i - s_0)
\end{aligned}
\tag{7.59}
$$

Minimization of MSE can be combined with $\sum_{i=1}^{n} w_i = 1$ condition in Eq. (7.56) to be the minimization of a new unconstrained Lagrange function (L) as below:

$$L = MSE - 2\lambda\left(\sum_{i=1}^{n} w_i - 1\right) = -\sum_{i=1}^{n}\sum_{j=1}^{n} w_i w_j \gamma(s_i - s_j) + 2\sum_{i=1}^{n} w_i \gamma(s_i - s_0)$$
$$- 2\lambda\left(\sum_{i=1}^{n} w_i - 1\right) \rightarrow \min$$

λ parameter in the above expression is called as Lagrange multiplier, and it serves as additional unknown to be determined. Minimum of L function can be obtained by taking its derivatives with respect to w_i and λ and then equating them to zero as below:

$$\frac{\partial L}{\partial w_i} = 0 \rightarrow -2\sum_{j=1}^{n} w_j \gamma(s_i - s_j) + 2\gamma(s_i - s_0) - 2\lambda = 0, \ (i = 1, \ldots, n)$$

$$\frac{\partial L}{\partial \lambda} = 0 \rightarrow \sum_{i=1}^{n} w_i = 1$$

Above equations can be rearranged as below:

$$\sum_{j=1}^{n} w_j \gamma(s_i - s_j) + \lambda = \gamma(s_i - s_0) \quad (i = 1, \ldots, n) \tag{7.60}$$

$$\sum_{i=1}^{n} w_i = 1 \tag{7.61}$$

Equations (7.60) and (7.61) can be written in expanded form as following:

$$w_1\gamma(s_i - s_1) + w_2\gamma(s_i - s_2) + w_3\gamma(s_i - s_3) + \ldots + w_n\gamma(s_i - s_n) + \lambda$$
$$= \gamma(s_i - s_0), \ (i = 1, \ldots, n)$$

$$w_1 + w_2 + w_3 + \ldots + w_n = 1$$

Above equations can be written in matrix form as below:

$$\underbrace{\begin{bmatrix} 0 & \gamma(s_1 - s_2) & \gamma(s_1 - s_3) & \ldots & \gamma(s_1 - s_n) & 1 \\ \gamma(s_2 - s_1) & 0 & \gamma(s_2 - s_3) & \ldots & \gamma(s_2 - s_n) & 1 \\ \gamma(s_3 - s_1) & \gamma(s_3 - s_2) & 0 & \ldots & \gamma(s_3 - s_n) & 1 \\ \ldots & \ldots & \ldots & \ldots & \ldots & \ldots \\ \gamma(s_n - s_1) & \gamma(s_n - s_2) & \gamma(s_n - s_3) & \ldots & 0 & 1 \\ 1 & 1 & 1 & \ldots & 1 & 0 \end{bmatrix}}_{A_{mat}} \underbrace{\begin{Bmatrix} w_1 \\ w_2 \\ w_3 \\ \ldots \\ w_n \\ \lambda \end{Bmatrix}}_{W_{mat}} = \underbrace{\begin{Bmatrix} \gamma(s_1 - s_0) \\ \gamma(s_2 - s_0) \\ \gamma(s_3 - s_0) \\ \ldots \\ \gamma(s_n - s_0) \\ 1 \end{Bmatrix}}_{B_{mat}} \tag{7.62}$$

Table 7.16 Function values at different points

s	1	2	3.5	4	5
Z	1	4	?	16	25

Solution of Eq. (7.62) using linear equation methods in Chap. 4 leads to w_i and λ values. Substituting the obtained w_i values in Eq. (7.54), approximate Z value at an s_0 point is obtained.

Note: Ordinary Kriging method assumes that variable has the same mean at all points. This may not be valid for a large number of data points. Therefore for accurate prediction of variable value at an intermediate point, it is suggested to consider up to 40–50 points around the intermediate point.

Note: MATLAB program was written to implement Kriging methods. Kriging program named as *MATLAB Program 7.2: Kriging Model* is given in appendix in this chapter. This program can be used to predict intermediate values in Examples 7.9 and 7.10.

Example 7.9: Function values in Table 7.16 are produced from $Z(s) = s^2$ function. Predict $Z(3.5)$ value using Ordinary Kriging method and compare it with the true value.

Indicating the distance between two points as $h = |s_i - s_j|$, Variogram values for Z function can be obtained using $\gamma(h) = \frac{1}{2N_h} \sum_{|s_i - s_j| = h} \left[Z(s_i) - Z(s_j)\right]^2$ equation, where

$\gamma(0)$ indicates the variogram value for points with zero distance from each other.
$\gamma(1)$ indicates the Variogram value for points with distance of 1 from each other.
$\gamma(2)$ indicates the Variogram value for points with distance of 2 from each other.
$\gamma(3)$ indicates the Variogram value for points with distance of 3 from each other.
$\gamma(4)$ indicates the Variogram value for points with distance of 4 from each other.

Point coordinates and their corresponding Z values in Table 7.16 can be written as below:

$$s_1 = 1, s_2 = 2, s_3 = 4, s_4 = 5 \rightarrow Z(s_1) = 1, Z(s_2) = 4, Z(s_3) = 16, Z(s_4) = 25$$

Using these values, Variogram values of $\gamma(0)$, $\gamma(1)$, $\gamma(2)$, $\gamma(3)$, $\gamma(4)$ can be calculated as following:

$$\gamma(0) = \frac{1}{2x4} \sum_{s_i - s_j = 1} [Z(s_1) - Z(s_1)]^2 + [Z(s_2) - Z(s_2)]^2 + [Z(s_3) - Z(s_3)]^2$$
$$+ [Z(s_4) - Z(s_4)]^2 = 0$$

$$\gamma(1) = \frac{1}{2x2} \left\{ [Z(s_1) - Z(s_2)]^2 + [Z(s_3) - Z(s_4)]^2 \right\} = \frac{1}{4} \left[(1-4)^2 + (16-25)^2 \right] = 22.5$$

$$\gamma(2) = \frac{1}{2x1}\left\{[Z(s_2) - Z(s_3)]^2\right\} = \frac{1}{2}\left[(4-16)^2\right] = 72$$

$$\gamma(3) = \frac{1}{2x2}\left\{[Z(s_1) - Z(s_3)]^2 + [Z(s_2) - Z(s_4)]^2\right\} = \frac{1}{4}\left[(1-16)^2 + (4-25)^2\right] = 166.5$$

$$\gamma(4) = \frac{1}{2x1}\left\{[Z(s_1) - Z(s_4)]^2\right\} = \frac{1}{2}\left[(1-25)^2\right] = 288$$

When intermediate variogram values are needed, they can be calculated from a theoretical Variogram model obtained by fitting a selected model to the available variogram values. Using least square method, linear, spherical, and exponential variogram models expressed in Eq. (7.53) can be fitted to the available data by taking $a = 4$ as following:

$$\text{Linear}: \gamma(h) = c\left(\frac{h}{a}\right) \rightarrow \gamma(h) = 242.4\left(\frac{h}{4}\right)$$

$$\text{Spherical}: \gamma(h) = c\left[\frac{3}{2}\left(\frac{h}{a}\right) - \frac{1}{2}\left(\frac{h}{a}\right)^2\right] \rightarrow \gamma(h) = 216.6966\left[\frac{3}{2}\left(\frac{h}{4}\right) - \frac{1}{2}\left(\frac{h}{4}\right)^2\right]$$

$$\text{Exponential}: \gamma(h) = c\left[1 - \exp\left(\frac{-h}{a}\right)\right] \rightarrow \gamma(h) = 343.8827\left[1 - \exp\left(\frac{-h}{4}\right)\right]$$

Linear, spherical, and exponential variogram models fitted to the variogram values are shown in Fig. 7.34.

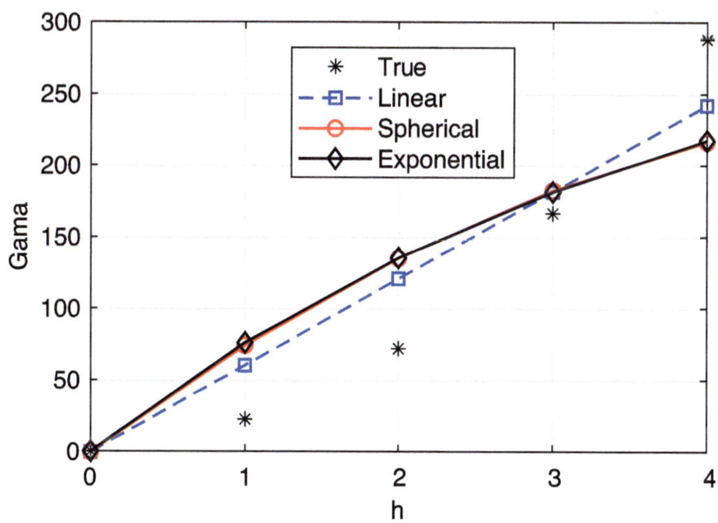

Fig. 7.34 Fitting linear, spherical, and exponential models to the variogram values

In this case, the set of equations giving the weight coefficients of Ordinary Kriging method can be expressed in terms of variogram values as following:

$$\begin{bmatrix} 0 & \gamma(s_1-s_2) & \gamma(s_1-s_3) & \gamma(s_1-s_4) & 1 \\ \gamma(s_2-s_1) & 0 & \gamma(s_2-s_3) & \gamma(s_2-s_4) & 1 \\ \gamma(s_3-s_1) & \gamma(s_3-s_2) & 0 & \gamma(s_3-s_4) & 1 \\ \gamma(s_4-s_1) & \gamma(s_4-s_2) & \gamma(s_4-s_3) & 0 & 1 \\ 1 & 1 & 1 & 1 & 1 \end{bmatrix} \begin{Bmatrix} w_1 \\ w_2 \\ w_3 \\ w_4 \\ \lambda \end{Bmatrix} = \begin{Bmatrix} \gamma(s_1-s_0) \\ \gamma(s_2-s_0) \\ \gamma(s_3-s_0) \\ \gamma(s_4-s_0) \\ 1 \end{Bmatrix} \rightarrow$$

$$\underbrace{\begin{bmatrix} 0 & \gamma(1) & \gamma(3) & \gamma(4) & 1 \\ \gamma(1) & 0 & \gamma(2) & \gamma(3) & 1 \\ \gamma(3) & \gamma(2) & 0 & \gamma(1) & 1 \\ \gamma(4) & \gamma(3) & \gamma(1) & 0 & 1 \\ 1 & 1 & 1 & 1 & 1 \end{bmatrix}}_{A_{mat}} \underbrace{\begin{Bmatrix} w_1 \\ w_2 \\ w_3 \\ w_4 \\ \lambda \end{Bmatrix}}_{W_{mat}} = \underbrace{\begin{Bmatrix} \gamma(2.5) \\ \gamma(1.5) \\ \gamma(0.5) \\ \gamma(1.5) \\ 1 \end{Bmatrix}}_{B_{mat}} \rightarrow$$

where intermediate values of $\gamma(2.5)$, $\gamma(1.5)$, and $\gamma(0.5)$ are calculated from the fitted Variogram model. Above equation will lead to different weight coefficients based on the selected Variogram model for $\gamma(h)$. For linear, spherical and exponential Variogram models, weight coefficients, Lagrange multiplier, and the predicted Z(3.5) value are given as below:

$$For\ Linear\ Variogram\ model: \gamma(h) = 242.4\left(\frac{h}{4}\right)$$

$$w_1 = 0.000, w_2 = 0.25, w_3 = 0.75, w_4\ 0.0000, \lambda = -0.0000$$

$$Z(3.5) = w_1Z_1 + w_2Z_2 + w_3Z_3 + w_4Z_4 = 13.0000$$

$$For\ Spherical\ Variogram\ model: \gamma(h) = 216.6966\left[\frac{3}{2}\left(\frac{h}{4}\right) - \frac{1}{2}\left(\frac{h}{4}\right)^2\right]$$

$$w_1 = 0.0000, w_2 = 0.25, w_3 = 0.75, w_4\ 0.0000, \lambda = 5.0788$$

$$Z(3.5) = w_1Z_1 + w_2Z_2 + w_3Z_3 + w_4Z_4 = 13.0000$$

$$For\ Exponential\ Variogram\ model: \gamma(h) = 343.8827\left[1 - \exp\left(\frac{-h}{4}\right)\right]$$

$$w_1 = 0.0086, w_2 = 0.2433, w_3 = 0.7395, w_4 = 0.0086, \lambda = 5.2659$$

$$Z(3.5) = w_1Z_1 + w_2Z_2 + w_3Z_3 + w_4Z_4 = 13.0284$$

True or actual Z value at $s_0 = 3.5$ is obtained as $Z(3.5) = 3.5^2 = 12.25$. From the results it is seen that there is a slight difference between Ordinary Kriging prediction and the actual value.

Note: For the solution of this problem with Ordinary Kriging Method, *MATLAB Program 7.2: Kriging Model* in appendix written in MATLAB programming language can be used.

7.7.4 Universal Kriging

Universal Kriging is also called as trend Kriging. In this method, it is assumed that mean values (i.e., $\mu(s)$) of a $Z(s)$ variable at different s_i points can be represented with a polynomial function as in Fig. 7.35. Additionally, Z_i values are assumed to have normal distribution around this polynomial. In Universal Kriging, trend function can be written in general form using basic functions $f_i(s)$ and weight coefficients β_i as following:

$$\mu(s) = \beta_0 f_0(s) + \beta_1 f_1(s) + \beta_2 f_2(s) + \ldots + \beta_p f_p(s) = \sum_{i=0}^{p} \beta_i f_i(s) \quad (7.63)$$

In Universal Kriging method, if values of a Z variable are known at different points such as $Z(s_1) = Z_1$, $Z(s_2) = Z_2$, $Z(s_3) = Z_3$, …, $Z(s_n) = Z_n$, approximate Z value (i.e., $\tilde{Z}(s_0)$) at a point s_0 can be obtained as the weighted sum of variable values as following:

$$\tilde{Z}_0 = \tilde{Z}(s_0) = w_1 Z_1 + w_2 Z_2 + w_3 Z_3 + \ldots + w_n Z_n = \sum_{i=1}^{n} w_i Z_i$$

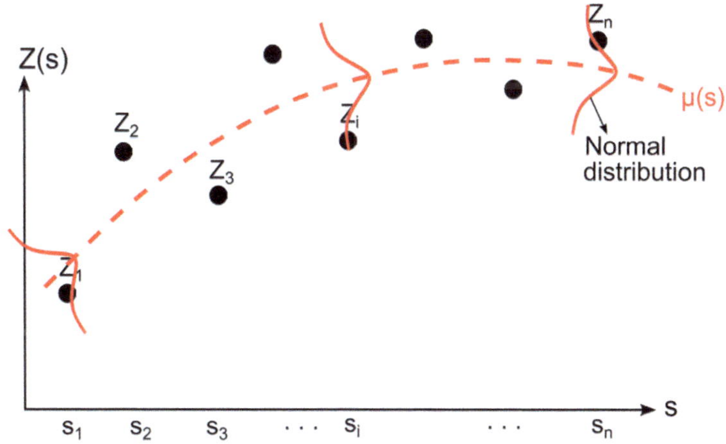

Fig. 7.35 Representing mean values with a function in Universal Kriging method

Expected values of true (Z_0) and approximate ($\tilde{Z}_0$) values of Z variable at s_0 point should be the same. This assumption can be expressed as $E\left[\tilde{Z}_0\right] = E[Z_0]$ where $E[Z_0]$ can be stated as below:

$$E[Z_0] = \mu(s_0) = \beta_0 + \beta_1 f_1(s_0) + \beta_2 f_2(s_0) + \ldots + \beta_p f_p(s_0) = \sum_{i=0}^{p} \beta_i f_i(s_0) \quad (7.64)$$

Using Eqs. (7.54) and (7.64), $E\left[\tilde{Z}_0\right] = E[Z_0]$ equation can be rewritten as following:

$$E\left[\sum_{i=1}^{n} w_i Z_i\right] = \sum_{i=0}^{p} \beta_i f_i(s_0) \rightarrow \sum_{i=1}^{n} w_i E[Z_i] = \sum_{i=0}^{p} \beta_i f_i(s_0) \quad (7.65)$$

where $E[Z_i]$ term in Eq. (7.65) can be expressed as below:

$$E[Z_i] = \mu(s_i) = \beta_0 + \beta_1 f_1(s_i) + \beta_2 f_2(s_i) + \ldots + \beta_p f_p(s_i) = \sum_{i=0}^{p} \beta_i f_i(s_i) \quad (7.66)$$

In this case, using Eqs. (7.64) and (7.66), Eq. (7.65) can be rewritten as following:

$$\sum_{i=1}^{n} w_i \left[\beta_0 + \beta_1 f_1(s_i) + \beta_2 f_2(s_i) + \ldots + \beta_p f_p(s_i)\right] = \beta_0 + \beta_1 f_1(s_0) + \beta_2 f_2(s_0) + \ldots + \beta_p f_p(s_0) \quad (7.67)$$

From the equivalency of the right and left sides in Eq. (7.67), following equations can be obtained for the weight coefficients:

$$\beta_0 \text{ term}: \sum_{i=1}^{n} w_i = 1 \rightarrow w_1 + w_2 + w_3 + \ldots w_n = 1 \quad (7.68)$$

$$\beta_1 \text{ term}: \sum_{i=1}^{n} w_i f_1(s_i) = f_1(s_0) \rightarrow w_1 f_1(s_1) + w_2 f_1(s_2) + w_3 f_1(s_3) + \ldots w_n f_1(s_n) = f_1(s_0) \quad (7.69)$$

$$\beta_2 \text{ term}: \sum_{i=1}^{n} w_i f_2(s_i) = f_2(s_0) \rightarrow w_1 f_2(s_1) + w_2 f_2(s_2) + w_3 f_2(s_3) + \ldots w_n f_2(s_n) = f_2(s_0) \quad (7.70)$$

...

$$\beta_p \text{ term} : \sum_{i=1}^{n} w_i f_p(s_i) = f_p(s_0) \rightarrow w_1 f_p(s_1) + w_2 f_p(s_2) + w_3 f_p(s_3)$$
$$+ \ldots w_n f_p(s_n) = f_p(s_0) \tag{7.71}$$

As in Ordinary Kriging method, closeness of the approximate ($\tilde{Z}_0$) and true (Z_0) values can be enforced with minimization of the variance of the difference between true and approximate values (shortly Mean Square Value = *MSE*) as following:

$$MSE = -\sum_{i=1}^{n}\sum_{j=1}^{n} w_i w_j \gamma(s_i - s_j) + 2\sum_{i=1}^{n} w_i \gamma(s_i - s_0) \rightarrow \min$$

Minimization of *MSE* can be combined with the conditions in Eqs. (7.68)–(7.71) to be the minimization of a new unconstrained Lagrange function (L) as below:

$$L = -\sum_{i=1}^{n}\sum_{j=1}^{n} w_i w_j \gamma(s_i - s_j) + 2\sum_{i=1}^{n} w_i \gamma(s_i - s_0) + 2\lambda_0\left(1 - \sum_{i=1}^{n} w_i\right)$$
$$+ 2\lambda_1\left(f_1(s_0) - \sum_{i=1}^{n} w_i f_1(s_i)\right) + +2\lambda_2\left(f_2(s_0) - \sum_{i=1}^{n} w_i f_2(s_i)\right) + .$$
$$\ldots + 2\lambda_p\left(f_p(s_0) - \sum_{i=1}^{n} w_i f_p(s_i)\right) \longrightarrow \min$$

λ_0, λ_1,..., λ_p parameters in the above expression are called as Lagrange multipliers, and they serve as additional unknowns to be determined. Minimum of L function can be obtained by taking its derivatives with respect to w_i and λ_i and then equating them to zero as below:

$$\frac{\partial L}{\partial w_i} = 0 \rightarrow -2\sum_{j=1}^{n} w_j \gamma(s_i - s_j) + 2\gamma(s_i - s_0) - 2\lambda_0 - 2\sum_{j=1}^{p} \lambda_j f_j(s_i) = 0 \ (i = 1, \ldots, n) \tag{7.72}$$

$$\frac{\partial L}{\partial \lambda_0} = 0 \rightarrow \sum_{i=1}^{n} w_i = 1 \tag{7.73}$$

$$\frac{\partial L}{\partial \lambda_k} = 0 : \sum_{i=1}^{n} w_i f_k(s_i) = f_k(s_0) \rightarrow (k = 1, \ldots, p) \tag{7.74}$$

Equations (7.72)–(7.74) can be rearranged as following:

$$\frac{\partial L}{\partial w_i}=0\rightarrow\sum_{j=1}^{n}w_j\gamma(s_i-s_j)+\lambda_0+\sum_{j=1}^{p}\lambda_j f_j(s_i)=\gamma(s_i-s_0)\ (i=1,\ldots,n) \tag{7.75}$$

$$\frac{\partial L}{\partial \lambda_0}=0\rightarrow\sum_{i=1}^{n}w_i=1 \tag{7.76}$$

$$\frac{\partial L}{\partial \lambda_k}=0:\sum_{i=1}^{n}w_i f_k(s_i)=f_k(s_0)\rightarrow(k=1,\ldots,p) \tag{7.77}$$

Equations (7.75)–(7.77) can be expressed in matrix form as below:

$$\underbrace{\begin{bmatrix} 0 & \gamma(s_1-s_2) & \gamma(s_1-s_3) & \ldots & \gamma(s_1-s_n) & 1 & f_1(s_1) & f_2(s_1) & \ldots & f_p(s_1) \\ \gamma(s_2-s_1) & 0 & \gamma(s_2-s_3) & \ldots & \gamma(s_2-s_n) & 1 & f_1(s_2) & f_2(s_2) & \ldots & f_p(s_2) \\ \gamma(s_3-s_1) & \gamma(s_3-s_2) & 0 & \ldots & \gamma(s_3-s_n) & 1 & f_1(s_3) & f_2(s_3) & \ldots & f_p(s_3) \\ \ldots & \ldots & \ldots & \ldots & \ldots & \ldots & \ldots & \ldots & \ldots & \ldots \\ \gamma(s_n-s_1) & \gamma(s_n-s_2) & \gamma(s_n-s_3) & \ldots & 0 & 1 & f_1(s_n) & f_2(s_n) & \ldots & f_p(s_n) \\ 1 & 1 & 1 & \ldots & 1 & 0 & 0 & 0 & \ldots & 0 \\ f_1(s_1) & f_1(s_2) & f_1(s_3) & \ldots & f_1(s_n) & 0 & 0 & 0 & \ldots & 0 \\ f_2(s_1) & f_2(s_2) & f_2(s_3) & \ldots & f_2(s_n) & 0 & 0 & 0 & \ldots & 0 \\ \ldots & \ldots & \ldots & \ldots & \ldots & \ldots & \ldots & \ldots & \ldots & \ldots \\ f_p(s_1) & f_p(s_2) & f_p(s_3) & \ldots & f_p(s_n) & 0 & 0 & 0 & \ldots & 0 \end{bmatrix}}_{A_{mat}} \cdot \underbrace{\begin{Bmatrix} w_1 \\ w_2 \\ w_3 \\ \ldots \\ w_n \\ \lambda_0 \\ \lambda_1 \\ \lambda_2 \\ \ldots \\ \lambda_p \end{Bmatrix}}_{W_{mat}} = \underbrace{\begin{Bmatrix} \gamma(s_0-s_1) \\ \gamma(s_0-s_2) \\ \gamma(s_0-s_3) \\ \ldots \\ \gamma(s_0-s_n) \\ 1 \\ f_1(s_0) \\ f_2(s_0) \\ \ldots \\ f_p(s_0) \end{Bmatrix}}_{B_{mat}} \tag{7.78}$$

Solution of Eq. (7.78) using linear equation methods in Chap. 4 leads to w_i and λ_i values. Substituting the obtained w_i values in Eq. (7.54), approximate Z value at s_0 point is obtained.

Example 7.10: Using function values in Table 7.16, predict $Z(3.5)$ value using Universal Kriging method and compare it with the true value.

With Universal Kriging method, mean function of the Z variable at points in Table 7.16 can be expressed with a parabolic polynomial function as following:

$$\mu(s) = \beta_0 \underbrace{f_0(s)}_{1} + \beta_1 \underbrace{f_1(s)}_{s} + \beta_2 \underbrace{f_2(s)}_{s^2} = \beta_0 + \beta_1 s + \beta_2 s^2$$

In this case, substitution of $f_0 = 1, f_1(s) = s$, and $f_2(s) = s^2$ into Eq. (7.78) leads to the following equation:

$$\begin{bmatrix} 0 & \gamma(s_1-s_2) & \gamma(s_1-s_3) & \gamma(s_1-s_4) & 1 & s_1 & s_1^2 \\ \gamma(s_2-s_1) & 0 & \gamma(s_2-s_3) & \gamma(s_2-s_4) & 1 & s_2 & s_2^2 \\ \gamma(s_3-s_1) & \gamma(s_3-s_2) & 0 & \gamma(s_3-s_4) & 1 & s_3 & s_3^2 \\ \gamma(s_4-s_1) & \gamma(s_4-s_2) & \gamma(s_4-s_3) & 0 & 1 & s_4 & s_4^2 \\ 1 & 1 & 1 & 1 & 0 & 0 & 0 \\ s_1 & s_2 & s_3 & s_4 & 0 & 0 & 0 \\ s_1^2 & s_2^2 & s_3^2 & s_4^2 & 0 & 0 & 0 \end{bmatrix} \begin{Bmatrix} w_1 \\ w_2 \\ w_3 \\ w_4 \\ \lambda_0 \\ \lambda_1 \\ \lambda_2 \end{Bmatrix} = \begin{Bmatrix} \gamma(s_0-s_1) \\ \gamma(s_0-s_2) \\ \gamma(s_0-s_3) \\ \gamma(s_0-s_4) \\ 1 \\ s_0 \\ s_0^2 \end{Bmatrix} \rightarrow$$

$$\underbrace{\begin{bmatrix} 0 & \gamma(1) & \gamma(3) & \gamma(4) & 1 & s_1 & s_1^2 \\ \gamma(1) & 0 & \gamma(2) & \gamma(3) & 1 & s_2 & s_2^2 \\ \gamma(3) & \gamma(2) & 0 & \gamma(1) & 1 & s_3 & s_3^2 \\ \gamma(4) & \gamma(3) & \gamma(1) & 0 & 1 & s_4 & s_4^2 \\ 1 & 1 & 1 & 1 & 0 & 0 & 0 \\ s_1 & s_2 & s_3 & s_4 & 0 & 0 & 0 \\ s_1^2 & s_2^2 & s_3^2 & s_4^2 & 0 & 0 & 0 \end{bmatrix}}_{A_{mat}} \underbrace{\begin{Bmatrix} w_1 \\ w_2 \\ w_3 \\ w_4 \\ \lambda_0 \\ \lambda_1 \\ \lambda_2 \end{Bmatrix}}_{W_{mat}} = \underbrace{\begin{Bmatrix} \gamma(2.5) \\ \gamma(1.5) \\ \gamma(0.5) \\ \gamma(1.5) \\ 1 \\ s_0 \\ s_0^2 \end{Bmatrix}}_{B_{mat}} \rightarrow$$

where intermediate values of $\gamma(2.5)$, $\gamma(1.5)$, and $\gamma(0.5)$ are calculated from the fitted Variogram model. Above equation will lead to different weight coefficients based on the selected Variogram model for $\gamma(h)$. For linear, spherical and exponential Variogram models, weight coefficients, Lagrange multipliers and predicted $Z(3.5)$ value are given as below:

$$\textit{For linear Variogram model}: \gamma(h) = 242.4\left(\frac{h}{4}\right)$$

$$w_1 = -0.1250, w_2 = 0.3750, w_3 = 0.8750, w_4 = -0.1250,$$

$$\lambda_0 = -25.25, \lambda_1 = 30.3, \lambda_2 = -5.05$$

$$Z(3.5) = w_1 Z_1 + w_2 Z_2 + w_3 Z_3 + w_4 Z_4 = 12.25$$

$$\textit{For spherical Variogram model}: \gamma(h) = 216.6966\left[\frac{3}{2}\left(\frac{h}{4}\right) - \frac{1}{2}\left(\frac{h}{4}\right)^2\right]$$

$$w_1 = -0.1250, w_2 = 0.3750, w_3 = 0.8750, w_4 = -0.1250,$$

$$\lambda_0 = -33.8588, \lambda_1 = 40.6306, \lambda_2 = -6.7718$$

$$Z(3.5) = w_1 Z_1 + w_2 Z_2 + w_3 Z_3 + w_4 Z_4 = 12.25$$

$$\textit{For exponential Variogram model}: \gamma(h) = 343.8827\left[1 - \exp\left(\frac{-h}{4}\right)\right]$$

$$w_1 = -0.1260, w_2 = 0.3769, w_3 = 0.8731, w_4 = -0.1240,$$

$$\lambda_0 = -36.5624, \lambda_1 = 43.4851, \lambda_2 = -7.2299$$

$$Z(3.5) = w_1 Z_1 + w_2 Z_2 + w_3 Z_3 + w_4 Z_4 = 12.25$$

True or actual Z value at $s_0 = 3.5$ is obtained as $Z(3.5) = 3.5^2 = 12.25$. From the results it is seen that Universal Kriging method yields the same as true value.

Note: For the solution of this problem with Universal Kriging Method, *MATLAB Program 7.2: Kriging Model* in appendix written in MATLAB programming language can be used.

7.8 Nonlinear Regression

So far, linear regression methods have been described. In linear regression, fitting models are linear in terms of unknown model constants. Unknown constants can be solved easily with linear equation methods in Chap. 4. Polynomial models are commonly used in linear regression. However, sometimes using nonlinear regression models a simpler and more accurate model compared to linear regression models can be fitted to the available data. In nonlinear regression, fitting models are nonlinear in terms of unknown model constants. For example, in engineering, tensile test is performed to determine the mechanical properties of materials. Tensile test results for metals can be expressed best with a nonlinear regression model of exponential type ($f(x) = ax^b$) as illustrated in Fig. 7.36.

Although they have simple expressions, nonlinear regression models present difficulties in the solution of model constants. Two approaches are used in the calculation of model constants in nonlinear regression:

- *Nonlinear regression approach*: Constants of the nonlinear regression model are solved iteratively using nonlinear equation methods in Chap. 5.
- *Linear regression approach*: Nonlinear regression model is converted into a linear regression model via variable transformation. Then, constants of the linear regression model are solved with linear equations methods in Chap. 4.

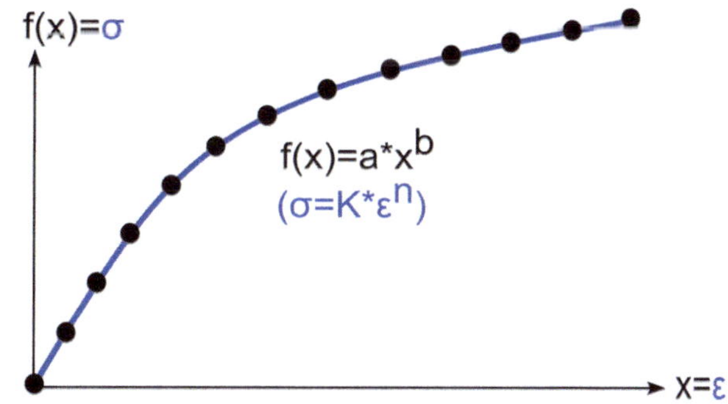

Fig. 7.36 Fitting exponential function to the tensile test results in mechanical engineering

These two approaches are explained in detail using the solution of the nonlinear regression model of $\tilde{f}(x) = ax^b$ as following:

Nonlinear Regression approach:

Assume that values of an $f(x)$ function are known at n points as $f_1, f_2, \ldots, f_n$. Best fitting of $\tilde{f}(x) = ax^b$ model to the function values can be achieved by minimizing the sum of squares of errors at all points (E) as following:

$$E = \sum_{i=1}^{n} (\tilde{f}_i - f_i)^2 = \sum_{i=1}^{n} (ax_i^b - f_i)^2 \rightarrow \min \tag{7.79}$$

Minimum of E function can be obtained by taking its derivatives with respect to a and b constants and equating them to zero as below:

$$\frac{\partial E}{\partial a} = 2\sum_{i=1}^{n} (ax_i^b - f_i)x_i^b = 0 \rightarrow a\sum_{i=1}^{n} x_i^b - \sum_{i=1}^{n} f_i x_i^b = 0 \tag{7.80}$$

$$\frac{\partial E}{\partial b} = 2\sum_{i=1}^{n} (ax_i^b - f_i)ax_i^b \log x_i = 0 \rightarrow a^2\sum_{i=1}^{n} x_i^{2b} \log x_i - a\sum_{i=1}^{n} f_i x_i^b \log x_i = 0 \tag{7.81}$$

Equations (7.80) and (7.81), which are nonlinear with respect to a and b constants, can be solved iteratively using the generalized Newton-Raphson method in Chap. 5.

Linear Regression approach:

In some cases, transforming a nonlinear regression model into a linear regression one using variable transformation can simplify the calculation of model constants. This approach is generally preferred where transformation is possible. For example, applying logarithmic transformation, $\tilde{f}(x) = ax^b$ model can be converted into linear regression model as below:

$$\ln\left(\tilde{f}(x)\right) = \ln\left(ax^b\right) \rightarrow \underbrace{\ln\left(\tilde{f}\right)}_{\tilde{F}} = \underbrace{\ln\left(a\right)}_{A} + b\underbrace{\ln\left(x\right)}_{y} \rightarrow \tilde{F} = A + by$$

Resulting model $\tilde{F} = A + by$ is a linear regression model. Coefficients or constants of this model can be obtained easily applying the least squares method. When function values upon logarithmic transformation are represented with F values ($F = \ln(f)$), sum of squares of errors at all points can be written as following:

$$E = \sum_{i=1}^{n} \left(\tilde{F}_i - F_i\right)^2 = \sum_{i=1}^{n} \left(A + by_i - f_i\right)^2 \tag{7.82}$$

Minimum of E function can be obtained by taking its derivatives with respect to A (new constant) and b constants and equating them to zero as below:

$$\frac{\partial E}{\partial A} = 2\sum_{i=1}^{n} \left(A + by_i - f_i\right) = 0 \rightarrow nA + b\sum_{i=1}^{n} y_i = \sum_{i=1}^{n} f_i \tag{7.83}$$

$$\frac{\partial E}{\partial b} = 2\sum_{i=1}^{n} \left(A + by_i - f_i\right)y_i = 0 \rightarrow A\sum_{i=1}^{n} y_i + b\sum_{i=1}^{n} y_i^2 = \sum_{i=1}^{n} f_i y_i \tag{7.84}$$

Equations (7.83) and (7.84) can be expressed in matrix form as following:

$$\begin{bmatrix} n & \sum\limits_{i=1}^{n} y_i \\ \sum\limits_{i=1}^{n} y_i & \sum\limits_{i=1}^{n} y_i^2 \end{bmatrix} \begin{Bmatrix} A \\ b \end{Bmatrix} = \begin{Bmatrix} \sum\limits_{i=1}^{n} f \\ \sum\limits_{i=1}^{n} f_i y_i \end{Bmatrix} \rightarrow A, b \tag{7.85}$$

Equation (7.85) corresponds to a linear equation set, and it can be solved easily with linear equation methods in Chap. 4.

7.9 Artificial Neural Networks

Artificial Neural Networks (ANNs) are commonly used as nonlinear regression methods. They consist of nested functions. Artificial Neural Networks simulate the functioning mechanism of a human brain [19–23]. Brain cells are made up of interconnected nerve cells called as neurons. Each neuron receives information from other neurons, processes it, and transmits to the other connected neurons. This process is simulated by artificial neurons in Artificial Neural Network models. Figure 7.37 shows functioning of a single neuron. In Fig. 7.37, a j-th neuron collects $x_1, x_2, \ldots, x_n$ signals (information) from previous neurons and amplifies them with a_{1j}, a_{2j}, and a_{nj} weight coefficients. Amplified information is then processed through

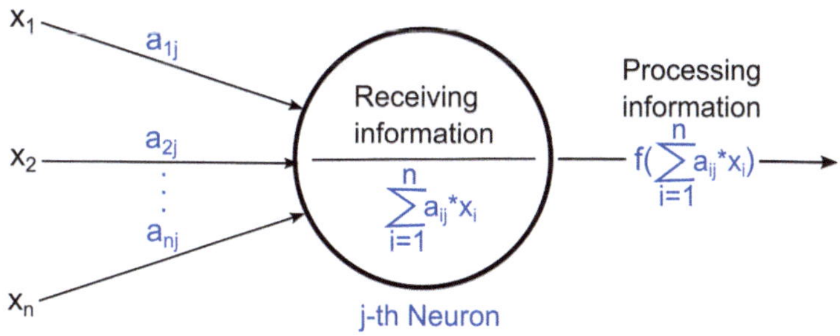

Fig. 7.37 Functioning of a neuron

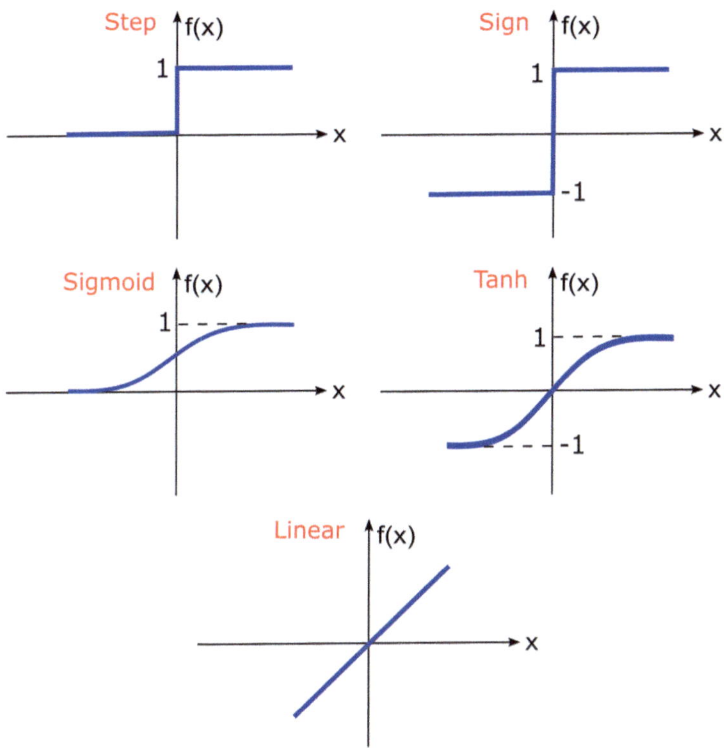

Fig. 7.38 Activation functions used in Artificial Neural Networks

an *f* function called as activation function and transmitted to the other connected neuron.

Step, Sign, Sigmoid, Tangent hyperbolic, and Linear functions shown in Fig. 7.38 are commonly used as activation functions.

Step, Sign, Sigmoid, Tangent hyperbolic, and Linear activation functions can be mathematically expressed as following:

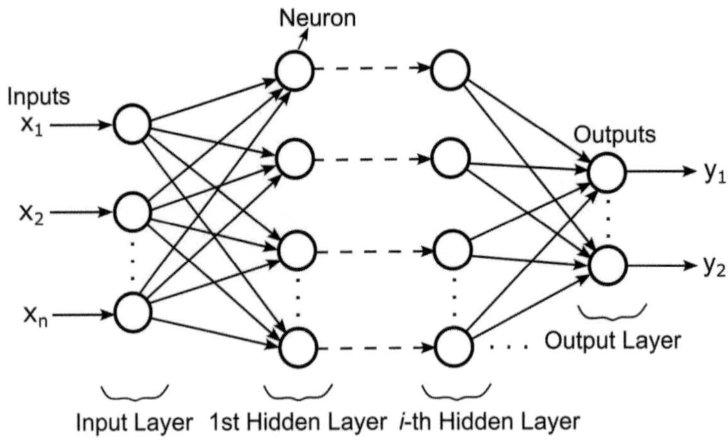

Fig. 7.39 Architecture of an Artificial Neural Network model

$$\text{Step function}: f(x) = \begin{cases} 1 & , \quad x \geq 0 \\ 0 & , \quad x < 0 \end{cases} \tag{7.86a}$$

$$\text{Sign function}: f(x) = \begin{cases} 1 & , \quad x \geq 0 \\ -1 & , \quad x < 0 \end{cases} \tag{7.86b}$$

$$\text{Sigmoid function}: f(x) = \frac{1}{1 + e^x} \tag{7.86c}$$

$$\text{Tangent hyperbolic function (Tanh)}: f(x) = \frac{e^x - e^{-x}}{e^x + e^{-x}} \tag{7.86d}$$

$$\text{Linear function}: f(x) = x \tag{7.86e}$$

An Artificial Neural Network model is created by using layers of neurons as shown in Fig. 7.39. Each layer consists of a certain number of neurons with selected activation function embedded in. Layers in a Neural Network are called as Input, Hidden (Intermediate), and Output layers. Intermediate (Hidden) layers can be more than one. In an Artificial Neural Network model, information enters the model through the Input layer, processed in Hidden layers and outputted through the Output layer.

7.10 Steps in Creating Artificial Neural Network Models

An Artificial Neural Network model can be systematically created using the following steps:

1. *Selection of layer and neuron numbers*:

   In an Artificial Neural Networks model, first, Input, Hidden, and Output layers are determined. Hidden layers can have multiple layers. Then, a number of neurons are determined in each layer. A number of neurons in Input and Output layers are determined from the problem information. For example, the number of neurons in Input layer is determined from the number of variables in the problem. The number of neurons in Output layer is determined from the number of outputs. For a single output, only a single neuron and for multiple outputs, multiple neurons will be utilized in Output layer. Although some intuitive approaches were developed for Hidden layers and their neuron numbers, there is no common rule valid everywhere. The number of Hidden layers and their neurons are usually determined by experience or trial-and-error approach depending on the nonlinearity of the data. In applications, a single Hidden layer is often used. A number of neurons in Hidden layers are usually chosen as the multiples of variables.
2. *Determination of neuron connections*:

   After creating the architecture of Artificial Neural Network model, second task is to determine neuron connections. In this task, which neuron will interact with which neuron is determined.
3. *Selection of activation functions in neurons:*

   Activation functions are selected in neurons of all layers except for Input layer to process the coming information. Activation functions are not used in Input layer since this layer corresponds to input variables. In Hidden layers, Step, Sign or Sigmoid functions are commonly used. In Output layer, Linear activation function is often used. With selection of activation functions in neurons, architecture of an Artificial Neural Network model is completed.
4. *Determination of weight coefficients between neurons:*

   How signals (information) between neurons will be amplified or strengthened is determined by weight coefficients in ANN models. Weight coefficients serve as the unknown model constants in ANN models. In Fig. 7.40, weight coefficients are indicated with a_{ij}, b_i, e_{ij}, and f_i terms. b_i (b_1, b_2,..., b_p) and f_i (f_1, f_2) coefficients are also called as bias. For the neurons having connections with b_i and f_i weight coefficients, input values are considered as one. Bias coefficients are actually used to create constant terms in ANN models. This situation is similar to using constant term c_1 in the response surface model of $f(x) = c_1 + c_2x + \ldots + c_{n+1}x^n$.
5. *Training of Neural Network model:*

   Calculation of weight coefficients a_{ij}, b_i, e_{ij}, and f_i in an ANN model is called as training of the ANN model. This situation is similar to the calculation of model constants of c_1, c_2, ..., c_{n+1} in the response surface model. Unlike response surface model, model constants (weight coefficients) in ANN models are determined solving nonlinear equations iteratively.

Because of the nested functions, mathematical expression of ANN model is very complex. However, this is very easy for response surface models. Complexity of ANN models is illustrated over a single neuron in the following examples:

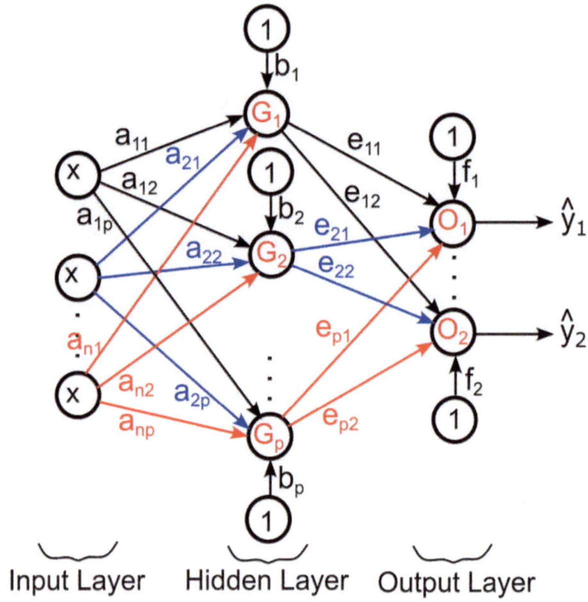

Fig. 7.40 Weight coefficients to be determined in an Artificial Neural Network model

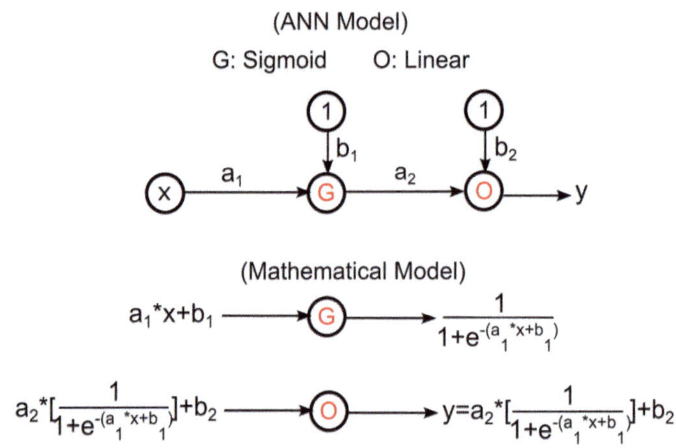

Fig. 7.41 Architecture and mathematical model of a simple ANN model for sigmoid and linear activation functions

Example 1: Consider the simple ANN model in Fig. 7.41a, where Input (x), Hidden (G), and Output (O) layers consist of single neurons. When activation functions are selected as Sigmoid for Hidden layer and linear for Output layer, output value y can be expressed with mathematical function of $y = a_2\left(\frac{1}{1+e^{-(a_1x+b_1)}}\right) + b_2$ as shown in Fig. 7.41b.

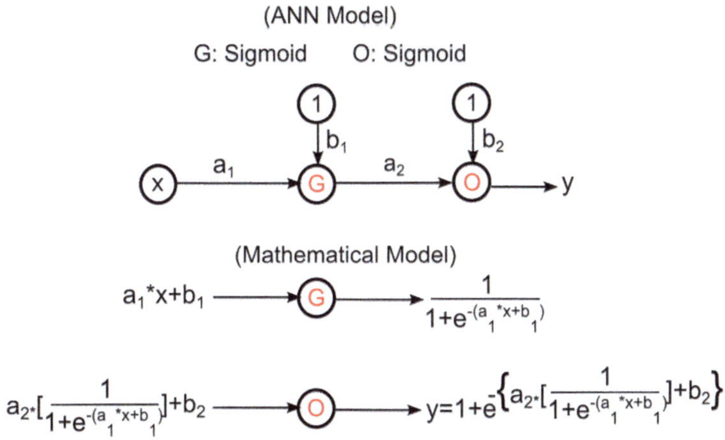

Fig. 7.42 Architecture and mathematical model of a simple ANN model for sigmoid activation functions

Example 2: Consider the simple ANN model in Fig. 7.42a, where Input (x), Hidden (G), and Output (O) layers consist of single neurons. When activation functions are selected as Sigmoid for Hidden and Output layers, output value y can be expressed with mathematical function of $y = 1 + e^{-\left\{a_2\left(\frac{1}{1+e^{-(a_1 x + b_1)}}\right)+b_2\right\}}$ as shown in Fig. 7.42b.

From the above examples, it is seen that mathematical expression is complicated even for a simple ANN model consisting of single neurons in layers. It can be easily thought that it will be much more complicated for a large number of neurons in layers. Because of this complexity, ANN models are not explicitly expressed mathematically. Rather, they are represented with an architecture expressed with a number of layers and neurons as well as their connections.

Because of highly nonlinear character of ANN models, weight coefficients are determined in an iterative way. First, initial values are selected for weight coefficients and output value of the ANN model is predicted using these values. This operation is called Feed Forwarding. Then, the difference between the actual and predicted output values is determined as the error. Using error value, weighting coefficients are corrected starting from the Output layer towards the Input layer. This process is called as Backpropagation of errors. Correction of weight coefficients are repeated using the values in the previous iteration as initial values in the current iteration until the error becomes sufficiently small. Training of an ANN model with a single hidden layer is explained in detail below.

7.11 Training of Artificial Neural Network Model

An Artificial Neural Network (ANN) model with one Hidden layer and one Output layer (layer) is shown in Fig. 7.43. Main unknowns in this model are the weight coefficients between layers. In Fig. 7.43, weight coefficients are shown with a_{ij}, b_i, e_i, and f_i constants. As stated before, determination of the weight coefficients in an ANN model is called as training of the ANN model. The data set used in the training of the ANN is called as Training Set. Weight coefficients are determined using optimization methods so that output values of the ANN model are sufficiently close to the actual output values in the data set. Training of an ANN model can be explained using the model in Fig. 7.43.

7.11.1 Feed Forwarding

Assume that there are m numbers of data in our data set to train the ANN model in Fig. 7.43. Thus, training process will include training of this model for all data set. To train the ANN model for s-th data set, we can consider $x_1^{(s)}, x_2^{(s)}, \ldots, x_n^{(s)}$ values are inputted in Input layer, then net input values coming to 1st, 2nd,…, pth neurons of Hidden layer (i.e., $\tilde{G}_1, \tilde{G}_2, \ldots, \tilde{G}_p$) can be expressed as the product of input values and weight coefficients which connect Input and Hidden layers as below:

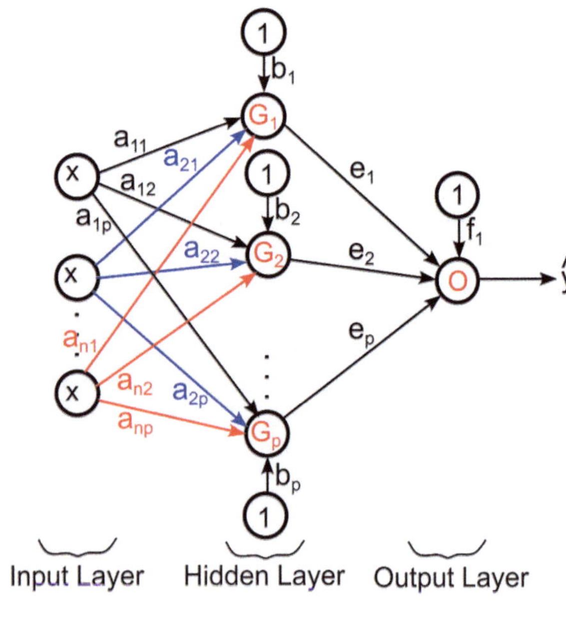

Fig. 7.43 An ANN model with one Hidden layer and their weight coefficients

$$
\begin{aligned}
\tilde{G}_1 &= b_1 + a_{11}x_1^{(s)} + a_{21}x_2^{(s)} + \ldots + a_{n1}x_n^{(s)} \\
\tilde{G}_2 &= b_2 + a_{12}x_1^{(s)} + a_{22}x_2^{(s)} + \ldots + a_{n2}x_n^{(s)} \\
&\ldots \\
\tilde{G}_p &= b_p + a_{1p}x_1^{(s)} + a_{2p}x_2^{(s)} + \ldots + a_{np}x_n^{(s)}
\end{aligned}
\tag{7.87}
$$

After processing net input values using activation functions, output values of 1st, 2nd,..., pth neurons in Hidden layer (G_1, G_2,..., G_p) can be expressed for *Sigmoid* activation functions as following:

$$
\begin{aligned}
G_1 &= \frac{1}{1+e^{-\tilde{G}_1}} \quad (G_1 : Sigmoid) \\
G_2 &= \frac{1}{1+e^{-\tilde{G}_2}} \quad (G_2 : Sigmoid) \\
&\ldots \\
G_p &= \frac{1}{1+e^{-\tilde{G}_p}} \quad (G_3 : Sigmoid)
\end{aligned}
\tag{7.88}
$$

Using Eq. (7.88), net input for the neuron in Output layer ($\tilde{y}$) can be expressed as the product of output values in Hidden layer and weight coefficients which connect Hidden and Output layers as below:

$$\tilde{y} = f_1 + G_1e_1 + G_2e_2 + \ldots + G_pe_p \tag{7.89}$$

After processing the net input value through the activation function, output value of the neuron in Output layer, which is also the output value ($\hat{y}_s$) of the ANN model, can be expressed for *Linear* activation function as following:

$$\hat{y}_s = \tilde{y} = f_1 + G_1e_1 + G_2e_2 + \ldots + G_pe_p \tag{7.90}$$

In this case, the difference between the actual value y_s and the predicted value $\hat{y}_s$ represents the error for the s-th data set. Error square for the s-th data set (E_S) can be expressed as following (1/2 is used for convenience in mathematical operations):

$$E_S = \frac{1}{2}(y_s - \hat{y}_s)^2 \tag{7.91}$$

Repeating above operations for all data set, sum of squares of errors (total error) can be written as below (m: number of data in training sets):

$$E = E_1 + E_2 + \ldots + E_m = \frac{1}{2}\left[(y_1 - \hat{y}_1)^2 + (y_2 - \hat{y}_2)^2 + \ldots + (y_m - \hat{y}_m)^2\right] = \sum_{s=1}^{m} E_S \tag{7.92}$$

7.11.2 Backpropagation

Correction of weight coefficients from Output layer to Hidden layer is called as backpropagation. In backpropagation, sensitivities of error with respect to weight coefficients are used. Sensitivity is expressed with derivative of error with respect to weight coefficients. Use of chain rule in derivatives simplifies sensitivity calculations. If activation functions in Hidden layer are chosen as Sigmoid, derivative of error with respect to weight coefficients can be expressed for the s-th data set as following (*Note*: if *f(x)* is *Sigmoid*, derivative is expressed as $\frac{\partial f}{\partial x}=f(1-f)$):

$$\left(\frac{\partial E}{\partial e_j}\right)_s=\frac{\partial E}{\partial \hat{y}}\underbrace{\frac{\partial \hat{y}}{\partial iy}}_{1}\underbrace{\frac{\partial iy}{\partial e_j}}_{G_j}=-(y_s-\hat{y}_s).1.G_j(j=1,\ldots,p) \tag{7.93}$$

$$\left(\frac{\partial E}{\partial f_1}\right)_s=\frac{\partial E}{\partial \hat{y}}\underbrace{\frac{\partial \hat{y}}{\partial iy}}_{1}\underbrace{\frac{\partial iy}{\partial f_1}}_{1}=-(y_s-\hat{y}_s).1.1 \tag{7.94}$$

$$\left(\frac{\partial E}{\partial a_{ij}}\right)_s=\frac{\partial E}{\partial \hat{y}}\underbrace{\frac{\partial \hat{y}}{\partial iy}}_{1}\underbrace{\frac{\partial iy}{\partial G_j}}_{e_j}\underbrace{\frac{\partial G_j}{\partial iG_j}}_{G_j(1-G_j)}\underbrace{\frac{\partial iG_j}{\partial a_{ij}}}_{x_i^s}$$

$$=-(y_s-\hat{y}_s).1.e_j.G_j(1-G_j).x_i^{(s)}(i=1,\ldots,n,j=1,\ldots,p) \tag{7.95}$$

$$\left(\frac{\partial E}{\partial b_j}\right)_s=\frac{\partial E}{\partial \hat{y}}\underbrace{\frac{\partial \hat{y}}{\partial iy}}_{1}\underbrace{\frac{\partial iy}{\partial G_j}}_{e_j}\underbrace{\frac{\partial G_j}{\partial iG_j}}_{G_j(1-G_j)}\underbrace{\frac{\partial iG_j}{\partial b_j}}_{1}=-(y_s-\hat{y}_s).1.e_j.G_j(1-G_j).1(j=1,\ldots,p) \tag{7.96}$$

Applying above calculations to all data sets, error derivatives for all data sets can be calculated.

7.11.3 Correction of Weight Coefficients

Summing error derivatives for all data sets, total error derivative can be obtained as following:

$$\frac{\partial E}{\partial e_j}=\sum_{s=1}^{m}\left(\frac{\partial E}{\partial e_j}\right)_s,\ \frac{\partial E}{\partial f_1}=\sum_{s=1}^{m}\left(\frac{\partial E}{\partial f_1}\right)_s,\ \frac{\partial E}{\partial a_{ij}}=\sum_{s=1}^{m}\left(\frac{\partial E}{\partial a_{ij}}\right)_s,\ \frac{\partial E}{\partial b_j}=\sum_{s=1}^{m}\left(\frac{\partial E}{\partial b_j}\right)_s$$

Using total error derivatives, weight coefficients can be updated as below (k: current iteration, k-1: previous iteration):

$$\left(a_{ij}\right)_k = \left(a_{ij}\right)_{k-1} - \eta \frac{\partial E}{\partial a_{ij}} \quad (i=1,\ldots,n, j=1,\ldots,p) \tag{7.97}$$

$$\left(b_j\right)_k = \left(b_j\right)_{k-1} - \eta \frac{\partial E}{\partial b_j} \quad (j=1,\ldots,p) \tag{7.98}$$

$$\left(e_j\right)_k = \left(e_j\right)_{k-1} - \eta \frac{\partial E}{\partial e_j} \quad (j=1,\ldots,p) \tag{7.99}$$

$$(f_1)_k = (f_1)_{k-1} - \eta \frac{\partial E}{\partial f_1} \tag{7.100}$$

where η is called as learning rate and it is taken as a small number such as $\eta = 0.01$. Equations (7.97)–(7.100) are applied repetitively until a_{ij}, b_j, e_j, and f_1 values converge to specific values. Training the ANN model using iterative approach can be summarized with a pseudo code in Table 7.17.

Example 7.11: Using the values of $y = 4 \sin x$ function at $x^{(1)} = 0$, $x^{(2)} = 0.7854$, $x^{(3)} = 1.5708$, $x^{(4)} = 2.3562$, $x^{(5)} = 3.1416$ points, predict the function value at $x = 2$ point with the ANN model in Fig. 7.44 and compare it with the true value. Use Sigmoid function as the activation function in the neurons of Hidden layer and use Linear function as the activation function in the neuron of Output layer. Take learning rate as $\eta = 0.01$.

For $x^{(1)} = 0$, $x^{(2)} = 0.7854$, $x^{(3)} = 1.5708$, $x^{(4)} = 2.3562$, $x^{(5)} = 3.1416$ points, values of $y = 4\sin x$ function are calculated as $y_1 = 0$, $y_2 = 2.8284$, $y_3 = 4.0$, $y_4 = 2.8284$, $y_5 = 0$, respectively. Using these values (5 training data), ANN model can be trained as following:

Training iteration 1 ($k = 1$):

Random initial values are widely used in ANN training since they give good results. Therefore, initial values of weight coefficients for the first iteration can be randomly selected in the range of [0, 1] as following:

$$a_{11} = (a_{11})_{k=0} = 0.4116, a_{12} = (a_{12})_{k=0} = 0.6026, b_1 = (b_1)_{k=0} = 0.7505, b_2 = (b_2)_{k=0} = 0.5835,$$

$$e_1 = (e_1)_{k=0} = 0.5518, \quad e_2 = (e_2)_{k=0} = 0.5836, f_1 = 0.5118.$$

For data point 1 ($s = 1$): $x^{(1)} = 0$, $y_1 = 0$

For this data point, net input and output values of the neurons in Hidden and Output layers as well as the error of the ANN model can be obtained as following:

Table 7.17 Training of the Artificial Neural Network model in Fig. 7.43

Selection of initial values: $(a_{ij})_0$, $(b_j)_0$, $(e_j)_0$, $(f_1)_0$
Loop 1: $k = 1,\ldots,$ *maxit*
  $a_{ij} = (a_{ij})_{k-1}$, $b_j = (b_j)_{k-1}$, $e_j = (e_j)_{k-1}$, $f_1 = (f_1)_{k-1}$
  Loop 2: $s = 1,\ldots, m$ (m: Training Set or Data Set)
    Error for s-th data: $E_s = \frac{1}{2}(y_s - \hat{y}_s)^2$
    Error derivatives for s-th data: $\left(\frac{\partial E}{\partial a_{ij}}\right)_s, \left(\frac{\partial E}{\partial b_j}\right)_s, \left(\frac{\partial E}{\partial e_j}\right)_s, \left(\frac{\partial E}{\partial f_1}\right)_s$
  End of Loop 2
  Total Error: $E = \sum_{s=1}^{m} E_s$
  Total Error Derivatives:
  $\frac{\partial E}{\partial a_{ij}} = \sum_{s=1}^{m}\left(\frac{\partial E}{\partial a_{ij}}\right)_s, \frac{\partial E}{\partial b_j} = \sum_{s=1}^{m}\left(\frac{\partial E}{\partial b_j}\right)_s, \frac{\partial E}{\partial e_j} = \sum_{s=1}^{m}\left(\frac{\partial E}{\partial e_j}\right)_s, \frac{\partial E}{\partial f_1} = \sum_{s=1}^{m}\left(\frac{\partial E}{\partial f_1}\right)_s$
  Updated weight coefficients:
  $(a_{ij})_k = (a_{ij})_{k-1} - \eta\frac{\partial E}{\partial a_{ij}}, (b_j)_k = (b_j)_{k-1} - \eta\frac{\partial E}{\partial b_j},$
  $(e_j)_k = (e_j)_{k-1} - \eta\frac{\partial E}{\partial e_j}, (f_1)_k = (f_1)_{k-1} - \eta\frac{\partial E}{\partial f_1}$
End of Loop 1

In Artificial Neural Networks, unlike Response Surface method, the number of experiments or test points does not need to be dependent on the number of variables. Reminding: at least $(n + 1)(n + 2)/2$ test points are required to create a quadratic response surface model with n variables. Also, test points in ANN models do not need to be evenly distributed. Lack of limitation on the number and distribution of data points forms the major advantage of the Artificial Neural Network method over Response Surface method. However, long training time due to iterative approach forms the disadvantage of ANN method.

In training of ANN models, initial values of weight coefficients are usually chosen randomly in the range of [0, 1] since it accelerates the training process. Therefore, convergence iterations of the ANN model may differ due to the use of random numbers when the algorithm in Table 7.17 is rerun.

Note: If convergence of an ANN model cannot be achieved in a large number of training iterations, this situation may be the indicator of insufficient number of neurons in Hidden layer. In this case, convergence can be ensured by increasing the number of neurons in Hidden layer.

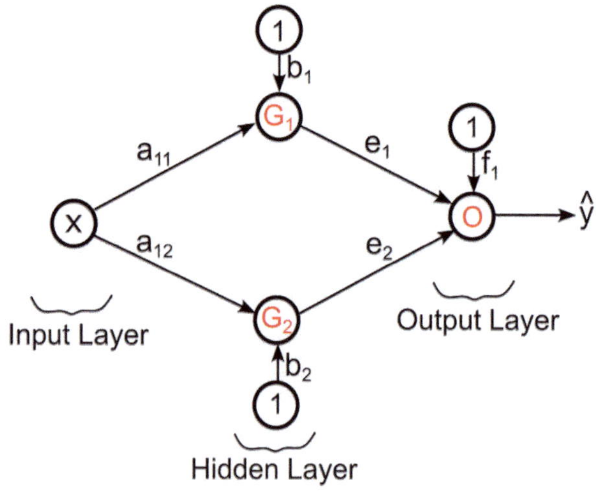

Fig. 7.44 Artificial Neural Network model for the $y = 4sinx$ function and the weighting coefficients to be determined

$$\tilde{G}_1 = b_1 + a_{11}x^{(1)} = 0.7505 \rightarrow G_1 = \frac{1}{1 + e^{-\tilde{G}_1}} = 0.6793$$

$$\tilde{G}_2 = b_2 + a_{12}x^{(1)} = 0.5835 \rightarrow G_2 = \frac{1}{1 + e^{-\tilde{G}_2}} = 0.6419$$

$$\hat{y}_1 = \tilde{y} = f_1 + G_1e_1 + G_2e_2 = 1.2612 \rightarrow E_1 = \frac{1}{2}(y_1 - \hat{y}_1)^2 = \frac{1}{2}(0 - 1.2612)^2 = 0.7954.$$

For this data point, derivative of the error with respect to the weight coefficients can be obtained as below:

$$\left(\frac{\partial E}{\partial e_1}\right)_1 = -(y_1 - \hat{y}_1)G_1 = 0.8567, \left(\frac{\partial E}{\partial e_2}\right)_1 = -(y_1 - \hat{y}_1)G_2 = 0.8096, \left(\frac{\partial E}{\partial f_1}\right)_1 = -(y_1 - \hat{y}_1) = 1.2612,$$

$$\left(\frac{\partial E}{\partial a_{11}}\right)_1 = -(y_1 - \hat{y}_1)e_1G_1(1 - G_1)x^{(1)} = 0, \left(\frac{\partial E}{\partial a_{12}}\right)_1 = -(y_1 - \hat{y}_1)e_2G_2(1 - G_2)x^{(1)} = 0,$$

$$\left(\frac{\partial E}{\partial b_1}\right)_1 = -(y_1 - \hat{y}_1)e_1G_1(1 - G_1) = 0.1516, \left(\frac{\partial E}{\partial b_2}\right)_1 = -(y_1 - \hat{y}_1)e_2G_2(1 - G_2) = 0.1692.$$

For data point 2 ($s = 2$): $x^{(2)} = 0.7854$, $y_2 = 2.8284$

For this data point, net input and output values of the neurons in Hidden and Output layers as well as the error of the ANN model can be obtained as following:

$$\tilde{G}_1 = b_1 + a_{11}x^{(2)} = 1.0738 \rightarrow G_1 = \frac{1}{1 + e^{-\tilde{G}_1}} = 0.7453$$

$$\tilde{G}_2 = b_2 + a_{12}x^{(2)} = 1.0568 \rightarrow G_2 = \frac{1}{1 + e^{-\tilde{G}_2}} = 0.7421$$

$$\hat{y}_2 = \tilde{y} = f_1 + G_1e_1 + G_2e_2 = 1.3561, E_2 = \frac{1}{2}(y_2 - \hat{y}_2)^2 = \frac{1}{2}(2.8284 - 1.3561)^2 = 1.0838.$$

For this *data* point, derivative of the error with respect to the weight coefficients can be obtained as below:

$$\left(\frac{\partial E}{\partial e_1}\right)_2 = -(y_2 - \hat{y}_2)G_1 = -1.0973, \left(\frac{\partial E}{\partial e_2}\right)_1 = -(y_2 - \hat{y}_2)G_2 = -1.0926, \left(\frac{\partial E}{\partial f_1}\right)_1 = -(y_2 - \hat{y}_2) = -1.4723,$$

$$\left(\frac{\partial E}{\partial a_{11}}\right)_2 = -(y_2 - \hat{y}_2)e_1 G_1(1 - G_1)x^{(2)} = -0.1211, \left(\frac{\partial E}{\partial a_{12}}\right)_2 = -(y_2 - \hat{y}_2)e_2 G_2(1 - G_2)x^{(2)} = -0.1292,$$

$$\left(\frac{\partial E}{\partial b_1}\right)_2 = -(y_2 - \hat{y}_2)e_1 G_1(1 - G_1) = -0.1542, \left(\frac{\partial E}{\partial b_2}\right)_2 = -(y_2 - \hat{y}_2)e_2 G_2(1 - G_2) = -0.1644.$$

For data point 3 ($s = 3$): $x^{(3)} = 1.5708$, $y_3 = 4.0$

For this data point, net input and output values of the neurons in Hidden and Output layers as well as the error of the ANN model can be obtained as following:

$$\tilde{G}_1 = b_1 + a_{11}x^{(3)} = 1.3970 \rightarrow G_1 = \frac{1}{1 + e^{-\tilde{G}_1}} = 0.8017$$

$$\tilde{G}_2 = b_2 + a_{12}x^{(3)} = 1.5302 \rightarrow G_2 = \frac{1}{1 + e^{-\tilde{G}_2}} = 0.8220$$

$$\hat{y}_3 = \tilde{y} = f_1 + G_1 e_1 + G_2 e_2 = 1.4339, E_3 = \frac{1}{2}(y_3 - \hat{y}_3)^2 = \frac{1}{2}(4.0 - 1.4339)^2 = 3.2924.$$

For this data point, derivative of the error with respect to the weight coefficients can be obtained as below:

$$\left(\frac{\partial E}{\partial e_1}\right)_3 = -(y_3 - \hat{y}_3)G_1 = -2.0573, \left(\frac{\partial E}{\partial e_2}\right)_3 = -(y_3 - \hat{y}_3)G_2 = -2.1094, \left(\frac{\partial E}{\partial f_1}\right)_3 = -(y_3 - \hat{y}_3) = -2.5661,$$

$$\left(\frac{\partial E}{\partial a_{11}}\right)_3 = -(y_3 - \hat{y}_3)e_1 G_1(1 - G_1)x^{(3)} = -0.3536, \left(\frac{\partial E}{\partial a_{12}}\right)_3 = -(y_3 - \hat{y}_3)e_2 G_2(1 - G_2)x^{(3)} = -0.3441,$$

$$\left(\frac{\partial E}{\partial b_1}\right)_3 = -(y_3 - \hat{y}_3)e_1 G_1(1 - G_1) = -0.2251, \left(\frac{\partial E}{\partial b_2}\right)_3 = -(y_3 - \hat{y}_3)e_2 G_2(1 - G_2) = -0.2191.$$

For data point 4 ($s = 4$): $x^{(4)} = 2.3562$, $y_4 = 2.8284$

For this data point, net input and output values of the neurons in Hidden and Output layers as well as the error of the ANN model can be obtained as following:

$$\tilde{G}_1 = b_1 + a_{11}x^{(4)} = 1.7203 \rightarrow G_1 = \frac{1}{1 + e^{-\tilde{G}_1}} = 0.8482$$

$$\tilde{G}_2 = b_2 + a_{12}x^{(4)} = 2.0035 \rightarrow G_2 = \frac{1}{1 + e^{-\tilde{G}_2}} = 0.8812$$

$$\hat{y}_4 = \tilde{y} = f_1 + G_1 e_1 + G_2 e_2 = 1.4941, E_4 = \frac{1}{2}(y_4 - \hat{y}_4)^2 = \frac{1}{2}(2.8284 - 1.4941)^2 = 0.8903.$$

For this data point, derivative of the error with respect to the weight coefficients can be obtained as below:

$$\left(\frac{\partial E}{\partial e_1}\right)_4 = -(y_4 - \hat{y}_4)G_1 = -1.1318, \left(\frac{\partial E}{\partial e_2}\right)_4 = -(y_4 - \hat{y}_4)G_2$$
$$= -1.1758, \left(\frac{\partial E}{\partial f_1}\right)_4 = -(y_4 - \hat{y}_4) = -1.3344,$$
$$\left(\frac{\partial E}{\partial a_{11}}\right)_4 = -(y_4 - \hat{y}_4)e_1 G_1(1 - G_1)x^{(4)} = -0.2234, \left(\frac{\partial E}{\partial a_{12}}\right)_4$$
$$= -(y_4 - \hat{y}_4)e_2 G_2(1 - G_2)x^{(4)} = -0.1921,$$
$$\left(\frac{\partial E}{\partial b_1}\right)_4 = -(y_4 - \hat{y}_4)e_1 G_1(1 - G_1) = -0.0948, \left(\frac{\partial E}{\partial b_2}\right)_4$$
$$= -(y_4 - \hat{y}_4)e_2 G_2(1 - G_2) = -0.0815.$$

For data point 5 ($s = 5$): $x^{(5)} = 3.1416$, $y_5 = 0$

For this data point, net input and output values of the neurons in Hidden and Output layers as well as the error of the ANN model can be obtained as following:

$$\tilde{G}_1 = b_1 + a_{11}x^{(5)} = 2.0436 \rightarrow G_1 = \frac{1}{1 + e^{-\tilde{G}_1}} = 0.8853$$

$$\tilde{G}_2 = b_2 + a_{12}x^{(5)} = 2.4768 \rightarrow G_2 = \frac{1}{1 + e^{-\tilde{G}_2}} = 0.9225$$

$$\hat{y}_5 = \tilde{y} = f_1 + G_1 e_1 + G_2 e_2 = 1.5387, E_5 = \frac{1}{2}(y_5 - \hat{y}_5)^2 = \frac{1}{2}(0 - 1.5387)^2 = 1.1837.$$

For this data point, derivative of the error with respect to the weight coefficients can be obtained as below:

$$\left(\frac{\partial E}{\partial e_1}\right)_5 = -(y_5 - \hat{y}_5)G_1 = 1.3622, \left(\frac{\partial E}{\partial e_2}\right)_5$$
$$= -(y_5 - \hat{y}_5)G_2 = 1.4194, \left(\frac{\partial E}{\partial f_1}\right)_5 = -(y_5 - \hat{y}_5) = 1.5387,$$

$$\left(\frac{\partial E}{\partial a_{11}}\right)_5 = -(y_5 - \hat{y}_5)e_1G_1(1-G_1)x^{(5)} = 0.2709, \left(\frac{\partial E}{\partial a_{12}}\right)_5$$
$$= -(y_5 - \hat{y}_5)e_2G_2(1-G_2)x^{(5)} = 0.12017,$$
$$\left(\frac{\partial E}{\partial b_1}\right)_5 = -(y_5 - \hat{y}_5)e_1G_1(1-G_1) = 0.0862, \left(\frac{\partial E}{\partial b_2}\right)_5$$
$$= -(y_5 - \hat{y}_5)e_2G_2(1-G_2) = 0.0642.$$

Thus, sum of the error derivatives for five data points can be obtained as following:

$$\frac{\partial E}{\partial a_{11}} = \left(\frac{\partial E}{\partial a_{11}}\right)_1 + \left(\frac{\partial E}{\partial a_{11}}\right)_2 + \left(\frac{\partial E}{\partial a_{11}}\right)_3 + \left(\frac{\partial E}{\partial a_{11}}\right)_4 + \left(\frac{\partial E}{\partial a_{11}}\right)_5 = -0.4273,$$
$$\frac{\partial E}{\partial a_{12}} = \left(\frac{\partial E}{\partial a_{12}}\right)_1 + \left(\frac{\partial E}{\partial a_{12}}\right)_2 + \left(\frac{\partial E}{\partial a_{12}}\right)_3 + \left(\frac{\partial E}{\partial a_{12}}\right)_4 + \left(\frac{\partial E}{\partial a_{12}}\right)_5 = -0.4638,$$
$$\frac{\partial E}{\partial b_1} = \left(\frac{\partial E}{\partial b_1}\right)_1 + \left(\frac{\partial E}{\partial b_1}\right)_2 + \left(\frac{\partial E}{\partial b_1}\right)_3 + \left(\frac{\partial E}{\partial b_1}\right)_4 + \left(\frac{\partial E}{\partial b_1}\right)_5 = -0.2363,$$
$$\frac{\partial E}{\partial b_2} = \left(\frac{\partial E}{\partial b_2}\right)_1 + \left(\frac{\partial E}{\partial b_2}\right)_2 + \left(\frac{\partial E}{\partial b_2}\right)_3 + \left(\frac{\partial E}{\partial b_2}\right)_4 + \left(\frac{\partial E}{\partial b_2}\right)_5 = -0.2317,$$
$$\frac{\partial E}{\partial e_1} = \left(\frac{\partial E}{\partial e_1}\right)_1 + \left(\frac{\partial E}{\partial e_1}\right)_2 + \left(\frac{\partial E}{\partial e_1}\right)_3 + \left(\frac{\partial E}{\partial e_1}\right)_4 + \left(\frac{\partial E}{\partial e_1}\right)_5 = -2.0674,$$
$$\frac{\partial E}{\partial e_2} = \left(\frac{\partial E}{\partial e_2}\right)_1 + \left(\frac{\partial E}{\partial e_2}\right)_2 + \left(\frac{\partial E}{\partial e_2}\right)_3 + \left(\frac{\partial E}{\partial e_2}\right)_4 + \left(\frac{\partial E}{\partial e_2}\right)_5 = -2.1487,$$
$$\frac{\partial E}{\partial f_1} = \left(\frac{\partial E}{\partial f_1}\right)_1 + \left(\frac{\partial E}{\partial f_1}\right)_2 + \left(\frac{\partial E}{\partial f_1}\right)_3 + \left(\frac{\partial E}{\partial f_1}\right)_4 + \left(\frac{\partial E}{\partial f_1}\right)_5 = -2.5728.$$

Using sum of the error derivatives, weight coefficients can be updated at the end of the first iteration (Epoch) as following:

$$(a_{11})_{k=1} = (a_{11})_{k=0} - \eta\frac{\partial E}{\partial a_{11}} = 0.4159, \quad (a_{12})_{k=1} = (a_{12})_{k=0} - \eta\frac{\partial E}{\partial a_{12}} = 0.6072,$$
$$(b_1)_{k=1} = (b_1)_{k=0} - \eta\frac{\partial E}{\partial b_1} = 0.7529, \quad (b_2)_{k=1} = (b_2)_{k=0} - \eta\frac{\partial E}{\partial b_2} = 0.5858$$
$$(e_1)_{k=1} = (e_1)_{k=0} - \eta\frac{\partial E}{\partial e_1} = 0.5725, \quad (e_2)_{k=1} = (e_2)_{k=0} - \eta\frac{\partial E}{\partial e_2} = 0.6051$$

$$(f_1)_{k=1} = (f_1)_{k=0} - \eta \frac{\partial E}{\partial f_1} = 0.5375$$

Training iteration 2 ($k = 2$):

Weight coefficient values at the end of first iteration are taken as the initial values for second iteration. Thus, initial values for second iteration are adopted as following:

$$a_{11} = (a_{11})_{k=1} = 0.4159, a_{12} = (a_{12})_{k=1} = 0.6072, b_1$$
$$= (b_1)_{k=1} = 0.7529, b_2 = (b_2)_{k=1} = 0.5858$$

$$e_1 = (e_1)_{k=1} = 0.5725, e_2 = (e_2)_{k=1} = 0.6051, f_1 = (f_1)_{k=1} = 0.5375$$

For data point 1 ($s = 1$): $x^{(1)} = 0$, $y_1 = 0$

For this data point, net input and output values of the neurons in Hidden and Output layers as well as the error of the ANN model can be obtained as following:

$$\tilde{G}_1 = b_1 + a_{11}x^{(1)} = 0.7529 \rightarrow G_1 = \frac{1}{1 + e^{-\tilde{G}_1}} = 0.6798$$

$$\tilde{G}_2 = b_2 + a_{12}x^{(1)} = 0.5858 \rightarrow G_2 = \frac{1}{1 + e^{-\tilde{G}_2}} = 0.6424$$

$$\hat{y}_1 = \tilde{y} = f_1 + G_1 e_1 + G_2 e_2 = 1.3154, E_1 = \frac{1}{2}(y_1 - \hat{y}_1)^2 = \frac{1}{2}(0 - 1.3154)^2 = 0.8652.$$

For this data point, derivative of the error with respect to the weight coefficients can be obtained as below:

$$\left(\frac{\partial E}{\partial e_1}\right)_1 = -(y_1 - \hat{y}_1)G_1 = 0.8942, \left(\frac{\partial E}{\partial e_2}\right)_1$$
$$= -(y_1 - \hat{y}_1)G_2 = 0.8450, \left(\frac{\partial E}{\partial f_1}\right)_1 = -(y_1 - \hat{y}_1) = 1.3154,$$

$$\left(\frac{\partial E}{\partial a_{11}}\right)_1 = -(y_1 - \hat{y}_1)e_1 G_1(1 - G_1)x^{(1)}$$
$$= 0, \left(\frac{\partial E}{\partial a_{12}}\right)_1 = -(y_1 - \hat{y}_1)e_2 G_2(1 - G_2)x^{(1)} = 0,$$

$$\left(\frac{\partial E}{\partial b_1}\right)_1 = -(y_1 - \hat{y}_1)e_1 G_1(1 - G_1) = 0.1639, \left(\frac{\partial E}{\partial b_2}\right)_1$$
$$= -(y_1 - \hat{y}_1)e_2 G_2(1 - G_2) = 0.1828.$$

For data point 2 ($s = 2$): $x^{(2)} = 0.7854$, $y_2 = 2.8284$

For this data point, net input and output values of the neurons in Hidden and Output layers as well as the error of the ANN model can be obtained as following:

$$\tilde{G}_1 = b_1 + a_{11}x^{(2)} = 1.0795 \rightarrow G_1 = \frac{1}{1+e^{-\tilde{G}_1}} = 0.7464$$

$$\tilde{G}_2 = b_2 + a_{12}x^{(2)} = 1.0627 \rightarrow G_2 = \frac{1}{1+e^{-\tilde{G}_2}} = 0.7432$$

$$\hat{y}_2 = \tilde{y} = f_1 + G_1e_1 + G_2e_2 = 1.4145, E_2 = \frac{1}{2}(y_2 - \hat{y}_2)^2 = \frac{1}{2}(2.8284 - 1.4145)^2 = 0.9996$$

For this data point, derivative of the error with respect to the weight coefficients can be obtained as below:

$$\left(\frac{\partial E}{\partial e_1}\right)_2 = -(y_2 - \hat{y}_2)G_1 = -1.0553, \left(\frac{\partial E}{\partial e_2}\right)_1$$
$$= -(y_2 - \hat{y}_2)G_2 = -1.0508, \left(\frac{\partial E}{\partial f_1}\right)_1 = -(y_2 - \hat{y}_2) = -1.4139,$$
$$\left(\frac{\partial E}{\partial a_{11}}\right)_2 = -(y_2 - \hat{y}_2)e_1G_1(1 - G_1)x^{(2)} = -0.1203, \left(\frac{\partial E}{\partial a_{12}}\right)_2$$
$$= -(y_2 - \hat{y}_2)e_2G_2(1 - G_2)x^{(2)} = -0.1282,$$
$$\left(\frac{\partial E}{\partial b_1}\right)_2 = -(y_2 - \hat{y}_2)e_1G_1(1 - G_1) = -0.1532, \left(\frac{\partial E}{\partial b_2}\right)_2$$
$$= -(y_2 - \hat{y}_2)e_2G_2(1 - G_2) = -0.1633.$$

For data point 3 ($s = 3$): $x^{(3)} = 1.5708$, $y_3 = 4.0$

For this data point, net input and output values of the neurons in Hidden and Output layers as well as the error of the ANN model can be obtained as following:

$$\tilde{G}_1 = b_1 + a_{11}x^{(3)} = 1.4061 \rightarrow G_1 = \frac{1}{1+e^{-\tilde{G}_1}} = 0.8032$$

$$\tilde{G}_2 = b_2 + a_{12}x^{(3)} = 1.5397 \rightarrow G_2 = \frac{1}{1+e^{-\tilde{G}_2}} = 0.8234$$

$$\hat{y}_3 = \tilde{y} = f_1 + G_1e_1 + G_2e_2 = 1.4956, E_3 = \frac{1}{2}(y_3 - \hat{y}_3)^2 = \frac{1}{2}(4.0 - 1.4956)^2 = 3.1361.$$

For this data point, derivative of the error with respect to the weight coefficients can be obtained as below:

$$\left(\frac{\partial E}{\partial e_1}\right)_3 = -(y_3 - \hat{y}_3)G_1 = -2.0115, \left(\frac{\partial E}{\partial e_2}\right)_3$$
$$= -(y_3 - \hat{y}_3)G_2 = -2.0622, \left(\frac{\partial E}{\partial f_1}\right)_3 = -(y_3 - \hat{y}_3) = -2.5044,$$
$$\left(\frac{\partial E}{\partial a_{11}}\right)_3 = -(y_3 - \hat{y}_3)e_1G_1(1 - G_1)x^{(3)} = -0.3561, \left(\frac{\partial E}{\partial a_{12}}\right)_3$$
$$= -(y_3 - \hat{y}_3)e_2G_2(1 - G_2)x^{(3)} = -0.3461,$$
$$\left(\frac{\partial E}{\partial b_1}\right)_3 = -(y_3 - \hat{y}_3)e_1G_1(1 - G_1) = -0.2267, \left(\frac{\partial E}{\partial b_2}\right)_3$$
$$= -(y_3 - \hat{y}_3)e_2G_2(1 - G_2) = -0.2203.$$

For data point 4 (s = 4): $x^{(4)} = 2.3562$, $y_4 = 2.8284$

For this data point, net input and output values of the neurons in Hidden and Output layers as well as the error of the ANN model can be obtained as following:

$$\tilde{G}_1 = b_1 + a_{11}x^{(4)} = 1.7327 \rightarrow G_1 = \frac{1}{1 + e^{-\tilde{G}_1}} = 0.8498$$

$$\tilde{G}_2 = b_2 + a_{12}x^{(4)} = 2.0166 \rightarrow G_2 = \frac{1}{1 + e^{-\tilde{G}_2}} = 0.8825$$

$$\hat{y}_4 = \tilde{y} = f_1 + G_1e_1 + G_2e_2 = 1.5580, E_4 = \frac{1}{2}(y_4 - \hat{y}_4)^2 = \frac{1}{2}(2.8284 - 1.5580)^2 = 0.8070.$$

For this data point, derivative of the error with respect to the weight coefficients can be obtained as below:

$$\left(\frac{\partial E}{\partial e_1}\right)_4 = -(y_4 - \hat{y}_4)G_1 = -1.0796, \left(\frac{\partial E}{\partial e_2}\right)_4$$
$$= -(y_4 - \hat{y}_4)G_2 = -1.1212, \left(\frac{\partial E}{\partial f_1}\right)_4 = -(y_4 - \hat{y}_4) = -1.2704,$$
$$\left(\frac{\partial E}{\partial a_{11}}\right)_4 = -(y_4 - \hat{y}_4)e_1G_1(1 - G_1)x^{(4)} = -0.2188, \left(\frac{\partial E}{\partial a_{12}}\right)_4$$
$$= -(y_4 - \hat{y}_4)e_2G_2(1 - G_2)x^{(4)} = -0.1878,$$
$$\left(\frac{\partial E}{\partial b_1}\right)_4 = -(y_4 - \hat{y}_4)e_1G_1(1 - G_1) = -0.0928, \left(\frac{\partial E}{\partial b_2}\right)_4$$
$$= -(y_4 - \hat{y}_4)e_2G_2(1 - G_2) = -0.0797.$$

For data point 5 (s = 5): $x^{(5)} = 3.1416$, $y_5 = 0$

For this data point, net input and output values of the neurons in Hidden and Output layers as well as the error of the ANN model can be obtained as following:

$$\tilde{G}_1 = b_1 + a_{11}x^{(5)} = 2.0594 \to G_1 = \frac{1}{1+e^{-\tilde{G}_1}} = 0.8869$$

$$\tilde{G}_2 = b_2 + a_{12}x^{(5)} = 2.4935 \to G_2 = \frac{1}{1+e^{-\tilde{G}_2}} = 0.9237$$

$$\hat{y}_5 = \tilde{y} = f_1 + G_1e_1 + G_2e_2 = 1.6042, E_5 = \frac{1}{2}(y_5 - \hat{y}_5)^2 = \frac{1}{2}(0 - 1.6042)^2 = 1.2867.$$

For this data point, derivative of the error with respect to the weight coefficients can be obtained as below:

$$\left(\frac{\partial E}{\partial e_1}\right)_5 = -(y_5 - \hat{y}_5)G_1 = 1.4227, \left(\frac{\partial E}{\partial e_2}\right)_5 = -(y_5 - \hat{y}_5)G_2 = 1.4817, \left(\frac{\partial E}{\partial f_1}\right)_5 = -(y_5 - \hat{y}_5) = 1.6042,$$

$$\left(\frac{\partial E}{\partial a_{11}}\right)_5 = -(y_5 - \hat{y}_5)e_1G_1(1 - G_1)x^{(5)} = 0.2894, \left(\frac{\partial E}{\partial a_{12}}\right)_5 = -(y_5 - \hat{y}_5)e_2G_2(1 - G_2)x^{(5)} = 0.2150,$$

$$\left(\frac{\partial E}{\partial b_1}\right)_5 = -(y_5 - \hat{y}_5)e_1G_1(1 - G_1) = 0.0921, \left(\frac{\partial E}{\partial b_2}\right)_5 = -(y_5 - \hat{y}_5)e_2G_2(1 - G_2) = 0.0684.$$

Thus, sum of the error derivatives for five data points can be obtained as following:

$$\frac{\partial E}{\partial a_{11}} = \left(\frac{\partial E}{\partial a_{11}}\right)_1 + \left(\frac{\partial E}{\partial a_{11}}\right)_2 + \left(\frac{\partial E}{\partial a_{11}}\right)_3 + \left(\frac{\partial E}{\partial a_{11}}\right)_4 + \left(\frac{\partial E}{\partial a_{11}}\right)_5 = -0.4057,$$

$$\frac{\partial E}{\partial a_{12}} = \left(\frac{\partial E}{\partial a_{12}}\right)_1 + \left(\frac{\partial E}{\partial a_{12}}\right)_2 + \left(\frac{\partial E}{\partial a_{12}}\right)_3 + \left(\frac{\partial E}{\partial a_{12}}\right)_4 + \left(\frac{\partial E}{\partial a_{12}}\right)_5 = -0.4472,$$

$$\frac{\partial E}{\partial b_1} = \left(\frac{\partial E}{\partial b_1}\right)_1 + \left(\frac{\partial E}{\partial b_1}\right)_2 + \left(\frac{\partial E}{\partial b_1}\right)_3 + \left(\frac{\partial E}{\partial b_1}\right)_4 + \left(\frac{\partial E}{\partial b_1}\right)_5 = -0.2167,$$

$$\frac{\partial E}{\partial b_2} = \left(\frac{\partial E}{\partial b_2}\right)_1 + \left(\frac{\partial E}{\partial b_2}\right)_2 + \left(\frac{\partial E}{\partial b_2}\right)_3 + \left(\frac{\partial E}{\partial b_2}\right)_4 + \left(\frac{\partial E}{\partial b_2}\right)_5 = -0.2120,$$

$$\frac{\partial E}{\partial e_1} = \left(\frac{\partial E}{\partial e_1}\right)_1 + \left(\frac{\partial E}{\partial e_1}\right)_2 + \left(\frac{\partial E}{\partial e_1}\right)_3 + \left(\frac{\partial E}{\partial e_1}\right)_4 + \left(\frac{\partial E}{\partial e_1}\right)_5 = -1.8294,$$

$$\frac{\partial E}{\partial e_2} = \left(\frac{\partial E}{\partial e_2}\right)_1 + \left(\frac{\partial E}{\partial e_2}\right)_2 + \left(\frac{\partial E}{\partial e_2}\right)_3 + \left(\frac{\partial E}{\partial e_2}\right)_4 + \left(\frac{\partial E}{\partial e_2}\right)_5 = -1.9074,$$

$$\frac{\partial E}{\partial f_1} = \left(\frac{\partial E}{\partial f_1}\right)_1 + \left(\frac{\partial E}{\partial f_1}\right)_2 + \left(\frac{\partial E}{\partial f_1}\right)_3 + \left(\frac{\partial E}{\partial f_1}\right)_4 + \left(\frac{\partial E}{\partial f_1}\right)_5 = -2.2692.$$

Using sum of the error derivatives, weight coefficients can be updated at the end of the second iteration (Epoch) as following:

$$(a_{11})_{k=2}=(a_{11})_{k=1}-\eta\frac{\partial E}{\partial a_{11}}=0.4199,\quad (a_{12})_{k=2}=(a_{12})_{k=1}-\eta\frac{\partial E}{\partial a_{12}}=0.6117$$

$$(b_1)_{k=2}=(b_1)_{k=1}-\eta\frac{\partial E}{\partial b_1}=0.7550,\quad (b_2)_{k=2}=(b_2)_{k=1}-\eta\frac{\partial E}{\partial b_2}=0.5879$$

$$(e_1)_{k=2}=(e_1)_{k=1}-\eta\frac{\partial E}{\partial e_1}=0.5908,\quad (e_2)_{k=2}=(e_2)_{k=1}-\eta\frac{\partial E}{\partial e_2}=0.6242$$

$$(f_1)_{k=2}=(f_1)_{k=1}-\eta\frac{\partial E}{\partial f_1}=0.5602$$

Calculations for other iterations are repeated similarly until convergence for weight coefficients can be achieved or maximum number of iterations is reached. Error Root Mean Square (*ERMS*) is commonly used as convergence criterion in ANN trainings. *ERMS* for n points in a data set can be expressed at an iteration as $ERMS=\sqrt{\frac{(E_1+E_2+\ldots+E_n)}{n}}$. For example, 7.11 (for 5 points) *ERMS* can be expressed at an iteration as $ERMS=\sqrt{\frac{(E_1+E_2+E_3+E_4+E_5)}{5}}$. Using the ANN program in appendix written in MATLAB programming language, weight coefficients were obtained after convergence (using the convergence tolerance of $\varepsilon=0.005$) as $a_{11}=2.3873$, $a_{12}=4.5406$, $b_1=-6.7872$, $b_2=-2.8251$, $e_1=-6.4948$, $e_2=4.6171$, $f_1=-0.2517$. Convergence history of ANN model is shown using *ERMS* values in Fig. 7.45a. Predicted and true values are compared at five data points in Fig. 7.45b.

At $x=2$ point, y value is predicted with ANN model as $y=3.5908$, whereas true value is calculated using $y=4\sin x$ function as $y=3.6372$. When the number of

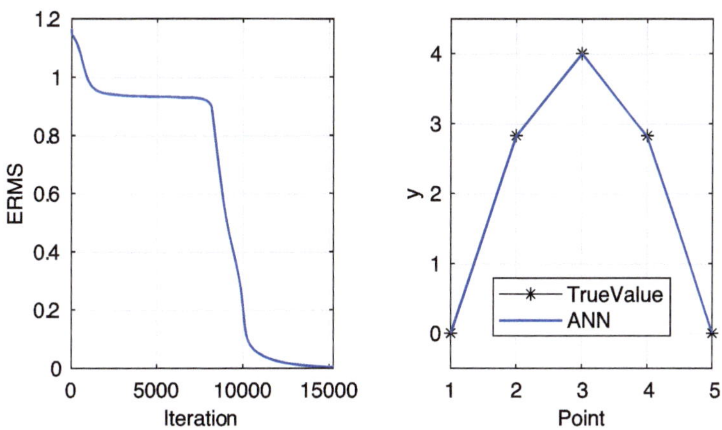

Fig. 7.45 Convergence history of the ANN model and comparison of the predicted (ANN) and true values (TrueValue) at five data points (Train Set)

neurons in Hidden layer is increased, ANN prediction will get closer to the true value. For example, when the number of neurons in Hidden layer is taken as 4, ANN prediction is obtained as $y = 3.6223$, which is very close to the true value.

Note: For the solution of this problem with Artificial Neural Networks method, MATLAB program in appendix entitled as *MATLAB Program 7.3: Artificial Neural Network (ANN)* can be used.

Reminding: Since the MATLAB program uses random numbers for the initial values of weight coefficients, converged results and the number of iterations for convergence might be different than the reported value here when the program is rerun.

7.12 Approximation of Complicated Functions with Simple Functions

In engineering, working with smooth and simple functions such as polynomial and trigonometric functions are desired to simplify the solution process. From mathematics, it is well-known that derivative or integral calculation of polynomial and trigonometric functions can be performed much more easily than other functions. Therefore, complicated functions in engineering problems are usually replaced or approximated with equivalent polynomial or trigonometric functions through Fourier series [1, 5, 24]. For example, in beam and plate bending problems in mechanics, load function is often approximated with a trigonometric function in the form of sine series to obtain a semi-analytical or approximate solution. In Fig. 7.46, a function and its approximation is illustrated. Figure 7.46 can be considered as fitting a function to another function similar to fitting a function to data points in curve fitting at beginning of this chapter.

To explain approximating a function with another function, let us indicate the real function with $f(x)$ and its approximation with $\tilde{f}(x)$. For both functions to be similar to each other in a [a, b] region, difference (error) between these functions should be small as much as possible at all points in this region. Similarity of $f(x)$ and $\tilde{f}(x)$

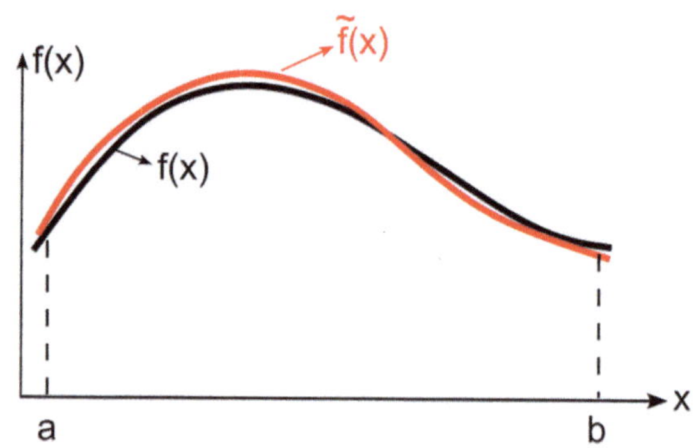

Fig. 7.46 Approximating a complicated $f(x)$ function with a simpler $\tilde{f}(x)$ function

functions can be enforced with minimization of sum of square of error (E) in [a, b] region through an optimization problem as following:

$$E=\int_a^b \left[\tilde{f}(x)-f(x)\right]^2 dx \rightarrow \min \tag{7.101}$$

Equation (7.101) can be seen as continuous form of the Least Square method in curve fitting and it can be called as Least Square expression.

Polynomial functions are widely used as approximate or approximation functions to replace more complicated functions. A polynomial function $P(x)$ can be written in general form as below:

$$\tilde{f}(x)=c_0+c_1x+c_2x^2+\ldots+c_nx^n \tag{7.102}$$

When polynomial function is substituted, Least Squares expression in Eq. (7.101) becomes as following:

$$E=\int_a^b \left[\left(c_0+c_1x+c_2x^2+\ldots+c_nx^n\right)-f(x)\right]^2 dx \rightarrow \min \rightarrow \tag{7.103}$$

Minimization of sum of error square E can alternatively be expressed with following equations:

$$\frac{\partial E}{\partial c_0}=0 \rightarrow 2\int_a^b \left[\left(c_0+c_1x+c_2x^2+\ldots+c_nx^n\right)-f(x)\right]dx=0 \tag{7.104a}$$

$$\frac{\partial E}{\partial c_1}=0 \rightarrow 2\int_a^b x\left[\left(c_0+c_1x+c_2x^2+\ldots+c_nx^n\right)-f(x)\right]dx=0 \tag{7.104b}$$

$$\frac{\partial E}{\partial c_2}=0 \rightarrow 2\int_a^b x^2\left[\left(c_0+c_1x+c_2x^2+\ldots+c_nx^n\right)-f(x)\right]dx=0 \tag{7.104c}$$

$$\ldots$$

$$\frac{\partial E}{\partial c_n}=0 \rightarrow 2\int_a^b x^n\left[\left(c_0+c_1x+c_2x^2+\ldots+c_nx^n\right)-f(x)\right]dx=0 \tag{7.104d}$$

Equations (7.104a)–(7.104d) can be written in matrix form as below:

$$\begin{bmatrix} \int_a^b 1dx & \int_a^b xdx & \int_a^b x^2dx & \cdots & \int_a^b x^ndx \\ \int_a^b xdx & \int_a^b x^2dx & \int_a^b x^3dx & \cdots & \int_a^b x^{n+1}dx \\ \int_a^b x^2dx & \int_a^b x^3dx & \int_a^b x^4dx & \cdots & \int_a^b x^{n+2}dx \\ \cdots & \cdots & \cdots & \cdots & \cdots \\ \int_a^b x^ndx & \int_a^b x^{n+1}dx & \int_a^b x^{n+2}dx & \cdots & \int_a^b x^{2n}dx \end{bmatrix} \begin{Bmatrix} c_0 \\ c_1 \\ c_2 \\ \cdots \\ c_n \end{Bmatrix}$$

$$= \begin{Bmatrix} \int_a^b f(x)dx \\ \int_a^b xf(x)dx \\ \int_a^b x^2f(x)dx \\ \cdots \\ \int_a^b x^nf(x)dx \end{Bmatrix} \rightarrow \begin{Bmatrix} c_0 \\ c_1 \\ c_2 \\ \cdots \\ c_n \end{Bmatrix} \tag{7.105}$$

Solution of Eq. (7.105) with linear equation methods in Chap. 4 yields polynomial coefficients.

Note: If general polynomial function is used, polynomial coefficients are calculated with the solution of a linear system as in Eq. (7.105). However, if *orthogonal* functions or orthogonal polynomials are used as approximate functions, solution of a linear system can be avoided. In this case, coefficients of approximate function can be obtained more easily with the solution of independent equations.

7.13 Approximation with Orthogonal Functions

Use of orthogonal functions as approximate functions simplifies the calculation of unknown coefficients. To explain orthogonality, we can consider $\Phi(x)$ and $\Psi(x)$ functions. If the integral of the product of $\Phi(x)$ and $\Psi(x)$ functions in an $[a, b]$

interval is zero as stated by $\int_a^b \Phi(x)\Psi(x)dx = 0$, then $\Phi(x)$ and $\Psi(x)$ are called orthogonal functions. Orthogonality for a family of functions can be expressed similarly. For example, for $\Phi_1(x)$, $\Phi_2(x)$,..., $\Phi_n(x)$ functions in a $\Phi(x)$ family, orthogonality in an $[a, b]$ interval can be expressed as following:

$$\int_a^b \Phi_m \Phi_n dx = \begin{cases} 0, & if \quad m \neq n \\ \lambda_n, & if \quad m = n \end{cases} \tag{7.106}$$

Orthogonal functions are independent functions. In an orthogonal family of functions, one function cannot be derived from others with mathematical operations. Commonly used orthogonal family of functions can be expressed as below:

- Trigonometric functions: $\cos 0 = 1$, $\cos x$, $\cos 2x$,..., $\cos nx$, $\sin x$, $\sin 2x$,...,$\sin nx$
- Legendre polynomials
- Chebyshev polynomials

These functions are explained in detail below.

7.13.1 Trigonometric Functions

Trigonometric functions are kinds of functions that repeat themselves with certain period. Among these functions, *sine* and *cosine* functions are commonly used. These functions repeat themselves with period of 2π. *Sine* and *cosine* functions retain their orthogonality in the interval of $[0, 2\pi]$ or $[-\pi, \pi]$. Fourier series include sine and cosine functions and it can be expressed as following:

$$f(x) = a_0 \underbrace{\cos 0}_{1} + a_1 \cos x + a_2 \cos 2x + \ldots + a_n \cos nx + b_1 \sin x + b_2 \sin 2x + \ldots + b_n \sin nx \tag{7.107}$$

Replacing a_0 with $\frac{a_0}{2}$ for convenience in calculations, above equation can be written briefly as below:

$$f(x) = \frac{a_0}{2} + \sum_{k=1}^{n} (a_k \cos kx + b_k \sin kx) \tag{7.108}$$

When sine and cosine functions have $2L$ period instead of 2π, Fourier series can be expressed as following:

$$f(x)=\frac{a_0}{2}+\sum_{k=1}^{n}\left(a_k\cos\frac{k\pi x}{L}+b_k\sin\frac{k\pi x}{L}\right) \tag{7.109}$$

Orthogonality of sine and cosine functions with $2L$ period in the interval of $[-L, L]$ can be expressed as below:

$$\int_{-L}^{L}\cos\frac{m\pi x}{L}\cos\frac{n\pi x}{L}dx=\begin{cases}0, & \text{if}\quad m\neq n\\ L, & \text{if}\quad m=n\neq 0\\ 2L, & \text{if}\quad m=n=0\end{cases} \tag{7.110a}$$

$$\int_{-L}^{L}\sin\frac{m\pi x}{L}\sin\frac{n\pi x}{L}dx=\begin{cases}0, & \text{if}\quad m\neq n\\ L, & \text{if}\quad m=n\neq 0\end{cases} \tag{7.110b}$$

$$\int_{-L}^{L}\cos\frac{m\pi x}{L}\sin\frac{n\pi x}{L}dx=\int_{-L}^{L}\sin\frac{m\pi x}{L}\cos\frac{n\pi x}{L}dx=0 \tag{7.110c}$$

An approximate function can be expressed using Fourier series expression in Eq. (7.109). a_0, a_k, and b_k coefficients in Fourier series can be determined using orthogonality condition of trigonometric functions as following:

a_0 calculation:

Fourier series in Eq. (7.109) can be written in expanded form as below:

$$f(x)=\frac{a_0}{2}+a_1\cos\frac{\pi x}{L}+a_2\cos\frac{2\pi x}{L}+\ldots+a_n\cos\frac{n\pi x}{L}+b_1\sin\frac{\pi x}{L}+b_2\sin\frac{2\pi x}{L}+\ldots$$
$$+b_n\sin\frac{n\pi x}{L} \tag{7.111}$$

Integration of Eq. (7.111) in $[-L, L]$ interval yields a_0 coefficient as following:

$$\int_{-L}^{L}f(x)dx=\int_{-L}^{L}\begin{bmatrix}\frac{a_0}{2}+a_1\cos\frac{\pi x}{L}+a_2\cos\frac{2\pi x}{L}+\ldots+a_k\cos\frac{k\pi x}{L}+\ldots+a_n\cos\frac{n\pi x}{L}\\ +b_1\sin\frac{\pi x}{L}+b_2\sin\frac{2\pi x}{L}+\ldots+b_k\sin\frac{k\pi x}{L}+\ldots+b_n\sin\frac{n\pi x}{L}\end{bmatrix}dx$$

$$=\frac{a_0}{2}\int_{-L}^{L}dx+a_1\underbrace{\int_{-L}^{L}\cos\frac{\pi x}{L}dx}_{=0}+a_2\underbrace{\int_{-L}^{L}\cos\frac{2\pi x}{L}dx}_{=0}+\ldots+a_k\underbrace{\int_{-L}^{L}\cos\frac{k\pi x}{L}dx}_{=0}+\ldots+a_n\underbrace{\int_{-L}^{L}\cos\frac{n\pi x}{L}dx}_{=0}+$$

$$b_1\underbrace{\int_{-L}^{L}\sin\frac{\pi x}{L}dx}_{=0}+b_2\underbrace{\int_{-L}^{L}\sin\frac{2\pi x}{L}dx}_{=0}+\ldots+b_n\underbrace{\int_{-L}^{L}\sin\frac{n\pi x}{L}dx}_{=0}$$

$$\rightarrow$$

$$\int_{-L}^{L} f(x)dx = a_0 L \rightarrow a_0 = \frac{1}{L} \int_{-L}^{L} f(x)dx$$

a_k *calculation*:

All $a_1 - a_n$ coefficients in Eq. (7.109) are obtained in the same way. Therefore, it will be more advantageous to derive a general expression for an a_k coefficient, from which all a_1- a_n coefficients can be obtained only by changing the k index. For this purpose, Fourier series can be written to include the general a_k coefficient as below:

$$f(x) = \frac{a_0}{2} + a_1 \cos\frac{\pi x}{L} + a_2 \cos\frac{2\pi x}{L} + \ldots + a_k \cos\frac{k\pi x}{L} + \ldots + a_n \cos\frac{n\pi x}{L}$$
$$+ b_1 \sin\frac{\pi x}{L} + b_2 \sin\frac{2\pi x}{L} + \ldots + b_k \sin\frac{k\pi x}{L} + \ldots + b_n \sin\frac{n\pi x}{L} \tag{7.112}$$

Multiplication of Eq. (7.112) by $\cos\frac{k\pi x}{L}$ and then integration in $[-L, L]$ interval using the orthogonality conditions in Eqs. (7.110a)–(7.110c) yield a_k coefficient as following:

$$\int_{-L}^{L} f(x)\cos\frac{k\pi x}{L}dx = \int_{-L}^{L} \cos\frac{k\pi x}{L} \begin{bmatrix} \frac{a_0}{2} + a_1\cos\frac{\pi x}{L} + a_2\cos\frac{2\pi x}{L} + \ldots + a_k\cos\frac{k\pi x}{L} + \ldots + a_n\cos\frac{n\pi x}{L} + \\ b_1\sin\frac{\pi x}{L} + b_2\sin\frac{2\pi x}{L} + \ldots + b_k\sin\frac{k\pi x}{L} + \ldots + b_n\sin\frac{n\pi x}{L} \end{bmatrix} dx$$

$$= \frac{a_0}{2}\underbrace{\int_{-L}^{L} \cos\frac{k\pi x}{L}dx}_{=0} + a_1\underbrace{\int_{-L}^{L} \cos\frac{k\pi x}{L}\cos\frac{\pi x}{L}dx}_{=0} + a_2\underbrace{\int_{-L}^{L} \cos\frac{k\pi x}{L}\cos\frac{2\pi x}{L}dx}_{=0} + \ldots a_k\underbrace{\int_{L}^{L} \cos\frac{k\pi x}{L}\cos\frac{k\pi x}{L}dx}_{=L} +$$

$$\ldots + a_n\underbrace{\int_{-L}^{L} \cos\frac{k\pi x}{L}\cos\frac{n\pi x}{L}dx}_{=0} + b_1\underbrace{\int_{-L}^{L} \cos\frac{k\pi x}{L}\sin\frac{\pi x}{L}dx}_{=0} + b_2\underbrace{\int_{-L}^{L} \cos\frac{k\pi x}{L}\sin\frac{2\pi x}{L}dx}_{=0} + \ldots$$

$$+ b_k\underbrace{\int_{-L}^{L} \cos\frac{k\pi x}{L}\sin\frac{k\pi x}{L}dx}_{=0} + \ldots + b_n\underbrace{\int_{-L}^{L} \cos\frac{k\pi x}{L}\sin\frac{n\pi x}{L}dx}_{=0}$$

$$\longrightarrow \int_{-L}^{L} f(x)\cos\frac{k\pi x}{L}dx = a_k L \longrightarrow a_k$$
$$= \frac{1}{L}\int_{-L}^{L} f(x)\cos\frac{k\pi x}{L}dx$$

a_k equation above serves as general equation. $a_1, a_2,\ldots, a_n$ coefficients can be easily obtained by changing k index in this equation.

b_k calculation:

All $b_1 - b_n$ coefficients in Eq. (7.109) are obtained in the same way. Therefore, it will be more advantageous to derive a general expression for b_k coefficient, from which all $b_1 - b_n$ coefficients can be obtained only by changing the k index. Multiplication of Eq. (7.112) by $\sin\frac{k\pi x}{L}$ and then integration in $[-L, L]$ interval using the orthogonality condition in Eqs. (7.110a)–(7.110c) yield b_k coefficient as following:

$$\int_{-L}^{L} f(x)\sin\frac{k\pi x}{L}dx = \int_{-L}^{L}\sin\frac{k\pi x}{L}\begin{bmatrix}\frac{a_0}{2}+a_1\cos\frac{\pi x}{L}+a_2\cos\frac{2\pi x}{L}+\ldots+a_k\cos\frac{k\pi x}{L}+\ldots+a_n\cos\frac{n\pi x}{L}+\\ b_1\sin\frac{\pi x}{L}+b_2\sin\frac{2\pi x}{L}+\ldots+b_k\sin\frac{k\pi x}{L}+\ldots+b_n\sin\frac{n\pi x}{L}\end{bmatrix}dx$$

$$=\frac{a_0}{2}\underbrace{\int_{-L}^{L}\sin\frac{k\pi x}{L}dx}_{=0}+a_1\underbrace{\int_{-L}^{L}\sin\frac{k\pi x}{L}\cos\frac{\pi x}{L}dx}_{=0}$$

$$+a_2\underbrace{\int_{-L}^{L}\sin\frac{k\pi x}{L}\cos\frac{2\pi x}{L}dx}_{=0}+\ldots a_k\underbrace{\int_{-L}^{L}\sin\frac{k\pi x}{L}\cos\frac{k\pi x}{L}dx}_{=0}$$

$$+\ldots+a_n\underbrace{\int_{-L}^{L}\sin\frac{k\pi x}{L}\cos\frac{n\pi x}{L}dx}_{=0}$$

$$+b_1\underbrace{\int_{-L}^{L}\sin\frac{k\pi x}{L}\sin\frac{\pi x}{L}dx}_{=0}+b_2\underbrace{\int_{-L}^{L}\sin\frac{k\pi x}{L}\sin\frac{2\pi x}{L}dx}_{=0}$$

$$+\ldots+b_k\underbrace{\int_{-L}^{L}\sin\frac{k\pi x}{L}\sin\frac{k\pi x}{L}dx}_{=L}+\ldots$$

$$+b_n\underbrace{\int_{-L}^{L}\sin\frac{k\pi x}{L}\sin\frac{n\pi x}{L}dx}_{=0}$$

$$\rightarrow \int_{-L}^{L} f(x)\sin\frac{k\pi x}{L}dx = b_kL \rightarrow b_k=\frac{1}{L}\int_{-L}^{L} f(x)\sin\frac{\pi x}{L}dx$$

Using the general b_k equation, $b_1, b_2,\ldots, b_n$ coefficients can be easily obtained by changing k index.

Expressions of a_0, a_k, b_k coefficients in Fourier series can be summarized as following:

$$a_0=\frac{1}{L}\int_{-L}^{L} f(x)dx, a_k=\frac{1}{L}\int_{-L}^{L} f(x)\cos\frac{k\pi x}{L}dx, b_k=\frac{1}{L}\int_{-L}^{L} f(x)\sin\frac{k\pi x}{L}dx \quad (7.113)$$

Example 7.12: An $f(x)$ is defined in $x = [-2, 2]$ interval as following:

$$f(x)=\begin{cases} 1+\frac{x}{2} \ , & -2\leq x\leq 0 \\ 1 \ , & 0\leq x\leq 2 \end{cases}$$

Obtain Fourier series approximation of $f(x)$ function.

Using Eq. (7.109), Fourier series approximation of $f(x)$ function in $x = [-2, 2]$ interval can be obtained as below (L: Half period, $L = 2$):

$$f(x)=\frac{a_0}{2}+\sum_{k=1}^{n}\left(a_k\cos\frac{k\pi x}{L}+b_k\sin\frac{k\pi x}{L}\right)\rightarrow f(x)=\frac{a_0}{2}+\sum_{k=1}^{n}\left(a_k\cos\frac{k\pi x}{2}+b_k\sin\frac{k\pi x}{2}\right)$$

When five terms are used, Fourier series approximation can be written as following:

$$f(x)=\frac{a_0}{2}+a_1\cos\frac{\pi x}{2}+a_2\cos\frac{2\pi x}{2}+b_1\sin\frac{\pi x}{2}+b_2\sin\frac{2\pi x}{2}$$

Using $a_0=\frac{1}{L}\int_{-L}^{L} f(x)dx$, $a_k=\frac{1}{L}\int_{-L}^{L} f(x)\cos\frac{k\pi x}{L}dx$, $b_k=\frac{1}{L}\int_{-L}^{L} f(x)\sin\frac{k\pi x}{L}dx$ expressions in Eq. (7.113), a_0, a_1, a_2, b_1, b_2 coefficients can be obtained as below:

$$a_0=\frac{1}{2}\int_{-2}^{2} f(x)dx=\frac{1}{2}\left[\int_{-2}^{0} f(x)dx+\int_{0}^{2} f(x)dx\right]=\frac{1}{2}\left[\int_{-2}^{0}\left(1+\frac{x}{2}\right)dx+\int_{0}^{2} 1dx\right]=1.5$$

$$a_1=\frac{1}{2}\int_{-2}^{2} f(x)\cos\frac{\pi x}{2}dx=\frac{1}{2}\left[\int_{-2}^{0} f(x)\cos\frac{\pi x}{2}dx+\int_{0}^{2} f(x)\cos\frac{\pi x}{2}dx\right]$$

$$=\frac{1}{2}\left[\int_{-2}^{0}\left(1+\frac{x}{2}\right)\cos\frac{\pi x}{2}dx+\int_{0}^{2} 1\cos\frac{\pi x}{2}dx\right]=0.2026$$

$$a_2 = \frac{1}{2}\int_{-2}^{2} f(x)\cos\frac{2\pi x}{2}dx = \frac{1}{2}\left[\int_{-2}^{0} f(x)\cos\frac{2\pi x}{2}dx + \int_{0}^{2} f(x)\cos\frac{2\pi x}{2}dx\right]$$

$$= \frac{1}{2}\left[\int_{-2}^{0}\left(1+\frac{x}{2}\right)\cos\frac{2\pi x}{2}dx + \int_{0}^{2} 1\cos\frac{2\pi x}{2}dx\right] = 0$$

$$b_1 = \frac{1}{2}\int_{-2}^{2} f(x)\sin\frac{\pi x}{2}dx = \frac{1}{2}\left[\int_{-2}^{0} f(x)\sin\frac{\pi x}{2}dx + \int_{0}^{2} f(x)\sin\frac{\pi x}{2}dx\right]$$

$$= \frac{1}{2}\left[\int_{-2}^{0}\left(1+\frac{x}{2}\right)\sin\frac{\pi x}{2}dx + \int_{0}^{2} 1\sin\frac{\pi x}{2}dx\right] = 0.3183$$

$$b_2 = \frac{1}{2}\int_{-2}^{2} f(x)\sin\frac{2\pi x}{2}dx = \frac{1}{2}\left[\int_{-2}^{0} f(x)\sin\frac{2\pi x}{2}dx + \int_{0}^{2} f(x)\sin\frac{2\pi x}{2}dx\right]$$

$$= \frac{1}{2}\left[\int_{-2}^{0}\left(1+\frac{x}{2}\right)\sin\frac{2\pi x}{2}dx + \int_{0}^{2} 1\sin\frac{2\pi x}{2}dx\right] = -0.1592$$

Thus, Fourier series approximation of $f(x)$ function is obtained as following:

$$f(x) = \frac{1.5}{2} + 0.2026\cos\frac{\pi x}{2} + 0.3183\sin\frac{\pi x}{2} - 0.1592\sin\frac{2\pi x}{2}$$

Comparison of $f(x)$ function and its Fourier series approximation with five terms is shown in Fig. 7.47. Approximation quality can be increased by including more terms in Fourier series approximation.

a_k, b_k coefficients in Fourier series approximation can be calculated using following commands in MATLAB:

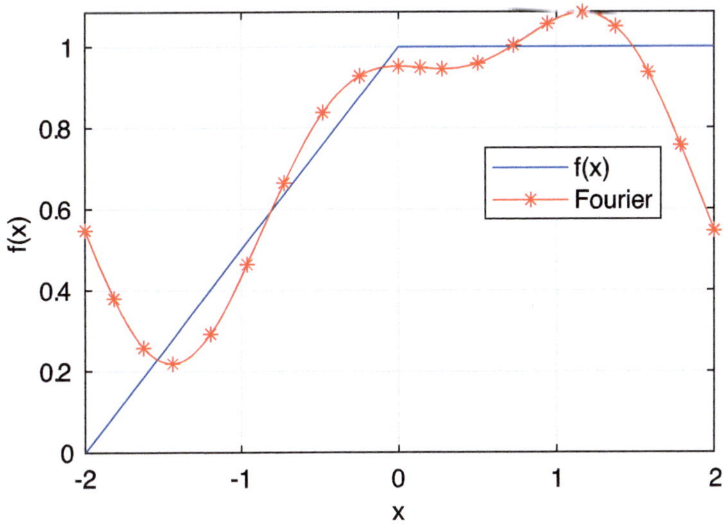

Fig. 7.47 Fourier series approximation of $f(x)$ function using five terms

```
% Calculation of ak and bk coefficients in MATLAB:
syms x

a0=int((1+x/2),x,-2,0)+int(1,x,0,2) ;
a1=int((1+x/2)*cos(pi*x/2),x,-2,0)+int(1*cos(pi*x/2),x,0,2) ;
a2=int((1+x/2)*cos(2*pi*x/2),x,-
2,0)+int(1*cos(2*pi*x/2),x,0,2) ;

b1=int((1+x/2)*sin(pi*x/2),x,-2,0)+int(1*sin(pi*x/2),x,0,2);
b2=int((1+x/2)*sin(2*pi*x/2),x,-
2,0)+int(1*sin(2*pi*x/2),x,0,2) ;

a0=(1/2)*eval(a0)
a1=(1/2)*eval(a1)
a2=(1/2)*eval(a2)

b1=(1/2)*eval(b1)
b2=(1/2)*eval(b2)

% Plotting functions
>> fplot(@(x) 1+x/2,[-2 0],'b'); hold on;
>> fplot(@(x) 1,[0 2],'b'); hold on;
>> xlabel('x'); ylabel('f(x)'); hold on;
>> fplot(@(x)
(1.5)/2+0.2026*cos(pi*x/2)+0*cos(2*pi*x/2)+0.3183*sin(pi*x/2)-
0.1592*sin(2*pi*x/2),[-2 2],'*-r');
>> legend('f(x)','','Fourier');
>> grid;
```

7.13.2 Fourier Series Expansions of Special Functions

Some functions have special features such as symmetry. Even and odd functions have these features. Symmetry feature simplifies Fourier series approximation of these functions as following.

(a) **Even Functions**

As shown in Fig. 7.48, if $f(-x) = f(x)$ relationship is valid for an $f(x)$ function, $f(x)$ function is called symmetric function with respect to $y = f(x)$ axis. This function is also called as even function. x^2, x^4, x^{2n}, $Cos(x)$, $Cos(nx)$, $Sec(x)$, and $Sec(nx)$ are examples of even functions.

Using $\int_a^b f(x)dx = -\int_b^a f(x)dx$ relationship in integrals, coefficients in Fourier series approximation of even functions can be calculated as below.

a_0 calculation:

Splitting the integral domain, a_0 coefficient can be expressed as following:

$$a_0 = \frac{1}{L}\int_{-L}^{L} f(x)dx = \frac{1}{L}\left[\int_{-L}^{0} f(x)dx + \int_{0}^{L} f(x)dx\right] = \frac{1}{L}\left[-\int_{0}^{-L} f(x)dx + \int_{0}^{L} f(x)dx\right]$$

Using variable transformation of $u = -x$ and $du = -dx$, $\int_0^{-L} f(x)dx$ integral can be stated as $\int_0^{-L} f(x)dx = \int_0^{L} f(-u)(-du)$. In this case, a_0 coefficient can be expressed as below:

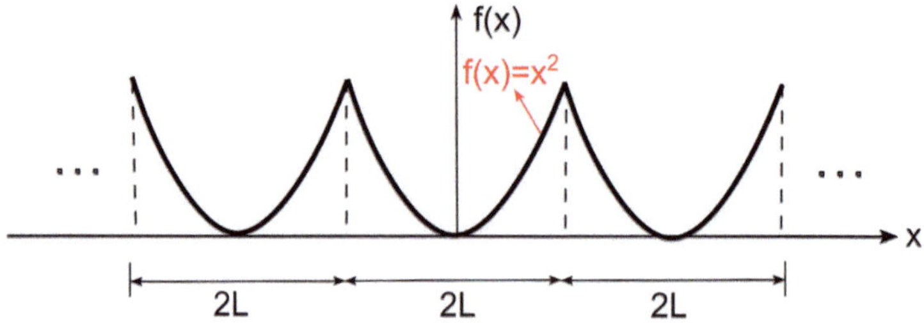

Fig. 7.48 An example of even function: $f(x) = x^2$ with $2L$ period

$$a_0 = \frac{1}{L}\left[-\int_0^{-L} f(x)dx + \int_0^{L} f(x)dx\right] = \frac{1}{L}\left[-\int_0^{L} \underbrace{f(-u)}_{Even=f(u)}(-du) + \int_0^{L} f(x)dx\right]$$

$$= \frac{1}{L}\left[\int_0^{L} f(u)du + \int_0^{L} f(x)dx\right]$$

If desired, $\int_0^L f(u)du$ integral in the above expression can be expressed with an alternative variable by replacing u with any other variable. In alternative form, integral result will not differ. For example, when u is replaced with x variable, $\int_0^L f(u)du$ integral can be expressed as $\int_0^L f(u)du = \int_0^L f(x)dx$. In this case, a_0 coefficient can be obtained as following:

$$a_0 = \frac{1}{L}\left[\int_0^{L} f(u)du + \int_0^{L} f(x)dx\right] = \frac{1}{L}\left[\int_0^{L} f(x)dx + \int_0^{L} f(x)dx\right] = \frac{2}{L}\int_0^{L} f(x)dx \tag{7.114}$$

a_k calculation:

Splitting the integral domain, a_k coefficient can be expressed as below:

$$a_k = \frac{1}{L}\int_{-L}^{L} f(x)\cos\frac{k\pi x}{L}dx = \frac{1}{L}\left[\int_{-L}^{0} f(x)\cos\frac{k\pi x}{L}dx + \int_0^{L} f(x)\cos\frac{k\pi x}{L}dx\right]$$

$$= \frac{1}{L}\left[-\int_0^{-L} f(x)\cos\frac{k\pi x}{L}dx + \int_0^{L} f(x)\cos\frac{k\pi x}{L}dx\right]$$

Using variable transformation of $u = -x$ and $du = -dx$, a_k coefficient can be expressed as following:

$$a_k = \frac{1}{L}\left[-\int_0^{-L} f(x)\cos\frac{k\pi x}{L}dx + \int_0^{L} f(x)\cos\frac{k\pi x}{L}dx\right]$$

$$= \frac{1}{L}\left[-\int_0^{L} \underbrace{f(-u)}_{Even=f(u)} \underbrace{\cos\frac{k\pi(-u)}{L}}_{Even=\cos\frac{k\pi u}{L}}(-du) + \int_0^{L} f(x)\cos\frac{k\pi x}{L}dx\right]$$

$$= \frac{1}{L}\left[\int_0^{L} f(u)\cos\frac{k\pi u}{L}du + \int_0^{L} f(x)\cos\frac{k\pi x}{L}dx\right]$$

Without changing the result, $\int_0^L f(u)\cos\frac{k\pi u}{L}du$ integral in the above expression can be expressed with an alternative variable by replacing u with any other variable. For example, when u is replaced with x variable, $\int_0^L f(u)\cos\frac{k\pi u}{L}du$ integral can be expressed as $\int_0^L f(u)\cos\frac{k\pi u}{L}du = \int_0^L f(x)\cos\frac{k\pi x}{L}dx$. In this case, a_k coefficient can be obtained as below:

$$a_k = \frac{1}{L}\left[\int_0^{L} f(u)\cos\frac{k\pi u}{L}du + \int_0^{L} f(x)\cos\frac{k\pi x}{L}dx\right]$$
$$= \frac{1}{L}\left[\int_0^{L} f(x)\cos\frac{k\pi x}{L}dx + \int_0^{L} f(x)\cos\frac{k\pi x}{L}dx\right] = \frac{2}{L}\int_0^{L} f(x)\cos\frac{k\pi x}{L}dx \tag{7.115}$$

b_k calculation:

Splitting the integral domain, b_k coefficient can be expressed as following:

$$b_k = \frac{1}{L}\int_{-L}^{L} f(x)\sin\frac{k\pi x}{L}dx = \frac{1}{L}\left[\int_{-L}^{0} f(x)\sin\frac{k\pi x}{L}dx + \int_0^{L} f(x)\sin\frac{k\pi x}{L}dx\right]$$
$$= \frac{1}{L}\left[-\int_0^{-L} f(x)\sin\frac{k\pi x}{L}dx + \int_0^{L} f(x)\sin\frac{k\pi x}{L}dx\right]$$

Using variable transformation of $u = -x$ and $du = -dx$, b_k coefficient can be expressed as below:

$$b_k = \frac{1}{L}\left[-\int_0^{-L} f(x)\sin\frac{k\pi x}{L}dx + \int_0^{L} f(x)\sin\frac{k\pi x}{L}dx\right]$$

$$= \frac{1}{L}\left[-\int_0^{L} \underbrace{f(-u)}_{Even=f(u)}\ \underbrace{\sin\frac{k\pi(-u)}{L}}_{Odd=-\sin\frac{k\pi u}{L}}(-du) + \int_0^{L} f(x)\sin\frac{k\pi x}{L}dx\right]$$

$$= \frac{1}{L}\left[-\int_0^{L} f(u)\sin\frac{k\pi u}{L}du + \int_0^{L} f(x)\sin\frac{k\pi x}{L}dx\right]$$

Without changing the result, $\int_0^L f(u)\sin\frac{k\pi u}{L}du$ integral in the above expression can be expressed with an alternative variable by replacing u with any other variable. For example, when u is replaced with x variable, $\int_0^L f(u)\sin\frac{k\pi u}{L}du$ integral can be expressed as $\int_0^L f(u)\sin\frac{k\pi u}{L}du = \int_0^L f(x)\sin\frac{k\pi x}{L}dx$. In this case, b_k coefficient can be obtained as following:

$$\begin{aligned} b_k &= \frac{1}{L}\left[-\int_0^{L} f(u)\sin\frac{k\pi u}{L}du + \int_0^{L} f(x)\sin\frac{k\pi x}{L}dx\right] \\ &= \frac{1}{L}\left[-\int_0^{L} f(x)\sin\frac{k\pi x}{L}dx + \int_0^{L} f(x)\sin\frac{k\pi x}{L}dx\right] - 0 \end{aligned} \tag{7.116}$$

Obtained coefficients for Fourier series approximation of even functions can be summarized as below:

$$a_0 = \frac{2}{L}\int_0^L f(x)dx, \quad a_k = \frac{2}{L}\int_0^L f(x)\cos\frac{k\pi x}{L}dx, \quad b_k = 0 \tag{7.117}$$

(b) **Odd Functions**

As shown in Fig. 7.49, if $f(-x) = -f(x)$ relationship is valid for an $f(x)$ function, $f(x)$ function is called symmetric function with respect to the origin. This function is also called as odd function. x, x^3, x^{2n+1}, Sin(x), Sin(nx), Cosec (x), and Cosec(nx) are examples of odd functions.

Using $\int_a^b f(x)dx = -\int_b^a f(x)dx$ relationship in integrals, coefficients in Fourier series approximation of odd functions can be calculated as below.

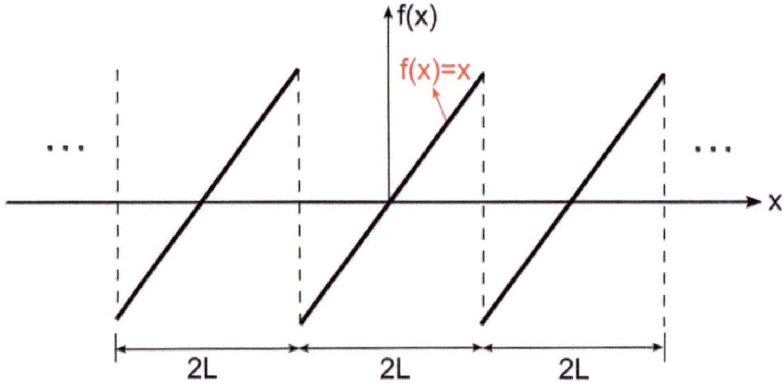

Fig. 7.49 An example of odd function: $f(x) = x$ with $2L$ period

a_0 *calculation:*

Splitting the integral domain, a_0 coefficient can be expressed as following:

$$a_0 = \frac{1}{L}\int_{-L}^{L} f(x)dx = \frac{1}{L}\left[\int_{-L}^{0} f(x)dx + \int_{0}^{L} f(x)dx\right] = \frac{1}{L}\left[-\int_{0}^{-L} f(x)dx + \int_{0}^{L} f(x)dx\right]$$

Using variable transformation of $u = -x$ and $du = -dx$, $\int_0^{-L} f(x)dx$ integral can be stated as $\int_0^{-L} f(x)dx = \int_0^{L} f(-u)(-du)$. In this case, a_0 coefficient can be expressed as below:

$$a_0 = \frac{1}{L}\left[-\int_{0}^{-L} f(x)dx + \int_{0}^{L} f(x)dx\right] = \frac{1}{L}\left[-\int_{0}^{L} \underbrace{f(-u)}_{Odd = -f(u)}(-du) + \int_{0}^{L} f(x)dx\right]$$

$$= \frac{1}{L}\left[-\int_{0}^{L} f(u)du + \int_{0}^{L} f(x)dx\right]$$

Without changing the result, $\int_0^L f(u)du$ integral in the above expression can be expressed with an alternative variable by replacing u with any other variable. For example, when u is replaced with x variable, $\int_0^L f(u)du$ integral can be expressed as $\int_0^L f(u)du = \int_0^L f(x)dx$. In this case, a_0 coefficient can be obtained as following:

$$a_0 = \frac{1}{L}\left[-\int_0^L f(u)du + \int_0^L f(x)dx\right] = \frac{1}{L}\left[-\int_0^L f(x)dx + \int_0^L f(x)dx\right] = 0 \tag{7.118}$$

a_k calculation:

Splitting the integral domain, a_k coefficient can be expressed as below

$$a_k = \frac{1}{L}\int_{-L}^{L} f(x)\cos\frac{k\pi x}{L}dx = \frac{1}{L}\left[\int_{-L}^{0} f(x)\cos\frac{k\pi x}{L}dx + \int_0^L f(x)\cos\frac{k\pi x}{L}dx\right]$$

$$= \frac{1}{L}\left[-\int_0^{-L} f(x)\cos\frac{k\pi x}{L}dx + \int_0^L f(x)\cos\frac{k\pi x}{L}dx\right]$$

Using variable transformation of $u = -x$ and $du = -dx$, a_k coefficient can be expressed as following:

$$a_k = \frac{1}{L}\left[-\int_0^{-L} f(x)\cos\frac{k\pi x}{L}dx + \int_0^L f(x)\cos\frac{k\pi x}{L}dx\right]$$

$$= \frac{1}{L}\left[-\int_0^L \underbrace{f(-u)}_{Odd = -f(u)}\ \underbrace{\cos\frac{k\pi(-u)}{L}}_{Even = \cos\frac{k\pi u}{L}}(-du) + \int_0^L f(x)\cos\frac{k\pi x}{L}dx\right]$$

$$= \frac{1}{L}\left[-\int_0^L f(u)\cos\frac{k\pi u}{L}du + \int_0^L f(x)\cos\frac{k\pi x}{L}dx\right]$$

Without changing the result, $\int_0^L f(u)\cos\frac{k\pi u}{L}du$ integral in the above expression can be expressed with an alternative variable by replacing u with any other variable. For example, when u is replaced with x variable, $\int_0^L f(u)\cos\frac{k\pi u}{L}du$ integral can be expressed as $\int_0^L f(u)\cos\frac{k\pi u}{L}du = \int_0^L f(x)\cos\frac{k\pi x}{L}dx$. In this case, a_k coefficient can be obtained as below:

$$a_k = \frac{1}{L}\left[-\int_0^L f(u)\cos\frac{k\pi u}{L}du + \int_0^L f(x)\cos\frac{k\pi x}{L}dx \right] \\ = \frac{1}{L}\left[-\int_0^L f(x)\cos\frac{k\pi x}{L}dx + \int_0^L f(x)\cos\frac{k\pi x}{L}dx \right] = 0 \tag{7.119}$$

b_k calculation:

Splitting the integral domain, b_k coefficient can be expressed as following:

$$b_k = \frac{1}{L}\int_{-L}^{L} f(x)\sin\frac{k\pi x}{L}dx = \frac{1}{L}\left[\int_{-L}^{0} f(x)\sin\frac{k\pi x}{L}dx + \int_0^L f(x)\sin\frac{k\pi x}{L}dx \right]$$

$$= \frac{1}{L}\left[-\int_0^{-L} f(x)\sin\frac{k\pi x}{L}dx + \int_0^L f(x)\sin\frac{k\pi x}{L}dx \right]$$

Using variable transformation of $u = -x$ and $du = -dx$, b_k coefficient can be expressed as below:

$$b_k = \frac{1}{L}\left[-\int_0^{-L} f(x)\sin\frac{k\pi x}{L}dx + \int_0^L f(x)\sin\frac{k\pi x}{L}dx \right]$$

$$= \frac{1}{L}\left[-\int_0^L \underbrace{f(-u)}_{Odd=-f(u)}\ \underbrace{\sin\frac{k\pi(-u)}{L}}_{Odd=-\sin\frac{k\pi u}{L}}(-du) + \int_0^L f(x)\sin\frac{k\pi x}{L}dx \right]$$

$$= \frac{1}{L}\left[\int_0^L f(u)\sin\frac{k\pi u}{L}du + \int_0^L f(x)\sin\frac{k\pi x}{L}dx \right]$$

Without changing the result, $\int_0^L f(u)\sin\frac{k\pi u}{L}du$ integral in the above expression can be expressed with an alternative variable by replacing u with any other variable. For example, when u is replaced with x variable, $\int_0^L f(u)\sin\frac{k\pi u}{L}du$ integral can be expressed as $\int_0^L f(u)\sin\frac{k\pi u}{L}du = \int_0^L f(x)\sin\frac{k\pi x}{L}dx$. In this case, b_k coefficient can be obtained as following:

$$
\begin{aligned}
b_k &= \frac{1}{L}\left[\int_0^L f(u)\sin\frac{k\pi u}{L}du + \int_0^L f(x)\sin\frac{k\pi x}{L}dx\right] \\
&= \frac{1}{L}\left[\int_0^L f(x)\sin\frac{k\pi x}{L}dx + \int_0^L f(x)\sin\frac{k\pi x}{L}dx\right] = \frac{2}{L}\int_0^L f(x)\sin\frac{k\pi x}{L}dx
\end{aligned} \tag{7.120}
$$

Obtained coefficients for Fourier series approximation of odd functions can be summarized as below:

$$
a_0 = 0, \quad a_k = 0, \quad b_k = \frac{2}{L}\int_0^L f(x)\sin\frac{k\pi x}{L}dx \tag{7.121}
$$

Summation and product of even and odd functions can be expressed as following:

Summation of even and odd functions:

- Odd+Odd = Odd
- Even+Even = Even

Product of even and odd functions:

- Odd*Odd = Even
- Even*Even = Even
- Odd*Even = Odd

Note: Calculation of a_k and b_k coefficients in Fourier series approximation requires integrals of $\int_{-1}^{1} f(x)dx$, $\frac{2}{L}\int_0^L f(x)\cos\frac{k\pi x}{L}dx$, $\int_0^L f(x)\sin\frac{k\pi x}{L}dx$. However, these integrals may not be calculated easily. For complicated functions, numerical methods that will be explained in Chap. 9 can be used for integration.

7.13.3 Fourier Series Expansion of Non-periodic Functions

In engineering applications, most of the functions are actually non-periodic. Non-periodic functions can also be approximated with Fourier series by virtually extending them beyond the application region. This can be done by assuming the non-periodic function as even or odd function in the extended region. For example, a general *f*(*x*) function defined in [0, *L*] region can be made periodic in [−*L*, *L*] region by extending it into [−*L*, 0] region in the form of even or odd function as shown in Fig. 7.50. Odd and even functions used for non-periodic functions are also known Half Range Fourier series.

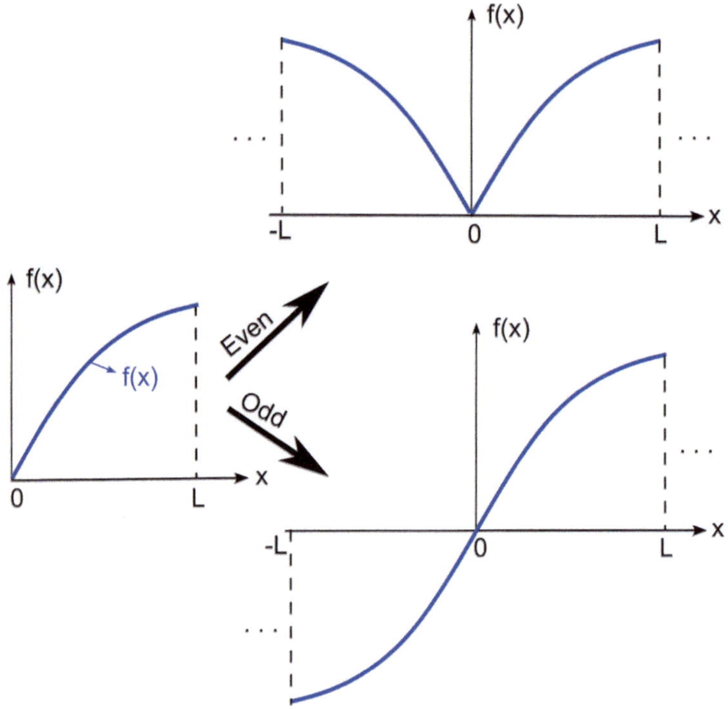

Fig. 7.50 Conversion of a general $f(x)$ function into periodic one by extending it as even or odd function outside of the defined region

Example 7.13: Obtain Fourier series approximation of $f(x) = e^x$ function in $x = [0, 2]$ interval by extending it as (a) even, (b) odd function

Fourier series approximation of $f(x) = e^x$ function can be obtained by converting it into an even or odd function as shown in Fig. 7.51. For this, $f(x) = e^x$ function is extended into $[-2, 0]$ interval to be an even (symmetric with respect to $y = f(x)$ axis) or odd (symmetric with respect to origin) function and have period of 4.

(a) *Assuming f(x) function even*:

$f(x) = e^x$ function can be converted into an even function by extending it into $[-2, 0]$ interval to be symmetric with respect to $y = f(x)$ axis. When $f(x)$ function defined in $[-2, 2]$ interval repeats itself periodically in even form, then its Fourier Series approximation can be expressed using Eq. (7.117) as following (L: Half period, $L = 2$):

$$f(x) = \frac{a_0}{2} + \sum_{k=1}^{n} a_k \cos\frac{k\pi x}{L} \rightarrow f(x) = \frac{a_0}{2} + \sum_{k=1}^{n} a_k \cos\frac{k\pi x}{2}$$

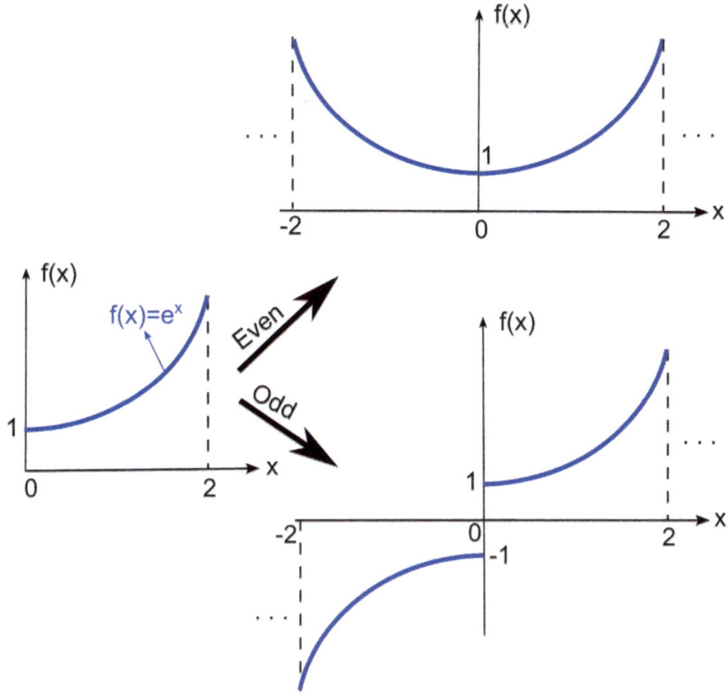

Fig. 7.51 Converting $f(x) = e^x$ function into an even or odd function

When the terms are used, Fourier Series approximation can be written as following:

$$f(x) = \frac{a_0}{2} + a_1 \cos\frac{\pi x}{2} + a_2 \cos\frac{2\pi x}{2}$$

Using $a_0 = \frac{2}{L}\int_0^L f(x)dx$, $a_k = \frac{2}{L}\int_0^L f(x)\cos\frac{k\pi x}{L}dx$ expressions in Eq. (7.117), a_0, a_1, a_2 coefficients can be obtained as below:

$$a_0 = \frac{2}{2}\int_0^L e^x dx = \int_0^2 e^x dx = 6.3891, a_1 = \frac{2}{2}\int_0^L e^x \cos\frac{\pi x}{2}dx = \int_0^2 e^x \cos\frac{\pi x}{2}dx = -2.4194,$$

$$a_2 = \frac{2}{2}\int_0^L e^x \cos\frac{2\pi x}{2}dx = \int_0^2 e^x \cos\frac{2\pi x}{2}dx = 0.5878$$

Thus, Fourier series approximation of $f(x) = e^x$ function is obtained as following:

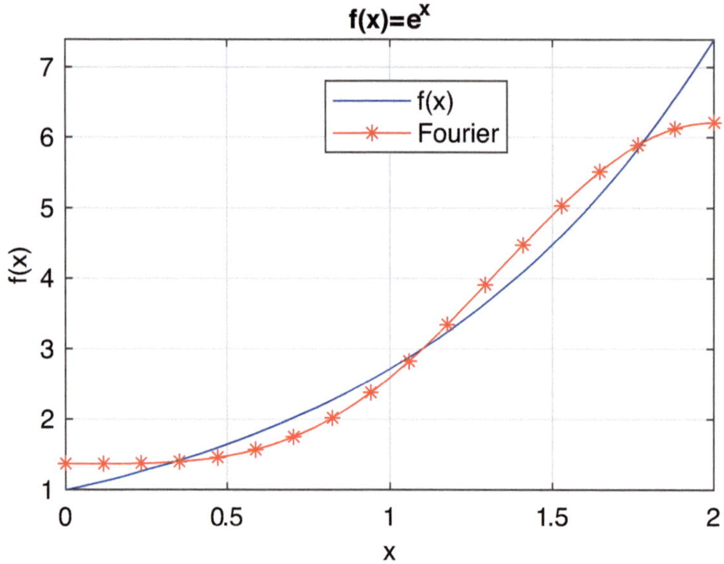

Fig. 7.52 Fourier series approximation of $f(x) = e^x$ function by assuming it even and periodic (three terms are used in Fourier series)

$$f(x) = \frac{6.3891}{2} - 2.4194\cos\left(\frac{\pi x}{2}\right) + 0.5978\cos\left(\frac{2\pi x}{2}\right)$$

Comparison of $f(x) = e^x$ function (assuming it as even function) and its Fourier series approximation using three terms is shown in Fig. 7.52. Approximation quality can be increased by including more terms in Fourier series approximation.

(b) *Assuming f(x) function odd*:

$f(x) = e^x$ function can be converted into an odd function by extending it into [−2, 0] interval to be symmetric with respect to the origin. When $f(x)$ function defined in [−2, 2] interval repeats itself periodically in odd form, then its Fourier Series approximation can be expressed using Eq. (7.121) as following (L: Half period, $L = 2$):

$$f(x) = \sum_{k=1}^{n} b_k \sin\frac{k\pi x}{L} \rightarrow f(x) = \sum_{k=1}^{n} b_k \sin\frac{k\pi x}{2}$$

When three terms are used, Fourier Series approximation can be written as following:

$$f(x) = b_1\sin\left(\frac{\pi x}{2}\right) + b_2\sin\left(\frac{2\pi x}{2}\right) + b_2\sin\left(\frac{3\pi x}{2}\right)$$

Using $b_k = \frac{2}{L}\int_0^L f(x)\sin\frac{k\pi x}{L}dx$ expression in Eq. (7.121), b_1, b_2, b_3 coefficients can be obtained as below:

$$b_1 = \frac{2}{L}\int_0^L e^x \sin\frac{\pi x}{L}dx = \frac{2}{2}\int_0^2 e^x \sin\frac{\pi x}{2}dx = 3.8004$$

$$b_2 = \frac{2}{L}\int_0^L e^x \sin\frac{2\pi x}{L}dx = \frac{2}{2}\int_0^2 e^x \sin\frac{2\pi x}{2}dx = -1.8466$$

$$b_3 = \frac{2}{L}\int_0^L e^x \sin\frac{3\pi x}{L}dx = \frac{2}{2}\int_0^2 e^x \sin\frac{3\pi x}{2}dx = 1.7035$$

Thus, Fourier Series approximation of $f(x) = e^x$ function is obtained as following $f(x) = 3.8004 \sin(\pi x/2) - 1.8466 \sin(2\pi x/2) + 1.7035 \sin(3\pi x/2)$.

Comparison of $f(x) = e^x$ function (assuming it as odd function) and its Fourier series approximation using three terms is shown in Fig. 7.53. Approximation quality can be increased by including more terms in Fourier series approximation.

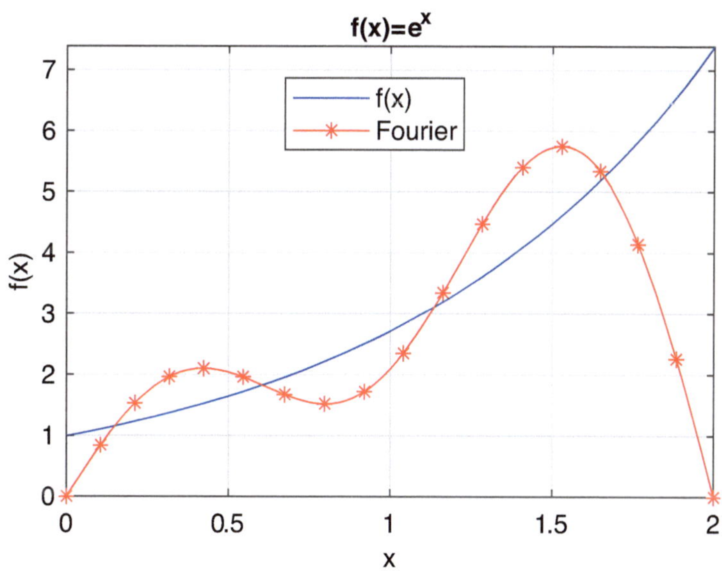

Fig. 7.53 Fourier series approximation of $f(x) = e^x$ function by assuming it odd and periodic (three terms are used in Fourier series)

7.13.4 Legendre Polynomials

Approximation of an *f*(*x*) function can be constructed using orthogonal Legendre polynomials as following:

$$f(x) \approx a_0 L_0(x) + a_1 L_1(x) + a_2 L_2(x) + \ldots + a_n L_n(x) = \sum_{k=0}^{n} a_k L_k(x) \qquad (7.122)$$

where $L_k(x)$ denotes Legendre polynomials and they can be obtained in two ways as below

1. *Rodrigue's method*:

   In this method, Legendre polynomials are obtained using Rodrigue's derivative expression of $L_k(x) = \frac{1}{2^k k!} \frac{d^k}{dx^k}\left[\left(x^2-1\right)^k\right]$. With Rodrigue's expression, Legendre polynomials are obtained as below:

$$L_0(x) = \frac{1}{2^0 0!} \frac{d^0}{dx^0}\left[\left(x^2-1\right)^0\right] = 1, L_1(x) = \frac{1}{2^1 1!} \frac{d^1}{dx^1}\left[\left(x^2-1\right)^1\right] = x, L_2(x) = \frac{1}{2^2 2!} \frac{d^2}{dx^2}\left[\left(x^2-1\right)^2\right] = \frac{3x^2-1}{2}, \ldots.,$$

   Rodrigues expression becomes complicated for $L_n(x)$ where n is a big number since higher order derivatives are needed.

2. *Recurrence method*:

   In this method, a higher order Legendre polynomial is obtained from the lower order Legendre polynomials using recurrence approach. For example, in recurrence method, $(k + 1)$-th Legendre polynomial L_{k+1} can be obtained from L_k and L_{k-1} as following:

$$L_{k+1}(x) = \left(\frac{2k+1}{k+1}\right) L_k(x) - \left(\frac{k}{k+1}\right) L_{k-1}(x) \qquad (7.123)$$

   Recurrence formula cannot start by itself since $L_0(x)$ and $L_1(x)$ initially need to be known. $L_0(x)$ and $L_1(x)$ can be obtained using Rodrigue's formula. With Recurrence formula, Legendre polynomials can be obtained as below:

$$L_0(x) = 1, L_1(x) = x, L_2(x) = \frac{3x^2-1}{2}, L_3(x) = \frac{5x^3-3x}{2},$$
$$L_4(x) = \frac{35x^4-30x^2+3}{8}, L_5(x) = \frac{63x^5-70x^3+15x}{8}, \ldots,$$

Legendre polynomials vary between −1 and 1 values. They are orthogonal in the range of [−1, 1]. Orthogonality condition for Legendre polynomials can be expressed as below:

$$\int_{-1}^{1} L_m(x)L_n(x)dx = \begin{cases} 0, & if \quad m \neq n \\ \dfrac{2}{2m+1}, & if \quad m = n \end{cases} \tag{7.124}$$

After obtaining Legendre polynomials, unknown coefficients a_k in the approximate function in Eq. (7.122) can be calculated using orthogonality condition as below:

a_k calculation:

All a_0–a_n coefficients are calculated similarly. Therefore, it will be more practical to obtain an expression for a general a_k coefficient from which other coefficients can be obtained by changing k index only.

Equation (7.122) can be rewritten to include a general a_k coefficient as below:

$$f(x) \approx a_0L_0(x) + a_1L_1(x) + a_2L_2(x) + \ldots + a_kL_k(x) + \ldots + a_nL_n(x) = \sum_{k=0}^{n} a_kL_k(x) \tag{7.125}$$

Multiplication of Eq. (7.125) by $L_k(x)$ and then integration in [−1, 1] interval using the orthogonality condition yield a_k coefficient as following:

$$\int_{-1}^{1} f(x)L_n(x)dx = \int_{-1}^{1} L_n(x)[a_0L_0(x) + a_1L_1(x) + a_2L_2(x) + \ldots + a_nL_n(x)]dx \rightarrow$$

$$\int_{-1}^{1} f(x)L_n(x)dx = a_0 \underbrace{\int_{-1}^{1} L_n(x)L_0(x)dx}_{=0} + a_1 \underbrace{\int_{-1}^{1} L_n(x)L_1(x)dx}_{=0} + \ldots + a_n \underbrace{\int_{-1}^{1} L_n(x)L_n(x)dx}_{=\frac{2}{2n+1}}$$

$$\rightarrow a_k = \frac{2k+1}{2} \int_{-1}^{1} f(x)L_k(x)dx \tag{7.126}$$

Changing the k index in Eq. (7.126), a_0, a_1,..., a_n coefficients can be easily obtained.

Note 1: Calculation of a_k coefficients in Legendre approximate function require integrals of $a_k = \frac{2k+1}{2} \int_{-1}^{1} f(x)L_k(x)dx$. However, these integrals may not be calculated easily. For complicated functions, numerical methods that will be explained in Chap. 9 can be used for integration.

Note 2: Legendre approximation function was constructed in the normalized interval of $[-1, 1]$. If the original interval is different, it can be converted into $[-1, 1]$ interval using variable transformation. For example, if the original interval is given as $y = [a, b]$, it can be converted into $x = [-1, 1]$ interval using the variable transformation as below:

$$y=[a,b]\rightarrow x=[-1,1]:y=\frac{(1-x)}{2}a+\frac{(1+x)}{2}b$$

Example 7.14: Obtain the approximation of $f(x) = 2|x|$ function in $x = [-1, 1]$ interval using Legendre polynomials.

When five terms are used in Eq. (7.122), approximate function of $f(x) = 2|x|$ with Legendre polynomials can be written as following:

$$f(x)=a_0L_0(x)+a_1L_1(x)+a_2L_2(x)+a_3L_3(x)+a_4L_4(x)$$

where a_0, a_1, …, a_4 coefficients can be calculated as below:

$$a_0=\frac{1}{2}\int_{-1}^{1}f(x)L_0(x)dx=\frac{1}{2}\int_{-1}^{1}2|x|1dx=1\quad(L_0(x)=1)$$

$$a_1=\frac{3}{2}\int_{-1}^{1}f(x)L_1(x)dx=\frac{3}{2}\int_{-1}^{1}2|x|xdx=0\quad(L_1(x)=x)$$

$$a_2=\frac{5}{2}\int_{-1}^{1}f(x)L_2(x)dx=\frac{5}{2}\int_{-1}^{1}2|x|\left(\frac{3x^2-1}{2}\right)dx=\frac{10}{8}\quad\left(L_2(x)=\frac{3x^2-1}{2}\right)$$

$$a_3=\frac{7}{2}\int_{-1}^{1}f(x)L_3(x)dx=\frac{7}{2}\int_{-1}^{1}2|x|\left(\frac{5x^3-3x}{2}\right)dx=0\quad\left(L_3(x)=\frac{5x^3-3x}{2}\right)$$

$$a_4=\frac{9}{2}\int_{-1}^{1}f(x)L_4(x)dx=\frac{9}{2}\int_{-1}^{1}2|x|\left(\frac{35x^4-30x^2+3}{8}\right)dx$$
$$=\frac{-6}{16}\quad\left(L_4(x)=\frac{35x^4-30x^2+3}{8}\right)$$

Thus, approximate function for $f(x) = 2|x|$ is obtained as following:

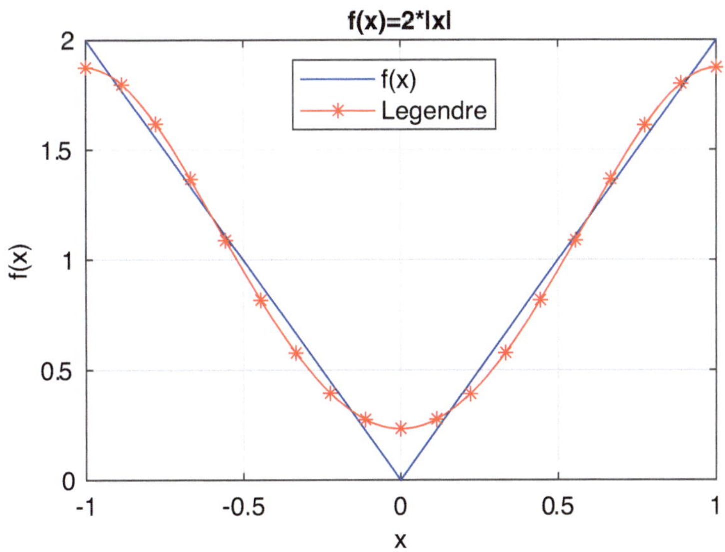

Fig. 7.54 Legendre approximation of $f(x) = 2|x|$ function using five terms

$$f(x) = 1 + \frac{10}{8}\left(\frac{3x^2 - 1}{2}\right) - \frac{6}{16}\left(\frac{35x^4 - 30x^2 + 3}{8}\right)$$

Above expression can be rearranged in power series form as $f(x) = \frac{30}{128} + \frac{420}{128}x^2 - \frac{210}{128}x^4$.

Comparison of $f(x) = 2|x|$ function and its Legendre approximation is shown in Fig. 7.54. Approximation quality can be increased by including more terms in Legendre approximation.

$f(x) = 2|x|$ function and its Legendre approximation can be plotted by using the following commands in MATLAB

```
>> fplot(@(x) 2*abs(x),[-1 1],'b');
>> xlabel('x'); ylabel('f(x)'); title('f(x)=2*|x|'); hold on;
>> fplot(@(x) (30/128)+(420/128)*x^2-(210/128)*x^4,[-1 1],'*-
r');
>> legend('f(x)','Legendre');
>> grid ;
```

Example 7.15: Obtain the approximation of $f(x) = \sin x$ function in $x = [-1, 1]$ interval using Legendre polynomials.

When three terms are used in Eq. (7.122), approximate function of $f(x) = \sin x$ with Legendre polynomials can be written as following:

$$f(x) = a_0L_0(x) + a_1L_1(x) + a_2L_2(x)$$

where a_0, a_1, a_2 coefficients can be calculated using $L_0(x) = 1$, $L_1(x) = x$, $L_2(x) = \frac{3x^2-1}{2}$ as below:

$$a_0 = \frac{1}{2}\int_{-1}^{1} f(x)L_0(x)dx = \frac{1}{2}\int_{-1}^{1} \sin x dx = 0$$

$$a_1 = \frac{3}{2}\int_{-1}^{1} f(x)L_1(x)dx = \frac{3}{2}\int_{-1}^{1} \sin x.(x)dx = 0.9035$$

$$a_2 = \frac{5}{2}\int_{-1}^{1} f(x)L_2(x)dx = \frac{5}{2}\int_{-1}^{1} \sin x\left(\frac{3x^2-1}{2}\right)dx = 0$$

Thus, approximate function for $f(x) = \sin x$ is obtained as $f(x) = 0.9035x$.

Comparison of $f(x) = \sin x$ function and its Legendre approximation is shown in Fig. 7.55. Approximation quality can be increased by including more terms in Legendre approximation.

$f(x) = \sin x$ function and its Legendre approximation can be plotted by using the following commands in MATLAB

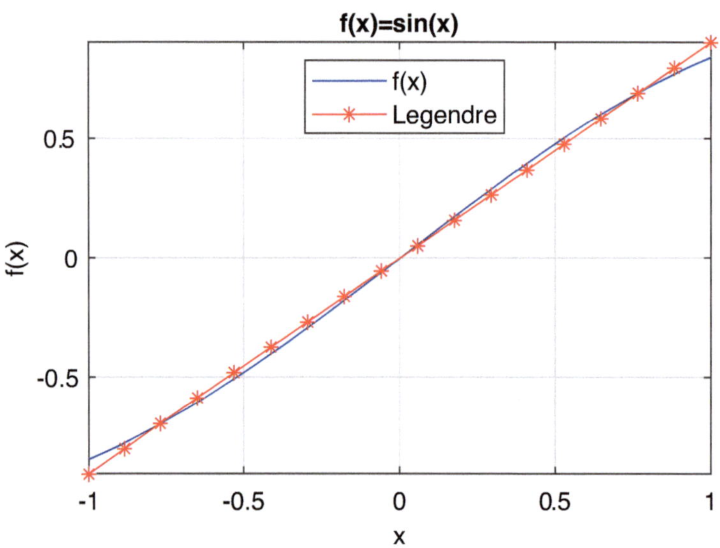

Fig. 7.55 Legendre approximation of $f(x) = \sin x$ function using three terms

```
>> fplot(@(x) sin(x),[-1 1],'b');
>> xlabel('x'); ylabel('f(x)'); title('f(x)=sin(x)'); hold on;
>> fplot(@(x) 0.9035*x,[-1 1],'*-r');
>> legend('f(x)','Legendre');
>> grid ;
```

7.13.5 Chebyshev Polynomials

Approximation of an $f(x)$ function can be generated using orthogonal Chebyshev polynomials as following:

$$f(x) \approx a_0T_0(x) + a_1T_1(x) + a_2T_2(x) + \ldots + a_kT_k(x) + \ldots + a_nT_n(x) = \sum_{k=0}^{n} a_kT_k(x) \tag{7.127}$$

where $T_k(x)$ denotes Chebyshev polynomials and they can be obtained in three ways as below:

1. *Rodrigue's method*:

   In this method, Chebyshev polynomials are obtained using Rodrigue's derivative expression of $T_k(x) = \frac{(-2)^k k!}{(2k)!}\sqrt{1-x^2}\frac{d^k}{dx^k}\left[(1-x^2)^{k-1/2}\right]$.

   With Rodrigue's expression, Chebyshev polynomials are obtained as below:

$$T_0(x) = \frac{(-2)^0 0!}{(2.0)!}\sqrt{1-x^2}\frac{d^0}{dx^0}\left[(1-x^2)^{0-1/2}\right] = 1, T_1(x) = \frac{(-2)^1 1!}{(2.1)!}\sqrt{1-x^2}\frac{d^1}{dx^1}\left[(1-x^2)^{1-1/2}\right] = x, T_2(x) = \frac{(-2)^2 2!}{(2.2)!}\sqrt{1-x^2}\frac{d^2}{dx^2}\left[(1-x^2)^{2-1/2}\right] = 2x^2 - 1, \ldots,$$

   Rodrigues expression becomes complicated for $T_n(x)$ where n is a big number since higher order derivatives are needed.

2. *Variable Transformation Method*:

   In this method, Chebyshev polynomials are obtained by transforming a general trigonometric $T_k(\theta) = \cos(k\theta)$ $(k \geq 1)$ function into a general cartesian $T_k(x)$ function. For this, a general trigonometric $T_k(\theta) = \cos(k\theta)$ function is expressed in terms of $\cos\theta$ first. Then, $\cos\theta$ is replaced with x leading to $T_k(x)$ function in x. This procedure can be explained in obtaining $T_0(x)$, $T_1(x)$, and $T_2(x)$ Chebyshev polynomials as following:

   Obtaining $T_0(x)$:

   For $k = 0$, $T_k(\theta) = \cos(k\theta)$ equation leads to $T_0(\theta) = \cos 0$. Using $x = \cos\theta$ transformation ($x = \cos 0 = 1$), $T_0(\theta)$ is converted into $T_0(x) = 1$.

Obtaining $T_1(x)$:

For $k = 1$, $T_k(\theta) = \cos(k\theta)$ equation leads to $T_1(\theta) = \cos(\theta)$. Using $x = \cos\theta$ transformation, $T_1(\theta)$ is converted into $T_1(x) = x$.

Obtaining $T_2(x)$:

For $k = 2$, $T_k(\theta) = \cos(k\theta)$ equation leads to $T_2(\theta) = \cos(2\theta)$. $T_2(\theta) = \cos(2\theta)$ can be further expressed in terms of $\cos\theta$ as $T_2(x) = \cos(2\theta) = 2(\cos\theta)^2 - 1$. Then, using $x = \cos\theta$ transformation, $T_2(\theta)$ is converted into $T_2(x) = 2x^2 - 1$.

Using similar approach, higher order Chebyshev polynomials can be obtained as $T_3(x) = 4x^3 - 3x$, $T_4(x) = 8x^4 - 8x^2 + 1$, $T_5(x) = 16x^5 - 20x^3 + 5x$, etc. Variable transformation method becomes complicated for $T_n(x)$ where n is a big number due to the expression of $\cos(n\theta)$ in terms of $\cos\theta$.

3. *Recurrence method*:

In this method, a higher order Chebyshev polynomial is obtained from the lower order Chebyshev polynomials using recurrence approach. For example, in recurrence method, $(k + 1)$-th Chebyshev polynomial T_{k+1} can be obtained from T_k and T_{k-1} as following:

$$T_{k+1}(x) = 2xT_k(x) - T_{k-1}(x) \tag{7.128}$$

Recurrence formula cannot start by itself since $T_0(x)$ and $T_1(x)$ initially need to be known. $T_0(x)$ and $T_1(x)$ can be obtained using Trigonometric transformation method. With Recurrence formula, Chebyshev polynomials are obtained as below:

$$T_2(x) = 2xT_1(x) - T_0(x) = 2x(x) - (1) = 2x^2 - 1$$

$$T_3(x) = 2xT_2(x) - T_1(x) = 2x\left(2x^2 - 1\right) - (x) = 4x^3 - 2x$$

$$T_4(x) = 2xT_3(x) - T_2(x) = 2x\left(4x^3 - 2x\right) - \left(2x^2 - 1\right) = 8x^4 - 8x^2 + 1$$

$$T_5(x) = 2xT_4(x) - T_3(x) = 2x\left(8x^4 - 8x^2 + 1\right) - \left(4x^3 - 2x\right) = 16x^5 - 20x^3 + 5x$$

Chebyshev polynomials vary between -1 and 1 values. They are orthogonal when multiplied with weight function of $w(x) = \frac{1}{\sqrt{1-x^2}}$ in the range of $[-1, 1]$. Orthogonality condition for Chebyshev polynomials can be expressed as below:

$$\int_{-1}^{1} \frac{1}{\sqrt{1-x^2}} T_m(x) T_n(x) dx = \begin{cases} 0, & if \quad m \neq n \\ \pi, & if \quad m = n = 0 \\ \frac{\pi}{2}, & if \quad m = n \neq 0 \end{cases} \tag{7.129}$$

Roots of k-th order Chebyshev polynomial can be expressed as following:

$$x_i = -\cos\left(\frac{(2i-1)\pi}{2k}\right) \quad i = 1, 2, .., k \tag{7.130}$$

Using roots of Chebyshev polynomials in Eq. (7.128), orthogonality condition in Eq. (7.129) can be written in *discrete* form as below:

$$\sum_{i=1}^{k} T_m(x_i)T_n(x_i) = \begin{cases} 0, & \text{if} \quad m \neq n \\ k, & \text{if} \quad m = n = 0 \\ \frac{k}{2}, & \text{if} \quad m = n \neq 0 \end{cases} \quad (m, n < k) \tag{7.131}$$

Discrete form of orthogonality condition allows the calculation of a_k coefficients in Chebyshev approximate function using numerical methods.

After obtaining Chebyshev polynomials, a_k coefficients can be obtained using orthogonality condition as following:

a_k calculation:

All a_0- a_n coefficients are calculated similarly. Therefore, it will be more practical to obtain an expression for a general a_k coefficient from which other coefficients can be obtained by changing the k index only.

Multiplication of Eq. (7.127) by $\frac{1}{\sqrt{1-x^2}}T_k(x)$ and then integration in [−1, 1] interval using the orthogonality condition yield a_k coefficient as following:

$$\int_{-1}^{1} f(x)\frac{1}{\sqrt{1-x^2}}T_k(x)dx$$

$$= \int_{-1}^{1} [a_0T_0(x) + a_1T_1(x) + a_2T_2(x) + \ldots + a_kT_k(x) + \ldots + a_nT_n(x)]\frac{1}{\sqrt{1-x^2}}T_k(x)dx$$

$$\int_{-1}^{1} f(x)\frac{1}{\sqrt{1-x^2}}T_k(x)dx = a_0 \underbrace{\int_{-1}^{1} \frac{1}{\sqrt{1-x^2}}T_0(x)T_k(x)dx}_{=0}$$

$$+ a_1 \underbrace{\int_{-1}^{1} \frac{1}{\sqrt{1-x^2}}T_1(x)T_k(x)dx}_{=0} + \ldots$$

$$+ a_k \underbrace{\int_{-1}^{1} \frac{1}{\sqrt{1-x^2}}T_k(x)T_k(x)dx}_{=\frac{\pi}{2}}$$

$$+ \ldots + a_n \underbrace{\int_{-1}^{1} \frac{1}{\sqrt{1-x^2}}T_n(x)T_k(x)dx}_{=0}$$

$$a_k = \frac{2}{\pi} \int_{-1}^{1} f(x) \frac{1}{\sqrt{1-x^2}} T_k(x) dx \tag{7.132}$$

$a_k = \frac{2}{\pi} \int_{-1}^{1} f(x) \frac{1}{\sqrt{1-x^2}} T_k(x) dx$ expression in Eq. (7.132) has discontinuity at $x = 1$. Discontinuity can be avoided with expression of Eq. (7.132) in terms of angle coordinate θ using the variable transformation of $\cos\theta = x$ as below:

$$a_k = \frac{2}{\pi} \int_{0}^{\pi} f(\cos\theta) \cos k\theta d\theta \tag{7.133}$$

Equation (7.133) is more suitable for integration since it does not contain any discontinuity.

Numerical calculation of a_k coefficient:

$a_k = \frac{2}{\pi} \int_{-1}^{1} f(x) \frac{1}{\sqrt{1-x^2}} T_k(x) dx$ integral in Eq. (7.132) is generally difficult to calculate in closed form. Another difficulty in integration is the discontinuity of the integrant function at $x = 1$. Discontinuity problem can be avoided with expression of the integral in terms of angle coordinate θ as $a_k = \frac{2}{\pi} \int_{0}^{\pi} f(\cos\theta) \cos(k\theta) d\theta$. $a_k = \frac{2}{\pi} \int_{0}^{\pi} f(\cos\theta) \cos(k\theta) d\theta$ integral can be calculated easily with numerical methods.

To calculate a_k coefficient numerically, both sides of Eq. (7.127) can be multiplied with $T_k(x)$ as below:

$$\begin{aligned} T_k(x)f(x) &= T_k(x)[a_0 T_0(x) + a_1 T_1(x) + a_2 T_2(x) + \ldots + a_k T_k(x) + \ldots + a_n T_n(x)] \\ &= a_0 T_k(x) T_0(x) + a_1 T_k(x) T_1(x) + a_2 T_k(x) T_2(x) + \ldots \\ &\quad + a_k T_k(x) T_k(x) + \ldots + a_n T_k(x) T_n(x) \end{aligned} \tag{7.134}$$

There are $n + 1$ unknown coefficients in Eq. (7.134). Therefore, $n + 1$ equations need to be generated to calculate them. These equations can be created rewriting Eq. (7.134) at roots of $(n + 1)$-th order Chebyshev polynomial. Indicating roots of $(n + 1)$-th order Chebyshev polynomial as $x_1, x_2, \ldots, x_{n+1}$, Eq. (7.134) can be expressed at these roots as following:

$$\begin{aligned} T_k(x_1)f(x_1) = a_0 T_k(x_1) T_0(x_1) + a_1 T_k(x_1) T_1(x_1) + a_2 T_k(x_1) T_2(x_1) + \ldots \\ + a_k T_k(x_1) T_k(x_1) + \ldots + a_n T_k(x_1) T_n(x_1) \end{aligned}$$

$$T_k(x_2)f(x_2)=a_0T_k(x_2)T_0(x_2)+a_1T_k(x_2)T_1(x_2)+a_2T_k(x_2)T_2(x_2)+\dots$$
$$+a_kT_k(x_2)T_k(x_2)+\dots+a_nT_k(x_2)T_n(x_2)$$

$$\dots$$

$$T_k(x_{n+1})f(x_{n+1})=a_0T_k(x_{n+1})T_0(x_{n+1})+a_1T_k(x_{n+1})T_1(x_{n+1})$$
$$+a_2T_k(x_{n+1})T_2(x_{n+1})+\dots$$
$$+a_kT_k(x_{n+1})T_k(x_{n+1})+\dots+a_nT_k(x_{n+1})T_n(x_{n+1})$$

If the above equations are summed up and discrete orthogonality condition is used, following expression is obtained:

$$\sum_{i=1}^{n+1}T_k(x_i)f(x_i)=a_0\underbrace{\sum_{i=1}^{n+1}T_k(x_i)T_0(x_i)}_{=0}+a_1\underbrace{\sum_{i=1}^{n+1}T_k(x_i)T_1(x_i)}_{=0}+\dots$$
$$+a_k\underbrace{\sum_{i=1}^{n+1}T_k(x_i)T_k(x_i)}_{n+1(if\ k=0),\ (n+1)/2(if\ k\neq0)}+\dots+a_n\underbrace{\sum_{i=1}^{n+1}T_k(x_i)T_n(x_i)}_{=0}$$

Thus, from the above equation, a_k coefficients are obtained as following:

$$a_0=\frac{1}{n+1}\sum_{i=1}^{n+1}T_0(x_i)f(x_i),\quad a_k=\frac{2}{n+1}\sum_{i=1}^{n+1}T_k(x_i)f(x_i)\quad(k=1,\dots,n) \tag{7.135}$$

Note: Chebyshev approximation function was constructed in the normalized interval of [−1, 1]. If the original interval is different, it can be converted into [−1, 1] interval using variable transformation. For example, if the original interval is given as [a, b], it can be converted into [−1, 1] interval using variable transformation as below:

$$y=[a,b]\rightarrow x=[-1,1]:y=\frac{(1-x)}{2}a+\frac{(1+x)}{2}b$$

Example 7.16: Obtain the approximation of $f(y) = \sin y$ function in the range of $y = [0, \pi]$ using Chebyshev polynomial function.

A simple approximation of $f(y) = \sin y$ function in $y = [0, \pi]$ interval can be obtained using three terms in Chebyshev approximate function in Eq. (7.127). As mentioned before, orthogonality of Chebyshev polynomials are valid in the normalized range of [−1, 1]. Therefore, $y = [0, \pi]$ interval should be converted into standard $x = [-1, 1]$ interval (x variable is used to denote normalized interval) using the variable transformation with the help of Lagrange interpolation method as following:

$$y = \frac{(x-1)}{(-1-1)}0 + \frac{(x-(-1))}{(1-(-1))}\pi = \frac{(x+1)\pi}{2}$$

Thus, $f(y) = \sin y$ function in $y = [0, \pi]$ range is expressed in terms of new x variable in $[-1, 1]$ range as $f(x) = \sin\frac{(x+1)\pi}{2}$. A simple approximation for $f(x) = \sin\frac{(x+1)\pi}{2}$ function in $x = [-1, 1]$ range can be obtained using three terms in Chebyshev approximate function in Eq. (7.127) as following:

$$\tilde{f}(x) = a_0T_0(x) + a_1T_1(x) + a_2T_2(x) = \sum_{k=0}^{n=2} a_kT_k(x)$$

There are three unknown coefficients in Chebyshev approximation. Therefore, three equations need to be generated to calculate them. These equations can be created rewriting Chebyshev approximation at roots of (n + 1)-th (n = 2) order Chebyshev polynomial. Roots of (n + 1 = 3)-th order Chebyshev polynomial can be expressed using general expression $x_i = -\cos\left(\frac{(2i-1)\pi}{2(n+1)}\right)$ as following:

$$x_1 = -\cos\left(\frac{(2x1-1)\pi}{2(3)}\right) = -\cos\left(\frac{\pi}{6}\right), x_2 = -\cos\left(\frac{(2x2-1)\pi}{2(3)}\right) = -\cos\left(\frac{3\pi}{6}\right), x_3 = -\cos\left(\frac{(2x3-1)\pi}{2(3)}\right) = -\cos\left(\frac{5\pi}{6}\right)$$

Using Chebyshev polynomials of $T_0(x) = 1$, $T_1(x) = x$, $T_2(x) = 2x^2 - 1$ and Eq. (7.135), a_0, a_1, and a_2 coefficients can be obtained by taking $n = 2$ as below:

$$a_0 = \frac{1}{n+1}\sum_{i=1}^{n+1} T_0(x_i)f(x_i) = \frac{1}{3}[T_0(x_1)f(x_1) + T_0(x_2)f(x_2) + T_0(x_3)f(x_3)] = 0.4726$$

$$a_1 = \frac{2}{n+1}\sum_{i=1}^{n+1} T_1(x_i)f(x_i) = \frac{2}{3}[T_1(x_1)f(x_1) + T_1(x_2)f(x_2) + T_1(x_3)f(x_3)] = 0$$

$$a_2 = \frac{2}{n+1}\sum_{i=1}^{n+1} T_2(x_i)f(x_i) = \frac{2}{3}[T_2(x_1)f(x_1) + T_2(x_2)f(x_2) + T_2(x_3)f(x_3)] = -0.5274$$

Thus, three-term Chebyshev approximation of $f(x) = \sin\frac{(x+1)\pi}{2}$ function is obtained as $f(x) = a_0T_0(x) + a_1T_1(x) + a_2T_2(x) = 0.4726 - 0.5274(2x^2 - 1)$.

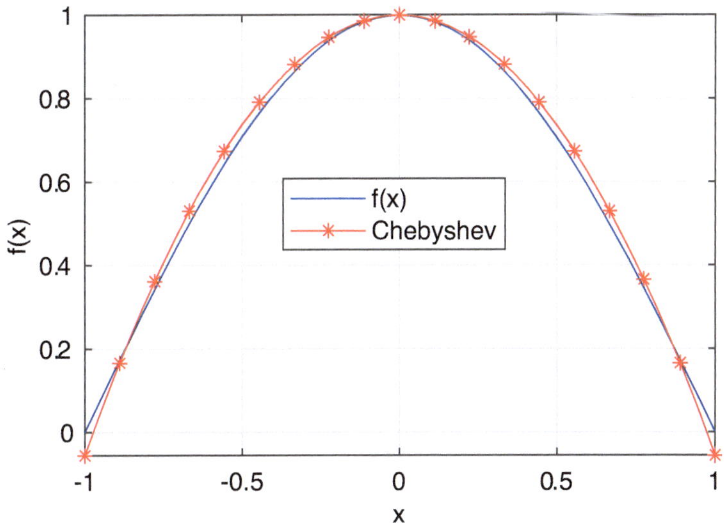

Fig. 7.56 Comparison of $f(x) = \sin\frac{(x+1)\pi}{2}$ function and its Chebyshev approximation in $x = [-1, 1]$ range

Chebyshev approximation and true function are compared in Fig. 7.56. From Fig. 7.56, it is seen that they are very similar to each other. Similarity can be increased by including more terms in the Chebyshev approximate function.

$f(x) = \sin\frac{(x+1)\pi}{2}$ function and its Chebyshev approximation can be plotted by using the following commands in MATLAB

```
>> fplot(@(x) sin((x+1)*pi/2),[-1 1],'b');
>> xlabel('x'); ylabel('f(x)'); hold on;
>> fplot(@(x) 0.4726-0.5274*(2*x^2-1),[-1 1],'*-r');
>> legend('f(x)','Chebyshev');
>> grid ;
```

7.14 Approximation of MultiVariable Functions

Multivariate functions such as two and three variables are often encountered in the expression of engineering problems. Therefore, approximation of complicated two- and three-variable functions as explained below is important in the efficient and effective solution of engineering problems.

7.15 Approximation of Two-Variable Functions

Approximation of a two-variable (bivariate) $f(x, y)$ function over a rectangular domain defined by $x = [0, a]$ and $y = [0, b]$ intervals as shown in Fig. 7.57 can be obtained with product of single-variable (univariate) approximate functions of $X(x)$ and $Y(y)$ as below:

$$f(x,y) = X(x) \otimes Y(y) = \underbrace{\left[a_0\Psi_{0x}(x) + a_1\Psi_{1x}(x) + a_2\Psi_{2x}(x) + \ldots + a_m\Psi_{mx}(x)\right]}_{X(x)}$$
$$\otimes \underbrace{\left[b_0\Psi_{0y}(y) + b_1\Psi_{1y}(y) + b_2\Psi_{2y}(y) + \ldots + b_n\Psi_{ny}(y)\right]}_{Y(y)} = \sum_{i=0}^{m}\sum_{j=0}^{n} c_{ij}\Psi_{ix}(x)\Psi_{jy}(y) \tag{7.136}$$

Note that $\otimes$ symbol in Eq. (7.136) means multiplication. In Eq. (7.136), $\Psi_{ix}(x)$ and $\Psi_{jy}(y)$ represent basic functions in trigonometric or polynomial forms, c_{ij} indicates unknown constants to be determined and $m + 1$ and $n + 1$ denote the number of terms in x and y directions.

Approximation of two-variable $f(x, y)$ function using Trigonometric functions, Legendre, and Chebyshev polynomials are explained below:

7.15.1 Approximation with Trigonometric Functions

Approximate function, which is valid over the rectangular domain in Fig. 7.57 can be constructed using cosine or sine functions.

Approximate function with cosine terms:

Selecting basic $\Psi_{ix}(x)$ and $\Psi_{jy}(y)$ functions in cosine form as $\Psi_{ix}(x) = \cos\frac{i\pi x}{a}$ and $\Psi_{jy}(y) = \cos\frac{j\pi y}{b}$, approximate function can be written as double cosine series as following:

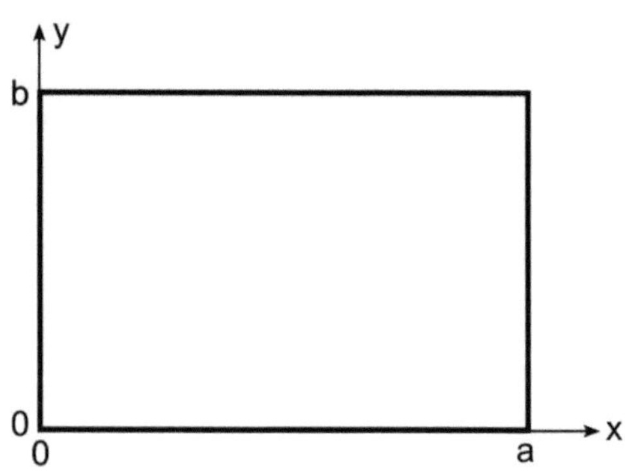

Fig. 7.57 Approximation of a two-variable $f(x, y)$ function in a rectangular domain defined by $x = [0, a]$ and $y = [0, b]$ intervals

$$f(x,y)=\sum_{i=0}^{m}\sum_{j=0}^{n}c_{ij}\Psi_{ix}(x)\Psi_{jy}(y)=\sum_{i=0}^{m}\sum_{j=0}^{n}c_{ij}\cos\frac{i\pi x}{a}\cos\frac{j\pi y}{b} \tag{7.137}$$

Approximate function with sine terms:

Selecting basic $\Psi_{ix}(x)$ and $\Psi_{jy}(y)$ functions in sine form as $\Psi_{ix}(x)=\sin\frac{i\pi x}{a}$ and $\Psi_{jy}(y)=\sin\frac{j\pi y}{b}$, approximate function can be written as double sine series as following:

$$f(x,y)=\sum_{i=0}^{m}\sum_{j=0}^{n}c_{ij}\Psi_{ix}(x)\Psi_{jy}(y)=\sum_{i=0}^{m}\sum_{j=0}^{n}c_{ij}\sin\frac{i\pi x}{a}\sin\frac{j\pi y}{b} \tag{7.138}$$

Since $\Psi_{0x}(x)=0$ and $\Psi_{0y}(y)=0$, *i* and *j* indices can be started from 1 and Eq. (7.138) can be rewritten as below:

$$f(x,y)=\sum_{i=1}^{m}\sum_{j=1}^{n}c_{ij}\Psi_{ix}(x)\Psi_{jy}(y)=\sum_{i=1}^{m}\sum_{j=1}^{n}c_{ij}\sin\frac{i\pi x}{a}\sin\frac{j\pi y}{b} \tag{7.139}$$

c_{ij} coefficients in Eq. (7.139) can be calculated using orthogonality condition of cosine and sine functions. Orthogonality condition for cosine and sine functions in $x=[0,a]$ and $y=[0,b]$ intervals can be expressed as following:

$$\int_0^a \cos\frac{m\pi x}{a}\cos\frac{n\pi x}{a}dx=\begin{cases}0, & \text{if} \quad m\neq n\\ a, & \text{if} \quad m=n=0\\ \frac{a}{2}, & \text{if} \quad m=n\neq 0\end{cases} \tag{7.140a}$$

$$\int_0^b \cos\frac{m\pi y}{b}\cos\frac{n\pi y}{b}dy=\begin{cases}0, & \text{if} \quad m\neq n\\ b, & \text{if} \quad m=n=0\\ \frac{b}{2}, & \text{if} \quad m=n\neq 0\end{cases} \tag{7.140b}$$

$$\int_0^a \sin\frac{m\pi x}{a}\sin\frac{n\pi x}{a}dx=\begin{cases}0, & \text{if} \quad m\neq n\\ \frac{a}{2}, & \text{if} \quad m=n\neq 0\end{cases} \tag{7.140c}$$

$$\int_0^b \sin\frac{m\pi y}{b}\sin\frac{n\pi y}{b}dy=\begin{cases}0, & \text{if} \quad m\neq n\\ \frac{b}{2}, & \text{if} \quad m=n\neq 0\end{cases} \tag{7.140d}$$

Using orthogonality condition, c_{ij} coefficients in Eq. (7.139) can be obtained as following:

c_{ij} *calculation (General coefficient)*:

Multiplication of Eq. (7.139) by $\sin\frac{k\pi x}{a}\sin\frac{l\pi y}{b}$ and then integration in $x = [0, a]$ and $y = [0, b]$ intervals as well as using orthogonality condition yield general c_{kl} coefficient as below:

$$\int_0^a\int_0^b \sin\frac{k\pi x}{a}\sin\frac{l\pi y}{b}f(x,y)dxdy$$
$$= \int_0^a\int_0^b \sin\frac{k\pi x}{a}\sin\frac{l\pi y}{b}\left[\sum_{i=1}^{2}\sum_{j=1}^{2}c_{ij}\sin\frac{i\pi x}{a}\sin\frac{j\pi y}{b}\right]dxdy$$
$$= \sum_{i=1}^{2}\sum_{j=1}^{2}c_{ij}\int_0^a\int_0^b \sin\frac{k\pi x}{a}\sin\frac{l\pi y}{b}\sin\frac{i\pi x}{a}\sin\frac{j\pi y}{b}dxdy$$
$$= \sum_{i=1}^{2}\sum_{j=1}^{2}c_{ij}\underbrace{\left[\int_0^a \sin\frac{k\pi x}{a}\sin\frac{i\pi x}{a}dx\right]}_{=a/2 \quad eger \quad i=k \quad ise}\underbrace{\left[\int_0^b \sin\frac{l\pi y}{b}\sin\frac{j\pi y}{b}dy\right]}_{=b/2 \quad eger \quad j=l \quad ise}$$
$$= c_{kl}\frac{ab}{4}$$

$$\rightarrow c_{kl} = \frac{4}{ab}\int_0^a\int_0^b \sin\frac{k\pi x}{a}\sin\frac{l\pi y}{b}f(x,y)dxdy$$

Changing indices, c_{kl} can also be expressed as c_{ij} as following:

$$c_{ij} = \frac{4}{ab}\int_0^a\int_0^b \sin\frac{i\pi x}{a}\sin\frac{j\pi y}{b}f(x,y)dxdy \tag{7.141}$$

Changing i, j indices in Eq. (7.141), all c_{ij} coefficients can be obtained easily.

Example 7.17: Obtain the approximation of $f(x, y) = x^2y^2 + xy$ function in the ranges of $x = [0, 2]$ and $y = [0, 2]$ using double Fourier Series based on sine function.

For simplicity, we can construct Fourier approximation based on double sine series using two terms in x direction and two terms in y direction in Eq. (7.139) as following:

$$f(x,y) = \sum_{i=1}^{2}\sum_{j=1}^{2}c_{ij}\Psi_{ix}(x)\Psi_{jy}(y) = \sum_{i=1}^{2}\sum_{j=1}^{2}c_{ij}\sin\frac{i\pi x}{a}\sin\frac{j\pi y}{b}$$
$$= c_{11}\sin\frac{\pi x}{a}\sin\frac{\pi y}{b} + c_{12}\sin\frac{\pi x}{a}\sin\frac{2\pi y}{b} + c_{21}\sin\frac{2\pi x}{a}\sin\frac{\pi y}{b} + c_{22}\sin\frac{2\pi x}{a}\sin\frac{2\pi y}{b}$$

where a and b values are taken as $a = 2$, $b = 2$. c_{11}, c_{12}, c_{21}, and c_{22} coefficients can be calculated using the general expression of $c_{ij} = \frac{4}{ab}\int_0^a\int_0^b \sin\frac{i\pi x}{a}\sin\frac{j\pi y}{b} f(x,y)dxdy$ in Eq. (7.141) as following:

$$c_{11} = \int_0^2\int_0^2 \sin\frac{\pi x}{2}\sin\frac{\pi y}{2}\left(x^2y^2 + xy\right)dxdy = 3.9146,$$

$$c_{12} = \int_0^2\int_0^2 \sin\frac{\pi x}{2}\sin\frac{2\pi y}{2}\left(x^2y^2 + xy\right)dxdy = -2.7388,$$

$$c_{21} = \int_0^2\int_0^2 \sin\frac{2\pi x}{2}\sin\frac{\pi y}{2}\left(x^2y^2 + xy\right)dxdy = -2.7388,$$

$$c_{22} = \int_0^2\int_0^2 \sin\frac{2\pi x}{2}\sin\frac{2\pi y}{2}\left(x^2y^2 + xy\right)dxdy = 2.0264.$$

Thus, approximation of $f(x, y) = x^2y^2 + xy$ function is obtained as below:

$$f(x,y) = 3.9146\sin\frac{\pi x}{2}\sin\frac{\pi y}{2} - 2.7388\sin\frac{\pi x}{2}\sin\frac{2\pi y}{2} - 2.7388\sin\frac{2\pi x}{2}\sin\frac{\pi y}{2}$$
$$+2.0264\sin\frac{2\pi x}{2}\sin\frac{2\pi y}{2}$$

Approximation quality can be increased by including more terms in the double sine series.

7.15.2 Approximation with Legendre Polynomials

Legendre approximation of a bivariate (two-variable) $f(x,y)$ function on a square domain defined by $x = [-1, 1]$ and $y = [-1, 1]$ intervals as shown in Fig. 7.58 can be created with product of univariate Legendre approximate functions of $X(x)$ and $Y(y)$ as below:

$$f(x,y) = X(x) \otimes Y(y) = [a_0L_{0x}(x) + a_1L_{1x}(x) + a_2L_{2x}(x) + \ldots + a_mL_{mx}(x)]$$
$$\otimes\left[b_0L_{0y}(y) + b_1L_{1y}(y) + b_2L_{2y}(y) + \ldots + b_nL_{ny}(y)\right] = \sum_{i=0}^{m}\sum_{j=0}^{n} c_{ij}L_{ix}(x)L_{jy}(y) \tag{7.142}$$

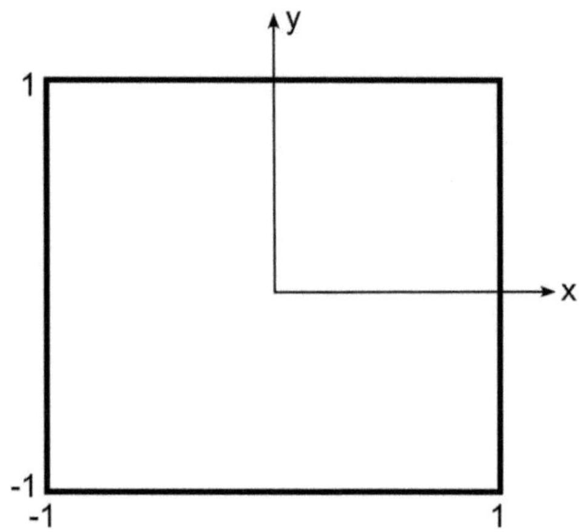

Fig. 7.58 Approximation of a bivariate $f(x,y)$ function in a square domain defined by $[-1, 1]$ intervals in x and y directions

In Eq. (7.142), L_x, L_y indicate Legendre polynomials and c_{ij} indicates unknown constants.

c_{ij} coefficients in Eq. (7.142) can be calculated using the orthogonality condition of Legendre polynomials. Multiplication of Eq. (7.142) by $L_{kx}(x)L_{ly}(y)$ and then integration in $x = [-1, 1]$ and $y = [-1, 1]$ intervals along with the orthogonality condition yield c_{kl} coefficient as following:

$$\begin{aligned}
&\int_{-1}^{1}\int_{-1}^{1} L_{kx}(x)L_{ly}(y)f(x,y)dxdy\\
&= \int_{-1}^{1}\int_{-1}^{1} L_{kx}(x)L_{ly}(y)\left[\sum_{i=0}^{m}\sum_{j=0}^{n} c_{ij}L_{ix}(x)L_{jy}(y)\right]dxdy\\
&= \sum_{i=0}^{m}\sum_{j=0}^{n} c_{ij}\int_{-1}^{1}\int_{-1}^{1} L_{kx}(x)L_{ly}(y)L_{ix}(x)L_{jy}(y)dxdy\\
&= \sum_{i=1}^{m}\sum_{j=1}^{n} c_{ij}\underbrace{\left[\int_{-1}^{1} L_{kx}(x)L_{ix}(x)dx\right]}_{=\frac{2k+1}{2},\quad if\quad i=k}\underbrace{\left[\int_{-1}^{1} L_{ly}(y)L_{jy}(y)dy\right]}_{=\frac{2l+1}{2},\quad if\quad j=l}\\
&= c_{kl}\left(\frac{2k+1}{2}\right)\left(\frac{2l+1}{2}\right)\\
&\rightarrow c_{kl} = \frac{2}{2k+1}\,\frac{2}{2l+1}\int_{-1}^{1}\int_{-1}^{1} L_{kx}(x)L_{ly}(y)f(x,y)dxdy
\end{aligned}$$

Changing indices, c_{kl} can also be expressed as c_{ij} as below:

$$c_{ij} = \frac{2}{2i+1}\frac{2}{2j+1}\int_{-1}^{1}\int_{-1}^{1} L_{ix}(x)L_{jy}(y)f(x,y)dxdy \tag{7.143}$$

Changing i, j indices in Eq. (7.143), all c_{ij} coefficients can be obtained easily.

Note: Legendre approximation function was constructed in normalized intervals of $[-1, 1]$. If original intervals are different, they can be converted into $[-1, 1]$ intervals using variable transformations. For example, if the original intervals are given as $x = [a, b]$ and $y = [c, d]$, they can be converted into standard $[-1, 1]$ intervals using variable transformation as below (New variables are defined for transformations, for example: ξ for x, η for y):

$$x = [a,b] \rightarrow \xi = [-1,1] : x = \frac{(1-\xi)}{2}a + \frac{(1+\xi)}{2}b$$

$$y = [c,d] \rightarrow \eta = [-1,1] : y = \frac{(1-\eta)}{2}c + \frac{(1+\eta)}{2}d$$

Example 7.18: Obtain the approximation of $f(x, y) = \sin x \sin y$ function in the ranges of $x = [-1, 1]$ and $y = [-1, 1]$ using Legendre polynomials.

For simplicity, we can construct approximate Legendre polynomial function using three terms in x and y directions in Eq. (7.142) as following:

$$\begin{aligned} f(x,y) &= \sum_{i=0}^{2}\sum_{j=0}^{2} c_{ij}L_{ix}(x)L_{jy}(y) \\ &= c_{00}L_{0x}(x)L_{0y}(y) + c_{01}L_{0x}(x)L_{1y}(y) + c_{02}L_{0x}(x)L_{2y}(y) \\ &\quad + c_{10}L_{1x}(x)L_{0y}(y) + c_{11}L_{1x}(x)L_{1y}(y) + c_{12}L_{1x}(x)L_{2y}(y) \\ &\quad + c_{20}L_{2x}(x)L_{0y}(y) + c_{21}L_{2x}(x)L_{1y}(y) + c_{22}L_{2x}(x)L_{2y}(y) \end{aligned}$$

where Legendre polynomials in $f(x, y)$ function are given as $L_{0x}(x) = 1$, $L_{1x}(x) = x$, $L_{2x}(x) = \frac{3x^2-1}{2}$, $L_{0y}(y) = 1$, $L_{1y}(y) = y$, $L_{2y}(y) = \frac{3y^2-1}{2}$.

c_{00}, c_{01}, c_{02}, c_{10}, c_{11}, c_{12}, c_{20}, c_{21}, c_{22} coefficients in $f(x, y)$ function can be calculated using the general expression $c_{ij} = \frac{2}{2i+1}\frac{2}{2j+1}\int_{-1}^{1} \times \int_{-1}^{1} L_{ix}(x)L_{jy}(y)f(x,y)dxdy$ in Eq. (7.143) as following:

$$c_{00} = \frac{2}{2.0+1}\frac{2}{2.0+1}\int_{-1}^{1}\int_{-1}^{1} L_{0x}(x)L_{0y}(y)f(x,y)dxdy = 4\int_{-1}^{1}\int_{-1}^{1} \sin x . \sin y dxdy = 0$$

$$c_{01}=\frac{2}{2.0+1}\frac{2}{2.1+1}\int_{-1}^{1}\int_{-1}^{1}L_{0x}(x)L_{1y}(y)f(x,y)dxdy=\frac{4}{3}\int_{-1}^{1}\int_{-1}^{1}(\sin x)y(\sin y)dxdy=0$$

$$c_{02}=\frac{2}{2.0+1}\frac{2}{2.2+1}\int_{-1}^{1}\int_{-1}^{1}L_{0x}(x)L_{2y}(y)f(x,y)dxdy=\frac{4}{5}\int_{-1}^{1}\int_{-1}^{1}(\sin x)\left(\frac{3y^2-1}{2}\right)(\sin y)dxdy=0$$

$$c_{10}=\frac{2}{2.1+1}\frac{2}{2.0+1}\int_{-1}^{1}\int_{-1}^{1}L_{1x}(x)L_{0y}(y)f(x,y)dxdy=\frac{4}{3}\int_{-1}^{1}\int_{-1}^{1}x(\sin x)(\sin y)dxdy=0$$

$$c_{11}=\frac{2}{2.1+1}\frac{2}{2.1+1}\int_{-1}^{1}\int_{-1}^{1}L_{1x}(x)L_{1y}(y)f(x,y)dxdy=\frac{4}{9}\int_{-1}^{1}\int_{-1}^{1}x(\sin x)y(\sin y)dxdy=0.1612$$

$$c_{12}=\frac{2}{2.1+1}\frac{2}{2.2+1}\int_{-1}^{1}\int_{-1}^{1}L_{1x}(x)L_{2y}(y)f(x,y)dxdy=\frac{4}{15}\int_{-1}^{1}\int_{-1}^{1}x(\sin x)\left(\frac{3y^2-1}{2}\right)(\sin y)dxdy=0$$

$$c_{20}=\frac{2}{2.2+1}\,\frac{2}{2.0+1}\int_{-1}^{1}\int_{-1}^{1}L_{2x}(x)L_{0y}(y)f(x,y)dxdy$$

$$=\frac{4}{5}\int_{-1}^{1}\int_{-1}^{1}\left(\frac{3x^2-1}{2}\right)(\sin x)(\sin y)dxdy=0$$

$$c_{21}=\frac{2}{2.2+1}\,\frac{2}{2.1+1}\int_{-1}^{1}\int_{-1}^{1}L_{2x}(x)L_{1y}(y)f(x,y)dxdy$$

$$=\frac{4}{15}\int_{-1}^{1}\int_{-1}^{1}\left(\frac{3x^2-1}{2}\right)(\sin x)y(\sin y)dxdy=0$$

$$c_{22}=\frac{2}{2.2+1}\,\frac{2}{2.2+1}\int_{-1}^{1}\int_{-1}^{1}L_{2x}(x)L_{2y}(y)f(x,y)dxdy$$

$$=\frac{4}{25}\int_{-1}^{1}\int_{-1}^{1}\left(\frac{3x^2-1}{2}\right)(\sin x)\left(\frac{3y^2-1}{2}\right)(\sin y)dxdy=0$$

Thus, approximation of $f(x, y) = \sin x.\sin y$ function with Legendre polynomials is obtained as below:

$$f(x,y)=c_{00}+c_{01}y+c_{02}\left(\frac{3y^2-1}{2}\right)+c_{10}x+c_{11}xy+c_{12}x\left(\frac{3y^2-1}{2}\right)+$$
$$c_{20}\left(\frac{3x^2-1}{2}\right)+c_{21}\left(\frac{3x^2-1}{2}\right)y+c_{22}\left(\frac{3x^2-1}{2}\right)\left(\frac{3y^2-1}{2}\right)=0.1612xy$$

Approximation quality can be increased by including more terms in the approximate function.

7.15.3 Approximation with Chebyshev Polynomials

Chebyshev approximation of a bivariate (two-variable) $f(x,y)$ function on the square domain in Fig. 7.58 can be created with product of univariate Chebyshev approximate functions of $X(x)$ and $Y(y)$ as below:

$$f(x,y)=X(x)\otimes Y(y)=\left[a_0T_{0x}(x)+a_1T_{1x}(x)+a_2T_{2x}(x)+\ldots+a_mT_{mx}(x)\right]$$
$$\otimes\left[b_0T_{0y}(y)+b_1T_{1y}(y)+b_2T_{2y}(y)+\ldots+b_nT_{ny}(y)\right]=\sum_{i=0}^{m}\sum_{j=0}^{n}c_{ij}T_{ix}(x)T_{jy}(y) \tag{7.144}$$

In Eq. (7.144), T_x, T_y indicate Chebyshev polynomials and c_{ij} indicates unknown constants. There are $(m + 1)(n + 1)$ unknown coefficients in Eq. (7.144). Necessary equations to obtain these coefficients can be created by expressing the approximate function at roots of $(m + 1)$-th order Chebyshev polynomials in x direction and $(n + 1)$-th order Chebyshev polynomials in y direction. Roots of $(m + 1)$-th order Chebyshev polynomials in x direction and $(n + 1)$-th order Chebyshev polynomials in y direction can be expressed as following:

$$x_i=-\cos\left(\frac{(2i-1)\pi}{2(m+1)}\right)\quad i=1,2,..,m+1 \tag{7.145a}$$

$$y_j=-\cos\left(\frac{(2j-1)\pi}{2(n+1)}\right)\quad j=1,2,..,n+1 \tag{7.145b}$$

Using root values, orthogonality condition of Chebyshev polynomials can be expressed in discrete form in x and y directions as following:

$$\sum_{i=1}^{m+1} T_{kx}(x_i)T_{lx}(x_i) = \begin{cases} 0, & \text{if} \quad k \neq l \\ m+1, & \text{if} \quad k=l=0 \\ \dfrac{m+1}{2}, & \text{if} \quad k=l\neq 0 \end{cases} \tag{7.146a}$$

$$\sum_{j=1}^{n+1} T_{ky}(y_j)T_{ly}(y_j) = \begin{cases} 0, & \text{if} \quad k \neq l \\ n+1, & \text{if} \quad k=l=0 \\ \dfrac{n+1}{2}, & \text{if} \quad k=l\neq 0 \end{cases} \tag{7.146b}$$

After expressing the Chebyshev approximate function in Eq. (7.144) at root values multiplied with $T_{kx}(x_i)T_{ly}(y_j)$ term and then summing them, following equation is obtained:

$$\begin{aligned} &\sum_{i=1}^{m+1}\sum_{j=1}^{n+1} T_{kx}(x_i)T_{ly}(y_j)f(x_i,y_j) = \sum_{i=1}^{m+1}\sum_{j=1}^{n+1} T_{kx}(x_i)T_{ly}(y_j) \\ &\times \begin{bmatrix} c_{00}T_{0x}(x_i)T_{0y}(y_j)+c_{01}T_{0x}(x_i)T_{1y}(y_j)+c_{02}T_{0x}(x_i)T_{2y}(y_j)+\ldots+c_{0n}T_{0x}(x_i)T_{ny}(y_j)+ \\ c_{10}T_{1x}(x_i)T_{0y}(y_j)+c_{11}T_{1x}(x_i)T_{1y}(y_j)+c_{12}T_{1x}(x_i)T_{2y}(y_j)+\ldots+c_{1n}T_{1x}(x_i)T_{ny}(y_j)+\ldots+ \\ c_{kl}T_{kx}(x_i)T_{ly}(y_j)+\ldots+c_{m0}T_{mx}(x_i)T_{0y}(y_j)+c_{m1}T_{mx}(x_i)T_{1y}(y_j)+c_{m2}T_{mx}(x_i)T_{2y}(y_j)+ \\ \ldots+c_{mn}T_{mx}(x_i)T_{ny}(y_j) \end{bmatrix} \end{aligned} \tag{7.147}$$

Equation (7.147) can be rearranged as below:

$$\sum_{i=1}^{m+1}\sum_{j=1}^{n+1} T_{kx}(x_i)T_{ly}(y_j)f(x_i,y_j) = c_{00}\underbrace{\sum_{i=1}^{m+1} T_{kx}(x_i)T_{0x}(x_i)}_{=0}\underbrace{\sum_{j=1}^{n+1} T_{ly}(y_j)T_{0y}(y_j)}_{=0}$$

$$+ c_{01}\underbrace{\sum_{i=1}^{m+1} T_{kx}(x_i)T_{0x}(x_i)}_{=0}\underbrace{\sum_{j=1}^{n+1} T_{ly}(y_j)T_{1y}(y_j)}_{=0} + c_{02}\underbrace{\sum_{i=1}^{m+1} T_{kx}(x_i)T_{0x}(x_i)}_{=0}\underbrace{\sum_{j=1}^{n+1} T_{ly}(y_j)T_{2y}(y_j)}_{=0} + \ldots$$

$$+ c_{0n}\underbrace{\sum_{i=1}^{m+1} T_{kx}(x_i)T_{0x}(x_i)}_{=0}\underbrace{\sum_{j=1}^{n+1} T_{ly}(y_j)T_{ny}(y_j)}_{=0} + c_{10}\underbrace{\sum_{i=1}^{m+1} T_{kx}(x_i)T_{1x}(x_i)}_{=0}\underbrace{\sum_{j=1}^{n+1} T_{ly}(y_j)T_{0y}(y_j)}_{=0}$$

$$+ c_{11}\underbrace{\sum_{i=1}^{m+1} T_{kx}(x_i)T_{1x}(x_i)}_{=0}\underbrace{\sum_{j=1}^{n+1} T_{ly}(y_j)T_{1y}(y_j)}_{=0} + c_{12}\underbrace{\sum_{i=1}^{m+1} T_{kx}(x_i)T_{1x}(x_i)}_{=0}\underbrace{\sum_{j=1}^{n+1} T_{ly}(y_j)T_{2y}(y_j)}_{=0} + \ldots$$

$$+ c_{1n}\underbrace{\sum_{i=1}^{m+1} T_{kx}(x_i)T_{1x}(x_i)}_{=0}\underbrace{\sum_{j=1}^{n+1} T_{ly}(y_j)T_{ny}(y_j)}_{=0} + \ldots$$

$$+ c_{kl}\underbrace{\sum_{i=1}^{m+1} T_{kx}(x_i)T_{kx}(x_i)}_{m+1\,(k=0),\ (m+1)/2\,(k\neq 0)}\underbrace{\sum_{j=1}^{n+1} T_{ly}(y_j)T_{ly}(y_j)}_{n+1\,(l=0),\ (n+1)/2\,(l\neq 0)}$$

$$+ \ldots c_{m0}\underbrace{\sum_{i=1}^{m+1} T_{kx}(x_i)T_{mx}(x_i)}_{=0}\underbrace{\sum_{j=1}^{n+1} T_{ly}(y_j)T_{0y}(y_j)}_{=0} + c_{m1}\underbrace{\sum_{i=1}^{m+1} T_{kx}(x_i)T_{mx}(x_i)}_{=0}\underbrace{\sum_{j=1}^{n+1} T_{ly}(y_j)T_{1y}(y_j)}_{=0}$$

$$+ c_{m2}\underbrace{\sum_{i=1}^{m+1} T_{kx}(x_i)T_{mx}(x_i)}_{=0}\underbrace{\sum_{j=1}^{n+1} T_{ly}(y_j)T_{2y}(y_j)}_{=0} + \ldots + c_{mn}\underbrace{\sum_{i=1}^{m+1} T_{kx}(x_i)T_{mx}(x_i)}_{=0}\underbrace{\sum_{j=1}^{n+1} T_{ly}(y_j)T_{ny}(y_j)}_{=0} \tag{7.148}$$

Using the discrete form of the orthogonality condition of Chebyshev polynomials in Eq. (7.148), c_{kl} coefficient can be obtained as following:

$$c_{kl} = \frac{2}{m+1}\frac{2}{n+1}\sum_{i=1}^{m+1}\sum_{j=1}^{n+1} T_{kx}(x_i)T_{ly}(y_j)f(x_i,y_j) \quad (k\neq 0, l\neq 0) \tag{7.149a}$$

c_{kl} coefficient can be specialized to other coefficients as following:

$$c_{00} = \frac{1}{m+1}\frac{1}{n+1}\sum_{i=1}^{m+1}\sum_{j=1}^{n+1} T_{0x}(x_i)T_{0y}(y_j)f(x_i, y_j) \tag{7.149b}$$

...

$$c_{0l} = \frac{1}{m+1}\frac{2}{n+1}\sum_{i=1}^{m+1}\sum_{j=1}^{n+1} T_{0x}(x_i)T_{ly}(y_j)f(x_i, y_j) \quad (k=0, l\neq 0) \tag{7.149c}$$

...

$$c_{k0} = \frac{2}{m+1}\frac{1}{n+1}\sum_{i=1}^{m+1}\sum_{j=1}^{n+1} T_{kx}(x_i)T_{0y}(y_j)f(x_i, y_j) \quad (k\neq 0, l=0) \tag{7.149d}$$

...

Note: Chebyshev approximation function was constructed in normalized interval of $[-1, 1]$. If original intervals are different, they can be converted into $[-1, 1]$ intervals using variable transformations. For example, if the original intervals are given as $x = [a, b]$ and $y = [c, d]$, they can be converted into standard $[-1, 1]$ intervals using variable transformation as below (New variables are defined for transformations, for example: ξ for x, η for y):

$$x=[a,b]\rightarrow\xi=[-1,1] : x=\frac{(1-\xi)}{2}a+\frac{(1+\xi)}{2}b$$

$$y=[c,d]\rightarrow\eta=[-1,1] : y=\frac{(1-\eta)}{2}c+\frac{(1+\eta)}{2}d$$

Example 7.19: Obtain the approximation of $f(r, s) = \sin r \sin s$ function in the ranges of $r = [0, \pi]$ and $s = [0, \pi]$ using Chebyshev polynomials.

Since the range of variables are different from the standard range of $[-1, 1]$, variable transformation is applied on r and s variables to convert them into the standard range. Indicating the transformed variables in the standard range with x and y, transformation between r, s and x, y variables can be expressed using Lagrange interpolation method as following:

$$r=[0,\pi]\rightarrow x=[-1,1] : r=\frac{(x-1)}{(-1-1)}0+\frac{(x-(-1))}{(1-(-1))}\pi=\frac{(x+1)\pi}{2},$$

$$s=[0,\pi]\rightarrow y=[-1,1] : s=\frac{(y-1)}{(-1-1)}0+\frac{(y-(-1))}{(1-(-1))}\pi=\frac{(y+1)\pi}{2}$$

In this case, original function $f(r, s) = \sin r \sin s$ can be expressed in terms of new x, y variables as below:

$$f(r,s) = \sin r \sin s \rightarrow f(x,y) = \sin\frac{(x+1)\pi}{2}\sin\frac{(y+1)\pi}{2}$$

A simple approximation can be created for $f(x,y) = \sin\frac{(x+1)\pi}{2}\sin\frac{(y+1)\pi}{2}$ function using three terms in x direction and three terms in y direction in Eq. (7.144) as following:

$$\begin{aligned} f(x,y) &= \sum_{i=0}^{m=2}\sum_{j=0}^{n=2} c_{ij}T_{ix}(x)T_j(y) \\ &= c_{00}T_{0x}(x)T_{0y}(y) + c_{01}T_{0x}(x)T_{1y}(y) + c_{02}T_{0x}(x)T_{2y}(y) \\ &\quad + c_{10}T_{1x}(x)T_{0y}(y) + c_{11}T_{1x}(x)T_{1y}(y) + c_{12}T_{1x}(x)T_{2y}(y) \\ &\quad + c_{20}T_{2x}(x)T_{0y}(y) + c_{21}T_{2x}(x)T_{1y}(y) + c_{22}T_{2x}(x)T_{2y}(y) \end{aligned}$$

To calculate coefficients, necessary equations can be obtained writing the approximate function at roots of third-order Chebyshev polynomials ($m + 1$, $m = 2$ and $n + 1$, $n = 2$) in x and y directions. Roots of third-order Chebyshev polynomials in x and y direction can be expressed as following:

$$x_i = -\cos\left(\frac{(2i-1)\pi}{2(m+1)}\right)\ (i = 1,2,3), \quad y_j = -\cos\left(\frac{(2j-1)\pi}{2(n+1)}\right)\ (j = 1,2,3),$$

Using Chebyshev polynomials $T_{0x}(x) = 1$, $T_{1x}(x) = x$, $T_{2x}(x) = 2x^2 - 1$, $T_{0y}(y) = 1$, $T_{1y}(y) = y$, $T_{2y}(y) = 2y^2 - 1$ in Eq. (7.149), c_{00}, c_{01}, c_{02}, c_{10}, c_{11}, c_{12}, c_{20}, c_{21}, c_{22} coefficients can be calculated as below:

$$c_{00} = \frac{1}{3}\frac{1}{3}\sum_{i=1}^{3}\sum_{j=1}^{3} T_{0x}(x_i)T_{0y}(y_j)f(x_i,y_j) = \frac{1}{9}\sum_{i=1}^{3}\sum_{j=1}^{3} f(x_i,y_j) = 0.2233$$

$$c_{01} = \frac{1}{3}\frac{2}{3}\sum_{i=1}^{3}\sum_{j=1}^{3} T_{0x}(x_i)T_{1y}(y_j)f(x_i,y_j) = \frac{2}{9}\sum_{i=1}^{3}\sum_{j=1}^{3} y_j f(x_i,y_j) = 0$$

$$c_{02} = \frac{1}{3}\frac{2}{3}\sum_{i=1}^{3}\sum_{j=1}^{3} T_{0x}(x_i)T_{2y}(y_j)f(x_i,y_j) = \frac{2}{9}\sum_{i=1}^{3}\sum_{j=1}^{3}\left(2y_j^2-1\right)f(x_i,y_j) = \text{-}0.2492$$

$$c_{10} = \frac{2}{3}\frac{1}{3}\sum_{i=1}^{3}\sum_{j=1}^{3} T_{1x}(x_i)T_{0y}(y_j)f(x_i,y_j) = \frac{2}{9}\sum_{i=1}^{3}\sum_{j=1}^{3} x_i f(x_i,y_j) = 0$$

$$c_{11} = \frac{2}{3}\frac{2}{3}\sum_{i=1}^{3}\sum_{j=1}^{3} T_{1x}(x_i)T_{1y}(y_j)f(x_i,y_j) = \frac{4}{9}\sum_{i=1}^{3}\sum_{j=1}^{3} x_i y_j f(x_i,y_j) = 0$$

$$c_{12}=\frac{2}{3}\frac{2}{3}\sum_{i=1}^{3}\sum_{j=1}^{3}T_{1x}(x_i)T_{2y}\left(y_j\right)f\left(x_i,y_j\right)=\frac{4}{9}\sum_{i=1}^{3}\sum_{j=1}^{3}x_i\left(2y_j^2-1\right)f\left(x_i,y_j\right)=0$$

$$c_{20}=\frac{2}{3}\frac{1}{3}\sum_{i=1}^{3}\sum_{j=1}^{3}T_{2x}(x_i)T_{0y}\left(y_j\right)f\left(x_i,y_j\right)=\frac{2}{9}\sum_{i=1}^{3}\sum_{j=1}^{3}\left(2x_i^2-1\right)f\left(x_i,y_j\right)=-0.2492$$

$$c_{21}=\frac{2}{3}\frac{2}{3}\sum_{i=1}^{3}\sum_{j=1}^{3}T_{2x}(x_i)T_{1y}\left(y_j\right)f\left(x_i,y_j\right)=\frac{4}{9}\sum_{i=1}^{3}\sum_{j=1}^{3}\left(2x_i^2-1\right)y_jf\left(x_i,y_j\right)=0$$

$$c_{22}=\frac{2}{3}\frac{2}{3}\sum_{i=1}^{3}\sum_{j=1}^{3}T_{2x}(x_i)T_{2y}\left(y_j\right)f\left(x_i,y_j\right)=\frac{4}{9}\sum_{i=1}^{3}\sum_{j=1}^{3}\left(2x_i^2-1\right)\left(2y_j^2-1\right)f\left(x_i,y_j\right)=0.2782$$

Thus, approximation of $f(r,s)=\sin r\sin s=\sin\frac{(x+1)\pi}{2}\sin\frac{(y+1)\pi}{2}$ function is obtained as following:

$$f(x,y)=0.2233-0.2492\left(2y^2-1\right)-0.2492\left(2x^2-1\right)+0.2782\left(2x^2-1\right)\left(2y^2-1\right)$$

Approximation quality can be increased by including more terms in the approximate function.

7.16 Approximation of Tri-Variable Functions

Approximation of a three-variable (trivariate) $f(x, y, z)$ function over a rectangular prism domain defined by $x = [0, a]$, $y = [0, b]$, and $z = [0, c]$ intervals as shown in Fig. 7.59 can be obtained with product of single-variable (univariate) approximate functions of $X(x)$, $Y(y)$, and $Z(z)$ as below:

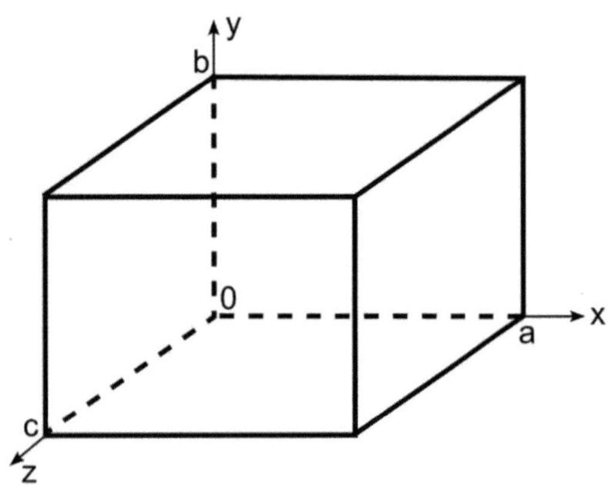

Fig. 7.59 Approximation of trivariate $f(x, y, z)$ function in a rectangular prism domain defined by $x = [0, a]$, $y = [0, b]$, and $z = [0, c]$ intervals

$$
\begin{aligned}
f(x,y,z) &= X(x) \bigotimes Y(y) \bigotimes Z(z) \\
&= \underbrace{[a_0\Psi_{0x}(x) + a_1\Psi_{1x}(x) + a_2\Psi_{2x}(x) + \ldots + a_m\Psi_{mx}(x)]}_{X(x)} \\
&\quad \bigotimes \underbrace{[b_0\Psi_{0y}(y) + b_1\Psi_{1y}(y) + b_2\Psi_{2y}(y) + \ldots + b_n\Psi_{ny}(y)]}_{Y(y)} \\
&\quad \bigotimes \underbrace{[c_0\Psi_{0z}(z) + c_1\Psi_{1z}(z) + c_2\Psi_{2z}(z) + \ldots + c_r\Psi_{rz}(z)]}_{Z(z)} \\
&= \sum_{i=0}^{m}\sum_{j=0}^{n}\sum_{k=0}^{r} d_{ijk}\Psi_{ix}(x)\Psi_{jy}(y)\Psi_{kz}(z)
\end{aligned}
\tag{7.150}
$$

$\otimes$ symbol means multiplication in Eq. (7.152). In Eq. (7.152), $\Psi_{ix}(x)$, $\Psi_{jy}(y)$, and $\Psi_{kz}(z)$ represent basic functions in trigonometric or polynomial forms, d_{ijk} indicates unknown constants to be determined, $m + 1$ and $n + 1$ and $r + 1$ denote the number of terms in x and y and z directions.

7.16.1 Approximation with Trigonometric Functions

Fourier approximation of a three-variable (trivariate) $f(x,y,z)$ function over the rectangular prism domain in Fig. 7.59 can be obtained by selecting basic functions in Eq. (7.150) in sine form as $\Psi_{ix}(x) = \sin\frac{i\pi x}{a}$, $\Psi_{jy}(y) = \sin\frac{j\pi y}{b}$ and $\Psi_{kz}(z) = \sin\frac{k\pi z}{c}$ as following:

$$
f(x,y,z) = \sum_{i=1}^{m}\sum_{j=1}^{n}\sum_{k=1}^{r} d_{ijk}\Psi_{ix}(x)\Psi_{jy}(y)\Psi_{kz}(z) = \sum_{i=1}^{m}\sum_{j=1}^{n}\sum_{k=1}^{r} d_{ijk}\sin\frac{i\pi x}{a}\sin\frac{j\pi y}{b}\sin\frac{k\pi z}{c}
\tag{7.151}
$$

d_{ijk} coefficients in Eq. (7.151) can be calculated using the orthogonality condition of sine functions. Multiplication of Eq. (7.151) by $\sin\frac{q\pi x}{a}\sin\frac{s\pi y}{b}\sin\frac{t\pi z}{c}$ and then integration in $x = [0,a]$, $y = [0,b]$, and $z = [0,c]$ intervals as well as using the orthogonality condition yield d_{qst} coefficient as following:

$$\begin{aligned}
&\int_0^a\int_0^b\int_0^c \sin\frac{q\pi x}{a}\sin\frac{s\pi y}{b}\sin\frac{t\pi z}{c}f(x,y,z)dxdydz\\
&=\int_0^a\int_0^b\int_0^c \sin\frac{q\pi x}{a}\sin\frac{s\pi y}{b}\sin\frac{t\pi z}{c}\left[\sum_{i=1}^{m}\sum_{j=1}^{n}\sum_{k=1}^{r}d_{ijk}\sin\frac{i\pi x}{a}\sin\frac{j\pi y}{b}\sin\frac{k\pi z}{c}\right]dxdydz\\
&=\sum_{i=1}^{m}\sum_{j=1}^{n}\sum_{k=1}^{r}d_{ijk}\int_0^a\int_0^b\int_0^c \sin\frac{q\pi x}{a}\sin\frac{s\pi y}{b}\sin\frac{t\pi z}{c}\sin\frac{i\pi x}{a}\sin\frac{j\pi y}{b}\sin\frac{k\pi z}{c}dxdydz\\
&=\sum_{i=1}^{m}\sum_{j=1}^{n}\sum_{k=1}^{r}d_{ijk}\underbrace{\left[\int_0^a \sin\frac{q\pi x}{a}\sin\frac{i\pi x}{a}dx\right]}_{=a/2\ \mathit{If}\ i=q}\underbrace{\left[\int_0^b \sin\frac{s\pi y}{b}\sin\frac{j\pi y}{b}dy\right]}_{=b/2\ \mathit{If}\ j=s}\underbrace{\left[\int_0^c \sin\frac{t\pi z}{c}\sin\frac{k\pi z}{c}dz\right]}_{=c/2\ \mathit{If}\ k=t}\\
&=d_{qst}\frac{abc}{8}
\end{aligned} \tag{7.152}$$

$$\rightarrow d_{qst}=\frac{8}{abc}\int_0^a\int_0^b\int_0^c \sin\frac{q\pi x}{a}\sin\frac{s\pi y}{b}\sin\frac{t\pi z}{c}f(x,y,z)dxdydz$$

Changing indices, d_{qst} can be written as d_{ijk} as below:

$$d_{ijk}=\frac{8}{abc}\int_0^a\int_0^b\int_0^c \sin\frac{i\pi x}{a}\sin\frac{j\pi y}{b}\sin\frac{k\pi z}{c}f(x,y,z)dxdydz \tag{7.153}$$

Changing i, j, k indices in Eq. (7.153), all d_{ijk} coefficients can be easily obtained.

Example 7.20: Obtain the approximation of $f(x, y, z) = x^2 + y^2 + z^2$ function in the ranges of $x = [0, a] = [0, 2]$, $y = [0, b] = [0, 2]$, and $z = [0, c] = [0, 2]$ using sine functions.

For simplicity, we can construct the approximate function based on triple sine series using two terms in x direction, two terms in y direction, and two terms in z direction in Eq. (7.151) as following:

$$f(x,y,z)=\sum_{i=1}^{2}\sum_{j=1}^{2}\sum_{k=1}^{2}d_{ijk}\Psi_{ix}(x)\Psi_{jy}(y)\Psi_{kz}(z)=\sum_{i=1}^{2}\sum_{j=1}^{2}\sum_{k=1}^{2}d_{ijk}\sin\frac{i\pi x}{a}\sin\frac{j\pi y}{b}\sin\frac{k\pi z}{c}$$

Taking $a = 2$, $b = 2$, and $c = 2$, approximate function $f(x, y, z)$ can be written in expanded form as below:

$$
\begin{aligned}
f(x,y,z) &= \sum_{i=1}^{2}\sum_{j=1}^{2}\sum_{k=1}^{2} d_{ijk}\sin\frac{i\pi x}{2}\sin\frac{j\pi y}{2}\sin\frac{k\pi z}{2} \\
&= d_{111}\sin\frac{\pi x}{2}\sin\frac{\pi y}{2}\sin\frac{\pi z}{2} + d_{112}\sin\frac{\pi x}{2}\sin\frac{\pi y}{2}\sin\frac{2\pi z}{2} \\
&\quad + d_{121}\sin\frac{\pi x}{2}\sin\frac{2\pi y}{2}\sin\frac{\pi z}{2} + d_{122}\sin\frac{\pi x}{2}\sin\frac{2\pi y}{2}\sin\frac{2\pi z}{2} \\
&\quad + d_{211}\sin\frac{2\pi x}{2}\sin\frac{\pi y}{2}\sin\frac{\pi z}{2} + d_{212}\sin\frac{2\pi x}{2}\sin\frac{\pi y}{2}\sin\frac{2\pi z}{2} \\
&\quad + d_{221}\sin\frac{2\pi x}{2}\sin\frac{2\pi y}{2}\sin\frac{\pi z}{2} + d_{222}\sin\frac{2\pi x}{2}\sin\frac{2\pi y}{2}\sin\frac{2\pi z}{2}
\end{aligned}
$$

Using general expression $d_{ijk} = \frac{8}{abc}\int_0^a\int_0^b\int_0^c \sin\frac{i\pi x}{a}\sin\frac{j\pi y}{b}\sin\frac{k\pi z}{c} f(x,y,z)dxdydz$ in Eq. (7.153), $d_{111}, d_{112}, d_{121}, d_{122}, d_{211}, d_{212}, d_{221}, d_{222}$ coefficients can be obtained as following:

$$d_{111} = \int_0^2\int_0^2\int_0^2 \sin\frac{\pi x}{2}\sin\frac{\pi y}{2}\sin\frac{\pi z}{2}\left(x^2+y^2+z^2\right)dxdydz = 7.3653,$$

$$d_{112} = \int_0^2\int_0^2\int_0^2 \sin\frac{\pi x}{2}\sin\frac{\pi y}{2}\sin\frac{2\pi z}{2}\left(x^2+y^2+z^2\right)dxdydz = -2.0641,$$

$$d_{121} = \int_0^2\int_0^2\int_0^2 \sin\frac{\pi x}{2}\sin\frac{2\pi y}{2}\sin\frac{\pi z}{2}\left(x^2+y^2+z^2\right)dxdydz = -2.0641,$$

$$d_{122} = \int_0^2\int_0^2\int_0^2 \sin\frac{\pi x}{2}\sin\frac{2\pi y}{2}\sin\frac{2\pi z}{2}\left(x^2+y^2+z^2\right)dxdydz = 0,$$

$$d_{211} = \int_0^2\int_0^2\int_0^2 \sin\frac{2\pi x}{2}\sin\frac{\pi y}{2}\sin\frac{\pi z}{2}\left(x^2+y^2+z^2\right)dxdydz = -2.0641,$$

$$d_{212} = \int_0^2\int_0^2\int_0^2 \sin\frac{2\pi x}{2}\sin\frac{\pi y}{2}\sin\frac{2\pi z}{2}\left(x^2+y^2+z^2\right)dxdydz = 0,$$

$$d_{221} = \int_0^2\int_0^2\int_0^2 \sin\frac{2\pi x}{2}\sin\frac{2\pi y}{2}\sin\frac{\pi z}{2}\left(x^2+y^2+z^2\right)dxdydz = 0,$$

$$d_{222} = \int_0^2 \int_0^2 \int_0^2 \sin\frac{2\pi x}{2}\sin\frac{2\pi y}{2}\sin\frac{2\pi z}{2}\left(x^2+y^2+z^2\right)dxdydz = 0.$$

Thus, approximation of $f(x, y, z) = x^2 + y^2 + z^2$ function is obtained as below:

$$f(x,y,z) = 7.3653\sin\frac{\pi x}{2}\sin\frac{\pi y}{2}\sin\frac{\pi z}{2} - 2.0641\sin\frac{\pi x}{2}\sin\frac{\pi y}{2}\sin\frac{2\pi z}{2}$$
$$-2.0641\sin\frac{\pi x}{2}\sin\frac{2\pi y}{2}\sin\frac{\pi z}{2} - 2.0641\sin\frac{2\pi x}{2}\sin\frac{\pi y}{2}\sin\frac{\pi z}{2}$$

Approximate function can be made closer to the true function by including more terms in triple sine series.

7.16.2 Approximation with Legendre Polynomials

Legendre approximation of a trivariate $f(x,y,z)$ function on a cube defined by $x = [-1, 1]$, $y = [-1, 1]$ and $z = [-1, 1]$ intervals as shown in Fig. 7.60 can be created with product of univariate Legendre approximate functions of $X(x)$, $Y(y)$, and $Z(z)$ as below:

$$f(x,y,z) = X(x)\otimes Y(y)\otimes Z(z) = \left[a_0L_{0x}(x)+a_1L_{1x}(x)+a_2L_{2x}(x)+\ldots+a_mL_{mx}(x)\right]$$
$$\otimes\left[b_0L_{0y}(y)+b_1L_{1y}(y)+b_2L_{2y}(y)+\ldots+b_nL_{ny}(y)\right]$$
$$\otimes\left[c_0L_{0z}(z)+c_1L_{1z}(z)+c_2L_{2z}(z)+\ldots+c_rL_{rz}(z)\right] = \sum_{i=0}^{m}\sum_{j=0}^{n}\sum_{k=0}^{r} d_{ijk}L_{ix}(x)L_{jy}(y)L_{kz}(z) \quad (7.154)$$

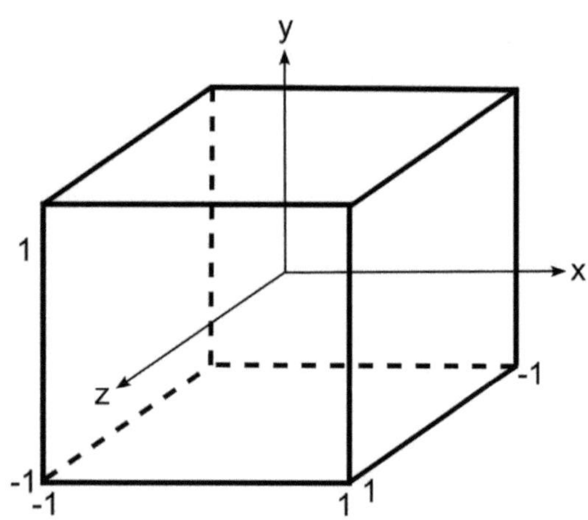

Fig. 7.60 Approximation of trivariate $f(x, y, z)$ function in a cube domain determined by $[-1, 1]$ intervals in x, y, and z directions

In Eq. (7.142), L_x, L_y, L_z indicate Legendre polynomials and d_{ijk} indicates unknown constants.

d_{ijk} coefficients in Eq. (7.154) can be calculated using the orthogonality condition of Legendre polynomials. Multiplication of Eq. (7.154) by $L_{qx}(x)L_{sy}(y)L_{tz}(z)$ and then integration in the ranges of $[-1, 1]$ in x direction, $[-1, 1]$ in y direction and $[-1, 1]$ in z direction along with the orthogonality condition yield d_{qst} coefficient as following:

$$\begin{aligned}
&\int_{-1}^{1}\int_{-1}^{1}\int_{-1}^{1} L_{qx}(x)L_{sy}(y)L_{tz}(z)f(x,y,z)dxdydz = \\
&\int_{-1}^{1}\int_{-1}^{1}\int_{-1}^{1} L_{qx}(x)L_{sy}(y)L_{tz}(z)\left[\sum_{i=1}^{m}\sum_{j=1}^{n}\sum_{k=1}^{r} d_{ijk}L_{ix}(x)L_{jy}(y)L_{kz}(z)\right]dxdydz \\
&= \sum_{i=1}^{m}\sum_{j=1}^{n}\sum_{k=1}^{r} d_{ijk}\int_{-1}^{1}\int_{-1}^{1}\int_{-1}^{1} L_{qx}(x)L_{sy}(y)L_{tz}(z)L_{ix}(x)L_{jy}(y)L_{kz}(z)dxdydz \\
&= \sum_{i=1}^{m}\sum_{j=1}^{n}\sum_{k=1}^{r} d_{ijk}\underbrace{\left[\int_{-1}^{1} L_{qx}(x)L_{ix}(x)dx\right]}_{=\frac{2q+1}{2}\quad If\quad i=q}\underbrace{\left[\int_{-1}^{1} L_{sy}(y)L_{jy}(y)dy\right]}_{=\frac{2s+1}{2}\quad If\quad j=s}\underbrace{\left[\int_{-1}^{1} L_{tz}(z)L_{kz}(z)dz\right]}_{=\frac{2t+1}{2}\quad If\quad k=t} \\
&= d_{qst}\left(\frac{2q+1}{2}\right)\left(\frac{2s+1}{2}\right)\left(\frac{2t+1}{2}\right)
\end{aligned} \tag{7.155}$$

$$\rightarrow d_{qst} = \left(\frac{2}{2q+1}\right)\left(\frac{2}{2s+1}\right)\left(\frac{2}{2t+1}\right)\int_{-1}^{1}\int_{-1}^{1} \times \int_{-1}^{1} L_{qx}(x)L_{sy}(y)L_{tz}(z)f(x,y,z)dxdydz$$

Changing indices d_{qst} can also be expressed as d_{ijk} as below:

$$d_{ijk} = \left(\frac{2}{2i+1}\right)\left(\frac{2}{2k+1}\right)\left(\frac{2}{2k+1}\right)\int_{-1}^{1}\int_{-1}^{1}\int_{-1}^{1} L_{ix}(x)L_{jy}(y)L_{kz}(z)f(x,y,z)dxdydz \tag{7.156}$$

Changing i, j, k indices in Eq. (7.156), all coefficients such as d_{111}, d_{112}, and d_{121} can be easily obtained.

Note: Legendre approximation function was constructed in normalized interval of $[-1, 1]$. If the original intervals are different, they can be converted into $[-1, 1]$

intervals using variable transformations. For example, if the original intervals are given as $x = [a, b]$, $y = [c, d]$, and $z = [e, g]$, they can be converted into $[-1, 1]$ intervals using variable transformation as below (New variables are defined for transformations, for example: ξ for x, η for y, and ζ for z):

$$x=[a,b]\rightarrow\xi=[-1,1]:x=\frac{(1-\xi)}{2}a+\frac{(1+\xi)}{2}b$$

$$y=[c,d]\rightarrow\eta=[-1,1]:y=\frac{(1-\eta)}{2}c+\frac{(1+\eta)}{2}d$$

$$z=[e,g]\rightarrow\zeta=[-1,1]:z=\frac{(1-\zeta)}{2}e+\frac{(1+\zeta)}{2}g$$

Example 7.21: Obtain the approximation of $f(x) = \sin x \sin y \sin z$ function in the ranges of $x = [-1, 1]$, $y = [-1, 1]$, and $z = [-1, 1]$ using Legendre polynomials.

For simplicity, we can construct the approximate Legendre polynomial function using two terms in x, y, and z directions in Eq. (7.154) as following:

$$\begin{aligned} f(x,y,z) &= \sum_{i=0}^{1}\sum_{j=0}^{1}\sum_{k=0}^{1} d_{ijk}L_{ix}(x)L_{jy}(y)L_{kz}(z) \\ &= d_{000}L_{0x}(x)L_{0y}(y)L_{0z}(z) + d_{001}L_{0x}(x)L_{0y}(y)L_{1z}(z) \\ &\quad + d_{010}L_{0x}(x)L_{1y}(y)L_{0z}(z) + d_{011}L_{0x}(x)L_{1y}(y)L_{1z}(z) \\ &\quad + d_{100}L_{1x}(x)L_{0y}(y)L_{0z}(z) + d_{101}L_{1x}(x)L_{0y}(y)L_{1z}(z) \\ &\quad + d_{110}L_{1x}(x)L_{1y}(y)L_{0z}(z) + d_{111}L_{1x}(x)L_{1y}(y)L_{1z}(z) \end{aligned}$$

where Legendre polynomials in $f(x, y)$ function are given as $L_{0x}(x) = 1$, $L_{1x}(x) = x$, $L_{0y}(y) = 1$, $L_{1y}(y) = y$, $L_{0z}(z) = 1$, $L_{1z}(z) = z$.

d_{000}, d_{001}, d_{010}, d_{011}, d_{100}, d_{101}, d_{110}, d_{111} coefficients in $f(x, y, z)$ function can be calculated using the general expression of $d_{ijk} = \left(\frac{2}{2i+1}\right)\left(\frac{2}{2k+1}\right)\left(\frac{2}{2k+1}\right)\int_{-1}^{1}\int_{-1}^{1} \times \int_{-1}^{1} L_{ix}(x)L_{jy}(y)L_{kz}(z)f(x,y,z)dxdydz$ in Eq. (7.156) as following:

$$d_{000} = 8\int_{-1}^{1}\int_{-1}^{1}\int_{-1}^{1}(\sin x)(\sin y)(\sin z)dxdydz = 0,$$

$$d_{001} = \left(\frac{8}{3}\right)\int_{-1}^{1}\int_{-1}^{1}\int_{-1}^{1}(\sin x)(\sin y)z(\sin z)dxdydz = 0,$$

$$d_{010} = \left(\frac{8}{3}\right) \int_{-1}^{1} \int_{-1}^{1} \int_{-1}^{1} (\sin x) y (\sin y)(\sin z) dxdydz = 0,$$

$$d_{011} = \left(\frac{8}{9}\right) \int_{-1}^{1} \int_{-1}^{1} \int_{-1}^{1} (\sin x) y (\sin y) z (\sin z) dxdydz = 0,$$

$$d_{100} = \left(\frac{8}{3}\right) \int_{-1}^{1} \int_{-1}^{1} \int_{-1}^{1} x (\sin x)(\sin y)(\sin z) dxdydz = 0,$$

$$d_{101} = \left(\frac{8}{9}\right) \int_{-1}^{1} \int_{-1}^{1} \int_{-1}^{1} x (\sin x)(\sin y) z (\sin z) dxdydz = 0,$$

$$d_{110} = \left(\frac{8}{9}\right) \int_{-1}^{1} \int_{-1}^{1} \int_{-1}^{1} x (\sin x) y (\sin y)(\sin z) dxdydz = 0,$$

$$d_{111} = \left(\frac{8}{27}\right) \int_{-1}^{1} \int_{-1}^{1} \int_{-1}^{1} x (\sin x) y (\sin y) z (\sin z) dxdydz = 0.0648.$$

Thus, approximation of $f(x, y, z) =$ sin x. sin y. sin z function is obtained as below:

$$f(x, y, z) = 0.0648xyz$$

Approximation quality can be increased by including more terms in the approximate function.

7.16.3 Approximation with Chebyshev Polynomials

Chebyshev approximation of a trivariate $f(x,y,z)$ function on a cube defined by $x = [-1, 1]$, $y = [-1, 1]$ and $z = [-1, 1]$ intervals as shown in Fig. 7.60 can be created with product of univariate Chebyshev approximate functions of $X(x)$, $Y(y)$, and $Z(z)$ as below:

$$
\begin{aligned}
f(x,y,z) &= X(x) \bigotimes Y(y) \bigotimes Z(z) \\
&= [a_0 T_{0x}(x) + a_1 T_{1x}(x) + a_2 T_{2x}(x) + \ldots + a_m T_{mx}(x)] \bigotimes \Big[b_0 T_{0y}(y) \\
&\quad + b_1 T_{1y}(y) + b_2 T_{2y}(y) + \ldots + b_n T_{ny}(y) \Big] \\
&\bigotimes [c_0 T_{0z}(z) + c_1 T_{1z}(z) + c_2 T_{2z}(z) + \ldots + c_r T_{rz}(z)] \\
&= \sum_{i=0}^{m} \sum_{j=0}^{n} \sum_{k=0}^{r} d_{ijk} T_{ix}(x) T_{jy}(y) T_{kz}(z)
\end{aligned}
\tag{7.157}
$$

In Eq. (7.157), T_x, T_y, T_z indicate Chebyshev polynomials and d_{ijk} indicates unknown constants. There are $(m + 1)(n + 1)(r + 1)$ unknown coefficients in Eq. (7.157). Necessary equations to find these coefficients can be created expressing the approximate function at roots of $(m + 1)$-th order Chebyshev polynomials in x direction, $(n + 1)$-th order Chebyshev polynomials in y direction and $(r + 1)$-th order Chebyshev polynomials in z direction. Roots of $(m + 1)$-th order Chebyshev polynomials in x direction, $(n + 1)$-th order Chebyshev polynomials in y direction and $(r + 1)$-th order Chebyshev polynomials in z direction can be expressed as following:

$$
x_i = -\cos\left(\frac{(2i-1)\pi}{2(m+1)}\right) \; i=1,2,..,m+1, \quad y_j = -\cos\left(\frac{(2j-1)\pi}{2(n+1)}\right) j=1,2,..,n+1,
$$

$$
z_k = -\cos\left(\frac{(2k-1)\pi}{2(r+1)}\right) \quad k=1,2,..,r+1 \tag{7.158}
$$

Using root values, orthogonality condition of Chebyshev polynomials can be expressed in discrete form in x, y, and z directions as below:

$$
\sum_{i=1}^{m+1} T_{px}(x_i) T_{qx}(x_i) = \begin{cases} 0, & \text{if} \quad q \neq p \\ m+1, & \text{if} \quad q = p = 0 \\ \dfrac{m+1}{2}, & \text{if} \quad q = p \neq 0 \end{cases}, \tag{7.159a}
$$

$$
\sum_{j=1}^{n+1} T_{py}(y_j) T_{qy}(y_j) = \begin{cases} 0, & \text{if} \quad q \neq p \\ n+1, & \text{if} \quad q = p = 0 \\ \dfrac{n+1}{2}, & \text{if} \quad q = p \neq 0 \end{cases}, \tag{7.159b}
$$

$$
\sum_{k=1}^{r+1} T_{pz}(z_k) T_{qz}(z_k) = \begin{cases} 0, & \text{if} \quad q \neq p \\ r+1, & \text{if} \quad q = p = 0 \\ \dfrac{r+1}{2}, & \text{if} \quad q = p \neq 0 \end{cases}, \tag{7.159c}
$$

After expressing the Chebyshev approximate function at root values multiplied with $T_{qx}(x_i)T_{sy}(y_j)T_{tz}(z_k)$ and then summing them, d_{qst} coefficient can be obtained as following:

$$d_{qst} = \frac{2}{m+1}\frac{2}{n+1}\frac{2}{r+1}\sum_{i=1}^{m+1}\sum_{j=1}^{n+1}\sum_{k=1}^{r+1} T_{qx}(x_i)T_{sy}(y_j)T_{tz}(z_k)f(x_i, y_j, z_k) \ (q, s, t \neq 0) \tag{7.160}$$

d_{qst} coefficient in Eq. (7.160) can be specialized to other coefficients as below:

$$d_{0st} = \frac{1}{m+1}\frac{2}{n+1}\frac{2}{r+1}\sum_{i=1}^{m+1}\sum_{j=1}^{n+1}\sum_{k=1}^{r+1} T_{0x}(x_i)T_{sy}(y_j)T_{tz}(z_k)f(x_i, y_j, z_k)(s, t \neq 0) \tag{7.161a}$$

$$d_{q0t} = \frac{2}{m+1}\frac{1}{n+1}\frac{2}{r+1}\sum_{i=1}^{m+1}\sum_{j=1}^{n+1}\sum_{k=1}^{r+1} T_{qx}(x_i)T_{0y}(y_j)T_{tz}(z_k)f(x_i, y_j, z_k) \ (q, t \neq 0) \tag{7.161b}$$

$$d_{qs0} = \frac{2}{m+1}\frac{2}{n+1}\frac{1}{r+1}\sum_{i=1}^{m+1}\sum_{j=1}^{n+1}\sum_{k=1}^{r+1} T_{qx}(x_i)T_{sy}(y_j)T_{0z}(z_k)f(x_i, y_j, z_k) \ (q, s \neq 0) \tag{7.161c}$$

Note: Chebyshev approximation function was constructed in the normalized interval of [−1, 1]. If the original intervals are different, they can be converted into [−1, 1] intervals using variable transformations. For example, if the original intervals are given as $x = [a, b]$, $y = [c, d]$, and $z = [e, g]$, they can be converted into standard [−1, 1] intervals using the variable transformation as below (New variables are defined for transformations, for example: ξ for x, η for y, and ζ for z)

$$x = [a, b] \rightarrow \xi = [-1, 1] : x = \frac{(1-\xi)}{2}a + \frac{(1+\xi)}{2}b$$

$$y = [c, d] \rightarrow \eta = [-1, 1] : y = \frac{(1-\eta)}{2}c + \frac{(1+\eta)}{2}d$$

$$z = [e, g] \rightarrow \zeta = [-1, 1] : z = \frac{(1-\zeta)}{2}e + \frac{(1+\zeta)}{2}g$$

Example 7.22: Obtain the approximation of $f(x, y, z) = \sin x \sin y \sin z$ function in the ranges of $x = [-1, 1]$, $y = [-1, 1]$, and $z = [-1, 1]$ using Chebyshev polynomials.

A simple approximation for $f(x, y, z) = \sin x \sin y \sin z$ function can be created by using two terms in x, y, and z directions in Eq. (7.157) as below:

$$
\begin{aligned}
f(x,y,z) &= \sum_{i=0}^{m=1}\sum_{j=0}^{n=1}\sum_{k=0}^{r=1} d_{ijk}T_{ix}(x)T_{jy}(y)T_{kz}(z)\\
&= d_{000}T_{0x}(x)T_{0y}(y)T_{0z}(z) + d_{001}T_{0x}(x)T_{0y}(y)T_{1z}(z)\\
&\quad + d_{010}T_{0x}(x)T_{1y}(y)T_{0z}(z) + d_{011}T_{0x}(x)T_{1y}(y)T_{1z}(z)\\
&\quad + d_{100}T_{1x}(x)T_{0y}(y)T_{0z}(z) + d_{101}T_{1x}(x)T_{0y}(y)T_{1z}(z)\\
&\quad + d_{110}T_{1x}(x)T_{1y}(y)T_{0z}(z) + d_{111}T_{1x}(x)T_{1y}(y)T_{1z}(z)
\end{aligned}
$$

There are $2 \times 2 \times 2 = 8$ unknown coefficients to be determined in $f(x, y, z)$ function. Necessary equations can be created by expressing the approximate function at roots of second-order Chebyshev polynomials in x, y, and z directions. Roots of second-order Chebyshev polynomials in x, y, and z directions can be expressed as following:

$$
x_i = -\cos\left(\frac{(2i-1)\pi}{2(m+1)}\right)(i=1,2),\quad y_j = -\cos\left(\frac{(2j-1)\pi}{2(n+1)}\right)(j=1,2),
$$
$$
z_k = -\cos\left(\frac{(2k-1)\pi}{2(r+1)}\right)(k=1,2)
$$

After calculating Chebyshev polynomials $T_{0x}(x) = 1$, $T_{1x}(x) = x$, $T_{0y}(y) = 1$, $T_{1y}(y) = y$, $T_{0z}(z) = 1$, $T_{1z}(z) = z$ at root values and then substituting them into Eqs. (7.161a)–(7.161c), $d_{000}, d_{001}, d_{010}, d_{011}, d_{100}, d_{101}, d_{110}, d_{111}$ coefficients can be obtained as below:

$$
d_{000} = \frac{1}{2}\frac{1}{2}\frac{1}{2}\sum_{i=1}^{2}\sum_{j=1}^{2}\sum_{k=1}^{2}\underbrace{(\sin x_i)(\sin y_j)(\sin z_k)}_{f(x_i,y_j,z_k)} = 0,
$$

$$
d_{001} = \frac{1}{2}\frac{1}{2}\frac{2}{2}\sum_{i=1}^{2}\sum_{j=1}^{2}\sum_{k=1}^{2}(z_k)\underbrace{(\sin x_i)(\sin y_j)(\sin z_k)}_{f(x_i,y_j,z_k)} = 0,
$$

$$
d_{010} = \frac{1}{2}\frac{2}{2}\frac{1}{2}\sum_{i=1}^{2}\sum_{j=1}^{2}\sum_{k=1}^{2}(y_j)\underbrace{(\sin x_i)(\sin y_j)(\sin z_k)}_{f(x_i,y_j,z_k)} = 0,
$$

$$
d_{011} = \frac{1}{2}\frac{2}{2}\frac{2}{2}\sum_{i=1}^{2}\sum_{j=1}^{2}\sum_{k=1}^{2}(y_j)(z_k)\underbrace{(\sin x_i)(\sin y_j)(\sin z_k)}_{f(x_i,y_j,z_k)} = 0,
$$

$$d_{100} = \frac{2}{2}\frac{1}{2}\frac{1}{2}\sum_{i=1}^{2}\sum_{j=1}^{2}\sum_{k=1}^{2}(x_i)\underbrace{(\sin x_i)(\sin y_j)(\sin z_k)}_{f(x_i, y_j, z_k)} = 0,$$

$$d_{101} = \frac{2}{2}\frac{1}{2}\frac{2}{2}\sum_{i=1}^{2}\sum_{j=1}^{2}\sum_{k=1}^{2}(x_i)(z_k)\underbrace{(\sin x_i)(\sin y_j)(\sin z_k)}_{f(x_i, y_j, z_k)} = 0,$$

$$d_{110} = \frac{2}{2}\frac{2}{2}\frac{1}{2}\sum_{i=1}^{2}\sum_{j=1}^{2}\sum_{k=1}^{2}(x_i)(y_j)\underbrace{(\sin x_i)(\sin y_j)(\sin z_k)}_{f(x_i, y_j, z_k)} = 0,$$

$$d_{111} = \frac{2}{2}\frac{2}{2}\frac{2}{2}\sum_{i=1}^{2}\sum_{j=1}^{2}\sum_{k=1}^{2}(x_i)(y_j)(z_k)\underbrace{(\sin x_i)(\sin y_j)(\sin z_k)}_{f(x_i, y_j, z_k)} = 0.7775.$$

Thus, approximation of $f(x, y, z) = \sin x. \sin y. \sin z$ function is obtained as below:

$$f(x, y, z) = 0.7755xyz$$

Approximation quality can be increased by including more terms in the approximate function.

7.17 Summary

- Prediction of an intermediate value in a data set by passing a function (which reflects the character of the data set in a balanced way is called as curve fitting. Curve fitting is also known as Regression.
- In curve fitting, data is implicitly considered to contain some errors such as measurement errors. Therefore, in curve fitting, prediction function is not passed through all data as in interpolation. In curve fitting, prediction function is passed through the data set in balanced manner using Least Squares method.
- Prediction functions in curve fitting vary widely, from simple polynomials to more complex functions (exponential, trigonometric, and nested functions).
- Use of polynomial functions in curve fitting is often called as Linear Regression. Coefficients of polynomial in linear regression are obtained by solving a set of linear equations.
- In linear regression, low-order polynomials such as linear or parabolic polynomials are often used. These polynomials are called as Response Surface models in multivariate cases.
- Response surface models can be easily created using design of experiment methods.

- Factorial design of experiment methods can be used to create all kinds of response surface models.
- Central Composite and Box-Behnken (when number of variables is higher than 2) designs can be used to create quadratic (parabolic) response surface models.
- Two-variable quadratic response surface model can be expressed in general form as $\widetilde{f}(x,y) = c_1 + c_2x + c_3y + c_4x^2 + c_5y^2 + c_6xy$.
- Model constants in quadratic response surface model can be easily obtained as $C = [X^TX]^{-1}X^TF$ where X and F are given as below:

$$X = \begin{bmatrix} 1 & x_1 & y_1 & x_1^2 & y_1^2 & x_1y_1 \\ 1 & x_2 & y_2 & x_2^2 & y_2^2 & x_2y_2 \\ 1 & x_3 & y_3 & x_3^2 & y_3^2 & x_3y_3 \\ . & . & . & . & . & . \\ 1 & x_n & y_n & x_n^2 & y_n^2 & x_ny_n \end{bmatrix}, F = \begin{Bmatrix} f_1 \\ f_2 \\ f_3 \\ . \\ f_n \end{Bmatrix}$$

R^2 criterion is widely used to determine fitting quality of regression models.

- Effect of variables on the fitting function or the data set can be determined quantitatively with Analysis of Variance (ANOVA) method.
- Effect of non-numeric variables (such as gender, color, taste) on the fitting function or the data set can be examined with ANOVA.
- In two-variable case, effect of variables (factors) on function can be expressed with ANOVA as following:

For x variable: $P_x = \frac{S_x}{S_t}$, for y variable: $P_y = \frac{S_y}{S_t}$, for error: $P_e = \frac{S_e}{S_t}$ where S_x, S_y, S_T, S_e are defined as below:

$$S_x = p\underbrace{\left(f_{x1,.m} - f_m\right)^2}_{x1 \quad level} + q\underbrace{\left(f_{x2,m} - f_m\right)^2}_{x2 \quad level} + \ldots + r\underbrace{\left(f_{xk,m} - f_m\right)^2}_{xk \quad level}$$

$$S_y = s\underbrace{\left(f_{y1,.m} - f_m\right)^2}_{y1 \quad level} + t\underbrace{\left(f_{y2,m} - f_m\right)^2}_{y2 \quad level} + \ldots + z\underbrace{\left(f_{yl,m} - f_m\right)^2}_{yl \quad level}$$

$$S_T = S_x + S_y + S_e \rightarrow S_e = S_T - \left(S_x + S_y\right)$$

Inverse distance method is one of the simplest methods to predict intermediate values. In this method, intermediate value is predicted with weighted sum of the function values in a narrow region which is often determined with a radius of R as following (d_j: distance between i and j points):

$$f_i = \frac{\frac{1}{d_1^p} f_1 + \frac{1}{d_2^p} f_2 + \frac{1}{d_3^p} f_3 + \frac{1}{d_4^p} f_4}{\frac{1}{d_1^p} + \frac{1}{d_2^p} + \frac{1}{d_3^p} + \frac{1}{d_4^p}} = \frac{\sum_{j=1}^{4} \frac{1}{d_j^p} f_j}{\sum_{j=1}^{4} \frac{1}{d_j^p}}$$

Kriging method combines regression and statistical methods to predict intermediate values. Several Kriging methods were developed such as Ordinary and Universal Kriging methods based on different simplifying assumptions.

- In Kriging methods, intermediate value $\tilde{Z}_0$ at an s_0 point can be obtained as the weighted sum of variable values multiplied with weight coefficients as following:

$$\tilde{Z}_0 = \tilde{Z}(s_0) = w_1 Z_1 + w_2 Z_2 + w_3 Z_3 + \ldots + w_n Z_n = \sum_{i=1}^{n} w_i Z_i$$

In Ordinary Kriging method, weight coefficients (w_i) are determined from the solution of the following matrix equation:

$$\underbrace{\begin{bmatrix} 0 & \gamma(s_1 - s_2) & \gamma(s_1 - s_3) & \ldots & \gamma(s_1 - s_n) & 1 \\ \gamma(s_2 - s_1) & 0 & \gamma(s_2 - s_3) & \ldots & \gamma(s_2 - s_n) & 1 \\ \gamma(s_3 - s_1) & \gamma(s_3 - s_2) & 0 & \ldots & \gamma(s_3 - s_n) & 1 \\ \ldots & \ldots & \ldots & \ldots & \ldots & \ldots \\ \gamma(s_n - s_1) & \gamma(s_n - s_2) & \gamma(s_n - s_3) & \ldots & 0 & 1 \\ 1 & 1 & 1 & \ldots & 1 & 0 \end{bmatrix}}_{A_{mat}} \underbrace{\begin{Bmatrix} w_1 \\ w_2 \\ w_3 \\ \ldots \\ w_n \\ \lambda \end{Bmatrix}}_{W_{mat}} = \underbrace{\begin{Bmatrix} \gamma(s_1 - s_0) \\ \gamma(s_2 - s_0) \\ \gamma(s_3 - s_0) \\ \ldots \\ \gamma(s_n - s_0) \\ 1 \end{Bmatrix}}_{B_{mat}}$$

In Universal Kriging method, weight coefficients (w_i) are determined from the solution of the following matrix equation:

$$\underbrace{\begin{bmatrix} 0 & \gamma(s_1-s_2) & \gamma(s_1-s_3) & \ldots & \gamma(s_1-s_n) & 1 & f_1(s_1) & f_2(s_1) & \ldots & f_p(s_1) \\ \gamma(s_2-s_1) & 0 & \gamma(s_2-s_3) & \ldots & \gamma(s_2-s_n) & 1 & f_1(s_2) & f_2(s_2) & \ldots & f_p(s_2) \\ \gamma(s_3-s_1) & \gamma(s_3-s_2) & 0 & \ldots & \gamma(s_3-s_n) & 1 & f_1(s_3) & f_2(s_3) & \ldots & f_p(s_3) \\ \ldots & \ldots & \ldots & \ldots & \ldots & \ldots & \ldots & \ldots & \ldots & \ldots \\ \gamma(s_n-s_1) & \gamma(s_n-s_2) & \gamma(s_n-s_3) & \ldots & 0 & 1 & f_1(s_n) & f_2(s_n) & \ldots & f_p(s_n) \\ 1 & 1 & 1 & \ldots & 1 & 0 & 0 & 0 & \ldots & 0 \\ f_1(s_1) & f_1(s_2) & f_1(s_3) & \ldots & f_1(s_n) & 0 & 0 & 0 & \ldots & 0 \\ f_2(s_1) & f_2(s_2) & f_2(s_3) & \ldots & f_2(s_n) & 0 & 0 & 0 & \ldots & 0 \\ \ldots & \ldots & \ldots & \ldots & \ldots & \ldots & \ldots & \ldots & \ldots & \ldots \\ f_p(s_1) & f_p(s_2) & f_p(s_3) & \ldots & f_p(s_n) & 0 & 0 & 0 & \ldots & 0 \end{bmatrix}}_{A_{mat}} \cdot \underbrace{\begin{Bmatrix} w_1 \\ w_2 \\ w_3 \\ \ldots \\ w_n \\ \lambda_0 \\ \lambda_1 \\ \lambda_2 \\ \ldots \\ \lambda_p \end{Bmatrix}}_{W_{mat}} = \underbrace{\begin{Bmatrix} \gamma(s_0-s_1) \\ \gamma(s_0-s_2) \\ \gamma(s_0-s_3) \\ \ldots \\ \gamma(s_0-s_n) \\ 1 \\ f_1(s_0) \\ f_2(s_0) \\ \ldots \\ f_p(s_0) \end{Bmatrix}}_{B_{mat}}$$

Fitting of non-polynomial functions that capture the character of the data set better is often called as Nonlinear Regression. In nonlinear regression, coefficients of fitting model are obtained using iterative methods.

- Artificial Neural Network (ANN) models are widely used in fitting models to complicated data sets. ANN models include nested functions and lead to nonlinear regression. Because of nested functions, ANN models cannot be expressed explicitly. Rather, ANN models are represented with an architecture consisting of layers and neurons.
- Determination of weight coefficients between neurons iteratively in ANN models is called as training of ANN model. Training of an ANN model can be time-consuming process due to iterative solution process.
- In the solution of engineering problems, complicated functions are often replaced with simpler ones. This procedure can be referred to fitting a simple function to a more complicated function, shortly fitting function to function.
- Fourier series, Legendre and Chebyshev polynomials are commonly used in the approximation of complicated functions.
- Fourier series approximation of an *f*(*x*) function is expressed as $f(x) = \frac{a_0}{2} + \sum_{k=1}^{n} (a_k \cos kx + b_k \sin kx)$. a_0, a_k, and b_k coefficients can be determined using orthogonality condition of trigonometric functions.
- Legendre approximation of an *f*(*x*) function can be constructed using orthogonal Legendre polynomials ($L_i(x)$) as following:

$$f(x) \approx a_0 L_0(x) + a_1 L_1(x) + a_2 L_2(x) + \ldots + a_n L_n(x) = \sum_{k=0}^{n} a_k L_k(x)$$

a_k coefficients can be determined using orthogonality properties of Legendre polynomials as $a_k = \frac{2k+1}{2} \int_{-1}^{1} f(x) L_k(x) dx$.

Chebyshev approximation of an *f*(*x*) function can be constructed using orthogonal Chebyshev polynomials ($T_i(x)$) as following:

$$f(x) \approx a_0 T_0(x) + a_1 T_1(x) + a_2 T_2(x) + \ldots + a_k T_k(x) + \ldots + a_n T_n(x) = \sum_{k=0}^{n} a_k T_k(x)$$

a_k coefficients can be determined using orthogonality properties of Chebyshev polynomials as $a_k = \frac{2}{\pi} \int_{-1}^{1} f(x) \frac{1}{\sqrt{1-x^2}} T_k(x) dx$.

Approximation of multivariable functions on regular domains can be constructed multiplying single variable approximate functions.

- Approximation of a two-variable $f(x,y)$ function over a rectangular domain can be obtained with product of single variable approximate functions in x and y variables as $f(x,y) = X(x)Y(y)$. $X(x)$ and $Y(y)$ functions can be created in series form as below:

$$X(x) = [a_0\Psi_{0x}(x) + a_1\Psi_{1x}(x) + a_2\Psi_{2x}(x) + \ldots + a_m\Psi_{mx}(x)]$$
$$Y(y) = [b_0\Psi_{0y}(y) + b_1\Psi_{1y}(y) + b_2\Psi_{2y}(y) + \ldots + b_n\Psi_{ny}(y)]$$

where $\Psi_{ix}(x)$ and $\Psi_{jy}(y)$ represent basic functions in trigonometric or polynomial (Legendre, Chebyshev) forms.

- Approximation of a trivariate $f(x,y,z)$ function over a rectangular prism domain can be obtained with product of single variable approximate functions in x, y, and z variables as $f(x,y,z) = X(x)Y(y)Z(z)$. $X(x)$, $Y(y)$, and $Z(z)$ functions can be created in series form as below:

$$X(x) = [a_0\Psi_{0x}(x) + a_1\Psi_{1x}(x) + a_2\Psi_{2x}(x) + \ldots + a_m\Psi_{mx}(x)]$$
$$Y(y) = [b_0\Psi_{0y}(y) + b_1\Psi_{1y}(y) + b_2\Psi_{2y}(y) + \ldots + b_n\Psi_{ny}(y)]$$
$$Z(z) = [c_0\Psi_{0z}(z) + c_1\Psi_{1z}(z) + c_2\Psi_{2z}(z) + \ldots + c_r\Psi_{rz}(z)]$$

where $\Psi_{ix}(x)$, $\Psi_{jy}(y)$, and $\Psi_{kz}(z)$ represent basic functions in trigonometric or polynomial (Legendre, Chebyshev) forms.

Appendix: MATLAB Programs

MATLAB Program 7.1: Response Surface Model

```
function ResponseSurfaceProgram_1st_and_2ndOrder

% NOTE: THIS PROGRAM FITS LINEAR AND QUADRATIC (PARABOLIC) RESPONSE
% SURFACE MODELS TO FUNCTION VALUES OF f(x) FUNCTION

% NOTE: After fitting Response Surface model, intermediate value f0
% is calculated at x0 point

% Selection of Order of Response Surface Model
% ModelOrder=1: LINEAR model: f=c1+c2*x1+c3*x2+...
% ModelOrder=2: QUADRATIC model:
F=c1+c2*x1+c3*x2+c4*x1^2+c5*x2^2+c6*x1*x2...
ModelOrder=2 ;

% DATA SET:
% Variable values at all points: xp
% xp=[x1 x2 .. xn;  % For Point 1
%     x1 x2 .. xn;  % For Point 2
%     ...........;
%     x1 x2 .. xn]; % For Point m
xp=[2   5 ;  % For Point 1
    6   5 ;  % For Point 2
    10  5 ;  % For Point 3
    2   10;  % For Point 4
    6   10;  % For Point 5
    10  10;  % For Point 6
    2   15;  % For Point 7
    6   15;  % For Point 8
    10  15]; % For Point 9

% Intermediate point: x0
x0=[5 9] ;

% Calculation of function values at xp points
% Note: Function values are calculated using f=FunctionValue(x) at the
% end of the program
npoint=size(xp,1) ;  %  Number of points
NumVar=size(xp,2) ;  %  Number of variables

f=zeros(npoint,1) ;
for s=1:npoint
    for i=1:NumVar
        x(i)=xp(s,i);
    end

    f(s)=FunctionValue(x) ;
end
% ========================================================================

% nterm: Number of constants in the Response Surface model
if ModelOrder==1
    nterm=NumVar+1 ;
else
    nterm=(NumVar+1)*(NumVar+2)/2 ;
end
```

```
% Values of model terms at xp points: Xmat
% Xmat=[1 x1 x2 x1^2 x1*x2  x2^2 ;   % For Point 1
%       1 x1 x2 x1^2 x1*x2  x2^2 ;   % For Point 2
%       ........................;
%       1 x1 x2 x1^2 x1*x2  x2^2 ] ; % For Point m
Xmat=zeros(npoint,nterm) ;
for k=1:npoint
   for i=1:NumVar
       x(i)=xp(k,i) ;
   end

   Xmat(k,1)=1 ; % 1. sütündaki değerler = 1

   % Lineer terms in Xmat
   num=1 ;
   for i=1:NumVar
       num=num+1;
       Xmat(k,num)=x(i);
   end

   if ModelOrder==2
       % Parabolic terms in Xmat
       num=NumVar+1;
       for i=1:NumVar
            num=num+1;
            Xmat(k,num)=x(i)^2;
       end

       % Interaction terms in Xmat
       for i=1:NumVar
           for j=i+1:NumVar
               num=num+1;
               Xmat(k,num)=x(i)*x(j);
           end
       end
   end
end

% Calculation of model constants
c=inv(Xmat'*Xmat)*(Xmat'*f) ;

% ---------------------------------------------------------------
% Function value at x0 point: f0
% Function value using LINEAR Response Surface Model
f0=c(1) ;

num=1 ;
for i=1:NumVar
    num=num+1;
    f0=f0+c(num)*x0(i);
end

% Function value using QUADRATIC Response Surface Model
% (Continuing LINEAR model)
if ModelOrder==2
    num=NumVar+1;
```

```
    for i=1:NumVar
        num=num+1;
        f0=f0+c(num)*x0(i)^2;
    end

    for i=1:NumVar
        for j=i+1:NumVar
            num=num+1;
            f0=f0+c(num)*x0(i)*x0(j);
        end
    end
end

% CHECKING FITTING QUALITY WITH R2 (R SQUARE) CRITERION
R2=R2_Calculation(npoint,NumVar,xp,f,c,ModelOrder) ;

% PRINTING RESULTS
if ModelOrder==1
    fprintf('SELECTED RESPONSE SURFACE MODEL: LINEAR \n') ;
else
    fprintf('SELECTED RESPONSE SURFACE MODEL: QUADRATIC \n') ;
end

fprintf('Model Coefficients: c \n') ;
c
x0
fprintf('f0=%f \n',f0) ;
fprintf('FITTING QUALITY (R2 VALUE)= %f \n',R2) ;


% -------------------------------------------------------------------------
% -------   DETERMINATION OF FITTING QUALITY USING R2 (R SQUARE)   ------
% -------------------------------------------------------------------------
function R2=R2_Calculation(npoint,NumVar,xp,f,c,ModelOrder)

% Mean of function values
Sumf=0 ;
for s=1:npoint
    Sumf=Sumf+f(s) ;
end
Avef=Sumf/npoint ;

% Sum of Squares of Error for Worst (Constant) Model
Ec=0 ;
E=0 ;
for s=1:npoint
    for i=1:NumVar
        x(i)=xp(s,i);
    end

    % Function value using LINEAR Response Surface Model
    % fa: function value
    fa=c(1) ;

    num=1 ;
    for i=1:NumVar
```

```
        num=num+1;
        fa=fa+c(num)*x(i);
    end

    % Function value using QUADRATIC Response Surface Model
    % (Continuing LINEAR model)
    if ModelOrder==2
        num=NumVar+1;
        for i=1:NumVar
             num=num+1;
             fa=fa+c(num)*x(i)^2;
        end

        for i=1:NumVar
            for j=i+1:NumVar
                num=num+1;
                fa=fa+c(num)*x(i)*x(j);
            end
        end
    end

    Ec=Ec+(Avef-f(s))^2 ;
    E=E+(fa-f(s))^2 ;
end
R2=(Ec-E)/Ec ;

% ------------------------------------------------
% --------------    FUNCTION: f(x)    ------------
% ------------------------------------------------
function  f=FunctionValue(x)
f=sqrt(x(1))+sqrt(x(2))+sqrt(x(1))*sqrt(x(2)) ;
```

MATLAB Program 7.2: Kriging Model

```
function Kriging_Regression

% NOTE: THIS PROGRAM FITS KRIGING MODELS TO FUNCTION VALUES OF
% f(x) FUNCTION

% NOTE: After fitting Kriging model, intermediate value f0
% is calculated at x0 point

% Selection of Kriging Method
% KrigingType=1: for Ordinary Kriging,
% KrigingType=2: for Universal Kriging,
KrigingType=2 ;

% Selection of Variogram Model
% VariogramFittingFunctionType=1: LINEAR model: Gama=c*h/a (h: uzaklık
değeri)
% VariogramFittingFunctionType=2: SPHERICAL model: Gama=c*[(3/2)*(h/a)-
(1/2)*(h/a)^2]
% VariogramFittingFunctionType=3. EXPONENTIAL model: Gama=c*[1-exp(-h/a)]
VariogramFittingFunctionType=3 ;

% DATA SET:
% Variable values at all points: xp
% xp=[x1 x2 .. xn;  % For Point 1
%     x1 x2 .. xn;  % For Point 2
%     ..........;
%     x1 x2 .. xn]; % For Point m
xp=[1 ;
    2 ;
    4 ;
    5 ] ;

% Intermediate point: x0
x0=[3.5] ;

% Calculation of function values at xp points
% Note: Function values are calculated using f=FunctionValue(x) at the
% end of the program
npoint=size(xp,1) ;  %  Number of points
NumVar=size(xp,2) ;  %  Number of variables

f=zeros(npoint,1) ;
for s=1:npoint
    for i=1:NumVar
        x(i)=xp(s,i);
    end

    f(s)=FunctionValue(x) ;
end
% ======================================================================

% FITTING OF VARIOGRAM MODEL AND CALCULATION OF MODEL CONSTANTS: a, c
[a,c]=VariogramCurveFitting(xp,f,npoint,NumVar,VariogramFittingFunctionTy
pe) ;

%  Number of terms for the Response Surface model in the Kriging model:
NumRSterms
if KrigingType==1
```

```
    NumRSterms=1 ;  % For Ordinary Kriging
else
    NumRSterms=(NumVar+1)*(NumVar+2)/2 ;  % For Universal Kriging
end

% ------------------------------------------------------------------
% Solution of Weight coefficients with Amat*Wmat=Bmat equation
Amat=zeros(npoint+NumRSterms,npoint+NumRSterms) ;
RSmat=zeros(npoint,NumRSterms) ;

% Amat calculation
for s=1:npoint
    for t=1:npoint
        % distance between points: h
        d=0 ;
        for i=1:NumVar
             d=d+(xp(s,i)-xp(t,i))^2 ;
        end

        h=sqrt(d) ;

        % For LINEAR model: Gama=c*h/a
        if VariogramFittingFunctionType==1
            Amat(s,t)=c*h/a ;
        end

        % For SPHERICAL model: Gama=c*[(3/2)*(h/a)-(1/2)*(h/a)^2]
        if VariogramFittingFunctionType==2
            Amat(s,t)=c*((3/2)*(h/a)-(1/2)*(h/a)^2) ;
        end

        % For EXPONENTIAL model: Gama=c*[1-exp(-h/a)]
        if VariogramFittingFunctionType==3
            Amat(s,t)=c*(1-exp(-h/a)) ;
        end
    end

    % Terms of Response Surface model in Kriging: RSmat
    % For Ordinary Kriging
    Amat(s,npoint+1)=1 ;
    RSmat(s,1)=Amat(s,npoint+1) ;

    % For Universal Kriging
    if KrigingType==2
        for i=1:NumVar
            Amat(s,npoint+1+i)=xp(s,i) ;
            RSmat(s,1+i)=Amat(s,npoint+1+i) ;
        end

        for i=1:NumVar
            Amat(s,npoint+NumVar+1+i)=xp(s,i)^2 ;
            RSmat(s,NumVar+1+i)=Amat(s,npoint+NumVar+1+i) ;
        end

        k=0 ;
        for i=1:NumVar
```

```
            for j=i+1:NumVar
                k=k+1 ;
                Amat(s,npoint+2*NumVar+1+k)=xp(s,i)*xp(s,j) ;
                RSmat(s,2*NumVar+1+k)=Amat(s,npoint+2*NumVar+1+k) ;
            end
        end
    end
end

trRSmat=RSmat'  ;

for i=1:NumRSterms
   for s=1:npoint
       Amat(npoint+i,s)=trRSmat(i,s) ;
   end
end

% ---------------------------------------------------------------------
% Bmat calculation
Bmat=zeros(npoint+NumRSterms,1) ;

for s=1:npoint
    % distance between points: h
    d=0 ;
    for i=1:NumVar
        d=d+(x0(i)-xp(s,i))^2 ;
    end

    h=sqrt(d) ;

    % For LINEAR model: Gama=c*h/a
    if VariogramFittingFunctionType==1
        Bmat(s)=c*h/a ;
    end

    % For SPHERICAL model: Gama=c*[(3/2)*(h/a)-(1/2)*(h/a)^2]
    if VariogramFittingFunctionType==2
        Bmat(s)=c*((3/2)*(h/a)-(1/2)*(h/a)^2) ;
    end

    % For EXPONENTIAL model: Gama=c*[1-exp(-h/a)]
    if VariogramFittingFunctionType==3
        Bmat(s)=c*(1-exp(-h/a)) ;
    end
end

% Quadratic Response Terms of Right Side matrix
% For Ordinary Kriging
Bmat(npoint+1)=1 ;

% for Universal Kriging
if KrigingType==2
    for i=1:NumVar
        Bmat(npoint+1+i)=x0(i) ;
    end
```

```
    for i=1:NumVar
       Bmat(npoint+NumVar+1+i)=x0(i)^2 ;
    end

    k=0 ;
    for i=1:NumVar
        for j=i+1:NumVar
            k=k+1 ;
            Bmat(npoint+2*NumVar+1+k)=x0(i)*x0(j) ;
        end
    end
end

% Solution of Weight coefficients (Wmat) using Amat*Wmat=Bmat equation
Wmat=Amat\Bmat ;

% Weight coefficients: w
for i=1:npoint
    w(i)=Wmat(i) ;
end

% Lagrange multipliers: Lam
if KrigingType==1
    Lam(1)=Wmat(npoint+1) ;
else
    for i=1:NumRSterms
        Lam(i)=Wmat(npoint+i) ;
    end
end

% Function value at x0 point: f0
f0=0 ;
for i=1:npoint
    f0=f0+w(i)*f(i) ;
end

if VariogramFittingFunctionType==1
    fprintf('Variogram Model: LINEAR: Gama=c*h/a \n') ;
end
if VariogramFittingFunctionType==2
    fprintf('Variogram Model: SPHERICAL: Gama=c*[(3/2)*(h/a)-
(1/2)*(h/a)^2] \n') ;
end
if VariogramFittingFunctionType==3
    fprintf('Variogram Model: EXPONENTIAL: Gama=c*[1-exp(-h/a)] \n') ;
end
fprintf('Variogram Model Coefficients: a=%f, c=%f \n',a,c) ;

x0
fprintf('f0=%f \n',f0) ;


% ------------------------------------------------------------------------
% --------------------     VARIOGRAM MODEL FITTING     --------------------
% ------------------------------------------------------------------------
function
[a,c]=VariogramCurveFitting(xp,f,npoint,NumVar,VariogramFittingFunctionTy
```

```
pe)
% CALCULATION OF VARIOGRAM VALUES: Gama
count=0 ;
for s=1:npoint
    for t=s:npoint
        count=count+1 ;

        % Distance between points: h
        d=0 ;
        for i=1:NumVar
            d=d+(xp(s,i)-xp(t,i))^2 ;
        end

        % Variogram value: Gama
        h(count)=sqrt(d) ;
        Gama(count)=((f(s)-f(t))^2)/2 ;
    end
end

% Sorting of h values: htemp
[htemp,index]=sort(h) ;

% Sorted h values
for i=1:count
    hSorted(i)=h(index(i)) ;
    GamaSorted(i)=Gama(index(i)) ;
end

% Determination of unique h values (removing repeating values)
uniqueh=unique(hSorted );

for i=1:length(uniqueh)
    k=0 ;
    Gamasum=0 ;

    for j=1:count
        if abs(hSorted(j)-uniqueh(i))<1e-10
             k=k+1 ;
             Gamasum=Gamasum+GamaSorted(j) ;
        end
    end

    hh(i)=uniqueh(i) ;
    Gamahh(i)=Gamasum/k ;

end

% CURVE FITTING TO VARIOGRAM VALUES
numh=length(hh) ;
a=hh(numh) ;

% Fitting LINEAR model: Gama=c*h/a
if VariogramFittingFunctionType==1
    Xi_2=0 ;
    fiXi=0 ;
    for i=1:numh
        Xi=hh(i)/a ;
```

```
        fi=Gamahh(i) ;
        fiXi=fiXi+fi*Xi ;
        Xi_2=Xi_2+Xi^2 ;
    end
    c=fiXi/Xi_2 ;
end

% ---------------------------------------------------------------------
% Fitting SPHERICAL model: Gama=c*[(3/2)*(h/a)-(1/2)*(h/a)^2]
if VariogramFittingFunctionType==2
    Xi_2=0 ;
    fiXi=0 ;
    for i=1:numh
        Xi=(3/2)*(hh(i)/a)-(1/2)*((hh(i)/a)^2) ;
        fi=Gamahh(i) ;
        fiXi=fiXi+fi*Xi ;
        Xi_2=Xi_2+Xi^2 ;
    end
    c=fiXi/Xi_2 ;
end

% ---------------------------------------------------------------------
% % Fitting EXPONENTIAL model: Gama=c*[1-exp(-h/a)]
if VariogramFittingFunctionType==3
    Xi_2=0 ;
    fiXi=0 ;
    for i=1:numh
        Xi=1-exp(-hh(i)/a)   ;
        fi=Gamahh(i) ;
        fiXi=fiXi+fi*Xi ;
        Xi_2=Xi_2+Xi^2 ;
    end
    c=fiXi/Xi_2 ;
end

% PLOTTING OF VARIOGRAM MODEL
if VariogramFittingFunctionType==1
    for i=1:numh
        Gamafit1(i)=c*hh(i)/a ;
        Gama1(i)=Gamahh(i) ;
    end
end

if VariogramFittingFunctionType==2
    for i=1:numh
        Gamafit2(i)=c*((3/2)*(hh(i)/a)-(1/2)*(hh(i)/a)^2) ;
        Gama2(i)=Gamahh(i) ;
    end
end

if VariogramFittingFunctionType==3
    for i=1:numh
        Gamafit3(i)=c*(1-exp(-hh(i)/a)) ;
        Gama3(i)=Gamahh(i) ;
    end
end
```

```
figure(1)
if VariogramFittingFunctionType==1
    plot(hh,Gamahh,'*',hh,Gamafit1,'r') ;
end
if VariogramFittingFunctionType==2
    plot(hh,Gamahh,'*',hh,Gamafit2,'r') ;
end
if VariogramFittingFunctionType==3
    plot(hh,Gamahh,'*',hh,Gamafit3,'r') ;
end

xlabel('h') ;
ylabel('Variogram Value') ;
legend('True','VariogramModel')
title('Variogram Model Fitting')
set(gcf,'color','w')

% ------------------------------------------------
% --------------    FUNCTION: f(x)    ------------
% ------------------------------------------------
function  f=FunctionValue(x)
f=x(1)^2 ;
```

MATLAB Program 7.3: Artificial Neural Network (ANN)

```
function ANN_1_Hidden_Layer_LinearOutput

% NOTE: THIS PROGRAM FITS ARTIFICIAL NEURAL NETWORK (ANN) MODEL TO
% FUNCTION VALUES OF f(x) FUNCTION

% NOTE 1: ANN MODEL IS TRAINED USING ALL DATA (SIMULTANEOUSLY)

% NOTE 2: Delta Rule is used for ANN training: Wij=Wij-ita*dError/dWij

% NOTE 3: After fitting ANN model, intermediate value f0 is calculated
% at x0 point

% Number of neurons in Hidden Layer: NumNeuron_Hiddenlayer
% Note: Number of neurons should be 4-5 times higher than number
% of variables n (NumNeuron_Hiddenlayer>=4*n)
NumNeuron_Hiddenlayer=2;

% Selection of Activation functions in neurons of Hidden Layer
% HiddenLayerActivationFunctionType==1: SIGMOID
% HiddenLayerActivationFunctionType==2: TANH
HiddenLayerActivationFunctionType=1  ;

ita=0.01 ;       % Learning rate
maxit=1e6;       % Maksimum iteration number
tol=0.005 ;      % Convergence tolerance

% DATA SET:
% Variable values at all points: xp
% xp=[x1 x2 .. xn;  % For Point 1
%     x1 x2 .. xn;  % For Point 2
%     ...........;
%     x1 x2 .. xn]; % For Point m
xp=[0        ;
    0.7854   ;
    1.5708   ;
    2.3562   ;
    3.1416   ] ;

% Intermediate point: x0
%x0=[x10 x20 .. xn0];
x0=[2 ] ;
%=======================================================================

% Calculation of function values at xp points
% Note: Function values are calculated using f=FunctionValue(x) at the
% end of the program
npoint=size(xp,1) ;  %  Number of points
NumVar=size(xp,2) ;  %  Number of variables

f=zeros(npoint,1) ;
for s=1:npoint
    for i=1:NumVar
        x(i)=xp(s,i) ;
    end

    f(s)=FunctionValue(x) ;
end
%
=======================================================================

p=NumNeuron_Hiddenlayer ;

% INITIAL VALUES FOR WEIGHT COEFFICIENTS: Random numbers can be selected
% in [0,1] interval
% Weight coefficients between Input and Hidden layers: a
```

```
a=zeros(NumVar,p) ;
for i=1:NumVar
   for j=1:p
       a(i,j)=rand ;
   end
end

% Bias coefficients for Hidden layer: b
b=zeros(p,1) ;
for j=1:p
   b(j)=rand ;
end

% Weight coefficients between Hidden and Output layers: e
e=zeros(p,1) ;
for j=1:p
    e(j)=rand ;
end

% Bias coefficients for Output layer:: f1
f1=rand ;

it=1 ;
ConVal=1e10 ; % Initial value for convergence criterion

while ((ConVal>tol) && (it<maxit))
    ErrorSum=0 ; % Sum of squares of error (Error Sum)

    % Derivative error with respect to weight coeffcients
    dEda=zeros(NumVar,p) ;
    dEdb=zeros(p,1) ;
    dEde=zeros(p,1) ;
    dEdf1=0 ;

    for s=1:npoint
           for i=1:NumVar
               x(i)=xp(s,i) ;
           end

           % Output value of Hidden layer: G
           iG=zeros(p,1) ;
           G=zeros(p,1) ;
           for j=1:p
                % Input
                iG(j)=b(j) ;

                for i=1:NumVar
                    iG(j)=iG(j)+a(i,j)*x(i) ;
                end

                % Output
                if HiddenLayerActivationFunctionType==1
                    G(j)=1/(1+exp(-iG(j))) ;  % SIGMOID
                else
                    G(j)=tanh(iG(j)) ;  % TANH
                end
           end

           % % Output value of Output layer: y
           % Input
           iy=f1 ;
           for j=1:p
                iy=iy+e(j)*G(j) ;
           end
```

```
            % Output
            y=iy ;

            % Error square
            Error=((f(s)-y)^2)/2 ;
            ErrorSum=ErrorSum+Error ;

            % DERIVATIVE OF TOTAL ERROR WITH RESPECT TO WEIGHT
COEFFICIENTS
            dEdy=-(f(s)-y) ;
            dydiy=1 ;
            dEdiy=dEdy*dydiy ;

            % Derivative of Total Error with respect to Hidden-Output
weight
            % coefficients
            for j=1:p
                dEde(j)=dEde(j)+dEdiy*G(j) ;
            end

            % Derivative of Total Error with respect to Bias coefficients
of
            % Output layer
            diydf1=1 ;
            dEdf1=dEdf1+dEdiy*diydf1 ;

            % Derivative of Total Error with respect to Input-Hidden
weight
            % coefficients
            for i=1:NumVar
                for j=1:p
                    if HiddenLayerActivationFunctionType==1
                        dGjdiGj=G(j)*(1-G(j)) ;  % SIGMOID
                    else
                        dGjdiGj=1-G(j)^2 ;  % TANH
                    end
                    dEda(i,j)=dEda(i,j)+dEdiy*e(j)*dGjdiGj*x(i) ;
                end
            end

            % Derivative of Total Error with respect to Bias coefficients
of
            % Hidden layer
            for j=1:p
                if HiddenLayerActivationFunctionType==1
                    dGjdiGj=G(j)*(1-G(j)) ;  % SIGMOID
                else
                    dGjdiGj=1-G(j)^2 ;  % TANH
                end
                dEdb(j)=dEdb(j)+dEdiy*e(j)*dGjdiGj*1 ;
            end
    end

    % New values of Input-Hidden weight coefficients
    for i=1:NumVar
        for j=1:p
            a(i,j)=a(i,j)-ita*dEda(i,j) ;
        end
    end

    % New values of Bias coefficients of Hidden layer
    for j=1:p
        b(j)=b(j)-ita*dEdb(j) ;
    end
```

```
    % New values of Hidden-Output weight coefficients
    for j=1:p
        e(j)=e(j)-ita*dEde(j) ;
    end

    % New values of Bias coefficients of Output layer
    f1=f1-ita*dEdf1 ;

    % Error Root Mean square (ERMS)
    ERMS=sqrt(ErrorSum/npoint) ;

    ERMS_iteration(it)=ERMS ;
    IterationCounter(it)=it ;

    % New values of convergence criterion
    %ConVal=ErrorSum ;
    ConVal=ERMS ;

    it=it+1 ;
end

% PREDICTION OF FUNCTION VALUES AT xp POINTS USING ANN MODEL
for s=1:npoint
    for i=1:NumVar
        x(i)=xp(s,i) ;
    end

    % Output value of Hidden layer: G
    iG=zeros(p,1) ;
    G=zeros(p,1) ;
    for j=1:p
        iG(j)=b(j) ;

        for i=1:NumVar
            iG(j)=iG(j)+a(i,j)*x(i) ;
        end

        if HiddenLayerActivationFunctionType==1
            G(j)=1/(1+exp(-iG(j))) ;  % SIGMOID
        else
            G(j)=tanh(iG(j)) ;  % TANH
        end
    end

    % Output value of Output layer: y
    iy=f1 ;
    for j=1:p
        iy=iy+e(j)*G(j) ;
    end

    y=iy ;

    fANN(s)=y ;
    Point(s)=s ;
end

% PREDICTION OF FUNCTION VALUES AT x0 POINT USING ANN MODEL
% Output value of Hidden layer: G
iG=zeros(p,1) ;
G=zeros(p,1) ;
for j=1:p
    iG(j)=b(j) ;

for i=1:NumVar
    iG(j)=iG(j)+a(i,j)*x0(i) ;
```

```
    end

    if HiddenLayerActivationFunctionType==1
         G(j)=1/(1+exp(-iG(j))) ;  % SIGMOID
    else
         G(j)=tanh(iG(j)) ;  % TANH
    end
end

% Output value of Output layer: y
iy=f1 ;
for j=1:p
    iy=iy+e(j)*G(j) ;
end

y=iy ;
f0=y ;

% PRINTING RESULTS
fprintf('COMPARISON OF TRUE VALUES AND ANN PREDICTIONS:\n ') ;
fprintf(' Point        f                    fANN
abs(Error)    \n ') ;
fprintf('--------   ----------         -----------           ---------
- \n ') ;
for s=1:npoint
    fprintf('%i         %e          %e            %f      \n
',s,f(s),fANN(s),abs(f(s)-fANN(s))) ;
end

fprintf(' \n ') ;
if HiddenLayerActivationFunctionType==1
    fprintf('Hidden Layer Activation Function = SIGMOID \n ') ;
else
    fprintf('Hidden Layer Activation Function = TANH \n ') ;
end

if it==maxit
    fprintf('ANN Training has NOT CONVERGED after %i iterations !!!
\n',maxit) ;
else
    fprintf('ANN Training has CONVERGED \n ') ;
end
fprintf(' \n ') ;
fprintf('ANN Prediction at x0 point: f0=%f \n',f0) ;

% PLOTS
% CONVERGENCE HISTORY
figure(1) ;
plot(IterationCounter,ERMS_iteration,'-b') ;
xlabel('Iteration') ;
ylabel('ERMS') ;
grid ;

% COMPARISON OF ANN PREDICTION AND FUNCTION (f(x)) VALUES AT DATA
% POINTS
figure(2) ;
plot(Point,f,'k-*',Point,fANN,'-b') ;
xlabel('Point') ;
legend('TrueValue','ANN') ;
ylabel('y') ;
grid ;

% ------------------------------------------------
% --------------     FUNCTION: f(x)    ------------
% ------------------------------------------------
```

```
function   f=FunctionValue(x)
f=4*sin(x(1)) ;
```

Problems

Problem 7.1: Function values at different points are given as $(x_1,f_1) = (0, -1)$, $(x_2f_2) = (2, 5)$, $(x_3,f_3) = (4, 15)$, $(x_4,f_4) = (6, 37)$. Fit a linear polynomial model to the function values using Least Squares method and predict the function value at $x = 5$.

Problem 7.2: Fit a parabolic polynomial model to the function values in Problem 7.1 using Least Squares method and predict the function value at $x = 5$.

Problem 7.3: Function values at different points are given as $(x_1,f_1) = (0, 0)$, $(x_2f_2) = (2, 2.46)$, $(x_3,f_3) = (4, 3.03)$, $(x_4,f_4) = (6, 3.42)$, $(x_5,f_5) = (8, 3.73)$. Fit $f(x) = ax^b$ model to the function values using Least Squares method and determine a, b constants.

Problem 7.4: Fit a two-variable quadratic (parabolic) response surface model to the Full Factorial experimental design data in Fig. 7.61 and predict the function value at $(x, y) = (4, 3)$.

Problem 7.5: Using Central Composite experimental design method, fit a quadratic (parabolic) response surface model to the values of $f(x, y) = 10 \sin x \sin y$ function within the ranges of $0 \leq x, y \leq 3$ (*Note*: Central composite design is given in normalized form in Table 7.10).

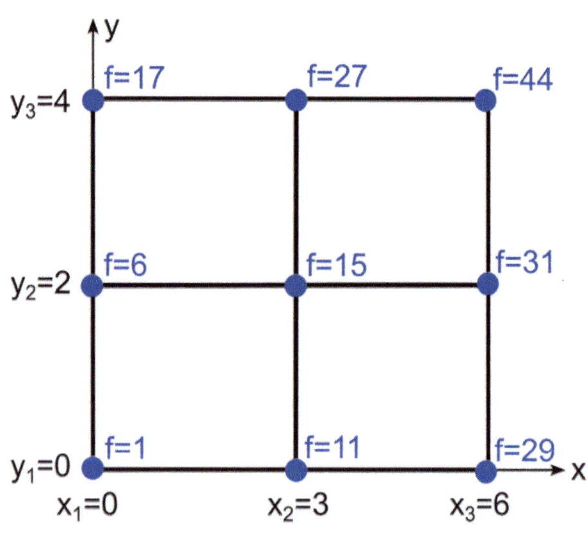

Fig. 7.61 Full Factorial experimental design data

Table 7.18 Factorial experimental design and corresponding function values

Point	x	y	$f(x, y)$
1	0	1	3
2	1	1	4
3	2	1	7
4	0	2	9
5	1	2	10
6	2	2	13
7	0	3	19
8	1	3	20
9	2	3	23

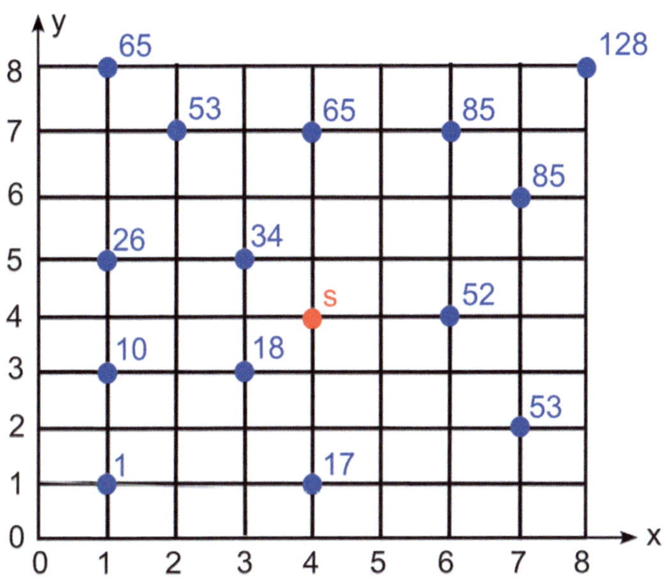

Fig. 7.62 Function values to be used in inverse distance method

Table 7.19 Function values at different points

s	2	4	5	6	8
Z	7	21	?	43	73

Problem 7.6: Perform ANOVA analysis using Table 7.18 and determine the effect of x and y variables on $f(x, y)$ function quantitatively.

Problem 7.7: Using the function values in Fig. 7.62, estimate the function value at intermediate point s with Inverse Distance method. Use $R = 3.5$ as the radius of domain of influence.

Problem 7.8: Using the data in Table 7.19, predict $Z(s = 5)$ value with Ordinary Kriging method.

Problem 7.9: Using the data in Table 7.19, predict $Z(s = 5)$ value with Universal Kriging method.

Problem 7.10: Function values at different points are given as $(x_1,f_1) = (0, -1)$, $(x_2 f_2) = (2, 5)$, $(x_3,f_3) = (4, 15)$, $(x_4,f_4) = (6, 37)$. Predict the function value at $x = 5$ using the ANN model in Fig. 7.44 (Single Hidden layer with two neurons). Use Sigmoid function as the activation function in the neurons of Hidden layer and use Linear function as the activation function in the neuron of Output layer. Take learning rate as $\eta = 0.01$.

Problem 7.11: An $f(x)$ function is defined in $x = [0, 2]$ interval as following:

$$f(x) = \begin{cases} 3x \ , & 0 \le x \le 1 \\ -3x + 6 \ , & 1 \le x \le 2 \end{cases}$$

Obtain Fourier series approximation of $f(x)$ function.

Problem 7.12: Obtain Fourier series approximation of $f(x) = e^{2x}$ function in $x = [0, 1]$ interval by extending it as (a) even, (b) odd function.

Problem 7.13: Obtain the approximation of $f(x) = \sin x + \cos x$ function in $x = [-1, 1]$ interval using Legendre polynomials. For simplicity, use three terms in Legendre approximation.

Problem 7.14: Obtain the approximation of $f(y) = \sin y + \cos y$ function in $y = [0, \pi]$ interval using Chebyshev polynomials. For simplicity, use three terms in y direction in Chebyshev approximation.

Problem 7.15: Obtain the approximation of $f(x, y) = 2 \sin x \sin y$ function in the ranges of $x = [0, 2]$ and $y = [0, 2]$ using double Fourier Series based on sine functions. For simplicity, use two terms in x and y directions.

Problem 7.16: Obtain the approximation of $f(x, y) = 2 \sin x \sin y$ function in the ranges of $x = [-1, 1]$ and $y = [-1, 1]$ using Legendre polynomials. For simplicity, use two terms in x and y directions in Legendre approximation.

Problem 7.17: Obtain the approximation of $f(r, s) = 3 \sin r \sin s$ function in the ranges of $r = [0, \pi]$ and $s = [0, \pi]$ using Chebyshev polynomials. For simplicity, use two terms in r and s directions in Chebyshev approximation.

Problem 7.18: Obtain the approximation of $f(x, y, z) = x^2 - y^2 + z^2 + 2xyz$ function in the ranges of $x = [0, 2]$, $y = [0, 2]$, and $z = [0, 2]$ using triple Fourier Series based on sine functions. For simplicity, use two terms in x, y, and z directions.

Problem 7.19: Obtain the approximation of $f(x, y, z) = \cos x \cos y \cos z$ function in the ranges of $x = [-1, 1]$, $y = [-1, 1]$, and $z = [-1, 1]$ using Legendre polynomials. For simplicity, use two terms in x, y, and z directions in Legendre approximation.

Problem 7.20: Obtain the approximation of $f(x, y, z) = \cos x \cos y \cos z$ function in the ranges of $x = [-1, 1]$, $y = [-1, 1]$, and $z = [-1, 1]$ using Chebyshev polynomials. For simplicity, use two terms in x, y, and z directions in Chebyshev approximation.

References

1. Chapra, S. C., & Canale, R. P. (2020). *Numerical methods for engineers*. McGraw-Hill.
2. Bernatz, R. A. (2010). *Fourier series and numerical methods for partial differential equations*. John Wiley & Sons.
3. Everstine G. C. (2010). *Numerical solution of partial differential equations* [Lecture notes, PDF]. George Washington University. Retrieved from https://www.academia.edu/31518755/Numerical_Solution_of_Partial_Differential_Equations.
4. Epperson, J. F. (2013). *An Introduction to numerical methods and analysis*. John Wiley & Sons.
5. Burden, R. L., & Faires, J. D. (2011). *Numerical analysis*. Cengage Learning.
6. Sert C. (2006). *ME310 Numerical Methods* [Lecture notes, PDF]. Middle East Technical University. Retrieved from https://users.metu.edu.tr/csert/me310/me310_1_introduction.pdf.
7. Grewal, B. S. (2019). *Numerical methods in engineering and science*. Mercury Learning and Information.
8. Sastry, S. S. (2012). *Introductory methods of numerical analysis*. PHI Learning.
9. Box, G. E. P., & Draper, N. R. (1987). *Empirical model-building and response surfaces*. John Wiley & Sons.
10. Myers, R. H., & Montgomery, D. C. (1995). *Response surface methodology process and product optimization using designed experiments*. John Wiley & Sons.
11. Free, J. W., Parkinson, A. R., Bryce, G. R., & Balling, R. J. (1987). Approximation of computationally expensive and noisy functions for constrained nonlinear optimization. *ASME Journal of Mechanisms, Transmissions, and Automation in Design, 109*, 528–532.
12. Draper, N. R., & Smith, H. (1981). *Applied regression analysis*. John Wiley & Sons.
13. Aguiar, P. F., Bourguignon, B., Khots, M. S., Massart, D. L., & Phan-Than-Luu, R. (1995). D-optimal designs. *Chemometrics and Intelligent Laboratory Systems, 30*, 199–210.
14. Valtrova M. (2009). *Computational aspects of Kriging in chosen engineering problems*. Master's thesis, Czech University of Life Sciences.
15. Journel, A. G., & Huijbregts, C. H. J. (1981). *Mining geostatistics*. Academic Press.
16. Isaaks, E. H., & Srivasta, R. M. (1989). *An introduction to applied geostatistics*. Oxford University Press.
17. Oliver, M. A., & Webster, R. (1990). Kriging: a method of interpolation for geographical information system. *International Journal of Geographical Information Systems, 4*(3), 313–332.
18. Cressie, N. A. C. (1991). *Statistics for spatial data*. Wiley-Interscience Publication.
19. Hassoun, M. H. (1995). *Fundamentals of artificial neural networks*. MIT Press.
20. Priddy, K. L., & Keller, P. E. (2005). *Artificial neural networks: An introduction*. SPIE-The International Society for Optical Engineering.
21. Sivanandam, S., & Paulraj, M. (2009). *Introduction to artificial neural networks*. Vikas Publishing House.
22. Schalkoff, R. J. (1997). *Artificial neural networks*. McGraw-Hill Education (India).
23. Patterson, D. W. (1996). *Artificial neural networks: Theory and applications*. Prentice-Hall.
24. Judd K. L. (2012). *Approximation Methods*. [Lecture notes, PDF]. Hoover Institution. Retrieved from https://kenjudd.org/wp-content/uploads/2017/03/Approximation_Methods_2012.pdf.

Chapter 8
Numerical Derivative

8.1 Introduction

Derivative expressions are frequently encountered in engineering problems that are expressed with differential equations. Some derivative terms are even given special names such as velocity, acceleration, slope, and curvature. Differential equations cannot be solved analytically under complex boundary conditions or in nonlinear situations. Numerical methods are preferred for such problems. In the numerical solution of differential equations, derivatives are expressed with numerical methods [1–12]. Therefore, numerical derivatives are explained in detail in this section.

In numerical derivatives, derivatives are expressed with weighted sum of function values at different points. For example, n-th order derivative of an $f(x)$ function at a point x_i can be expressed with numerical methods as following (Fig. 8.1):

$$\begin{aligned}\frac{d^n f}{dx^n}(x_i) = f^{(n)}(x_i) = \ldots + w_{i-2} f(x_{i-2}) + w_{i-1} f(x_{i-1}) \\ + w_i f(x_i) + w_{i+1} f(x_{i+1}) + w_{i+2} f(x_{i+2}) + \ldots \end{aligned} \quad (8.1)$$

In Eq. 8.1, w_i shows the weight coefficient for the function value of $f(x_i)$. A number of points and weight values are the most effective parameters on the accuracy and computational cost of the numerical derivative. Numerical derivative methods differ from each other according to the number of points and the methods to obtain the weight coefficients. They can be divided into two groups:

1. Methods based on Taylor series
2. Methods based on interpolation

In Taylor series-based methods, derivatives are obtained using function values which are expressed with truncated Taylor series expansions. Finite Difference method is such a method. In interpolation-based methods, a Lagrange interpolation function is passed through the function values at different points. Then, taking

H. Kurtaran, *Engineering Mathematics with MATLAB Applications: Vol. 1, Fundamentals*, https://doi.org/10.1007/978-3-031-81869-1_8

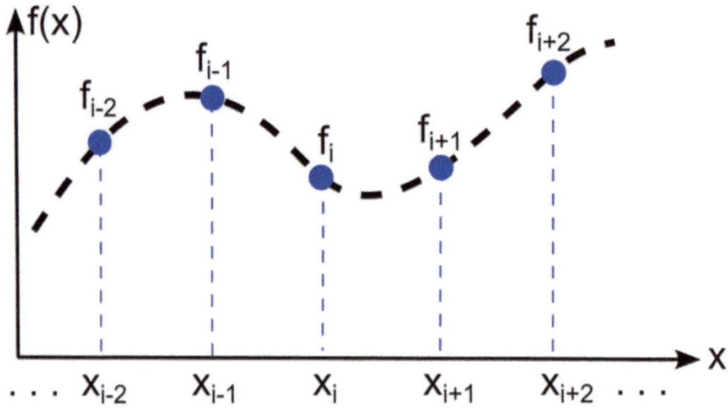

Fig. 8.1 Values of an $f(x)$ function at discrete points

derivative of Lagrange interpolation function, numerical derivative expression is obtained. An example of such method is the Differential Quadrature Method.

Numerical derivative methods such as Finite Difference and Differential Quadrature methods were actually developed to obtain derivatives of polynomial functions. They implicitly assume that functions are polynomials. Among numerical derivative methods, Differential Quadrature is more general and flexible than Finite Difference method. Finite Difference method can be considered as a special case of Differential Quadrature Method. Finite Difference and Differential Quadrature methods are explained in detail as following.

8.2 Derivative Calculation with Finite Difference Method

Taylor series expansion can be used to approximate an $f(x)$ function with polynomial series or shortly to express an $f(x)$ function in polynomial form. Most of the numerical derivative formulas are based on Taylor series expansion [1–8]. Therefore, derivation of Taylor series expansion is explained in detail below:

An $f(x)$ function around a general x_0 point can be expressed in polynomial form as following:

$$f(x)=a_0+a_1(x-x_0)+a_2(x-x_0)^2+a_3(x-x_0)^3+a_4(x-x_0)^4+\ldots+a_n(x-x_0)^n+\ldots \tag{8.2}$$

Derivatives of $f(x)$ function in Eq. 8.2 can be expressed as below:

$$f'(x)=\underbrace{1}_{1!}a_1+2a_2(x-x_0)+3a_3(x-x_0)^2+4a_4(x-x_0)^3+\ldots+na_n(x-x_0)^{n-1}+\ldots \tag{8.3}$$

$$f''(x)=\underbrace{2}_{2!}a_2+(3)(2)a_3(x-x_0)+(4)(3)a_4(x-x_0)^2+\ldots+n(n-1)a_n(x-x_0)^{n-2}+\ldots \tag{8.4}$$

$$f'''(x)=\underbrace{(3)(2)}_{3!}a_3+(4)(3)(2)a_4(x-x_0)+\ldots+n(n-1)(n-2)a_n(x-x_0)^{n-3}+\ldots \tag{8.5}$$

$$\ldots$$

$$f^{(n)}(x)=\underbrace{n(n-1)(n-2)\ldots.(1)}_{n!}a_n \tag{8.6}$$

If function value and its derivatives are known at x_0 point as $f(x_0)$, $f'(x_0)$, $f''(x_0)$, $f'''(x_0)$,..., $f^{(n)}(x_0)$, then a_0, a_1, a_2,..., a_n coefficients in Eq. 8.2 can be determined simply as below:

$$\text{For } x=x_0: \ f(x)=f(x_0)\rightarrow f(x_0)=a_0\rightarrow a_0=f(x_0) \tag{8.7}$$

$$\text{For } x=x_0: \ f'(x)=f'(x_0)\rightarrow f'(x_0)=a_1\rightarrow a_1=f'(x_0) \tag{8.8}$$

$$\text{For } x=x_0: \ f''(x)=f''(x_0)\rightarrow f''(x_0)=\underbrace{2}_{2!}a_2\rightarrow a_2=\frac{f''(x_0)}{2!} \tag{8.9}$$

$$\text{For } x=x_0: \ f'''(x)=f'''(x_0)\rightarrow f'''(x_0)=\underbrace{(3)(2)}_{3!}a_3\rightarrow a_3=\frac{f'''(x_0)}{3!} \tag{8.10}$$

$$\ldots$$

$$\text{For } x=x_0: f^{(n)}(x)=f^{(n)}(x_0)\rightarrow f^{(n)}(x_0)=\underbrace{n(n-1)(n-2)\ldots.(1)}_{n!}a_n\rightarrow a_n=\frac{f^{(n)}(x_0)}{n!} \tag{8.11}$$

When a_0, a_1, a_2,..., a_n expressions in Eqs. 8.7, 8.8, 8.9, 8.10, and 8.11 are substituted into Eq. 8.2, $f(x)$ function can be converted into the following form:

$$\begin{aligned} f(x) &= f(x_0)+f'(x_0)(x-x_0)+\frac{f''(x_0)}{2!}(x-x_0)^2+\ldots+\frac{f^{(n)}(x_0)}{n!}(x-x_0)^n+\ldots \\ &= \sum_{n=0}^{\infty}\frac{f^{(n)}(x_0)}{n!}(x-x_0)^n \end{aligned} \tag{8.12}$$

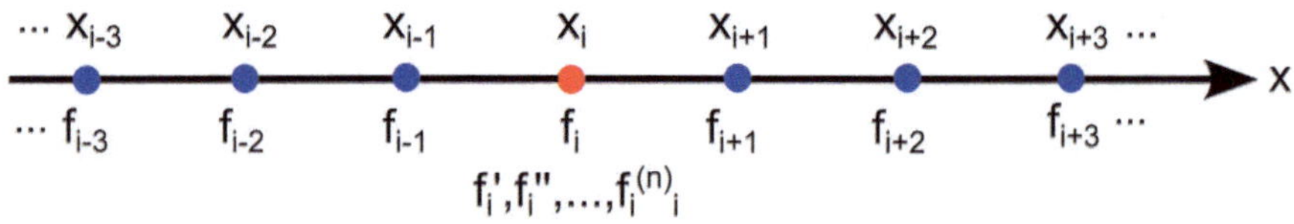

Fig. 8.2 Prediction of function values at further points (…, $i-2$, $i-1$, $i+1$, $i+2$, etc.) with Taylor series expansion using function value and its derivatives at point i

Equation 8.12 is also known as Taylor series expansion of $f(x)$ function around x_0 point.

Finite Difference derivative formulas are based on Taylor series expansion of an $f(x)$ function at different points [1–8]. Using Taylor series expansion, function and derivative values at a point can be used to estimate the function value at another point. For example, consider $i-3$, $i-2$, $i-1$, i, $i+1$, $i+2$, $i+3$ points in Fig. 8.2.

With help of Eq. 8.12, function values of $f(x_{i-3})$, $f(x_{i-2})$, $f(x_{i-1})$, $f(x_{i+1})$, $f(x_{i+2})$, $f(x_{i+3})$ at $i-3$, $i-2$, $i-1$, $i+1$, $i+2$, $i+3$ points can be expressed in terms of function and derivative values at point i using Taylor series expansion as below:

$$f(x_{i-3}) = f(x_i) + \frac{f'(x_i)}{1!}(x_{i-3}-x_i) + \frac{f''(x_i)}{2!}(x_{i-3}-x_i)^2 + \ldots + \frac{f^{(n)}(x_i)}{n!}(x_{i-3}-x_i)^n + \frac{f^{(n+1)}(x_i)}{(n+1)!}(x_{i-3}-x_i)^{n+1} + \ldots \tag{8.13a}$$

$$f(x_{i-2}) = f(x_i) + \frac{f'(x_i)}{1!}(x_{i-2}-x_i) + \frac{f''(x_i)}{2!}(x_{i-2}-x_i)^2 + \ldots + \frac{f^{(n)}(x_i)}{n!}(x_{i-2}-x_i)^n + \frac{f^{(n+1)}(x_i)}{(n+1)!}(x_{i-2}-x_i)^{n+1} + \ldots \tag{8.13b}$$

$$f(x_{i-1}) = f(x_i) + \frac{f'(x_i)}{1!}(x_{i-1}-x_i) + \frac{f''(x_i)}{2!}(x_{i-1}-x_i)^2 + \ldots + \frac{f^{(n)}(x_i)}{n!}(x_{i-1}-x_i)^n + \frac{f^{(n+1)}(x_i)}{(n+1)!}(x_{i-1}-x_i)^{n+1} + \ldots \tag{8.13c}$$

$$f(x_{i+1}) = f(x_i) + \frac{f'(x_i)}{1!}(x_{i+1}-x_i) + \frac{f''(x_i)}{2!}(x_{i+1}-x_i)^2 + \ldots + \frac{f^{(n)}(x_i)}{n!}(x_{i+1}-x_i)^n + \frac{f^{(n+1)}(x_i)}{(n+1)!}(x_{i+1}-x_i)^{n+1} + \ldots \tag{8.13d}$$

$$f(x_{i+2}) = f(x_i) + \frac{f'(x_i)}{1!}(x_{i+2} - x_i) + \frac{f''(x_i)}{2!}(x_{i+2} - x_i)^2 + \ldots + \frac{f^{(n)}(x_i)}{n!}(x_{i+2} - x_i)^n + \frac{f^{(n+1)}(x_i)}{(n+1)!}(x_{i+2} - x_i)^{n+1} + \ldots \tag{8.13e}$$

$$f(x_{i+3}) = f(x_i) + \frac{f'(x_i)}{1!}(x_{i+3} - x_i) + \frac{f''(x_i)}{2!}(x_{i+3} - x_i)^2 + \ldots + \frac{f^{(n)}(x_i)}{n!}(x_{i+3} - x_i)^n + \frac{f^{(n+1)}(x_i)}{(n+1)!}(x_{i+3} - x_i)^{n+1} + \ldots \tag{8.13f}$$

When sum of the terms including n-th and higher order derivatives in Taylor series expansion is indicated with R_n, Eq. 8.13a can be rewritten as following:

$$f(x_{i-3}) = f(x_i) + \frac{f'(x_i)}{1!}(x_{i-3} - x_i) + \frac{f''(x_i)}{2!}(x_{i-3} - x_i)^2 + \ldots + \underbrace{\frac{f^{(n)}(\xi_{i-3})}{n!}(x_{i-3} - x_i)^n}_{R_n(x_{i-3})} \tag{8.14a}$$

$$f(x_{i-2}) = f(x_i) + \frac{f'(x_i)}{1!}(x_{i-2} - x_i) + \frac{f''(x_i)}{2!}(x_{i-2} - x_i)^2 + \ldots + \underbrace{\frac{f^{(n)}(\xi_{i-2})}{n!}(x_{i-2} - x_i)^n}_{R_n(x_{i-2})} \tag{8.14b}$$

$$f(x_{i-1}) = f(x_i) + \frac{f'(x_i)}{1!}(x_{i-1} - x_i) + \frac{f''(x_i)}{2!}(x_{i-1} - x_i)^2 + \ldots + \underbrace{\frac{f^{(n)}(\xi_{i-1})}{n!}(x_{i-1} - x_i)^n}_{R_n(x_{i-1})} \tag{8.14c}$$

$$f(x_{i+1}) = f(x_i) + \frac{f'(x_i)}{1!}(x_{i+1} - x_i) + \frac{f''(x_i)}{2!}(x_{i+1} - x_i)^2 + \ldots + \underbrace{\frac{f^{(n)}(\xi_{i+1})}{n!}(x_{i+1} - x_i)^n}_{R_n(x_{i+1})} \tag{8.14d}$$

$$f(x_{i+2}) = f(x_i) + \frac{f'(x_i)}{1!}(x_{i+2} - x_i) + \frac{f''(x_i)}{2!}(x_{i+2} - x_i)^2 + \ldots + \underbrace{\frac{f^{(n)}(\xi_{i+2})}{n!}(x_{i+2} - x_i)^n}_{R_n(x_{i+2})} \tag{8.14e}$$

$$f(x_{i+3}) = f(x_i) + \frac{f'(x_i)}{1!}(x_{i+3} - x_i) + \frac{f''(x_i)}{2!}(x_{i+3} - x_i)^2 + \ldots + \underbrace{\frac{f^{(n)}(\xi_{i+3})}{n!}(x_{i+3} - x_i)^n}_{R_n(x_{i+3})} \tag{8.14f}$$

R_n terms in Eq. 8.14a can also be expressed as below:

$R_n(x_{i-3}) = \frac{f^{(n)}(\xi_{i-3})}{n!}(x_{i-3} - x_i)^n, x_{i-3} \leq \xi_{i-3} \leq x_i$ (ξ_{i-3} is a value between x_i and x_{i-3})
$R_n(x_{i-2}) = \frac{f^{(n)}(\xi_{i-2})}{n!}(x_{i-2} - x_i)^n, x_{i-2} \leq \xi_{i-2} \leq x_i$ (ξ_{i-2} is a value between x_i and x_{i-2})
$R_n(x_{i-1}) = \frac{f^{(n)}(\xi_{i-1})}{n!}(x_{i-1} - x_i)^n, x_{i-1} \leq \xi_{i-1} \leq x_i$ (ξ_{i-1} is a value between x_i and x_{i-1})
$R_n(x_{i+1}) = \frac{f^{(n)}(\xi_{i+1})}{n!}(x_{i+1} - x_i)^n, x_i \leq \xi_{i+1} \leq x_{i+1}$ (ξ_{i+1} is a value between x_i and x_{i+1})
$R_n(x_{i+2}) = \frac{f^{(n)}(\xi_{i+2})}{n!}(x_{i+2} - x_i)^n, x_i \leq \xi_{i+2} \leq x_{i+2}$ (ξ_{i+2} is a value between x_i and x_{i+2})
$R_n(x_{i+3}) = \frac{f^{(n)}(\xi_{i+3})}{n!}(x_{i+3} - x_i)^n, x_i \leq \xi_{i+3} \leq x_{i+3}$ (ξ_{i+3} is a value between x_i and x_{i+3})

By manipulating the expressions in Eq. 8.14a, desired derivative expression can be obtained in terms of function values of $f(x_{i-3})$, $f(x_{i-2})$, $f(x_{i-1})$, $f(x_{i+1})$, $f(x_{i+2})$, $f(x_{i+3})$.

8.2.1 *First-Order Derivative Expression with Finite Difference Method*

First-Order Derivative Expression with Forward Difference Method

Forward difference method uses points *i* (current point) and *i*+1 (forward point) in obtaining first-order derivative expression. To derive the expression, we can consider Eq. 8.14d. When sum of the terms including second- and higher order derivatives in Taylor series expansion is indicated with R_2, Eq. 8.14d can be written as following by taking: $x_{i+1} - x_i = \Delta x$

$$f(x_{i+1}) = f(x_i) + \frac{f'(x_i)}{1!}(\Delta x) + \underbrace{\frac{f^{(2)}(\xi_{i+1})}{2!}(\Delta x)^2}_{R_2} \tag{8.15}$$

R_2 term in Eq. 8.15 represents truncation error when second- and higher order derivative terms are discarded from the Taylor series expansion. From Eq. 8.15, first-order derivative term can be left alone as below:

$$f'(x_i)(\Delta x) = f(x_{i+1}) - f(x_i) - \underbrace{\frac{f^{(2)}(\xi_{i+1})}{2!}(\Delta x)^2}_{R_2} \tag{8.16}$$

After dividing both sides with Δx, first-order derivative expression is obtained as below:

$$f'(x_i) = \frac{f(x_{i+1}) - f(x_i)}{\Delta x} - \underbrace{\frac{f^{(2)}(\xi_{i+1})}{2!}(\Delta x)}_{R_2 = \text{Truncation Error}} \tag{8.17}$$

For practical use, R_2 term in Eq. 8.17 can be neglected and thereby first-order derivative expression can be written in terms of just function values as following:

$$f'(x_i) = \frac{f(x_{i+1}) - f(x_i)}{\Delta x} = \underbrace{\frac{1}{\Delta x}}_{w_{i+1}} f(x_{i+1}) + \underbrace{\frac{-1}{\Delta x}}_{w_i} f(x_i) \tag{8.18}$$

Neglecting R_2 term will cause a truncation error in Eq. 8.18 in the order of Δx (i.e. $R_2 = \frac{f^{(2)}(\xi_{i+1})}{2!}(\Delta x) = O(\Delta x)$, O indicates order of error). In order to reduce the truncation error in Eq. 8.18, Δx should be chosen very small. Note that Forward Difference derivative expression can be used only if there are further points.

More Accurate First-Order Derivative Expressions with Forward Difference

Using more points in forward direction and their Taylor series expansions, more accurate first derivative expressions can be developed. Below is a Forward Difference first derivative expression based on three points (i, i+1, i+2) leading to the order of error $O(\Delta x^2)$

$$f'(x_i) = \frac{-3f(x_i) + 4f(x_{i+1}) - f(x_{i+2})}{2\Delta x} + O(\Delta x^2) \tag{8.19}$$

First-Order Derivative Expression with Backward Difference Method

Backward difference method uses points i and $i-1$ (backward point) in obtaining first-order derivative expression. To derive the expression we can consider Eq. 8.14c. When sum of the terms including second- and higher order derivative terms in Taylor series expansion is indicated with R_2, Eq. 8.14c can be written as following by taking $x_i - x_{i-1} = \Delta x$

$$f(x_{i-1}) = f(x_i) - \frac{f'(x_i)}{1!}(\Delta x) + \underbrace{\frac{f^{(2)}(\xi_{i-1})}{2!}(\Delta x)^2}_{R_2} \tag{8.20}$$

R_2 term in Eq. 8.20 represents truncation error when second- and higher order derivative terms are discarded from the Taylor series expansion. From Eq. 8.20, first derivative term can be left alone as below:

$$f'(x_i)(\Delta x) = f(x_i) - f(x_{i-1}) + \underbrace{\frac{f^{(2)}(\xi_{i-1})}{2!}(\Delta x)^2}_{R_2} \tag{8.21}$$

After dividing both sides with Δx, first derivative expression is obtained as below:

$$f'(x_i) = \frac{f(x_i) - f(x_{i-1})}{\Delta x} + \underbrace{\frac{f^{(2)}(\xi_{i-1})}{2!}(\Delta x)}_{R_2 = TruncationError} \tag{8.22}$$

For practical use, R_2 term in Eq. 8.22 can be neglected and thereby first-order derivative expression can be written in terms of just function values as following:

$$f'(x_i) = \frac{f(x_i) - f(x_{i-1})}{\Delta x} = \underbrace{\frac{1}{\Delta x}}_{w_i} f(x_i) + \underbrace{\frac{-1}{\Delta x}}_{w_{i-1}} f(x_{i-1}) \tag{8.23}$$

Neglecting R_2 term will cause a truncation error in Eq. 8.23 in the order of Δx ($R_2 = \frac{f^{(2)}(\xi_{i-1})}{2!}(\Delta x) = O(\Delta x)$, O indicates order of error). In order to reduce the truncation error in Eq. 8.23, Δx should be chosen very small. Note that Backward Difference derivative expression can be used only if there are backward points.

More Accurate First-Order Derivative Expressions with Backward Difference

Using more points in backward direction and their Taylor series expansions, more accurate first derivative expressions can be developed. Below is a Backward Difference first derivative expression based on three points (i, $i-1$, $i-2$) leading to the order of error $O(\Delta x^2)$

$$f'(x_i) = \frac{3f(x_i) - 4f(x_{i-1}) + f(x_{i-2})}{2\Delta x} + O(\Delta x^2) \tag{8.24}$$

First-Order Derivative Expression with Central Difference Method

Central Difference method uses points $i-1$ and $i+1$ (backward and forward points) in obtaining first-order derivative expression. To derive the expression, we can consider Eqs. 8.14c and 8.14d. When sum of the terms including third- and higher order derivative terms in Taylor series expansion is indicated with R_3, Eqs. 8.14c and 8.14d can be written as following by taking $x_i - x_{i-1} = x_{i+1} - x_i = \Delta x$

$$f(x_{i-1}) = f(x_i) - \frac{f'(x_i)}{1!}(\Delta x) + \frac{f^{(2)}(x_i)}{2!}(\Delta x)^2 - \underbrace{\frac{f^{(3)}(\xi_{i-1})}{3!}(\Delta x)^3}_{R_3} \tag{8.25}$$

$$f(x_{i+1}) = f(x_i) + \frac{f'(x_i)}{1!}(\Delta x) + \frac{f^{(2)}(x_i)}{2!}(\Delta x)^2 + \underbrace{\frac{f^{(3)}(\xi_{i+1})}{3!}(\Delta x)^3}_{R_3} \tag{8.26}$$

If $f(x_{i-1})$ equation is subtracted from $f(x_{i+1})$ equation, following expression is obtained:

$$f(x_{i+1}) - f(x_{i-1}) = 2\frac{f'(x_i)}{1!}(\Delta x) + \underbrace{\frac{\left[f^{(3)}(\xi_{i+1}) + f^{(3)}(\xi_{i-1})\right]}{3!}(\Delta x)^3}_{R_3} \tag{8.27}$$

By rearranging the above expression, following Central Difference first derivative expression can be obtained:

$$f'(x_i) = \frac{f(x_{i+1}) - f(x_{i-1})}{2\Delta x} - \underbrace{\frac{\left[f^{(3)}(\xi_{i+1}) + f^{(3)}(\xi_{i-1})\right]}{2\times 3!}(\Delta x)^2}_{R_3 = TruncationError} \tag{8.28}$$

R_3 term in Eq. 8.28 represents truncation error when third- and higher order derivative terms are discarded from Taylor series expansions. When R_3 term is neglected, Eq. 8.28 can be written as following:

$$f'(x_i) = \frac{f(x_{i+1}) - f(x_{i-1})}{2\Delta x} = \underbrace{\frac{1}{2\Delta x}}_{w_{i+1}} f(x_{i+1}) + \underbrace{\frac{-1}{2\Delta x}}_{w_{i-1}} f(x_{i-1}) \tag{8.29}$$

Neglecting R_3 term will cause a truncation error in Eq. 8.29 in the order of $(\Delta x)^2$ ($R_3 = \frac{\left[f^{(3)}(\xi_{i+1})+f^{(3)}(\xi_{i-1})\right]}{2\times 3!}(\Delta x)^2 = O(\Delta x)^2$, O indicates order of error). When finite difference methods are compared for first-order derivative, it is seen that Central Difference method is more accurate (i.e., order of error is smaller) than Forward and Backward Difference methods since $O(\Delta x)^2 < O(\Delta x)$ for the selection of $0 < \Delta x < 1$. Note that Central Difference expression can be used when there are both further and backward points.

8.2.2 Second-Order Derivative with Finite Difference Method

Second-order derivative expression with Finite Difference method can be obtained using only forward points (Forward Difference), only backward points (Back Difference), or both forward and backward points (Central Difference) similar to the first derivative.

Second-Order Derivative with Central Finite Difference Method

Central Difference method uses points of $i-1$ and $i+1$ (backward and forward points) in obtaining second-order derivative expression. To derive the expression, we can consider Eqs. 8.14c and 8.14d again. When sum of the terms including fourth- and higher order derivative terms in Taylor series expansion is indicated with R_4, Eqs. 8.14c and 8.14d can be written as following by taking $x_i - x_{i-1} = x_{i+1} - x_i = \Delta x$

$$f(x_{i-1}) = f(x_i) - \frac{f'(x_i)}{1!}(\Delta x) + \frac{f^{(2)}(x_i)}{2!}(\Delta x)^2 - \frac{f^{(3)}(x_i)}{3!}(\Delta x)^3 + \underbrace{\frac{f^{(4)}(\xi_{i-1})}{4!}(\Delta x)^4}_{R_4} \tag{8.30}$$

$$f(x_{i+1}) = f(x_i) + \frac{f'(x_i)}{1!}(\Delta x) + \frac{f^{(2)}(x_i)}{2!}(\Delta x)^2 + \frac{f^{(3)}(x_i)}{3!}(\Delta x)^3 + \underbrace{\frac{f^{(4)}(\xi_{i+1})}{4!}(\Delta x)^4}_{R_4} \tag{8.31}$$

If $f(x_{i-1})$ and $f(x_{i+1})$ equations are summed up, following expression is obtained:

$$f(x_{i-1}) + f(x_{i+1}) = 2f(x_i) + f^{(2)}(x_i)(\Delta x)^2 + \underbrace{\frac{\left[f^{(4)}(\xi_{i+1}) + f^{(4)}(\xi_{i-1})\right]}{4!}(\Delta x)^4}_{R_4} \tag{8.32}$$

By rearranging Eq. 8.32, following second derivative Finite Difference expression can be obtained:

$$f^{('')}(x_i) = \frac{f(x_{i-1}) - 2f(x_i) + f(x_{i+1})}{(\Delta x)^2} - \underbrace{\frac{\left[f^{(4)}(\xi_{i+1}) + f^{(4)}(\xi_{i-1})\right]}{4!}(\Delta x)^2}_{R_4 = \text{Truncation Error}} \tag{8.33}$$

R_4 term in Eq. 8.33 represents truncation error when fourth- and higher order derivative terms are discarded from Taylor series expansions. When R_4 term is neglected, Eq. 8.33 can be written as following:

$$f^{('')}(x_i) = \frac{f(x_{i-1}) - 2f(x_i) + f(x_{i+1})}{(\Delta x)^2} = \underbrace{\frac{1}{(\Delta x)^2}}_{w_{i-1}} f(x_{i-1}) + \underbrace{\frac{-2}{(\Delta x)^2}}_{w_i} f(x_i) + \underbrace{\frac{1}{(\Delta x)^2}}_{w_{i+1}} f(x_{i+1}) \tag{8.34}$$

Neglecting R_4 term will cause a truncation error in Eq. 8.34 in the order of $(\Delta x)^2$ ($R_4 = \frac{\left[f^{(4)}(\xi_{i+1}) + f^{(4)}(\xi_{i-1})\right]}{4!}(\Delta x)^2 = O(\Delta x)^2$, O indicates order of error). Central Difference expression for second derivative can be used when there are both further and backward points.

Second-Order Derivative with Forward Difference Method

Forward Difference method uses forward points such as i+1 and i+2 in obtaining second-order derivative expressions. Using Eqs. 8.13d and 8.13e, Taylor series expansions for i+1 and i+2 points can be written as following (taking $x_{i+1} - x_i = \Delta x$ and $x_{i+2} - x_i = 2\Delta x$):

$$f(x_{i+1}) = f(x_i) + \frac{f'(x_i)}{1!}(\Delta x) + \frac{f''(x_i)}{2!}(\Delta x)^2 + \frac{f^{(3)}(x_i)}{3!}(\Delta x)^3 + \ldots \quad (8.35)$$

$$f(x_{i+2}) = f(x_i) + \frac{f'(x_i)}{1!}(2\Delta x) + \frac{f''(x_i)}{2!}(2\Delta x)^2 + \frac{f^{(3)}(x_i)}{3!}(2\Delta x)^3 + \ldots \quad (8.36)$$

If Eq. 8.35 is multiplied by (−2) and added to Eq. 8.36, following expression is obtained:

$$-2f(x_{i+1}) + f(x_{i+2}) = -f(x_i) + f^{('')}(x_i)(\Delta x)^2 + f^{(3)}(x_i)(\Delta x)^3 + \ldots \quad (8.37)$$

If the terms after second-order derivative are omitted in the above expression, it can be written as below:

$$-2f(x_{i+1}) + f(x_{i+2}) = -f(x_i) + f^{('')}(x_i)(\Delta x)^2 + O(\Delta x)^3 \quad (8.38)$$

If Eq. 8.38 is rearranged, following second-order derivative Forward Difference expression can be obtained:

$$f^{('')}(x_i) = \frac{f(x_i) - 2f(x_{i+1}) + f(x_{i+2})}{(\Delta x)^2} - O(\Delta x) \quad (8.39)$$

From Eq. 8.39, it is seen that order of error is $O(\Delta x)$ when second-order derivative is expressed as $f^{('')}(x_i) = \frac{f(x_i) - 2f(x_{i+1}) + f(x_{i+2})}{(\Delta x)^2}$. Since order of error is $O(\Delta x)$, Forward Difference method is less accurate than Central Difference method. Forward Difference expression can be used when there are further points.

More Accurate Second Derivative Expressions with Forward Difference: Using more points in forward direction and their Taylor series expansions, more accurate second-order derivative expressions can be developed. Below are examples of second-order derivative expressions based on a different number of forward points leading to higher accuracies.

$$f^{('')}(x_i) = \frac{2f(x_i) - 5f(x_{i+1}) + 4f(x_{i+2}) - f(x_{i+3})}{(\Delta x)^2} + O(\Delta x^2)$$
$$f^{('')}(x_i) = \frac{35f(x_i) - 104f(x_{i+1}) + 114f(x_{i+2}) - 56f(x_{i+3}) + 11f(x_{i+4})}{12(\Delta x)^2} + O(\Delta x^3)$$
$$f^{('')}(x_i) = \frac{45f(x_i) - 154f(x_{i+1}) + 214f(x_{i+2}) - 156f(x_{i+3}) + 61f(x_{i+4}) - 10f(x_{i+5})}{12(\Delta x)^2}$$
$$+ O(\Delta x^4) \tag{8.40}$$

When second derivative expressions in Eq. 8.34 and Eq. 8.40 are compared, it is seen that more points should be used with Forward Difference method in order to have the same accuracy with Central Difference method. From this, it is concluded that derivative with Forward Difference method is less efficient than derivative with Central Difference method.

Second-Order Derivative with Backward Difference Method

Backward Difference method uses backward points such as $i-1$ and $i-2$ in obtaining second-order derivative expressions. Using Eqs. 8.13b and 8.13c, Taylor series expansions for $i-1$ and $i-2$ points can be written as following (taking $x_i - x_{i-1} = \Delta x$ and $x_i - x_{i-2} = 2\Delta x$):

$$f(x_{i-1}) = f(x_i) - \frac{f'(x_i)}{1!}(\Delta x) + \frac{f''(x_i)}{2!}(\Delta x)^2 - \frac{f^{(3)}(x_i)}{3!}(\Delta x)^3 + \ldots \tag{8.41}$$

$$f(x_{i-2}) = f(x_i) - \frac{f'(x_i)}{1!}(2\Delta x) + \frac{f''(x_i)}{2!}(2\Delta x)^2 - \frac{f^{(3)}(x_i)}{3!}(2\Delta x)^3 + \ldots \tag{8.42}$$

If Eq. 8.41 is multiplied by (−2) and added to Eq. 8.42, following expression is obtained.

$$-2f(x_{i-1}) + f(x_{i-2}) = -f(x_i) + f^{('')}(x_i)(\Delta x)^2 - f^{(3)}(x_i)(\Delta x)^3 + \ldots \tag{8.43}$$

If the terms after second derivative are omitted in the above expression, it can be written as below:

$$-2f(x_{i-1}) + f(x_{i-2}) = -f(x_i) + f^{('')}(x_i)(\Delta x)^2 - O(\Delta x)^3 \tag{8.44}$$

If Eq. 8.44 is rearranged, following second derivative Backward Difference expression can be obtained:

$$f^{('')}(x_i) = \frac{f(x_i) - 2f(x_{i-1}) + f(x_{i-2})}{(\Delta x)^2} + O(\Delta x) \tag{8.45}$$

From Eq. 8.45, it is seen that order of error will be $O(\Delta x)$ when second-order derivative is expressed as $f^{('')}(x_i) = \frac{f(x_i) - 2f(x_{i-1}) + f(x_{i-2})}{(\Delta x)^2}$. Since order of error is $O(\Delta x)$, Backward Difference method is less accurate than Central Difference method. Backward Difference expression can be used when there are backward points.

More Accurate Second Derivative Expressions with Backward Difference: Using more points in backward direction and their Taylor series expansions, more accurate second-order derivative expressions can be developed. Below are examples of second-order derivative expressions based on a different number of backward points leading to higher accuracies.

$$f^{('')}(x_i) = \frac{2f(x_i) - 5f(x_{i-1}) + 4f(x_{i-2}) - f(x_{i-3})}{(\Delta x)^2} + O(\Delta x^2)$$
$$f^{('')}(x_i) = \frac{35f(x_i) - 104f(x_{i-1}) + 114f(x_{i-2}) - 56f(x_{i-3}) + 11f(x_{i-4})}{12(\Delta x)^2} + O(\Delta x^3)$$
$$f^{('')}(x_i) = \frac{45f(x_i) - 154f(x_{i-1}) + 214f(x_{i-2}) - 156f(x_{i-3}) + 61f(x_{i-4}) - 10f(x_{i-5})}{12(\Delta x)^2} + O(\Delta x^4) \tag{8.46}$$

When second-order derivative expressions in Eq. 8.34 and Eq. 8.46 are compared, it is seen that more points should be used with Backward Differences in order to have the same accuracy with Central Difference method. From this, it turns out that derivative with Backward Difference is less efficient than derivative with Central Difference.

8.3 General Method (Systematic Approach) for Obtaining Finite Difference Derivatives

When a large number of points are used, Finite Difference derivatives become complicated to obtain. A general method (systematic approach) can be devised to facilitate this process. In general method (systematic approach), the number and location of points that will be used in derivative expression are decided first. Then, Finite Difference derivative $\frac{d^n f}{dx^n}(x_i)$ is expressed as the weighted sum of the function values as in Eq. 8.1 (repeated as following):

$$\frac{d^n f}{dx^n}(x_i) = f^{(n)}(x_i) = \ldots + w_{i-2} f(x_{i-2}) + w_{i-1} f(x_{i-1}) + w_i f(x_i)$$
$$+ w_{i+1} f(x_{i+1}) + w_{i+2} f(x_{i+2}) + \ldots$$

For convenience, derivative expression above can be alternatively written by dividing with $(\Delta x)^n$ as below:

$$\frac{d^n f}{dx^n}(x_i) = f^{(n)}(x_i)$$
$$= \frac{\ldots + w_{i-2} f(x_{i-2}) + w_{i-1} f(x_{i-1}) + w_i f(x_i) + w_{i+1} f(x_{i+1}) + w_{i+2} f(x_{i+2}) + ..}{(\Delta x)^n} \tag{8.47}$$

Where $\ldots, w_{i-2}, w_{i-1}, w_i, w_{i+1}, w_{i+2}, \ldots$, in Eq. 8.47 indicate unknown weight coefficients to be determined.

Finally, Taylor series expansions of, $\ldots, f(x_{i-2}), f(x_{i-1}), f(x_{i+1}), f(x_{i+2}), \ldots$, are substituted into Eq. 8.47 and using equivalency of the left and right sides of Eq. 8.47, algebraic equations are obtained to determine weight coefficients. The process is shown in detail in the following example.

***Note*:** A number of terms in Taylor series expansion must be determined to produce a sufficient number of equations to determine the weight coefficients in Eq. 8.47. In short, it should not be less than the number of points used in Eq. 8.47.

Example 8.1: Obtain second-order Finite Difference derivative expression with general method (systematic approach) using $i-1$, i, $i+1$ points.

Second-order Finite Difference derivative can be expressed as the weighted sum of the function values at $i-1$, i, $i+1$ points as below:

$$f''(x_i) = w_{i-1} f(x_{i-1}) + w_i f(x_i) + w_{i+1} f(x_{i+1}) \tag{8.48}$$

In Eq. 8.48, w_{i-1}, w_i, w_{i+1} indicate weight coefficients to be determined. Taylor series expansions for $i-1$ and $i+1$ points can be written by discarding fourth- and higher order derivative terms as following:

$$f(x_{i-1}) = f(x_i) - \frac{f'(x_i)}{1!}(\Delta x) + \frac{f''(x_i)}{2!}(\Delta x)^2 - \frac{f^{(3)}(x_i)}{3!}(\Delta x)^3 + \underbrace{\frac{f^{(4)}(\xi_{i-1})}{4!}(\Delta x)^4}_{R_4}$$
$$f(x_{i+1}) = f(x_i) + \frac{f'(x_i)}{1!}(\Delta x) + \frac{f''(x_i)}{2!}(\Delta x)^2 + \frac{f^{(3)}(x_i)}{3!}(\Delta x)^3 + \underbrace{\frac{f^{(4)}(\xi_{i+1})}{4!}(\Delta x)^4}_{R_4} \tag{8.49}$$

Substituting Eq. 8.49 into the general derivative expression in Eq. 8.48, following expression is obtained:

$$f''(x_i) = w_{i-1}\left[f(x_i) - \frac{f'(x_i)}{1!}(\Delta x) + \frac{f''(x_i)}{2!}(\Delta x)^2 - \frac{f^{(3)}(x_i)}{3!}(\Delta x)^3 + \frac{f^{(4)}(\xi_{i-1})}{4!}(\Delta x)^4\right]$$
$$+ w_i f(x_i) + w_{i+1}\left[f(x_i) + \frac{f'(x_i)}{1!}(\Delta x) + \frac{f''(x_i)}{2!}(\Delta x)^2 + \frac{f^{(3)}(x_i)}{3!}(\Delta x)^3 + \frac{f^{(4)}(\xi_{i+1})}{4!}(\Delta x)^4\right] \tag{8.50}$$

By rearranging Eq. 8.50, it can be written as below:

$$\begin{aligned} f''(x_i) = {} & (w_{i-1} + w_i + w_{i+1})f(x_i) + (-w_{i-1} + w_{i+1})\Delta x f'(x_i) \\ & + (w_{i-1} + w_{i+1})\frac{(\Delta x)^2}{2!}f''(x_i) + (-w_{i-1} + w_{i+1})\frac{(\Delta x)^3}{3!}f^{(3)}(x_i) \\ & + \left(w_{i-1}f^{(4)}(\xi_{i-1}) + w_{i+1}f^{(4)}(\xi_{i+1})\right)\frac{(\Delta x)^4}{4!} \end{aligned} \tag{8.51}$$

Using equivalency of left and right sides of Eq. 8.51, following algebraic equations can be obtained for weight coefficients:

$$\begin{aligned} &\text{Equation 1}: w_{i-1} + w_i + w_{i+1} = 0 \\ &\text{Equation 2}: (-w_{i-1} + w_{i+1})\Delta x = 0 \\ &\text{Equation 3}: (w_{i-1} + w_{i+1})\frac{(\Delta x)^2}{2!} = 1 \\ &\text{Equation 4}: (-w_{i-1} + w_{i+1})\frac{(\Delta x)^3}{3!} = 0 \end{aligned}$$

Actually, Eq. 2 and Eq. 4 yield the same equation. Therefore, Eqs. 1–4 can be reduced to three independent equations as below:

$$\begin{aligned} w_{i-1} + w_i + w_{i+1} &= 0 \\ -w_{i-1} + w_{i+1} &= 0 \\ (w_{i-1} + w_{i+1}) &= \frac{2}{(\Delta x)^2} \end{aligned}$$

From solution of the above equations, weight coefficients can be determined as following:

$$\begin{aligned} &-w_{i-1} + w_{i+1} = 0 \Rightarrow w_{i-1} = w_{i+1} \\ &(w_{i-1} + w_{i+1}) = \frac{2}{(\Delta x)^2} \Rightarrow w_{i-1} = w_{i+1} = \frac{1}{(\Delta x)^2} \\ &w_{i-1} + w_i + w_{i+1} = 0 \Rightarrow w_i = -\frac{2}{(\Delta x)^2} \end{aligned}$$

Equation 8.48 and Eq. 8.51 are just different expressions of $f''(x_i)$. Therefore, using equivalency of Eq. 8.48 and Eq. 8.51, following equation can be written:

$$\begin{aligned} w_{i-1}f(x_{i-1})+w_if(x_i)+w_{i+1}f(x_{i+1}) &= \underbrace{(w_{i-1}+w_i+w_{i+1})}_{=0}f(x_i)+\underbrace{(-w_{i-1}+w_{i+1})}_{=0}\Delta x f'(x_i) \\ &+(w_{i-1}+w_{i+1})\frac{(\Delta x)^2}{2!}f''(x_i)+\underbrace{(-w_{i-1}+w_{i+1})}_{=0}\frac{(\Delta x)^3}{3!}f^{(3)}(x_i) \\ &+\left(w_{i-1}f^{(4)}(\xi_{i-1})+w_{i+1}f^{(4)}(\xi_{i+1})\right)\frac{(\Delta x)^4}{4!} \end{aligned} \quad (8.52)$$

Substituting the obtained weight coefficients into Eq. 8.52, following equation can be obtained:

$$\begin{aligned} &\frac{1}{(\Delta x)^2}f(x_{i-1})-2\frac{1}{(\Delta x)^2}f(x_i)+\frac{1}{(\Delta x)^2}f(x_{i+1}) \\ &=f''(x_i)+\left(\frac{1}{(\Delta x)^2}f^{(4)}(\xi_{i-1})+\frac{1}{(\Delta x)^2}f^{(4)}(\xi_{i+1})\right)\frac{(\Delta x)^4}{4!} \end{aligned} \quad (8.53)$$

Leaving second derivative alone in Eq. 8.53, following Finite Difference derivative expression can be obtained:

$$f''(x_i)=\frac{f(x_{i-1})-2f(x_i)+f(x_{i+1})}{(\Delta x)^2}\underbrace{-\left(f^{(4)}(\xi_{i-1})+f^{(4)}(\xi_{i+1})\right)\frac{(\Delta x)^2}{4!}}_{R\ =\ \text{Truncation Error}} \quad (8.54)$$

In Eq. 8.54, R shows the truncation error which can be controlled or reduced by reducing Δx. As can be seen from Eq. 8.54, second-order derivative expression $f''(x_i)=\frac{f(x_{i-1})-2f(x_i)+f(x_{i+1})}{(\Delta x)^2}$ using $i-1$, i, $i+1$ points has the order of error $O(\Delta x)^2$. As can be seen from Eq. 8.39 and Eq. 8.54, general method (systematic approach) and central difference methods yield identical Finite Difference second derivative formulas.

Example 8.2: Obtain first-order Finite Difference derivative expression with general method (systematic approach) using i, $i+1$, $i+2$ points.

First-order Finite Difference derivative can be expressed as the weighted sum of the function values at i, $i+1$, $i+2$ points as below:

$$f'(x_i)=w_if(x_i)+w_{i+1}f(x_{i+1})+w_{i+2}f(x_{i+2}) \quad (8.55)$$

In Eq. 8.55, w_i, w_{i+1}, w_{i+2} indicate weight coefficients to be determined. Taylor series expansions at $i+1$, $i+2$ points can be written as following:

$$f(x_{i+1}) = f(x_i) + f'(x_i)\Delta x + \frac{f''(x_i)}{2!}(\Delta x)^2 + \frac{f^{(3)}(x_i)}{3!}(\Delta x)^3 + \frac{f^{(4)}(x_i)}{4!}(\Delta x)^4 + \ldots + \frac{f^{(n)}(x_i)}{n!}(\Delta x)^n \tag{8.56a}$$

$$f(x_{i+2}) = f(x_i) + f'(x_i)(2\Delta x) + \frac{f''(x_i)}{2!}(2\Delta x)^2 + \frac{f^{(3)}(x_i)}{3!}(2\Delta x)^3 + \frac{f^{(4)}(x_i)}{4!}(2\Delta x)^4 + \ldots + \frac{f^{(n)}(x_i)}{n!}(2\Delta x)^n \tag{8.56b}$$

Taylor series expansions can be written by discarding third- and higher order derivative terms as following:

$$\begin{aligned} f(x_{i+1}) &= f(x_i) + f'(x_i)\Delta x + \frac{f''(x_i)}{2!}(\Delta x)^2 + \underbrace{\frac{f^{(3)}(\xi_{i+1})}{3!}(\Delta x)^3}_{R_3(x_{i+1})} \\ f(x_{i+2}) &= f(x_i) + f'(x_i)(2\Delta x) + \frac{f''(x_i)}{2!}(2\Delta x)^2 + \underbrace{\frac{f^{(3)}(\xi_{i+2})}{3!}(2\Delta x)^3}_{R_3(x_{i+2})} \end{aligned} \tag{8.57}$$

Substituting Eq. 8.57 into Eq. 8.55, following expression is obtained:

$$\begin{aligned} f'(x_i) &= w_i f(x_i) + w_{i+1}\left[f(x_i) + f'(x_i)\Delta x + \frac{f''(x_i)}{2!}(\Delta x)^2 + \frac{f^{(3)}(\xi_{i+1})}{3!}(\Delta x)^3\right] \\ &\quad + w_{i+2}\left[f(x_i) + f'(x_i)(2\Delta x) + \frac{f''(x_i)}{2!}(2\Delta x)^2 + \frac{f^{(3)}(\xi_{i+2})}{3!}(2\Delta x)^3\right] \end{aligned} \tag{8.58}$$

By rearranging Eq. 8.58, it can be written as below:

$$\begin{aligned} f'(x_i) &= (w_i + w_{i+1} + w_{i+2})f(x_i) + (w_{i+1} + 2w_{i+2})\Delta x f'(x_i) \\ &\quad + (w_{i+1} + 4w_{i+2})\frac{(\Delta x)^2}{2!}f''(x_i) + \left(w_{i+1}f^{(3)}(\xi_{i+1}) + 8w_{i+2}f^{(3)}(\xi_{i+2})\right)\frac{(\Delta x)^3}{3!} \end{aligned} \tag{8.59}$$

Using equivalency of left and right sides of Eq. 8.59, following algebraic equations can be written for weight coefficients:

$$\begin{array}{ll} \text{Equation 1}: w_i + w_{i+1} + w_{i+2} = 0 & w_i + w_{i+1} + w_{i+2} = 0 \\ \text{Equation 2}: (w_{i+1} + 2w_{i+2})\Delta x = 1 & \rightarrow \quad (w_{i+1} + 2w_{i+2}) = \dfrac{1}{\Delta x} \\ \text{Equation 3}: (w_{i+1} + 4w_{i+2})\dfrac{(\Delta x)^2}{2!} = 0 & w_{i+1} + 4w_{i+2} = 0 \end{array}$$

From Eq. 3, it can be written as $w_{i+1} = -4w_{i+2}$. By substituting $w_{i+1} = -4w_{i+2}$ into Eq. 2, $w_{i+2} = \frac{-1}{2\Delta x}$ is obtained. Thus w_{i+1} is obtained as $w_{i+1} = \frac{4}{2\Delta x}$. Substituting $w_{i+1} = \frac{4}{2\Delta x}$ and $w_{i+2} = \frac{-1}{2\Delta x}$ into Eq. 1, w_i is obtained as $w_i = \frac{-3}{2\Delta x}$.

Equation 8.55 and Eq. 8.59 are just different expressions of $f'(x_i)$. Therefore, using equivalency of Eq. 8.55 and Eq. 8.59, following equation can be written:

$$\begin{aligned} w_i f(x_i) + w_{i+1} f(x_{i+1}) + w_{i+2} f(x_{i+2}) &= (w_i + w_{i+1} + w_{i+2}) f(x_i) + (w_{i+1} + 2w_{i+2})\Delta x f'(x_i) \\ &+ (w_{i+1} + 4w_{i+2})\frac{(\Delta x)^2}{2!} f''(x_i) \\ &+ \left(w_{i+1} f^{(3)}(\xi_{i+1}) + 8w_{i+2} f^{(3)}(\xi_{i+2})\right)\frac{(\Delta x)^3}{3!} \end{aligned} \tag{8.60}$$

Substituting the weight coefficients into Eq. 8.60, following equation can be obtained:

$$\frac{-3f(x_i) + 4f(x_{i+1}) - f(x_{i+2})}{2\Delta x} = f'(x_i) + \left(\frac{4}{2\Delta x} f^{(3)}(\xi_{i+1}) + 8\frac{-1}{2\Delta x} f^{(3)}(\xi_{i+2})\right)\frac{(\Delta x)^3}{3!}$$

Leaving the first-order term alone in the above equation, following first-order Finite Difference derivative expression can be obtained:

$$f'(x_i) = \frac{-3f(x_i) + 4f(x_{i+1}) - f(x_{i+2})}{2\Delta x} - \underbrace{\left(2f^{(3)}(\xi_{i+1}) - 4f^{(3)}(\xi_{i+2})\right)\frac{(\Delta x)^2}{3!}}_{R = \text{Truncation Error}} \tag{8.61}$$

In Eq. 8.61, R shows the truncation error which can be controlled or reduced by reducing Δx. As can be seen from Eq. 8.61, first-order derivative expression $f'(x_i) = \frac{-3f(x_i) + 4f(x_{i+1}) - f(x_{i+2})}{2\Delta x}$ using i, $i + 1$, $i + 2$ points has the order of error as $O(\Delta x)^2$.

***Note*:** In order to calculate derivative expressions correctly with Finite Difference method, Δx corresponding to the distance between points should be chosen very small if a small number of points are used in derivative expression.

8.4 High-Order Derivatives with Finite Difference Method

One of the following approaches can be used to obtain higher order derivatives:

1. Using general method (systematic approach).
2. Using low-order Finite Difference derivative expressions (e.g., fourth-order finite difference expression can be obtained by taking second derivative of second-order finite difference expression). This approach can sometimes be easier and more practical than systematic approach.

General methodology (systematic approach) was explained in the previous section. In this section, second approach of taking derivative of low-order derivative expressions to obtain higher derivatives will be explained.

Third-Order Finite Difference Derivative Expression:
Third-order Finite Difference derivative expression can be obtained by taking derivative of second-order finite difference expression as below:

$$
\begin{aligned}
f^{(3)}(x_i) = f'''(x_i) &= \frac{d}{dx}\left[\frac{d^2 f(x_i)}{dx^2}\right] \\
&= \frac{d}{dx}\left[\frac{f(x_{i-1}) - 2f(x_i) + f(x_{i+1})}{(\Delta x)^2}\right] \\
&= \frac{1}{(\Delta x)^2}\left[\frac{df(x_{i-1})}{dx} - 2\frac{df(x_i)}{dx} + \frac{df(x_{i+1})}{dx}\right] \\
&= \frac{1}{(\Delta x)^2}\left[\left(\frac{-f(x_{i-2}) + f(x_i)}{2\Delta x}\right) - 2\left(\frac{-f(x_{i-1}) + f(x_{i+1})}{2\Delta x}\right) + \left(\frac{-f(x_i) + f(x_{i+2})}{2\Delta x}\right)\right] \\
&= \frac{1}{2(\Delta x)^3}\left[-f(x_{i-2}) + 2f(x_{i-1}) - 2f(x_{i+1}) + f(x_{i+2})\right]
\end{aligned}
\tag{8.62}
$$

In Eq. 8.62, $\frac{df(x_{i-1})}{dx}$, $\frac{df(x_i)}{dx}$, and $\frac{df(x_{i+1})}{dx}$ denote first-order derivative expressions at $i-1$, i, $i+1$ points, respectively.

Fourth-Order Finite Difference Derivative Expression:
Fourth-order Finite Difference derivative expression can be obtained by taking derivative of second-order finite difference expression twice as below:

$$\begin{aligned}
f^{(4)}(x_i) &= \frac{d^2}{dx^2}\left[\frac{d^2 f(x_i)}{dx^2}\right] \\
&= \frac{d^2}{dx^2}\left[\frac{f(x_{i-1}) - 2f(x_i) + f(x_{i+1})}{(\Delta x)^2}\right] \\
&= \frac{1}{(\Delta x)^2}\left[\frac{d^2 f(x_{i-1})}{dx^2} - 2\frac{d^2 f(x_i)}{dx^2} + \frac{d^2 f(x_{i+1})}{dx^2}\right] \\
&= \frac{1}{(\Delta x)^2}\left[\begin{array}{l}\left(\frac{f(x_{i-2}) - 2f(x_{i-1}) + f(x_i)}{(\Delta x)^2}\right) - 2\left(\frac{f(x_{i-1}) - 2f(x_i) + f(x_{i+1})}{(\Delta x)^2}\right) \\ + \left(\frac{f(x_i) - 2f(x_{i+1}) + f(x_{i+2})}{(\Delta x)^2}\right)\end{array}\right] \\
&= \frac{1}{(\Delta x)^4}\left[f(x_{i-2}) - 4f(x_{i-1}) + 6f(x_i) - 4f(x_{i+1}) + f(x_{i+2})\right]
\end{aligned} \tag{8.63}$$

In Eq. 8.63, $\frac{d^2 f(x_{i-1})}{dx^2}$, $\frac{d^2 f(x_i)}{dx^2}$, and $\frac{d^2 f(x_{i+1})}{dx^2}$ denote second-order derivative expressions at $i-1$, i, $i+1$ points, respectively.

Example 8.3: $f(x)$ function is given as $f(x) = 2 \sin x \cos x$. Calculate $\frac{df}{dx}(x=0.5)$ $\frac{d^2 f}{dx^2}(x=0.5)$, $\frac{d^3 f}{dx^3}(x=0.5)$, $\frac{d^4 f}{dx^4}(x=0.5)$ derivatives using Finite Difference method by taking the distance between adjacent points (step length) as $\Delta x = 0.1$. Compare finite difference results with analytical derivatives.

If coordinate of point i is chosen as $x_i = 0.5$, then coordinates of $i-2$, $i-1$, $i+1$, $i+2$ points can be specified as $x_{i-2} = 0.3$, $x_{i-1} = 0.4$, $x_{i+1} = 0.6$, $x_{i+2} = 0.7$ using step length of $\Delta x = 0.1$. In this case, $\frac{df}{dx}(x=0.5)$, $\frac{d^2 f}{dx^2}(x=0.5)$, $\frac{d^3 f}{dx^3}(x=0.5)$, $\frac{d^4 f}{dx^4} \times (x=0.5)$ derivatives can be calculated with Central Difference-based Finite Difference methods as following:

- *First derivative calculation*:

$$f'(x_i) = \frac{-f(x_{i-1}) + f(x_{i+1})}{2\Delta x} \rightarrow f'(x_i = 0.5) = \frac{-f(0.4) + f(0.6)}{2(0.1)} = 1.0734$$

- *Second derivative calculation*:

$$f''(x_i) = \frac{f(x_{i-1}) - 2f(x_i) + f(x_{i+1})}{(\Delta x)^2} \rightarrow f''(x_i = 0.5) = \frac{f(0.4) - 2f(0.5) + f(0.6)}{(0.1)^2}$$
$$= -3.3547$$

- *Third derivative calculation*:

$$f^{(3)}(x_i) = f'''(x_i) = \frac{1}{2(\Delta x)^3}[-f(x_{i-2}) + 2f(x_{i-1}) - 2f(x_{i+1}) + f(x_{i+2})]$$

$$f^{(3)}(x_i = 0.5) = f'''(x_i = 0.5) = \frac{1}{2(0.1)^3}[-f(0.3) + 2f(0.4) - 2f(0.6) + f(0.7)] = -4.279$$

- *Fourth derivative calculation*:

$$f^{(4)}(x_i) = \frac{1}{(\Delta x)^4}[f(x_{i-2}) - 4f(x_{i-1}) + 6f(x_i) - 4f(x_{i+1}) + f(x_{i+2})]$$

$$f^{(4)}(x_i = 0.5) = \frac{1}{(0.1)^4}[f(0.3) - 4f(0.4) + 6f(0.5) - 4f(0.6) + f(0.7)] = 13.38$$

- *Analytical Derivative Calculation*:

$$f(x) = 2\sin x \cos x \rightarrow$$

$$\frac{df}{dx}(x) = 2\left[(\cos x)^2 - (\sin x)^2\right] \rightarrow \frac{df}{dx}(0.5) = 2\left[(\cos 0.5)^2 - (\sin 0.5)^2\right] = 1.0806$$

$$\frac{d^2f}{dx^2}(x) = -8\cos x \sin x \rightarrow \frac{d^2f}{dx^2}(0.5) = -8\cos 0.5 \sin 0.5 = -3.3659$$

$$\frac{d^3f}{dx^3}(x) = -8\left[(\cos x)^2 - (\sin x)^2\right] \rightarrow \frac{d^3f}{dx^3}(0.5) = -8\left[(\cos 0.5)^2 - (\sin 0.5)^2\right] = -4.3224$$

$$\frac{d^4f}{dx^4}(x) = 32\sin x \cos x \rightarrow \frac{d^4f}{dx^4}(0.5) = 32\sin 0.5 \cos 0.5 = 13.4635$$

From the aforementioned results, it is seen that Finite Difference method based on the step length of $\Delta x = 0.1$ yields very close results to analytical derivatives.

Solution with MATLAB Command:
Derivatives of $f(x) = 2\sin x \cos x$ function can be calculated with MATLAB using a two-step calculation.

Step 1: With *diff(f,x,n)* command in MATLAB, *n*-th order derivative expression of *f*(*x*) function can be obtained. If first-, second-, third-, and fourth-order derivatives are indicated with df1, df2, df3, and df4 terms, respectively, they can be obtained in MATLAB symbolically as following:

```
>>syms f(x)  % to indicate that f(x) is symbolic
>>f(x)=2*sin(x)*cos(x);
>>df1=diff(f,x,1); df2=diff(f,x,2); df3=diff(f,x,3); df4=diff(f,
x,4);
```

Step 2: Symbolic derivative expressions can be converted into numerical values by specifying x value in the derivative expressions and using *double* command as below:

```
>> v1=double(df1(0.5))
v1 =
1.0806
>> v2=double(df2(0.5))
v2 =
-3.3659
>> v3=double(df3(0.5))
v3 =
-4.3224
>> v4=double(df4(0.5))
v4 =
13.4635
```

***Note*:** To calculate derivatives using Central Finite Difference Method, *MATLAB Program 8.1: Derivative Calculation with Central Finite Difference Method* in Appendix can be used.

Example 8.4: $f(x)$ function is given as $f(x) = 2 \sin x \cos x$. Calculate derivative $\frac{d^2f}{dx^2}(x=0.5)$ using: (a) Central Difference method, (b) Forward Difference method, (c) Backward Difference method. Compare finite difference results with analytical derivatives. Take the distance between adjacent points (step length) as $\Delta x = 0.1$.

Analytical derivative value for $\frac{d^2f}{dx^2}(x=0.5)$ was calculated in Example 8.3 as $\frac{d^2f}{dx^2}(0.5) = -8\cos 0.5 \sin 0.5 = -3.3659$. This value can be used to compare numerical derivative results.

(a) Second derivative with Central Difference method

In Example 8.3, it was calculated as following:

$$f''(x_i) = \frac{f(x_{i-1}) - 2f(x_i) + f(x_{i+1})}{(\Delta x)^2} \rightarrow f''(0.5) = \frac{f(0.4) - 2f(0.5) + f(0.6)}{(0.1)^2} = -3.3547$$

(b) Second derivative with Forward Difference method

Derivative value using three points:

$$f''(x_i) = \frac{f(x_i) - 2f(x_{i+1}) + f(x_{i+2})}{(\Delta x)^2} \rightarrow f''(0.5) = \frac{f(0.5) - 2f(0.6) + f(0.7)}{(0.1)^2} = -3.7157$$

Derivative value using four points:

$$f^{('')}(x_i) = \frac{2f(x_i) - 5f(x_{i+1}) + 4f(x_{i+2}) - f(x_{i+3})}{(\Delta x)^2} \rightarrow$$

$$f^{('')}(0.5) = \frac{2f(0.5) - 5f(0.6) + 4f(0.7) - f(0.8)}{(0.1)^2} = -3.5028$$

Derivative value using five points:

$$f^{('')}(x_i) = \frac{35f(x_i) - 104f(x_{i+1}) + 114f(x_{i+2}) - 56f(x_{i+3}) + 11f(x_{i+4})}{12(\Delta x)^2} \rightarrow$$

$$f^{('')}(0.5) = \frac{35f(0.5) - 104f(0.6) + 114f(0.7) - 56f(0.8) + 11f(0.9)}{12(0.1)^2} = -3.3592$$

Derivative value using six points:

$$f^{('')}(x_i) = \frac{45f(x_i) - 154f(x_{i+1}) + 214f(x_{i+2}) - 156f(x_{i+3}) + 61f(x_{i+4}) - 10f(x_{i+5})}{12(\Delta x)^2} \rightarrow$$

$$f^{('')}(0.5) = \frac{45f(0.5) - 154f(0.6) + 214f(0.7) - 156f(0.8) + 61f(0.9) - 10f(1)}{12(0.1)^2} = -3.3611$$

(c) Second derivative with Backward Difference method

Derivative value using three points:

$$f''(x_i) = \frac{f(x_i) - 2f(x_{i-1}) + f(x_{i-2})}{(\Delta x)^2} \rightarrow f''(0.5) = \frac{f(0.5) - 2f(0.4) + f(0.3)}{(0.1)^2} = -2.8599$$

Derivative value using four points:

$$f^{('')}(x_i) = \frac{2f(x_i) - 5f(x_{i-1}) + 4f(x_{i-2}) - f(x_{i-3})}{(\Delta x)^2} \rightarrow$$

$$f^{('')}(0.5) = \frac{2f(0.5) - 5f(0.4) + 4f(0.3) - f(0.2)}{(0.1)^2} = -3.4687$$

Derivative value using five points:

$$f^{('')}(x_i) = \frac{35f(x_i) - 104f(x_{i-1}) + 114f(x_{i-2}) - 56f(x_{i-3}) + 11f(x_{i-4})}{12(\Delta x)^2} \rightarrow$$

$$f^{('')}(0.5) = \frac{35f(0.5) - 104f(0.4) + 114f(0.3) - 56f(0.2) + 11f(0.1)}{12(0.1)^2} = -3.3864$$

Derivative value using six points:

$$f^{('')}(x_i) = \frac{45f(x_i) - 154f(x_{i-1}) + 214f(x_{i-2}) - 156f(x_{i-3}) + 61f(x_{i-4}) - 10f(x_{i-5})}{12(\Delta x)^2} \rightarrow$$

$$f^{('')}(0.5) = \frac{45f(0.5) - 154f(0.4) + 214f(0.3) - 156f(0.2) + 61f(0.1) - 10f(0)}{12(0.1)^2} = -3.3632$$

From above results, it is seen that Central Difference method is more efficient than Forward and Backward Difference methods. It gives close results to analytical derivatives with fewer points.

8.5 Partial Derivatives with Finite Difference Method

Partial derivative expressions based on Finite Difference method can be easily obtained using Finite Difference expressions of ordinary derivatives. In partial derivative expression, all variables are kept constant except for one variable and Finite Difference derivative expression is obtained for non-constant variable.

To obtain partial derivatives of an $f(x,y)$ function, x-y plane can be discretized into nodes or points as in Fig. 8.3. In Fig. 8.3, i index indicates x direction and j index indicates y direction. Using i and j indices, center point where partial derivative will be calculated can be represented with (i, j). In this case, forward and backward points can be referred to as $(i+1, j)$ and $(i, j-1)$ in x direction. Similarly, forward and backward points can be referred to as $(i, j+1)$ and $(i-1, j)$ in y direction. For simplicity in derivative expressions, we can assume uniform distances between adjacent points (step length) such as Δx in x direction and Δy in y direction. To obtain partial derivative of $f(x,y)$ function with respect to x variable, x coordinate (i index) is varied while y coordinate (j index) is kept constant. Similarly, to obtain partial derivative with respect to y variable, y coordinate (j index) is varied while

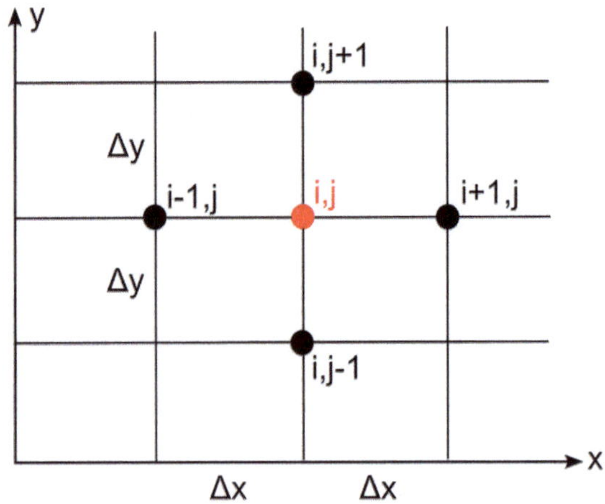

Fig. 8.3 Points used in the calculation of the partial derivative of an $f(x,y)$ function using finite difference method

x coordinate (i index) is kept constant. First-, second-, and fourth-order partial derivatives are explained below.

8.5.1 First-Order Partial Derivative Expressions with Forward, Backward, and Central Finite Difference Methods

By using first-order central finite difference expressions of ordinary derivatives, second-order partial derivative expressions can be obtained as following:

Forward Difference method:

$$\begin{aligned} \frac{\partial f}{\partial x}(x_i, y_j) &= \frac{f(x_{i+1}, y_j) - f(x_i, y_j)}{\Delta x} \quad (j = \text{fixed}) \\ \frac{\partial f}{\partial y}(x_i, y_j) &= \frac{f(x_i, y_{j+1}) - f(x_i, y_j)}{\Delta y} \quad (i = \text{fixed}) \end{aligned} \tag{8.64}$$

Backward Difference method:

$$\begin{aligned} \frac{\partial f}{\partial x}(x_i, y_j) &= \frac{f(x_i, y_j) - f(x_{i-1}, y_j)}{\Delta x} \quad (j = \text{fixed}) \\ \frac{\partial f}{\partial y}(x_i, y_j) &= \frac{f(x_i, y_j) - f(x_i, y_{j-1})}{\Delta y} \quad (i = \text{fixed}) \end{aligned} \tag{8.65}$$

Central Difference method:

$$\begin{aligned} \frac{\partial f}{\partial x}(x_i, y_j) &= \frac{f(x_{i+1}, y_j) - f(x_{i-1}, y_j)}{2\Delta x} \quad (j = \text{fixed}) \\ \frac{\partial f}{\partial y}(x_i, y_j) &= \frac{f(x_i, y_{j+1}) - f(x_i, y_{j-1})}{2\Delta y} \quad (i = \text{fixed}) \end{aligned} \tag{8.66}$$

8.5.2 Second-Order Partial Derivative Expressions with Finite Difference Method

By using first-order central finite difference expressions of ordinary derivatives, second-order partial derivative expressions can be obtained as following:

$$
\begin{aligned}
\frac{\partial^2 f}{\partial x^2}(x_i, y_j) &= \frac{f(x_{i-1}, y_j) - 2f(x_i, y_j) + f(x_{i+1}, y_j)}{(\Delta x)^2} \\
\frac{\partial^2 f}{\partial y^2}(x_i, y_j) &= \frac{f(x_i, y_{j-1}) - 2f(x_i, y_j) + f(x_i, y_{j+1})}{(\Delta y)^2} \\
\frac{\partial^2 f}{\partial x \partial y}(x_i, y_j) &= \frac{\partial}{\partial x}\left(\frac{\partial f}{\partial y}(x_i, y_j)\right) = \frac{\partial}{\partial x}\left(\frac{f(x_i, y_{j+1}) - f(x_i, y_{j-1})}{2\Delta y}\right) \\
&= \frac{1}{2\Delta y}\left(\frac{\partial f(x_i, y_{j+1})}{\partial x} - \frac{\partial f(x_i, y_{j-1})}{\partial x}\right) \\
&= \frac{1}{2\Delta y}\left(\frac{f(x_{i+1}, y_{j+1}) - f(x_{i-1}, y_{j+1})}{2\Delta x} - \frac{f(x_{i+1}, y_{j-1}) - f(x_{i-1}, y_{j-1})}{2\Delta x}\right) \\
&= \frac{1}{4\Delta x \Delta y}\left(-f(x_{i-1}, y_{j+1}) + f(x_{i-1}, y_{j-1}) + f(x_{i+1}, y_{j+1}) - f(x_{i+1}, y_{j-1})\right)
\end{aligned}
\tag{8.67}
$$

8.5.3 *Fourth-Order Partial Derivative Expression with Finite Difference Method*

By using second-order central finite difference expressions of ordinary derivatives, fourth-order partial derivative expressions can be obtained as following:

$$
\begin{aligned}
\frac{\partial^4 f}{\partial x^4}(x_i, y_j) &= \frac{f(x_{i-2}, y_j) - 4f(x_{i-1}, y_j) + 6f(x_i, y_j) - 4f(x_{i+1}, y_j) + f(x_{i+2}, y_j)}{(\Delta x)^4} \\
\frac{\partial^4 f}{\partial y^4}(x_i, y_j) &= \frac{f(x_i, y_{j-2}) - 4f(x_i, y_{j-1}) + 6f(x_i, y_j) - 4f(x_i, y_{j+1}) + f(x_i, y_{j+2})}{(\Delta y)^4} \\
\frac{\partial^4 f}{\partial x^2 \partial y^2}(x_i, y_j) &= \frac{\begin{bmatrix} f(x_{i-1}, y_{j+1}) - 2f(x_{i-1}, y_j) + f(x_{i-1}, y_{j-1}) - 2f(x_i, y_{j+1}) + \\ 4f(x_i, y_j) - 2f(x_i, y_{j-1}) + f(x_{i+1}, y_{j+1}) - 2f(x_{i+1}, y_j) + f(x_{i+1}, y_{j-1}) \end{bmatrix}}{(\Delta x)^2 (\Delta y)^2}
\end{aligned}
\tag{8.68}
$$

8.6 Improvement of Derivative Results with Richardson Extrapolation

As mentioned before, accuracy of derivative results can be increased by using small step length of $\Delta x = h_1$ between adjacent points. Another alternative to increase accuracy is calculating derivatives for different D_1 values and then extrapolating these results to obtain more accurate derivative value. This approach is known as Richardson Extrapolation. Richardson Extrapolation method can be easily explained using second-order finite difference derivative expression. To explain the method, second-order derivative result and the error for step length of $\Delta x = h_1$ can be indicated with D_1 and E_1, respectively. Similarly, second-order derivative result and the error for step length of $\Delta x = h_2$ can be indicated with D_2 and E_2, respectively. Orders of errors for E_1 and E_2 can be expressed as $E_1 = O(h_1^2)$ and $E_2 = O(h_2^2)$. In this case, accurate derivative result (D) using D_1, E_1 and D_2, E_2 can be expressed as below:

for $\Delta x = h_1$: $D = D_1 + E_1 = D_1 + O(h_1^2)$
for $\Delta x = h_2$: $D = D_2 + E_2 = D_2 + O(h_2^2)$

Relationship between error expressions of E_1 and E_2 and step lengths of h_1 and h_2 can be stated as following:

$$\frac{E_1}{E_2} \approx \frac{h_1^2}{h_2^2} \rightarrow E_1 \approx E_2 \frac{h_1^2}{h_2^2}$$

Using previous equation, error E_2 can be written in terms of derivative results and step lengths as below:

$$D = D_1 + E_1, \quad D = D_2 + E_2 \rightarrow D_1 + E_1 = D_2 + E_2 \rightarrow E_2 = D_1 + E_1 - D_2$$
$$\rightarrow E_2 = D_1 - D_2 + E_2 \frac{h_1^2}{h_2^2} \rightarrow E_2 = \frac{D_1 - D_2}{\left(1 - \frac{h_1^2}{h_2^2}\right)}$$

Then, using previous equation, accurate derivative result can be expressed as below:

$$D = D_2 + E_2 = D_2 + \frac{D_1 - D_2}{\left(1 - \frac{h_1^2}{h_2^2}\right)}$$

In the special case of $h_2 = \frac{h_1}{2}$, accurate derivative result can be expressed as following:

$$D = \frac{4}{3} D_2 - \frac{1}{3} D_1 \tag{8.69}$$

Equation 8.69 leads to more accurate derivative value compared to D_1 and D_2 values.

Example 8.5: $f(x)$ function is given as $f(x) = 2 \sin x \cos x$. Calculate second-order derivative $\frac{d^2 f}{dx^2}\left(x = \frac{\pi}{6}\right)$ using Finite Difference method and then apply Richardson Extrapolation to improve the result. Compare the improved result with analytical one.

To apply Richardson Extrapolation method, second-order derivative values can be calculated with Finite Difference method using two different step lengths such as $\Delta x = h_1 = \frac{\pi}{6}$ and $\Delta x = h_2 = \frac{\pi}{12}$..

For $\Delta x = h_1 = \frac{\pi}{6}$:

$$f''(x_i)=\frac{f(x_{i-1})-2f(x_i)+f(x_{i+1})}{(\Delta x)^2}\rightarrow D_1=f''\left(\frac{\pi}{6}\right)=\frac{f(0)-2f\left(\frac{\pi}{6}\right)+f\left(\frac{2\pi}{6}\right)}{\left(\frac{\pi}{6}\right)^2}=-3.1589$$

For $\Delta x=h_2=\frac{\pi}{12}$:

$$f''(x_i)=\frac{f(x_{i-1})-2f(x_i)+f(x_{i+1})}{(\Delta x)^2}\rightarrow D_2=f''\left(\frac{\pi}{6}\right)=\frac{f\left(\frac{\pi}{12}\right)-2f\left(\frac{\pi}{6}\right)+f\left(\frac{3\pi}{12}\right)}{\left(\frac{\pi}{12}\right)^2}=-3.3857$$

Using derivative values of D_1 and D_2, corrected (improved) value can be obtained with Richardson extrapolation as following:

$$D=\frac{4}{3}D_2-\frac{1}{3}D_1=-3.4613$$

Analytical Derivative Calculation:

$$f(x)=2\sin x\cos x\rightarrow\frac{df}{dx}(x)=2\left[(\cos x)^2-(\sin x)^2\right]\rightarrow\frac{d^2f}{dx^2}(x)=-8\cos x\sin x$$

$$\frac{d^2f}{dx^2}\left(\frac{\pi}{6}\right)=-8\cos\left(\frac{\pi}{6}\right)\sin\left(\frac{\pi}{6}\right)=-3.4641$$

From the above results, it is seen that Richardson extrapolation improves Finite Difference results. Improved result based on selected step lengths is very close to the analytical derivative value.

8.7 Derivative Calculation with Interpolation Methods

8.7.1 Derivative Calculation Using Lagrange Interpolation Polynomials

Another derivative calculation method is based on interpolation. In this approach, first, an interpolation polynomial is passed through the selected points and then derivative is obtained by taking the derivative of this interpolation function. As explained in Chap. 6 (Interpolation), Lagrange interpolation method is an efficient method that allows the interpolation polynomial to be written directly. Therefore, this method can be used in obtaining derivative expressions as well. Since accuracy of the interpolation polynomial will also affect accuracy of derivative result, interpolation function should be obtained correctly. In Chap. 6 (Interpolation), it was mentioned that non-uniform points lead to more accurate interpolation functions. Therefore, in generating Lagrange interpolation polynomials in this section, non-uniform points such as those generated by Chebyshev and Gauss-Lobatto methods will be used.

In creating interpolation polynomials to be used in derivatives, the number of points must be equal to or greater than the order of derivative. If the order of derivative is m and the number of points is n, this relationship can be expressed as $n \geq m + 1$. Using interpolation methods, first and second derivative expressions can be obtained as below:

First-Order Derivative Expression with Interpolation Method:
Parabolic Lagrange interpolation polynomial passing through $i-1$, i, $i+1$ points can be expressed as below:

$$\begin{aligned} f(x) &= \frac{(x-x_i)(x-x_{i+1})}{(x_{i-1}-x_i)(x_{i-1}-x_{i+1})} f(x_{i-1}) + \frac{(x-x_{i-1})(x-x_{i+1})}{(x_i-x_{i-1})(x_i-x_{i+1})} f(x_i) \\ &+ \frac{(x-x_{i-1})(x-x_i)}{(x_{i+1}-x_{i-1})(x_{i+1}-x_i)} f(x_{i+1}) \end{aligned} \tag{8.70}$$

Interpolation polynomial can be written more clearly as below:

$$\begin{aligned} f(x) &= \frac{x^2-(x_i+x_{i+1})x+x_i x_{i+1}}{(x_{i-1}-x_i)(x_{i-1}-x_{i+1})} f(x_{i-1}) + \frac{x^2-(x_{i-1}+x_{i+1})x+x_{i-1}x_{i+1}}{(x_i-x_{i-1})(x_i-x_{i+1})} f(x_i) \\ &+ \frac{x^2-(x_{i-1}+x_i)x+x_{i-1}x_i}{(x_{i+1}-x_{i-1})(x_{i+1}-x_i)} f(x_{i+1}) \end{aligned}$$

Derivative of the above function can be written as following:

$$\begin{aligned} f'(x) = &\frac{2x-(x_i+x_{i+1})}{(x_{i-1}-x_i)(x_{i-1}-x_{i+1})} f(x_{i-1}) + \frac{2x-(x_{i-1}+x_{i+1})}{(x_i-x_{i-1})(x_i-x_{i+1})} f(x_i) \\ &+ \frac{2x-(x_{i-1}+x_i)}{(x_{i+1}-x_{i-1})(x_{i+1}-x_i)} f(x_{i+1}) \end{aligned}$$

Using above expression, first-order derivative at point $x = x_i$ can be written in terms of distance or step length Δx ($\Delta x = x_i - x_{i-1} = x_{i+1} - x_i$) as below:

$$\begin{aligned} f'(x_i) = &\frac{2x_i-(x_i+x_{i+1})}{(x_{i-1}-x_i)(x_{i-1}-x_{i+1})} f(x_{i-1}) + \frac{2x_i-(x_{i-1}+x_{i+1})}{(x_i-x_{i-1})(x_i-x_{i+1})} f(x_i) \\ &+ \frac{2x_i-(x_{i-1}+x_i)}{(x_{i+1}-x_{i-1})(x_{i+1}-x_i)} f(x_{i+1}) \end{aligned}$$

$$\begin{aligned} f'(x_i) = &\frac{(x_i-x_{i+1})}{(x_{i-1}-x_i)(x_{i-1}-x_{i+1})} f(x_{i-1}) + \frac{(x_i-x_{i-1})+(x_i-x_{i+1})}{(x_i-x_{i-1})(x_i-x_{i+1})} f(x_i) \\ &+ \frac{(x_i-x_{i-1})}{(x_{i+1}-x_{i-1})(x_{i+1}-x_i)} f(x_{i+1}) \end{aligned}$$

$$f'(x_i)=\frac{-\Delta x}{(-\Delta x)(-2\Delta x)}f(x_{i-1})+\frac{(\Delta x)+(-\Delta x)}{(\Delta x)(-\Delta x)}f(x_i)+\frac{(\Delta x)}{(2\Delta x)(\Delta x)}f(x_{i+1})$$

$$f'(x_i)=\frac{-\Delta x}{2(\Delta x)^2}f(x_{i-1})+\frac{(\Delta x)+(-\Delta x)}{-(\Delta x)^2}f(x_i)+\frac{(\Delta x)}{2(\Delta x)^2}f(x_{i+1})$$

$$f'(x_i)=\underbrace{\frac{-1}{2(\Delta x)}}_{w_{i-1}}f(x_{i-1})+\underbrace{\frac{0}{-(\Delta x)^2}}_{w_i}f(x_i)+\underbrace{\frac{1}{2(\Delta x)}}_{w_{i+1}}f(x_{i+1})$$

$$=\frac{-f(x_{i-1})+f(x_{i+1})}{2(\Delta x)} \tag{8.71}$$

First-order derivative expression obtained by taking derivative of parabolic interpolation polynomial in Eq. 8.71 is the same with first-order derivative expression obtained using central finite difference method.

Second-Order Derivative Expression with Interpolation Method:
Taking derivative of parabolic interpolation polynomial in Eq. 8.69 twice, second-order derivative expression can be obtained as following:

$$f''(x)=\frac{2}{(x_{i-1}-x_i)(x_{i-1}-x_{i+1})}f(x_{i-1})+\frac{2}{(x_i-x_{i-1})(x_i-x_{i+1})}f(x_i)+\frac{2}{(x_{i+1}-x_{i-1})(x_{i+1}-x_i)}f(x_{i+1})$$

Using above expression, second-order derivative at point $x=x_i$ can be written in terms of step length Δx ($\Delta x=x_i-x_{i-1}=x_{i+1}-x_i$) as below:

$$f''(x_i)=\frac{2}{(-\Delta x)(-2\Delta x)}f(x_{i-1})+\frac{2}{(\Delta x)(-\Delta x)}f(x_i)+\frac{2}{(2\Delta x)(\Delta x)}f(x_{i+1})$$

$$f''(x_i)=\frac{2}{2(\Delta x)^2}f(x_{i-1})+\frac{2}{-(\Delta x)^2}f(x_i)+\frac{2}{2(\Delta x)^2}f(x_{i+1})$$

$$f''(x_i)=\frac{2}{2(\Delta x)^2}f(x_{i-1})+\frac{2}{-(\Delta x)^2}f(x_i)+\frac{2}{2(\Delta x)^2}f(x_{i+1})$$

$$f''(x_i)=\underbrace{\frac{1}{(\Delta x)^2}}_{w_{i-1}}f(x_{i-1})-\underbrace{\frac{2}{(\Delta x)^2}}_{w_i}f(x_i)+\underbrace{\frac{1}{(\Delta x)^2}}_{w_{i+1}}f(x_{i+1})=\frac{f(x_{i-1})-2f(x_i)+f(x_{i+1})}{(\Delta x)^2} \tag{8.72}$$

Second-order derivative expression obtained by taking the derivative of parabolic interpolation polynomial in Eq. 8.72 is the same with second-order derivative expression obtained using finite difference method.

8.7.2 Derivative Calculation with Differential Quadrature Method

Creating an interpolation polynomial and then taking its derivative provides a direct and systematic approach in obtaining derivative expressions. However, this approach loses its efficiency for higher order derivatives due to the difficulty in generation and derivation of higher order interpolation polynomials. To alleviate this process, several methods were proposed. Differential Quadrature (DQ) method is one of them. Differential Quadrature and its more advanced form known as Generalized Differential Quadrature (GDQ) method are increasingly used in numerical derivative calculations. Differential quadrature method was first proposed by Bellman and Casti in 1971. Later, it was improved by Shu and Richards for easy usage. Comprehensive information about Differential Quadrature method can be found in references of [9–12].

Inspiration for Differential Quadrature method is quadrature methods in numerical integration. In numerical integration with quadrature methods, integral of a function is expressed as the weighted sum of function values at different points. Similar approach is used in calculating derivatives with Differential Quadrature method. In Differential Quadrature method, *m*-th order derivative of a function is also expressed as the weighted sum of function values at different points as following (Fig. 8.4):

$$\frac{d^m f}{dx^m}(x_i) = f^{(m)}(x_i) = w_{i1} f(x_1) + w_{i2} f(x_2) + w_{i3} f(x_3) + \ldots + w_{in} f(x_n) = \sum_{j=1}^{n} w_{ij} f(x_j) \tag{8.73}$$

***Note*:** To calculate derivatives with Eq. 8.73, the number of points n must be equal to or greater than derivative order m as expressed with $n \geq m + 1$ relationship.

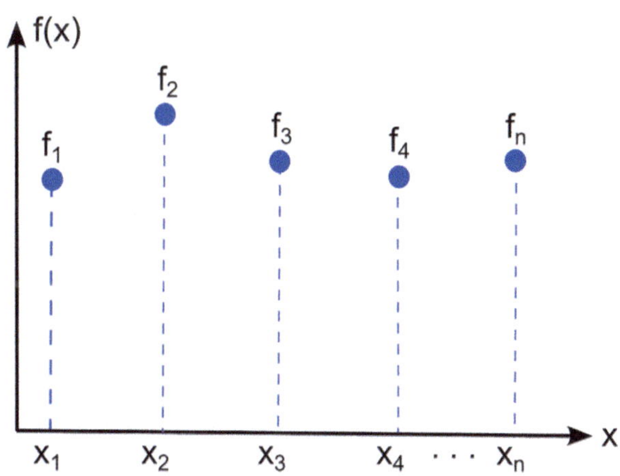

Fig. 8.4 Function values used in the numerical derivative calculation of an *f*(*x*) function with Differential Quadrature method

In the early version of the Differential Quadrature method, interpolation polynomial function $f(x)$ was expressed as below:

$$f(x) = c_0.1 + c_1 x + c_2 x^2 + c_3 x^3 + \ldots + c_n x^n \tag{8.74}$$

From Eq. 8.74, it is seen that the n-th order interpolation polynomial consists of basic terms of 1, x, x^2, x^3,..., x^n. Therefore, derivative of interpolation function will also consist of derivatives of basic terms of 1, x, x^2, x^3,..., x^n.

A quadrature method that correctly calculates derivatives of basic terms will also correctly calculate the derivative of interpolation function. Therefore, weight coefficients in Eq. 8.73 can be determined using derivatives of basic terms. For example, Eq. 8.73 can be utilized to express the first-order derivative of an n-th order polynomial function at x_i point as following:

$$\frac{df}{dx}(x_i) = w_{i1} f(x_1) + w_{i2} f(x_2) + w_{i3} f(x_3) + \ldots + w_{in} f(x_n)$$

Weight coefficients in the above equation can be obtained by creating and solving necessary equations. Necessary equations can be generated as below by substituting basic functions 1, x, x^2, x^3,..., x^n into the above equation as in early version of differential quadrature method by Bellman and Casti

$$f(x) = 1 \rightarrow \frac{df(x)}{dx} = 0 \rightarrow \frac{df(x_i)}{dx} = 0 \rightarrow 0 = w_{i1} 1 + w_{i2} 1 + w_{i3} 1 + \ldots + w_{in} 1$$

$$f(x) = x \rightarrow \frac{df(x)}{dx} = 1 \rightarrow \frac{df(x_i)}{dx} = 1 \rightarrow 1 = w_{i1} x_1 + w_{i2} x_2 + w_{i3} x_3 + \ldots + w_{in} x_n$$

$$f(x) = x^2 \rightarrow \frac{df(x)}{dx} = 2x \rightarrow \frac{df(x_i)}{dx} = 2x_i \rightarrow 2x_i = w_{i1} x_1^2 + w_{i2} x_2^2 + w_{i3} x_3^2 + \ldots + w_{in} x_n^2$$

$$f(x) = x^3 \rightarrow \frac{df(x)}{dx} = 3x^2 \rightarrow \frac{df(x_i)}{dx} = 3x_i^2 \rightarrow 3x_i^2 = w_{i1} x_1^3 + w_{i2} x_2^3 + w_{i3} x_3^3 + \ldots + w_{in} x_n^3$$

$$\ldots$$

$$f(x) = x^n \rightarrow \frac{df(x)}{dx} = nx^{n-1} \rightarrow \frac{df(x_i)}{dx} = nx_i^{n-1} \rightarrow nx_i^{n-1} = w_{i1} x_1^n + w_{i2} x_2^n + w_{i3} x_3^n + \ldots + w_{in} x_n^n$$

Obtained equations can be expressed in matrix form as below:

$$\underbrace{\begin{bmatrix} 1 & 1 & 1 & \ldots & 1 \\ x_1 & x_2 & x_3 & \ldots & x_n \\ x_1^2 & x_2^2 & x_3^2 & \ldots & x_n^2 \\ x_1^3 & x_2^3 & x_3^3 & \ldots & x_n^3 \\ \ldots & \ldots & \ldots & \ldots & \ldots \\ x_1^n & x_2^n & x_3^n & \ldots & x_n^n \end{bmatrix}}_{A} \begin{Bmatrix} w_{i1} \\ w_{i2} \\ w_{i3} \\ w_{i4} \\ \ldots \\ w_{in} \end{Bmatrix} = \begin{Bmatrix} 0 \\ 1 \\ 2x_i \\ 3x_i^2 \\ \ldots \\ nx_i^{n-1} \end{Bmatrix} \tag{8.75}$$

From solution of Eq. 8.75, weight coefficients w_{ij} that correctly calculate first-order derivative of n-th order polynomial function at x_i point are obtained.

Coefficient matrix A in Eq. 8.75 is called Vandermonde matrix. If the order of polynomial is high or equivalently if the number of points used in creating interpolation polynomial is high, Vandermonde matrix becomes ill-conditioned and solution of Eq. 8.75 becomes difficult and prone to error. For example, for higher than 13 points (i.e., 12th order polynomial), weight coefficients cannot be calculated correctly.

Richard and Shu developed an alternative approach to calculate weight coefficients accurately and easily. In the method of Richard and Shu, polynomial function is expressed with Legendre or Lagrange interpolation polynomials instead of power series as in Eq. 8.74. For example, a general polynomial function can be expressed using Lagrange interpolation method as following:

$$f(x) = L_1(x)f(x_1) + L_2(x)f(x_2) + \ldots + L_j(x)f(x_j) + \ldots + L_n(x)f(x_n) \tag{8.76}$$

Derivative expression can be obtained by taking derivative of Eq. 8.76. For example, first-order derivative expression can be written as below:

$$\frac{df(x)}{dx} = \frac{dL_1(x)}{dx}f(x_1) + \frac{dL_2(x)}{dx}f(x_2) + \ldots + \frac{dL_j(x)}{dx}f(x_j) + \ldots + \frac{dL_n(x)}{dx}(x)f(x_n) \tag{8.77}$$

In Eq. 8.77, $L_1(x)$, $L_2(x)$, $L_3(x)$,..., $L_n(x)$ show basic Lagrange functions. Basic Lagrange functions can be written in general form as following:

$$L_j(x) = \prod_{k=1,\,k\neq j}^{n} \frac{x - x_k}{x_j - x_k} \tag{8.78}$$

Alternatively, basic Lagrange functions can be expressed using a new function $\varphi(x)$, which is expressed as $\varphi(x) = \prod_{k=1}^{n} (x - x_k)$, and its derivative $\varphi'(x)$ as following:

$$L_j(x) = \frac{\varphi(x)}{(x - x_j)\varphi'(x_j)} \quad \left(\varphi'(x_j) = \frac{d\varphi(x_j)}{dx} = \prod_{k=1(k\neq j)}^{n} (x_j - x_k)\right) \tag{8.79}$$

where $\varphi(x)$ and its derivative are expressed as $\varphi(x) = \prod_{k=1}^{n} (x - x_k)$ and $\varphi'(x) = \frac{d\varphi(x)}{dx} = \prod_{k=1}^{n} (x - x_k)$, respectively.

Using Eq. 8.77, first-order derivative of $f(x)$ function at x_i point can be written as below:

$$\frac{df(x_i)}{dx} = \underbrace{\frac{dL_1(x_i)}{dx}}_{w_{i1}} f_1 + \underbrace{\frac{dL_2(x_i)}{dx}}_{w_{i2}} f_2 + \ldots + \underbrace{\frac{dL_j(x_i)}{dx}}_{w_{ij}} f_j + \ldots + \underbrace{\frac{dL_n(x_i)}{dx}}_{w_{in}} f_n \tag{8.80}$$

Derivatives of basic Lagrange functions give weight coefficients in Eq. 8.73. Therefore, weight coefficients can be expressed using $\varphi(x)$ function and its derivative as following:

$$w_{ij} = \frac{dL_j}{dx}(x_i) = \frac{\varphi'(x_i)}{(x_i - x_j)\varphi'(x_j)} \tag{8.81}$$

Similarly, weight coefficients for higher order derivatives can be obtained by taking higher order derivatives of basic Lagrange functions. However, this process becomes complex as the number of points and order of derivative increase. Shu and Richards developed a simple method to obtain weight coefficients of higher order derivatives from low-order ones in a successive way. To explain the method, *1*-st and *m*-th order derivatives of an $f(x)$ function at $i = 1,2,3,\ldots\ n$ points can be expressed in matrix form as following:

First-order derivative of $f(x)$ function at $i = 1,2,3,\ldots, n$ points:

$$\left.\begin{aligned} \frac{df(x_1)}{dx} &= A_{1,1}f_1 + A_{1,2}f_2 + A_{1,3}f_3 + \ldots + A_{1,n}f_n \\ \frac{df(x_2)}{dx} &= A_{2,1}f_1 + A_{2,2}f_2 + A_{2,3}f_3 + \ldots + A_{2,n}f_n \\ \frac{df(x_3)}{dx} &= A_{3,1}f_1 + A_{3,2}f_2 + A_{3,3}f_3 + \ldots + A_{3,n}f_n \\ &\ldots \\ \frac{df(x_n)}{dx} &= A_{n,1}f_1 + A_{n,2}f_2 + A_{n,3}f_3 + \ldots + A_{n,n}f_n \end{aligned}\right\} \Rightarrow$$

$$\begin{Bmatrix} \frac{df(x_1)}{dx} \\ \frac{df(x_2)}{dx} \\ \frac{df(x_3)}{dx} \\ \ldots \\ \frac{df(x_n)}{dx} \end{Bmatrix} = \underbrace{\begin{bmatrix} A_{1,1} & A_{1,2} & A_{1,3} & \ldots & A_{1,n} \\ A_{2,1} & A_{2,2} & A_{2,3} & \ldots & A_{2,n} \\ A_{3,1} & A_{3,2} & A_{3,3} & \ldots & A_{3,n} \\ \ldots & \ldots & \ldots & \ldots & \ldots \\ A_{n,1} & A_{n,2} & A_{n,3} & \ldots & A_{n,n} \end{bmatrix}}_{A} \begin{Bmatrix} f_1 \\ f_2 \\ f_3 \\ \ldots \\ f_n \end{Bmatrix} \tag{8.82}$$

where $A_{i,\,j}$ terms in Eq. 8.82 can be expressed as following:

$$A_{i,j} = \frac{1}{x_j - x_i} \prod_{k=1(k \neq i,j)}^{n} \frac{x_i - x_k}{x_j - x_k} \quad (i \neq j), \quad A_{i,i} = -\sum_{j=1, j \neq i}^{n} A_{i,j} \tag{8.83}$$

m-th order derivative of $f(x)$ function at $i = 1,2,3,\ldots\ n$ points:

$$\left.\begin{aligned}
\frac{d^m f(x_1)}{dx^m} &= C_{1,1}^{(m)} f_1 + C_{1,2}^{(m)} f_2 + C_{1,3}^{(m)} f_3 + \ldots + C_{1,n}^{(m)} f_n \\
\frac{d^m f(x_2)}{dx^m} &= C_{2,1}^{(m)} f_1 + C_{2,2}^{(m)} f_2 + C_{2,3}^{(m)} f_3 + \ldots + C_{2,n}^{(m)} f_n \\
\frac{d^m f(x_3)}{dx^m} &= C_{3,1}^{(m)} f_1 + C_{3,2}^{(m)} f_2 + C_{3,3}^{(m)} f_3 + \ldots + C_{3,n}^{(m)} f_n \\
\ldots \\
\frac{d^m f(x_n)}{dx^m} &= C_{n,1}^{(m)} f_1 + C_{n,2}^{(m)} f_2 + C_{n,3}^{(m)} f_3 + \ldots + C_{n,n}^{(m)} f_n
\end{aligned}\right\} \Rightarrow$$

$$\begin{Bmatrix} \frac{d^m f(x_1)}{dx^m} \\ \frac{d^m f(x_2)}{dx^m} \\ \frac{d^m f(x_3)}{dx^m} \\ \ldots \\ \frac{d^m f(x_n)}{dx^m} \end{Bmatrix} = \underbrace{\begin{bmatrix} C_{1,1}^{(m)} & C_{1,2}^{(m)} & C_{1,3}^{(m)} & \ldots & C_{1,n}^{(m)} \\ C_{2,1}^{(m)} & C_{2,2}^{(m)} & C_{2,3}^{(m)} & \ldots & C_{2,n}^{(m)} \\ C_{3,1}^{(m)} & C_{3,2}^{(m)} & C_{3,3}^{(m)} & \ldots & C_{3,n}^{(m)} \\ \ldots & \ldots & \ldots & \ldots & \ldots \\ C_{n,1}^{(m)} & C_{n,2}^{(m)} & C_{n,3}^{(m)} & \ldots & C_{n,n}^{(m)} \end{bmatrix}}_{C^{(m)}} \begin{Bmatrix} f_1 \\ f_2 \\ f_3 \\ \ldots \\ f_n \end{Bmatrix} \tag{8.84}$$

In Eqs. 8.82 and 8.84, A and $C^{(m)}$ matrices denote matrices of weight coefficients for first-order and m-th order derivatives, respectively. In these matrices, each row represents weight coefficients for the derivative of function at a point. For example, first row denotes weight coefficients of derivative at first point, second row denotes weight coefficients of derivative at second point, and so on. $C^{(m)}$ coefficients can be obtained using A and $C^{(m-1)}$ coefficients as following:

$$C^{(m)} = AC^{(m-1)} \tag{8.85}$$

Differential Quadrature method based on Eq. 8.85 is called as Generalized Differential Quadrature method in the literature. From now on, Differential Quadrature method will be used to mean Generalized Differential Quadrature method.

It was previously mentioned that accuracy of numerical derivative decreases when more than 13 uniformly distributed points are used in interpolation-based derivative methods. To keep accuracy high, *non-uniform* points are used in interpolation-based derivative methods such as Differential Quadrature method. Indicating the number of points with n, n number of non-uniform points in an [a, b] interval can be determined by using Gauss-Lobatto-Chebyshev or shortly Chebyshev equation as below:

$$x_i = a + \frac{1}{2}\left(1 - \cos\frac{(i-1)\pi}{n-1}\right)(b-a) \quad (i = 1, 2, \ldots, n) \tag{8.86}$$

When Chebyshev points are used in derivative calculations, it is observed that Differential Quadrature method maintains its accuracy for the number of points >13. Since more points can be used easily in Differential Quadrature method, accuracy of Differential Quadrature method is higher than Finite Difference methods. Also, with the method proposed by Shu and Richards, higher order derivatives can be expressed more flexibly and easily with Differential Quadrature method than with Finite Difference method.

***Note*:** Differential Quadrature method can be considered as an easier and more systematic way of derivatives of Lagrange interpolation polynomial.

8.7.3 Partial Derivative with Differential Quadrature Method

Partial derivative with Differential Quadrature method is obtained in a similar way with ordinary derivative. All variables are kept constant except for one variable. Then, partial derivative expression is obtained by applying ordinary derivative expression for non-constant variable. To obtain partial derivative expressions using Differential Quadrature method, x-y plane can be discretized into nodes or points as in Fig. 8.5.

Using Fig. 8.5, partial derivative of a bivariate $f(x, y)$ function at a general (i,j) point can be obtained. Here, i index is used to represent x direction and j index is used to represent y direction. For simplicity, we can show the value of $f(x, y)$ function at (i,j) point with $f_{i,\ j}$. To calculate derivative of $f(x,y)$ function with respect to x variable points on the i line and to calculate derivative with respect to y variable points on the j line are used.

For convenience, we can indicate weight coefficients for first-order derivative of $f(x,y)$ function with respect to x variable with $Ax_{i,\ j}$ and with respect to y variable with $Ay_{i,\ j}$ coefficients. Similarly, we can indicate weight coefficients for second-order

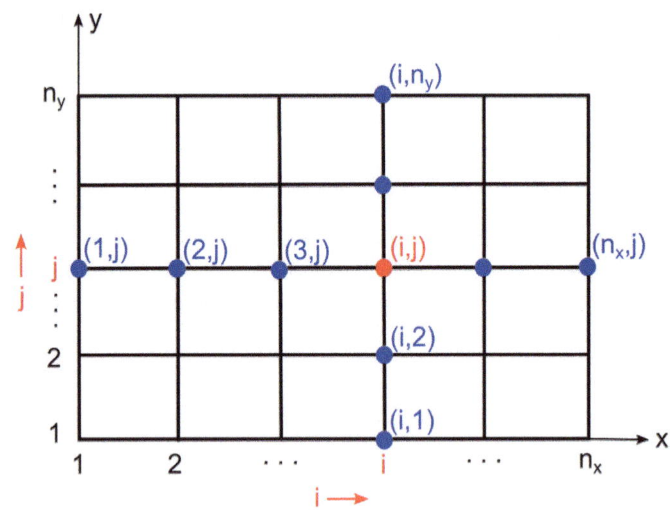

Fig. 8.5 Discretization of x-y plane for the calculation of partial derivatives with Differential Quadrature method at (i,j) point

derivative of $f(x,y)$ function with respect to x variable with $Bx_{i,j}$ and with respect to y variable with $By_{i,j}$ coefficients. Finally, we can indicate weight coefficients for fourth-order derivative of $f(x,y)$ function with respect to x variable with $Dx_{i,j}$ and with respect to y variable with $Dy_{i,j}$ coefficients. In this case, partial derivatives of an $f(x,y)$ function with respect x and y variables at (i,j) point can be expressed as following:

First-order partial derivative expression:

$$\begin{aligned} \frac{\partial f(x_i,y_j)}{\partial x} &= Ax_{i,1}f_{1,j} + Ax_{i,2}f_{2,j} + Ax_{i,3}f_{3,j} + \ldots + Ax_{i,n_x}f_{nx,j} \quad (j=\text{fixed}) \\ \frac{\partial f(x_i,y_j)}{\partial y} &= Ay_{j,1}f_{i,1} + Ay_{j,2}f_{i,2} + Ay_{j,3}f_{i,3} + \ldots + Ay_{j,n_y}f_{i,ny} \quad (i=\text{fixed}) \end{aligned} \tag{8.87}$$

Second-order partial derivative expression:

$$\begin{aligned} \frac{\partial^2 f(x_i,y_j)}{\partial x^2} &= Bx_{i,1}f_{1,j} + Bx_{i,2}f_{2,j} + Bx_{i,3}f_{3,j} + \ldots + Bx_{i,n_x}f_{nx,j} \quad (j=\text{fixed}) \\ \frac{\partial^2 f(x_i,y_j)}{\partial y^2} &= By_{j,1}f_{i,1} + By_{j,2}f_{i,2} + By_{j,3}f_{i,3} + \ldots + By_{j,n_y}f_{i,ny} \quad (i=\text{fixed}) \end{aligned} \tag{8.88}$$

$$\begin{aligned} \frac{\partial^2 f}{\partial x \partial y}(x_i,y_j) &= \frac{\partial}{\partial x}\left(\frac{\partial f(x_i,y_j)}{\partial y}\right) = \frac{\partial}{\partial x}\left(Ay_{j,1}f_{i,1} + Ay_{j,2}f_{i,2} + Ay_{j,3}f_{i,3} + \ldots + Ay_{j,n_y}f_{i,ny}\right) \\ &= Ay_{j,1}\frac{\partial f_{i,1}}{\partial x} + Ay_{j,2}\frac{\partial f_{i,2}}{\partial x} + Ay_{j,3}\frac{\partial f_{i,3}}{\partial x} + \ldots + Ay_{j,n_y}\frac{\partial f_{i,ny}}{\partial x} \\ &= Ay_{j,1}\underbrace{\left[Ax_{i,1}f_{1,1} + Ax_{i,2}f_{2,1} + Ax_{i,3}f_{3,1} + \ldots + Ax_{i,n_x}f_{nx,1}\right]}_{\frac{\partial f_{i,1}}{\partial x}} \\ &\quad + Ay_{j,2}\underbrace{\left[Ax_{i,1}f_{1,2} + Ax_{i,2}f_{2,2} + Ax_{i,3}f_{3,2} + \ldots + Ax_{i,n_x}f_{nx,2}\right]}_{\frac{\partial f_{i,2}}{\partial x}} \\ &\quad + Ay_{j,3}\underbrace{\left[Ax_{i,1}f_{1,3} + Ax_{i,2}f_{2,3} + Ax_{i,3}f_{3,3} + \ldots + Ax_{i,n_x}f_{nx,3}\right]}_{\frac{\partial f_{i,3}}{\partial x}} \\ &\quad \ldots \\ &\quad + Ay_{j,n_y}\underbrace{\left[Ax_{i,1}f_{1,ny} + Ax_{i,2}f_{2,ny} + Ax_{i,3}f_{3,ny} + \ldots + Ax_{i,n_x}f_{nx,ny}\right]}_{\frac{\partial f_{i,ny}}{\partial x}} \\ &= \sum_{l=1}^{n_y} Ay_{j,l}\sum_{k=1}^{n_x} Ax_{i,k}.f_{k,l} = \sum_{k=1}^{n_x}\sum_{l=1}^{n_y} Ax_{i,k}.Ay_{j,l}.f_{k,l} \end{aligned} \tag{8.89}$$

Fourth-order partial derivative expression:

$$\frac{\partial^4 f(x_i, y_j)}{\partial x^4} = Dx_{i,1} f_{1,j} + Dx_{i,2} f_{2,j} + Dx_{i,3} f_{3,j} + \ldots + Dx_{i,n_x} f_{nx,j} \quad (j = \text{fixed})$$
$$\frac{\partial^4 f(x_i, y_j)}{\partial y^4} = Dy_{j,1} f_{i,1} + Dy_{j,2} f_{i,2} + Dy_{j,3} f_{i,3} + \ldots + Dy_{j,n_y} f_{i,ny} \quad (i = \text{fixed})$$
(8.90)

$$\begin{aligned}
\frac{\partial^4 f}{\partial x^2 \partial y^2}(x_i, y_j) &= \frac{\partial^2}{\partial x^2}\left(\frac{\partial^2 f(x_i, y_j)}{\partial y^2}\right) = \frac{\partial}{\partial x^2}\left(By_{j,1} f_{i,1} + By_{j,2} f_{i,2} + By_{j,3} f_{i,3} + \ldots + By_{j,n_y} f_{i,ny}\right) \\
&= By_{j,1} \frac{\partial^2 f_{i,1}}{\partial x^2} + By_{j,2} \frac{\partial^2 f_{i,2}}{\partial x^2} + By_{j,3} \frac{\partial^2 f_{i,3}}{\partial x^2} + \ldots + By_{j,n_y} \frac{\partial^2 f_{i,ny}}{\partial x^2} \\
&= By_{j,1} \underbrace{\left[Bx_{i,1} f_{1,1} + Bx_{i,2} f_{2,1} + Bx_{i,3} f_{3,1} + \ldots + Bx_{i,n_x} f_{nx,1}\right]}_{\frac{\partial^2 f_{i,1}}{\partial x^2}} \\
&+ By_{j,2} \underbrace{\left[Bx_{i,1} f_{1,2} + Bx_{i,2} f_{2,2} + Bx_{i,3} f_{3,2} + \ldots + Bx_{i,n_x} f_{nx,2}\right]}_{\frac{\partial^2 f_{i,2}}{\partial x^2}} \\
&+ By_{j,3} \underbrace{\left[Bx_{i,1} f_{1,3} + Bx_{i,2} f_{2,3} + Bx_{i,3} f_{3,3} + \ldots + Bx_{i,n_x} f_{nx,3}\right]}_{\frac{\partial^2 f_{i,3}}{\partial x^2}} \\
&\ldots \\
&+ By_{j,n_y} \underbrace{\left[Bx_{i,1} f_{1,ny} + Bx_{i,2} f_{2,ny} + Bx_{i,3} f_{3,ny} + \ldots + Bx_{i,n_x} f_{nx,ny}\right]}_{\frac{\partial^2 f_{i,ny}}{\partial x^2}} \\
&= \sum_{l=1}^{n_y} By_{j,l} \sum_{k=1}^{n_x} Bx_{i,k} . f_{k,l} = \sum_{k=1}^{n_x} \sum_{l=1}^{n_y} Bx_{i,k} . By_{j,l} . f_{k,l}
\end{aligned}$$
(8.91)

Example 8.6: $f(x)$ function is given as $f(x) = \sin 2x$. Calculate $f'(0.5)$, $f''(0.5)$, and $f'''(0.5)$ derivatives with Differential Quadrature method using uniformly distributed five points in x direction. Compare Differential Quadrature results with analytical ones.

Since derivative is to be calculated at $x = 0.5$, an interval of [0,1] can be selected to include $x = 0.5$ point at the center. Five uniformly distributed points in [0,1] interval can be illustrated as in Fig. 8.6.

Coordinates of uniform five points can be expressed as below:

$x_1 = 0, x_2 = 0.25, x_3 = 0.5, x_4 = 0.75, x_5 = 1$

Function values corresponding to these positions can be written as following:

$f_1 = \sin(2x_1) = 0$, $f_2 = \sin(2x_2) = 0.4794$, $f_3 = \sin(2x_3) = 0.8415$, $f_4 = \sin(2x_4) = 0.9975$, $f_5 = \sin(2x_5) = 0.9093$

Indicating weight coefficients for first-, second-, and third-order derivatives with $A_{i,j}$, $B_{i,j}$, and $C_{i,j}$, respectively, weight coefficients for five points can be obtained as following:

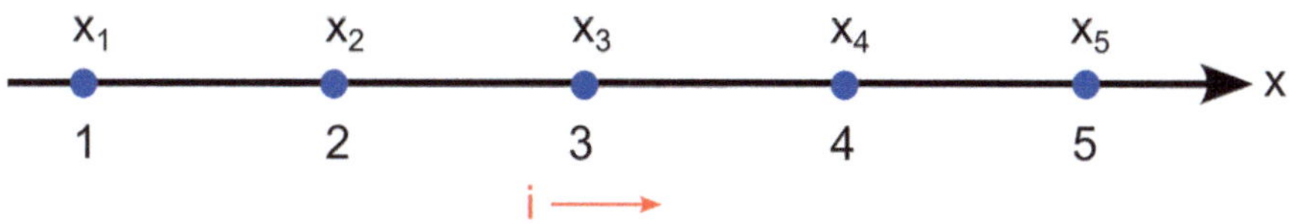

Fig. 8.6 Five uniformly distributed points in [0,1] interval to be used in derivative calculation with Differential Quadrature method

Weight coefficients for first-order derivative ($A_{i,\,j}$):

General expression: $A_{i,j} = \frac{1}{x_j - x_i} \prod\limits_{k=1(k\neq i,j)}^{n} \frac{x_i - x_k}{x_j - x_k} \quad (i \neq j)\,,\, A_{i,i} = -\sum\limits_{j=1,j\neq i}^{n} A_{i,j}$

$A_{1,\,1}, A_{1,\,2}, A_{1,\,3}, A_{1,\,4}, A_{1,\,5}$ calculation:

$$A_{1,2} = \frac{1}{x_2 - x_1}\left[\frac{(x_1 - x_3)(x_1 - x_4)(x_1 - x_5)}{(x_2 - x_3)(x_2 - x_4)(x_2 - x_5)}\right] = 16,$$

$$A_{1,3} = \frac{1}{x_3 - x_1}\left[\frac{(x_1 - x_2)(x_1 - x_4)(x_1 - x_5)}{(x_3 - x_2)(x_3 - x_4)(x_3 - x_5)}\right] = -12,$$

$$A_{1,4} = \frac{1}{x_4 - x_1}\left[\frac{(x_1 - x_2)(x_1 - x_3)(x_1 - x_5)}{(x_4 - x_2)(x_4 - x_3)(x_4 - x_5)}\right] = 5.3333,$$

$$A_{1,5} = \frac{1}{x_5 - x_1}\left[\frac{(x_1 - x_2)(x_1 - x_3)(x_1 - x_4)}{(x_5 - x_2)(x_5 - x_3)(x_5 - x_4)}\right] = -1,$$

$$A_{1,1} = -(A_{1,2} + A_{1,3} + A_{1,4} + A_{1,5}) = -8.3333$$

$A_{2,\,1}, A_{2,\,2}, A_{2,\,3}, A_{2,\,4}, A_{2,\,5}$ calculation:

$$A_{2,1} = \frac{1}{x_1 - x_2}\left[\frac{(x_2 - x_3)(x_2 - x_4)(x_2 - x_5)}{(x_1 - x_3)(x_1 - x_4)(x_1 - x_5)}\right] = -1,$$

$$A_{2,3} = \frac{1}{x_3 - x_2}\left[\frac{(x_2 - x_1)(x_2 - x_4)(x_2 - x_5)}{(x_3 - x_1)(x_3 - x_4)(x_3 - x_5)}\right] = 6,$$

$$A_{2,4} = \frac{1}{x_4 - x_2}\left[\frac{(x_2 - x_1)(x_2 - x_3)(x_2 - x_5)}{(x_4 - x_1)(x_4 - x_3)(x_4 - x_5)}\right] = -2,$$

$$A_{2,5} = \frac{1}{x_5 - x_2}\left[\frac{(x_2 - x_1)(x_2 - x_3)(x_2 - x_4)}{(x_5 - x_1)(x_5 - x_3)(x_5 - x_4)}\right] = 0.3333,$$

$$A_{2,2} = -(A_{2,1} + A_{2,3} + A_{2,4} + A_{2,5}) = -3.3333$$

$A_{3,\,1}, A_{3,\,2}, A_{3,\,3}, A_{3,\,4}, A_{3,\,5}$ calculation:

$$A_{3,1}=\frac{1}{x_1-x_3}\left[\frac{(x_3-x_2)(x_3-x_4)(x_3-x_5)}{(x_1-x_2)(x_1-x_4)(x_1-x_5)}\right]=0.3333,$$

$$A_{3,2}=\frac{1}{x_2-x_3}\left[\frac{(x_3-x_1)(x_3-x_4)(x_3-x_5)}{(x_2-x_1)(x_2-x_4)(x_2-x_5)}\right]=-2.6667,$$

$$A_{3,4}=\frac{1}{x_4-x_3}\left[\frac{(x_3-x_1)(x_3-x_2)(x_3-x_5)}{(x_4-x_1)(x_4-x_2)(x_4-x_5)}\right]=2.6667,$$

$$A_{3,5}=\frac{1}{x_5-x_3}\left[\frac{(x_3-x_1)(x_3-x_2)(x_3-x_4)}{(x_5-x_1)(x_5-x_2)(x_5-x_4)}\right]=-0.3333,$$

$$A_{3,3}=-(A_{3,1}+A_{3,2}+A_{3,4}+A_{3,5})=0$$

$A_{4,\,1}, A_{4,\,2}, A_{4,\,3}, A_{4,\,4}, A_{4,\,5}$ calculation:

$$A_{4,1}=\frac{1}{x_1-x_4}\left[\frac{(x_4-x_2)(x_4-x_3)(x_4-x_5)}{(x_1-x_2)(x_1-x_3)(x_1-x_5)}\right]=-0.3333,$$

$$A_{4,2}=\frac{1}{x_2-x_4}\left[\frac{(x_4-x_1)(x_4-x_3)(x_4-x_5)}{(x_2-x_1)(x_2-x_3)(x_2-x_5)}\right]=2,$$

$$A_{4,3}=\frac{1}{x_3-x_4}\left[\frac{(x_4-x_1)(x_4-x_2)(x_4-x_5)}{(x_3-x_1)(x_3-x_2)(x_3-x_5)}\right]=-6,$$

$$A_{4,5}=\frac{1}{x_5-x_4}\left[\frac{(x_4-x_1)(x_4-x_2)(x_4-x_3)}{(x_5-x_1)(x_5-x_2)(x_5-x_3)}\right]=1,$$

$$A_{4,4}=-(A_{4,1}+A_{4,2}+A_{4,3}+A_{4,5})=3.3333$$

$A_{5,\,1}, A_{5,\,2}, A_{5,\,3}, A_{5,\,4}, A_{5,\,5}$ calculation:

$$A_{5,1}=\frac{1}{x_1-x_5}\left[\frac{(x_5-x_2)(x_5-x_3)(x_5-x_4)}{(x_1-x_2)(x_1-x_3)(x_1-x_4)}\right]=1,$$

$$A_{5,2}=\frac{1}{x_2-x_5}\left[\frac{(x_5-x_1)(x_5-x_3)(x_5-x_4)}{(x_2-x_1)(x_2-x_3)(x_2-x_4)}\right]=-5.3333,$$

$$A_{5,3}=\frac{1}{x_3-x_5}\left[\frac{(x_5-x_1)(x_5-x_2)(x_5-x_4)}{(x_3-x_1)(x_3-x_2)(x_3-x_4)}\right]=12,$$

$$A_{5,4}=\frac{1}{x_4-x_5}\left[\frac{(x_5-x_1)(x_5-x_2)(x_5-x_3)}{(x_4-x_1)(x_4-x_2)(x_4-x_3)}\right]=-16,$$

$$A_{5,5}=-(A_{5,1}+A_{5,2}+A_{5,3}+A_{5,4})=8.3333$$

Weight coefficients obtained for first-order derivative can be written in matrix form as below:

$$A=\begin{bmatrix} A_{1,1} & A_{1,2} & A_{1,3} & A_{1,4} & A_{1,5} \\ A_{2,1} & A_{2,2} & A_{2,3} & A_{2,4} & A_{2,5} \\ A_{3,1} & A_{3,2} & A_{3,3} & A_{3,4} & A_{3,5} \\ A_{4,1} & A_{4,2} & A_{4,3} & A_{4,4} & A_{4,5} \\ A_{5,1} & A_{5,2} & A_{5,3} & A_{5,4} & A_{5,5} \end{bmatrix}=\begin{bmatrix} 8.3333 & 16 & -12 & 5.3333 & -1 \\ -1 & -3.3333 & 6 & -2 & 0.3333 \\ 0.3333 & -2.6667 & 0 & 2.6667 & -0.3333 \\ -0.3333 & 2 & -6 & 3.3333 & 1 \\ 1 & -5.3333 & 12 & -16 & 8.3333 \end{bmatrix}$$

Weight Coefficients for Second-Order Derivative ($B_{i,j}$):
Weight coefficient matrix of second derivative (B) can be obtained using weight coefficient matrix of first derivative (A) as following:

$$B=\begin{bmatrix} B_{1,1} & B_{1,2} & B_{1,3} & B_{1,4} & B_{1,5} \\ B_{2,1} & B_{2,2} & B_{2,3} & B_{2,4} & B_{2,5} \\ B_{3,1} & B_{3,2} & B_{3,3} & B_{3,4} & B_{3,5} \\ B_{4,1} & B_{4,2} & B_{4,3} & B_{4,4} & B_{4,5} \\ B_{5,1} & B_{5,2} & B_{5,3} & B_{5,4} & B_{5,5} \end{bmatrix}=A.A$$

$$=\begin{bmatrix} -8.3333 & 16 & -12 & 5.3333 & -1 \\ -1 & -3.3333 & 6 & -2 & 0.3333 \\ 0.3333 & -2.6667 & 0 & 2.6667 & -0.3333 \\ -0.3333 & 2 & -6 & 3.3333 & 1 \\ 1 & -5.3333 & 12 & -16 & 8.3333 \end{bmatrix}\begin{bmatrix} -8.3333 & 16 & -12 & 5.3333 & -1 \\ -1 & -3.3333 & 6 & -2 & 0.3333 \\ 0.3333 & -2.6667 & 0 & 2.6667 & -0.3333 \\ -0.3333 & 2 & -6 & 3.3333 & 1 \\ 1 & -5.3333 & 12 & -16 & 8.3333 \end{bmatrix}$$

$$=\begin{bmatrix} 46.6667 & -138.6667 & 152 & -74.6667 & 14.6667 \\ 14.6667 & -26.6667 & 8 & 5.3333 & -1.3333 \\ -1.3333 & 21.3333 & -40 & 21.3333 & -1.3333 \\ -1.3333 & 5.3333 & 8 & -26.6667 & 14.6667 \\ 14.6667 & -74.6667 & 152 & -138.6667 & 46.6667 \end{bmatrix}$$

Weight Coefficients for Third-Order Derivative ($C_{i,j}$):
Weight coefficient matrix C for third derivative can be obtained using first (A) and second derivative (B) weight coefficient matrices as following:

$$C=\begin{bmatrix} C_{1,1} & C_{1,2} & C_{1,3} & C_{1,4} & C_{1,5} \\ C_{2,1} & C_{2,2} & C_{2,3} & C_{2,4} & C_{2,5} \\ C_{3,1} & C_{3,2} & C_{3,3} & C_{3,4} & C_{3,5} \\ C_{4,1} & C_{4,2} & C_{4,3} & C_{4,4} & C_{4,5} \\ C_{5,1} & C_{5,2} & C_{5,3} & C_{5,4} & C_{5,5} \end{bmatrix}=A.B$$

$$=\begin{bmatrix} -8.3333 & 16 & -12 & 5.3333 & -1 \\ -1 & -3.3333 & 6 & -2 & 0.3333 \\ 0.3333 & -2.6667 & 0 & 2.6667 & -0.3333 \\ -0.3333 & 2 & -6 & 3.3333 & 1 \\ 1 & -5.3333 & 12 & -16 & 8.3333 \end{bmatrix}\begin{bmatrix} 46.6667 & -138.6667 & 152 & -74.6667 & 14.6667 \\ 14.6667 & -26.6667 & 8 & 5.3333 & -1.3333 \\ -1.3333 & 21.3333 & -40 & 21.3333 & -1.3333 \\ -1.3333 & 5.3333 & 8 & -26.6667 & 14.6667 \\ 14.6667 & -74.6667 & 152 & -138.6667 & 46.6667 \end{bmatrix}$$

$$=\begin{bmatrix} -160 & 576 & -768 & 448 & -96 \\ -96 & 320 & -384 & 192 & -32 \\ -32 & 64 & 0 & -64 & 32 \\ 32 & -192 & 384 & -320 & 96 \\ 96 & -448 & 768 & -576 & 160 \end{bmatrix}$$

In derivative calculations, $x = 0.5$ corresponds to the third point among five uniform points. Therefore, weight coefficients in the third row shown with $A_{3,\,j}$ $(j = 1,\ldots,5)$, $B_{3,\,j}$ $(j = 1,\ldots,5)$, and $C_{3,\,j}$ $(j = 1,\ldots,5)$ are used in derivative calculation with Differential Quadrature method. These weight coefficients can be expressed as below:

$A_{3,\,1}$=0.3333, $A_{3,\,2} = -2.6667$, $A_{3,\,3} = -0.0000$, $A_{3,\,4}$ =2.6667, $A_{3,\,5} = -$ 0.3333.

$B_{3,\,1}= -1.3333$, $B_{3,\,2} = 21.3333$, $B_{3,\,3} = -40.0000$, $B_{3,\,4} = 21.3333$, $B_{3,\,5} = -$ 1.3333.

$C_{3,\,1} = -32$, $C_{3,\,2} = 64$, $C_{3,\,3} = 0$, $C_{3,\,4} = -64$, $C_{3,\,5} = 32$.

Using above weight coefficients, derivative values at $x = 0.5$ can be obtained as following:

$$f'(x_3=0.5)=A_{3,1}f_1+A_{3,2}f_2+A_{3,3}f_3+A_{3,4}f_4+A_{3,5}f_5=1.0784$$

$$f''(x_3=0.5)=B_{3,1}f_1+B_{3,2}f_2+B_{3,3}f_3+B_{3,4}f_4+B_{3,5}f_5= -3.3636$$

$$f'''(x_3=0.5)=C_{3,1}f_1+C_{3,2}f_2+C_{3,3}f_3+C_{3,4}f_4+C_{3,5}f_5= -4.0589$$

Analytical derivative results for first, second, and third derivatives are obtained as following:

$$f'(x) = 2\cos(2x) \rightarrow f'(0.5) = 2\cos(2x0.5) = 1.0806$$
$$f''(x) = -4\sin(2x) \rightarrow f''(0.5) = -4\sin(2x0.5) = -3.3659$$
$$f'''(x) = -8\cos(2x) \rightarrow f'''(0.5) = -8\cos(2x0.5) = -4.3224$$

From the above results, it is seen that 5-point Differential Quadrature method yields very close values to analytical ones for low-order derivatives.

***Note*:** To calculate derivative values using Differential Quadrature Method, *MATLAB Program 8.2: Derivative Calculation with Differential Quadrature Method* in Appendix can be used.

Example 8.7: Function $f(x)$ is given as $f(x, y) = \sin x \cos y$. Calculate partial derivative $\frac{\partial^2 f}{\partial x \partial y}(0.5, 0.5)$ with Differential Quadrature method using non-uniform five points obtained with Gauss-Lobatto-Chebyshev method. Compare the Differential Quadrature result with analytical one.

Since partial derivative is to be calculated at $(x = 0.5, y = 0.5)$ point, intervals of [0,1] can be selected in x and y directions to locate $(x = 0.5, y = 0.5)$ point at the center. Five non-uniform points in x and y directions can be illustrated in [0,1] intervals as in Figs. 8.7 and 8.8.

Two-dimensional mesh points in x-y plane can be generated as shown in Fig. 8.9 using one-dimensional non-uniform mesh points in x and y directions. These mesh points can be utilized in the calculation of partial derivative.

Coordinates of five non-uniform points in x and y directions can be determined using Gauss-Lobatto-Chebyshev expression. Indicating lower limit of interval with a, upper limit of interval with b and the number of points with n, general expression for Gauss-Lobatto-Chebyshev points can be written as below:

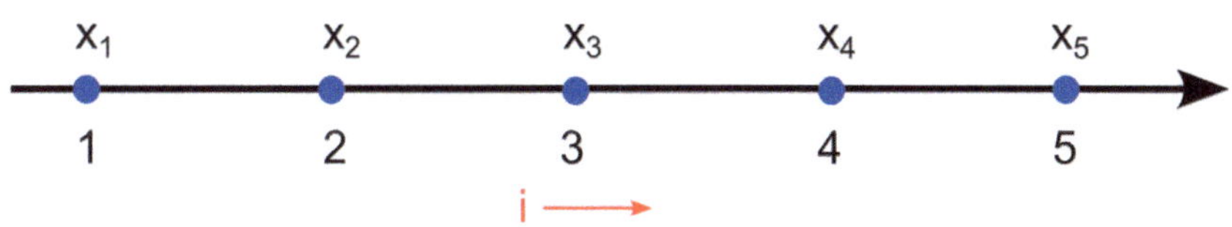

Fig. 8.7 Points in x direction for the calculation of partial derivative with Differential Quadrature method

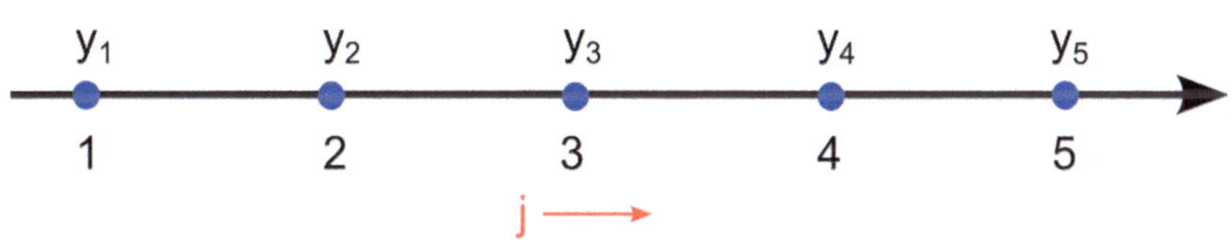

Fig. 8.8 Points in y direction for the calculation of partial derivative with Differential Quadrature method

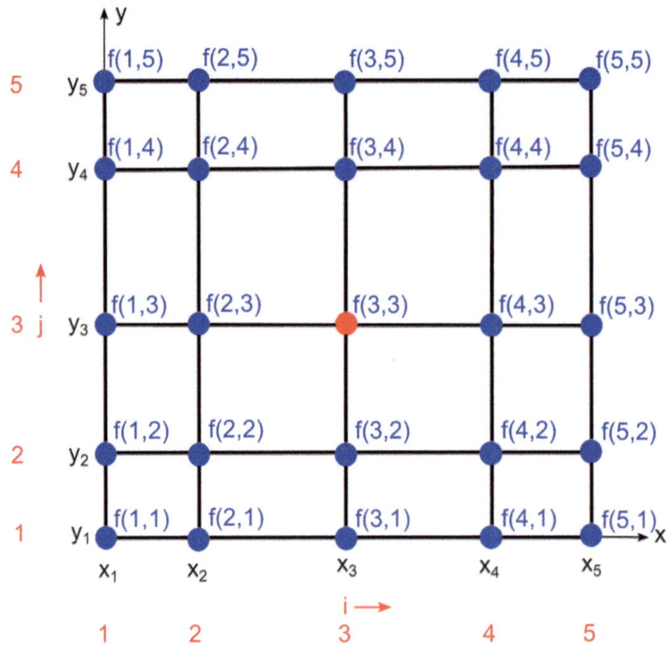

Fig. 8.9 Mesh points in the calculation of partial derivative using Differential Quadrature method

$$x_i = a + \frac{1}{2}\left(1 - \cos\frac{(i-1)\pi}{n-1}\right)(b-a) = 0$$
$$+ \frac{1}{2}\left(1 - \cos\frac{(i-1)\pi}{5-1}\right)(1-0) = \frac{1}{2}\left(1 - \cos\frac{(i-1)\pi}{4}\right)$$
$$y_j = a + \frac{1}{2}\left(1 - \cos\frac{(j-1)\pi}{n-1}\right)(b-a)$$
$$= 0 + \frac{1}{2}\left(1 - \cos\frac{(i-1)\pi}{5-1}\right)(1-0) = \frac{1}{2}\left(1 - \cos\frac{(j-1)\pi}{4}\right)$$

Using aforementioned equations, coordinates of five non-uniform points in x and y directions can be obtained as following:

$x_1 = 0$, $x_2 = 0.1464$, $x_3 = 0.5$, $x_4 = 0.8536$, $x_5 = 1$

$y_1 = 0$, $y_2 = 0.1464$, $y_3 = 0.5$, $y_4 = 0.8536$, $y_5 = 1$

$(x, y) = (0.5, 0.5)$ point corresponds to indices of $(i, j) = (3, 3)$. Therefore, weight coefficients shown with $Ax_{3,\,j}\,(j = 1,\ldots,5)$ and $Ay_{3,\,j}\,(j = 1,\ldots,5)$ are used in partial derivative calculation with Differential Quadrature method. These weight coefficients can be expressed as below:

$Ax_{3,\,1} = 1.0000$, $Ax_{3,\,2} = -2.8284$, $Ax_{3,\,3} = 0.0000$, $Ax_{3,\,4} = 2.8284$, $Ax_{3,\,5} = -1.0000$

$Ay_{3,1} = 1.0000, Ay_{3,2} = 2.8284, Ay_{3,3} = 0.0000, Ay_{3,4} = 2.8284, Ay_{3,5} = -1.0000$

Using the weight coefficients above, partial derivative $\frac{\partial^2 f}{\partial x \partial y}(x_3, y_3)$ can be expressed as below:

$$
\begin{aligned}
\frac{\partial^2 f}{\partial x \partial y}(x_3, y_3) &= \frac{\partial}{\partial x}\left(Ay_{3,1} f_{3,1} + Ay_{3,2} f_{3,2} + Ay_{3,3} f_{3,3} + Ay_{3,4} f_{3,4} + Ay_{3,5} f_{3,5}\right) \\
&= Ay_{3,1}\frac{\partial f_{3,1}}{\partial x} + Ay_{3,2}\frac{\partial f_{3,2}}{\partial x} + Ay_{3,3}\frac{\partial f_{3,3}}{\partial x} + Ay_{3,4}\frac{\partial f_{3,4}}{\partial x} + Ay_{3,5}\frac{\partial f_{3,5}}{\partial x} \\
&= Ay_{3,1}\underbrace{\left[Ax_{3,1} f_{1,1} + Ax_{3,2} f_{2,1} + Ax_{3,3} f_{3,1} + Ax_{3,4} f_{4,1} + Ax_{3,5} f_{5,1}\right]}_{\frac{\partial f_{3,1}}{\partial x}} \\
&+ Ay_{3,2}\underbrace{\left[Ax_{3,1} f_{1,2} + Ax_{3,2} f_{2,2} + Ax_{3,3} f_{3,2} + Ax_{3,4} f_{4,2} + Ax_{3,5} f_{5,2}\right]}_{\frac{\partial f_{3,1} 2}{\partial x}} \\
&+ Ay_{3,3}\underbrace{\left[Ax_{3,1} f_{1,3} + Ax_{3,2} f_{2,3} + Ax_{3,3} f_{3,3} + Ax_{3,4} f_{4,3} + Ax_{3,5} f_{5,3}\right]}_{\frac{\partial f_{3,3}}{\partial x}} \\
&+ Ay_{3,4}\underbrace{\left[Ax_{3,1} f_{1,4} + Ax_{3,2} f_{2,4} + Ax_{3,3} f_{3,4} + Ax_{3,4} f_{4,4} + Ax_{3,5} f_{5,4}\right]}_{\frac{\partial f_{3,4}}{\partial x}} \\
&+ Ay_{3,5}\underbrace{\left[Ax_{3,1} f_{1,5} + Ax_{3,2} f_{2,5} + Ax_{3,3} f_{3,5} + Ax_{3,4} f_{4,5} + Ax_{3,5} f_{5,5}\right]}_{\frac{\partial f_{3,5}}{\partial x}} \\
&= \sum_{l=1}^{5} Ay_{j,l} \sum_{k=1}^{5} Ax_{i,k} f_{k,l} = \sum_{k=1}^{5}\sum_{l=1}^{5} Ax_{i,k}.Ay_{j,l} f_{k,l}
\end{aligned}
$$

Substituting numerical values into the above expression, partial derivative value at $(x, y) = (0.5, 0.5)$ point is obtained as $\frac{\partial^2 f}{\partial x \partial y}(x_3, y_3) = \frac{\partial^2 f}{\partial x \partial y}(0.5, 0.5) = -0.4205$.

Analytical derivative value can be calculated as below:

$$f(x, y) = \sin x \cos y \rightarrow \frac{\partial^2 f}{\partial x \partial y}(x, y) = -\cos x.\sin y$$

$$\frac{\partial^2 f}{\partial x \partial y}(0.5, 0.5) = -\cos(0.5).\sin(0.5) = -0.4207$$

From above results, it is seen that Differential Quadrature method using non-uniform grid points gives very close result to the analytical derivative.

8.8 Comparison of Finite Difference and Differential Quadrature Methods

In Finite Difference method, derivative expression must be rederived for an increasing number of points. However, in differential quadrature method this is not so. Method can easily adapt itself for a varying number of points. Finite Difference method can be accepted as a special case of Differential Quadrature method. It can be thought as a different version of Differential Quadrature method that uses uniform grid points. This can be shown by comparison of Finite Difference and Differential Quadrature formulations for second-order derivative of an $f(x)$ function. Assume that three points ($i-1$, i, $i+1$) with step length (distance between adjacent points) of $\Delta x = 0.5$ are given. In this case, second derivative of $f(x)$ function at point i using Central Finite Difference and Differential Quadrature methods can be expressed as below:

$$\text{Central Difference}: \frac{d^2f(x_i)}{dx^2} = \frac{f_{i-1} - 2f_i + f_{i+1}}{(\Delta x)^2} \rightarrow$$

$$\text{Using } \Delta x = 0.5: \frac{d^2f(x_i)}{dx^2} = 4f_{i-1} - 8f_i + 4f_{i+1}$$

$$\text{Differential Quadrature method}: \frac{d^2f(x_i)}{dx^2} = B_{i,i-1}f_{i-1} + B_{i,i}f_i + B_{i,i+1}f_{i+1}$$

where $B_{i,\,i-1}, B_{i,\,i}, B_{i,\,i+1}$ indicate weight coefficients for second-order derivative at $i-1$, i, $i+1$ points. When three points are uniformly distributed in the interval of [−0.5,0.5], coordinates of the points can be obtained as $x_{i-1} = -0.5$, $x_i = 0$, $x_{i+1} = 0.5$. Using Eqs. 8.83 and 8.85, weight coefficients $B_{i,\,i-1}, B_{i,\,i}, B_{i,\,i+1}$ are obtained as $B_{i,\,i-1} = 4$, $B_{i,\,i} = -8$, $B_{i,\,i+1} = 4$. In this case, second-order derivative expression with Differential Quadrature method is obtained as $\frac{d^2f(x_i)}{dx^2} = 4f_{i-1} - 8f_i + 4f_{i+1}$. This expression is the same as that of Finite Difference method. This proves that Differential Quadrature method can degenerate (transform) to Finite Difference method for the same number of uniform points.

8.9 Summary

- Derivative expressions are frequently encountered in engineering problems that are expressed with differential equations. Some derivative terms are given special names such as velocity, acceleration, slope, and curvature.
- Taking derivative is generally difficult for functions other than polynomial and trigonometric functions. For this reason, numerical derivatives are used to calculate derivatives of such functions.

- In numerical derivative, derivative of a function is expressed as weighted sum of function values at different points as below:

$$\frac{d^n f}{dx^n}(x_i) = f^{(n)}(x_i) = \ldots + w_{i-2} f(x_{i-2}) + w_{i-1} f(x_{i-1})$$
$$+ w_i f(x_i) + w_{i+1} f(x_{i+1}) + w_{i+2} f(x_{i+2}) + \ldots$$

- Derivative formulas are derived from Taylor series expansion or interpolation expressions of functions.
- Finite Difference as a classical method and Differential Quadrature as a recent method are used in calculation of numerical derivative values.
- Finite Difference method is obtained by manipulating Taylor series expansion of $f(x)$ function at different points to lead to the desired derivative term.
- In Finite Difference method, derivatives are expressed using a limited number of predetermined uniform points as following:

  First-order derivative (Forward Difference): $f'(x_i) = \frac{f(x_{i+1}) - f(x_i)}{\Delta x}$

  First-order derivative (Backward Difference): $f'(x_i) = \frac{f(x_i) - f(x_{i-1})}{\Delta x}$

  First-order derivative (Central Difference): $f'(x_i) = \frac{f(x_{i+1}) - f(x_{i-1})}{2\Delta x}$

  Second-order derivative: $f^{('')}(x_i) = \frac{f(x_{i-1}) - 2f(x_i) + f(x_{i+1})}{(\Delta x)^2}$
- Partial derivative expressions based on Finite Difference method can be easily obtained using Finite Difference expressions of ordinary derivatives.
- In partial derivative expression, all variables are kept constant except for one variable. In Finite Difference method, partial derivatives can be expressed as following:

  First-order derivatives:

$$\frac{\partial f}{\partial x}(x_i, y_j) = \frac{f(x_{i+1}, y_j) - f(x_{i-1}, y_j)}{2\Delta x} \quad (j = \text{fixed})$$
$$\frac{\partial f}{\partial y}(x_i, y_j) = \frac{f(x_i, y_{j+1}) - f(x_i, y_{j-1})}{2\Delta y} \quad (i = \text{fixed})$$

  Second-order derivatives:

$$\frac{\partial^2 f}{\partial x^2}(x_i, y_j) = \frac{f(x_{i-1}, y_j) - 2f(x_i, y_j) + f(x_{i+1}, y_j)}{(\Delta x)^2}$$
$$\frac{\partial^2 f}{\partial y^2}(x_i, y_j) = \frac{f(x_i, y_{j-1}) - 2f(x_i, y_j) + f(x_i, y_{j+1})}{(\Delta y)^2}$$
$$\frac{\partial^2 f}{\partial x \partial y}(x_i, y_j) = \frac{1}{4\Delta x \Delta y}\left(-f(x_{i-1}, y_{j+1}) + f(x_{i-1}, y_{j-1}) + f(x_{i+1}, y_{j+1}) - f(x_{i+1}, y_{j-1})\right)$$

- For high accuracy in Finite Difference method, step lengths of Δx and Δy (distance between adjacent points in x and y directions) should be small values.
- Another alternative to increase accuracy in Finite Difference method is calculating derivatives for different Δx values and then extrapolating these results to obtain more accurate derivative value. This approach is known as Richardson Extrapolation.
- Using derivative results (D_1 and D_2) for two different Δx values (D_1 taking $\Delta x = h_1$ and D_2 taking $\Delta x = h_1/2$), corrected derivative value with Richardson Extrapolation can be calculated as $D = \frac{4}{3}D_2 - \frac{1}{3}D_1$.
- Differential Quadrature method is obtained by implicitly creating a global interpolation polynomial passing through the desired number of points and then taking derivative of this function.
- In Differential Quadrature method, derivatives can be calculated using the desired number of non-uniform points.
- In Differential Quadrature method, derivative of a function is expressed as weighted sum of function values at non-uniform points as following:

$$\frac{d^m f}{dx^m}(x_i) = f^{(m)}(x_i) = C_{i1}^{(m)} f(x_1) + C_{i2}^{(m)} f(x_2) + C_{i3}^{(m)} f(x_3) + \ldots + C_{in}^{(m)} f(x_n) = \sum_{j=1}^{n} C_{ij}^{(m)} f(x_j)$$

- Non-uniformity of points is very critical in accurate calculation of derivatives with Differential Quadrature method.
- Indicating a number of points with n, non-uniform points in an [a,b] interval can be determined by using Gauss-Lobatto-Chebyshev or shortly Gauss-Lobatto equation as below:

$$x_i = a + \frac{1}{2}\left(1 - \cos\frac{(i-1)\pi}{n-1}\right)(b-a) \quad (i = 1, 2, \ldots, n)$$

- If first-order derivative coefficients $A_{ij} = C_{ij}^{(1)}$ are known, higher order derivative coefficients $C_{ij}^{(m)}$ can be obtained from lower order derivative coefficients recursively as below:

$$C_{ij}^{(m)} = m\left[A_{ij} C_{ii}^{(m-1)} - \frac{C_{ij}^{(m-1)}}{x_j - x_i}\right] \quad (i \neq j)$$

$$C_{ii}^{(m)} = -\sum_{j=1, j\neq i}^{n} C_{ij}^{(m)}$$

- Partial derivative with Differential Quadrature method is obtained in a similar way with ordinary derivative. All variables are kept constant except for one variable. Then, partial derivative expression is obtained by applying ordinary derivative expression for non-constant variable.

- Differential Quadrature method yields more accurate derivative results compared to Finite Difference method for the same number of points.
- Differential Quadrature method can be specialized to Finite Difference method by using uniform points.
- In Finite Difference method, derivative expression must be rederived for increasing the number of points. However, in differential quadrature method this is not so. Method can easily adapt itself for the varying number of points.

Appendix: MATLAB Programs

MATLAB Program 8.1: Derivative Calculation with Central Finite Difference Method

```
function DerivativeCalculationwithCentralDifference_1D

% NOTE: THIS PROGRAM CALCULATES NUMERICAL DERIVATIVES AT A x0 POINT
% USING FINITE DIFFERENCE METHOD

x0=5 ;      % Derivative point
h=0.001 ;   % Step length

% NOTE: h should be selected very small for accurate derivative
calculation
%
==========================================================================

% Backward and Forward points to be used in derivative calculation
x_1=x0-h ;
x_2=x0-2*h ;
x1=x0+h ;
x2=x0+2*h ;

% Function values at Backward and Forward points
f_2=FunctionValue(x_2) ;
f_1=FunctionValue(x_1) ;
f0=FunctionValue(x0) ;
f1=FunctionValue(x1) ;
f2=FunctionValue(x2) ;

% DERIVATIVE CALCULATIONS
% 1-st Order Derivative: df1
df1=(-f_1+f1)/(2*h) ;

% 2-nd Order Derivative: df2
df2=(f_1-2*f0+f1)/(h^2) ;

% 3-rd Order Derivative: df3
df3=(-f_2+2*f_1-2*f1+f2)/(2*h^3) ;

% 4-th Order Derivative: df4
df4=(f_2-4*f_1+6*f0-4*f1+f2)/(h^4) ;

fprintf('x0=%f , 1-st Order Derivative Value: df1=%f \n',x0,df1) ;
fprintf('x0=%f , 2-nd Order Derivative Value: df2=%f \n',x0,df2) ;
fprintf('x0=%f , 3-rd Order Derivative Value: df3=%f \n',x0,df3) ;
fprintf('x0=%f , 4-th Order Derivative Value: df4=%f \n',x0,df4) ;

%=====================================================
%================     FUNCTION: f(x)     ==============
%=====================================================
function f=FunctionValue(x)
f=sin(x) ;
```

MATLAB Program 8.2: Derivative Calculation with Differential Quadrature Method

```
function DerivativeCalculationwithDifferentialQuadrature_1D

% NOTE: THIS PROGRAM CALCULATES NUMERICAL DERIVATIVES AT A x0 POINT
% USING FINITE DIFFERENCE METHOD

% NOTE: Derivative will be calculated using non-uniform n points
% in [x0-L/2,x0+L/2] interval

L=1 ;  % Interval length to be used in derivative calculation
x0=5 ; % Derivative point
n=5 ;  % Number of points to be used in derivative calculation

% NOTE: n should be even number to locate x0 at the center of
% [x0-L/2,x0+L/2] interval
%
========================================================================
m=round(n/2) ; % center grid number

% Gauss-Lobatto-Chebyshev points (grids) in normalized [-1,1] interval:
% ksi(i)
for i=1:n
    ksi(i)=-cos((i-1)*pi/(n-1)) ;
end

% Transformation of Gauss-Lobatto-Chebyshev point into true values in
% [x0-L/2,x0+L/2] interval: xc
for i=1:n
    xc(i)=x0-(1-ksi(i))*L/4+(1+ksi(i))*L/4 ;
end

% DERIVATIVE CALCULATIONS
% A: 1-st Order Derivative Weight Coefficients
% B: 2-nd Order Derivative Weight Coefficients
% C: 3-rd Order Derivative Weight Coefficients
% D: 4-th Order Derivative Weight Coefficients
A=FirstOrderDerivativeCoefficients(n,xc);
B=A*A;
C=A*B;
D=A*C;

% 1-st Order Derivative: df1
df1=0 ;
for i=1:n
    f=FunctionValue(xc(i)) ;
    df1=df1+A(m,i)*f ;
end

% 2-nd Order Derivative: df2
df2=0 ;
for i=1:n
```

```
    f=FunctionValue(xc(i)) ;
    df2=df2+B(m,i)*f ;
end

% 3-rd Order Derivative: df3
df3=0 ;
for i=1:n
    f=FunctionValue(xc(i)) ;
    df3=df3+C(m,i)*f ;
end

% 4-th Order Derivative: df4
df4=0 ;
for i=1:n
    f=FunctionValue(xc(i)) ;
    df4=df4+D(m,i)*f ;
end

fprintf('x0=%f , 1-st Order Derivative Value: df1=%f \n',x0,df1) ;
fprintf('x0=%f , 2-nd Order Derivative Value: df2=%f \n',x0,df2) ;
fprintf('x0=%f , 3-rd Order Derivative Value: df3=%f \n',x0,df3) ;
fprintf('x0=%f , 4-th Order Derivative Value: df4=%f \n',x0,df4) ;

%=================================================================
%=================================================================
function A=FirstOrderDerivativeCoefficients(n,xc)
% A(i,j): 1-st Order Derivative Weight Coefficients
A=zeros(n,n) ;

for i=1:n
    for j=1:n
            if j~=i
                    mult1=1 ;
                    for k=1:n
                         if ((k~=i) && (k~=j))
                             mult1=mult1*((xc(i)-xc(k))/(xc(j)-xc(k))) ;

                         end
                    end
                    A(i,j)=mult1/(xc(j)-xc(i)) ;
            end
    end
end

for i=1:n
        A(i,i)=0 ;
        for j=1:n
             if j~=i
                   A(i,i)=A(i,i)-A(i,j) ;
             end
        end
end

%====================================================
%================    FUNCTION: f(x)     ==============
%====================================================
function f=FunctionValue(x)
f=sin(x) ;
```

Problems

Problem 8.1: $f(x)$ function is given as $f(x) = 2\sqrt{\sin x}$. Calculate $\frac{df}{dx}\left(x = \frac{\pi}{2}\right)$ derivative using (a) Forward Finite Difference method, (b) Backward Finite Difference method, (c) Central Finite Difference method by taking the distance between adjacent points (step length) as $\Delta x = 0.1$. Compare finite difference results with analytical derivative value.

Problem 8.2: $f(x)$ function is given as $f(x) = 2\sqrt{\sin x}$. Calculate $\frac{d^2f}{dx^2}\left(x = \frac{\pi}{2}\right)$ derivative using Central Finite Difference method by taking the distance between adjacent points (step length) as $\Delta x = 0.1$. Compare finite difference result with analytical derivative value.

Problem 8.3: $f(x)$ function is given as $f(x) = 2\sqrt{\sin x}$. Calculate $\frac{d^3f}{dx^3}\left(x = \frac{\pi}{2}\right)$ derivative using Central Finite Difference method by taking the distance between adjacent points (step length) as $\Delta x = 0.1$. Compare finite difference result with analytical derivative value.

Problem 8.4: $f(x)$ function is given as $f(x) = 2\sqrt{\sin x}$. Calculate $\frac{d^4f}{dx^4}\left(x = \frac{\pi}{2}\right)$ derivative using Central Finite Difference method by taking the distance between adjacent points (step length) as $\Delta x = 0.1$. Compare finite difference result with analytical derivative value.

Problem 8.5: Obtain first-order Finite Difference derivative expression of an $f(x)$ function with general method (systematic approach) using $i-1$, i, $i+2$ points.

Problem 8.6: Obtain first-order Finite Difference derivative expression of an $f(x)$ function with general method (systematic approach) using i, $i-1$, $i-2$ points.

Problem 8.7: Obtain second-order Finite Difference derivative expression of an $f(x)$ function with general method (systematic approach) using i, $i+1$, $i+2$ points.

Problem 8.8: Obtain second-order Finite Difference derivative expression of an $f(x)$ function with general method (systematic approach) using i, $i-1$, $i-2$ points.

Problem 8.9: Derive numerical derivative expression for $\frac{\partial^4 f}{\partial x^2 \partial y^2}\left(x_i, y_j\right)$ using lower order Central Finite Difference derivative expressions.

Problem 8.10: $f(x,y)$ function is given as $f(x, y) = 3\sin x\sqrt{\cos y}$. Calculate second-order derivative of $\frac{\partial^2 f}{\partial x^2}\left(x = \frac{\pi}{4}, y = \frac{\pi}{4}\right)$ using Finite Difference method by taking the distance between adjacent points (step length) in x and y directions as $\Delta x = 0.01$, $\Delta y = 0.01$. Compare finite difference result with analytical derivative value.

Problem 8.11: $f(x,y)$ function is given as $f(x, y) = 3\sin x\sqrt{\cos y}$. Calculate second-order derivative of $\frac{\partial^2 f}{\partial y^2}\left(x = \frac{\pi}{4}, y = \frac{\pi}{4}\right)$ using Finite Difference method by taking the distance between adjacent points (step length) in x and y directions as $\Delta x = 0.01$, $\Delta y = 0.01$. Compare finite difference result with analytical derivative value.

Problem 8.12: $f(x,y)$ function is given as $f(x,y) = 3\sin x\sqrt{\cos y}$. Calculate fourth-order derivative of $\frac{\partial^4 f}{\partial x^2 \partial y^2}\left(x = \frac{\pi}{4}, y = \frac{\pi}{4}\right)$ using Finite Difference method by taking the distance between adjacent points (step length) in x and y directions as $\Delta x = 0.01$, $\Delta y = 0.01$. Compare finite difference result with analytical derivative value.

Problem 8.13: $f(x)$ function is given as $f(x) = 2\sin x + 3\cos x$. Calculate $\frac{df}{dx} \times (x = 0.5)$ derivative with Differential Quadrature method using five non-uniform points (*Hint*: five non-uniform points can be created centered around $x = 0.5$ point in [0,1] interval using Chebyshev equation). Compare the Differential Quadrature result with analytical one.

Problem 8.14: $f(x)$ function is given as $f(x) = 2\sin x + 3\cos x$. Calculate $\frac{d^2 f}{dx^2} \times (x = 0.5)$ derivative with Differential Quadrature method using five non-uniform points (*Hint*: five non-uniform points can be created centered around $x = 0.5$ point in [0,1] interval using Chebyshev equation). Compare the Differential Quadrature result with analytical one.

Problem 8.15: $f(x,y)$ function is given as $f(x,y) = 2\sin x\cos y$. Calculate the partial derivative of $\frac{\partial^2 f}{\partial x \partial y}\left(x = \frac{\pi}{6}, y = \frac{\pi}{3}\right)$ with Differential Quadrature method using three non-uniform points in x and y directions (*Hint*: three non-uniform points can be created centered around $x = \frac{\pi}{6}$ point in $[0, \frac{2\pi}{6}]$ interval and $y = \frac{\pi}{3}$ points in $[\frac{\pi}{6}, \frac{3\pi}{6}]$ interval using Chebyshev equation). Compare the Differential Quadrature result with analytical one.

References

1. Chapra, S. C., & Canale, R. P. (2020). *Numerical methods for engineers*. McGraw-Hill.
2. Bernatz, R. A. (2010). *Fourier series and numerical methods for partial differential equations*. John Wiley & Sons.
3. Everstine, G. C. (2010). *Numerical solution of partial differential equations* [Lecture notes, PDF]. George Washington University. https://www.academia.edu/31518755/Numerical_Solution_of_Partial_Differential_Equations
4. Epperson, J. F. (2013). *An introduction to numerical methods and analysis*. John Wiley & Sons.
5. Burden, R. L., & Faires, J. D. (2011). *Numerical analysis*. Cengage Learning.
6. Sert, C. (2006). *ME310 numerical methods* [Lecture notes, PDF]. Middle East Technical University. https://users.metu.edu.tr/csert/me310/me310_1_introduction.pdf
7. Grewal, B. S. (2019). *Numerical methods in engineering and science*. Mercury Learning and Information.
8. Sastry, S. S. (2012). *Introductory methods of numerical analysis*. PHI Learning.
9. Chang, S. (2000). *Differential quadrature and its application in engineering*. Springer.

10. Zong, Z., & Zhang, Y. (2009). *Advanced differential quadrature methods*. CRC Press.
11. Bert, C. W., & Malik, M. (1996). Differential quadrature method in computational mechanics: A review. *Applied Mechanics Review, 29*(1), 1–28.
12. Wu, T. Y., & Liu, G. R. (1999). A differential quadrature as a numerical method to solve differential equations. *Computational Mechanics, 24*, 197–205.

Chapter 9
Numerical Integral

9.1 Introduction

Integral expressions are frequently encountered in engineering problems. Integral expressions are used either to express a part of the problem or the whole problem. For example, Finite Element method (it will be explained in Chap. 6 of Vol. 2) is widely used as an integral method in the solution of engineering problems. In Finite Element method, engineering problem is expressed in the $KU = F$ form where K matrix is called as stiffness matrix and is calculated with $K = \int_L \underbrace{B^T CB}_{f(x)} dx$ line integral, $K = \int_A \underbrace{B^T CB}_{f(x,y)} \underbrace{dA}_{dxdy}$ area integral or $K = \int_V \underbrace{B^T CB}_{f(x,y,z)} \underbrace{dV}_{dxdydz}$ volume integral depending on the dimension of the solution domain [1–3].

Similarly, equilibrium equations for elasticity problems can be obtained with energy methods (Total potential energy or virtual work) that use integral equations as following:

Total potential energy minimization:

$$\text{Total potential energy minimization} : \Pi = \int_V \underbrace{\varepsilon^T \sigma}_{f(x,y,z)} \underbrace{dV}_{dxdydz} - u^T F \rightarrow \min$$

$$\text{Virtual work} : \int_V \underbrace{\delta\varepsilon^T \sigma}_{f(x,y,z)} \underbrace{dV}_{dxdydz} = \delta u^T F.$$

These integrals can be calculated analytically (in closed from) if integrant function $f(x)$ is simple such as polynomial, trigonometric, or exponential function and integral domain is regular as shown in Fig. 9.1a. However, when the integrant function is complicated such as $f(x) = \sqrt{\tanh(e^{x^2})}$ or the integral domain is irregular as shown in Fig. 9.1b, these integrals can only be calculated numerically.

H. Kurtaran, *Engineering Mathematics with MATLAB Applications: Vol. 1, Fundamentals*, https://doi.org/10.1007/978-3-031-81869-1_9

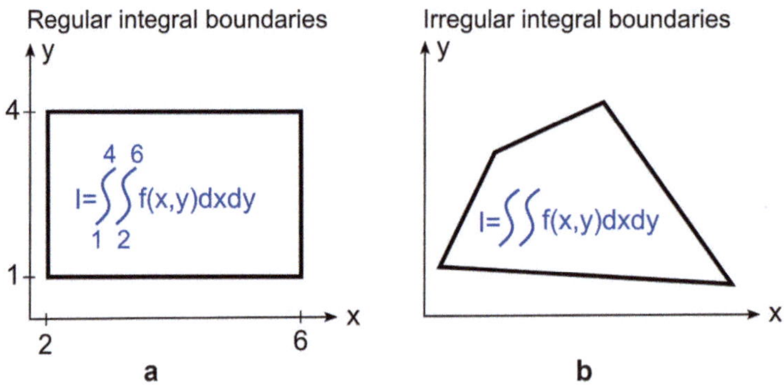

Fig. 9.1 Integral boundaries: (**a**) regular boundaries (**b**) irregular boundaries

Numerical integral methods can be divided into two main groups [4–11]:

1. Newton-Cotes methods: Integral of an $f(x)$ function is calculated using its polynomial approximation ($\tilde{f}(x)$) as below:

$$I=\int_a^b f(x)dx=\int_a^b \tilde{f}(x)dx=\int_a^b \left(c_1+c_2x+c_3x^2+c_4x^3+\ldots+c_{n+1}x^n\right)dx \tag{9.1}$$

2. Quadrature methods: Integral of an $f(x)$ function is calculated as the weighted sum of function values at different points as below:

$$I=\int_a^b f(x)dx=w_1f(x_1)+w_2f(x_2)+\ldots+w_nf(x_n)=\sum_{i=1}^{p} w_if(x_i) \tag{9.2}$$

Both groups of methods were actually developed to calculate integral of polynomial functions. In both groups of methods, it is assumed that a real function can be represented with an equivalent polynomial function and thereby integral of the real function can be approximated with integral of the equivalent polynomial function. In Newton-Cotes methods, polynomials, which are approximations of real function, are obtained using interpolation methods. Numerical integral methods are described in detail below.

9.2 Newton-Cotes Methods

In Newton-Cotes methods, real function is approximated with an interpolation polynomial and integral of polynomial function is calculated to represent the integral of the real function. Depending on the order of the polynomial function Newton-Cotes methods take different names [4–11].

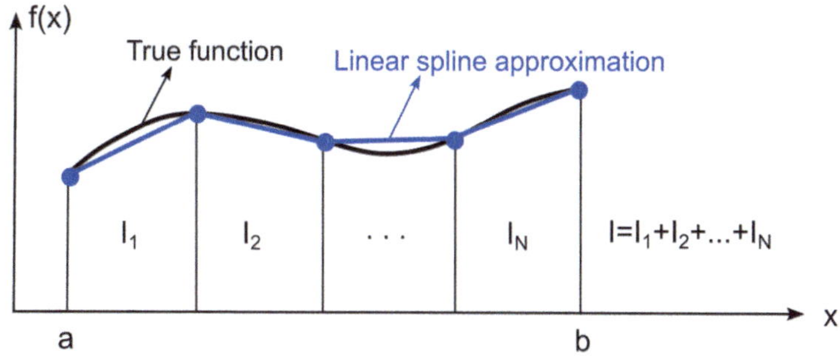

Fig. 9.2 Integral of an $f(x)$ function using integrals of linear piecewise polynomial functions

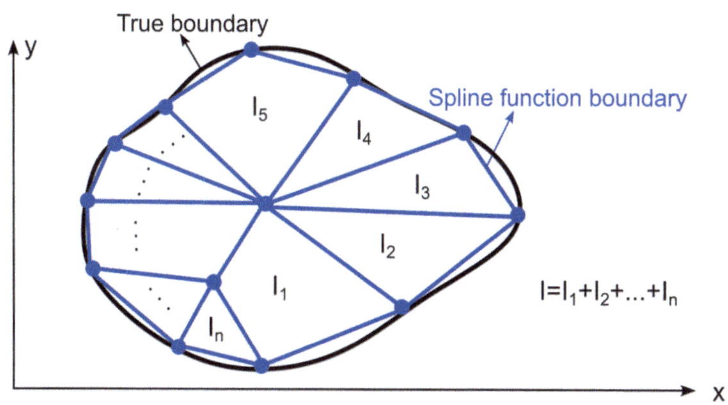

Fig. 9.3 Integral of an $f(x,y)$ function over a complex two-dimensional domain using integrals over simpler sub-domains

It was explained in Chap. 6 (Interpolation) that global higher order polynomials generated from a large number of data points will have oscillatory character and interpolation with these functions will be erroneous. Using such functions in integral calculations will also yield inaccurate integral results. Therefore, common approach in Newton-Cotes methods is to use piecewise polynomials. Whole integral range is divided into sub-ranges where piecewise polynomial functions using lower orders are created in. Then, integral value is calculated by summing integrals of piecewise functions as shown in Figs. 9.2 and 9.3.

In Newton-Cotes methods, interpolation polynomials are generated using uniformly distributed data points. This simplifies resulting integral formulations. Numerical integral methods in Newton-Cotes group are explained below:

9.2.1 Trapezoidal Method

In trapezoidal method, real function is approximated with a linear interpolation polynomial within the integral range of [a,b] as shown in Fig. 9.4. Linear polynomial function $f(x)$ can be expressed with Lagrange interpolation method using function values at $x_1 = a$ and $x_2 = b$ as following:

$$f(x) = \frac{x - x_2}{x_1 - x_2} f_1 + \frac{x - x_1}{x_2 - x_1} f_2 \tag{9.3}$$

Integral of the linear polynomial can be expressed as below:

$$I = \int_{x_1 = a}^{x_2 = b} f(x)dx = \int_{x_1 = a}^{x_2 = b} \left(\frac{x - x_2}{x_1 - x_2} f_1 + \frac{x - x_1}{x_2 - x_1} f_2 \right) dx = \frac{f_1 + f_2}{2} (x_2 - x_1) \tag{9.4}$$

When real function is a non-polynomial one, a single linear polynomial will not approximate it. Therefore, integral of a non-polynomial function using that of a single linear polynomial will not be accurate. As a remedy, a number of linear polynomials are increased in the integral range of [a,b] as shown in Fig. 9.5 to approximate the real function better. Indicating the number of intervals with N, the number of points with n ($n = N + 1$), and interval length (distance between adjacent points) with $h = \frac{b-a}{N}$, integral of a general $f(x)$ function can be expressed as the sum of the integrals of piecewise linear polynomials as below:

$$\begin{aligned} I &= \int_{x_1 = a}^{x_n = b} f(x)dx = \int_{x_1 = a}^{x_2} f(x)dx + \int_{x_2}^{x_3} f(x)dx + \ldots + \int_{x_{n-1}}^{x_n = b} f(x)dx \\ &= \frac{h}{2} \left[(f_1 + f_2) + (f_2 + f_3) + \ldots + \left(f_{n-1=N} + f_{n=N+1} \right) \right] \end{aligned} \tag{9.5}$$

Integral expression in Eq. 9.5 can be rearranged as following

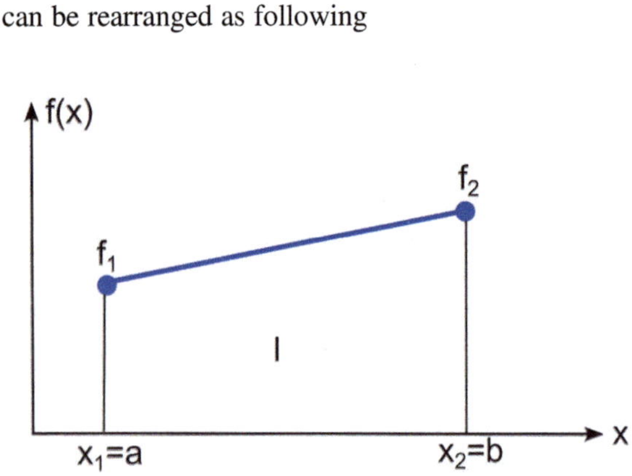

Fig. 9.4 Integral of the linear polynomial function

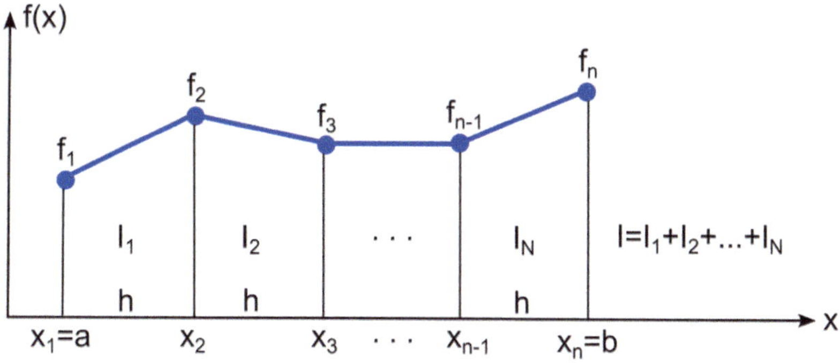

Fig. 9.5 Integral of a general *f*(*x*) function using piecewise linear functions

Table 9.1 Obtaining correct integral value by increasing the number of linear polynomials (i.e., reducing h by half) in the integral range $[a,b]$

	Number of trapezoidal	Integral
h	$N = \frac{b-a}{h}$	I_1
$\frac{h}{2}$	$2xN$	I_2
$\frac{h}{2^2}$	$(2^2)xN$	I_3
$\frac{h}{2^3}$	$(2^3)xN$	I_4
…	…	…
$\frac{h}{2^i}$	$(2^i)xN$	I_{i+1}

$$I = \frac{h}{2}\left[f_1 + 2(f_2 + f_3 + \ldots + f_{n-1=N}) + f_{n=N+1}\right] \tag{9.6}$$

Difficulty with trapezoidal method as in all Newton-Cotes methods is the determination of the appropriate number of piecewise polynomials that will yield accurate integral result. Unfortunately, there is no rule for this. Trial-and-error approach is used for this purpose. A number of linear polynomials are increased successively until the last two integral results do not differ much or the convergence condition is satisfied. Table 9.1 shows integral results when the number of linear polynomials (N) is doubled successively.

Convergence condition for the accurate integral result can be expressed in terms of integral results based on an increasing number of piecewise polynomials as following:

$$\left|\frac{I_{i+1} - I_i}{I_i}\right| \leq \varepsilon \tag{9.7}$$

In Eq. 9.7, ε denotes the convergence tolerance which is chosen as a small number such as 0.001.

Determination of Error Level in Trapezoidal Method

Although trial-and-error approach is used in practical applications, order of error study can guide how fast the number of piecewise polynomials in integral range should be increased to converge to the accurate integral value. As an example, order of error study is given below for piecewise linear polynomials.

Order of Error for a Single Linear Function:
As mentioned in Chap. 6 (Interpolation), an interpolation function $f(x)$ can be created in polynomial form with Taylor series, Divided Difference, and Newton-Gregory methods using function values. For example, using $h = (x_2 - x_1)$ and $\alpha = \frac{(x - x_1)}{h}$ definitions, Newton-Gregory polynomial can be written as below:

$$\begin{aligned} f(\alpha) =& f_1 + \Delta f_1 \alpha + \frac{\Delta^2 f_1}{2!}\alpha(\alpha-1) + \frac{\Delta^3 f_1}{3!}\alpha(\alpha-1)(\alpha-2) + \ldots \\ &+ \frac{\Delta^{n+1} f_1}{(n+1)!h^{n+1}}\alpha(\alpha-1)(\alpha-2)\ldots(\alpha-n) \end{aligned} \tag{9.8}$$

Considering only linear terms, Eq. 9.8 can be written as following:

$$\begin{aligned} f(\alpha) =& f_1 + \Delta f_1 \alpha + \underbrace{\frac{\Delta^2 f_1}{2!}\alpha(\alpha-1) + \frac{\Delta^3 f_1}{3!}\alpha(\alpha-1)(\alpha-2) + \ldots}_{R = \text{Truncation Error}} \\ &\underbrace{+ \frac{\Delta^{n+1} f_1}{(n+1)!h^{n+1}}\alpha(\alpha-1)(\alpha-2)\ldots(\alpha-n)}_{R = \text{Truncation Error}} \end{aligned}$$

Sum of the truncated terms (R) can be expressed as below:

$$\begin{aligned} R =& \frac{\Delta^2 f_1}{2!}\alpha(\alpha-1) + \frac{\Delta^3 f_1}{3!}\alpha(\alpha-1)(\alpha-2) + \ldots + \frac{\Delta^{n+1} f_1}{(n+1)!h^{n+1}}\alpha(\alpha-1)(\alpha-2)\ldots(\alpha-n) \\ \approx& \frac{f''(\xi)}{2!}\alpha(\alpha-1)h^2 \end{aligned}$$

where ξ represents a value in the integral range of [0, 1]. Considering the sum of the truncated terms, Newton-Gregory series can be rewritten as below:

$$f(\alpha) = f_1 + \Delta f_1 \alpha + \underbrace{\frac{f''(\xi)}{2!}\alpha(\alpha-1)h^2}_{R}$$

In this case, integral of $f(x)$ function in $[x_1 = a, x_2 = b]$ interval can be expressed using $dx = h d\alpha$ transformation as below:

$$
\begin{aligned}
I &= \int_{x_1=a}^{x_2=b} f(x)dx = \int_0^1 \left[f_1 + \Delta f_1 \alpha + \frac{f''(\xi)}{2!}\alpha(\alpha-1)h^2\right] h d\alpha \\
&= \int_0^1 [f_1 + \Delta f_1 \alpha] h d\alpha + \underbrace{\int_0^1 \left[\frac{f''(\xi)}{2!}\alpha(\alpha-1)h^2\right] h d\alpha}_{E:\text{Truncation Error}} \\
&= \int_0^1 [f_1 + \Delta f_1 \alpha] h d\alpha + \underbrace{\int_0^1 \left[\frac{f''(\xi)}{2!}\alpha(\alpha-1)h^2\right] h d\alpha}_{E:\text{Truncation Error}} \\
&= h\left[f_1\alpha + \Delta f_1 \frac{\alpha^2}{2}\right]_0^1 + h\left[\frac{f''(\xi)}{2!}\left(\frac{\alpha^3}{3} - \frac{\alpha^2}{2}\right)h^2\right]_0^1 \\
&= h\left[f_1 + \Delta f_1 \frac{1}{2}\right] - \frac{f''(\xi)}{12}h^3 = \underbrace{h\left(\frac{f_1+f_2}{2}\right)}_{\text{Trapezoidal Method}} \underbrace{-\frac{f''(\xi)}{12}h^3}_{E:\text{Truncation Error}}
\end{aligned}
\tag{9.9}
$$

From Eq. 9.9, it is seen that if a single linear polynomial function is used in the integral range of $[a, b]$, order of error due to truncation of terms becomes $O(h^3)$.

Order of Error for Multi-Linear Function:

If n number of linear polynomial functions are used in the integral range of $[a, b]$, total error in integral can be expressed as the sum of the error for each linear polynomial as below:

$$
\text{Total error}: -\frac{h^3}{12}[f''(\xi_1) + f''(\xi_2) + f''(\xi_3) + \ldots + f''(\xi_n)]
$$

where $\xi_1, \xi_2, \xi_3 \ldots, \xi_n$ represent values in sub-ranges ($x_i + (i-1)h \le \xi_i \le x_i + ih$, $i = 1,2,\ldots,n$). If $\xi_1, \xi_2, \xi_3 \ldots, \xi_n$ values are taken as ξ, total error can be written as following:

$$
\begin{aligned}
\text{Total Error} &= -\frac{h^3}{12}[f''(\xi_1) + f''(\xi_2) + f''(\xi_3) + \ldots + f''(\xi_n)] \\
&= -\frac{h^3}{12}[f''(\xi) + f''(\xi) + f''(\xi) + \ldots + f''(\xi)] \\
&= -\frac{h^3}{12}\underbrace{n}_{\frac{(b-a)}{h}} f''(\xi) = -\frac{(b-a)}{12}h^2 f''(\xi) = O(h^2)
\end{aligned}
\tag{9.10}
$$

From Eq. 9.10, it is seen that if multi-linear polynomial functions are used in the integral range of $[a,b]$, order of error due to truncation of terms becomes $O(h^2)$ which is lower than that of a single polynomial function.

Acceleration of Convergence in Trapezoidal Method

As stated before accurate integral value can be calculated by increasing the number of piecewise polynomials successively. However, this strategy is not an efficient strategy especially for the trapezoidal method. Therefore, more effective strategies were developed to increase the convergence rate for methods with low accuracy such as the trapezoidal method. One of them is Richardson extrapolation method adapted to the integral calculation. Applying Richardson extrapolation successively, convergence to the accurate integral value can be accelerated.

To explain Richardson extrapolation method, we can calculate an integral value using two different interval values between adjacent points such as $h = h_1$ and $h = h_2$. For $h = h_1$, we can represent integral value with I_1 and order of error with E_1. Similarly, for $h = h_2$, we can represent integral value with I_2 and order of error with E_2. In this case, true integral value I can be written for $h = h_1$ and $h = h_2$ as following:

For $h = h_1$: $I = I_1 + E_1 = I_1 + O\left(h_1^2\right)$

For $h = h_2$: $I = I_2 + E_2 = I_2 + O\left(h_2^2\right)$

Also, following equations can be written for error terms of E_1 and E_2

$$E_1 = -\frac{(b-a)}{12} h_1^2 f''(\xi), \quad E_2 = -\frac{(b-a)}{12} h_2^2 f''(\xi) \rightarrow \frac{E_1}{E_2} \approx \frac{h_1^2}{h_2^2} \rightarrow E_1 \approx E_2 \frac{h_1^2}{h_2^2}$$

In this case, following relationships can be written between integral and error expressions

$$\begin{aligned} I &= I_1 + E_1, \quad I = I_2 + E_2 \rightarrow I_1 + E_1 = I_2 + E_2 \rightarrow E_2 = I_1 + E_1 - I_2 \\ \rightarrow E_2 &= I_1 - I_2 + E_2 \frac{h_1^2}{h_2^2} \rightarrow E_2 = \frac{I_1 - I_2}{\left(1 - \frac{h_1^2}{h_2^2}\right)} \end{aligned} \tag{9.11}$$

Thus, actual integral result I can be expressed in terms of I_1, I_2, h_1 and h_2 as below:

$$I = I_2 + E_2 = I_2 + \frac{I_1 - I_2}{\left(1 - \frac{h_1^2}{h_2^2}\right)}$$

In the special case of $h_2 = \frac{h_1}{2}$, integral result I can be expressed as following

$$I = \frac{4}{3} I_2 - \frac{1}{3} I_1 \tag{9.12}$$

Example 9.1: Calculate $I = \int_0^{\pi/2} x^2 \sin(x)dx$ integral using trapezoidal method and then apply correction with Richardson extrapolation method.

Integral of $f(x) = x^2 \sin(x)$ function can be roughly calculated using 1 and 2 linear polynomial functions (i.e., line) as below (h: distance between adjacent points)

Using 1 linear polynomial function ($h = \frac{\pi}{2}$): $I_1 = \frac{h}{2}\left[f(0) + f\left(\frac{\pi}{2}\right)\right] = 1.9379$

Using 2 linear polynomial functions ($h = \frac{\pi}{4}$): $I_2 = \frac{h}{2}\left[f(0) + 2f\left(\frac{\pi}{4}\right) + f\left(\frac{\pi}{2}\right)\right] = 1.3115$

Using I_1 and I_2 results, corrected integral value can be obtained using Richardson extrapolation method as following:

$$I = \frac{4}{3}I_2 - \frac{1}{3}I_1 = 1.1027$$

If desired, more corrections can be applied successively as in Table 9.2.

From Table 9.2, it is seen that Richardson Extrapolation method accelerates convergence to the true integral value.

Integral Calculation with MATLAB Command:

Integral of $f(x) = x^2 \sin(x)$ function in the interval of $[0, \pi/2]$ can be calculated using a general *int* command or a special *trapz* (Trapezoidal method) command in MATLAB as below:

Using *int* command:

```
>> syms x; f=(x^2)*sin(x); I=int(f,x,0,pi/2)
I =
pi - 2
>> I=double(I)
I =
1.1416
```

Table 9.2 Integral values after successive corrections with Richardson extrapolation method

		Integral		
Number of line (linear segment)	h	Trapezoidal method	Richardson extrapolation $I = \frac{4}{3}I_{i+1}(h_{i+1}) - \frac{1}{3}I_i(h_i)$	Analytical
1	$h_1 = \pi/2$	$I_1 = 1.9379$	–	1.1416
2	$h_2 = h_1/2 = \pi/4$	$I_2 = 1.3115$	$I = \frac{4}{3}I_2 - \frac{1}{3}I_1 = 1.1027$	1.1416
4	$h_3 = h_2/2 = \pi/8$	$I_3 = 1.1825$	$I = \frac{4}{3}I_3 - \frac{1}{3}I_2 = 1.1395$	1.1416
8	$h_4 = h_3/2 = \pi/16$	$I_4 = 1.1517$	$I = \frac{4}{3}I_4 - \frac{1}{3}I_3 = 1.1414$	1.1416
16	$h_5 = h_4/2 = \pi/32$	$I_5 = 1.1441$	$I = \frac{4}{3}I_5 - \frac{1}{3}I_4 = 1.1416$	1.1416
32	$h_6 = h_5/2 = \pi/64$	$I_6 = 1.1416$	$I = \frac{4}{3}I_6 - \frac{1}{3}I_5 = 1.1416$	1.1416

Using *trapz* command:

Integral value can be calculated with *trapz* command using 33 points (32 equally spaced points) in the integral range of [0,π/2] as following:

```
>> x=linspace(0,pi/2,33);
>> f=(x.^2).*sin(x);
>> I=trapz(x,f)
I =
1.1422
```

Note: For the solution of this problem with Trapezoidal Method, *MATLAB Program 9.1: Integral Calculation with the Trapezoid Method* in Appendix can also be used.

9.2.2 1/3 Simpson Method

In 1/3 Simpson method, real function is approximated with a parabolic interpolation polynomial in the integral range of [a,b] as shown in Fig. 9.6. Parabolic polynomial function $f(x)$ can be expressed with Lagrange interpolation method using function values at three points ($x_1 = a$, $x_2 = \frac{a+b}{2}$, $x_3 = b$) as following:

$$f(x) = \frac{(x-x_2)(x-x_3)}{(x_1-x_2)(x_1-x_3)}f_1 + \frac{(x-x_1)(x-x_3)}{(x_2-x_1)(x_2-x_3)}f_2 + \frac{(x-x_1)(x-x_2)}{(x_3-x_1)(x_3-x_2)}f_3 \quad (9.13)$$

If equidistant points ($x_2 - x_1 = x_3 - x_2 = h$) are used, in this special case, integral value can be expressed as below:

$$I = \frac{h}{3}(f_1 + 4f_2 + f_3) \quad (9.14)$$

Fig. 9.6 Integral of parabolic function

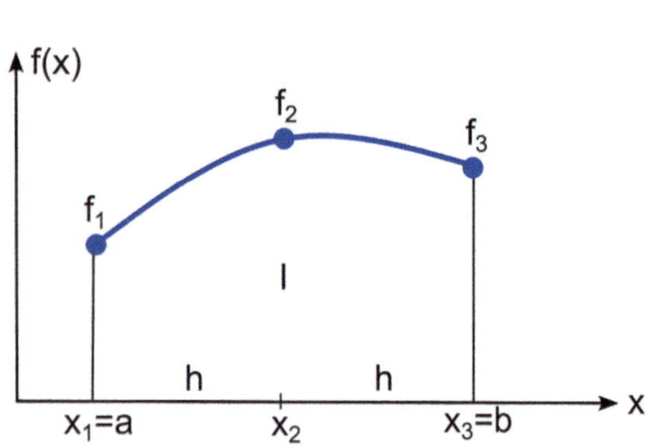

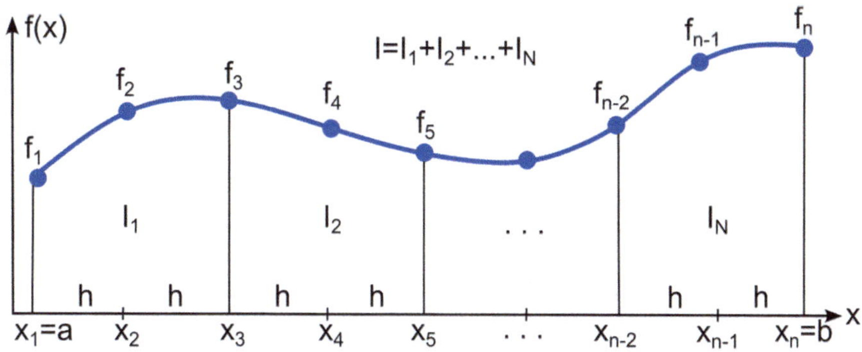

Fig. 9.7 Integration with piecewise parabolic functions

Fraction value of 1/3 in Eq. 9.14 gives Simpson method its name known as 1/3 Simpson method.

Single parabolic function usually cannot represent a general function exactly. In such a case, the number of piecewise parabolas can be increased as shown in Fig. 9.7 for better approximation. Indicating the number of intervals with N, the number of points with n ($n = 2N + 1$), and interval length with $h = \frac{b-a}{2N}$ in the integral range $[a, b]$, integral of a general $f(x)$ function can be calculated as the sum of the integrals of piecewise parabolic polynomials as below:

$$
\begin{aligned}
I = \int_{x_1=a}^{x_n=b} f(x)dx &= \int_{x_1=a}^{x_3} f(x)dx + \int_{x_3}^{x_5} f(x)dx + \ldots + \int_{x_{n-2}}^{x_n=b} f(x)dx \\
&= \frac{h}{3}(f_1 + 4f_2 + f_3) + \frac{h}{3}(f_3 + 4f_4 + f_5) + \ldots \\
&\quad + \frac{h}{3}\left(f_{n-2=2N-1} + 4f_{n-1=2N} + f_{n=2N+1}\right) \\
&= \frac{h}{3}\Big[f_1 + 4(f_2 + f_4 + f_6 + \ldots + f_{n-1=2N}) + 2(f_3 + f_5 + f_7 + \ldots + f_{n-2=2N-1}) \\
&\quad + f_{n=2N+1}\Big]
\end{aligned}
\tag{9.15}
$$

A number of piecewise parabolas leading to accurate integral result in an $[a, b]$ interval can be determined with trial-and-error approach. For this, the number of piecewise parabolas is increased successively until last two integral results do not differ much or the convergence condition is satisfied.

Table 9.3 Integral values with 1/3 Simpson method by an increasing number of parabolic splines in the interval of $[0, \frac{\pi}{2}]$

Number of parabola	h	Integral	
		1/3 Simpson method	Analytical
1	$h_1 = \pi/4$	$I_1 = 1.1027$	1.1416
2	$h_2 = h_1/2 = \pi/8$	$I_2 = 1.1395$	1.1416
4	$h_3 = h_2/2 = \pi/16$	$I_3 = 1.1415$	1.1416
8	$h_4 = h_3/2 = \pi/32$	$I_4 = 1.1416$	1.1416

Example 9.2: Calculate $I = \int_0^{\pi/2} x^2 \sin(x)dx$ integral using 1/3 Simpson method.

Integral of $f(x) = x^2 \sin(x)$ function in $[0, \frac{\pi}{2}]$ interval can be approximately calculated using 1 and 2 parabolic polynomial functions as below (h: distance between adjacent points).

Using 1 parabolic polynomial function ($h = \frac{\pi}{4}$):

$$I_1 = \frac{h}{3}\left[f(0) + 4f\left(\frac{\pi}{4}\right) + f\left(\frac{\pi}{2}\right)\right] = 1.1027$$

Using 2 parabolic polynomial functions ($h = \frac{\pi}{8}$):

$$I_2 = \frac{h}{3}\left[f(0) + 4f\left(\frac{\pi}{8}\right) + f\left(\frac{2\pi}{8}\right)\right] + \frac{h}{3}\left[f\left(\frac{2\pi}{8}\right) + 4f\left(\frac{3\pi}{8}\right) + f\left(\frac{\pi}{2}\right)\right] = 1.1395$$

Increasing the number of parabolic splines in the specified interval, integral values can be calculated as in Table 9.3.

Integral Calculation with MATLAB Command:
With 1/3 Simpson method, integral of $f(x) = x^2 \sin(x)$ function in the interval of $[0, \pi/2]$ can be calculated using *quad* command in MATLAB as following:

```
>> f='(x.^2).*sin(x)';
>> I=quad(f,0,pi/2)
I =
1.1416
```

Note: For the solution of this problem with 1/3 Simpson Method, *MATLAB Program 9.2: Integral Calculation* program with 1/3 Simpson Method in Appendix can also be used.

9.2.3 3/8 Simpson Method

In 3/8 Simpson method, real function is approximated with a cubic interpolation polynomial in the integral range of $[a,b]$ as shown in Fig. 9.8. Cubic polynomial function $f(x)$ can be expressed with Lagrange interpolation method using function values at four points ($x_1 = a$, $x_2, x_3, x_4 = b$) as following:

$$\begin{aligned} f(x) &= \frac{(x-x_2)(x-x_3)(x-x_4)}{(x_1-x_2)(x_1-x_3)(x_1-x_4)}f_1 + \frac{(x-x_1)(x-x_3)(x-x_4)}{(x_2-x_1)(x_2-x_3)(x_2-x_4)}f_2 \\ &+ \frac{(x-x_1)(x-x_2)(x-x_4)}{(x_3-x_1)(x_3-x_2)(x_3-x_4)}f_3 + \frac{(x-x_1)(x-x_2)(x-x_3)}{(x_4-x_1)(x_4-x_2)(x_4-x_3)}f_4 \end{aligned} \tag{9.16}$$

If equidistant points ($x_2 - x_1 = x_3 - x_2 = x_4 - x_3 = h$) are used, in this special case, integral value can be expressed as below:

$$I = \int_{x_1=a}^{x_4=b} f(x)dx$$

$$= \int_{x_1=a}^{x_4=b} \begin{bmatrix} \tilde{f}(x) = \frac{(x-x_2)(x-x_3)(x-x_4)}{(x_1-x_2)(x_1-x_3)(x_1-x_4)}f_1 + \frac{(x-x_1)(x-x_3)(x-x_4)}{(x_2-x_1)(x_2-x_3)(x_2-x_4)}f_2 + \\ \frac{(x-x_1)(x-x_2)(x-x_4)}{(x_3-x_1)(x_3-x_2)(x_3-x_4)}f_3 + \frac{(x-x_1)(x-x_2)(x-x_3)}{(x_4-x_1)(x_4-x_2)(x_4-x_3)}f_4 \end{bmatrix} dx$$

$$= \frac{3h}{8}(f_1 + 3f_2 + 3f_3 + f_4) \tag{9.17}$$

Fraction value of 3/8 in Eq. 9.17 gives the Simpson method its name known as 3/8 Simpson method.

Single cubic function usually cannot represent a general function exactly. In such a case, the number of piecewise parabolas can be increased as shown in Fig. 9.9 for

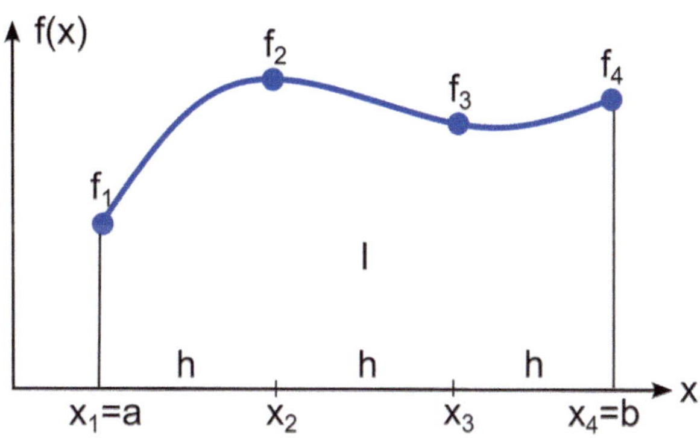

Fig. 9.8 Integral of the cubic function

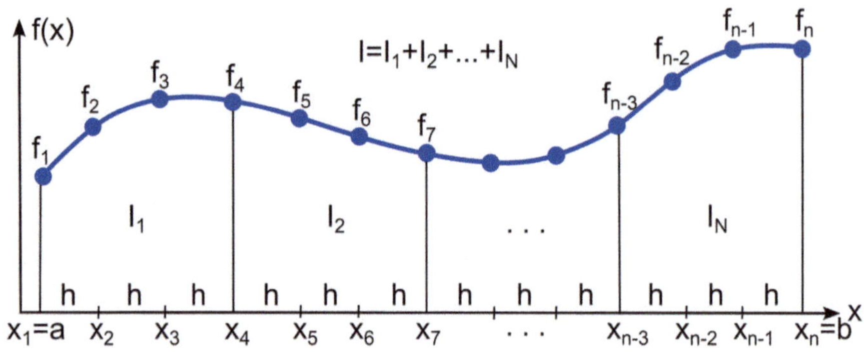

Fig. 9.9 Integration with piecewise cubic polynomial functions

better approximation. Indicating the number of intervals with N, the number of points with n ($n = 3N + 1$) and interval length with $h = \frac{b-a}{3N}$ in the integral range $[a,b]$, integral of a general $f(x)$ function can be calculated as the sum of the integrals of piecewise cubic polynomials as below:

$$\begin{aligned} I &= \int_{x_1=a}^{x_n=b} f(x)dx = \int_{x_1=a}^{x_4} f(x)dx + \int_{x_4}^{x_7} f(x)dx + \ldots + \int_{x_{n-3}}^{x_n=b} f(x)dx \\ &= \frac{3h}{8}(f_1 + 3f_2 + 3f_3 + f_4) + \frac{3h}{8}(f_4 + 3f_5 + 3f_6 + f_7) + \ldots \\ &\quad + \frac{h}{3}\left(f_{n-3=3N-2} + 3f_{n-2=3N-1} + 3f_{n-1=3N} + f_{n=3N+1}\right) \end{aligned} \tag{9.18}$$

To calculate integrals with 3/8 Simpson method accurately, the number of cubic functions is increased successively within the integral region $[a, b]$ until the last two integral results do not differ much or the convergence condition is satisfied.

Example 9.3: Calculate $I = \int_0^{\pi/2} x^2 \sin(x)dx$ integral using 3/8 Simpson method.

Integral of $f(x) = x^2 \sin(x)$ function in $[0, \frac{\pi}{2}]$ interval can be roughly calculated using one and two cubic polynomial functions as below (h: distance between adjacent points).

Using one cubic polynomial function ($h = \frac{\pi}{6}$):

$$I_1 = \frac{3h}{8}\left[f(0) + 3f\left(\frac{\pi}{6}\right) + 3f\left(\frac{2\pi}{6}\right) + f\left(\frac{\pi}{2}\right)\right] = 1.1246$$

Using two cubic polynomial functions ($h = \frac{\pi}{12}$):

Table 9.4 Integral values with 3/8 Simpson method by increasing the number of cubic splines in the interval of [0, $\pi/2$]

Number of cubic splines	h	Integral	
		3/8 Simpson method	Analytical
1	$h_1 = \pi/6$	$I_1 = 1.1246$	1.1416
2	$h_2 = h_1/2 = \pi/12$	$I_2 = 1.1407$	1.1416
4	$h_3 = h_2/2 = \pi/24$	$I_3 = 1.1415$	1.1416

$$\begin{aligned} I_2 &= \frac{3h}{8}\left[f(0) + 3f\left(\frac{\pi}{12}\right) + 3f\left(\frac{2\pi}{12}\right) + f\left(\frac{3\pi}{12}\right)\right] \\ &+ \frac{3h}{8}\left[f\left(\frac{3\pi}{12}\right) + 3f\left(\frac{4\pi}{12}\right) + 3f\left(\frac{5\pi}{12}\right) + f\left(\frac{\pi}{2}\right)\right] \\ &= 1.1407 \end{aligned}$$

Increasing the number of cubic splines in the specified interval, integral values can be calculated as in Table 9.4.

As seen in Table 9.4, 3/8 Simpson method converges to the analytical result very quickly (faster than 1/3 Simpson method).

Note**:** For the solution of this problem with 3/8 Simpson Method, *MATLAB Program 9.3: Integral Calculation Program with 3/8 Simpson Method* in Appendix can be used.

9.3 Numerical Integration with Quadrature Methods

Quadrature methods were developed as alternative to Newton-Cotes methods such as Trapezoid, 1/3 Simpson, 3/8 Simpson methods [4–11]. Quadrature methods are more efficient in calculating integrals than Newton-Cotes methods. Their efficiency comes from the use of non-uniform and less number of integration points. Due to their efficiency, quadrature methods are widely used in the calculation of stiffness matrices which are expressed with integrals in popular Finite Element Method.

In quadrature methods, original function is approximated in the range of integral limits [a,b] with a global polynomial function $f(x)$. Then, integral of global polynomial function $f(x)$ in [a,b] interval is equivalently expressed as the weighted sum of the function values at different points as in Eq. 9.2 expressed before:

$$I = \int_a^b f(x)dx = w_1 f(x_1) + w_2 f(x_2) + \ldots + w_n f(x_n) = \sum_{i=1}^{p} w_i f(x_i)$$

where x_1, x_2,…,x_n indicate integration points and w_1, w_2,…,w_n denote their corresponding weight coefficients. A number of integration points and weight coefficients vary with the order of polynomial function. Quadrature methods have different names based on the number and location of integration points such as Classical Gauss, Gauss-Legendre, and Gauss-Lobatto.

For convenience, numerical methods including quadrature methods use integral limits of [−1,1]. If actual integral range is different from this such as [a,b], it can be converted into the normalized limits of [−1,1] using mapping techniques. Use of [−1,1] limits allows development of standard formulations in table form for easy integration. Quadrature methods are explained in detail below:

9.3.1 Numerical Integration with Gauss Quadrature Method

A general polynomial function *f*(*x*) with order *n* can be expressed in terms of basic functions as following:

$$f(x)=c_1\underbrace{1}_{f_1(x)}+c_2\underbrace{x}_{f_2(x)}+c_3\underbrace{x^2}_{f_3(x)}+c_4\underbrace{x^3}_{f_4(x)}+\ldots+c_{n+1}\underbrace{x^n}_{f_{n+1}(x)} \tag{9.19}$$

where $f_1(x)=1$, $f_2(x)=x$, $f_3(x)=x^2$, $f_4(x)=x^3$, …,$f_{n+1}(x)=x^n$ functions in Eq. 9.19 show basic polynomial functions. Integral of *f*(*x*) function in [−1,1] interval can be expressed with Classical Gauss (or Gauss) Quadrature method using *p* number of integration points as below:

$$\begin{aligned}&\int_{-1}^{1} f(x)dx = w_1 f(x_1)+w_2 f(x_2)+\ldots+w_p f(x_p) \Rightarrow\\ &\int_{-1}^{1}\left(c_1\underbrace{1}_{f_1(x)}+c_2\underbrace{x}_{f_2(x)}+c_3\underbrace{x^2}_{f_3(x)}+c_4\underbrace{x^3}_{f_4(x)}+\ldots+c_{n+1}\underbrace{x^n}_{f_{n+1}(x)}\right)dx\\ &\quad = w_1 f(x_1)+w_2 f(x_2)+\ldots+w_p f(x_p)\end{aligned} \tag{9.20}$$

Two approaches can be used to calculate integration points and weight coefficients in Eq. 9.20:

- First approach: Using integral of global function
- Second approach: Using integrals of basic functions

First Approach (Integral of Global Function)

In this approach, necessary equations to determine the location/position of integration points and their weight coefficients are obtained by integrating global function. This approach is explained below for integration of linear and cubic functions.

Integral of Linear Polynomial Function:
Linear polynomial function can be written in general form as $f(x) = c_1 + c_2x$. Since degree of linear function is one ($n = 1$), the number of Gauss integration points needed for accurate integral value will be $p = 1$ ($p = (n + 1)/2 = 1$). We can denote location and weight coefficient of this single point as x_1 and w_1, respectively, as shown in Fig. 9.10. In this case, integral of linear function in the range of [−1,1] can be expressed as below:

$$I = \int_{-1}^{1} f(x)dx = w_1 f(x_1) \Rightarrow I = \int_{-1}^{1} \underbrace{(c_1 + c_2x)}_{f(x)}dx = w_1\underbrace{(c_1 + c_2x_1)}_{f(x_1)} \tag{9.21}$$

After integrating left side of Eq. 9.21, following equation can be obtained.

$$2c_1 = w_1c_1 + w_1x_1c_2$$

Using equivalency of right and left sides of the above equation, w_1 and x_1 can be determined as following:

$$2c_1 = w_1c_1 + w_1x_1c_2 \rightarrow \begin{array}{l} w_1 = 2 \\ w_1x_1 = 0 \Rightarrow x_1 = 0 \end{array}$$

Thus, integral of linear polynomial function with a single integration point using Gauss Quadrature method can be expressed as below:

$$I = \int_{-1}^{1} \underbrace{(c_1 + c_2x)}_{f(x)}dx = 2f(0) \tag{9.22}$$

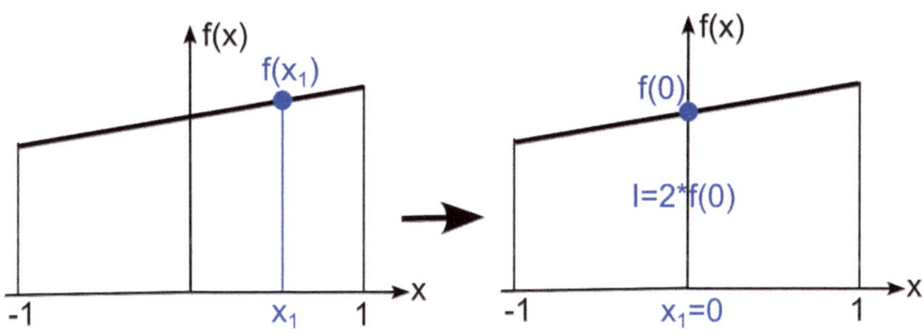

Fig. 9.10 Integral of linear function with Gauss Quadrature method

Integral of Cubic Polynomial Function:

Cubic polynomial function can be written in general form as $f(x) = c_1 + c_2x + c_3x^2 + c_4x^3$. Since degree of cubic function is three ($n = 3$), the number of Gauss integration points needed for accurate integral value will be $p = 2$ ($p = (n + 1)/2 = 2$). We can denote these points as x_1, x_2 and their weight coefficients as w_1, w_2 as shown in Fig. 9.11. In this case, integral of cubic function in $[-1,1]$ range can be expressed as below:

$$I = \int_{-1}^{1} f(x)dx = w_1 f(x_1) + w_2 f(x_2) \Rightarrow$$

$$\begin{aligned} I &= \int_{-1}^{1} \underbrace{\left(c_1 + c_2x + c_3x^2 + c_4x^3\right)}_{f(x)} dx \\ &= w_1 \underbrace{\left(c_1 + c_2x_1 + c_3x_1^2 + c_4x_1^3\right)}_{f(x_1)} + w_2 \underbrace{\left(c_1 + c_2x_2 + c_3x_2^2 + c_4x_2^3\right)}_{f(x_2)} \end{aligned} \tag{9.23}$$

After integrating left side of Eq. 9.23, following equation can be obtained

$$2c_1 + \frac{2}{3}c_3 = (w_1 + w_2)c_1 + (w_1x_1 + w_2x_2)c_2 + \left(w_1x_1^2 + w_2x_2^2\right)c_3 + \left(w_1x_1^3 + w_2x_2^3\right)c_4$$

From equivalency of left and right sides of the above equation, integration points and their weight coefficients can be obtained as following:

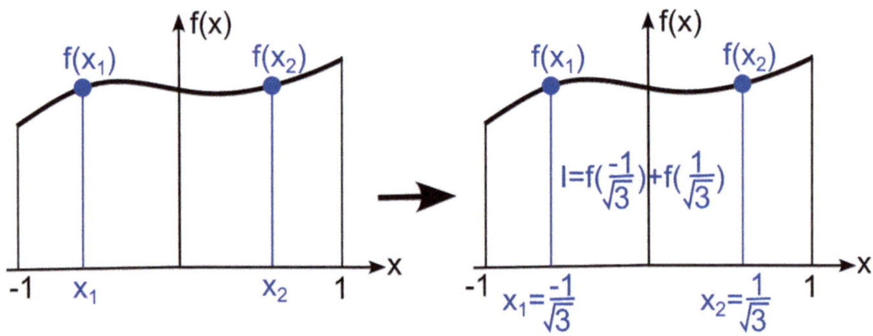

Fig. 9.11 Integral of cubic function with Gauss Quadrature method

$$\left.\begin{aligned} w_1 + w_2 &= 2 \\ w_1x_1 + w_2x_2 &= 0 \\ w_1x_1^2 + w_2x_2^2 &= \frac{2}{3} \\ w_1x_1^3 + w_2x_2^3 &= 0 \end{aligned}\right\} \Rightarrow w_1 = 1, w_2 = 1, x_1 = -\frac{1}{\sqrt{3}} = -0.5774, x_2 = \frac{1}{\sqrt{3}} = 0.5774$$

***Note*:** Above four equations form a nonlinear equation system. They can be solved using the Generalized Newton-Raphson method in Chap. 5.

Thus, integral of cubic polynomial function with Gauss Quadrature method in the range of $[-1,1]$ can be expressed as below:

$$I = \int_{-1}^{1} \underbrace{\left(c_1 + c_2x + c_3x^2 + c_4x^3\right)}_{f(x)} dx = f\left(-\frac{1}{\sqrt{3}}\right) + f\left(\frac{1}{\sqrt{3}}\right) \tag{9.24}$$

Integral expression in Eq. 9.24 can also be used in integration of polynomial functions with lower orders such as linear and parabolic polynomials.

Second Approach (Integrals of Basic Functions)

Second approach is easier and more general to implement than first approach. Therefore, it is preferred more in determining integration points and weight coefficients for higher order polynomial functions. In this approach, necessary equations are produced from the integrals of basic functions. For example, a general polynomial function with order n can be expressed in terms of basic functions as $f(x) = c_1 \underbrace{1}_{f_1(x)} + c_2 \underbrace{x}_{f_2(x)} + \ldots + c_{n+1} \underbrace{x^n}_{f_{n+1}(x)}$. Integral of this function can be expressed using p integration points as following:

$$\int_{-1}^{1} \underbrace{\left(c_1 f_1(x) + c_2 f_2(x) + \ldots + c_{n+1} f_{n+1}(x)\right)}_{f(x)} dx = w_1 f(x_1) + w_2 f(x_2) + \ldots + w_p f(x_p) \tag{9.25}$$

Second approach uses the principle of *a method that correctly calculates integrals of basic polynomial functions also correctly calculates the integral of a general function constructed from them.* Using this principle, integrals of basic functions ($f_1(x), f_2(x), \ldots, f_{n+1}(x)$) can be expressed with Gauss Quadrature as below:

$$\left.\begin{array}{l}\int\limits_{-1}^{1} f_1(x)dx = w_1 f_1(x_1) + w_2 f_1(x_2) + \ldots + w_p f_1(x_p)\\ \int\limits_{-1}^{1} f_2(x)dx = w_1 f_2(x_1) + w_2 f_2(x_2) + \ldots + w_p f_2(x_p)\\ \int\limits_{-1}^{1} f_3(x)dx = w_1 f_3(x_1) + w_2 f_3(x_2) + \ldots + w_p f_3(x_p)\\ \ldots\\ \int\limits_{-1}^{1} f_{n+1}(x)dx = w_1 f_{n+1}(x_1) + w_2 f_{n+1}(x_2) + \ldots + w_p f_{n+1}(x_p)\end{array}\right\} \Rightarrow \begin{array}{ll} n+1 & \text{equations,}\\ 2p & \text{unknowns}\end{array} \tag{9.26}$$

From Eq. 9.26, it is seen that there are $n + 1$ equations available to solve p number of weight coefficients and p number of integration points (total $2p$ unknowns). Therefore, $n + 1 = 2p$ condition should be satisfied for the solution of unknowns, Using $n + 1 = 2p$ relationship, the number of integration points to integrate a polynomial function of order n is determined as $p = (n + 1)/2$. Second approach can be applied to the integration of linear and cubic polynomial functions as following:

Integral of Linear Polynomial Function:
A general linear polynomial function can be written in terms of basic functions as $f(x) = c_1 \underbrace{1}_{f_1(x)} + c_2 \underbrace{x}_{f_2(x)} = c_1 f_1(x) + c_2 f_2(x)$, where $f_1(x)$ and $f_2(x)$ indicate basic functions. Using $p = (n + 1)/2$ equation, a single integration point ($p = (1 + 1)/2 = 1$) can be determined to integrate a general linear function accurately. Thus, integral of a general linear function with Gauss Quadrature method can be expressed as $\int_{-1}^{1} f(x)dx = w_1 f(x_1)$. Equations to find w_1 and x_1 unknowns can be obtained from the integrals of basic functions ($f(x)$ is replaced with $f_1(x)$ and $f_2(x)$ in $\int_{-1}^{1} f(x)dx = w_1 f(x_1)$ equation) as following:

$$f_1(x) = 1: \int\limits_{-1}^{1} f_1(x)dx = w_1 f_1(x_1) \rightarrow \int\limits_{-1}^{1} 1dx = w.1 \rightarrow 2 = w$$

$$f_2(x) = x: \int_{-1}^{1} f_2(x)dx = w_1 f_2(x_1) \rightarrow \int_{-1}^{1} xdx = w.x_1 \rightarrow 0 = w.x_1$$

From above equations, $w = 2$ and $x_1 = 0$ are obtained. Thus, integral of a linear polynomial function with Gauss Quadrature method in the range of $[-1,1]$ can be expressed as below:

$$\int_{-1}^{1} \underbrace{(c_1 + c_2 x)}_{f(x)} dx = 2f(0)$$

Gauss Quadrature and Trapezoidal method can be compared in terms of efficiency (i.e., number of points) in integration of linear polynomial function. For this, trapezoidal method can be written in quadrature form as below:

$$I = \frac{f_a + f_b}{2}(b - a) = \frac{f(-1) + f(1)}{2}(1 - (-1)) = f(-1) + f(1)$$

From the above expression, it is seen that for integral of a linear function, two integration points are used in Trapezoidal method while just one point is utilized in Gauss Quadrature method. This shows that Gauss Quadrature method is more efficient than Trapezoidal method due to the use of fewer integration points in integral calculations.

Integral of Cubic Polynomial Function

A general cubic polynomial function can be written in terms of basic functions as

$$f(x) = c_1 \underbrace{1}_{f_1(x)} + c_2 \underbrace{x}_{f_2(x)} + c_3 \underbrace{x^2}_{f_3(x)} + c_4 \underbrace{x^3}_{f_4(x)} = c_1 f_1(x) + c_2 f_2(x) + c_3 f_3(x) + c_4 f_4(x),$$

where $f_1(x)$, $f_2(x)$, $f_3(x)$, and $f_4(x)$ represent basic functions. Using $p = (n + 1)/2$ equation, two integration points ($p = (3 + 1)/2 = 2$) can be determined to integrate a general cubic function accurately. Using two integration points, integral of cubic function with Gauss Quadrature method can be expressed as $\int_{-1}^{1} f(x)dx = w_1 f(x_1) + w_2 f(x_2)$. Equations to find w_1, w_2 and x_1, x_2 unknowns can be obtained from the integrals of basic functions ($f(x)$ is replaced with $f_1(x)$, $f_2(x)$, $f_3(x)$ and $f_4(x)$ in $\int_{-1}^{1} f(x)dx = w_1 f(x_1)$ equation) as following:

$$f_1(x)=1: \int_{-1}^{1} f_1(x)dx = w_1 f_1(x_1) + w_2 f_1(x_2) \rightarrow \int_{-1}^{1} 1dx = w_1.1 + w_2.1 \rightarrow 2$$
$$= w_1 + w_2$$

$$f_2(x)=x: \int_{-1}^{1} f_2(x)dx = w_1 f_2(x_1) + w_2 f_2(x_2) \rightarrow \int_{-1}^{1} xdx = w_1.x_1 + w_2.x_2 \rightarrow 0$$
$$= w_1 x_1 + w_2 x_2$$

$$f_3(x)=x^2: \int_{-1}^{1} f_3(x)dx = w_1 f_3(x_1) + w_2 f_3(x_2) \rightarrow \int_{-1}^{1} x^2dx = w_1.x_1^2 + w_2.x_2^2 \rightarrow \frac{2}{3}$$
$$= w_1 x_1^2 + w_2 x_2^2$$

$$f_3(x)=x^3: \int_{-1}^{1} f_4(x)dx = w_1 f_4(x_1) + w_2 f_4(x_2) \rightarrow \int_{-1}^{1} x^3dx = w_1.x_1^3 + w_2.x_2^3 \rightarrow 0$$
$$= w_1 x_1^3 + w_2 x_2^3$$

From above equations, $w_1 = 1$, $w_2 = 1$, $x_1 = -\frac{1}{\sqrt{3}} = -0.5774$, $x_2 = \frac{1}{\sqrt{3}} = 0.5774$ are obtained. Thus, integral of cubic polynomial function with Gauss Quadrature method in the range of [−1,1] can be expressed as below:

$$\int_{-1}^{1} \underbrace{\left(c_1 + c_2 x + c_3 x^2 + c_4 x^3\right)}_{f(x)} dx = f\left(\frac{-1}{\sqrt{3}}\right) + f\left(\frac{1}{\sqrt{3}}\right)$$

Comparison of Gauss Quadrature and 1/3 Simpson Methods

Gauss Quadrature and 1/3 Simpson methods can be compared in terms of their efficiency (i.e., number of points) in integration of cubic polynomial function. For this, 1/3 Simpson method can be written in quadrature form for the range of [−1,1] as below:

$$I = \frac{h}{3}(f_1 + 4f_2 + f_3) = \frac{1}{3}f(-1) + \frac{4}{3}f(0) + \frac{1}{3}f(1)$$

From the above expression, it is seen that for integral of a cubic function, three integration points are used in 1/3 Simpson method while just two points are utilized in Gauss Quadrature method. This shows that Gauss Quadrature method is more efficient than 1/3 Simpson method due to the use of fewer integration points in integral calculations.

In Gauss Quadrature method, using p number of integration points, polynomials with order of 2p-1 or less can be integrated exactly. For example, a Gauss Quadrature method, which can correctly calculate integral of a cubic function, also correctly calculates integrals of linear and parabolic functions. In Gauss Quadrature method, integration points are symmetric with respect to the center and they are located in the inner region of [−1,1] interval.

In the classical Gauss Quadrature method as explained before, integration points and weight coefficients are obtained by solving a nonlinear equation set. Because of this, determination of unknown coefficients is highly complicated when a large number of integration points are used in integration. Therefore, in practical applications, integration points and weight coefficients are tabulated before their usage as shown in Table 9.5.

Table 9.5 Integration points and weight coefficients for integration of polynomial functions with different orders in the range of [−1,1] using Gauss Quadrature method

Gauss integration points

$I=\int_{-1}^{1} f(x)dx=\sum_{i=1}^{p} w_i f(x_i) \quad (-1\leq x\leq 1)$

Polynomial order (n)	Number of points (p)	Integration point (x_i)	Weight (w_i)
1	1	0	2
3	2	−0.5774	1
		0.5774	1
5	3	−0.7746	0.5555
		0	0.8888
		0.7746	0.5555
7	4	−0.8611	0.3479
		−0.3400	0.6521
		0.3400	0.6521
		0.8611	0.3479
9	5	−0.9062	0.2369
		−0.5385	0.4786
		0	0.5689
		0.5385	0.4786
		0.9062	0.2369
11	6	−0.9325	0.1713
		−0.6612	0.3608
		−0.2386	0.4679
		0.23861	0.4679
		0.6612	0.3608
		0.9325	0.1713

9.3.2 Numerical Integration with Gauss-Legendre Quadrature Method

Gauss-Legendre quadrature method was developed as an effective alternative to classical Gauss quadrature method. In Gauss-Legendre quadrature method, integration points and weight coefficients are directly obtained from Legendre polynomials without solving a set of equations. Gauss-Legendre quadrature method uses the same integration points and weight coefficients with classical Gauss quadrature method. For this reason, Gauss-Legendre quadrature method can be viewed as a more systematic and simplified version of Classical Gauss quadrature method. In Gauss-Legendre quadrature method, a general function $f(x)$ can be expressed in terms of orthogonal Legendre polynomials as below:

$$f(x) = c_0 L_0(x) + c_1 L_1(x) + \ldots + c_n L_n(x) \tag{9.27}$$

where $L_0(x)$, $L_1(x)$,…,$L_n(x)$ represent orthogonal Legendre polynomials which were explained in Chap. 7. Integration points can be obtained from roots of Legendre polynomials as following:

First-order Legendre polynomial: $L_1(x) = x \rightarrow \text{Root} : x_1 = 0$
Second-order Legendre polynomial: $L_2(x) = \frac{3x^2 - 1}{2} \rightarrow \text{Roots} : x_1 = \frac{-1}{\sqrt{3}}, \quad x_2 = \frac{1}{\sqrt{3}}$
Third-order Legendre polynomial: $L_3(x) = \frac{5x^3 - 3x}{2} \rightarrow \text{Roots} : x_1 = -\sqrt{\frac{3}{5}}, \quad x_2 = 0,$ $x_3 = \sqrt{\frac{3}{5}}$

Weight coefficients corresponding to n number of integration points can be obtained by using derivatives of Legendre polynomials as below:

$$w_i = \frac{2}{\left(1 - x_i^2\right)\left[\frac{dL_n}{dx}(x_i)\right]^2} \tag{9.28}$$

For example, for one and two integration points, weight coefficients can be expressed as below:

For one integration point:

$$L_1(x) = x \rightarrow \text{Roots} : x_1 = 0 \rightarrow w_1 = \frac{2}{\left(1 - x_1^2\right)\left[\frac{dL_1}{dx}(x_1)\right]^2} = \frac{2}{\left(1 - 0^2\right)[1]^2} = 2$$

For two integration points:

$$L_2(x) = \frac{3x^2 - 1}{2} \rightarrow \text{Roots} : x_1 = \frac{-1}{\sqrt{3}}, \quad x_2 = \frac{1}{\sqrt{3}} : \rightarrow$$

$$w_1 = \frac{2}{\left(1-x_1^2\right)\left[\frac{dL_2}{dx}(x_1)\right]^2} = \frac{2}{\left(1-\left(\frac{-1}{\sqrt{3}}\right)^2\right)\left[3\left(\frac{-1}{\sqrt{3}}\right)\right]^2} = 1,$$

$$w_2 = \frac{2}{\left(1-x_2^2\right)\left[\frac{dL_2}{dx}(x_2)\right]^2} = \frac{2}{\left(1-\left(\frac{1}{\sqrt{3}}\right)^2\right)\left[3\left(\frac{1}{\sqrt{3}}\right)\right]^2} = 1,$$

From above results, it is seen that Gauss-Legendre quadrature method yields the same integration points and weight coefficients with Classical Gauss quadrature method.

9.3.3 Numerical Integration with Gauss-Lobatto Quadrature Method

Gauss-Lobatto quadrature method is another alternative to classical Gauss Quadrature method. It is an efficient integration method that uses integration points both at boundaries and interior regions of the interval. Integral of a general *f*(*x*) function in the range of [−1,1] can be expressed with Gauss-Lobatto method using *p* integration point as following:

$$\begin{aligned}\int_{-1}^{1} f(x)dx &= \underbrace{\frac{2}{p(p-1)}}_{w_1} f\left(\underbrace{-1}_{x_1}\right) + w_2 f(x_2) + w_3 f(x_3) + \ldots + w_{p-1} f(x_{p-1}) + \underbrace{\frac{2}{p(p-1)}}_{w_p} f\left(\underbrace{1}_{x_p}\right) \\ &= \underbrace{\frac{2}{p(p-1)}}_{w_1} f\left(\underbrace{-1}_{x_1}\right) + \sum_{i=2}^{p-1} w_i f(x_i) + \underbrace{\frac{2}{p(p-1)}}_{w_p} f\left(\underbrace{1}_{x_p}\right)\end{aligned} \tag{9.29}$$

From Eq. 9.29 it is seen that first ($x_1 = -1$) and last ($x_p = 1$) integration points correspond to integral limits. Internal integration points ($x_2, x_3, \ldots, x_{p-1}$) are obtained from roots of derivatives of Legendre polynomials. Regarding weight coefficients, first and last weight coefficients are obtained as $w_1 = w_p = \frac{2}{p(p-1)}$, while internal weight coefficients are calculated using the following equation:

$$w_i = \frac{2}{p(p-1)\left[L_{p-1}(x_i)\right]^2} \quad (i = 2, 3, \ldots, p-1) \tag{9.30}$$

Gauss-Lobatto integration points and weight coefficients for *p*=3 and *p*=4 point numbers can be obtained in more detail as below:

Number of integration point for p = 3:

x_1 and x_3 are taken as integral limits and are expressed as $x_1 = -1, x_3 = 1$. Internal point x_2 is obtained from root of derivative of Legendre polynomial with order of $p-1 = 2$ as following:

Legendre polynomial: $L_2(r) = \frac{3r^2-1}{2} \rightarrow \text{Derivative}: \frac{dL_2}{dr} = 3r \rightarrow \text{Root}: r = 0 \rightarrow x_2 = r = 0$

Weight coefficients:

First and last points: $w_1 = w_3 = \frac{2}{p(p-1)} = \frac{2}{3(3-1)} = \frac{1}{3}$,

Internal point: $w_2 = \frac{2}{p(p-1)\left[L_{p-1}(x_2)\right]^2} = \frac{2}{3x2[L_2(0)]^2} = \frac{2}{3x2\left[\frac{-1}{2}\right]^2} = \frac{4}{3}$

Number of integration point for p = 4:

x_1 and x_4 are taken as integral limits and are expressed as $x_1 = -1, x_4 = 1$. Internal points x_2 and x_3 are obtained from roots of derivative of Legendre polynomial with order of $p-1 = 3$ as following:

Legendre polynomial: $L_3(r) = \frac{5r^3-3r}{2} \rightarrow \text{Derivative}: \frac{dL_3}{dr} = \frac{15r^2-3}{2} \rightarrow$

Roots: $r_1 = -\frac{1}{\sqrt{5}}, \quad r_2 = \frac{1}{\sqrt{5}} \rightarrow x_2 = r_1 = -\frac{1}{\sqrt{5}}, \quad x_3 = r_2 = \frac{1}{\sqrt{5}}$

Weight coefficients:

$$w_1 = w_4 = \frac{2}{p(p-1)} = \frac{2}{4(4-1)} = \frac{1}{6},$$

$$w_2 = \frac{2}{p(p-1)\left[L_{p-1}(x_2)\right]^2} = \frac{2}{4x3\left[L_3\left(-\frac{1}{\sqrt{5}}\right)\right]^2} = \frac{5}{6},$$

$$w_3 = \frac{2}{p(p-1)\left[L_{p-1}(x_3)\right]^2} = \frac{2}{4x3\left[L_3\left(\frac{1}{\sqrt{5}}\right)\right]^2} = \frac{5}{6}$$

Integration points and weight coefficients for integration of polynomial functions with different orders in the range of [−1,1] using Gauss-Lobatto Quadrature method are given in Table 9.6.

In Gauss-Lobatto Quadrature method, integration points and weight coefficients can be obtained directly without solving a set of equations as in Gauss-Legendre method. In Gauss-Lobatto Quadrature method, using *p* number of integration point, polynomials with order of 2*p*-3 or less can be integrated exactly. In Gauss-Lobatto Quadrature method, integration points are symmetric with respect to the center.

Table 9.6 Integration points and weight coefficients for integration of polynomial functions in the range of [−1,1] with Gauss-Lobatto Quadrature method

Gauss-Lobatto integration points

$I=\int_{-1}^{1} f(x)dx=\sum_{i=1}^{p} w_i f(x_i) \quad (-1\leq x\leq 1)$

Polynomial order (n)	Number of points (p)	Integration point (x_i)	Weight (w_i)
3	3	−1	0.3333
		0	1.3333
		1	0.3333
5	4	−1	0.1667
		−0.4472	0.8333
		0.4472	0.8333
		1	0.1667
7	5	−1	0.1000
		−0.6547	0.5444
		0	0.7111
		0.6547	0.5444
		1	0.1000

9.4 Conversion of Arbitrary Integral Limits to [−1,1]

Quadrature formulas developed above calculate integrals in the standard range of [−1,1]. But in real applications, actual integral ranges differ. Therefore, to use quadrature formulas, actual integral range is converted into the standard range of [−1,1]. To explain the conversion, we can show our variable in the range of $[a,b]$ with x and the transformed variable in the range of [−1,1] with ξ. Corresponding values between x and ξ can be shown as in Table 9.7.

Using Table 9.7, x can be expressed in terms of ξ through Lagrange interpolation as below:

$$\begin{aligned} x &= \frac{\xi-\xi_2}{\xi_1-\xi_2}x_1+\frac{\xi-\xi_1}{\xi_2-\xi_1}x_2=\frac{\xi-1}{(-1)-1}x_1+\frac{\xi-(-1)}{1-(-1)}x_2=\frac{\xi-1}{-2}a+\frac{\xi+1}{2}b \\ &= \left(\frac{b+a}{2}\right)+\left(\frac{b-a}{2}\right)\xi \end{aligned} \tag{9.31}$$

In this case, dx conversion can be expressed in terms of ξ variable as $dx=\left(\frac{b-a}{2}\right)d\xi$. After variable transformation, $\int_a^b f(x)dx$ integral can be expressed in terms of ξ in the

Table 9.7 Corresponding values between x and ξ variables

x	ξ
a	-1
b	1

Table 9.8 Corresponding values for x and ξ variables

x	ξ
2	-1
6	1

standard range of $[-1,1]$ as $\int_a^b f(x)dx = \int_{-1}^{1} \underbrace{f(\xi)\left(\frac{b-a}{2}\right)}_{F(\xi)} d\xi = \int_{-1}^{1} F(\xi)d\xi$. Quadrature methods can be easily applied to the calculation of $\int_{-1}^{1} F(\xi)d\xi$ integral.

Example 9.4: Calculate $I = \int_2^6 [x^2 + 5x + 3]dx$ integral using classical Gauss and Gauss-Lobatto quadrature methods.

x variable defined in [2, 6] interval can be expressed in terms of a new ξ variable defined in $[-1,1]$ interval. In this case, corresponding values for x and ξ variables can be shown as in Table 9.8.

Using Table 9.8, x and dx can be expressed in terms of ξ through Lagrange interpolation as below:

$$x = \frac{\xi - \xi_2}{\xi_1 - \xi_2}x_1 + \frac{\xi - \xi_1}{\xi_2 - \xi_1}x_2 = \frac{\xi - 1}{(-1) - 1}x_1 + \frac{\xi - (-1)}{1 - (-1)}x_2 = \frac{\xi - 1}{-2}2 + \frac{\xi + 1}{2}6$$
$$= 4 + 2\xi \rightarrow dx = 2d\xi$$

Thus, $I = \int_2^6 [x^2 + 5x + 3]dx$ integral can be expressed in terms of ξ as following:

$$I = \int_2^6 \left[x^2 + 5x + 3\right]dx = \int_{-1}^{1} \left[(4 + 2\xi)^2 + 5(4 + 2\xi) + 3\right]2d\xi$$
$$= \int_{-1}^{1} \left[\underbrace{8\xi^2 + 52\xi + 78}_{F(\xi) = \text{Parabolic}}\right] d\xi$$

Integrant function in new integral expression is $F(\xi) = 8\xi^2 + 52\xi + 78$ which is a parabolic polynomial. Integral of this function using classical Gauss (equivalent of Gauss-Legendre) and Gauss-Lobatto methods can be calculated as below.

Integration with Classical Gauss Quadrature Method:
In classical Gauss quadrature method, using p number of integration points, polynomials functions with order of $2p-1$ or less can be integrated exactly. Since integrant function $F(\xi) = 8\xi^2 + 52\xi + 78$ is a parabolic polynomial (i.e., $n = 2$), the number of integration points can be determined using the relationship of $2p-1 = n$ leading to $p = 2$ after rounding up. Then, using Table 9.5, integration points are determined as $\xi_1 = \frac{-1}{\sqrt{3}}$ and $\xi_2 = \frac{1}{\sqrt{3}}$ for $p = 2$ integration points. In this case, integral value can be calculated as below:

$$I = \int_{-1}^{1} \left[\underbrace{8\xi^2 + 52\xi + 78}_{F(\xi) = \text{Parabolic}} \right] d\xi = F\left(\frac{-1}{\sqrt{3}}\right) + F\left(\frac{1}{\sqrt{3}}\right) = 161.3333$$

Integration with Gauss-Lobatto Quadrature method:
In Gauss-Lobatto method, using p number of integration points, polynomials functions with order of $2p\text{-}3$ or less can be integrated exactly. Since integrant function $F(\xi) = 8\xi^2 + 52\xi + 78$ is a parabolic polynomial (i.e., $n = 2$), the number of integration points can be determined using the relationship of $2p-3 = n$ leading to $p = 3$ after rounding up. Then, using Table 9.6, integration points are determined as $\xi_1 = -1$, $\xi_2 = 0$, and $\xi_3 = 1$ for $p = 3$ integration points. In this case, integral value can be calculated as below:

$$I = \int_{-1}^{1} \left[\underbrace{8\xi^2 + 52\xi + 78}_{F(\xi) = Parabolic} \right] d\xi = \frac{1}{3}F(-1) + \frac{4}{3}F(0) + \frac{1}{3}F(1) = 161.3333$$

Example 9.5: Calculate $I = \int_0^{\pi} [x \sin x] dx$ integral using classical Gauss and Gauss-Lobatto quadrature methods.

x variable defined in $[0, \pi]$ interval can be expressed in terms of a new ξ variable defined in $[-1,1]$ interval. In this case, corresponding values for x and ξ variables can be shown as in Table 9.9.

Using Table 9.9, x and dx can be expressed in terms of ξ through Lagrange interpolation as below:

Table 9.9 Corresponding values for x and ξ variables

x	ξ
0	−1
π	1

$$x = \frac{\xi - \xi_2}{\xi_1 - \xi_2} x_1 + \frac{\xi - \xi_1}{\xi_2 - \xi_1} x_2 = \frac{\xi - 1}{(-1) - 1} x_1 + \frac{\xi - (-1)}{1 - (-1)} x_2 = \frac{\xi - 1}{-2} 0 + \frac{\xi + 1}{2} \pi$$
$$= \frac{\pi}{2}(1 + \xi) \rightarrow dx = \frac{\pi}{2} d\xi$$

Thus, $I = \int_0^{\pi} [x\sin x]dx$ integral can be expressed in terms of ξ as following:

$$I = \int_0^{\pi} [x\sin x]dx = \int_{-1}^{1} \left[\frac{\pi}{2}(1 + \xi) \sin \frac{\pi}{2}(1 + \xi)\right] \frac{\pi}{2} d\xi$$
$$= \int_{-1}^{1} \underbrace{\left[\left(\frac{\pi}{2}\right)^2 (1 + \xi) \sin \frac{\pi}{2}(1 + \xi)\right]}_{F(\xi)} d\xi$$

Integrant function in new integral expression is $F(\xi) = \left(\frac{\pi}{2}\right)^2 (1 + \xi) \sin \frac{\pi}{2}(1 + \xi)$, which is a non-polynomial (trigonometric) function. However, quadrature methods were developed to integrate polynomial functions. In this case, integrant function of $F(\xi) = \left(\frac{\pi}{2}\right)^2 (1 + \xi) \sin \frac{\pi}{2}(1 + \xi)$ needs to be approximated with a polynomial function. Approximation is usually done with trial-and-error approach starting from a low order. In trial-and-error approach, first, integrant function is assumed as a low-order polynomial and its integral is calculated. Then, order of approximate polynomial is increased and its integral is calculated again. Old and new integral values are compared with each other. If they are close to each other, it is considered that correct integral result has been reached with assumed polynomial order. If not, order of the approximate polynomial is increased again followed by the calculation if it is integral. Note that in numerical integration, increasing the number of integration points means increasing the order of approximate polynomial since they are related.

Using classical Gauss and Gauss-Lobatto methods, integral values can be calculated for varying number of integration points (i.e., varying order of polynomials) as in Table 9.10. From Table 9.10, it is seen that integral value converges 3.14.

It can be seen from Table 9.10 that classical Gauss method (equivalent of Gauss-Legendre) leads to analytical integral result with three integration points while

Table 9.10 Integral values using classical Gauss and Gauss-Lobatto methods for varying number of integration points

	Integral		
Number of Points (p)	Classical Gauss (Gauss-Legendre)	Gauss-Lobatto	Analytical
1	4.9348	–	3.1416
2	3.0408	–	3.1416
3	3.1438	3.2899	3.1416
4	3.1416	3.1387	3.1416
5	3.1416	3.1416	3.1416

Gauss-Lobatto method does this with five points. Table 9.10 proves that Gauss method is more efficient than Gauss-Lobatto method.

***Note*:** For the solution of this problem with Gauss Quadrature Method, *MATLAB Program 9.4: Gauss Quadrature Method* program in Appendix can be used

9.5 Integration of Multi-Variable Functions

In this section, numerical integrals of two-variable *f*(*x*, *y*) and three-variable *f*(*x*, *y*, *z*) functions will be explained over two- and three-dimensional domains using multiple integrals. In multiple integrals, integral boundaries can be regular or irregular. Regular integral domains corresponding to rectangular area and prism are shown in Fig. 9.12. Integral over a rectangular area can be calculated using double integrals, while integral over rectangular prism can be calculated using triple integrals. Double integrals of an *f*(*x*,*y*) function can be expressed as $I = \int_{x=a}^{x=b} \int_{y=c}^{y=d} f(x,y)dxdy$ and triple integrals of an *f*(*x*,*y*,*z*) function can be stated as $I = \int_{x=a}^{x=b} \int_{y=c}^{y=d} \int_{z=g}^{z=h} f(x,y,z)dxdydz$

For regular domains, multiple integrals can be calculated by applying single integrals successively as below:

$$I = \int_{x=a}^{x=b} \int_{y=c}^{y=d} f(x,y)dxdy = \int_{x=a}^{x=b} \underbrace{\left[\int_{y=c}^{y=d} f(x,y)dy \right]}_{I_y} dx = \int_{x=a}^{x=b} I_y dx \quad (9.32)$$

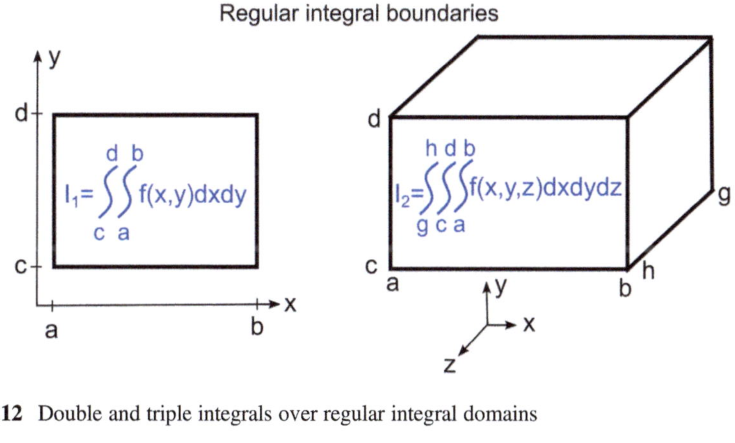

Fig. 9.12 Double and triple integrals over regular integral domains

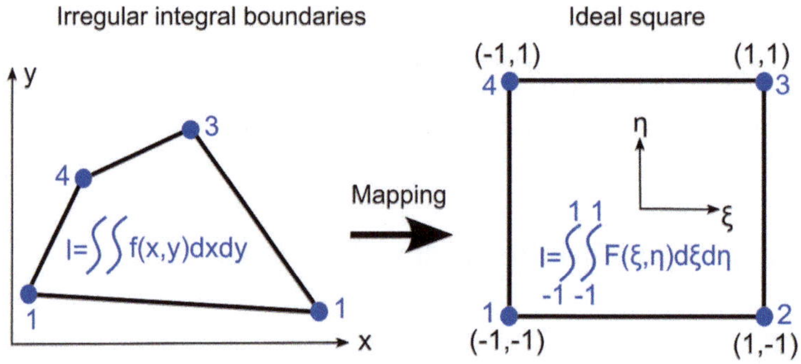

Fig. 9.13 Transformation of an irregular areal domain into a regular square domain for numerical integration

$$I = \int_{x=a}^{x=b}\int_{y=c}^{y=d}\int_{z=g}^{z=h} f(x,y,z)dxdydz = \int_{x=a}^{x=b}\int_{y=c}^{y=d}\underbrace{\left[\int_{z=g}^{z=h} f(x,y,z)dz\right]}_{I_z} dydx$$
$$= \int_{x=a}^{x=b}\underbrace{\left[\int_{y=c}^{y=d} I_z dy\right]}_{I_y} dx = \int_{x=a}^{x=b} I_y dx \tag{9.33}$$

When integral domain is irregular as shown in Fig. 9.13a and Fig. 9.14a, integrals cannot be calculated so easily. In calculation of integrals over irregular domains, irregular domain is converted into a regular domain by using *mapping* concepts. For example, in Fig. 9.13, an irregular areal domain can be converted into a regular square domain. Representing irregular areal domain with x, y variables and regular square domain with ξ, η variables, integral of an $f(x,y)$ function can be expressed in terms of ξ, η variables as below:

$$I = \int\int f(x,y)dxdy = \int_{\xi=-1}^{\xi=1}\int_{\eta=-1}^{\eta=1} F(\xi,\eta)d\xi d\eta \tag{9.34}$$

Similarly, in Fig. 9.14, an irregular volumetric domain can be converted into a regular cubic domain. Representing irregular volumetric domain with x, y, z variables and cubic domain with ξ, η, ζ variables, integral of an $f(x,y,z)$ function can be expressed in terms of ξ, η, ζ variables as below:

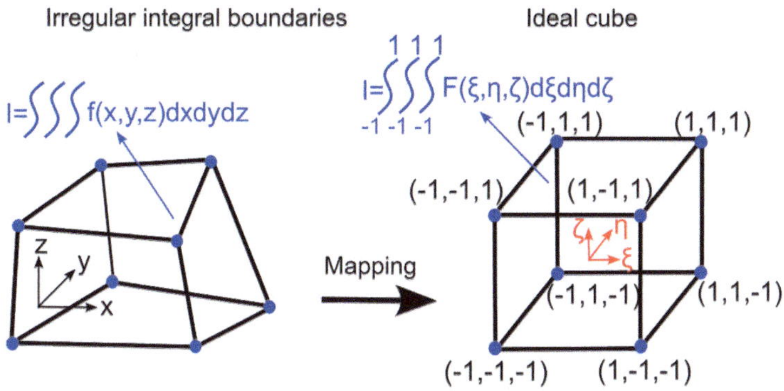

Fig. 9.14 Transformation of an irregular volumetric domain into a regular cubic domain for numerical integration

$$I = \int\int\int f(x, y, z)dxdydz = \int_{\xi=-1}^{\xi=1}\int_{\eta=-1}^{\eta=1}\int_{\zeta=-1}^{\zeta=1} F(\xi, \eta, \zeta)d\xi d\eta d\zeta \quad (9.35)$$

In the following sections, numerical integrals of two- and three-variable functions on different domains using Gauss Quadrature method are explained in detail.

9.6 Integration of Bivariable Functions with Gauss Quadrature Method

Integration of functions with two variables (bivariate functions) over two-dimensional domains such as triangular and quadrilateral areas is frequently encountered in solving engineering problems with popular Finite Element method. For this reason, double integral calculations on triangular and quadrilateral areas will be explained in detail in this section.

9.6.1 Integral Calculation on Quadrilateral Area

In this section, integral calculations with Gauss Quadrature method on square and quadrilateral areas or domains are explained.

Integral Calculation on Square Area

Square area can be considered as a special form of quadrilateral area. For simplicity in the development of numerical integration method, boundaries of square are normalized to be in the range of [−1,1] as in Fig. 9.15. Integral of a polynomial $f(x, y)$ function with Gauss quadrature method in ranges of [−1,1] in x and y directions can be obtained by applying single integrals in x and y directions successively as following:

$$I = \int_{-1}^{1}\int_{-1}^{1} f(x,y)dxdy = \int_{-1}^{1}\underbrace{\left[\sum_{i=1}^{p_x} w_i f(x_i,y)\right]}_{\int_{-1}^{1} f(x)dx} dy$$
$$= \sum_{j=1}^{p_y} w_j \underbrace{\left[\sum_{i=1}^{p_x} w_i f\left(x_i,y_j\right)\right]}_{f(y_j)} = \sum_{i=1}^{p_x}\sum_{j=1}^{p_y} w_i w_j f\left(x_i,y_j\right) \tag{9.36}$$

In Eq. 9.36, i index indicates x direction and j index indicates y direction. p_x and p_y denote the number of integration points in x and y directions.

Integral of a general *linear* bivariate function $f(x, y) = c_1 + c_2x + c_3y$ over square area can be obtained with Gauss Quadrature method using 1x1 integration points (i.e., 1 point in x direction and 1 point in y direction) as below:

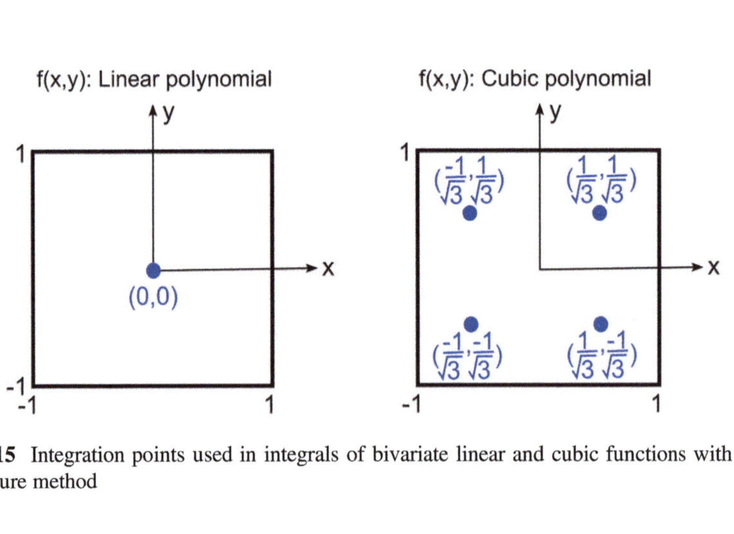

Fig. 9.15 Integration points used in integrals of bivariate linear and cubic functions with Gauss quadrature method

$$\int_{-1}^{1}\int_{-1}^{1} f(x,y)dxdy = \int_{-1}^{1}\left[\int_{-1}^{1} f(x,y)dx\right]dy = \int_{-1}^{1}\left[\underbrace{2f(0,y)}_{F(y)}\right]dy = 2\int_{-1}^{1}\left[\underbrace{f(0,y)}_{F(y)}\right]dy$$
$$= 2x2\underbrace{f(0,0)}_{F(0)} = 4f(0,0) \tag{9.37}$$

Similarly, integral of a general bivariate *cubic* function $f(x, y) = c_1 + c_2x + c_3y + c_4x^2 + c_5y^2 + c_6xy + c_7x^3 + c_8y^3 + c_9x^2y + c_{10}xy^2$ over square area can be obtained with Gauss Quadrature method using 2×2 integration points (i.e., two points in x direction and two points in y direction) as below:

$$\int_{-1}^{1}\int_{-1}^{1} f(x,y)dxdy = \int_{-1}^{1}\left[\int_{-1}^{1} f(x,y)dx\right]dy = \int_{-1}^{1}\left[\underbrace{f\left(\frac{-1}{\sqrt{3}},y\right)+f\left(\frac{1}{\sqrt{3}},y\right)}_{F(y)}\right]dy$$
$$= \underbrace{f\left(\frac{-1}{\sqrt{3}},\frac{-1}{\sqrt{3}}\right)+f\left(\frac{1}{\sqrt{3}},\frac{-1}{\sqrt{3}}\right)}_{F\left(\frac{-1}{\sqrt{3}}\right)} + \underbrace{f\left(\frac{-1}{\sqrt{3}},\frac{1}{\sqrt{3}}\right)+f\left(\frac{1}{\sqrt{3}},\frac{-1}{\sqrt{3}}\right)}_{F\left(\frac{1}{\sqrt{3}}\right)} \tag{9.38}$$

Integration points used in integrals of bivariate linear and cubic functions with Gauss quadrature method are shown in Fig. 9.15.

Integral Calculation on Quadrilateral Area

Quadrilateral area may have an irregular shape as in Fig. 9.16a. In such a case, integral of $I = \int\int f(x, y)dxdy$ cannot be calculated directly. Common approach to simplify the integration process is to convert the quadrilateral domain into ideal square area in the range of [−1,1] as shown in Fig. 9.16. If quadrilateral area is described with x, y variables and ideal square is described with ξ, η variables, relationships between these variables can be defined using Lagrange interpolation method as below:

$$x = \frac{(1-\xi)(1-\eta)}{4}x_1 + \frac{(1+\xi)(1-\eta)}{4}x_2 + \frac{(1+\xi)(1+\eta)}{4}x_3 + \frac{(1-\xi)(1+\eta)}{4}x_4$$
$$y = \frac{(1-\xi)(1-\eta)}{4}y_1 + \frac{(1+\xi)(1-\eta)}{4}y_2 + \frac{(1+\xi)(1+\eta)}{4}y_3 + \frac{(1-\xi)(1+\eta)}{4}y_4 \tag{9.39}$$

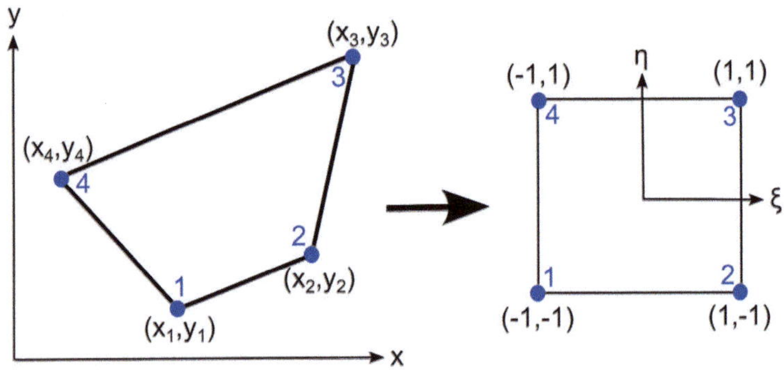

Fig. 9.16 Conversion of quadrilateral domain into ideal square

Using information in Eq. 9.39, $I = \int\int f(x, y)dxdy$ integral can be expressed in terms of ξ and η variables for easy calculation. For this: (1) $f(x,y)$ term in the integral expression I is converted into $f(\xi, \eta)$ term by expressing x and y variables in terms of ξ and η variables using Eq. 9.39. (2) $dxdy$ term in the integral expression I is also expressed in terms of ξ and η variables as $dxdy = |J(\xi, \eta)|d\xi d\eta$ using the determinant of Jacobian matrix J given below:

$$\text{Jacobian matrix}: J(\xi, \eta) = \begin{bmatrix} \frac{\partial x}{\partial \xi} & \frac{\partial y}{\partial \xi} \\ \frac{\partial x}{\partial \eta} & \frac{\partial y}{\partial \eta} \end{bmatrix} \tag{9.40}$$

Using above information, integral $I = \int\int f(x, y)dxdy$ can be expressed in terms of ξ, η variables as following:

$$I = \int\int f(x, y)dxdy = \int_{-1}^{1}\int_{-1}^{1} \underbrace{f(\xi, \eta)|J(\xi, \eta)|}_{F(\xi,\eta)} d\xi d\eta = \int_{-1}^{1}\int_{-1}^{1} F(\xi, \eta)d\xi d\eta \tag{9.41}$$

Integral expression in Eq. 9.41 is now defined in the standard form over the square domain. Therefore, integral value can be easily calculated using values of integrant function $F(\xi, \eta)$ at integration points as below:

$$I = \int_{-1}^{1}\int_{-1}^{1} F(\xi, \eta)d\xi d\eta = \sum_{i=1}^{p_\xi}\sum_{j=1}^{p_\eta} w_i w_j F(\xi_i, \eta_j) \tag{9.42}$$

Integration points can be selected from Table 9.5 according to the order of integrant function. Equation 9.42 can be implemented in a computer program using nested loops in ξ and η directions.

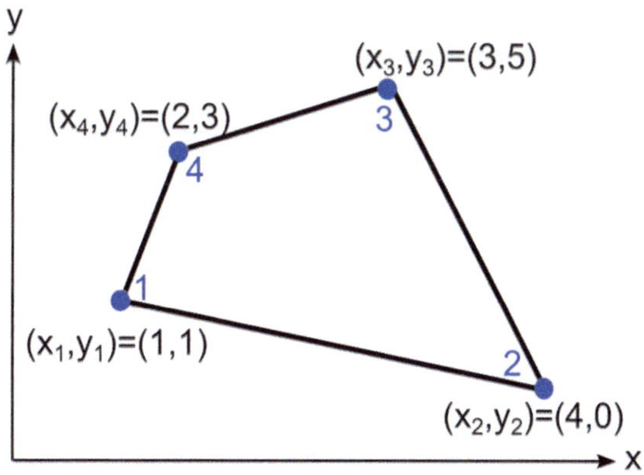

Fig. 9.17 Quadrilateral domain

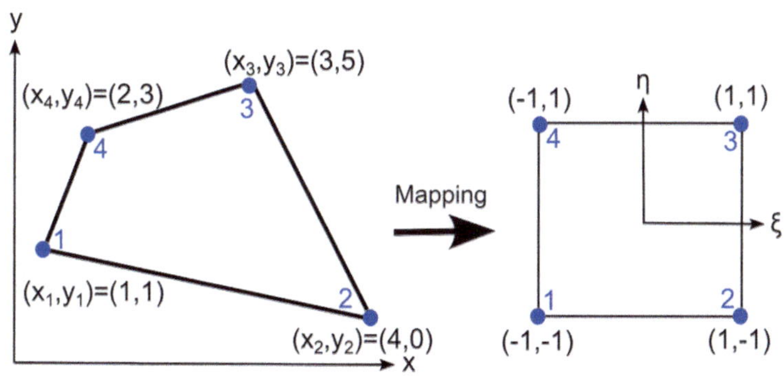

Fig. 9.18 Conversion of a quadrilateral domain into an ideal square area for integral calculation

Example 9.6: Calculate $I = \int\int [x^2 + y^2 + 2xy]dxdy$ integral over the quadrilateral area/domain shown in Fig. 9.17 using (a) classical Gauss or Gauss-Legendre quadrature method (b) Gauss-Lobatto quadrature method.

Integral of $I = \int\int [x^2 + y^2 + 2xy]dxdy$ over the quadrilateral area in Fig. 9.17 can be calculated by converting it into an ideal square area as in Fig. 9.18. Defining quadrilateral area with x, y variables and ideal square with ξ, η variables, relationships between these variables can be defined using Lagrange interpolation method as below:

$$x = \frac{(1-\xi)(1-\eta)}{4}x_1 + \frac{(1+\xi)(1-\eta)}{4}x_2 + \frac{(1+\xi)(1+\eta)}{4}x_3 + \frac{(1-\xi)(1+\eta)}{4}x_4 = \frac{10 + 4\xi - 2\xi\eta}{4}$$

$$y = \frac{(1-\xi)(1-\eta)}{4}y_1 + \frac{(1+\xi)(1-\eta)}{4}y_2 + \frac{(1+\xi)(1+\eta)}{4}y_3 + \frac{(1-\xi)(1+\eta)}{4}y_4 = \frac{9+\xi+7\eta+3\xi\eta}{4}$$

Transformation of $dxdy$ term can be expressed as $dxdy = |J(\xi,\eta)|d\xi d\eta$ using the determinant of Jacobian matrix J as given below:

$$\text{Jacobian matrix}: J = \begin{bmatrix} \frac{\partial x}{\partial \xi} & \frac{\partial y}{\partial \xi} \\ \frac{\partial x}{\partial \eta} & \frac{\partial y}{\partial \eta} \end{bmatrix} = \begin{bmatrix} \frac{4-2\eta}{4} & \frac{1+3\eta}{4} \\ -\frac{2\xi}{4} & \frac{7+3\xi}{4} \end{bmatrix}$$

$$\text{Determinant}: |J(\xi,\eta)| = \frac{28+14\xi-20\eta+6\xi\eta}{16} \rightarrow dxdy = \frac{28+14\xi-20\eta+6\xi\eta}{16}d\xi d\eta$$

In this case, using above information, integral $I = \int\int [x^2 + y^2 + 2xy]dxdy$ can be expressed in terms of variables ξ,η variables as following:

$$I = \int\int \left[x^2+y^2\right]dxdy$$

$$= \int_{-1}^{1}\int_{-1}^{1}\underbrace{\begin{bmatrix} \left(\frac{10+4\xi-2\xi\eta}{4}\right)^2 + \left(\frac{9+\xi+7\eta+3\xi\eta}{4}\right)^2 + \\ 2\left(\frac{10+4\xi-2\xi\eta}{4}\right)\left(\frac{9+\xi+7\eta+3\xi\eta}{4}\right) \end{bmatrix}\left(\frac{28+14\xi-20\eta+6\xi\eta}{16}\right)}_{F(\xi,\eta)}d\xi d\eta$$

$$= \int_{-1}^{1}\int_{-1}^{1} F(\xi,\eta)d\xi d\eta$$

From the integral expression above, integrant function $F(\xi,\eta)$ can be expressed as below:

$$F(\xi,\eta) = \left[\left(\frac{10+4\xi-2\xi\eta}{4}\right)^2 + \left(\frac{9+\xi+7\eta+3\xi\eta}{4}\right)^2 + 2\left(\frac{10+4\xi-2\xi\eta}{4}\right)\left(\frac{9+\xi+7\eta+3\xi\eta}{4}\right)\right] x\left(\frac{28+14\xi-20\eta+6\xi\eta}{16}\right)$$

The highest order of the integrant function $F(\xi,\eta)$ is 3 corresponding to ξ^3, η^3 terms. Integral $I(\xi,\eta)$ can be calculated using Gauss (Gauss-Legendre) and Gauss-Lobatto quadrature methods as following:

(a) *Integration with Gauss or Gauss-Legendre Method*:

Using $2p - 1 = n$ relationship (n is the highest order of integrant function, here $n = 3$), the number of integration points for Gauss Quadrature method is determined as $p_\xi = p_\eta = 2$ in ξ and η directions. Locations and weights of these two integration points can be read from Table 9.5 as below:

In ξ direction (for $p_\xi = 2$): $\xi_1 = -\frac{1}{\sqrt{3}}$, $\xi_2 = \frac{1}{\sqrt{3}}$, $w_1 = 1$, $w_2 = 1$

In η direction (for $p_\eta = 2$): $\eta_1 = -\frac{1}{\sqrt{3}}$, $\eta_2 = \frac{1}{\sqrt{3}}$, $w_1 = 1$, $w_2 = 1$

Using integration points and their weights, integral value can be calculated as below:

$$
\begin{aligned}
I &= \int_{-1}^{1}\int_{-1}^{1} F(\xi,\eta)d\xi d\eta \\
&= \sum_{i=1}^{2}\sum_{j=1}^{2} w_i w_j F(\xi_i,\eta_j) \\
&= w_1w_1F(\xi_1,\eta_1) + w_1w_2F(\xi_1,\eta_2) + w_2w_1F(\xi_2,\eta_1) + w_2w_2F(\xi_2,\eta_2) \\
&= F\left(-\frac{1}{\sqrt{3}}, -\frac{1}{\sqrt{3}}\right) + F\left(-\frac{1}{\sqrt{3}}, \frac{1}{\sqrt{3}}\right) + F\left(\frac{1}{\sqrt{3}}, -\frac{1}{\sqrt{3}}\right) + F\left(\frac{1}{\sqrt{3}}, \frac{1}{\sqrt{3}}\right) = 163.3333
\end{aligned}
$$

(b) *Integration with Gauss-Lobatto Method*:

Using $2p - 3 = n$ relationship (n is the highest order of integrant function, here $n = 3$), the number of integration points for Gauss-Lobatto method is determined as $p_\xi = p_\eta = 3$ in ξ and η directions. Locations and weights of these three integration points can be read from Table 9.5 as below:

In ξ direction (for $p_\xi = 3$): $\xi_1 = -1$, $\xi_2 = 0$, $\xi_3 = 1$ $w_1 = \frac{1}{3}$, $w_2 = \frac{4}{3}$, $w_3 = \frac{1}{3}$

In η direction (for $p_\eta = 3$): $\eta_1 = -1$, $\eta_2 = 0$, $\eta_3 = 1$ $w_1 = \frac{1}{3}$, $w_2 = \frac{4}{3}$, $w_3 = \frac{1}{3}$

Using integration points and their weights, integral value can be calculated as below:

$$
\begin{aligned}
I &= \int_{-1}^{1}\int_{-1}^{1} F(\xi,\eta)d\xi d\eta \\
&= \sum_{i=1}^{3}\sum_{j=1}^{3} w_i w_j F(\xi_i,\eta_j) \\
&= w_1w_1F(\xi_1,\eta_1) + w_1w_2F(\xi_1,\eta_2) + w_1w_3F(\xi_1,\eta_3) \\
&\quad + w_2w_1F(\xi_2,\eta_1) + w_2w_2F(\xi_2,\eta_2) + w_2w_3F(\xi_2,\eta_3) \\
&\quad + w_3w_1F(\xi_3,\eta_1) + w_3w_2F(\xi_3,\eta_2) + w_3w_3F(\xi_3,\eta_3) \\
&= \frac{11}{33}F(-1,-1) + \frac{14}{33}F(-1,0) + \frac{11}{33}F(-1,1) + \frac{41}{33}F(0,-1) + \frac{44}{33}F(0,0) + \frac{41}{33}F(0,1) \\
&\quad + \frac{11}{33}F(1,-1) + \frac{14}{33}F(1,0) + \frac{11}{33}F(1,1) \\
&= 163.3333
\end{aligned}
$$

***Note*:** In this example, to calculate integral value using Gauss Quadrature Method, *MATLAB Program 9.5: Integration Over Quadrilateral Area Using Gauss Quadrature Method* in Appendix can be used.

9.6.2 Integral Calculation on Triangular Area

In this section, integral calculations on ideal right triangular and general triangular areas using Gauss Quadrature method are explained.

Integral Calculation on an Ideal Right Triangle Area

Ideal right triangle can be described as the right triangle with side lengths of 1 as shown in Fig. 9.19. Since any general triangular can be converted into this ideal right (unit) triangle for integral calculations, integration on this element is important.

In ideal right triangle as shown in Fig. 9.19, hypotenuse can be expressed with equation of $x + y = 1$. From this equation, relationship of $x = 1 - y$ can be obtained. Then integral of an $f(x,y)$ function over the ideal right triangle can be expressed as $\int_{y=0}^{1} \int_{x=0}^{1-y} f(x, y)dxdy$. This integral can be calculated numerically using Gauss Quadrature method as below:

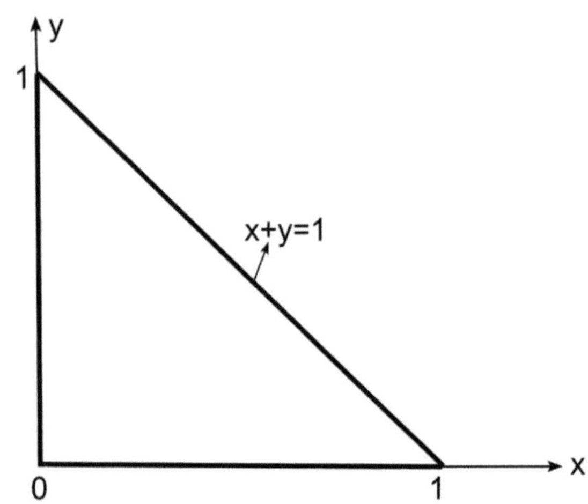

Fig. 9.19 Ideal right (unit) triangular area

Integral of Linear *f*(*x*,*y*) Function:
If *f*(*x*,*y*) function is linear, it can be written in general form as below:

$$f(x,y)=c_1\underbrace{1}_{f_1(x,y)}+c_2\underbrace{x}_{f_2(x,y)}+c_3\underbrace{y}_{f_3(x,y)}=c_1f_1(x,y)+c_2f_2(x,y)+c_3f_3(x,y) \quad (9.43)$$

From the above expression, it is seen that a general two-variable linear $f(x, y)$ function consists of $f_1(x, y) = 1, f_2(x, y) = x$, and $f_3(x, y) = y$ basic functions. Integral of linear $f(x, y)$ function can be calculated using Gauss Quadrature method with a single integration point as following:

$$\int_{y=0}^{1}\int_{x=0}^{1-y} f(x,y)dxdy = w_1 f(x_1,y_1) \quad (9.44)$$

where x_1, y_1 describe coordinates of the integration point and w_1 denotes its weight. Necessary equations to determine x_1, y_1, and w_1 can be obtained by applying Eq. 9.44 in integrals of basic functions as below:

$$f(x,y)=f_1(x,y)=1:\ \int_{y=0}^{1}\int_{x=0}^{1-y} f_1(x,y)dxdy = w_1 f_1(x_1,y_1)\rightarrow$$

$$\int_{y=0}^{1}\int_{x=0}^{1-y} 1dxdy = w_1 \rightarrow \frac{1}{2}=w_1$$

$$f(x,y)=f_2(x,y)=x:\ \int_{y=0}^{1}\int_{x=0}^{1-y} f_2(x,y)dxdy = w_1 f_2(x_1,y_1)\rightarrow$$

$$\int_{y=0}^{1}\int_{x=0}^{1-y} xdxdy = w_1.x_1 \rightarrow \frac{1}{6}=\underbrace{w_1}_{\frac{1}{2}}.x_1\rightarrow x_1=\frac{1}{3}$$

$$f(x,y)=f_3(x,y)=y:\ \int_{y=0}^{1}\int_{x=0}^{1-y} f_3(x,y)dxdy = w_1 f_3(x_1,y_1)\rightarrow$$

$$\int_{y=0}^{1}\int_{x=0}^{1-y} ydxdy = w_1.y_1 \rightarrow \frac{1}{6}=\underbrace{w_1}_{\frac{1}{2}}.y_1\rightarrow y_1=\frac{1}{3}$$

Thus, from the solution of the above equations, following Gauss Quadrature expression is obtained for the integration of the general linear $f(x, y) = c_1 + c_2x + c_3y$ function.

$$I = \int_{y=0}^{1} \int_{x=0}^{1-y} f(x, y)dxdy = \int_{y=0}^{1} \int_{x=0}^{1-y} (c_1 + c_2x + c_3y)dxdy = \frac{1}{2} f\left(\frac{1}{3}, \frac{1}{3}\right) \tag{9.45}$$

Integral of Parabolic *f*(*x*,*y*) Function:
If $f(x,y)$ function is parabolic, it can be written in general form as below:

$$\begin{aligned} f(x, y) &= c_1 \underbrace{1}_{f_1(x,y)} + c_2 \underbrace{x}_{f_2(x,y)} + c_3 \underbrace{y}_{f_3(x,y)} + c_4 \underbrace{x^2}_{f_4(x,y)} + c_5 \underbrace{y^2}_{f_5(x,y)} + c_6 \underbrace{xy}_{f_6(x,y)} \\ &= c_1 f_1(x, y) + c_2 f_2(x, y) + c_3 f_3(x, y) + c_4 f_4(x, y) + c_5 f_5(x, y) + c_6 f_6(x, y) \end{aligned} \tag{9.46}$$

From the above expression, it is seen that a general two-variable parabolic $f(x, y)$ function consists of $f_1(x, y) = 1$, $f_2(x, y) = x$, $f_3(x, y) = y$, $f_4(x, y) = x^2$, $f_5(x, y) = y^2$, and $f_6(x, y) = xy$ basic functions. Integral of a parabolic $f(x, y)$ function with Gauss Quadrature method can be calculated using three integration points as following:

$$\int_{y=0}^{1} \int_{x=0}^{1-y} f(x, y)dxdy = w_1 f(x_1, y_1) + w_2 f(x_2, y_2) + w_3 f(x_3, y_3) \tag{9.47}$$

From Eq. 9.47, it is seen that there are nine unknown constants which are x_1, y_1, w_1, x_2, y_2, w_2, x_3, y_3, w_3. But we can only generate six equations from integrals of basic functions. In this case, to reduce the number of unknowns to that of basic functions, we can impose some constraints such as the symmetry of integration points with respect to $y = x$ line and equality of weight coefficients for symmetric points. If points 1 and 3 are selected to be symmetrical about $y = x$ line, then Eq. 9.47 can be written as following:

$$\int_{y=0}^{1} \int_{x=0}^{1-y} f(x, y)dxdy = w_1 f(x_1, y_1) + w_2 f(x_2, y_2) + w_1 f(y_1, x_1) \tag{9.48}$$

Now in the new expression in Eq. 9.48, unknowns are x_1, y_1, w_1, x_2, y_2, w_2 and they can be can be obtained by applying Eq. 9.48 in integrals of basic functions as below:

$$f(x,y) = f_1(x,y) = 1 : \int\limits_{y=0}^{1} \int\limits_{x=0}^{1-y} f_1(x,y)dxdy$$
$$= w_1 f_1(x_1, y_1) + w_2 f_1(x_2, y_2) + w_1 f_1(y_1, x_1) \rightarrow$$
$$\int\limits_{y=0}^{1} \int\limits_{x=0}^{1-y} 1dxdy = 2w_1 + w_2 \rightarrow \frac{1}{2} = 2w_1 + w_2 \tag{9.49a}$$

$$f(x,y) = f_2(x,y) = x : \int\limits_{y=0}^{1} \int\limits_{x=0}^{1-y} f_2(x,y)dxdy$$
$$= w_1 f_2(x_1, y_1) + w_2 f_2(x_2, y_2) + w_1 f_2(y_1, x_1) \rightarrow$$
$$\int\limits_{y=0}^{1} \int\limits_{x=0}^{1-y} xdxdy = w_1.x_1 + w_2.x_2 + w_1.y_1 \rightarrow \frac{1}{6} = w_1(x_1 + y_1) + w_2.x_2 \tag{9.49b}$$

$$f(x,y) = f_3(x,y) = y : \int\limits_{y=0}^{1} \int\limits_{x=0}^{1-y} f_3(x,y)dxdy$$
$$= w_1 f_3(x_1, y_1) + w_2 f_3(x_2, y_2) + w_1 f_3(y_1, x_1) \rightarrow$$
$$\int\limits_{y=0}^{1} \int\limits_{x=0}^{1-y} ydxdy = w_1.y_1 + w_2.y_2 + w_1.x_1 \rightarrow \frac{1}{6} = w_1(x_1 + y_1) + w_2.y_2 \tag{9.49c}$$

$$f(x,y) = f_4(x,y) = x^2 : \int\limits_{y=0}^{1} \int\limits_{x=0}^{1-y} f_4(x,y)dxdy$$
$$= w_1 f_4(x_1, y_1) + w_2 f_4(x_2, y_2) + w_1 f_4(y_1, x_1) \rightarrow$$
$$\int\limits_{y=0}^{1} \int\limits_{x=0}^{1-y} x^2dxdy = w_1.x_1^2 + w_2.x_2^2 + w_1.y_1^2 \rightarrow \frac{1}{12} = w_1\left(x_1^2 + y_1^2\right) + w_2.x_2^2 \tag{9.49d}$$

$$f(x,y) = f_5(x,y) = y^2 : \int\limits_{y=0}^{1} \int\limits_{x=0}^{1-y} f_5(x,y)dxdy = w_1 f_5(x_1, y_1) + w_2 f_5(x_2, y_2)$$
$$+ w_1 f_5(y_1, x_1) \rightarrow$$

$$\int_{y=0}^{1}\int_{x=0}^{1-y} y^2 dxdy = w_1.y_1^2 + w_2.y_2^2 + w_1.x_1^2 \rightarrow \frac{1}{12} = w_1\left(x_1^2 + y_1^2\right) + w_2.y_2^2 \quad (9.49e)$$

$$f(x,y) = f_6(x,y) = xy : \int_{y=0}^{1}\int_{x=0}^{1-y} f_6(x,y)dxdy$$

$$= w_1 f_6(x_1, y_1) + w_2 f_6(x_2, y_2) + w_1 f_6(y_1, x_1) \rightarrow$$

$$\int_{y=0}^{1}\int_{x=0}^{1-y} xydxdy = 2w_1.x_1y_1 + w_2.x_2y_2 \rightarrow \frac{1}{24} = 2w_1.x_1y_1 + w_2.x_2y_2 \quad (9.49f)$$

All six equations obtained in Eqs. 9.49a–9.49f are nonlinear and have a degree of 2. Therefore, two solution sets are obtained from the solution of these equations.

First solution set: $w_1 = \frac{1}{6}, x_1 = \frac{1}{2}, y_1 = 0, w_2 = \frac{1}{6}, x_2 = \frac{1}{2}, y_2 = \frac{1}{2}$
Second solution set: $w_1 = \frac{1}{6}, x_1 = \frac{2}{3}, y_1 = \frac{1}{6}, w_2 = \frac{1}{6}, x_2 = \frac{1}{6}, y_2 = \frac{1}{6}$

Thus, two different expressions are obtained for the integral of parabolic functions using Gauss Quadrature method on the ideal right triangle as following:

For solution set 1: $I = \int_{y=0}^{1}\int_{x=0}^{1-y} f(x,y)dxdy = \frac{1}{6}f\left(\frac{1}{2}, 0\right) + \frac{1}{6}f\left(\frac{1}{2}, \frac{1}{2}\right) + \frac{1}{6}f\left(0, \frac{1}{2}\right)$

For solution set 2: $I = \int_{y=0}^{1}\int_{x=0}^{1-y} f(x,y)dxdy = \frac{1}{6}f\left(\frac{2}{3}, \frac{1}{6}\right) + \frac{1}{6}f\left(\frac{1}{6}, \frac{1}{6}\right) + \frac{1}{6}f\left(\frac{1}{6}, \frac{2}{3}\right)$

Aforementioned methodology along with the symmetry of integration points about $y = x$ line can be used in the development of integral expression for higher order polynomials. Table 9.11 and Fig. 9.20 show integration points and their weights in the integration of various two-variable polynomial functions over the ideal right triangular area.

***Note*:** Since there are more than one solution set for the same number of points, integration points, and weight coefficients are given only for one solution set in Table 9.11.

***Note*:** Integration points and weight coefficients cannot be solved for two integration points. Therefore, information for two integration point is not included in Table 9.11.

Integral Calculation on General Triangular Area

Integration of an $f(x,y)$ function over a general (arbitrary) triangular domain ($I = \int\int f(x, y)dxdy$) is difficult due to irregular integral limits. However, integral over a general triangle can be calculated easily by transforming it into ideal right triangle using

Table 9.11 Integration points and their weights in the integration of various two-variable polynomial functions over ideal right triangular area using Gauss Quadrature method

Gauss integration points for a right unit triangle

$$I = \int_{x=0}^{x=1} \int_{y=0}^{y=1-x} f(x,y)dxdy = \sum_{i=1}^{p} w_i f(x_i, y_i) \quad (0 \leq x \leq 1, \quad 0 \leq y \leq 1)$$

Polynomial order (n)	Number of points (p)	Integration point (x_i)	Integration point (y_i)	Weight (w_i)
1	1	0.3333	0.3333	0.5000
2	3	0.6667	0.1667	0.1667
		0.1667	0.1667	0.1667
		0.1667	0.6667	0.1667
3	4	0.3333	0.3333	−0.2812
		0.6000	0.2000	0.2604
		0.2000	0.2000	0.2604
		0.2000	0.6000	0.2604
4	7	0	0	0.0250
		0.5000	0	0.0667
		1	0	0.0250
		0.5000	0.5000	0.0667
		0	1	0.0250
		0	0.5000	0.0667
		0.3333	0.3333	0.2250

variable transformation as shown in Fig. 9.21. If general triangle is described with x, y variables and ideal right triangle is described with ξ, η variables, then, relationships between these variables can be defined as below:

$$\begin{aligned} x &= (1-\xi-\eta)x_1 + \xi x_2 + \eta x_3 \\ y &= (1-\xi-\eta)y_1 + \xi y_2 + \eta y_3 \end{aligned} \tag{9.50}$$

Using Eq. 9.50, $I = \int \int f(x,y)dxdy$ integral can be expressed in terms of ξ and η variables for easy calculation. For this: (1) $f(x,y)$ term in the integral expression I is converted into $f(\xi,\eta)$ term by expressing x and y variables in terms of ξ and η variables using Eq. 9.50, (2) $dxdy$ term in the integral expression I is also expressed in terms of ξ and η variables as $dxdy = |J(\xi,\eta)|d\xi d\eta$ using the determinant of Jacobian matrix J given below:

$$\text{Jacobian matrix}: \quad J(\xi,\eta) = \begin{bmatrix} \frac{\partial x}{\partial \xi} & \frac{\partial y}{\partial \xi} \\ \frac{\partial x}{\partial \eta} & \frac{\partial y}{\partial \eta} \end{bmatrix} = \begin{bmatrix} x_2 - x_1 & y_2 - y_1 \\ x_3 - x_1 & y_3 - y_1 \end{bmatrix} \tag{9.51}$$

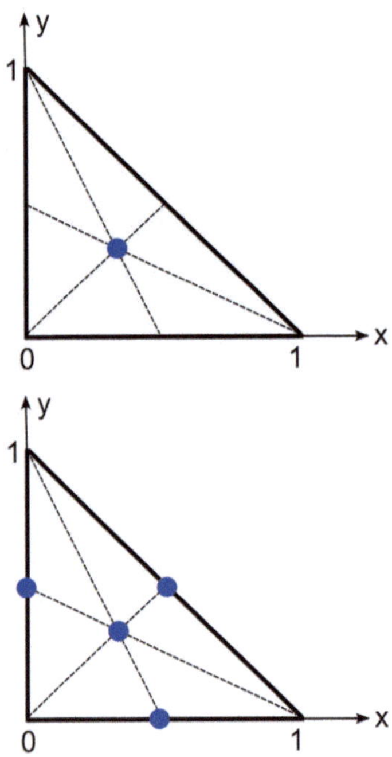

Fig. 9.20 Symmetric integration points in the integration of various two-variable polynomial functions over ideal right triangular area using Gauss Quadrature method

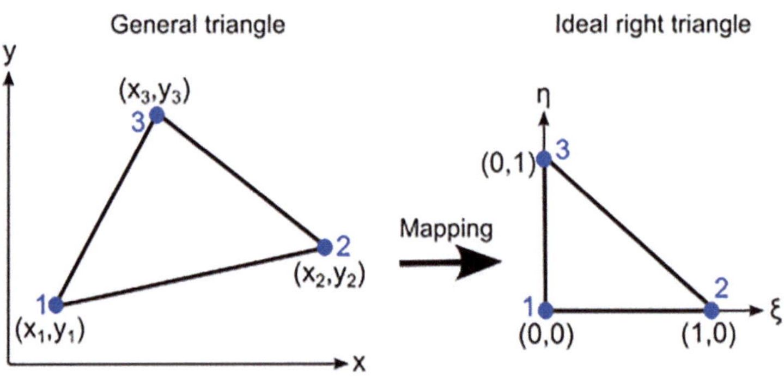

Fig. 9.21 Transformation of a general triangle into ideal right triangle

$$\text{Determinant of Jacobian}: |J(\xi,\eta)| = [(x_2 - x_1)(y_3 - y_1) - (x_3 - x_1)(y_2 - y_1)]$$

In this case, using above information, integral $I = \int\int f(x,y)dxdy$ can be expressed in terms ξ and η variables as following:

$$\begin{aligned} I &= \int_{\eta=0}^{\eta=1} \int_{\xi=0}^{\xi=1-\eta} f(\xi,\eta)|J(\xi,\eta)|d\xi d\eta \\ &= \int_{\eta=0}^{\eta=1} \int_{\xi=0}^{\xi=1-\eta} \underbrace{f(\xi,\eta)|J(\xi,\eta)|}_{F(\xi,\eta)} d\xi d\eta \\ &= \int_{\eta=0}^{\eta=1} \int_{\xi=0}^{\xi=1-\eta} F(\xi,\eta)d\xi d\eta \end{aligned} \tag{9.52}$$

In the new integral expression in Eq. 9.52, integrant function is $F(\xi,\eta) = f(\xi,\eta)|J(\xi,\eta)|$. Integral of this function can be calculated as below using integrant values at different integration points given in Table 9.11.

$$I = \int_{\eta=0}^{\eta=1} \int_{\xi=0}^{\xi=1-\eta} F(\xi,\eta)d\xi d\eta = \sum_{i=1}^{p} w_i F(\xi_i,\eta_i) \tag{9.53}$$

If integrant function $F(\xi,\eta)$ is a non-polynomial function, trial-and-error approach is used in the selection of the number of integration points (p) in Eq. 9.53. Integration points are increased successively until the last two integral values are close to each other.

Example 9.7: Calculate $I = \int \int (x^3 + y^3)dxdy$ integral over the general triangle with corner coordinates of $(x_1, y_1) = (-1,0)$, $(x_2, y_2) = (1,0)$ and $(x_3, y_3) = (0, \sqrt{3})$ using Gauss Quadrature method.

Since the triangle is not an ideal right triangle, it can be transformed into an ideal right triangle using variable transformation for easy integration. If general triangle is described with x, y variables and ideal right triangle is described with ξ, η variables, then, relationships between these variables can be defined as below:

$$\begin{aligned} x &= (1-\xi-\eta)x_1 + \xi x_2 + \eta x_3 = -1 + 2\xi + \eta \\ y &= (1-\xi-\eta)y_1 + \xi y_2 + \eta y_3 = \sqrt{3}\eta \end{aligned}$$

Transformation of $dxdy$ term can be obtained using the determinant of Jacobian matrix $|J|$ as below:

$$\text{Jacobian matrix}: \ J = \begin{bmatrix} \dfrac{\partial x}{\partial \xi} & \dfrac{\partial y}{\partial \xi} \\ \dfrac{\partial x}{\partial \eta} & \dfrac{\partial y}{\partial \eta} \end{bmatrix} = \begin{bmatrix} 2 & 0 \\ 1 & \sqrt{3} \end{bmatrix} \rightarrow |J| = 2\sqrt{3}, \quad dxdy = |J|d\xi d\eta$$
$$= 2\sqrt{3}d\xi d\eta$$

In this case, $I = \int \int (x^3 + y^3)dxdy$ integral over the general triangle can be expressed in terms of the integral over the ideal right triangle as following:

$$I = \int_{\eta=0}^{\eta=1} \int_{\xi=0}^{\xi=1-\eta} \left[(-1+2\xi+\eta)^3 + \left(\sqrt{3}\eta\right)^3 \right] 2\sqrt{3} d\xi d\eta$$
$$= \int_{\eta=0}^{\eta=1} \int_{\xi=0}^{\xi=1-\eta} \underbrace{\left[(-1+2\xi+\eta)^3 + \left(\sqrt{3}\eta\right)^3 \right] 2\sqrt{3}}_{F(\xi,\eta)} d\xi d\eta$$

Integrant function in terms of new variables is $F(\xi,\eta) = \left[(-1+2\xi+\eta)^3 + (\sqrt{3}\eta)^3\right] 2\sqrt{3}$, which is a cubic polynomial ($n = 3$). Integral of this two-variable cubic function can be calculated using four integration points corresponding to $n = 3$ as below:

$$I = \frac{-27}{96} F\left(\frac{1}{3}, \frac{1}{3}\right) + \frac{25}{96} F\left(\frac{3}{5}, \frac{1}{5}\right) + \frac{25}{96} F\left(\frac{1}{5}, \frac{1}{5}\right) + \frac{25}{96} F\left(\frac{1}{5}, \frac{3}{5}\right) = 0.9$$

Note: In this example, to calculate integral value using Gauss Quadrature Method, *MATLAB Program 9.6: Integration Over General Triangular Area Using Gauss Quadrature Method* in Appendix can be used.

Integral Calculation on General Triangular Area (Alternative Method)

Integration with Gauss quadrature method over square domain is quite mature. Therefore, advantage of such integration can also be used in integration over general triangular domain through an alternative approach. In an alternative approach, integration over a general triangle can be carried out by transforming it first into an ideal right triangle and then into a square following a two-stage process as shown in Fig. 9.22.

This two-stage transformation can be explained in more detail as below:

1. Conversion of general triangle into ideal right triangle
2. Conversion of ideal right triangle into ideal square

In order to explain these transformations, we can represent general triangle with x, y coordinates and ideal right triangle with r, s coordinates and ideal square with ξ, η coordinates.

1. Conversion of General Triangle into Ideal Right Triangle:

Integral of an $f(x,y)$ function over a general triangle ($I = \int \int f(x,y)dxdy$) can be calculated by transforming general triangle into ideal right triangle as shown in Fig. 9.23. Transformation of general triangle (defined by x and y coordinates) into ideal right (unit) triangle (defined by r and s coordinates) can be expressed using corner coordinates (x_1, y_1), (x_2, y_2) and (x_3, y_3) of general triangle as below:

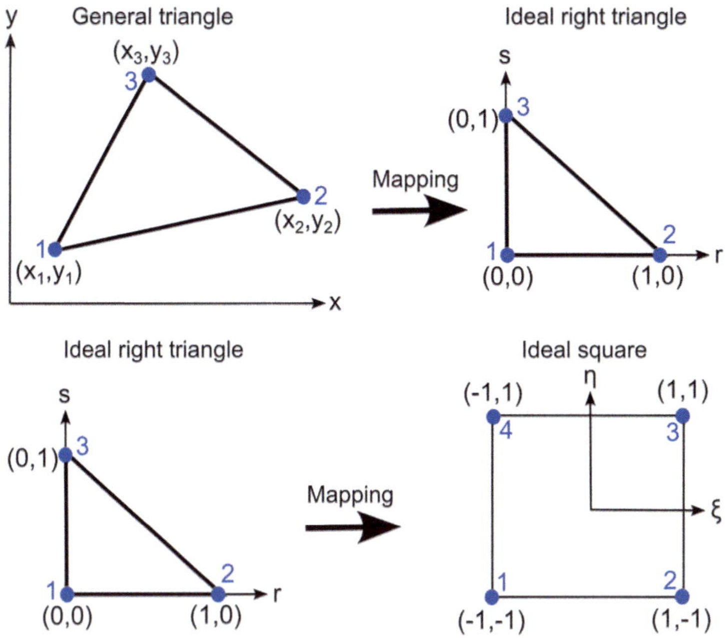

Fig. 9.22 Integration over a general triangle by transforming it first into an ideal right triangle and then into a square

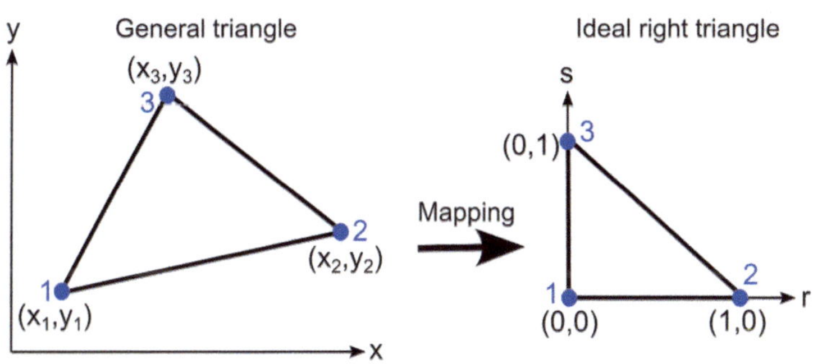

Fig. 9.23 Conversion of general triangle to ideal right triangle

$$
\begin{aligned}
x &= (1 - r - s)x_1 + rx_2 + sx_3 \\
y &= (1 - r - s)y_1 + ry_2 + sy_3
\end{aligned}
\tag{9.54}
$$

Using Eq. 9.54, $I = \int \int f(x, y)dxdy$ integral can be expressed in terms of r and s variables for easy calculation. For this: (1) $f(x,y)$ term in the integral expression I is converted into $f(r,s)$ term by expressing x and y variables in terms of r and s variables using Eq. 9.54, (2) *dxdy* term in the integral expression I is expressed in terms of r and s variables.

dxdy term can be expressed in terms of r and s variables as $dxdy = |J_{rs}(r, s)|drds$ using the determinant of Jacobian matrix J_{rs} given below:

$$\text{Jacobian matrix}: J_{rs}(r,s) = \begin{bmatrix} \dfrac{\partial x}{\partial r} & \dfrac{\partial y}{\partial r} \\ \dfrac{\partial x}{\partial s} & \dfrac{\partial y}{\partial s} \end{bmatrix} = \begin{bmatrix} x_2 - x_1 & y_2 - y_1 \\ x_3 - x_1 & y_3 - y_1 \end{bmatrix} \tag{9.55}$$

$$\text{Determinant of Jacobian}: |J_{rs}(r,s)| = [(x_2 - x_1)(y_3 - y_1) - (x_3 - x_1)(y_2 - y_1)]$$

In this case, using Eqs. 9.54 and 9.55, integral $I = \int\int f(x, y)dxdy$ can be expressed in terms of r and s variables as following:

$$I = \int\int f(x,y)dxdy = \int_{s=0}^{s=1}\int_{r=0}^{r=1-s} \underbrace{f(r,s)|J_{rs}|}_{F(r,s)}drds$$

$$= \int_{s=0}^{s=1}\int_{r=0}^{r=1-s} F(r,s)drds \tag{9.56}$$

2. Conversion of Ideal Right Triangle into Ideal Square:

Transformation of ideal right triangle into square can be illustrated as in Fig. 9.24. From Fig. 9.24, it is seen that to enable such conversion, a new point is created on ideal right triangle in addition to the corner points. This extra point is needed to have an equal number of points on ideal right triangle and ideal square and thereby to provide one-to-one correspondence between points of ideal right triangle and ideal square. Using Fig. 9.24, transformation of ideal right triangle into square can then be expressed with Lagrange interpolation method using $(r_1, s_1) = (0, 0)$, $(r_2, s_2) = (1, 0)$, $(r_3, s_3) = \left(\frac{1}{2}, \frac{1}{2}\right)$, $(r_4, s_4) = (1, 0)$ coordinates of reference points on ideal right triangle as following (See Chap. 6 for two-dimensional Lagrange interpolation).

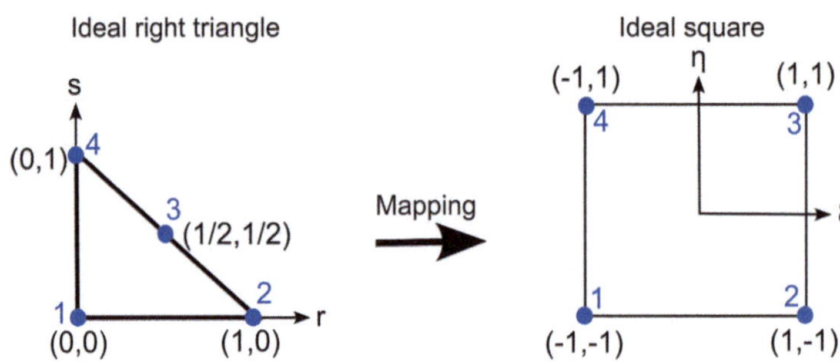

Fig. 9.24 Conversion of ideal right triangle into ideal square

$$r = \frac{(1-\xi)(1-\eta)}{4} r_1 + \frac{(1+\xi)(1-\eta)}{4} r_2 + \frac{(1+\xi)(1+\eta)}{4} r_3 + \frac{(1-\xi)(1+\eta)}{4} r_4$$
$$s = \frac{(1-\xi)(1-\eta)}{4} s_1 + \frac{(1+\xi)(1-\eta)}{4} s_2 + \frac{(1+\xi)(1+\eta)}{4} s_3 + \frac{(1-\xi)(1+\eta)}{4} s_4 \quad (9.57)$$

Using Eq. 9.57, $I = \int_{s=0}^{s=1}\int_{r=0}^{r=1-s} F(r,s)drds$ integral can be expressed in terms of ξ and η variables for easy calculation. For this: (1) $F(r,s)$ term in the integral expression I is converted into $F(\xi, \eta)$ term by expressing r and s variables in terms of ξ and η variables using Eq. 9.57. (2) $drds$ term in the integral expression I is expressed in terms of ξ and η variables.

$drds$ term can be expressed in terms of ξ and η variables as $drds = |J_{\xi\eta}(\xi,\eta)|d\xi d\eta$ using the determinant of Jacobian matrix $J_{\xi\eta}$ given below:

$$\text{Jacobian matrix}: J_{\xi\eta}(\xi,\eta) = \begin{bmatrix} \frac{\partial r}{\partial \xi} & \frac{\partial s}{\partial \xi} \\ \frac{\partial r}{\partial \eta} & \frac{\partial s}{\partial \eta} \end{bmatrix} \rightarrow |J_{\xi\eta}(\xi,\eta)| = \frac{2-\xi-\eta}{16} \quad (9.58)$$

In this case, using above information, integral $I = \int_{s=0}^{s=1}\int_{r=0}^{r=1-s} F(r,s)drds$ can be expressed in terms of ξ, η coordinates as following:

$$I = \int_{s=0}^{s=1}\int_{r=0}^{r=1-s} F(r,s)drds = \int_{-1}^{1}\int_{-1}^{1} \underbrace{F(r(\xi,\eta), s(\xi,\eta))|J_{\xi\eta}|d\xi d\eta}_{G(\xi,\eta)}$$
$$= \int_{-1}^{1}\int_{-1}^{1} G(\xi,\eta)d\xi d\eta \quad (9.59)$$

Integral expression in Eq. 9.59 is now defined in standard form over the ideal square domain. Integral of new integrant function $G(\xi, \eta)$ can be easily calculated using appropriate integration points and their weights from Table 9.5 as below:

$$I = \int_{-1}^{1}\int_{-1}^{1} G(\xi,\eta)d\xi d\eta = \sum_{i=1}^{p_\xi}\sum_{j=1}^{p_\eta} w_i w_j G(\xi_i, \eta_j) \quad (9.60)$$

In Eq. 9.60, p_ξ, p_η denote numbers of integration points and w_i, w_j indicate their weights in ξ and η directions. For non-polynomial integrant functions, p_ξ, p_η in Eq. 9.60 are determined by trial-and-error approach. A predetermined number of integration points are increased successively until corresponding integral results do not differ much.

Example 9.8: Calculate $I = \int\int (x^3 + y^3)dxdy$ integral over the general triangular area in Example 9.7 by transforming it into ideal square using (a) classical Gauss Quadrature (Gauss-Legendre) method, (b) Gauss-Lobatto method.

Integral $I = \int\int (x^3 + y^3)dxdy$ over the general triangular area can be calculated by transforming the general triangle first into an ideal right triangle and then into a square as below.

1. *Conversion of general triangle into ideal right triangle*:

   Transformation of the general triangle into ideal right triangle can be expressed using the corner coordinates (x_1, y_1), (x_2, y_2), and (x_3, y_3) of the general triangle as below:

$$x = (1 - r - s)x_1 + rx_2 + sx_3 = -1 + 2r + s$$
$$y = (1 - r - s)y_1 + ry_2 + sy_3 = \sqrt{3}s$$

   Transformation of $dxdy$ term can be obtained as $dxdy = |J_{rs}(r, s)|drds$ using the determinant of Jacobian matrix given below:

$$\text{Jacobian matrix} : J_{rs} = \begin{bmatrix} \frac{\partial x}{\partial r} & \frac{\partial y}{\partial r} \\ \frac{\partial x}{\partial s} & \frac{\partial y}{\partial s} \end{bmatrix} = \begin{bmatrix} 2 & 0 \\ 1 & \sqrt{3} \end{bmatrix} \rightarrow |J_{rs}(r, s)| = 2\sqrt{3} \rightarrow dxdy$$
$$= 2\sqrt{3}drds$$

   Using above information, integral $I = \int\int (x^3 + y^3)dxdy$ can be expressed in terms of r, s coordinates as following:

$$I = \int\int (x^3 + y^3)dxdy = \int_{s=0}^{s=1}\int_{r=0}^{r=1-s} \underbrace{\left[(-1 + 2r + s)^3 + \left(\sqrt{3}s\right)^3\right]2\sqrt{3}drds}_{F(r,s)}$$

2. *Conversion of ideal right triangle into ideal square*:

   Transformation of ideal right triangle into ideal square can be expressed with Lagrange interpolation method using $(r_1, s_1) = (0, 0)$, $(r_2, s_2) = (1, 0)$, $(r_3, s_3) = \left(\frac{1}{2}, \frac{1}{2}\right)$, $(r_4, s_4) = (1, 0)$ coordinates of reference points on ideal right triangle as following:

$$r = \frac{(1-\xi)(1-\eta)}{4}r_1 + \frac{(1+\xi)(1-\eta)}{4}r_2 + \frac{(1+\xi)(1+\eta)}{4}r_3 + \frac{(1-\xi)(1+\eta)}{4}r_4$$
$$s = \frac{(1-\xi)(1-\eta)}{4}s_1 + \frac{(1+\xi)(1-\eta)}{4}s_2 + \frac{(1+\xi)(1+\eta)}{4}s_3 + \frac{(1-\xi)(1+\eta)}{4}s_4$$

   Transformation of $drds$ term can be obtained as $drds = |J_{\xi\eta}(\xi, \eta)|d\xi d\eta$ using the determinant of Jacobian matrix given below:

$$\text{Jacobian matrix}: J_{\xi\eta} = \begin{bmatrix} \frac{\partial r}{\partial \xi} & \frac{\partial s}{\partial \xi} \\ \frac{\partial r}{\partial \eta} & \frac{\partial s}{\partial \eta} \end{bmatrix} \rightarrow$$

$$\text{Determinant}: \left|J_{\xi\eta}(\xi,\eta)\right| = \frac{2-\xi-\eta}{16} \rightarrow drds = \left|J_{\xi\eta}(\xi,\eta)\right| d\xi d\eta = \left(\frac{2-\xi-\eta}{16}\right) d\xi d\eta$$

Using above information, integral $I = \int_{s=0}^{s=1}\int_{r=0}^{r=1-s}$ $\underbrace{\left[(-1+2r+s)^3 + \left(\sqrt{3}s\right)^3\right]}_{F(r,s)} 2\sqrt{3}drds$ can be expressed in terms of ξ, η coordinates by substituting r, s, $|J_{\xi\eta}|$, and $drds$ expressions as following:

$$I = \int_{s=0}^{s=1}\int_{r=0}^{r=1-s} F(r,s)drds = \int_{-1}^{1}\int_{-1}^{1} \underbrace{F(\xi,\eta)\left|J_{\xi\eta}\right|}_{G(\xi,\eta)} d\xi d\eta = \int_{-1}^{1}\int_{-1}^{1} G(\xi,\eta)d\xi d\eta$$

Final integral expression above is now defined in standard form over the ideal square domain. Therefore, it can be easily calculated using integration points and their weights from Table 9.5 as below:

$$I = \int_{-1}^{1}\int_{-1}^{1} G(\xi,\eta)d\xi d\eta = \sum_{i=1}^{p_\xi}\sum_{j=1}^{p_\eta} w_i w_j G(\xi_i,\eta_j)$$

New integrant function $G(\xi,\eta)$ is polynomial with order of 4 (ξ^4, η^4) in ξ and η directions. Selecting appropriate numbers of integration points, integral I can be calculated using Gauss quadrature and Gauss-Lobatto methods as below:

(a) *Gauss quadrature method*:
Using $2p-1 = n$ ($n = 4$) relationship, numbers of integration points in ξ and η directions are determined as $p_\xi = p_\eta = 3$. Selecting three integration points in ξ and η directions using Table 9.5, integral value is obtained as $I = \int_{-1}^{1}\int_{-1}^{1} G(\xi,\eta)d\xi d\eta = \sum_{i=1}^{3}\sum_{j=1}^{3} w_i w_j G(\xi_i,\eta_j) = 0.9$.

(b) *Gauss-Lobatto method*:
Using $2p-3 = n$ ($n = 4$) relationship, numbers of integration points in ξ and η directions are determined as $p_\xi = p_\eta = 4$. Selecting four integration points in ξ and η directions using Table 9.6, integral value is obtained as $I = \int_{-1}^{1}\int_{-1}^{1} G(\xi,\eta)d\xi d\eta = \sum_{i=1}^{4}\sum_{j=1}^{4} w_i w_j G(\xi_i,\eta_j) = 0.9$.

From above results, it is seen that Gauss quadrature method is more efficient than Gauss-Lobatto method. The same integral value is calculated with fewer integration points using Gauss quadrature method compared to Gauss-Lobatto method.

Note: In this example, to calculate integral value using Gauss Quadrature Method, *MATLAB Program 9.7: Integration Over General Triangular Area Using Gauss Quadrature Method (Converting General Triangular Area First Into Ideal Right Triangle and Then Into Ideal Square)* in Appendix can be used.

Note: After comparison of results in Example 9.7 and Example 9.8, it is seen that integral over a general triangular area is calculated more efficiently with conversion of general triangular area into ideal right triangle than conversion of general triangular area first into right triangular area and then into ideal square. In Example 9.7, integral $I = \int\int (x^3 + y^3)dxdy$ is calculated over general triangular area by converting it into ideal right triangle using four integration points. However, the same integral value is calculated converting general triangular area first into right triangle and then into ideal square in Example 9.8 with $3 \times 3 = 9$ integration points using classical Gauss method and with $4 \times 4 = 16$ integration points using Gauss-Lobatto method.

9.7 Integration of Tri-Variable Functions with Gauss Quadrature Method

In the solution of engineering problems such as in elasticity, heat transfer, fluid mechanics, integrals of three-variable/trivariate $f(x,y,z)$ functions are often encountered. For example, integral-based popular Finite Element method uses such integrals in the solution. In this section, integrals on hexahedral and tetrahedral volumes as examples of trivariate function integrals are covered.

9.7.1 Integral Calculation on Hexahedral Volume

In this section, integral of tri-variable $f(x,y,z)$ function over hexahedral volume is explained. In this context, integrals over cubic volume and general hexahedral volume are considered.

Integral Calculation on Cubic Volume

Cubic volume with normalized boundary limits of $[-1,1]$ are often used in derivation of numerical integral methods. Such cube is sometimes referred to as ideal cubic volume. Integral of a general tri-variable $f(x,y,z)$ function over the ideal cubic volume can be calculated by applying single integrals in x, y, and z directions successively. For example, integral of a *linear* three-variable polynomial $f(x, y, z) = c_1 + c_2x + c_3y + c_4z$ function over the ideal cubic volume can be calculated with Gauss Quadrature method by applying one-dimensional integrals successively as below:

$$
\begin{aligned}
\int_{-1}^{1}\int_{-1}^{1}\int_{-1}^{1} f(x,y,z)dxdydz &= \int_{-1}^{1}\int_{-1}^{1} \underbrace{\left[\int_{-1}^{1} f(x,y,z)dx\right]}_{2f(0,y,z)} dydz \\
&= \int_{-1}^{1}\int_{-1}^{1} [2f(0,y,z)]dydz = 2\int_{-1}^{1} \underbrace{\int_{-1}^{1} [f(0,y,z)]dy}_{2f(0,0,z)} dz \\
&= 2\int_{-1}^{1} 2f(0,0,z)dz = 4\int_{-1}^{1} f(0,0,z)dz = 8f(0,0,0)
\end{aligned}
\tag{9.61}
$$

Similarly, integral of a *cubic* three-variable polynomial *f(x,y,z)* function over cubic volume can be calculated with Gauss Quadrature method by applying one-dimensional integrals successively as following:

$$
\begin{aligned}
&\int_{-1}^{1}\int_{-1}^{1}\int_{-1}^{1} f(x,y,z)dxdydz \\
&= \int_{-1}^{1}\int_{-1}^{1} \left[\int_{-1}^{1} f(x,y,z)dx\right] dydz \\
&= \int_{-1}^{1}\int_{-1}^{1} \left[f\left(\frac{1}{\sqrt{3}},y,z\right) + f\left(\frac{1}{\sqrt{3}},y,z\right)\right] dydz \\
&= \int_{-1}^{1} \left[f\left(\frac{-1}{\sqrt{3}},\frac{-1}{\sqrt{3}},z\right) + f\left(\frac{1}{\sqrt{3}},\frac{-1}{\sqrt{3}},z\right) + f\left(\frac{-1}{\sqrt{3}},\frac{1}{\sqrt{3}},z\right) + f\left(\frac{1}{\sqrt{3}},\frac{1}{\sqrt{3}},z\right)\right] dz \\
&= f\left(\frac{-1}{\sqrt{3}},\frac{-1}{\sqrt{3}},\frac{-1}{\sqrt{3}}\right) + f\left(\frac{1}{\sqrt{3}},\frac{-1}{\sqrt{3}},\frac{-1}{\sqrt{3}}\right) + f\left(\frac{-1}{\sqrt{3}},\frac{1}{\sqrt{3}},\frac{-1}{\sqrt{3}}\right) \\
&\quad + f\left(\frac{1}{\sqrt{3}},\frac{1}{\sqrt{3}},\frac{-1}{\sqrt{3}}\right) + f\left(\frac{-1}{\sqrt{3}},\frac{-1}{\sqrt{3}},\frac{1}{\sqrt{3}}\right) + f\left(\frac{1}{\sqrt{3}},\frac{-1}{\sqrt{3}},\frac{1}{\sqrt{3}}\right) \\
&\quad + f\left(\frac{-1}{\sqrt{3}},\frac{1}{\sqrt{3}},\frac{1}{\sqrt{3}}\right) + f\left(\frac{1}{\sqrt{3}},\frac{1}{\sqrt{3}},\frac{1}{\sqrt{3}}\right)
\end{aligned}
\tag{9.62}
$$

Similar to the integrals of linear and cubic functions, integral of a general non-polynomial three-variable $f(x, y, z)$ function can be calculated with Gauss Quadrature method using single integrals successively as below:

$$
\begin{aligned}
&\int_{-1}^{1}\int_{-1}^{1}\int_{-1}^{1} f(x,y,z)dxdydz \\
&= \int_{-1}^{1}\int_{-1}^{1}\left[\int_{-1}^{1} f(x,y,z)dx\right] dydz \\
&= \int_{-1}^{1}\int_{-1}^{1}\left[\underbrace{\sum_{i=1}^{p_x} w_i f(x_i,y,z)}_{\int_{-1}^{1} f(x)dx}\right] dydz \\
&= \int_{-1}^{1}\underbrace{\sum_{j=1}^{p_y} w_j\left[\left(\sum_{i=1}^{p_x} w_i f\left(x_i,y_j,z\right)\right)\right]}_{\int_{-1}^{1} f(y)dy} dz \\
&= \int_{-1}^{1}\left[\sum_{i=1}^{p_x}\sum_{j=1}^{p_y} w_i w_j f\left(x_i,y_j,z\right)\right] dz \\
&= \underbrace{\sum_{k=1}^{p_z} w_k\left[\sum_{i=1}^{p_x}\sum_{j=1}^{p_y} w_i w_j f\left(x_i,y_j,z_k\right)\right]}_{\int_{-1}^{1} f(z)dz} \\
&= \sum_{i=1}^{p_x}\sum_{j=1}^{p_y}\sum_{k=1}^{p_z} w_i w_j w_k f\left(x_i,y_j,z_k\right)
\end{aligned}
\tag{9.63}
$$

In Eq. 9.63, p_x, p_y, and p_z indicate numbers of Gauss integration points in x, y, and z directions. They are determined considering the equivalent polynomial order of the function in variables. For non-polynomial function, trial-and-error approach is used since prediction of equivalent polynomial function is difficult. For non-polynomial functions, Eq. 9.63 is applied successively by increasing the number of integration points until last two integral values do not differ much.

Integral Calculation on General Hexahedral Volume

General hexahedral volume may have irregular shape as shown in Fig. 9.25a. Because of the irregularity, integral $I = \int\int\int f(x, y, z)dxdydz$ may not be calculated directly. Common approach in integration of functions over a general hexahedral volume is to convert it into ideal cubic volume as shown in Fig. 9.25. If general hexahedral volume is described with x, y, z variables and ideal cubic volume is described with ξ, η, ζ variables, relationships between these variables can be defined using Lagrange interpolation method (see Chap. 6) as below:

$$\begin{aligned} x(\xi,\eta,\zeta) &= N_1(\xi,\eta,\zeta)x_1 + N_2(\xi,\eta,\zeta)x_2 + N_3(\xi,\eta,\zeta)x_3 + N_4(\xi,\eta,\zeta)x_4 \\ &+ N_5(\xi,\eta,\zeta)x_5 + N_6(\xi,\eta,\zeta)x_6 + N_7(\xi,\eta,\zeta)x_7 + N_8(\xi,\eta,\zeta)x_8 \end{aligned} \tag{9.64a}$$

$$\begin{aligned} y(\xi,\eta,\zeta) &= N_1(\xi,\eta,\zeta)y_1 + N_2(\xi,\eta,\zeta)y_2 + N_3(\xi,\eta,\zeta)y_3 + N_4(\xi,\eta,\zeta)y_4 \\ &+ N_5(\xi,\eta,\zeta)y_5 + N_6(\xi,\eta,\zeta)y_6 + N_7(\xi,\eta,\zeta)y_7 + N_8(\xi,\eta,\zeta)y_8 \end{aligned} \tag{9.64b}$$

$$\begin{aligned} z(\xi,\eta,\zeta) &= N_1(\xi,\eta,\zeta)z_1 + N_2(\xi,\eta,\zeta)z_2 + N_3(\xi,\eta,\zeta)z_3 + N_4(\xi,\eta,\zeta)z_4 \\ &+ N_5(\xi,\eta,\zeta)z_5 + N_6(\xi,\eta,\zeta)z_6 + N_7(\xi,\eta,\zeta)z_7 + N_8(\xi,\eta,\zeta)z_8 \end{aligned} \tag{9.64c}$$

where (x_1, y_1, z_1), (x_2, y_2, z_2),…, (x_8, y_8, z_8) denote vertex coordinates of cubic volume and ξ, η, ζ variables change in limits of $[-1,1]$. In Eq. 9.64a, N_1,… N_8 are called as Lagrange or shape functions and are expressed as following:

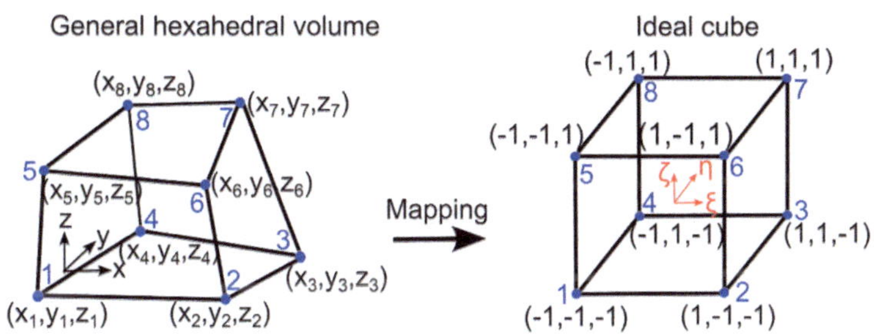

Fig. 9.25 Conversion of a general hexagonal volume into ideal cubic volume

$$
\begin{aligned}
N_1(\xi,\eta,\zeta) &= \frac{(1-\xi)(1-\eta)(1-\zeta)}{8}, & N_2(\xi,\eta,\zeta) &= \frac{(1+\xi)(1-\eta)(1-\zeta)}{8} \\
N_3(\xi,\eta,\zeta) &= \frac{(1+\xi)(1+\eta)(1-\zeta)}{8}, & N_4(\xi,\eta,\zeta) &= \frac{(1-\xi)(1+\eta)(1-\zeta)}{8} \\
N_5(\xi,\eta,\zeta) &= \frac{(1-\xi)(1-\eta)(1+\zeta)}{8}, & N_6(\xi,\eta,\zeta) &= \frac{(1+\xi)(1-\eta)(1+\zeta)}{8} \\
N_7(\xi,\eta,\zeta) &= \frac{(1+\xi)(1+\eta)(1+\zeta)}{8}, & N_8(\xi,\eta,\zeta) &= \frac{(1-\xi)(1+\eta)(1+\zeta)}{8}
\end{aligned}
\tag{9.65}
$$

Using Eq. 9.65, $I = \int \int \int f(x,y,z)dxdydz$ integral can be expressed in terms of ξ,η ζ variables for easy calculation. For this: (1) $f(x,y,z)$ term in the integral expression I is converted into $f(\xi,\eta,\zeta)$ term by expressing x, y, z variables in terms of ξ,η ζ variables using Eq. 9.64a. (2) $dxdydz$ term in the integral expression I is expressed in terms of ξ,η ζ variables.

$dxdydz$ term can be expressed in terms of ξ, η, ζ variables as $dxdydz = |J(\xi,\eta,\zeta)|\, d\xi d\eta d\zeta$ using the determinant of Jacobian matrix J given below:

$$
\text{Jacobian matrix}: J(\xi,\eta,\zeta) = \begin{bmatrix} \frac{\partial x}{\partial \xi} & \frac{\partial y}{\partial \xi} & \frac{\partial z}{\partial \xi} \\ \frac{\partial x}{\partial \eta} & \frac{\partial y}{\partial \eta} & \frac{\partial z}{\partial \eta} \\ \frac{\partial x}{\partial \zeta} & \frac{\partial y}{\partial \zeta} & \frac{\partial z}{\partial \zeta} \end{bmatrix}
\tag{9.66}
$$

In this case, using above information, integral $I = \int \int \int f(x,y,z)dxdydz$ can be expressed in terms of ξ,η ζ variables as below:

$$
I = \int \int \int f(x,y,z)dxdydz = \int_{-1}^{1} \int_{-1}^{1} \int_{-1}^{1} \underbrace{f(\xi,\eta,\zeta)|J|d\xi d\eta d\zeta}_{F(\xi,\eta,\zeta)}
\tag{9.67}
$$

Integral value in Eq. 9.67 can be easily calculated using an appropriate number of integration points and their weights in Table 9.5 for ξ,η, and ζ directions as below:

$$
I = \int_{-1}^{1} \int_{-1}^{1} \int_{-1}^{1} F(\xi,\eta,\zeta)d\xi d\eta d\zeta = \sum_{i=1}^{p_\xi} \sum_{j=1}^{p_\eta} \sum_{k=1}^{p_\zeta} w_i w_j w_k F(\xi_i,\eta_j,\zeta_k)
\tag{9.68}
$$

Equation 9.68 can be easily calculated with the help of a computer program that includes nested loops in ξ,η, and ζ directions. For non-polynomial integrant functions, p_ξ, p_η, p_ζ in Eq. 9.68 are determined by trial-and-error approach. A predetermined number of integration points are increased successively until last two integral results do not differ much.

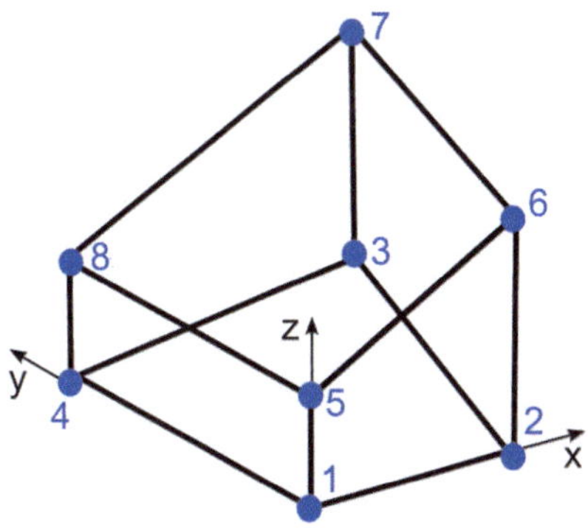

Corner point	x	y	z
1	0	0	0
2	2	0	0
3	3	2	0
4	0	2	0
5	0	0	2
6	2	0	4
7	3	2	4
8	0	2	2

Fig. 9.26 General hexahedral volume

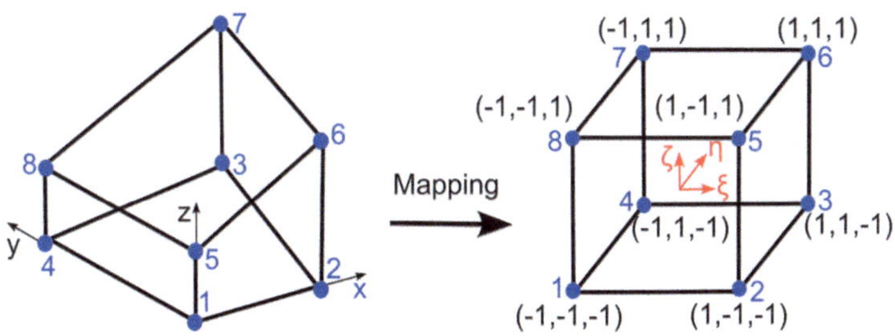

Fig. 9.27 Conversion of general hexahedral volume into ideal cubic volume

Example 9.9: Calculate $I = \int\int\int 10 \sin x \sin y \sin z dxdydz$ integral over the general hexahedral volume shown in Fig. 9.26 using (a) classical Gauss quadrature method, (b) Gauss-Lobatto quadrature method.

Integral $I = \int\int\int 10 \sin x \sin y \sin z dxdydz$ over the general hexahedral volume can be calculated by converting the general hexahedral volume into ideal cubic volume as shown in Fig. 9.27. Defining the general hexahedral volume with x, y, z variables and ideal cubic volume with ξ, η, ζ variables, relationships between these variables can be defined using Lagrange interpolation method as below:

$$
\begin{aligned}
x(\xi,\eta,\zeta) = & \frac{(1-\xi)(1-\eta)(1-\zeta)}{8}x_1 + \frac{(1+\xi)(1-\eta)(1-\zeta)}{8}x_2 + \frac{(1+\xi)(1+\eta)(1-\zeta)}{8}x_3 \\
& + \frac{(1-\xi)(1+\eta)(1-\zeta)}{8}x_4 + \frac{(1-\xi)(1-\eta)(1+\zeta)}{8}x_5 + \frac{(1+\xi)(1-\eta)(1+\zeta)}{8}x_6 \\
& + \frac{(1+\xi)(1+\eta)(1+\zeta)}{8}x_7 + \frac{(1-\xi)(1+\eta)(1+\zeta)}{8}x_8
\end{aligned}
$$

$$y(\xi,\eta,\zeta)=\frac{(1-\xi)(1-\eta)(1-\zeta)}{8}y_1+\frac{(1+\xi)(1-\eta)(1-\zeta)}{8}y_2+\frac{(1+\xi)(1+\eta)(1-\zeta)}{8}y_3$$
$$+\frac{(1-\xi)(1+\eta)(1-\zeta)}{8}y_4+\frac{(1-\xi)(1-\eta)(1+\zeta)}{8}y_5+\frac{(1+\xi)(1-\eta)(1+\zeta)}{8}y_6$$
$$+\frac{(1+\xi)(1+\eta)(1+\zeta)}{8}y_7+\frac{(1-\xi)(1+\eta)(1+\zeta)}{8}y_8$$

$$z(\xi,\eta,\zeta)=\frac{(1-\xi)(1-\eta)(1-\zeta)}{8}z_1+\frac{(1+\xi)(1-\eta)(1-\zeta)}{8}z_2+\frac{(1+\xi)(1+\eta)(1-\zeta)}{8}z_3$$
$$+\frac{(1-\xi)(1+\eta)(1-\zeta)}{8}z_4+\frac{(1-\xi)(1-\eta)(1+\zeta)}{8}z_5+\frac{(1+\xi)(1-\eta)(1+\zeta)}{8}z_6$$
$$+\frac{(1+\xi)(1+\eta)(1+\zeta)}{8}z_7+\frac{(1-\xi)(1+\eta)(1+\zeta)}{8}z_8$$

Transformation of $dxdydz$ term in integral expression I can be obtained as $dxdydz = |J(\xi,\eta,\zeta)|d\xi d\eta d\zeta$ using the determinant of Jacobian matrix $|J(\xi,\eta,\zeta)|$. In this case, using above information, integral $I = \int\int\int 10 \sin x \sin y \sin z dxdydz$ can be expressed in terms of ξ,η, ζ variables as following:

$$I=\int_{-1}^{1}\int_{-1}^{1}\int_{-1}^{1}\underbrace{10[\sin x(\xi,\eta,\zeta)][\sin y(\xi,\eta,\zeta)][\sin z(\xi,\eta,\zeta)]|J(\xi,\eta,\zeta)|}_{F(\xi,\eta,\zeta)}d\xi d\eta d\zeta$$

From the new integral expression, integrant function can be expressed as below:

$$F(\xi,\eta,\zeta)=10[\sin x(\xi,\eta,\zeta)][\sin y(\xi,\eta,\zeta)][\sin z(\xi,\eta,\zeta)]|J(\xi,\eta,\zeta)|$$

Integral of this function can be calculated using quadrature methods as below:

$$I=\int_{-1}^{1}\int_{-1}^{1}\int_{-1}^{1}F(\xi,\eta,\zeta)d\xi d\eta d\zeta=\sum_{i=1}^{p_\xi}\sum_{j=1}^{p_\eta}\sum_{k=1}^{p_\zeta}w_iw_jw_kF(\xi_i,\eta_j,\zeta_k)$$
$$=\sum_{i=1}^{p_\xi}\sum_{j=1}^{p_\eta}\sum_{k=1}^{p_\zeta}w_iw_jw_k10[\sin x(\xi_i,\eta_j,\zeta_k)][\sin y(\xi_i,\eta_j,\zeta_k)][\sin z(\xi_i,\eta_j,\zeta_k)]|J(\xi_i,\eta_j,\zeta_k)|$$

In above expression, the number of integration points p_ξ, p_η, p_z, location of points ξ_i, η_j, ζ_k and weight coefficients w_i, w_j, w_k differ according to the quadrature method. Since integrant function $F(\xi,\eta,\zeta)$ here is non-polynomial, integral of this function is calculated by applying quadrature formula successively for an increasing number of integration points as in Table 9.12.

In integral calculations in Table 9.12, integration points and weight coefficients are selected from Table 9.5 for Gauss or Gauss-Legendre method and from Table 9.6 for Gauss-Lobatto method. From Table 9.12, it is seen that the integral result

Table 9.12 Variation of integral results with a number of integration points

Number of Points ($p = p_\xi \times p_\eta \times p_\zeta$)	Integral: Classical Gauss (Gauss-Legendre)	Integral: Gauss-Lobatto
1×1×1	119.4814	–
2×2×2	45.18	–
3×3×3	49.4464	54.9271
4×4×4	49.3103	49.1371
5×5×5	49.3131	49.3164

converges to the value of 49. *Note*: Integration points and weight coefficients are the same for the classical Gauss and Gauss-Legendre methods.

***Note*:** In this example, to calculate integral value using Gauss Quadrature Method, *MATLAB Program 9.8: Integration Over General Hexahedral Volume Using Gauss Quadrature Method* in Appendix can be used

9.7.2 Integral Calculation on Tetrahedral Volume

Simplest volume among three-dimensional domains is the tetrahedral volume. Tetrahedral volume can be classified as ideal tetrahedral volume and general tetrahedral volume. Integrals over such volumes can be calculated as following:

Calculation of Integral on Ideal Tetrahedral Volume

Ideal tetrahedral volume with unit lengths in *x*, *y*, and *z* directions can be represented as in Fig. 9.28.

Integral of a trivariate *f*(*x,y,z*) function over the ideal tetrahedral volume can be expressed as below:

$$I = \int_{x=0}^{x=1} \int_{y=0}^{y=1-x} \int_{z=0}^{z=1-x-y} f(x, y, z) dxdydz \tag{9.69}$$

Integral in Eq. 9.69 can be calculated with Gauss Quadrature method by selecting a sufficient number of integration points and their weights from Table 9.13 considering the order of *f*(*x*, *y*, *z*) function as below:

$$I = \int_{x=0}^{x=1} \int_{y=0}^{y=1-x} \int_{z=0}^{z=1-x-y} f(x, y, z) dxdydz = \sum_{i=1}^{p} w_i f(x_i, y_i, z_i) \tag{9.70}$$

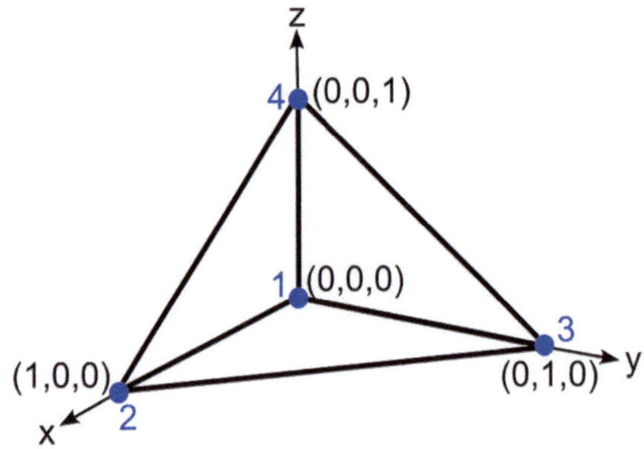

Fig. 9.28 Ideal tetrahedral volume with unit lengths in x, y, and z directions

Table 9.13 Integration points and weight coefficients for numerical integration with Gauss Quadrature method over ideal tetrahedral volume

		Gauss integration points for ideal tetrahedral volume $I = \int_{x=0}^{x=1} \int_{y=0}^{y=1-x} \int_{z=0}^{\zeta=1-x-y} f(x, y, z)dxdydz = \sum_{i=1}^{p} w_i f(x_i, y_i, z_i)$ $(0 \leq x \leq 1, 0 \leq y \leq 1, 0 \leq z \leq 1)$			
Polynomial degree (n)	Number of points (p)	Integration point (x_i)	Integration point (y_i)	Integration point (z_i)	Weight (w_i)
1	1	0.2500	0.2500	0.2500	0.1667
2	4	0.5854	0.1382	0.1382	0.0417
		0.1382	0.5854	0.1382	0.0417
		0.1382	0.1382	0.5854	0.0417
		0.1382	0.1382	0.1382	0.0417
3	5	0.2500	0.2500	0.2500	−0.1333
		0.5000	0.1667	0.1667	0.0750
		0.1667	0.5000	0.1667	0.0750
		0.1667	0.1667	0.5000	0.0750
		0.1667	0.1667	0.1667	0.0750
4	11	0.2500	0.2500	0.2500	−0.0132
		0.7857	0.0714	0.0714	0.0076
		0.0714	0.7857	0.0714	0.0076
		0.0714	0.0714	0.7857	0.0076
		0.0714	0.0714	0.0714	0.0076
		0.3994	0.3994	0.1006	0.0249
		0.3994	0.1006	0.3994	0.0249
		0.3994	0.1006	0.1006	0.0249
		0.1006	0.3994	0.3994	0.0249
		0.1006	0.3994	0.1006	0.0249
		0.1006	0.1006	0.3994	0.0249

where p indicates the number of integration points in ideal tetrahedral volume. For non-polynomial functions, p is often determined with trial-and-error process by increasing it and calculating integrals successively.

Integral Calculation on General Tetrahedral Volume

A general tetrahedral volume is shown in Fig. 9.29. Integral of an $f(x,y,z)$ function over general tetrahedral volume ($I = \int \int \int f(x, y, z)dxdydz$) cannot be calculated directly due to irregular integral limits.

In these cases, $I = \int \int \int f(x, y, z)dxdydz$ integral can be calculated easily by transforming general tetrahedral volume into ideal tetrahedral volume as shown in Fig. 9.30.

Representing general tetrahedral volume with x, y, z variables and ideal tetrahedral volume with ξ, η, ζ variables, conversion of general tetrahedral volume into ideal tetrahedral volume can be expressed as following:

Fig. 9.29 General tetrahedral volume

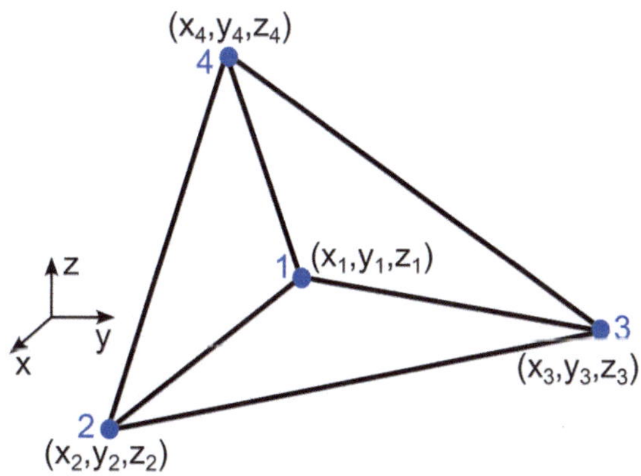

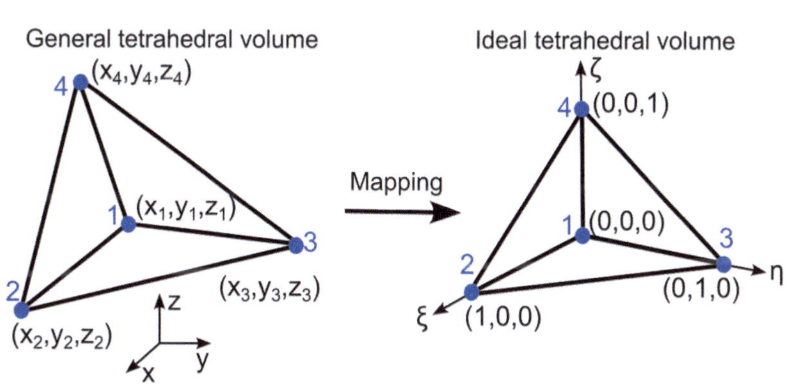

Fig. 9.30 Conversion of a general tetrahedral volume into ideal tetrahedral volume

$$\begin{aligned} x &= (1-\xi-\eta-\zeta)x_1 + \xi x_2 + \eta x_3 + \zeta x_4 \\ y &= (1-\xi-\eta-\zeta)y_1 + \xi y_2 + \eta y_3 + \zeta y_4 \\ z &= (1-\xi-\eta-\zeta)z_1 + \xi z_2 + \eta z_3 + \zeta z_4 \end{aligned} \tag{9.71}$$

$I = \int \int \int f(x,y,z)dxdydz$ integral can be expressed in terms of ξ,η ζ variables for easy calculation. For this: (1) $f(x,y,z)$ term in the integral expression I is converted into $f(\xi,\eta,\zeta)$ term by expressing x, y, z variables in terms of ξ,η, ζ variables using Eq. 9.71, (2) $dxdydz$ term in the integral expression I is expressed in terms of ξ,η, ζ variables.

$dxdydz$ term can be expressed in terms of ξ, η, ζ variables as $dxdydz = |J(\xi,\eta,\zeta)|$ $d\xi d\eta d\zeta$ using the determinant of Jacobian matrix J given below:

$$\text{Jacobian matrix}: J(\xi,\eta,\zeta) = \begin{bmatrix} \frac{\partial x}{\partial \xi} & \frac{\partial y}{\partial \xi} & \frac{\partial z}{\partial \xi} \\ \frac{\partial x}{\partial \eta} & \frac{\partial y}{\partial \eta} & \frac{\partial z}{\partial \eta} \\ \frac{\partial x}{\partial \zeta} & \frac{\partial y}{\partial \zeta} & \frac{\partial z}{\partial \zeta} \end{bmatrix} = \begin{bmatrix} x_2 - x_1 & y_2 - y_1 & z_2 - z_1 \\ x_3 - x_1 & y_3 - y_1 & z_3 - z_1 \\ x_4 - x_1 & y_4 - y_1 & z_4 - z_1 \end{bmatrix} \tag{9.72}$$

In this case, using above information, integral $I = \int \int \int f(x,y,z)dxdydz$ can be expressed in terms of ξ,η, ζ variables as below:

$$\begin{aligned} I &= \int \int \int f(x,y,z)dxdydz = \int_{\xi=0}^{\xi=1} \int_{\eta=0}^{\eta=1-\xi} \int_{\zeta=0}^{\zeta=1-\xi-\eta} \underbrace{f(\xi,\eta,\zeta)|J(\xi,\eta,\zeta)|}_{F(\xi,\eta,\zeta)} d\xi d\eta d\zeta \\ &= \int_{\xi=0}^{\xi=1} \int_{\eta=0}^{\eta=1-\xi} \int_{\zeta=0}^{\zeta=1-\xi-\eta} F(\xi,\eta,\zeta)d\xi d\eta d\zeta \end{aligned} \tag{9.73}$$

After transformation, integral is now defined in standard form as $I = \int_{\xi=0}^{\xi=1} \times \int_{\eta=0}^{\eta=1-\xi} \int_{\zeta=0}^{\zeta=1-\xi-\eta} F(\xi,\eta,\zeta)d\xi d\eta d\zeta$ over the ideal tetrahedral volume. Integral in Eq. 9.73 can be easily calculated with Gauss Quadrature method by selecting appropriate integration points in Table 9.13 according to the order of function as below:

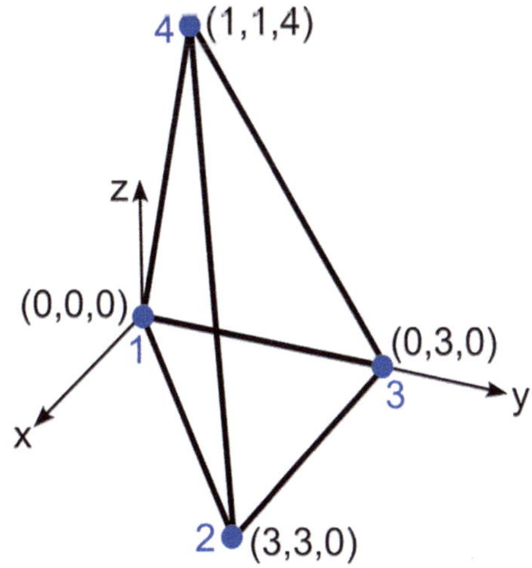

Fig. 9.31 General tetrahedral volume

$$I = \int_{\xi=0}^{\xi=1} \int_{\eta=0}^{\eta=1-\xi} \int_{\zeta=0}^{\zeta=1-\xi-\eta} F(\xi,\eta,\zeta) d\xi d\eta d\zeta = \sum_{i=1}^{p} w_i F(\xi_i,\eta_i,\zeta_i) \tag{9.74}$$

The number of integration points p is determined considering the order of integrant function $F(\xi,\eta,\zeta)$. For non-polynomial integrant functions, p value in Eq. 9.74 is determined by trial-and-error approach. A predetermined number of integration points are increased successively until last two integral results do not differ much.

Example 9.10: Calculate $I = \int \int \int 10 \sin x \sin y \sin z dxdydz$ integral over the general tetrahedral volume shown in Fig. 9.31 using Gauss Quadrature method.

Representing general tetrahedral volume with x, y, z variables and ideal tetrahedral volume with ξ, η, ζvariables, conversion of the general tetrahedral volume into ideal tetrahedral volume can be expressed as following:

$$x = (1-\xi-\eta-\zeta)x_1 + \xi x_2 + \eta x_3 + \zeta x_4 = 3\xi + \zeta$$
$$y = (1-\xi-\eta-\zeta)y_1 + \xi y_2 + \eta y_3 + \zeta y_4 = 3\xi + 3\eta + \zeta$$
$$z = (1-\xi-\eta-\zeta)z_1 + \xi z_2 + \eta z_3 + \zeta z_4 = 4\zeta$$

Jacobian matrix J and its determinant $|J|$ can be calculated as below:

$$\text{Jacobian matrix} : J(\xi,\eta,\zeta) = \begin{bmatrix} \frac{\partial x}{\partial \xi} & \frac{\partial y}{\partial \xi} & \frac{\partial z}{\partial \xi} \\ \frac{\partial x}{\partial \eta} & \frac{\partial y}{\partial \eta} & \frac{\partial z}{\partial \eta} \\ \frac{\partial x}{\partial \zeta} & \frac{\partial y}{\partial \zeta} & \frac{\partial z}{\partial \zeta} \end{bmatrix} = \begin{bmatrix} 3 & 3 & 0 \\ 0 & 3 & 0 \\ 1 & 1 & 4 \end{bmatrix}$$

$$\text{Determinant}: |J(\xi,\eta,\zeta)| = 36$$

Using above information, integral $I = \int\int\int 10 \sin x \sin y \sin z dxdydz$ can be expressed in terms of ξ, η, ζ variables as below:

$$
\begin{aligned}
I &= \int\int\int f(x,y,z)dxdydz \\
&= \int_{\xi=0}^{\xi=1}\int_{\eta=0}^{\eta=1-\xi}\int_{\zeta=0}^{\zeta=1-\xi-\eta} \underbrace{10[\sin(3\xi+\zeta)][\sin(3\xi+3\eta+\zeta)][\sin(4\zeta)]36d\xi d\eta d\zeta}_{F(\xi,\eta,\zeta)} \\
&= \int_{\xi=0}^{\xi=1}\int_{\eta=0}^{\eta=1-\xi}\int_{\zeta=0}^{\zeta=1-\xi-\eta} F(\xi,\eta,\zeta)d\xi d\eta d\zeta
\end{aligned}
$$

After conversion of the integral expression into the standard form over ideal tetrahedral volume, integral value can be easily calculated with Gauss Quadrature method selecting appropriate integration points in Table 9.13 as below:

$$I = \int_{\xi=0}^{\xi=1}\int_{\eta=0}^{\eta=1-\xi}\int_{\zeta=0}^{\zeta=1-\xi-\eta} F(\xi,\eta,\zeta)d\xi d\eta d\zeta = \sum_{i=1}^{p} w_i F(\xi_i,\eta_i,\zeta_i)$$

Since integrant function $F(\xi,\eta,\zeta)$ is non-polynomial (sinusoidal) function, number of integration points p is determined with trial and error. p is increased successively until last two integral values do not differ much. Variation of integral values with number of integration points for $F(\xi,\eta,\zeta)$ function is given in Table 9.14.

***Note*:** In this example, to calculate integral value using Gauss Quadrature Method, *MATLAB Program 9.9: Integration Over General Tetrahedral Volume Using Gauss Quadrature Method* in Appendix can be used

Integral Calculation on General Tetrahedral Volume (Alternative Method)

Integration with Gauss quadrature method over ideal cubic volume is quite mature. Therefore, advantage of such integration can also be used in integration over general tetrahedral volume through an alternative approach. In an alternative approach,

Table 9.14 Variation of integral values with number of integration points for $F(\xi,\eta,\zeta)$ 9.function

Number of Point (p)	Integral
1	41.8041
4	20.9308
5	18.9568
11	22.7461

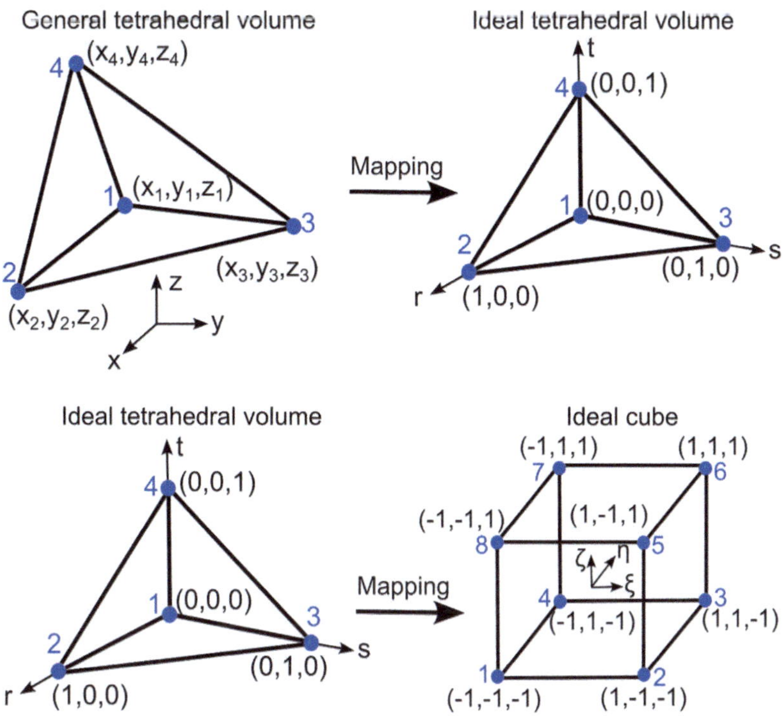

Fig. 9.32 Integration over a general tetrahedral volume by transforming it first into an ideal tetrahedral volume and then into ideal cubic volume

integration over a general tetrahedral volume can be carried out by transforming it first into an ideal tetrahedral volume and then into ideal cubic volume following a two-stage process as shown in Fig. 9.32.

This two-stage transformation can be explained in more detail as below:

1. Conversion of general tetrahedral volume into ideal tetrahedral volume.
2. Conversion of ideal tetrahedral volume into ideal cubic volume.

In order to explain these transformations, we can represent general tetrahedral volume with x, y, z coordinates, ideal tetrahedral volume with r, s, t coordinates and ideal cubic volume with ξ, η, ζ coordinates.

1. *Conversion of general tetrahedral volume into ideal tetrahedral volume*:

   Conversion of the general tetrahedral volume into ideal tetrahedral volume can be illustrated as in Fig. 9.33. Using Fig. 9.33. this conversion can be expressed using corner coordinates of the general tetrahedral volume as below:

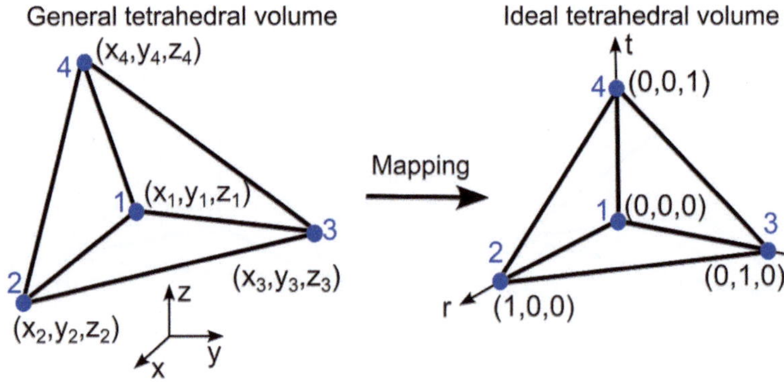

Fig. 9.33 Conversion of general tetrahedral volume into ideal tetrahedral volume

$$
\begin{aligned}
x &= (1-r-s-t)x_1 + rx_2 + sx_3 + tx_4 \\
y &= (1-r-s-t)y_1 + ry_2 + sy_3 + ty_4 \\
z &= (1-r-s-t)z_1 + rz_2 + sz_3 + tz_4
\end{aligned} \tag{9.75}
$$

In this case, $I = \int\int f(x, y, z)dxdydz$ integral can be expressed in terms of r, s, t variables for easy calculation. For this: (1) $f(x,y,z)$ term in the integral expression I is converted into $f(r,s,t)$ term by expressing x, y, z variables in terms of r, s, t variables using Eq. 9.75. (2) $dxdydz$ term in the integral expression I is expressed in terms of r, s, t variables.

$dxdydz$ term can be expressed in terms of r, s, t variables as $dxdydz = |J_{rst}(r, s, t)|drdsdt$ using the determinant of Jacobian matrix J_{rst} given below:

$$
\text{Jacobian matrix} : J_{rst}(r,s,t) = \begin{bmatrix} \frac{\partial x}{\partial r} & \frac{\partial y}{\partial r} & \frac{\partial z}{\partial r} \\ \frac{\partial x}{\partial s} & \frac{\partial y}{\partial s} & \frac{\partial z}{\partial s} \\ \frac{\partial x}{\partial t} & \frac{\partial y}{\partial t} & \frac{\partial z}{\partial t} \end{bmatrix} = \begin{bmatrix} x_2 - x_1 & y_2 - y_1 & z_2 - z_1 \\ x_3 - x_1 & y_3 - y_1 & z_3 - z_1 \\ x_4 - x_1 & y_4 - y_1 & z_4 - z_1 \end{bmatrix} \tag{9.76}
$$

Using Eqs. 9.75 and 9.76, integral $I = \int\int\int f(x, y, z)dxdydz$ can be expressed in terms r, s, t coordinates as following:

$$
\begin{aligned}
I &= \int\int\int f(x,y,z)dxdydz = \int_{r=0}^{r=1}\int_{s=0}^{s=1-r}\int_{t=0}^{t=1-r-s} \underbrace{f(r,s,t)|J_{rst}|}_{F(r,s,t)}drdsdt \\
&= \int_{r=0}^{r=1}\int_{s=0}^{s=1-r}\int_{t=0}^{t=1-r-s} F(r,s,t)drdsdt
\end{aligned} \tag{9.77}
$$

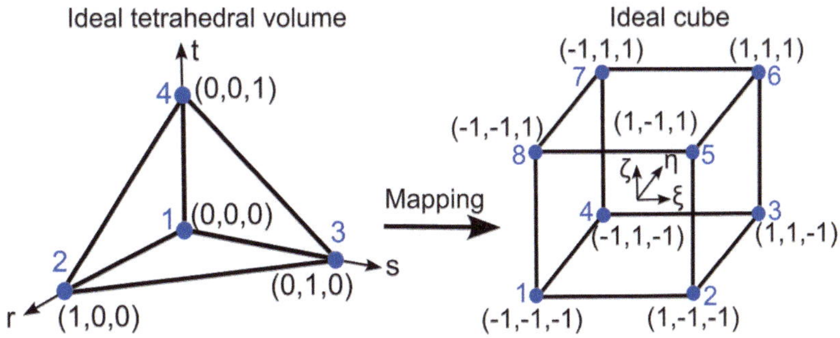

Fig. 9.34 Conversion of ideal tetrahedral volume into ideal cubic volume

2. *Conversion of ideal tetrahedral volume into ideal cubic volume*:

Integral of an *F(r,s,t)* function over ideal tetrahedral volume ($I = \int_{r=0}^{r=1} \int_{s=0}^{s=1-r} \int_{t=0}^{t=1-r-s} F(r,s,t)drdsdt$) can be calculated by transforming ideal tetrahedral volume into ideal cube as shown in Fig. 9.34. This transformation can be expressed using corner coordinates of ideal tetrahedral volume as below:

$$r = \frac{1+\xi}{2}, \quad s = \left(\frac{1-\xi}{2}\right)\left(\frac{1+\eta}{2}\right), \quad t = \left(\frac{1-\xi}{2}\right)\left(\frac{1-\eta}{2}\right)\left(\frac{1+\zeta}{2}\right) \tag{9.78}$$

In this case, $I = \int_{r=0}^{r=1} \int_{s=0}^{s=1-r} \int_{t=0}^{t=1-r-s} F(r,s,t)drdsdt$ integral can be expressed in terms of ξ, η, ζ variables for easy calculation. For this: (1) *F(r,s,t)* term in the integral expression *I* is converted into $F(\xi,\eta,\zeta)$ term by expressing *r*, *s*, *t* variables in terms of ξ, η, ζ variables using Eq. 9.57, (2) *drdsdt* term in the integral expression *I* is expressed in terms of ξ, η, ζ variables.

drdsdt term can be expressed in terms of ξ, η, ζ variables as $drdsdt = |J_{\xi\eta\zeta}(\xi,\eta,\zeta)| d\xi d\eta d\zeta$ using the determinant of Jacobian matrix $J_{\xi\eta\zeta}$ given below:

$$\text{Jacobian matrix}: J_{\xi\eta\zeta} = \begin{bmatrix} \frac{\partial r}{\partial \xi} & \frac{\partial s}{\partial \xi} & \frac{\partial t}{\partial \xi} \\ \frac{\partial r}{\partial \eta} & \frac{\partial s}{\partial \eta} & \frac{\partial t}{\partial \eta} \\ \frac{\partial r}{\partial \zeta} & \frac{\partial s}{\partial \zeta} & \frac{\partial t}{\partial \zeta} \end{bmatrix} = \begin{bmatrix} \frac{1}{2} & -\frac{1}{2}\left(\frac{1+\eta}{2}\right) & -\frac{1}{2}\left(\frac{1-\eta}{2}\right)\left(\frac{1+\zeta}{2}\right) \\ 0 & \frac{1}{2}\left(\frac{1-\xi}{2}\right) & -\frac{1}{2}\left(\frac{1-\xi}{2}\right)\left(\frac{1+\zeta}{2}\right) \\ 0 & 0 & \frac{1}{2}\left(\frac{1-\xi}{2}\right)\left(\frac{1-\eta}{2}\right) \end{bmatrix} \tag{9.79}$$

$$\text{Determinant}: \left|J_{\xi\eta\zeta}\right| = \frac{1}{8}\left(\frac{1-\xi}{2}\right)^2\left(\frac{1-\eta}{2}\right)$$

In this case, using above information, integral $I = \int_{r=0}^{r=1}\int_{s=0}^{s=1-r} \times \int_{t=0}^{t=1-r-s} F(r,s,t)drdsdt$ can be expressed in terms of ξ, η, ζ variables as following:

$$\begin{aligned} I &= \int_{r=0}^{r=1}\int_{s=0}^{s=1-r}\int_{t=0}^{t=1-r-s} F(r,s,t)drdsdt \\ &= \int_{-1}^{1}\int_{-1}^{1}\int_{-1}^{1} \underbrace{F(r(\xi,\eta,\zeta), s(\xi,\eta,\zeta), t(\xi,\eta,\zeta))\left|J_{\xi\eta\zeta}\right|d\xi d\eta d\zeta}_{G(\xi,\eta,\zeta)} \\ &= \int_{-1}^{1}\int_{-1}^{1}\int_{-1}^{1} G(\xi,\eta,\zeta)d\xi d\eta d\zeta \end{aligned} \tag{9.80}$$

Final integral expression of $I = \int_{-1}^{1}\int_{-1}^{1}\int_{-1}^{1} G(\xi,\eta,\zeta)d\xi d\eta d\zeta$ in Eq. 9.80 is now defined in standard form over the ideal cubic volume. Integral value can be easily calculated selecting appropriate integration points and their weights in Table 9.5 as below:

$$I = \int_{-1}^{1}\int_{-1}^{1}\int_{-1}^{1} G(\xi,\eta,\zeta)d\xi d\eta d\zeta = \sum_{i=1}^{p_\xi}\sum_{j=1}^{p_\eta}\sum_{k=1}^{p_\zeta} w_i w_j w_k G(\xi_i,\eta_j,\zeta_k) \tag{9.81}$$

In Eq. 9.81, p_ξ, p_η, p_ζ denote numbers of integration points in ξ, η, and ζ directions and w_i, w_j, w_k indicate their weights. For non-polynomial integrant functions, p_ξ, p_η, p_ζ in Eq. 9.81 are determined by trial-and-error approach. A predetermined number of integration points are increased successively until last two integral results do not differ much.

Example 9.11: Calculate $I = \int\int\int 10 \sin x \sin y \sin z dxdydz$ integral over the general tetrahedral volume given in Example 9.10 (corner coordinates: $(x_1, y_1, z_1) = (0,0,0)$, $(x_2, y_2, z_2) = (3,3,0)$, $(x_3, y_3, z_3) = (0,3,0)$, $(x_4, y_4, z_4) = (1,1,4)$) with (a) Classical Gauss or Gauss-Legendre method (b) Gauss-Lobatto method by using alternative approach (i.e., by first converting the general tetrahedral volume into ideal tetrahedral volume and then converting ideal tetrahedral volume into ideal cubic volume).

Integral $I = \int \int \int 10 \sin x \sin y \sin z dxdydz$ over the general tetrahedral volume can be calculated by transforming the general tetrahedral volume first into an ideal tetrahedral volume and then into ideal cubic volume as below:

1. *Conversion of general tetrahedral volume into ideal tetrahedral volume*:
   Conversion of general tetrahedral volume into ideal tetrahedral volume as illustrated in Fig. 9.33 can be expressed using Eq. 9.75 as below:

$$x = (1 - r - s - t)x_1 + rx_2 + sx_3 + tx_4 = 3r + t$$
$$y = (1 - r - s - t)y_1 + ry_2 + sy_3 + ty_4 = 3r + 3s + t$$
$$z = (1 - r - s - t)z_1 + rz_2 + sz_3 + tz_4 = 4t$$

Jacobian matrix and its determinant can be given as following:

$$J_{rst} = \begin{bmatrix} \frac{\partial x}{\partial r} & \frac{\partial y}{\partial r} & \frac{\partial z}{\partial r} \\ \frac{\partial x}{\partial s} & \frac{\partial y}{\partial s} & \frac{\partial z}{\partial s} \\ \frac{\partial x}{\partial t} & \frac{\partial y}{\partial t} & \frac{\partial z}{\partial t} \end{bmatrix} = \begin{bmatrix} 3 & 3 & 0 \\ 0 & 3 & 0 \\ 1 & 1 & 4 \end{bmatrix} \rightarrow \text{Determinant} : |J_{rst}| = 36$$

In this case, integral $I = \int \int \int 10 \sin x \sin y \sin z dxdydz$ can be expressed in terms r, s, t variables as following:

$$\begin{aligned} I &= \int \int \int 10 \sin x \sin y \sin z dxdydz \\ &= \int_{r=0}^{r=1} \int_{s=0}^{s=1-r} \int_{t=0}^{t=1-r-s} \underbrace{10[\sin(3r+t)][\sin(3r+3s+t)][\sin(4t)]36}_{F(r,s,t)} drdsdt \\ &= \int_{r=0}^{r=1} \int_{s=0}^{s=1-r} \int_{t=0}^{t=1-r-s} F(r,s,t) drdsdt \end{aligned}$$

2. *Conversion of ideal tetrahedral volume into ideal cubic volume*:
   Conversion of ideal tetrahedral volume into ideal cubic volume as illustrated in Fig. 9.34 can be expressed using corner coordinates of ideal tetrahedral volume as below:

$$r = \frac{1+\xi}{2}, \quad s = \left(\frac{1-\xi}{2}\right)\left(\frac{1+\eta}{2}\right), \quad t = \left(\frac{1-\xi}{2}\right)\left(\frac{1-\eta}{2}\right)\left(\frac{1+\zeta}{2}\right)$$

Transformation of *drdsdt* term can be obtained as $drdsdt = |J_{\xi\eta\zeta}(\xi,\eta,\zeta)|d\xi d\eta d\zeta$ using the determinant of Jacobian matrix $J_{\xi\eta\zeta}$ given below:

$$\text{Jacobian matrix}: J_{\xi\eta\zeta} = \begin{bmatrix} \frac{\partial r}{\partial \xi} & \frac{\partial s}{\partial \xi} & \frac{\partial t}{\partial \xi} \\ \frac{\partial r}{\partial \eta} & \frac{\partial s}{\partial \eta} & \frac{\partial t}{\partial \eta} \\ \frac{\partial r}{\partial \zeta} & \frac{\partial s}{\partial \zeta} & \frac{\partial t}{\partial \zeta} \end{bmatrix} = \begin{bmatrix} \frac{1}{2} & -\frac{1}{2}\left(\frac{1+\eta}{2}\right) & -\frac{1}{2}\left(\frac{1-\eta}{2}\right)\left(\frac{1+\zeta}{2}\right) \\ 0 & \frac{1}{2}\left(\frac{1-\xi}{2}\right) & -\frac{1}{2}\left(\frac{1-\xi}{2}\right)\left(\frac{1+\zeta}{2}\right) \\ 0 & 0 & \frac{1}{2}\left(\frac{1-\xi}{2}\right)\left(\frac{1-\eta}{2}\right) \end{bmatrix}$$

$$\text{Determinant}: |J_{\xi\eta\zeta}| = \frac{1}{8}\left(\frac{1-\xi}{2}\right)^2\left(\frac{1-\eta}{2}\right)$$

In this case, integral $I = \int_{r=0}^{r=1}\int_{s=0}^{s=1-r}\int_{t=0}^{t=1-r-s} F(r,s,t)drdsdt$ can be expressed in terms of ξ, η, ζ variables as following:

$$I = \int_{r=0}^{r=1}\int_{s=0}^{s=1-r}\int_{t=0}^{t=1-r-s} F(r,s,t)drdsdt$$

$$= \int_{-1}^{1}\int_{-1}^{1}\int_{-1}^{1} 10[\sin(1-r+2s-t)][\sin(1-r+2s+2t)][\sin(4-4r-4s-4t)]36drdsdt$$

$$= \int_{-1}^{1}\int_{-1}^{1}\int_{-1}^{1} \underbrace{\left\{ \begin{array}{l} 10\left[\sin\left(1-\underbrace{\left(\frac{1+\xi}{2}\right)}_{r}+2\underbrace{\left(\frac{1-\xi}{2}\right)\left(\frac{1+\eta}{2}\right)}_{s}-\underbrace{\left(\frac{1-\xi}{2}\right)\left(\frac{1-\eta}{2}\right)\left(\frac{1+\zeta}{2}\right)}_{t}\right)\right] \\ \otimes\left[\sin\left(1-\underbrace{\left(\frac{1+\xi}{2}\right)}_{r}+2\underbrace{\left(\frac{1-\xi}{2}\right)\left(\frac{1+\eta}{2}\right)}_{s}+2\underbrace{\left(\frac{1-\xi}{2}\right)\left(\frac{1-\eta}{2}\right)\left(\frac{1+\zeta}{2}\right)}_{t}\right)\right] \\ \otimes\left[\sin\left(4-4\underbrace{\left(\frac{1+\xi}{2}\right)}_{r}-4\underbrace{\left(\frac{1-\xi}{2}\right)\left(\frac{1+\eta}{2}\right)}_{s}-4\underbrace{\left(\frac{1-\xi}{2}\right)\left(\frac{1-\eta}{2}\right)\left(\frac{1+\zeta}{2}\right)}_{t}\right)\right] \\ \otimes 36\frac{1}{8}\left(\frac{1-\xi}{2}\right)^2\left(\frac{1-\eta}{2}\right) \end{array} \right\}}_{G(\xi,\eta,\zeta)} d\xi d\eta d\zeta$$

Final integral expression $I = \int_{-1}^{1} \int_{-1}^{1} \int_{-1}^{1} G(\xi, \eta, \zeta) d\xi d\eta d\zeta$is now defined in standard form over cubic volume. Therefore, integral value can be easily calculated selecting appropriate integration points and their weights in Table 9.5 as below:

$$I = \int_{-1}^{1} \int_{-1}^{1} \int_{-1}^{1} G(\xi, \eta, \zeta) d\xi d\eta d\zeta = \sum_{i=1}^{p_\xi} \sum_{j=1}^{p_\eta} \sum_{k=1}^{p_\zeta} w_i w_j w_k G(\xi_i, \eta_j, \zeta_k)$$

A number of integration points p_ξ, p_η, and p_ζ in ξ, η, and ζ directions are chosen according to the order of integrant function $G(\xi, \eta, \zeta)$. Since integrant function is non-polynomial (sinusoidal), its polynomial equivalent is predicted with trial-and-error approach. Integrals are calculated by increasing the number of integration points until last two integral values are close to each other as in Table 9.15.

In calculation of integrals in Table 9.15, integration points and weight coefficients are selected from Table 9.5 for Gauss or Gauss-Legendre method and from Table 9.6 for Gauss-Lobatto method. From Table 9.15, it is seen that integral result converges to the value of 22.

***Note*:** Integration points and weight coefficients are the same for the classical Gauss and Gauss-Legendre methods.

***Note*:** In this example, to calculate integral value using Gauss Quadrature Method, *MATLAB Program 9.10: Integration Over General Tetrahedral Volume Using Gauss Quadrature Method (Converting General Tetrahedral Volume First Into Ideal Tetrahedral Volume and Then Into Ideal Cube)* in Appendix can be used.

***Note*:** After comparison of results in Example 9.10 and Example 9.11, it is seen that integral over a general tetrahedral volume is calculated more efficiently with conversion of general tetrahedral volume into ideal tetrahedral volume than conversion of general tetrahedral volume first into ideal tetrahedral volume and then into

Table 9.15 Variation of integral values with a number of integration points

Number of integration points ($p = p_\xi \times p_\eta \times p_\zeta$)	Integral	
	Classical Gauss (Gauss-Legendre)	Gauss-Lobatto
1×1×1	14.9437	–
2×2×2	27.3309	–
3×3×3	22.0936	17.3170
4×4×4	22.8517	23.6965
5×5×5	22.7945	22.7945

ideal cubic volume. In Example 9.10, integral $I = \int\int\int 10 \sin x \sin y \sin z dxdydz$ is calculated over the tetrahedral volume by converting it into ideal tetrahedral volume using 11 integration points. However, the same integral value is calculated in this example by converting general tetrahedral volume first into ideal tetrahedral volume and then into ideal cubic volume with 4×4×4 = 64 integration points using classical Gauss method and with 5×5×5 = 125 integration points using Gauss-Lobatto method.

9.8 Summary

- Integrals of non-polynomial functions are generally difficult to calculate analytically (in closed form).
- Also, integral of functions over irregular domains cannot be calculated analytically (in closed form).
- When function is complicated or integral domain is irregular, numerical methods can be used to calculate integrals easily.
- In numerical integration, integral is calculated by summing weighted function values at predetermined points (x_i) called as integration points as following:

$$I = \int_a^b f(x)dx = w_1 f(x_1) + w_2 f(x_2) + \ldots + w_n f(x_n) = \sum_{i=1}^{p} w_i f(x_i)$$

- Weight coefficients (w_i) and integration points (x_i) in I expression are determined using either Newton-Cotes or Quadrature methods.
- In Newton-Cotes methods, Integral of an $f(x)$ function is calculated using its polynomial approximation ($\tilde{f}(x)$) as below:

$$I = \int_a^b f(x)dx = \int_a^b \tilde{f}(x)dx = \int_a^b \left(c_1 + c_2 x + c_3 x^2 + c_4 x^3 + \ldots + c_{n+1} x^n\right)dx$$

- Newton-Cotes methods use uniform integration points. That is integral domain is divided into equal sub-intervals.
- Trapezoidal, 1/3 and 3/8 Simpson methods are common numerical integral methods in Newton-Cotes group.
- Trapezoidal method uses piecewise linear approximations of original $f(x)$ function in integral domain represented with [a,b] interval.

- When integral domain is divided into N sub-intervals ($N+1$ points), integral of a general $f(x)$ function can be expressed with Trapezoidal method as below:

$$I = \int_{x_1=a}^{x_n=b} f(x)dx = \frac{h}{2}\left[f_1 + 2(f_2 + f_3 + \ldots + f_{n-1=N}) + f_{n=N+1}\right] \quad \left(h = \frac{b-a}{N}\right)$$

- 1/3 Simpson method uses piecewise parabolic approximations of original $f(x)$ function in integral domain represented with [a,b] interval.
- When integral domain is divided into N sub-intervals ($2N+1$ points), integral of a general $f(x)$ function can be expressed with 1/3 Simpson method as below:

$$I = \int_{x_1=a}^{x_n=b} f(x)dx$$

$$= \frac{h}{3}[f_1 + 4(f_2 + f_4 + f_6 + \ldots + f_{n-1=2N}) + 2(f_3 + f_5 + f_7 + \ldots + f_{n-2=2N-1}) + f_{n=2N+1}] \left(h = \frac{b-a}{2N}\right)$$

- 3/8 Simpson method uses piecewise cubic approximations of original $f(x)$ function in integral domain represented with [a,b] interval.
- When integral domain is divided into N sub-intervals ($3N+1$ points), integral of a general $f(x)$ function can be expressed with 3/8 Simpson method as below:

$$I = \int_{x_1=a}^{x_n=b} f(x)dx = = \frac{3h}{8}(f_1 + 3f_2 + 3f_3 + f_4) + \frac{3h}{8}(f_4 + 3f_5 + 3f_6 + f_7) + \ldots \left(h = \frac{b-a}{3N}\right)$$

$$+ \frac{h}{3}\left(f_{n-3=3N-2} + 3f_{n-2=3N-1} + 3f_{n-1=3N} + f_{n=3N+1}\right)$$

- For accurate integral results, h should be small in Newton-Cotes methods.
- Another alternative to increase accuracy in Newton-Cotes methods is calculating integrals for different h values and then extrapolating these results to obtain more accurate integral value. This approach is known as Richardson Extrapolation.
- Using integral results (I_1 and I_2) for two different h values (I_1 taking $h = h_1$ and I_2 taking $h = h_1/2$), corrected integral value with Richardson Extrapolation can be calculated as $I = \frac{4}{3}I_2 - \frac{1}{3}I_1$.
- Numerical integral methods based on non-uniform points are often referred to as Quadrature methods.
- Using different distribution of non-uniform points in integral domain, quadrature methods of Gauss, Gauss-Legendre, Gauss-Lobatto, etc. were developed.

- Gauss and Gauss-Legendre methods use only inner integration points, while Gauss-Lobatto uses both inner and boundary (end) integration points.
- In Gauss Quadrature method, following formulations are used in integrating linear and cubic $f(x)$ functions in integral domain represented with $[-1,1]$ interval:

$$f(x): \text{linear} \rightarrow I = \int_{-1}^{1} \underbrace{(c_1 + c_2 x)}_{f(x)} dx = 2f(0)$$

$$f(x): \text{cubic} \rightarrow I = \int_{-1}^{1} \underbrace{(c_1 + c_2 x + c_3 x^2 + c_4 x^3)}_{f(x)} dx = f\left(-\frac{1}{\sqrt{3}}\right) + f\left(\frac{1}{\sqrt{3}}\right)$$

- Gauss-Legendre method yields identical integral formulations with Gauss Quadrature method.
- Gauss-Lobatto method uses more integration points compared to Gauss and Gauss-Legendre methods. Therefore, it is less efficient than these methods.
- Gauss-Lobatto method uses following formulation in integrating cubic $f(x)$ functions in integral domain represented with $[-1,1]$ interval:

$$f(x): \text{cubic} \rightarrow I = \int_{-1}^{1} \underbrace{(c_1 + c_2 x + c_3 x^2 + c_4 x^3)}_{f(x)} dx = \frac{1}{3}f(-1) + \frac{4}{3}f(0) + \frac{1}{3}f(1)$$

- Quadrature methods are more efficient than Newton-Cotes methods due to use of non-uniform points.
- Due to their efficiency, quadrature methods are more commonly used in the solution of engineering problems. For example, Finite Element method, which is an integral-based method, uses quadrature methods in area and volume integrals of finite elements.

Appendix: MATLAB Programs

MATLAB Program 9.1: Integral Calculation with Trapezoid Method

```
function TrapezIntegralMethod

% THIS PROGRAM CALCULATES INTEGRAL OF AN f(x) FUNCTION IN [a,b] INTERVAL
% USING TRAPEZOIDAL METHOD

a=0 ;      % Lower bound of interval
b=pi/2 ;   % Upper bound of interval

N=32 ;     % Number of Linear segments

% ----------------------------------------------------------------------
--
h=(b-a)/N ;  % Length between adjacent integration points
n=N+1 ;      % Number of integration points

x=zeros(n,1) ;
f=zeros(n,1) ;

% Integration points
x(1)=0 ;
for i=1:n-1
    x(i+1)=x(i)+h ;
end

% Function values at integration points
for i=1:n
     f(i)=FunctionValue(x(i)) ;
end

% Integral calculation
Int=0 ;
for i=1:N
    Int=Int+(h/2)*(f(i)+f(i+1)) ;
end

fprintf('Integral Value: Int = %f \n',Int) ;

% ------------------------------------------------------
% ------------------   FUNCTION: f(x)   -----------------
% ------------------------------------------------------
function f=FunctionValue(x)
% f(x) function
f=(x^2)*sin(x) ;
```

MATLAB Program 9.2: Integral Calculation with 1/3 Simpson's Method

```
function Simpson1_3IntegralMethod

% THIS PROGRAM CALCULATES INTEGRAL OF AN f(x) FUNCTION IN [a,b] INTERVAL
% USING 1/3 SIMPSON METHOD

a=0 ;      % Lower bound of interval
b=pi/2 ;   % Upper bound of interval

N=4 ; % Number of Parabolic segments

% -----------------------------------------------------------------------
--
h=(b-a)/(2*N) ;  % Length between adjacent integration points
n=2*N+1 ;        % Number of integration points

x=zeros(n,1) ;
f=zeros(n,1) ;

% Integration points
x(1)=0 ;
for i=1:n-1
    x(i+1)=x(i)+h ;
end

% Function values at integration points
for i=1:n
     f(i)=FunctionValue(x(i)) ;
end

% Integral calculation
Int=0 ;
for i=1:N
    j=2*i-1 ;
    Int=Int+(h/3)*(f(j)+4*f(j+1)+f(j+2)) ;
end

fprintf('Integral Value: Int = %f \n',Int) ;

% ---------------------------------------------------
% ------------------   FUNCTION: f(x)   -----------------
% ---------------------------------------------------
function f=FunctionValue(x)
% f(x) function
f=(x^2)*sin(x) ;
```

MATLAB Program 9.3: Integration with 3/8 Simpson's Method

```
function Simpson3_8IntegralMethod

% THIS PROGRAM CALCULATES INTEGRAL OF AN f(x) FUNCTION IN [a,b] INTERVAL
% USING 3/8 SIMPSON METHOD

a=0 ;      % Lower bound of interval
b=pi/2 ;   % Upper bound of interval

N=3 ; % Number of Cubic segments

% -----------------------------------------------------------------------
--
h=(b-a)/(3*N) ; % Length between adjacent integration points
n=3*N+1 ;       % Number of integration points

x=zeros(n,1) ;
f=zeros(n,1) ;

% Integration points
x(1)=0 ;
for i=1:n-1
    x(i+1)=x(i)+h ;
end

% Function values at integration points
for i=1:n
     f(i)=FunctionValue(x(i)) ;
end

% Integral calculation
Int=0 ;
for i=1:N
    j=3*i-2 ;
    Int=Int+(3*h/8)*(f(j)+3*f(j+1)+3*f(j+2)+f(j+3)) ;
end

fprintf('Integral Value: Int = %f \n',Int) ;

% -------------------------------------------------------
% ------------------   FUNCTION: f(x)   -----------------
% -------------------------------------------------------
function f=FunctionValue(x)
% f(x) function
f=(x^2)*sin(x) ;
```

MATLAB Program 9.4: Gauss Quadrature Method

```
function GaussQuadratureIntegration_1D

% THIS PROGRAM CALCULATES INTEGRAL OF AN f(x) FUNCTION IN [a,b] INTERVAL
% USING GAUSS QUADRATURE METHOD

a=0 ;      % Lower bound of interval
b=pi/2 ;   % Upper bound of interval

n=5 ;      % Number of Gauss integration points

% Gauss integration points in [-1,1] interval and therir weights
if n==1
    ksi(1)=0 ;    w(1)=2 ;
end

if n==2
    ksi(1)=-1/sqrt(3)   ;   ksi(2)=1/sqrt(3) ;
    w(1)=1              ;   w(2)=1 ;
end

if n==3
    ksi(1)=-sqrt(3/5)   ;  ksi(2)=0   ;      ksi(3)=sqrt(3/5)   ;
    w(1)=5/9            ;  w(2)=8/9   ;      w(3)=5/9 ;
end

if n==4
    ksi(1)=-0.86113631   ;  ksi(2)=-0.33998104   ;      ksi(3)=0.33998104   ;
ksi(4)=0.86113631   ;
    w(1)=0.34785485      ;  w(2)=0.65214515      ;      w(3)=0.65214515     ;
w(4)=0.34785485     ;
end

if n==5
    ksi(1)=-0.90617985   ;  ksi(2)=-0.53846931   ;      ksi(3)=0          ;
ksi(4)=0.53846931   ;     ksi(5)=0.90617985   ;
    w(1)=0.23692689      ;  w(2)=0.47862867      ;      w(3)=0.56888889 ;
w(4)=0.47862867     ;     w(5)=0.23692689     ;
end


Int=0 ;
for i=1:n
    N(1)=(1-ksi(i))/2 ;  N(2)=(1+ksi(i))/2 ;
    dNdksi(1)=-1/2    ;  dNdksi(2)=1/2     ;

    % Conversion from [-1,1] interval to [a,b] interval
    x=N(1)*a+N(2)*b ;
    dxdksi=dNdksi(1)*a+dNdksi(2)*b ;

    J=dxdksi ;

    % function value
    f=FunctionValue(x) ;

    fpoint=w(i)*f*det(J) ;
    Int=Int+fpoint ;
end

fprintf('Number of Gauss Integration Points: n = %i \n',n) ;
fprintf('Integral Value: Int = %f \n',Int) ;

% -------------------------------------------------------
% ------------------   FUNCTION: f(x)   -----------------
% -------------------------------------------------------
function f=FunctionValue(x)
% f(x) function
f=x*sin(x) ;
```

MATLAB Program 9.5: Integration Over Quadrilateral Area Using Gauss Quadrature Method

```
function GaussQuadratureIntegration_2D_Quadrilateral

% THIS PROGRAM CALCULATES INTEGRAL OF f(x,y) FUNCTION OVER QUADRILATERAL
% AREA USING GASUSS QUADRATURE METHOD

% NOTE: Quadrilateral area is transformed into Ideal Square area

% n: Number of integration points in Ideal Square in ksi, ita directions
n=2 ;

% Corner points of quadrilateral
x1=1  ; y1=1  ;
x2=4  ; y2=0  ;
x3=3  ; y3=5  ;
x4=2  ; y4=3  ;

% Integration points and weights
if n==1
    ksi(1)=0 ;    ita(1)=0 ;    w(1)=2 ;
end

if n==2
    ksi(1)=-1/sqrt(3)  ;   ksi(2)=1/sqrt(3) ;
    ita(1)=-1/sqrt(3)  ;   ita(2)=1/sqrt(3) ;
    w(1)=1             ;   w(2)=1 ;
end

if n==3
    ksi(1)=-sqrt(3/5)  ;  ksi(2)=0  ;     ksi(3)=sqrt(3/5)   ;
    ita(1)=-sqrt(3/5)  ;  ita(2)=0  ;     ita(3)=sqrt(3/5)   ;
    w(1)=5/9           ;  w(2)=8/9  ;     w(3)=5/9 ;
end

if n==4
    ksi(1)=-0.86113631  ;  ksi(2)=-0.33998104  ;     ksi(3)=0.33998104  ;
ksi(4)=0.86113631  ;
    ita(1)=-0.86113631  ;  ita(2)=-0.33998104  ;     ita(3)=0.33998104  ;
ita(4)=0.86113631  ;
    w(1)=0.34785485     ;  w(2)=0.65214515     ;     w(3)=0.65214515    ;
w(4)=0.34785485    ;
end

if n==5
    ksi(1)=-0.90617985  ;  ksi(2)=-0.53846931  ;     ksi(3)=0          ;
ksi(4)=0.53846931  ;    ksi(5)=0.90617985  ;
```

```
    ita(1)=-0.90617985  ;  ita(2)=-0.53846931  ;     ita(3)=0        ;
ita(4)=0.53846931  ;     ita(5)=0.90617985  ;
    w(1)=0.23692689     ;  w(2)=0.47862867     ;     w(3)=0.56888889 ;
w(4)=0.47862867    ;     w(5)=0.23692689     ;
end


IntValue=0 ;
for i=1:n
    for j=1:n

        N(1)=(1-ksi(i))*(1-ita(j))/4 ; N(2)=(1+ksi(i))*(1-ita(j))/4 ;
        N(3)=(1+ksi(i))*(1+ita(j))/4 ; N(4)=(1-ksi(i))*(1+ita(j))/4 ;

        dNdksi(1)=-(1-ita(j))/4 ; dNdksi(2)=(1-ita(j))/4 ;
        dNdksi(3)=(1+ita(j))/4  ; dNdksi(4)=-(1+ita(j))/4 ;

        dNdita(1)=-(1-ksi(i))/4 ; dNdita(2)=-(1+ksi(i))/4 ;
        dNdita(3)=(1+ksi(i))/4  ; dNdita(4)=(1-ksi(i))/4 ;

        x=N(1)*x1+N(2)*x2+N(3)*x3+N(4)*x4 ;
        y=N(1)*y1+N(2)*y2+N(3)*y3+N(4)*y4 ;

        dxdksi=dNdksi(1)*x1+dNdksi(2)*x2+dNdksi(3)*x3+dNdksi(4)*x4 ;
        dxdita=dNdita(1)*x1+dNdita(2)*x2+dNdita(3)*x3+dNdita(4)*x4 ;

        dydksi=dNdksi(1)*y1+dNdksi(2)*y2+dNdksi(3)*y3+dNdksi(4)*y4 ;
        dydita=dNdita(1)*y1+dNdita(2)*y2+dNdita(3)*y3+dNdita(4)*y4 ;

        J=[dxdksi  dydksi ;
           dxdita  dydita ] ;

        % function value: f
        f=FunctionValue(x,y) ;

        fpoint=w(i)*w(j)*f*det(J) ;
        IntValue=IntValue+fpoint ;
    end
end

fprintf('Number of integration points=%ix%i,  Integral=%f
\n',n,n,IntValue) ;

% ========================================================
% ============    FUNCTION: f(x,y)   =====================
% ========================================================
function f=FunctionValue(x,y)
f=x^2+y^2+2*x*y ;
```

MATLAB Program 9.6: Integration Over General Triangular Area Using Gauss Quadrature Method

```
function GaussQuadratureIntegration_2D_Triangle

% THIS PROGRAM CALCULATES INTEGRAL OF f(x,y) FUNCTION OVER GENERAL
% TRIANGULAR AREA USING GASUSS QUADRATURE METHOD

% NOTE: General Triangle is transformed into Ideal Right Triangle

% n: Number of integration points in Ideal Right Triangle
% n should be n=1, 3, 4, 7
n=4 ;

% Corner points of general Triangle (Counter clockwise node numbering)
x1=-1 ;  y1=0   ;
x2=1  ;  y2=0 ;
x3=0  ;  y3=sqrt(3) ;

% Integration points and weights
if n==1
    ksi(1)=1/3   ;    ita(1)=1/3   ;   w(1)=1/2 ;
end

if n==3
    ksi(1)=2/3   ;    ksi(2)=1/6   ;     ksi(3)=1/6   ;
    ita(1)=1/6   ;    ita(2)=1/6   ;     ita(3)=2/3   ;
    w(1)=1/6     ;    w(2)=1/6     ;     w(3)=1/6     ;
end

if n==4
    ksi(1)=1/3   ;    ksi(2)=3/5   ;     ksi(3)=1/5   ;   ksi(4)=1/5   ;
    ita(1)=1/3   ;    ita(2)=1/5   ;     ita(3)=1/5   ;   ita(4)=3/5   ;
    w(1)=-27/96 ;     w(2)=25/96   ;     w(3)=25/96   ;   w(4)=25/96   ;
end

if n==7
    ksi(1)=0    ;     ksi(2)=1/2   ;     ksi(3)=1     ;   ksi(4)=1/2   ;
ksi(5)=0    ;    ksi(6)=0     ;     ksi(7)=1/3   ;
    ita(1)=0    ;     ita(2)=0     ;     ita(3)=0     ;   ita(4)=1/2   ;
ita(5)=1    ;    ita(6)=1/2   ;     ita(7)=1/3   ;
    w(1)=1/40   ;     w(2)=1/15    ;     w(3)=1/40    ;   w(4)=1/15    ;
w(5)=1/40   ;    w(6)=1/15    ;     w(7)=9/40    ;
end


IntValue=0 ;
for i=1:n

        N(1)=1-ksi(i)-ita(i)   ; N(2)=ksi(i)  ;   N(3)=ita(i) ;

        dNdksi(1)=-1 ; dNdksi(2)=1 ; dNdksi(3)=0   ;
        dNdita(1)=-1 ; dNdita(2)=0 ; dNdita(3)=1   ;
```

```
        x=N(1)*x1+N(2)*x2+N(3)*x3 ;
        y=N(1)*y1+N(2)*y2+N(3)*y3 ;

        dxdksi=dNdksi(1)*x1+dNdksi(2)*x2+dNdksi(3)*x3 ;
        dxdita=dNdita(1)*x1+dNdita(2)*x2+dNdita(3)*x3 ;

        dydksi=dNdksi(1)*y1+dNdksi(2)*y2+dNdksi(3)*y3 ;
        dydita=dNdita(1)*y1+dNdita(2)*y2+dNdita(3)*y3 ;

       J=[dxdksi  dydksi ;
          dxdita  dydita ] ;

        % function value: f
        f=FunctionValue(x,y) ;

        fpoint=w(i)*f*det(J) ;
        IntValue=IntValue+fpoint ;
end

fprintf('Number of integration points=%i,  Integral=%f \n',n,IntValue) ;

% =========================================================
% ===========    FUNCTION: f(x,y)   =====================
% =========================================================
function f=FunctionValue(x,y)
f=(x^3)+(y^3) ;
```

MATLAB Program 9.7: Integration Over General Triangular Area Using Gauss Quadrature Method (Converting General Triangular Area First into Ideal Right Triangle and Then into Ideal Square)

```
function GaussQuadratureIntegration_2D_Triangle_Square_Transformation

% THIS PROGRAM CALCULATES INTEGRAL OF f(x,y) FUNCTION OVER GENERAL
% TRIANGULAR AREA USING GASUSS QUADRATURE METHOD

% NOTE: Integral is calculated using two steps as following:
% 1- General Traingle is transformed into Ideal Right Triangle
% 2- Ideal Right Triangle is transformed into Ideal Square

% n: Number of integration points in Ideal Square in ksi, ita directions
n=4 ;

% Corner points of general Triangle
x1=-1 ;  y1=0   ;
x2=1  ;  y2=0 ;
x3=0  ;  y3=sqrt(3) ;

% Integration points and weights
if n==1
    ksi(1)=0 ;    ita(1)=0 ;    w(1)=2 ;
end

if n==2
    ksi(1)=-1/sqrt(3)   ;   ksi(2)=1/sqrt(3) ;
    ita(1)=-1/sqrt(3)   ;   ita(2)=1/sqrt(3) ;
    w(1)=1              ;   w(2)=1 ;
end

if n==3
    ksi(1)=-sqrt(3/5)   ;  ksi(2)=0   ;     ksi(3)=sqrt(3/5)   ;
    ita(1)=-sqrt(3/5)   ;  ita(2)=0   ;     ita(3)=sqrt(3/5)   ;
    w(1)=5/9            ;  w(2)=8/9   ;     w(3)=5/9 ;
end

if n==4
    ksi(1)=-0.86113631  ;  ksi(2)=-0.33998104  ;     ksi(3)=0.33998104  ;
ksi(4)=0.86113631  ;
    ita(1)=-0.86113631  ;  ita(2)=-0.33998104  ;     ita(3)=0.33998104  ;
ita(4)=0.86113631  ;
    w(1)=0.34785485     ;  w(2)=0.65214515     ;     w(3)=0.65214515    ;
w(4)=0.34785485    ;
end

if n==5
    ksi(1)=-0.90617985  ;  ksi(2)=-0.53846931  ;     ksi(3)=0         ;
ksi(4)=0.53846931  ;    ksi(5)=0.90617985  ;
    ita(1)=-0.90617985  ;  ita(2)=-0.53846931  ;     ita(3)=0         ;
ita(4)=0.53846931  ;    ita(5)=0.90617985  ;
    w(1)=0.23692689     ;  w(2)=0.47862867     ;     w(3)=0.56888889 ;
w(4)=0.47862867    ;    w(5)=0.23692689    ;
end
```

```
IntValue=0 ;
for i=1:n
    for j=1:n

            % Transformation of General Triangle into Ideal Right
Triangle
            % Ideal Right Triangle: length is 1 with straight sides
            r=(3+3*ksi(i)-ita(j)-ksi(i)*ita(j))/8 ;
            s=(3-ksi(i)+3*ita(j)-ksi(i)*ita(j))/8 ;
            detJ=(2-ksi(i)-ita(j))/16 ;

            % Transformation of Ideal Right Triangle into Ideal Square
            % Corner points of Ideal Right Triangle
            % Point 1:  (r,s,t)=(0,0)
            % Point 2:  (r,s,t)=(1,0)
            % Point 3:  (r,s,t)=(1/2,1/2)
            % Point 4:  (r,s,t)=(0,1)

            N1=1-r-s ;  N2=r ;  N3=s ;
            x=N1*x1+N2*x2+N3*x3 ;
            y=N1*y1+N2*y2+N3*y3 ;

            Jxyz=[x2-x1 y2-y1 ;
                  x3-x1 y3-y1 ]  ;

            detJxyz=abs(det(Jxyz)) ;

            % function value: f
            f=FunctionValue(x,y) ;

            f=f*detJxyz ;
            fpoint=w(i)*w(j)*f*detJ ;
            IntValue=IntValue+fpoint ;
    end
end

fprintf('Number of integration points=%ix%i,  Integral=%f
\n',n,n,IntValue) ;

% ==========================================================
% ============    FUNCTION: f(x,y)   ======================
% ==========================================================
function f=FunctionValue(x,y)
f=(x^3)+(y^3) ;
```

MATLAB Program 9.8: Integration Over General Hexahedral Volume Using Gauss Quadrature Method

```
function GaussQuadratureIntegration_3D_Hexahedral

% THIS PROGRAM CALCULATES INTEGRAL OF f(x,y) FUNCTION OVER GENERAL
% HEXAHEDRAL VOLUME USING GASUSS QUADRATURE METHOD

% NOTE: General Hexahedral volume is transformed into Ideal Cube volume

% n: Number of integration points in ideal cube in ksi,ita,zeta
directions
n=5 ;

% Corner points of Hexahedral Volume (Hexahedral Cubic Volume)
x1=0 ; y1=0; z1=0 ;
x2=2 ; y2=0; z2=0 ;
x3=3 ; y3=2; z3=0 ;
x4=0 ; y4=2; z4=0 ;
x5=0 ; y5=0; z5=2 ;
x6=2 ; y6=0; z6=4 ;
x7=3 ; y7=2; z7=4 ;
x8=0 ; y8=2; z8=2 ;

% Integration points and weights
if n==1
    ksi(1)=0 ;    ita(1)=0 ;    zeta(1)=0 ;  w(1)=2 ;
end

if n==2
    ksi(1)=-1/sqrt(3)   ;    ksi(2)=1/sqrt(3) ;
    ita(1)=-1/sqrt(3)   ;    ita(2)=1/sqrt(3) ;
    zeta(1)=-1/sqrt(3)  ;    zeta(2)=1/sqrt(3) ;
    w(1)=1              ;    w(2)=1 ;
end

if n==3
    ksi(1)=-sqrt(3/5)   ;  ksi(2)=0   ;     ksi(3)=sqrt(3/5)   ;
    ita(1)=-sqrt(3/5)   ;  ita(2)=0   ;     ita(3)=sqrt(3/5)   ;
    zeta(1)=-sqrt(3/5)  ;  zeta(2)=0  ;     zeta(3)=sqrt(3/5)  ;
    w(1)=5/9            ;  w(2)=8/9   ;     w(3)=5/9 ;
end

if n==4
    ksi(1)=-0.86113631  ;  ksi(2)=-0.33998104  ;     ksi(3)=0.33998104   ;
ksi(4)=0.86113631   ;
    ita(1)=-0.86113631  ;  ita(2)=-0.33998104  ;     ita(3)=0.33998104   ;
ita(4)=0.86113631   ;
    zeta(1)=-0.86113631 ;  zeta(2)=-0.33998104 ;     zeta(3)=0.33998104  ;
zeta(4)=0.86113631  ;
    w(1)=0.34785485     ;  w(2)=0.65214515     ;     w(3)=0.65214515     ;
w(4)=0.34785485     ;
end
```

```
if n==5
    ksi(1)=-0.90617985  ; ksi(2)=-0.53846931  ;     ksi(3)=0        ;
ksi(4)=0.53846931  ;    ksi(5)=0.90617985  ;
    ita(1)=-0.90617985  ; ita(2)=-0.53846931  ;     ita(3)=0        ;
ita(4)=0.53846931  ;    ita(5)=0.90617985  ;
    zeta(1)=-0.90617985 ; zeta(2)=-0.53846931 ;     zeta(3)=0       ;
zeta(4)=0.53846931 ;    zeta(5)=0.90617985 ;
    w(1)=0.23692689     ; w(2)=0.47862867     ;     w(3)=0.56888889 ;
w(4)=0.47862867    ;    w(5)=0.23692689    ;
end


IntValue=0 ;
for i=1:n
    for j=1:n
        for k=1:n

            N(1)=(1-ksi(i))*(1-ita(j))*(1-zeta(k))/8 ;
N(2)=(1+ksi(i))*(1-ita(j))*(1-zeta(k))/8 ;
            N(3)=(1+ksi(i))*(1+ita(j))*(1-zeta(k))/8 ; N(4)=(1-
ksi(i))*(1+ita(j))*(1-zeta(k))/8 ;
            N(5)=(1-ksi(i))*(1-ita(j))*(1+zeta(k))/8 ;
N(6)=(1+ksi(i))*(1-ita(j))*(1+zeta(k))/8 ;
            N(7)=(1+ksi(i))*(1+ita(j))*(1+zeta(k))/8 ; N(8)=(1-
ksi(i))*(1+ita(j))*(1+zeta(k))/8 ;

            dNdksi(1)=-(1-ita(j))*(1-zeta(k))/8 ; dNdksi(2)=(1-
ita(j))*(1-zeta(k))/8  ;
            dNdksi(3)=(1+ita(j))*(1-zeta(k))/8  ; dNdksi(4)=-
(1+ita(j))*(1-zeta(k))/8 ;
            dNdksi(5)=-(1-ita(j))*(1+zeta(k))/8 ; dNdksi(6)=(1-
ita(j))*(1+zeta(k))/8  ;
            dNdksi(7)=(1+ita(j))*(1+zeta(k))/8  ; dNdksi(8)=-
(1+ita(j))*(1+zeta(k))/8 ;

            dNdita(1)=-(1-ksi(i))*(1-zeta(k))/8 ; dNdita(2)=-
(1+ksi(i))*(1-zeta(k))/8 ;
            dNdita(3)=(1+ksi(i))*(1-zeta(k))/8  ; dNdita(4)=(1-
ksi(i))*(1-zeta(k))/8  ;
            dNdita(5)=-(1-ksi(i))*(1+zeta(k))/8 ; dNdita(6)=-
(1+ksi(i))*(1+zeta(k))/8 ;
            dNdita(7)=(1+ksi(i))*(1+zeta(k))/8  ; dNdita(8)=(1-
ksi(i))*(1+zeta(k))/8  ;

            dNdzeta(1)=-(1-ksi(i))*(1-ita(j))/8 ; dNdzeta(2)=-
(1+ksi(i))*(1-ita(j))/8 ;
            dNdzeta(3)=-(1+ksi(i))*(1+ita(j))/8 ; dNdzeta(4)=-(1-
ksi(i))*(1+ita(j))/8 ;
            dNdzeta(5)=(1-ksi(i))*(1-ita(j))/8  ;
dNdzeta(6)=(1+ksi(i))*(1-ita(j))/8  ;
            dNdzeta(7)=(1+ksi(i))*(1+ita(j))/8  ; dNdzeta(8)=(1-
ksi(i))*(1+ita(j))/8  ;
```

```
x=N(1)*x1+N(2)*x2+N(3)*x3+N(4)*x4+N(5)*x5+N(6)*x6+N(7)*x7+N(8)*x8 ;

y=N(1)*y1+N(2)*y2+N(3)*y3+N(4)*y4+N(5)*y5+N(6)*y6+N(7)*y7+N(8)*y8 ;

z=N(1)*z1+N(2)*z2+N(3)*z3+N(4)*z4+N(5)*z5+N(6)*z6+N(7)*z7+N(8)*z8 ;

dxdksi=dNdksi(1)*x1+dNdksi(2)*x2+dNdksi(3)*x3+dNdksi(4)*x4+dNdksi(5)*x5+d
Ndksi(6)*x6+dNdksi(7)*x7+dNdksi(8)*x8 ;

dxdita=dNdita(1)*x1+dNdita(2)*x2+dNdita(3)*x3+dNdita(4)*x4+dNdita(5)*x5+d
Ndita(6)*x6+dNdita(7)*x7+dNdita(8)*x8 ;

dxdzeta=dNdzeta(1)*x1+dNdzeta(2)*x2+dNdzeta(3)*x3+dNdzeta(4)*x4+dNdzeta(5
)*x5+dNdzeta(6)*x6+dNdzeta(7)*x7+dNdzeta(8)*x8 ;


dydksi=dNdksi(1)*y1+dNdksi(2)*y2+dNdksi(3)*y3+dNdksi(4)*y4+dNdksi(5)*y5+d
Ndksi(6)*y6+dNdksi(7)*y7+dNdksi(8)*y8 ;

dydita=dNdita(1)*y1+dNdita(2)*y2+dNdita(3)*y3+dNdita(4)*y4+dNdita(5)*y5+d
Ndita(6)*y6+dNdita(7)*y7+dNdita(8)*y8 ;

dydzeta=dNdzeta(1)*y1+dNdzeta(2)*y2+dNdzeta(3)*y3+dNdzeta(4)*y4+dNdzeta(5
)*y5+dNdzeta(6)*y6+dNdzeta(7)*y7+dNdzeta(8)*y8 ;


dzdksi=dNdksi(1)*z1+dNdksi(2)*z2+dNdksi(3)*z3+dNdksi(4)*z4+dNdksi(5)*z5+d
Ndksi(6)*z6+dNdksi(7)*z7+dNdksi(8)*z8 ;

dzdita=dNdita(1)*z1+dNdita(2)*z2+dNdita(3)*z3+dNdita(4)*z4+dNdita(5)*z5+d
Ndita(6)*z6+dNdita(7)*z7+dNdita(8)*z8 ;

dzdzeta=dNdzeta(1)*z1+dNdzeta(2)*z2+dNdzeta(3)*z3+dNdzeta(4)*z4+dNdzeta(5
)*z5+dNdzeta(6)*z6+dNdzeta(7)*z7+dNdzeta(8)*z8 ;

            J=[dxdksi   dydksi   dzdksi   ;
               dxdita   dydita   dzdita   ;
               dxdzeta  dydzeta  dzdzeta ] ;

            % function value: f
            f=FunctionValue(x,y,z) ;

            fpoint=w(i)*w(j)*w(k)*f*det(J) ;
            IntValue=IntValue+fpoint ;
        end
    end
end

fprintf('Number of integration points=%ix%ix%i,  Integral=%f
\n',n,n,n,IntValue) ;

% ======================================================
% ============    FUNCTION: f(x,y,z)   ==================
% ======================================================
function f=FunctionValue(x,y,z)
f=10*sin(x)*sin(y)*sin(z) ;
```

MATLAB Program 9.9: Integration Over General Tetrahedral Volume Using Gauss Quadrature Method

```
function GaussQuadratureIntegration_3D_Tetrahedron

% THIS PROGRAM CALCULATES INTEGRAL OF f(x,y) FUNCTION OVER GENERAL
% TETRAHEDRAL VOLUME USING GASUSS QUADRATURE METHOD

% NOTE: General Tetrahedral Volume is transformed into Ideal Tetrahedron
% Volume

% n: Number of integration points in tetrahedral volume
% n should be equal to n=1,4,5,11
n=5 ; % number of integration points

% Corner points of General Tetrahedral Volume
%x1=2  ; x2=0 ; x3=0 ;  x4=0 ;
%y1=0  ; y2=2 ; y3=0 ;  y4=0 ;
%z1=0  ; z2=0 ; z3=0 ;  z4=2 ;

x1=0 ; y1=0; z1=0 ;
x2=3 ; y2=3; z2=0 ;
x3=0 ; y3=3; z3=0 ;
x4=1 ; y4=1; z4=4 ;

% Integration points and weights
if n==1
    ksi(1)=1/4 ;   ita(1)=1/4 ;   zeta(1)=1/4 ;  w(1)=1/6 ;
end

if n==4
    a=(5+3*sqrt(5))/20 ;  b=(5-3*sqrt(5))/20 ;

    ksi(1)=a  ;  ksi(2)=b  ;      ksi(3)=b  ;   ksi(4)=b  ;
    ita(1)=b  ;  ita(2)=a  ;      ita(3)=b  ;   ita(4)=b  ;
    zeta(1)=b ;  zeta(2)=b ;      zeta(3)=a ;   zeta(4)=b ;
    w(1)=1/24      ;  w(2)=1/24      ;      w(3)=1/24     ;   w(4)=1/24     ;
end

if n==5
    ksi(1)=1/4   ;  ksi(2)=1/2   ;     ksi(3)=1/6        ;    ksi(4)=1/6  ;
ksi(5)=1/6 ;
    ita(1)=1/4   ;  ita(2)=1/6   ;     ita(3)=1/2        ;    ita(4)=1/6 ;
ita(5)=1/6  ;
    zeta(1)=1/4 ;  zeta(2)=1/6 ;     zeta(3)=1/6        ;    zeta(4)=1/2 ;
zeta(5)=1/6 ;
    w(1)=-4/30      ;  w(2)=9/120      ;      w(3)=9/120 ;     w(4)=9/120
;    w(5)=9/120     ;
end


if n==11
     a=(1+sqrt(5/14))/4 ;  b=(1-sqrt(5/14))/4 ;
```

```
    ksi(1)=1/4  ;  ksi(2)=11/14  ;     ksi(3)=1/14         ;
ksi(4)=1/14  ;    ksi(5)=1/14 ;  ksi(6)=a  ;  ksi(7)=a  ;     ksi(8)=a
;   ksi(9)=b  ;    ksi(10)=b ;    ksi(11)=b ;
    ita(1)=1/4  ;  ita(2)=1/14  ;    ita(3)=11/14        ;
ita(4)=1/14 ;     ita(5)=1/14  ; ita(6)=a  ;  ita(7)=b  ;     ita(8)=b
;   ita(9)=a ;     ita(10)=a  ;   ita(11)=b  ;
    zeta(1)=1/4 ;  zeta(2)=1/14 ;    zeta(3)=1/14        ;
zeta(4)=11/14 ;    zeta(5)=1/14 ; zeta(6)=b ;  zeta(7)=a ;     zeta(8)=b
;   zeta(9)=a ;    zeta(10)=b ;   zeta(11)=a ;
    w(1)=-74/5625     ;  w(2)=343/45000    ;    w(3)=343/45000 ;
w(4)=343/45000    ;    w(5)=343/45000    ;  w(6)=56/2250     ;
w(7)=56/2250     ;     w(8)=56/2250 ;     w(9)=56/2250    ;
w(10)=56/2250    ;  w(11)=56/2250    ;
end

IntValue=0 ;
for i=1:n

            % Transformation of General Tetrahedral Volume into Ideal
            % Tetrahedral Volume
            % Corner points of Ideal Tetrahedral Volume:
            % Point 1:  (ksi,ita,zeta)=(1,0,0)
            % Point 2:  (ksi,ita,zeta)=(0,1,0)
            % Point 3:  (ksi,ita,zeta)=(0,0,0)
            % Point 4:  (ksi,ita,zeta)=(0,0,1)

            N1=1-ksi(i)-ita(i)-zeta(i) ;  N2=ksi(i) ;  N3=ita(i) ;
N4=zeta(i) ;
            x=N1*x1+N2*x2+N3*x3+N4*x4 ;
            y=N1*y1+N2*y2+N3*y3+N4*y4 ;
            z=N1*z1+N2*z2+N3*z3+N4*z4 ;

            Jxyz=[x2-x1 y2-y1 z2-z1 ;
                  x3-x1 y3-y1 z3-z1 ;
                  x4-x1 y4-y1 z4-z1 ]  ;

            detJxyz=det(Jxyz) ;

            % function value: f
            f=FunctionValue(x,y,z) ;

            fpoint=w(i)*f*detJxyz ;
            IntValue=IntValue+fpoint ;

end

fprintf('Number of integration points=%i,  Integral=%f \n',n,IntValue) ;

% ==========================================================
% ============    FUNCTION: f(x,y,z)   ====================
% ==========================================================
function f=FunctionValue(x,y,z)
f=10*sin(x)*sin(y)*sin(z) ;
```

MATLAB Program 9.10: Integration Over General Tetrahedral Volume Using Gauss Quadrature Method (Converting General Tetrahedral Volume First into Ideal Tetrahedral Volume and Then into Ideal Cube)

```
function GaussQuadratureIntegration_3D_Tetrahedron_transformed_to_Cube

% THIS PROGRAM CALCULATES INTEGRAL OF f(x,y) FUNCTION OVER GENERAL
% TETRAHEDRAL VOLUME USING GASUSS QUADRATURE METHOD

% NOTE: Integral is calculated using two steps as following:
% 1- General Tetrahedral Volume is transformed into Ideal Tetrahedron
Volume
% 2- Ideal Tetrahedron Volume is transformed into Ideal Cube Volume

% n: Number of integration points in Ideal Cube in ksi,ita,zeta
directions
n=5 ;

% Corner points of Arbitrary (General) Tetrahedron Volume
%x1=2  ; x2=0 ; x3=0 ;  x4=0 ;
%y1=0  ; y2=2 ; y3=0 ;  y4=0 ;
%z1=0  ; z2=0 ; z3=0 ;  z4=2 ;

x1=0 ; y1=0; z1=0 ;
x2=3 ; y2=3; z2=0 ;
x3=0 ; y3=3; z3=0 ;
x4=1 ; y4=1; z4=4 ;

% Integration points and weights
if n==1
    ksi(1)=0 ;    ita(1)=0 ;    zeta(1)=0 ;  w(1)=2 ;
end

if n==2
    ksi(1)=-1/sqrt(3)   ;   ksi(2)=1/sqrt(3) ;
    ita(1)=-1/sqrt(3)   ;   ita(2)=1/sqrt(3) ;
    zeta(1)=-1/sqrt(3)  ;   zeta(2)=1/sqrt(3) ;
    w(1)=1              ;   w(2)=1 ;
end

if n==3
    ksi(1)=-sqrt(3/5)   ;  ksi(2)=0   ;      ksi(3)=sqrt(3/5)   ;
    ita(1)=-sqrt(3/5)   ;  ita(2)=0   ;      ita(3)=sqrt(3/5)   ;
    zeta(1)=-sqrt(3/5)  ;  zeta(2)=0  ;      zeta(3)=sqrt(3/5)  ;
    w(1)=5/9            ;  w(2)=8/9   ;      w(3)=5/9 ;
end

if n==4
    ksi(1)=-0.86113631   ;  ksi(2)=-0.33998104   ;     ksi(3)=0.33998104   ;
ksi(4)=0.86113631   ;
    ita(1)=-0.86113631   ;  ita(2)=-0.33998104   ;     ita(3)=0.33998104   ;
ita(4)=0.86113631   ;
    zeta(1)=-0.86113631  ;  zeta(2)=-0.33998104  ;     zeta(3)=0.33998104  ;
zeta(4)=0.86113631  ;
    w(1)=0.34785485      ;  w(2)=0.65214515      ;     w(3)=0.65214515     ;
w(4)=0.34785485     ;
end

if n==5
    ksi(1)=-0.90617985   ;  ksi(2)=-0.53846931   ;     ksi(3)=0            ;
ksi(4)=0.53846931   ;     ksi(5)=0.90617985   ;
    ita(1)=-0.90617985   ;  ita(2)=-0.53846931   ;     ita(3)=0            ;
ita(4)=0.53846931   ;     ita(5)=0.90617985   ;
```

```
    zeta(1)=-0.90617985 ;  zeta(2)=-0.53846931 ;     zeta(3)=0        ;
zeta(4)=0.53846931 ;    zeta(5)=0.90617985 ;
    w(1)=0.23692689     ;  w(2)=0.47862867     ;     w(3)=0.56888889 ;
w(4)=0.47862867    ;    w(5)=0.23692689    ;
end


IntValue=0 ;
for i=1:n
    for j=1:n
        for k=1:n

            % Transformation of General Tetrahedral to Ideal Tetrahedral
            % Ideal Tetrahedral: length is 1 with straight sides
            r=(1+ksi(i))/2 ;
            s=(1-ksi(i))*(1+ita(j))/4 ;
            t=(1-ksi(i))*(1-ita(j))*(1+zeta(k))/8 ;
            detJ=(1/64)*((1-ksi(i))^2)*(1-ita(j)) ;

            % Transformation of Ideal Tetrahedral into Ideal Cube
            % Corner points of Ideal Tetrahedral Volume
            % Point 1:  (r,s,t)=(1,0,0)
            % Point 2:  (r,s,t)=(0,1,0)
            % Point 3:  (r,s,t)=(0,0,0)
            % Point 4:  (r,s,t)=(0,0,1)

            N1=1-r-s-t ;  N2=r ;  N3=s ;  N4=t ;
            x=N1*x1+N2*x2+N3*x3+N4*x4 ;
            y=N1*y1+N2*y2+N3*y3+N4*y4 ;
            z=N1*z1+N2*z2+N3*z3+N4*z4 ;

            Jxyz=[x2-x1  y2-y1  z2-z1 ;
                  x3-x1  y3-y1  z3-z1 ;
                  x4-x1  y4-y1  z4-z1 ]  ;

            detJxyz=abs(det(Jxyz)) ;

            % function value: f
            f=FunctionValue(x,y,z) ;

            f=f*detJxyz ;
            fpoint=w(i)*w(j)*w(k)*f*detJ ;
            IntValue=IntValue+fpoint ;
        end
    end
end

fprintf('Number of integration points=%ix%ix%i,  Integral=%f
\n',n,n,n,IntValue) ;

% ==========================================================
% ============     FUNCTION: f(x,y,z)   ====================
% ==========================================================
function f=FunctionValue(x,y,z)
f=10*sin(x)*sin(y)*sin(z) ;
```

Problems

Problem 9.1: Calculate $I = \int_1^2 [x^3 + 2x^2 + 3x]dx$ integral using trapezoidal method.

Problem 9.2: Calculate $I = \int_0^{\pi/2} \sin(2x)dx$ integral using trapezoidal method.

Problem 9.3: Calculate $I = \int_0^2 \sqrt{e^x \cos x}dx$ integral using trapezoidal method and then apply correction with Richardson extrapolation method.

Problem 9.4: Calculate $I = \int_0^{\pi/3} 3\tan x dx$ integral using trapezoidal method and then apply correction with Richardson extrapolation method.

Problem 9.5: Calculate $I = \int_1^2 [x^3 + 2x^2 + 3x]dx$ integral using 1/3 Simpson method.

Problem 9.6: Calculate $I = \int_0^{\pi/3} (x+1)\cos x dx$ integral using 1/3 Simpson method.

Problem 9.7: Calculate $I = \int_0^2 \sqrt{e^x \cos x}dx$ integral using 1/3 Simpson method.

Problem 9.8: Calculate $I = \int_1^2 [x^3 + 2x^2 + 3x]dx$ integral using 3/8 Simpson method.

Problem 9.9: Calculate $I = \int_0^{\pi/3} (x+1)\cos x dx$ integral using 3/8 Simpson method.

Problem 9.10: Calculate $I = \int_0^2 \sqrt{e^x \cos x}dx$ integral using 3/8 Simpson method.

Problem 9.11: Calculate $I = \int_2^4 [x^3 - 2x^2 + 5x + 3]dx$ integral using (a) classical Gauss quadrature (i.e., Gauss-Legendre method), (b) Gauss-Lobatto quadrature method.

Problem 9.12: Calculate $I = \int_0^{\pi/6} x\cos x dx$ integral using (a) classical Gauss quadrature, (b) Gauss-Lobatto quadrature method.

Problem 9.13: Calculate $I = \int_1^2 \int_1^2 [x^3 + y^2 + 2xy] dxdy$ double integral using (a) classical Gauss quadrature, (b) Gauss-Lobatto quadrature method.

Problem 9.14: Calculate $I = \int_0^{\pi/2} \int_0^{\pi/2} 2 \sin x \sin y dxdy$ double integral using (a) classical Gauss quadrature, (b) Gauss-Lobatto quadrature method.

Problem 9.15: Calculate $I = \int \int 2(x + y) \cos x \cos y dxdy$ double integral over the quadrilateral domain shown in Fig. 9.17 using (a) classical Gauss quadrature, (b) Gauss-Lobatto quadrature method.

Problem 9.16: Calculate $I = \int \int (x^3 + 2y^3 + 2xy) dxdy$ double integral over the general triangle with corner coordinates of $(x_1, y_1) = (0,1)$, $(x_2, y_2) = (2,1)$, and $(x_3, y_3) = (1,2\sqrt{3})$ using (a) classical Gauss quadrature, (b) Gauss-Lobatto quadrature method.

Problem 9.17: Calculate $I = \int \int xy \cos x \cos y dxdy$ double integral over the general triangle with corner coordinates of $(x_1, y_1) = (0,1)$, $(x_2, y_2) = (2,1)$ and $(x_3, y_3) = (1, 2\sqrt{3})$ using (a) classical Gauss quadrature, (b) Gauss-Lobatto quadrature method.

Problem 9.18: Calculate $I = \int_0^3 \int_0^3 \int_0^3 [2x^3 + y^2 + 3z^3 + 2xyz] dxdydz$ triple integral using (a) classical Gauss quadrature, (b) Gauss-Lobatto quadrature method.

Problem 9.19: Calculate $I = \int_1^2 \int_1^2 \int_1^2 (3 \sin x \sin y \sin z) dxdydz$ triple integral using (a) classical Gauss quadrature, (b) Gauss-Lobatto quadrature method.

Problem 9.20: Calculate $I = \int \int \int (x^3 + 2y^3 + z^3 + 2xyz) dxdydz$ triple integral over the general hexahedral volume shown in Fig. 9.26 using (a) classical Gauss, (b) Gauss-Lobatto quadrature method.

Problem 9.21: Calculate $I = \int \int \int (x^3 + 2y^3 + z^3 + 2xyz) dxdydz$ triple integral over the general tetrahedral volume defined with corner coordinates of $(x_1, y_1, z_1) = (0, 0, 0)$, $(x_2, y_2, z_2) = (4, 4, 0)$, $(x_3, y_3, z_3) = (0, 4, 0)$, $(x_4, y_4, z_4) = (1, 1, 3)$ using (a) classical Gauss quadrature, (b) Gauss-Lobatto quadrature method.

Problem 9.22: Calculate $I = \int \int \int (x^3 + 2y^3 + z^3 + 2xyz) dxdydz$ integral over the general tetrahedral volume defined with corner coordinates of $(x_1, y_1, z_1) = (0, 0, 0)$, $(x_2, y_2, z_2) = (4, 4, 0)$, $(x_3, y_3, z_3) = (0, 4, 0)$, $(x_4, y_4, z_4) = (1, 1, 3)$ using (a) Classical Gauss or Gauss-Legendre method (b) Gauss-Lobatto method by using alternative approach (i.e., by first converting the general tetrahedral volume into ideal tetrahedral volume and then converting ideal tetrahedral volume into ideal cubic volume).

References

1. Bhavikatti, S. S. (2005). *Finite element analysis*. New Age International.
2. Hutton, D. (2003). *Fundamentals of finite element analysis*. McGraw-Hill.
3. Bathe, K. J. (2006). *Finite element procedures*. Prentice Hall.
4. Chapra, S. C., & Canale, R. P. (2020). *Numerical methods for engineers*. McGraw-Hill.
5. Bernatz, R. A. (2010). *Fourier series and numerical methods for partial differential equations*. John Wiley & Sons.
6. Everstine, G. C. (2010). *Numerical solution of partial differential equations* [Lecture notes, PDF]. George Washington University. https://www.academia.edu/31518755/Numerical_Solution_of_Partial_Differential_Equations
7. Epperson, J. F. (2013). *An introduction to numerical methods and analysis*. John Wiley & Sons.
8. Burden, R. L., & Faires, J. D. (2011). *Numerical analysis*. Cengage Learning.
9. Sert, C. (2006). *ME310 numerical methods* [lecture notes, PDF]. Middle East Technical University. https://users.metu.edu.tr/csert/me310/me310_1_introduction.pdf
10. Grewal, B. S. (2019). *Numerical methods in engineering and science*. Mercury Learning and Information.
11. Sastry, S. S. (2012). *Introductory methods of numerical analysis*. PHI Learning.

Chapter 10
Solution of Eigenvalue Problems

10.1 Introduction

Eigenvalue problems are used to investigate fundamental characteristics of engineering systems. For example, finding natural frequencies and vibration patterns (vibration modes) in free vibration problems require eigenvalue solution [1, 2]. Also, finding critical buckling loads and buckling shapes (buckling modes) in buckling problems require eigenvalue solution. Eigenvalue problems are usually expressed in mathematical form as $AX = \lambda X$ or $AX = \lambda BX$ where λ and X indicate eigenvalues and eigenvectors, respectively. A and B denote properties of the system such as stiffness and mass. Mathematical expression of some eigenvalue problems in mechanics are given below.

Free vibration of spring-mass system:
Spring-mass-damper system is often used in engineering to create simple dynamic models of complex structures. This way dynamic analysis of the structure is simplified extensively and carried out efficiently. Consider the spring-mass system in Fig. 10.1 which is subjected to the external force F. Assuming all springs are stretched (assuming $x_4(t) > x_3(t) > x_2(t) > x_1(t)$), equation of motion of spring-mass system can be expressed as following:

$$
\begin{aligned}
&k_2(x_2 - x_1) - k_1x_1 = m_1\ddot{x}_1 \\
&k_3(x_3 - x_2) - k_2(x_2 - x_1) = m_2\ddot{x}_2 \\
&k_4(x_4 - x_3) - k_3(x_3 - x_2) = m_3\ddot{x}_3 \\
&F - k_4(x_4 - x_3) = m_4\ddot{x}_4
\end{aligned}
\tag{10.1}
$$

Equation (10.1) can be expressed in matrix form as below:

H. Kurtaran, *Engineering Mathematics with MATLAB Applications: Vol. 1, Fundamentals*, https://doi.org/10.1007/978-3-031-81869-1_10

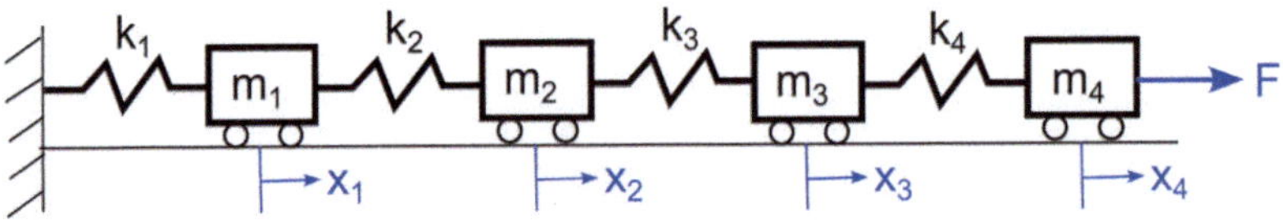

Fig. 10.1 Spring-mass system subjected to force F

$$\begin{Bmatrix} 0 \\ 0 \\ 0 \\ F \end{Bmatrix} + \begin{bmatrix} -(k_1+k_2) & k_2 & 0 & 0 \\ k_2 & -(k_2+k_3) & k_3 & 0 \\ 0 & k_3 & -(k_3+k_4) & k_4 \\ 0 & 0 & k_4 & -k_4 \end{bmatrix} \begin{Bmatrix} x_1 \\ x_2 \\ x_3 \\ x_4 \end{Bmatrix}$$

$$= \begin{bmatrix} m_1 & 0 & 0 & 0 \\ 0 & m_2 & 0 & 0 \\ 0 & 0 & m_3 & 0 \\ 0 & 0 & 0 & m_4 \end{bmatrix} \begin{Bmatrix} \ddot{x}_1 \\ \ddot{x}_2 \\ \ddot{x}_3 \\ \ddot{x}_4 \end{Bmatrix} \tag{10.2}$$

If external force in Eq. (10.2) is taken as zero ($F = 0$), free vibration equation of the spring-mass system is obtained. Free vibration equation can be solved using the solution function $x_i(t) = X_i \sin(\omega t + \phi)$ corresponding to mass displacements where X_i indicates displacement amplitude, ω indicates natural frequency, and ϕ indicates phase angle. After substituting the solution function in Eq. (10.2) and cancelling common terms on both sides, following eigenvalue problem can be obtained:

$$\underbrace{\begin{bmatrix} -(k_1+k_2) & k_2 & 0 & 0 \\ k_2 & -(k_2+k_3) & k_3 & 0 \\ 0 & k_3 & -(k_3+k_4) & k_4 \\ 0 & 0 & k_4 & -k_4 \end{bmatrix}}_{A} \underbrace{\begin{Bmatrix} X_1 \\ X_2 \\ X_3 \\ X_4 \end{Bmatrix}}_{X}$$

$$= \underbrace{\omega^2}_{\lambda} \underbrace{(-1) \begin{bmatrix} m_1 & 0 & 0 & 0 \\ 0 & m_2 & 0 & 0 \\ 0 & 0 & m_3 & 0 \\ 0 & 0 & 0 & m_4 \end{bmatrix}}_{B} \underbrace{\begin{Bmatrix} X_1 \\ X_2 \\ X_3 \\ X_4 \end{Bmatrix}}_{X} \tag{10.3}$$

Above eigenvalue problem can be briefly expressed as $AX = \lambda BX$ where $\lambda = \omega^2$ and X denote eigenvalues and eigenvectors, respectively. If $k_1 = k_2 = k_3 = k_4 = 1$ and $m_1 = m_2 = m_3 = m_4 = 1$ values are used, eigenvalues of the spring-mass system are obtained as $\lambda_1 = 3.5321$, $\lambda_2 = 2.3473$, $\lambda_3 = 1$, $\lambda_4 = 0.1206$, and eigenvectors (mode shapes) are obtained as in Fig. 10.2.

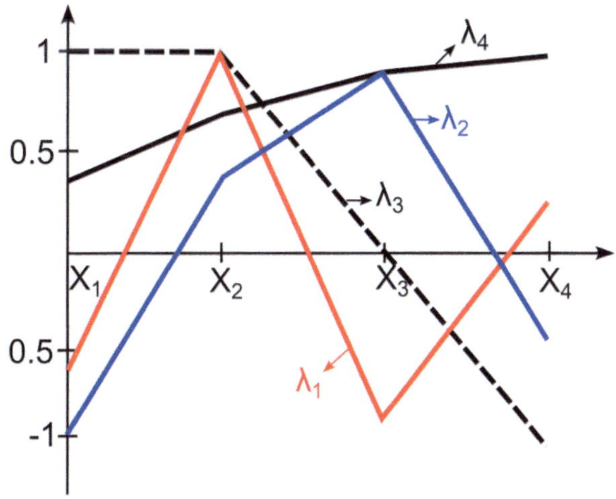

Fig. 10.2 Eigenvectors (free vibration mode shapes) of the spring-mass system

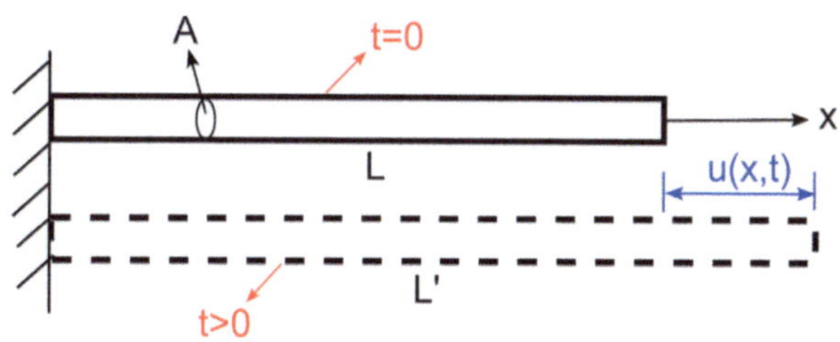

Fig. 10.3 Axial vibration of bar

Bar axial vibration:

Bar models are used to analyze slender structures in engineering. Consider the bar in Fig. 10.3 which has cross-sectional area A, length L, density ρ, and modulus of elasticity E. Axial vibration equation of the bar can be expressed as following (u: axial displacement, t: time, σ: stress, ε: strain):

$$EA\frac{\partial^2 u(x,t)}{\partial x^2} = \rho A\frac{\partial^2 u(x,t)}{\partial t^2} \rightarrow \frac{\partial^2 u(x,t)}{\partial x^2} = \frac{1}{c^2}\frac{\partial^2 u(x,t)}{\partial t^2} \quad \left(c = \sqrt{\frac{E}{\rho}}\right) \tag{10.4}$$

Substituting the solution function $u(x, t) = U(x) \sin(\omega t + \phi)$ in Eq. (10.4), following eigenvalue problem is obtained:

$$\frac{d^2U(x)}{dx^2} + \left(\frac{\omega}{c}\right)^2 U = 0 \rightarrow -\frac{d^2U(x)}{dx^2} = \alpha^2 U \quad \left(\alpha = \frac{\omega}{c}\right) \tag{10.5}$$

Eigenvalue problem expressed as a differential equation in Eq. (10.5) can be solved using numerical methods such as Finite Difference method (it will be explained in Chap. 4 of Vol. 2). For the solution with Finite Difference method, the bar can be discretized into nodes as in Fig. 10.4. If boundary conditions are expressed with

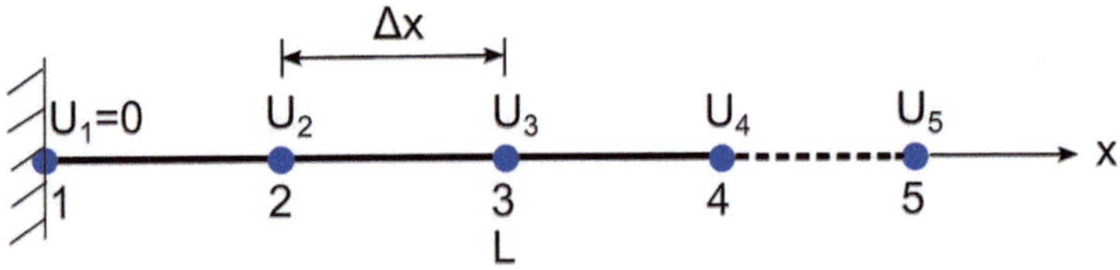

Fig. 10.4 Discretization of the bar for solution with Finite Difference method

derivatives, virtual nodes are needed outside the bar. Node 5 in Fig. 10.4 serves as the virtual node for the application of the stress boundary condition at the right end. First and second derivatives with Finite Difference method at *i*-th node can be expressed in terms of function values at adjacent nodes as $\frac{dU_i}{dx} = \frac{U_{i+1} - U_{i-1}}{2\Delta x}$ and $\frac{d^2U_i}{dx^2} = \frac{U_{i-1} - 2U_i + U_{i+1}}{(\Delta x)^2}$, respectively. In this case, boundary conditions for the bar in Fig. 10.3 can be written using Finite Difference method for the left end (first node) as $U_1 = 0$ and for the right end (fourth node) as $\sigma_4 = E\varepsilon_4 = E\frac{dU_4}{dx} = E\frac{U_5 - U_3}{2\Delta x} = 0 \Rightarrow U_5 = U_3$. On the other hand, differential equation at second, third, and fourth nodes can be expressed using Finite Difference method as following (*Note:* Solution of differential equations with Finite Difference method will be explained in detail in Chap. 4 of Vol. 2).

$$\text{For Node 2}: -\frac{\left(\underbrace{U_1}_{0} - 2U_2 + U_3\right)}{(\Delta x)^2} = \alpha^2 U_2, \quad \text{For Node 3}$$

$$: -\frac{(U_2 - 2U_3 + U_4)}{(\Delta x)^2} = \alpha^2 U_3, \quad \text{For node 4}$$

$$: -\frac{\left(U_3 - 2U_4 + \underbrace{U_5}_{U_3}\right)}{(\Delta x)^2} = \alpha^2 U_4$$

After substituting boundary conditions in the above equations, following eigenvalue problem for bar vibration is obtained.

$$\underbrace{-\frac{1}{(\Delta x)^2}\begin{bmatrix} -2 & 1 & 0 \\ 1 & -2 & 1 \\ 0 & 2 & -2 \end{bmatrix}}_{A} \underbrace{\begin{Bmatrix} U_2 \\ U_3 \\ U_4 \end{Bmatrix}}_{X} = \underbrace{(\alpha)^2}_{\lambda} \underbrace{\begin{Bmatrix} U_2 \\ U_3 \\ U_4 \end{Bmatrix}}_{X}$$

$$\rightarrow \underbrace{-\frac{1}{(\Delta x)^2}\begin{bmatrix} -2 & 1 & 0 \\ 1 & -2 & 1 \\ 0 & 2 & -2 \end{bmatrix}}_{A} \underbrace{\begin{Bmatrix} U_2 \\ U_3 \\ U_4 \end{Bmatrix}}_{X} = \lambda \underbrace{\begin{Bmatrix} U_2 \\ U_3 \\ U_4 \end{Bmatrix}}_{X}$$

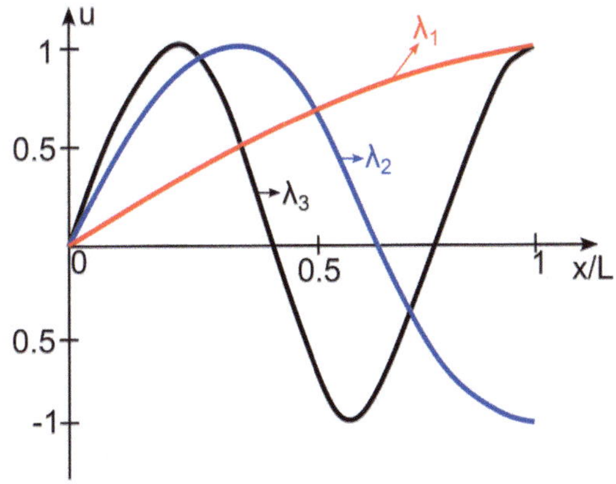

Fig. 10.5 Eigenvectors $X = X_1, X_2, X_3$ corresponding to eigenvalues $\lambda = \lambda_1, \lambda_2, \lambda_3$ ($\lambda_3 > \lambda_2 > \lambda_1$)

Aforementioned eigenvalue problem can be expressed shortly as $AX = \lambda X$ where λ and X denote eigenvalues (natural frequencies) and eigenvectors (vibration mode shapes), respectively. Solution of the eigenvalue problem yields eigenvectors shown in Fig. 10.5.

Column buckling:
Stability analyses of slender structures under compressive loading are investigated using column buckling. Columns or bars can buckle as in Fig. 10.6b under the axial compressive load if the load exceeds a critical value referred to as critical buckling load. Buckling of a column can be expressed with the following differential equation:

$$EI\frac{d^2v(x)}{dx^2} + Pv(x) = 0 \rightarrow -\frac{d^2v}{dx^2} = \frac{P}{EI}v \tag{10.6}$$

Eigenvalue problem expressed as a differential equation in Eq. (10.6) can be solved with numerical methods such as Finite Difference method. For solution with Finite Difference method, column can be discretized into nodes as in Fig. 10.6b. First and sixth nodes on the column serve as virtual nodes to apply derivative boundary conditions. Boundary conditions can be written at node 2 and node 5 with Finite Difference method as following (v: lateral displacement, M: moment):

$$\text{For node 2}: v_2 = 0, \quad M_2 = EI\frac{d^2v_2}{dx^2} = EI\frac{v_1 - 2\underbrace{v_2}_{0} + v_3}{(\Delta x)^2} = 0 \Rightarrow (v_1 = -v_3)$$

$$\text{For node 5}: v_5 = 0, \quad M_5 = EI\frac{d^2v_4}{dx^2} = EI\frac{v_4 - 2\underbrace{v_5}_{0} + v_6}{(\Delta x)^2} = 0 \Rightarrow (v_6 = -v_4)$$

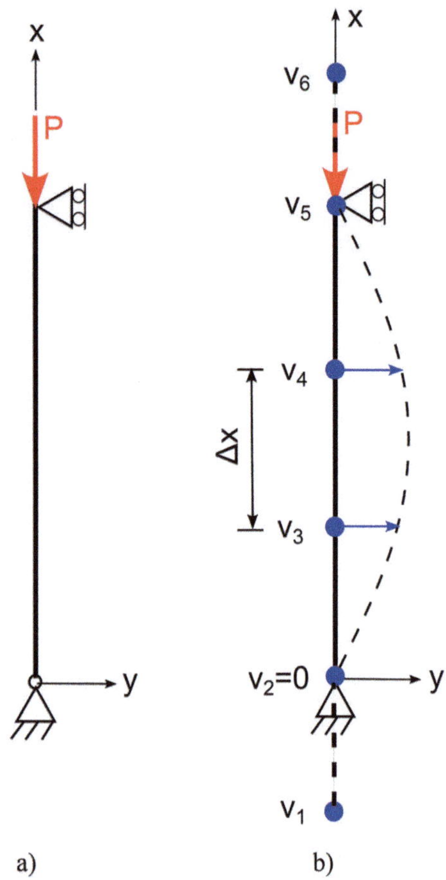

Fig. 10.6 (**a**) Buckling of a column with one end simply supported and other end subjected to the axial load, (**b**) Discretization of the column for numerical solution with Finite Difference method

On the other hand, differential equation can be expressed with Finite Difference method at third and fourth nodes as below:

$$\text{For node 3}: -\frac{\left(\underbrace{v_2}_{0} - 2v_3 + v_4\right)}{(\Delta x)^2} = \frac{P}{EI}v_3, \quad \text{For node 4}$$

$$: -\frac{\left(v_3 - 2v_4 + \underbrace{v_5}_{0}\right)}{(\Delta x)^2} = \frac{P}{EI}v_4$$

After substituting boundary conditions in above equations, following eigenvalue problem is obtained.

Fig. 10.7 Three smallest eigenvectors ($X = X_1, X_2, X_3$) and their corresponding eigenvalues ($\lambda = \lambda_1, \lambda_2, \lambda_3$) for the buckling of column with sufficient number of nodes

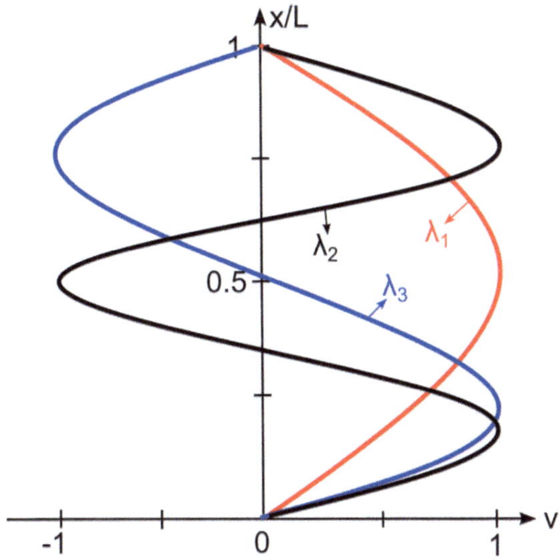

$$\underbrace{-\frac{1}{(\Delta x)^2}\begin{bmatrix} -2 & 1 \\ 1 & -2 \end{bmatrix}}_{A}\underbrace{\begin{Bmatrix} v_3 \\ v_4 \end{Bmatrix}}_{X} = \underbrace{\left(\frac{P}{EI}\right)}_{\lambda}\underbrace{\begin{Bmatrix} v_3 \\ v_4 \end{Bmatrix}}_{X} \rightarrow \underbrace{-\frac{1}{(\Delta x)^2}\begin{bmatrix} -2 & 1 \\ 1 & -2 \end{bmatrix}}_{A}\underbrace{\begin{Bmatrix} v_3 \\ v_4 \end{Bmatrix}}_{X} = \lambda\underbrace{\begin{Bmatrix} v_3 \\ v_4 \end{Bmatrix}}_{X}$$

Eigenvalue problem above can be expressed shortly as $AX = \lambda X$ where λ and X denote eigenvalues (buckling loads) and eigenvectors (buckled shapes), respectively. Smallest eigenvalue $\lambda_{\min}$ is called as critical buckling load. With a sufficient number of nodes, eigenvectors (three smallest ones) can be obtained as in Fig. 10.7.

In the solution of eigenvalue problems expressed with $AX = \lambda X$ equation, the number of eigenvalues and eigenvectors will be equal to the size of A matrix (row or column number). But in practice, not all eigenvalues and eigenvectors need to be calculated. Sometimes, solution of just minimum or maximum eigenvalue can be sufficient. For example, in column buckling problem smallest eigenvalue and the corresponding eigenvector is enough to see the onset of buckling that will put the column into dangerous position.

10.2 Expression of Eigenvalue Problems in Different Forms

As stated before, eigenvalue problems can be expressed with $AX = \lambda X$ equation. But sometimes expressing the eigenvalue problem in an alternative form can simplify the solution process. Therefore, for convenience, eigenvalue problems are expressed in one of the following forms.

1. Original eigenvalue problem: $AX = \lambda X$
2. Transformed eigenvalue problem: $\underbrace{(Q^T AQ)}_{B} Y = \lambda Y \rightarrow BY = \lambda Y$

Transformed eigenvalue problem:
Original eigenvalue equation $AX = \lambda X$ can be converted into a simpler new form as $BY = \lambda Y$ by applying orthogonal transformation such that eigenvalues do not change. For example, when X is expressed as the product of an orthogonal vector Q and new vector Y in the form of $X = QY$, then eigenvalue equation $AX = \lambda X$ can be written as below:

$$AX = \lambda X \rightarrow AQY = \lambda QY \tag{10.7}$$

When Eq. (10.7) is premultiplied with Q^{-1} (inverse), it can be converted into $BY = \lambda Y$ form as following (*Note*: inverse of an orthogonal matrix Q is equal to its transpose, i.e., $Q^{-1} = Q^T$).

$$\underbrace{Q^{-1}}_{Q^T}AQY = \lambda \underbrace{Q^{-1}Q}_{I}Y \rightarrow \underbrace{Q^T AQ}_{B}Y = \lambda Y \rightarrow BY = \lambda Y \tag{10.8}$$

Since the same λ term appears in $AX = \lambda X$ and $BY = \lambda Y$ expressions, their solution will lead to the same eigenvalues. However, transformed equation $BY = \lambda Y$ will be usually easier to solve than $AX = \lambda X$ in calculating λ eigenvalues. For example, if B matrix is converted into diagonal form as below:

$$B = \begin{bmatrix} b_{11} & 0 & 0 \\ 0 & b_{22} & 0 \\ 0 & 0 & b_{33} \end{bmatrix}$$

then, eigenvalues of $BY = \lambda Y$ will be simply equal to the diagonal terms of B matrix as $\lambda_1 = b_{11}, \lambda_2 = b_{22}, \lambda_3 = b_{33}$. These values will also be the eigenvalues of $AX = \lambda X$ equation. Due to advantage of orthogonal rotation (transformation), A matrix is usually rotated (converted) into diagonal, lower or upper triangular B matrix for the efficient calculation of eigenvalues. However, this transformation generally requires computational effort and is usually carried out iteratively. Transformation methods to convert A matrix into B matrix differ from each other in the way of constructing Q matrix.

10.3 Solution of Eigenvalue Problems

Solution of eigenvalue problems involves calculation of: (1) Eigenvalues λ and (2) Eigenvectors X. Depending on the size and property of eigenvalue problem, several methods have been developed to solve it efficiently and effectively [3–10]. Some methods calculate just a single eigenvalue while others calculate all eigenvalues. Also, some methods can calculate eigenvalues and eigenvectors simultaneously, while others calculate just eigenvalues. Solution of eigenvalue problems can be summarized as below:

1. *Calculation of eigenvalues*:

   Eigenvalue equation $AX = \lambda X$ can be expressed as $(A - \lambda I)X = 0$. For the solution, either $X = 0$ or $(A - \lambda I) = 0$ condition must be satisfied. $X = 0$ condition will not correspond to a meaningful solution (a vibrating system or a buckled column). Therefore, $(A - \lambda I) = 0$ condition is used for a non-trivial solution. $(A - \lambda I) = 0$ equation can be solved using one of the following methods to yield eigenvalues of $\lambda = \{\lambda_1, \lambda_2, \ldots, \lambda_n\}$:

   (a) Polynomial methods
   (b) Iterative methods

2. *Calculation of eigenvectors*:

   After eigenvalues $\lambda = \{\lambda_1, \lambda_2, \ldots, \lambda_n\}$ are obtained, each λ_i is substituted into $(A - \lambda_i I)X = 0$ equation to obtain its corresponding eigenvector ofX_i.

10.4 Calculation of Eigenvalues

10.4.1 Polynomial Methods

In polynomial methods, eigenvalue problem $(A - \lambda I) = 0$ (size of A matrix is assumed as $n \times n$) is converted into a root finding problem. For this, determinant of $(A - \lambda I)$ matrix is zeroed as $\det(A - \lambda I) = 0$. $\det(A - \lambda I) = 0$ yields characteristic equation of $c_1 + c_2\lambda + \ldots + c_{n+1}\lambda^n = 0$ which is a root finding problem. Roots of characteristic equation gives eigenvalues of $\lambda = \{\lambda_1, \lambda_2, \ldots, \lambda_n\}$. After eigenvalues $\lambda = \{\lambda_1, \lambda_2, \ldots, \lambda_n\}$ are obtained, each λ_i is substituted into the equation $(A - \lambda I)X = 0$ to calculate its corresponding eigenvector ofX_i.

Polynomial methods require explicit expression of the characteristic equation which becomes difficult for big-size eigenvalue problems. Therefore, these methods are mostly utilized for solving small-size eigenvalue problems. Also, since these methods calculate eigenvalues and eigenvectors successively, they are generally less efficient than other methods that calculate eigenvalues and eigenvectors simultaneously.

Example 10.1: Solve the following eigenvalue problem with polynomial method.

$$A = \begin{bmatrix} 4 & -2 & 0 \\ -2 & 4 & -2 \\ 0 & -2 & 4 \end{bmatrix}$$

Using A matrix above, eigenvalue problem can be expressed as below:

$$(A-\lambda I)X=0\rightarrow$$

$$\left\{\underbrace{\begin{bmatrix}4 & -2 & 0\\ -2 & 4 & -2\\ 0 & -2 & 4\end{bmatrix}}_{A}-\lambda\underbrace{\begin{bmatrix}1 & 0 & 0\\ 0 & 1 & 0\\ 0 & 0 & 1\end{bmatrix}}_{I}\right\}\underbrace{\begin{Bmatrix}x_1\\ x_2\\ x_3\end{Bmatrix}}_{X}$$

$$=\underbrace{\begin{Bmatrix}0\\ 0\\ 0\end{Bmatrix}}_{0}\rightarrow\underbrace{\begin{bmatrix}4-\lambda & -2 & 0\\ -2 & 4-\lambda & -2\\ 0 & -2 & 4-\lambda\end{bmatrix}}_{A-\lambda I}\underbrace{\begin{Bmatrix}x_1\\ x_2\\ x_3\end{Bmatrix}}_{X}=\underbrace{\begin{Bmatrix}0\\ 0\\ 0\end{Bmatrix}}_{0}$$

For $X \neq 0$ solution, condition $\det(A-\lambda I)=|A-\lambda I|=0$ must be satisfied. Satisfaction of this condition yields the following characteristics equation:

$$|A-\lambda I|=\begin{vmatrix}4-\lambda & -2 & 0\\ -2 & 4-\lambda & -2\\ 0 & -2 & 4-\lambda\end{vmatrix}=0\rightarrow-\lambda^3+12\lambda^2-40\lambda+32=0$$

Roots of characteristics equation $-\lambda^3+12\lambda^2-40\lambda+32=0$ are obtained as $\lambda_1=6.8284$, $\lambda_2=4.0$, $\lambda_3=1.1716$. In this case, eigenvectors corresponding to these eigenvalues can be obtained as below:

Eigenvector calculation for $\lambda_1=6.8284$:

$$\underbrace{\begin{bmatrix}4-\lambda_1 & -2 & 0\\ -2 & 4-\lambda_1 & -2\\ 0 & -2 & 4-\lambda_1\end{bmatrix}}_{A-\lambda I}\underbrace{\begin{Bmatrix}x_1\\ x_2\\ x_3\end{Bmatrix}}_{X}=\underbrace{\begin{Bmatrix}0\\ 0\\ 0\end{Bmatrix}}_{0}\rightarrow\begin{matrix}-2.8284x_1-2x_2=0\\ -2x_1-2.8284x_2-2x_3=0\\ -2.8284x_2-2x_3=0\end{matrix}$$

In the solution of above equations, one unknown is assigned a predetermined value and then remaining unknowns are solved. For example, by assigning x_1 two different values, eigenvector problem can be solved as following:

Assigning $x_1=1$:

Substituting $x_1=1$ in equations and using any two of the equations, x_2 and x_3 can be solved as below:

$$\begin{matrix}-2.8284x_1-2x_2=0\\ -2x_1-2.8284x_2-2x_3=0\end{matrix}\rightarrow\begin{matrix}-2.8284-2x_2=0\\ -2-2.8284x_2-2x_3=0\end{matrix}\rightarrow\begin{matrix}x_2=-1.4142\\ x_3=1\end{matrix}$$

In this case, eigenvector is obtained as $X=\begin{Bmatrix}x_1\\ x_2\\ x_3\end{Bmatrix}=\begin{Bmatrix}1\\ -1.4142\\ 1\end{Bmatrix}$.

Assigning $x_1 = 2$:

Substituting $x_1 = 1$ in equations and using any two of the equations, x_2 and x_3 can be solved as below:

$$\begin{matrix} -2.8284x_1 - 2x_2 = 0 \\ -2x_1 - 2.8284x_2 - 2x_3 = 0 \end{matrix} \rightarrow \begin{matrix} -2.8284 - 2x_2 = 0 \\ -2 - 2.8284x_2 - 2x_3 = 0 \end{matrix} \rightarrow \begin{matrix} x_2 = -2.8284 \\ x_3 = 1.9999 \end{matrix}$$

In this case, eigenvector is obtained as $X = \begin{Bmatrix} x_1 \\ x_2 \\ x_3 \end{Bmatrix} = \begin{Bmatrix} 2 \\ -2.8482 \\ 2 \end{Bmatrix}$

If X vector above is divided by 2 as $X = \frac{1}{2}\begin{Bmatrix} 2 \\ -2.8482 \\ 2 \end{Bmatrix} = \begin{Bmatrix} 1 \\ -1.4142 \\ 1 \end{Bmatrix}$, the same eigenvector in the case of Assigning $x_1 = 1$ is obtained. This shows that assigned value is not important in eigenvector calculation and dividing/multiplying an eigenvector (often referred to as normalization) does not change its character. Therefore, eigenvectors are normalized in several ways to keep their size or norm limited. For example, eigenvector $X = \begin{Bmatrix} 1 \\ -1.4142 \\ 1 \end{Bmatrix}$ can be normalized by dividing its elements with -1.4142 as following:

$$X = \frac{1}{-1.4142}\begin{Bmatrix} 1 \\ -1.4142 \\ 1 \end{Bmatrix} = \underbrace{\begin{Bmatrix} -0.7070 \\ 1 \\ -0.7070 \end{Bmatrix}}_{\substack{Normalized \\ Form}}$$

Alternatively, eigenvector X can be normalized by dividing with its norm as following:

$$X = \underbrace{\frac{1}{\sqrt{1^2 + (-1.4142)^2 + 1^1}}}_{Norm(X)}\begin{Bmatrix} 1 \\ -1.4142 \\ 1 \end{Bmatrix} = \begin{Bmatrix} 0.5 \\ -0.7071 \\ 0.5 \end{Bmatrix}$$

Eigenvector calculation for $\lambda_2 = 4$:

$$\underbrace{\begin{bmatrix} 4-\lambda_2 & -2 & 0 \\ -2 & 4-\lambda_2 & -2 \\ 0 & -2 & 4-\lambda_2 \end{bmatrix}}_{A-\lambda I}\underbrace{\begin{Bmatrix} x_1 \\ x_2 \\ x_3 \end{Bmatrix}}_{X} = \underbrace{\begin{Bmatrix} 0 \\ 0 \\ 0 \end{Bmatrix}}_{0} \rightarrow \begin{aligned} -2x_2 &= 0 \\ -2x_1 - 2x_3 &= 0 \\ -2x_2 &= 0 \end{aligned}$$

Substituting $x_1 = 1$ in equations and using any two of the equations, x_2 and x_3 can be solved as below:

$$\begin{aligned} 2x_2 &= 0 \\ -2x_1 - 2x_3 &= 0 \end{aligned} \rightarrow \begin{aligned} x_2 &= 0 \\ x_3 &= -1 \end{aligned}$$

In this case, eigenvector is obtained as $X = \begin{Bmatrix} x_1 \\ x_2 \\ x_3 \end{Bmatrix} = \begin{Bmatrix} 1 \\ 0 \\ -1 \end{Bmatrix}$.

Eigenvector calculation for $\lambda_3 = 1.1716$:

$$\underbrace{\begin{bmatrix} 4-\lambda_3 & -2 & 0 \\ -2 & 4-\lambda_3 & -2 \\ 0 & -2 & 4-\lambda_3 \end{bmatrix}}_{A-\lambda I}\underbrace{\begin{Bmatrix} x_1 \\ x_2 \\ x_3 \end{Bmatrix}}_{X} = \underbrace{\begin{Bmatrix} 0 \\ 0 \\ 0 \end{Bmatrix}}_{0} \rightarrow \begin{aligned} 2.8284x_1 - 2x_2 &= 0 \\ -2x_1 + 2.8284x_2 - 2x_3 &= 0 \\ -2x_2 + 2.8284x_3 &= 0 \end{aligned}$$

Substituting $x_1 = 1$ in equations and using any two of the equations, x_2 and x_3 can be solved as below:

$$\begin{aligned} 2.8284x_1 - 2x_2 &= 0 \\ -2x_1 + 2.8284x_2 - 2x_3 &= 0 \end{aligned} \rightarrow \begin{aligned} x_2 &= 1.4142 \\ x_3 &= 1 \end{aligned}$$

In this case, eigenvector is obtained as $X = \begin{Bmatrix} x_1 \\ x_2 \\ x_3 \end{Bmatrix} = \begin{Bmatrix} 1 \\ 1.4142 \\ 1 \end{Bmatrix} = \underbrace{\begin{Bmatrix} 0.7070 \\ 1 \\ 0.7070 \end{Bmatrix}}_{\substack{\textit{Normalized} \\ \textit{Form}}}$

Solution with MATLAB command:

Eigenvalue problem in this example can be solved using *eig* command in MATLAB as following:

```
>> A = [4 -2 0 ; -2 4 -2; 0 -2 4];
>> [Xmat,Lam] = eig (A)
Xmat =
0.5000 -0.7071 -0.5000
```

```
0.7071 0.0000 0.7071
0.5000 0.7071 -0.5000
Lam =
1.1716 0 0
0 4.0000 0
0 0 6.8284
```

Lam matrix give eigenvalues and columns of *Xmat* matrix gives eigenvectors. Note that eigenvectors in MATLAB solution look different from ours since MATLAB applies different normalization technique. However, MATLAB eigenvectors can be easily converted into our solution by applying scaling. For example, eigenvector $X = \begin{Bmatrix} 0.5 \\ 0.7071 \\ 0.5 \end{Bmatrix}$ corresponding to the eigenvalue of $\lambda_1 = 1.1716$ can be normalized by multiplying it with 1/0.7071 to be the same of our solution as $X = \begin{Bmatrix} 0.7070 \\ 1 \\ 0.7070 \end{Bmatrix}$.

10.4.2 Iterative Methods

Iterative methods solve eigenvalue problems in iterative way. Some of them give only eigenvalues at the end of the solution, while some others give both eigenvalues and eigenvectors simultaneously. Iterative methods are generally preferred for solving big size eigenvalue problems. There are various iterative methods in the literature. Some of them are general and they can be used to solve any eigenvalue problem. Some are more efficient and effective in solving special eigenvalue problems where *A* matrix is symmetric or banded. Some methods were developed to find the largest or smallest eigenvalue and its corresponding eigenvector instead of finding all eigenvalues. Such methods can also be used with some modifications to find smaller or larger eigenvalues and their corresponding eigenvectors. Common iterative methods to solve eigenvalue problems are explained in detail below.

(a) **Power Method**

Power method was developed to find the biggest eigenvalue and its corresponding eigenvector in eigenvalue problems expressed with $AX = \lambda X$ equation. However, if *A* matrix is replaced with its inverse A^{-1}, then power method can also be used to find the smallest eigenvalue and its corresponding eigenvector. Using Power method, smaller eigenvalues and their corresponding eigenvectors can also be obtained by subtracting the eigenvalues from *A* matrix. Power method may not always lead to the biggest eigenvalue. It may converge to the second biggest eigenvalue instead of the biggest one as well.

It can be proved as following that power method gives the biggest eigenvalue. In the solution of eigenvalue problem $AX = \lambda X$, eigenvalues can be indicated with λ_1, λ_2,... λ_n (assuming $\lambda_1 > \lambda_2 \geq \ldots \geq \lambda_n$) and eigenvectors with V_1, V_2,... V_n. Any vector V can be represented with combination of these eigenvectors as following:

$$V = c_1 V_1 + c_2 V_2 + \ldots + c_n V_n \tag{10.9}$$

Replacing X with V_i in eigenvalue equation of $AX = \lambda X$, $AV_i = \lambda V_i$ equation is obtained. Using this equation, product of V vector with A matrix can be expressed as below:

$$AV = c_1 \underbrace{AV_1}_{\lambda_1 V_1} + c_2 \underbrace{AV_2}_{\lambda_2 V_2} + \ldots + c_n \underbrace{AV_n}_{\lambda_n V_n} = c_1 \lambda_1 V_1 + c_2 \lambda_2 V_2 + \ldots + c_n \lambda_n V_n$$

Similarly, product of V vector with A^2, A^3,..., A^k matrices can be obtained as following:

$$A^2 V = c_1 \lambda_1 \underbrace{AV_1}_{\lambda_1 V_1} + c_2 \lambda_2 \underbrace{AV_2}_{\lambda_2 V_2} + \ldots + c_n \lambda_n \underbrace{AV_n}_{\lambda_n V_n} = c_1 \lambda_1^2 V_1 + c_2 \lambda_2^2 V_2 + \ldots$$
$$+ c_n \lambda_2^2 V_n \text{ (second iteration)}$$
$$A^3 V = c_1 \lambda_1^2 \underbrace{AV_1}_{\lambda_1 V_1} + c_2 \lambda_2^2 \underbrace{AV_2}_{\lambda_2 V_2} + \ldots + c_n \lambda_n^2 \underbrace{AV_n}_{\lambda_n V_n} = c_1 \lambda_1^3 V_1 + c_2 \lambda_2^3 V_2 + \ldots$$
$$+ c_n \lambda_2^3 V_n \text{ (third iteration)}$$

$$\ldots$$

$$A^k V = c_1 \lambda_1^k \underbrace{AV_1}_{\lambda_1 V_1} + c_2 \lambda_2^k \underbrace{AV_2}_{\lambda_2 V_2} + \ldots + c_n \lambda_n^k \underbrace{AV_n}_{\lambda_n V_n} = c_1 \lambda_1^k V_1 + c_2 \lambda_2^k V_2 + \ldots$$
$$+ c_n \lambda_{2n}^k \text{ } (k - \text{th iteration})$$

$A^k V$ equation above can be rearranged as below:

$$\begin{aligned} A^k V &= c_1 \lambda_1^k V_1 + c_2 \lambda_2^k V_2 + \ldots + c_n \lambda_2^k V_n \\ &= \lambda_1^k \left[c_1 V_1 + c_2 \left(\frac{\lambda_2}{\lambda_1} \right)^k V_2 + \ldots + c_n \left(\frac{\lambda_n}{\lambda_1} \right)^k V_n \right] \end{aligned} \tag{10.10}$$

When biggest eigenvalue (λ_1) dominates, Eq. (10.10) can be written in the limit case (for $k \to \infty$) as below:

$$\lim_{k\to\infty} A^k V = \lambda_1^k \left[c_1 V_1 + c_2 \underbrace{\left(\frac{\lambda_2}{\lambda_1}\right)^k}_{\approx 0} V_2 + \ldots + c_n \underbrace{\left(\frac{\lambda_n}{\lambda_1}\right)^k}_{\approx 0} V_n \right] = c_1 \lambda_1^k V_1 \quad (10.11)$$

From Eq. (10.11), it is seen that as iterations are continued, biggest eigenvalue λ_1 and its corresponding eigenvector are converged.

In the solution of eigenvalue problems with Power method, $AX = \lambda X$ equation is solved iteratively. To start iterations, an initial value $X^{(0)}$ is selected and substituted into $AX^{(0)}$ expression. Then, $AX^{(0)}$ product is decomposed into the updated values of $\lambda^{(1)}$ and $X^{(1)}$ as $AX^{(0)} = \lambda^{(1)} X^{(1)}$ (Note: $\lambda^{(1)}$ can be taken as the biggest value in $AX^{(0)}$ product). New iterations are repeated in the same way. Solution iterations with Power method can be expressed symbolically as below:

$$\text{Calculate} : AX^{(0)} \to \text{Decompose}$$

$$: AX^{(0)} = \lambda^{(1)} X^{(1)}, \left(\lambda^{(1)} : \text{absolute biggest value in } AX^{(0)}\right)$$

$$\text{Calculate} : AX^{(1)} \to \text{Decompose}$$

$$: AX^{(1)} = \lambda^{(2)} X^{(2)}, \left(\lambda^{(2)} : \text{absolute biggest value in } AX^{(1)}\right)$$

$$\ldots$$

$$\text{Calculate} : AX^{(k-1)} \to \text{Decompose}$$

$$: AX^{(k-1)} = \lambda^{(k)} X^{(k)}, \left(\lambda^{(k)} : \text{absolute biggest value in } AX^{(k-1)}\right)$$

Iterations are stopped when the convergence criterion $\left|\frac{\lambda^{(k)} - \lambda^{(k-1)}}{\lambda^{(k-1)}}\right| \leq \varepsilon$ is satisfied. ε indicates convergence tolerance corresponding to a small value such as $\varepsilon = 0.001$.

Advantages and disadvantages of power method can be stated as below:

Advantages:

1. Method is simple and easy to implement.
2. Method yields both eigenvalue and eigenvector simultaneously.

Disadvantages:

1. Method yields only the biggest eigenvalue and its corresponding eigenvector.
2. Depending on the chosen initial vector, it may converge to a smaller eigenvalue. For example, if the chosen initial vector corresponds to any eigenvalue other than the biggest one, it will immediately converge to that eigenvalue.

Example 10.2: Symmetric A matrix is given as $A = \begin{bmatrix} 4 & -2 & 0 \\ -2 & 4 & -2 \\ 0 & -2 & 4 \end{bmatrix}$. Find the biggest eigenvalue and it corresponding eigenvector of this matrix using Power method.

Biggest eigenvalue and its corresponding eigenvector can be calculated iteratively with Power method by applying product ($AX^{(k-1)}$) and decomposition steps ($AX^{(k-1)} = \lambda^{(k)}X^{(k)}$). When initial value for the eigenvector is selected as $X^{(0)} = \begin{Bmatrix} 1 \\ 1 \\ 1 \end{Bmatrix}$, iteration results can be presented as following:

Iteration 1:

$$\text{Calculate}: AX^{(0)} \rightarrow \underbrace{\begin{bmatrix} 4 & -2 & 0 \\ -2 & 4 & -2 \\ 0 & -2 & 4 \end{bmatrix}}_{A} \underbrace{\begin{Bmatrix} 1 \\ 1 \\ 1 \end{Bmatrix}}_{X^{(0)}} = \begin{Bmatrix} 2 \\ 0 \\ 2 \end{Bmatrix},$$

$$\text{Decompose}: AX^{(0)} = \lambda^{(1)}X^{(1)} \rightarrow \begin{Bmatrix} 2 \\ 0 \\ 2 \end{Bmatrix} = 2\begin{Bmatrix} 1 \\ 0 \\ 1 \end{Bmatrix} \rightarrow \lambda^{(1)} = 2, X^{(1)} = \begin{Bmatrix} 1 \\ 0 \\ 1 \end{Bmatrix}$$

Iteration 2:

$$\text{Calculate}: AX^{(1)} \rightarrow \underbrace{\begin{bmatrix} 4 & -2 & 0 \\ -2 & 4 & -2 \\ 0 & -2 & 4 \end{bmatrix}}_{A} \underbrace{\begin{Bmatrix} 1 \\ 0 \\ 1 \end{Bmatrix}}_{X^{(1)}} = \begin{Bmatrix} 4 \\ -4 \\ 4 \end{Bmatrix},$$

$$\text{Decompose}: AX^{(1)} = \lambda^{(2)}X^{(2)} \rightarrow \begin{Bmatrix} 4 \\ -4 \\ 4 \end{Bmatrix} = 4\begin{Bmatrix} 1 \\ -1 \\ 1 \end{Bmatrix} \rightarrow \lambda^{(2)} = 4\ X^{(2)} = \begin{Bmatrix} 1 \\ -1 \\ 1 \end{Bmatrix}$$

Iteration 3:

$$\text{Calculate}: AX^{(2)} \rightarrow \underbrace{\begin{bmatrix} 4 & -2 & 0 \\ -2 & 4 & -2 \\ 0 & -2 & 4 \end{bmatrix}}_{A} \underbrace{\begin{Bmatrix} 1 \\ -1 \\ 1 \end{Bmatrix}}_{X^{(2)}} = \begin{Bmatrix} 6 \\ -8 \\ 6 \end{Bmatrix},$$

$$\text{Decompose}: AX^{(2)} = \lambda^{(3)} X^{(3)} \rightarrow \begin{Bmatrix} 6 \\ -8 \\ 6 \end{Bmatrix} = 8 \begin{Bmatrix} 0.75 \\ -1 \\ 0.75 \end{Bmatrix} \rightarrow \lambda^{(3)} = 8, X^{(3)} = \begin{Bmatrix} 0.75 \\ -1 \\ 0.75 \end{Bmatrix}$$

Iteration 4:

$$\text{Calculate}: AX^{(3)} \rightarrow \underbrace{\begin{bmatrix} 4 & -2 & 0 \\ -2 & 4 & -2 \\ 0 & -2 & 4 \end{bmatrix}}_{A} \underbrace{\begin{Bmatrix} 0.75 \\ -1 \\ 0.75 \end{Bmatrix}}_{X^{(3)}} = \begin{Bmatrix} 5 \\ -7 \\ 5 \end{Bmatrix}$$

$$\text{Decompose}: AX^{(3)} = \lambda^{(4)} X^{(4)} \rightarrow \begin{Bmatrix} 5 \\ -7 \\ 5 \end{Bmatrix} = 7 \begin{Bmatrix} 0.7143 \\ -1 \\ 0.7143 \end{Bmatrix} \rightarrow \lambda^{(4)} = 7, X^{(4)}$$

$$= \begin{Bmatrix} 0.7143 \\ -1 \\ 0.7143 \end{Bmatrix}$$

Iteration 5:

$$\text{Calculate}: AX^{(4)} \rightarrow \underbrace{\begin{bmatrix} 4 & -2 & 0 \\ -2 & 4 & -2 \\ 0 & -2 & 4 \end{bmatrix}}_{A} \underbrace{\begin{Bmatrix} 0.7143 \\ -1 \\ 0.7143 \end{Bmatrix}}_{X^{(4)}} = \begin{Bmatrix} 4.8571 \\ -6.8571 \\ 4.8571 \end{Bmatrix},$$

$$\text{Decompose}: AX^{(4)} = \lambda^{(5)} X^{(5)} \rightarrow \begin{Bmatrix} 4.8571 \\ -6.8571 \\ 4.8571 \end{Bmatrix} = 6.8571 \begin{Bmatrix} 0.7083 \\ -1 \\ 0.7083 \end{Bmatrix} \rightarrow$$

$$\lambda^{(5)} = 6.8571, X^{(5)} = \begin{Bmatrix} 0.7083 \\ -1 \\ 0.7083 \end{Bmatrix}$$

Iteration 6:

$$\text{Calculate}: AX^{(5)} \rightarrow \underbrace{\begin{bmatrix} 4 & -2 & 0 \\ -2 & 4 & -2 \\ 0 & -2 & 4 \end{bmatrix}}_{A} \underbrace{\begin{Bmatrix} 0.7083 \\ -1 \\ 0.7083 \end{Bmatrix}}_{X^{(5)}} = \begin{Bmatrix} 4.8333 \\ -6.8333 \\ 4.8333 \end{Bmatrix},$$

$$\text{Decompose}: AX^{(5)} = \lambda^{(6)}X^{(6)} \rightarrow \begin{Bmatrix} 4.8333 \\ -6.8333 \\ 4.8333 \end{Bmatrix} = 6.8333 \begin{Bmatrix} 0.7073 \\ -1 \\ 0.7073 \end{Bmatrix} \rightarrow$$

$$\lambda^{(6)} = 6.8333, X^{(6)} = \begin{Bmatrix} 0.7073 \\ -1 \\ 0.7073 \end{Bmatrix}$$

After six iterations, convergence condition $\varepsilon \leq 0.005$ is satisfied for the biggest eigenvalue. In this case, the biggest eigenvalue and its corresponding eigenvector is obtained as below:

$$\lambda_{\max} = 6.8333, X_{\max} = \begin{Bmatrix} -0.7073 \\ 1 \\ -0.7073 \end{Bmatrix}$$

Note: Eigenvalues and eigenvectors can also be calculated using *MATLAB Program 10.1: Eigenvalue Calculation with Power Method* in appendix.

(b) **Jacobi Method**

Jacobi method is one of the oldest methods in iterative solution of eigenvalue problems. It was developed to find all eigenvalues and eigenvectors simultaneously where A matrix is symmetric. In Jacobi method, instead of the original eigenvalue problem of $AX = \lambda X$, its transformed form which is expressed as $BY = \lambda Y$ is solved. B matrix is obtained with iterative rotation of A matrix until it transforms to a diagonal matrix. Using rotation matrix Q, this transformation is expressed as $B = Q^{-1}AQ$. Transformed eigenvalue problem is assumed to have the same eigenvalues with the original problem. In the diagonalization of A matrix, at each iteration an off-diagonal term of A matrix is set to zero using a new rotation matrix Q. Zeroing off-diagonal terms can be done column by column or selecting the biggest off-diagonal term.

To explain Jacobi method, we can express A matrix symbolically as below:

$$A = \begin{bmatrix} a_{11} & a_{12} & \cdots & \overbrace{a_{1p}}^{Column:p} & \cdots & \overbrace{a_{1q}}^{Column:q} & \cdots & a_{1n} \\ a_{21} & a_{22} & \cdots & a_{2p} & \cdots & a_{2q} & \cdots & a_{2n} \\ \cdots & \cdots & \cdots & \cdots & \cdots & \cdots & \cdots & \cdots \\ a_{p1} & a_{p2} & \cdots & a_{pp} & \cdots & a_{pq} & \cdots & a_{pn} \\ \cdots & \cdots & \cdots & \cdots & \cdots & \cdots & \cdots & \cdots \\ a_{q1} & a_{q2} & \cdots & a_{qp} & \cdots & a_{qq} & \cdots & a_{qn} \\ \cdots & \cdots & \cdots & \cdots & \cdots & \cdots & \cdots & \cdots \\ a_{n1} & a_{n2} & \cdots & a_{np} & \cdots & a_{nq} & \cdots & a_{nn} \end{bmatrix} \begin{matrix} \\ \\ \\ Row:p \\ \\ Row:q \\ \\ \\ \end{matrix}$$

Since A matrix is assumed symmetric, $A_{ij} = A_{ji}$ relation can be used for its terms. For zeroing an $A(p, q)$ term in A matrix, rotation submatrix Q_R can be generated in terms of rotation angle θ as following:

$$Q_R = \begin{bmatrix} Q_{pp} & Q_{pq} \\ Q_{qp} & Q_{qq} \end{bmatrix} = [\cos\,\theta - \ \sin\theta \sin\theta \cos\theta] \tag{10.12}$$

Substituting the rotation submatrix Q_R into $(p,\ q)$ columns-rows of a unit matrix I that has the same size with A matrix, rotation matrix Q can be obtained as below:

$$I = \begin{bmatrix} 1 & 0 & \ldots & 0 & \ldots & 0 & \ldots & 0 \\ 0 & 1 & \ldots & 0 & \ldots & 0 & \ldots & 0 \\ \ldots & \ldots & \ldots & \ldots & \ldots & \ldots & \ldots & \ldots \\ 0 & 0 & \ldots & 1 & \ldots & 0 & \ldots & 0 \\ \ldots & \ldots & \ldots & \ldots & \ldots & \ldots & \ldots & \ldots \\ 0 & 0 & \ldots & 0 & \ldots & 1 & \ldots & 0 \\ \ldots & \ldots & \ldots & \ldots & \ldots & \ldots & \ldots & \ldots \\ 0 & 0 & \ldots & 0 & \ldots & 0 & \ldots & 1 \end{bmatrix} \rightarrow$$

$$Q = \left[\begin{array}{cccccccc|l} 1 & 0 & \ldots & \overbrace{0}^{Column:p} & \ldots & \overbrace{0}^{Column:q} & \ldots & 0 & \\ 0 & 1 & \ldots & 0 & \ldots & 0 & \ldots & 0 & \\ \ldots & \ldots & \ldots & \ldots & \ldots & \ldots & \ldots & \ldots & \\ 0 & 0 & \ldots & \cos\theta & \ldots & -\sin\theta & \ldots & 0 & Row : p \\ \ldots & \ldots & \ldots & \ldots & \ldots & \ldots & \ldots & \ldots & \\ 0 & 0 & \ldots & \sin\theta & \ldots & \cos\theta & \ldots & 0 & Row : q \\ \ldots & \ldots & \ldots & \ldots & \ldots & \ldots & \ldots & \ldots & \\ 0 & 0 & \ldots & 0 & \ldots & 0 & \ldots & 1 & \end{array}\right] \tag{10.13}$$

In this case, rotated form of A matrix can be expressed as $B = Q^{-1}AQ$ as following:

$$
\begin{aligned}
B &= Q^{-1}AQ \\
&= \begin{bmatrix}
1 & 0 & \dots & \overbrace{0}^{Column:p} & \dots & \overbrace{0}^{Column:q} & \dots & 0 \\
0 & 1 & \dots & 0 & \dots & 0 & \dots & 0 \\
\dots & \dots & \dots & \dots & \dots & \dots & \dots & \dots \\
0 & 0 & \dots & \cos\theta & \dots & -\sin\theta & \dots & 0 \\
\dots & \dots & \dots & \dots & \dots & \dots & \dots & \dots \\
0 & 0 & \dots & \sin\theta & \dots & \cos\theta & \dots & 0 \\
\dots & \dots & \dots & \dots & \dots & \dots & \dots & \dots \\
0 & 0 & \dots & 0 & \dots & 0 & \dots & 1
\end{bmatrix}^{-1} \\
&\otimes \begin{bmatrix}
a_{11} & a_{12} & \dots & \overbrace{a_{1p}}^{Column:p} & \dots & \overbrace{a_{1q}}^{Column:q} & \dots & a_{1n} \\
a_{21} & a_{22} & \dots & a_{2p} & \dots & a_{2q} & \dots & a_{2n} \\
\dots & \dots & \dots & \dots & \dots & \dots & \dots & \dots \\
a_{p1} & a_{p2} & \dots & a_{pp} & \dots & a_{pq} & \dots & a_{pn} \\
\dots & \dots & \dots & \dots & \dots & \dots & \dots & \dots \\
a_{q1} & a_{q2} & \dots & a_{qp} & \dots & a_{qq} & \dots & a_{qn} \\
\dots & \dots & \dots & \dots & \dots & \dots & \dots & \dots \\
a_{n1} & a_{n2} & \dots & a_{np} & \dots & a_{nq} & \dots & a_{nn}
\end{bmatrix} \\
&\otimes \begin{bmatrix}
1 & 0 & \dots & \overbrace{0}^{Column:p} & \dots & \overbrace{0}^{Column:q} & \dots & 0 \\
0 & 1 & \dots & 0 & \dots & 0 & \dots & 0 \\
\dots & \dots & \dots & \dots & \dots & \dots & \dots & \dots \\
0 & 0 & \dots & \cos\theta & \dots & -\sin\theta & \dots & 0 \\
\dots & \dots & \dots & \dots & \dots & \dots & \dots & \dots \\
0 & 0 & \dots & \sin\theta & \dots & \cos\theta & \dots & 0 \\
\dots & \dots & \dots & \dots & \dots & \dots & \dots & \dots \\
0 & 0 & \dots & 0 & \dots & 0 & \dots & 1
\end{bmatrix} \\
&= \underbrace{\begin{bmatrix}
a_{11} & a_{12} & \dots & \overbrace{a_{1p}}^{Column:p} & \dots & \overbrace{a_{1q}}^{Column:q} & \dots & a_{1n} \\
a_{21} & a_{22} & \dots & a_{2p} & \dots & a_{2q} & \dots & a_{2n} \\
\dots & \dots & \dots & \dots & \dots & \dots & \dots & \dots \\
a_{p1} & a_{p2} & \dots & b_{pp} & \dots & b_{pq} & \dots & a_{pn} \\
\dots & \dots & \dots & \dots & \dots & \dots & \dots & \dots \\
a_{q1} & a_{q2} & \dots & b_{qp} & \dots & b_{qq} & \dots & a_{qn} \\
\dots & \dots & \dots & \dots & \dots & \dots & \dots & \dots \\
a_{n1} & a_{n2} & \dots & a_{np} & \dots & a_{nq} & \dots & a_{nn}
\end{bmatrix}}_{B} \begin{matrix} Row : p \\ Row : q \end{matrix}
\end{aligned}
\tag{10.14}
$$

All terms in B matrix will be the same as A matrix except for b_{pp}, b_{pq}, b_{qp}, b_{qq} terms. These terms can be expressed in terms of rotation angle θ as following:

$$\begin{aligned} b_{pp} &= a_{pp}\cos^2\theta + a_{qq}\sin^2\theta - 2a_{pq}\sin\theta\cos\theta \\ b_{qq} &= a_{pp}\sin^2\theta + a_{qq}\cos^2\theta + 2a_{pq}\sin\theta\cos\theta \\ b_{pq} &= b_{qp} = \left(a_{qq} - a_{pp}\right)\sin\theta\cos\theta + a_{pq}\left(\cos^2\theta - \sin^2\theta\right) \end{aligned} \tag{10.15}$$

b_{pp} and b_{qq} terms in B matrix correspond to diagonal terms whereas b_{pq} and b_{qp} correspond to off-diagonal terms. Rotation angle θ can be determined to make any off-diagonal term in B matrix zero. For example, angle θ to make b_{pq} term zero ($b_{pq} = 0$) can be obtained as below:

$$0 = \left(a_{qq} - a_{pp}\right)\sin\theta\cos\theta + a_{pq}\left(\cos^2\theta - \sin^2\theta\right) \rightarrow \tan 2\theta = \frac{2a_{pq}}{\left(a_{pp} - a_{qq}\right)} \tag{10.16}$$

Zeroing an off-diagonal term in Jacobi method is referred to as an iteration. By changing p, q indices in iterations, all off-diagonal terms can be zeroed. In Jacobi method, at each iteration *maximum* off-diagonal term is zeroed. That is in a iteration, (p, q) corresponds to row and columns of maximum off-diagonal coefficient. Rotation of A matrix into B matrix where $B(p, q) = 0$ can be expressed in iterative form as $B = Q^{-1}AQ \rightarrow B_{i+1} = Q_i^{-1}B_iQ_i$. In this case, zeroing of all off-diagonal terms after k iteration can be expressed in terms of previous iterations as following:

$$B_k = \underbrace{Q_{k-1}^{-1}\ldots Q_2^{-1}Q_1^{-1}Q_0^{-1}}_{Q^{-1}=Q^T}A\underbrace{Q_0Q_1Q_2\ldots Q_{k-1}}_{Q} = Q^TAQ \tag{10.17}$$

Q matrix expressed as multiplication of previous Q_i matrices ($Q = Q_1Q_2Q_3\ldots Q_k$) in Eq. (10.17) give eigenvectors. Each column in Q matrix gives an eigenvector corresponding to the associated eigenvalue. For example, first column represents the eigenvector for the first eigenvalue, second column represents the eigenvector for second eigenvalue, etc.

Although Jacobi method can calculate all eigenvalues and eigenvectors simultaneously, its computational efficiency is generally lower than other methods. Advantages and disadvantages of Jacobi method can be stated as following:

Advantage:

1. Method is easy to implement.
2. It returns all eigenvalues and eigenvectors simultaneously.

Disadvantage:

1. Convergence rate is usually low since only one off-diagonal term is set to zero at an iteration.
2. Previously zeroed coefficients can again become non-zero in a new iteration. This can cause extra iterations in zeroing process reducing the efficiency of the method.

Example 10.3: Symmetric A matrix is given as $A=\begin{bmatrix} 4 & -2 & 0 \\ -2 & 4 & -2 \\ 0 & -2 & 4 \end{bmatrix}$. Find eigenvalues and eigenvectors of this matrix using Jacobi method.

With Jacobi method, iterative solution of the eigenvalue problem can be obtained using the iterative schema of $B_{i+1}=Q_i^{-1}B_iQ_i$. Jacobi iterations can be conducted by taking the initial value of B_0 as $B_0=A$ as following:

Iteration 1:

$$B_0=A=\begin{bmatrix} 4 & -2 & 0 \\ -2 & 4 & -2 \\ 0 & -2 & 4 \end{bmatrix} \rightarrow \text{Maximum off}-\text{diagonal coefficient (absolute)}$$

$$: |a_{12}|=2$$

Since off-diagonal coefficient $|a_{12}|$ is maximum, this coefficient is zeroed. Angle θ to make $a_{12}=0$ can be calculated using Eq. (10.16) (taking $p=1$, $q=2$) as following:

$$\tan 2\theta=\frac{2a_{12}}{(a_{11}-a_{22})}=\frac{2(-2)}{(4-4)}=\frac{-4}{0}=\infty \rightarrow \theta=-0.7854 \text{ rad}$$

Using $\theta=-0.7854$, rotation submatrix matrix Q_R can be obtained as below:

$$Q_R=\begin{bmatrix} Q_{11} & Q_{12} \\ Q_{21} & Q_{22} \end{bmatrix}=[\cos(\theta) - \sin(\theta) \sin(\theta) \cos(\theta)]=\begin{bmatrix} 0.7071 & 0.7071 \\ -0.7071 & 0.7071 \end{bmatrix}$$

Substituting the rotation submatrix matrix Q_R into ($p=1$, $q=2$) rows/columns of unit matrix I of size 3×3 (size of $B_0=A$), rotation matrix Q_0 can be obtained as following:

$$Q_0=\begin{bmatrix} 0.7071 & 0.7071 & 0 \\ -0.7071 & 0.7071 & 0 \\ 0 & 0 & 1 \end{bmatrix}$$

Then, a_{12} coefficient in $B_0 = A$ matrix can be zeroed by rotating $B_0 = A$ using $B_1 = (Q_0)^{-1} \underbrace{B_0}_{A} Q_0 = tr(Q_0) \underbrace{B_0}_{A} Q_0$ equation as following:

$$B_1 = tr(Q_0) \underbrace{A_0}_{A} Q_0 = \begin{bmatrix} 6.0000 & 0 & 1.4142 \\ -0.0000 & 2.0000 & -1.4142 \\ 1.4142 & -1.4142 & 4.0000 \end{bmatrix}$$

Iteration 2:

$$B_1 = \begin{bmatrix} 6 & 0 & 1.4142 \\ 0 & 2 & -1.4142 \\ 1.4142 & -1.4142 & 4 \end{bmatrix} \rightarrow \text{Maximum off}-\text{diagonal coefficient}$$

$: |b_{23}| = 1.4142$

Since the off-diagonal coefficient $|b_{23}|$ is maximum, this coefficient is zeroed. Angle θ to make $b_{23} = 0$ can be calculated using Eq. (10.16) (taking $p = 2, q = 3$) as following:

$$\tan 2\theta = \frac{2b_{23}}{(b_{22} - b_{33})} = \frac{2(-1.4142)}{(2-4)} = 1.4142 \rightarrow \theta = 0.4777 \text{ rad}$$

Using $\theta = 0.4777$, rotation submatrix matrix Q_R can be obtained as below:

$$Q_R = \begin{bmatrix} Q_{22} & Q_{23} \\ Q_{32} & Q_{33} \end{bmatrix} = [\cos(\theta) - \sin(\theta) \sin(\theta) \cos(\theta)] = \begin{bmatrix} 0.8881 & -0.4597 \\ 0.4597 & 0.8881 \end{bmatrix}$$

Substituting the rotation submatrix matrix Q_R into ($p = 2$, $q = 3$) rows/ columns of unit matrix I of size 3 × 3 (size of B_1), rotation matrix Q_1 can be obtained as following:

$$Q_1 = \begin{bmatrix} 1.0000 & 0 & 0 \\ 0 & 0.8881 & -0.4597 \\ 0 & 0.4597 & 0.8881 \end{bmatrix}$$

Then, b_{23} coefficient in B_1 matrix can be zeroed by rotating B_1 using $B_2 = (Q_1)^{-1} B_1 Q_1 = tr(Q_1) B_1 Q_1$ equation as following:

$$B_2 = (Q_1)^{-1} B_1 Q_1 = tr(Q_1) B_1 Q_1 = \begin{bmatrix} 6.0000 & 0.6501 & 1.2559 \\ 0.6501 & 1.2679 & -0.0000 \\ 1.2559 & 0 & 4.7321 \end{bmatrix}$$

Iteration 3:

$$B_2 = \begin{bmatrix} 6.0000 & 0.6501 & 1.2559 \\ 0.6501 & 1.2679 & -0.0000 \\ 1.2559 & 0 & 4.7321 \end{bmatrix} \rightarrow \text{Maximum off}-\text{diagonal coefficient}$$

$$: |b_{13}| = 1.2559$$

Since the off-diagonal coefficient $|b_{13}|$ is maximum, this coefficient is zeroed. Angle θ to make $b_{13} = 0$ can be calculated using Eq. (10.16) (taking $p = 1, q = 3$) as following:

$$\tan 2\theta = \frac{2b_{13}}{(b_{11} - b_{33})} = \frac{2(1.2559)}{(6.0000 - 4.7321)} \rightarrow \theta = 0.5517 \text{ rad}$$

Using $\theta = -0.5517$, rotation submatrix matrix Q_R can be obtained as below:

$$Q_R = \begin{bmatrix} Q_{11} & Q_{13} \\ Q_{31} & Q_{33} \end{bmatrix} = [\cos(\theta) - \sin(\theta) \sin(\theta) \cos(\theta)] = \begin{bmatrix} 0.8517 & -0.5241 \\ 0.5241 & 0.8517 \end{bmatrix}$$

Substituting the rotation submatrix matrix Q_R into ($p = 1$, $q = 3$) rows/columns of unit matrix I of size 3 × 3 (size of B_2), rotation matrix Q_2 can be obtained as following:

$$Q_2 = \begin{bmatrix} 0.8517 & 0 & -0.5241 \\ 0 & 1.0000 & 0 \\ 0.5241 & 0 & 0.8517 \end{bmatrix}$$

Then, b_{31} coefficient in B_2 matrix can be zeroed by rotating B_2 using $B_3 = (Q_2)^{-1} B_2 Q_2 = tr(Q_2) B_2 Q_2$ equation as following:

$$B_3 = (Q_2)^{-1} B_2 Q_2 = tr(Q_2) B_2 Q_2 = \begin{bmatrix} 6.7729 & 0.5537 & -0.0000 \\ 0.5537 & 1.2679 & -0.3407 \\ -0.0000 & -0.3407 & 3.9592 \end{bmatrix}$$

Other iterations can be carried out similarly. After eight iterations, diagonalized B matrix is obtained (using convergence condition of $\varepsilon \leq 0.001$) as below:

$$\text{Diagonalized } B = B_8 = \begin{bmatrix} 6.8284 & 0.0000 & 0.0000 \\ 0.0000 & 1.1716 & 0.0000 \\ 0.0000 & -0.0000 & 4.0000 \end{bmatrix}$$

Thus, diagonal terms of $B = B_8$ matrix yield eigenvalues of $\lambda_1 = 6.8284$, $\lambda_2 = 1.1716$, $\lambda_3 = 4$. Multiplication of all Q matrices yield eigenvectors as following (each column corresponds to an eigenvector).

$$Q = Q_0 Q_1 Q_2 \ldots Q_7 = \begin{bmatrix} \overbrace{0.5000}^{\lambda_1 = 6.8284} & \overbrace{0.5000}^{\lambda_2 = 1.1716} & \overbrace{-0.7071}^{\lambda_3 = 4} \\ -0.7071 & 0.7071 & -0.0000 \\ 0.5000 & 0.5000 & 0.7071 \end{bmatrix}$$

Note: Eigenvalues and eigenvectors can also be calculated using *MATLAB Program 10.2: Eigenvalue and Eigenvector Calculation with Jacobi Method* in appendix.

(c) ***QR* Method**

QR method like Jacobi method is used in the solution of eigenvalue problem $AX = \lambda X$ where A matrix is symmetric. In *QR* method, A matrix is decomposed into the product of an orthogonal Q matrix and an upper triangular R matrix (all elements below diagonal terms are zero) as $A = QR$. *QR* method calculates only eigenvalues. Therefore, eigenvectors need to be calculated separately.

In *QR* method, orthogonal rotation is applied to A matrix using $B = Q^T AQ$ equation to convert it into an upper triangular matrix B. In *QR* method, to rotate A matrix, alternative form of rotation equation is used. In alternative form, A matrix is decomposed into the product of Q and R matrices first. Then, rotated form of A matrix is obtained with product of R and Q matrices as $B = RQ$.

Equivalency of $B = Q^T AQ$ and $B = RQ$ equations can be shown as below:

$$A = QR \rightarrow B = Q^T AQ \rightarrow B = Q^T (QR)Q \rightarrow B = \underbrace{\underbrace{Q^T}_{Q^{-1}} Q}_{I} RQ = RQ \qquad (10.18)$$

In *QR* method, rotation of A matrix is performed iteratively until it transforms to an upper triangular matrix B using decomposition and rotation steps as following:

$$\text{Decomposition} : B_i = A_i = Q_i R_i \rightarrow Q_i, R_i, \text{Rotation}$$
$$: B_{i+1} = R_i Q_i \ \left(\text{or } B_{i+1} = Q_i^T A_i Q_i\right)$$

Decomposition and rotation steps can be written for iterations more explicitly as following:

$$\text{Iteration 1 : Decompose : } B_0 = A = Q_0R_0 \rightarrow Q_0, R_0, \qquad \text{Rotate : } B_1 = R_0Q_0$$

$$\text{Iteration 2 : Decompose : } B_1 = Q_1R_1 \rightarrow Q_1, R_1, \qquad \text{Rotate : } B_2 = R_1Q_1$$

$$\ldots$$

$$\text{Iteration } k \text{ : Decompose : } B_{k-1} = Q_{k-1}R_{k-1} \rightarrow Q_{k-1}, R_{k-1}, \qquad \text{Rotate : } B_k = R_{k-1}Q_{k1}$$

Iterations can be combined in a single equation using original A matrix and Q matrix as below:

$$B_k = \underbrace{Q_{k-1}^T \ldots Q_3^T Q_2^T Q_1^T}_{Q^T} A \underbrace{Q_1 Q_2 Q_3 \ldots Q_{k-1}}_{Q} = Q^T A Q = Q^T A Q \tag{10.19}$$

After iterations are completed, A matrix is converted (rotated) into an upper triangular B matrix (under diagonal terms are zero), where *diagonal* terms of B matrix gives eigenvalues of A matrix.

$$B = QR \textit{ decomposition :}$$

In QR method, critical step is the decomposition of original A (at the beginning) or B matrix (in further iterations) into Q and R matrices using $B = QR$ equation at each iteration. There are different ways to do this decomposition. One of the alternatives is to use Householder method. To explain Householder method, consider $B = QR$ expression. R can be expressed from this equation as $R = \underbrace{Q^{-1}}_{Q^T} B = Q^T B$. In Householder method, Q is expressed with product of *orthogonal* Householder matrices (H_i) as following:

$$Q = H_1 H_2 H_3 \ldots H_{n-1} \qquad (n \times n \text{ is size of } B) \tag{10.20}$$

In this case, in Householder method, R matrix can be obtained by converting B matrix into upper triangular form by premultiplying it with Householder matrix H_i as following:

$$R = H_{n-1} \ldots H_3 H_2 H_1 B \tag{10.21}$$

Note that Q and R matrices in Eqs. (10.20) and (10.21) are calculated successively since H_i requires H_{i-1} information. For calculation of R, Eq. (10.21) can be expressed in successive form as following:

$$R = \underbrace{H_{n-1} \ldots \underbrace{H_3 \underbrace{H_2 \underbrace{H_1 B}_{R_1}}_{R_2}}_{R_3}}_{R_{n-1}} : H_1 \underbrace{B}_{R_0} = R_1, H_2R_1 = R_2, \ldots, H_iR_{i-1} = R_i, \ldots, H_{n-1}R_{n-2} = R_{n-1} \tag{10.22}$$

In Eq. (10.22), R_i indicates a matrix where all under diagonal coefficients up to i-th diagonal term are zero as below:

$$R_i = \begin{bmatrix} a''_{11} & a''_{12} & \ldots & \overbrace{a''_{1i}}^{Column:i} & \ldots & a''_{1n} \\ 0 & a''_{22} & \ldots & a''_{2i} & \ldots & a''_{2n} \\ \ldots & \ldots & \ldots & \ldots & \ldots & \ldots \\ 0 & 0 & \ldots & a''_{pi} & \ldots & a''_{pn} \\ \ldots & \ldots & \ldots & \ldots & \ldots & \ldots \\ 0 & 0 & \ldots & 0 & \ldots & a''_{nn} \end{bmatrix}$$

H_i matrix is calculated before using it in $H_iR_{i-1} = R_i$ equation in Eq. (10.22) so as to zero under diagonal terms in the i-th column of R_i matrix. That is, H_i is determined such that when premultiplied with R_{i-1} will produce the matrix with zero values under diagonal terms up to and including i-th column as following:

$$H_i \underbrace{\begin{bmatrix} a'_{11} & \ldots & \overbrace{a'_{1i-1}}^{Column:i-1} & \overbrace{a'_{1i}}^{Column:i} & \ldots & a'_{1n} \\ \ldots & \ldots & \ldots & \ldots & \ldots & \ldots \\ 0 & \ldots & a'_{i-1i-1} & \ldots & \ldots & \ldots \\ 0 & \ldots & 0 & a'_{ii} & \ldots & a'_{in} \\ \ldots & \ldots & \ldots & \ldots & \ldots & \ldots \\ 0 & \ldots & 0 & a'_{ni} & \ldots & a'_{nn} \end{bmatrix}}_{R_{i-1}}$$

$$= \underbrace{\begin{bmatrix} a''_{11} & \ldots & \overbrace{a''_{1i-1}}^{Column:i-1} & \overbrace{a''_{1i}}^{Column:i} & \ldots & a''_{1n} \\ \ldots & \ldots & \ldots & \ldots & \ldots & \ldots \\ 0 & \ldots & a''_{i-1i-1} & \ldots & \ldots & \ldots \\ 0 & \ldots & 0 & a''_{ii} & \ldots & a''_{in} \\ \ldots & \ldots & \ldots & \ldots & \ldots & \ldots \\ 0 & \ldots & 0 & 0 & \ldots & a''_{nn} \end{bmatrix}}_{R_i} \tag{10.23}$$

To obtain H_i, a new vector c can be created from the i-th column of R_{i-1} matrix upon zeroing above diagonal terms as following:

$$c = \begin{Bmatrix} 0 \\ 0 \\ \cdots \\ a'_{ii} \Leftarrow Diagonal \\ \cdots \\ a'_{ni} \end{Bmatrix} \tag{10.24}$$

Then from c vector, a new unit vector e in size of c vector where only i-th element is non-zero ($e_i = \pm 1$) can be created as following:

$$e = \begin{Bmatrix} 0 \\ 0 \\ \cdots \\ e_i = \pm 1 \Leftarrow sign(Diagonal) \\ \cdots \\ 0 \end{Bmatrix} \tag{10.25}$$

In vector e, $e_i = 1$ if a'_{ii} is positive and $e_i = -1$ if a'_{ii} is negative.
Later, using c and e vectors a new vector v can be created as below:

$$v = c + c_{norm}e \left(c_{norm} = \sqrt{c_1^2 + c_2^2 + c_3^2 + \ldots + c_n^2} \right) \tag{10.26}$$

Finally, using v vector, Householder matrix H_i can be obtained as following:

$$H_i = I - \frac{2(vv^T)}{(v^Tv)} \quad (I : \text{unit matrix with } n \times n \text{ size}) \tag{10.27}$$

In the following example, obtaining Householder matrix H for an A matrix with 3×3 size is explained in detail.

Advantages and disadvantages of *QR* method can be stated as following:

Advantages:

1. *QR* method gives all eigenvalues.
2. Diagonalization is performed column by column in contrast to term by term in Jacobi method. Therefore, *QR* method is more efficient than Jacobi method.

Disadvantages:

1. QR method does not give eigenvectors. They should be calculated separately.
2. Since QR method is computationally intensive method, it requires programming to perform mathematical operations.

Example 10.4: Symmetric matrix A is given as $A = \begin{bmatrix} a_{11} & a_{12} & a_{13} \\ a_{12} & a_{22} & a_{23} \\ a_{13} & a_{23} & a_{33} \end{bmatrix}$. Obtain Householder matrices H_i symbolically to decompose it into $A = QR$ form where Q is orthogonal and R is upper triangular matrix.

From $H_i R_{i-1} = R_i$ expression in Eq. (10.22), it is known that upper triangular matrix R can be obtained using Householder matrices H_i successively and iteratively. $H_i R_{i-1} = R_i$ iterations can be implemented as following:

Calculation of H_1 matrix:

Using $H_1 = I - \frac{2(vv^T)}{(v^T v)}$ expression, H_1 matrix can be calculated so as to zero all under diagonal terms in the first column of $R_0 = A$ matrix. In H_1 expression, v vector is obtained using the first column of A matrix as below:

$$c = \begin{Bmatrix} a_{11} \Leftarrow Diagonal \\ a_{21} \\ a_{31} \end{Bmatrix} \rightarrow c_{norm} = \sqrt{a_{11}^2 + a_{21}^2 + a_{31}^2}), e$$

$$= \begin{Bmatrix} \pm 1 \Leftarrow sign(Diagonal) \\ 0 \\ 0 \end{Bmatrix}$$

$$\rightarrow v = c + c_{norm} e = \begin{Bmatrix} a_{11} \\ a_{21} \\ a_{31} \end{Bmatrix} + c_{norm} \begin{Bmatrix} \pm 1 \\ 0 \\ 0 \end{Bmatrix}$$

Product of H_1 and $R_0 = A$ matrices leads to a new R_1 matrix ($H_1 \underbrace{R_0}_{A} = R_1$) where all under diagonal terms in the *first* column are zero as following:

$$R_1 = H_1 \underbrace{R_0}_{A} = \begin{bmatrix} a'_{11} & a'_{12} & a'_{13} \\ 0 & a'_{22} & a'_{23} \\ 0 & a'_{32} & a'_{33} \end{bmatrix}$$

Calculation of H_2 matrix:

Using $H_2 = I - \frac{2(vv^T)}{(v^T v)}$ expression, H_2 matrix can be calculated so as to zero all under diagonal terms in the second column of R_1 matrix. In H_2 expression, v vector is obtained using the second column of R_1 matrix as below:

$$c = \begin{Bmatrix} 0 \\ a'_{22} \Leftarrow Diagonal \\ a'_{32} \end{Bmatrix} \rightarrow c_{norm} = \sqrt{(a'_{22})^2 + (a'_{32})}, e$$

$$= \begin{Bmatrix} 0 \\ \pm 1 \Leftarrow sign(Diagonal) \\ 0 \end{Bmatrix}$$

$$\rightarrow v = c + c_{norm} e = \begin{Bmatrix} 0 \\ a'_{22} \\ a'_{32} \end{Bmatrix} + c_{norm} \begin{Bmatrix} 0 \\ \pm 1 \\ 0 \end{Bmatrix}$$

Product of H_2 and R_1 matrices leads to a new R_2 matrix ($H_2 R_1 = R_2$) where all under diagonal terms in the *second* column are zero as following:

$$R_2 = H_2 R_1 = H_2 H_1 \underbrace{R_0}_{A} = \begin{bmatrix} a''_{11} & a''_{12} & a''_{13} \\ 0 & a''_{22} & a''_{23} \\ 0 & 0 & a''_{33} \end{bmatrix}$$

Thus, iterations of zeroing under diagonal terms are completed for A matrix of 3×3 size. R matrix is obtained as $R = R_2$ and Q matrix is obtained with product of Householder matrices as $Q = H_1 H_2$. In this case, rotated A matrix is obtained as $B = RQ$.

Note: In QR method, zero values under diagonal terms are preserved during decomposition process increasing the efficiency of the method in calculating eigenvalues.

Example 10.5: Symmetric A matrix is given as $A = \begin{bmatrix} 4 & -2 & 0 \\ -2 & 4 & -2 \\ 0 & -2 & 4 \end{bmatrix}$. Obtain eigenvalues of this matrix using QR method.

In eigenvalue calculation with QR method, A matrix is first decomposed into special Q and R matrices. Then, using product of these matrices, A matrix is rotated using $A = RQ$ product. In QR method, Q and R matrices are obtained successively and iteratively using Householder matrices H_i. When iterations are completed, A matrix is converted into an upper triangular matrix (equivalently diagonal matrix since A is symmetric). From upper triangular form of A matrix, eigenvalues are directly obtained as diagonal coefficients.

Note: Since size of A matrix is 3×3, decomposition of A matrix can be achieved using two successive Householder matrices of $H1$ and $H2$ that have the same size of A.

Solution iterations with QR method can be carried out as following:

Iteration 1:

Step 1: Decomposition of A matrix

- *Calculation of H1 matrix*:

  Using $H_1 = I - \frac{2(vv^T)}{(v^T v)}$ expression, H_1 matrix can be calculated so as to zero all under diagonal terms in the first column of $R_0 = A$ matrix. In H_1 expression, v vector is obtained using the first column of $R_0 = A$ matrix as below:

$$\text{Using first column of } R_0 = A \rightarrow c = \begin{Bmatrix} 4 \Leftarrow Diagonal \\ -2 \\ 0 \end{Bmatrix} \rightarrow c_{norm} = 4.4721$$

e vector is obtained as the unit vector in size of c vector (zeroing non-diagonal terms in *c vector*) as following:

$$\text{Unit vector : } e = \begin{Bmatrix} 1 \Leftarrow sign(Diagonal) > 0 \\ 0 \\ 0 \end{Bmatrix}$$

In this case, using obtained information, v vector and Householder matrix H_1 can be obtained as below:

$$v = c + c_{norm} e \rightarrow v = \begin{Bmatrix} 8.4721 \\ -2.0000 \\ 0 \end{Bmatrix} \rightarrow H_1 = I - \frac{2(vv^T)}{(v^T v)}$$

$$= \begin{bmatrix} -0.8944 & 0.4472 & 0 \\ 0.4472 & 0.8944 & 0 \\ 0 & 0 & 1.0000 \end{bmatrix}$$

Product of H_1 and $R_0 = A$ matrices leads to a new R_1 matrix where all under diagonal terms in the first column are zero as following:

$$H_1 \underbrace{R_0}_{A} = R_1 = \begin{bmatrix} -4.4721 & 3.5777 & -0.8944 \\ 0 & 2.6833 & -1.7889 \\ 0 & -2.0000 & 4.0000 \end{bmatrix}$$

– *Calculation of H2 matrix*:

Using $H_2 = I - \frac{2(vv^T)}{(v^Tv)}$ expression, H_2 matrix can be calculated so as to zero all under diagonal terms in the second column of R_1 matrix. In H_2 expression, v vector is obtained using the second column of R_1 matrix as below:

$$\text{Using second column of } R_1 : c = \left\{ \begin{matrix} 0 \\ 2.6833 \Leftarrow Diagonal \\ -2.0000 \end{matrix} \right\}, c_{norm} = 3.3466$$

e vector is obtained as the unit vector in size of c vector (zeroing non-diagonal terms in c *vector*) as following:

$$\text{Unit vector} : e = \left\{ \begin{matrix} 0 \\ 1 \Leftarrow sign(Diagonal) > 0 \\ 0 \end{matrix} \right\},$$

In this case, using obtained information, v vector and Householder matrix H_2 can be obtained as below:

$$v = c + c_{norm}e \rightarrow v = \left\{ \begin{matrix} 0 \\ 6.0299 \\ -2.0000 \end{matrix} \right\} \rightarrow H_2 = I - \frac{2(vv^T)}{(v^Tv)}$$

$$= \begin{bmatrix} 1.0000 & 0 & 0 \\ 0 & -0.8018 & 0.5976 \\ 0 & 0.5976 & 0.8018 \end{bmatrix}$$

Product of H_2 and R_1 matrices leads to a new R_2 matrix ($H_2R_1 = R_2$) where all under diagonal terms in the second column are zero as following:

$$R_2 = H_2R_1 = H_2H_1\underbrace{R_0}_{A} = \begin{bmatrix} -4.4721 & 3.5777 & -0.8944 \\ 0 & -3.3466 & 3.8247 \\ 0 & -0.0000 & 2.1381 \end{bmatrix}$$

For iteration 1, decomposition process is completed after calculating H_1 and H_2 matrices. After decomposition, Q and R matrices are obtained as $Q = H_1H_2$ and $R = R_2$ as below:

$$R = R_2 = \begin{bmatrix} -4.4721 & 3.5777 & -0.8944 \\ 0 & -3.3466 & 3.8247 \\ 0 & -0.0000 & 2.1381 \end{bmatrix}, Q = H_1 H_2$$

$$= \begin{bmatrix} -0.8944 & -0.3586 & 0.2673 \\ 0.4472 & -0.7171 & 0.5345 \\ -0.0000 & 0.5976 & 0.8018 \end{bmatrix}$$

Step 2: Rotation of A matrix

Rotated A matrix is obtained using $B = R^*Q$ equation as below:

$$B = RQ = \begin{bmatrix} 5.6000 & -1.4967 & 0.0000 \\ -1.4967 & 4.6857 & 1.2778 \\ -0.0000 & 1.2778 & 1.7143 \end{bmatrix}$$

Iteration 2:

Step 1: Decomposition of B matrix (B replaces A)

- *Calculation of H1 matrix*:

  Using $H_1 = I - \frac{2(vv^T)}{(v^T v)}$ expression, H_1 matrix can be calculated so as to zero all under diagonal terms in the first column of $R_0 = B$ matrix. In H_1 expression, v vector is obtained using the first column of $R_0 = B$ matrix as below:

$$\text{Using first column of } R_0 = B \rightarrow c = \begin{Bmatrix} 5.6000 \Leftarrow Diagonal \\ -1.4967 \\ -0.0000 \end{Bmatrix}, c_{norm} = 5.7966$$

  e vector is obtained as the unit vector in size of c vector (zeroing non-diagonal terms in *c vector*) as following:

$$\text{Unit vector} : e = \begin{Bmatrix} 1 \Leftarrow sign(Diagonal) > 0 \\ 0 \\ 0 \end{Bmatrix}$$

  In this case, using obtained information, v vector and Householder matrix H_2 can be obtained as below:

$$v = c + c_{norm} e \rightarrow v = \begin{Bmatrix} 11.3966 \\ -1.4967 \\ -0.0000 \end{Bmatrix} \rightarrow H_1 = I - \frac{2(vv^T)}{(v^T v)}$$

$$= \begin{bmatrix} -0.9661 & 0.2582 & 0.0000 \\ 0.2582 & 0.9661 & -0.0000 \\ 0.0000 & -0.0000 & 1.0000 \end{bmatrix}$$

Product of H_1 and $R_0 = B$ matrices leads to a new R_1 matrix where all under diagonal terms in the first column are zero as following:

$$H_1 \underbrace{R_0}_{B} = R_1 = \begin{bmatrix} -5.7966 & 2.6558 & 0.3299 \\ -0.0000 & 4.1404 & 1.2344 \\ -0.0000 & 1.2778 & 1.7143 \end{bmatrix}$$

- *Calculation of H2 matrix*:

  Using $H_2 = I - \frac{2(vv^T)}{(v^T v)}$ expression, H_2 matrix can be calculated so as to zero all under diagonal terms in the second column of R_1 matrix. In H_2 expression, v vector is obtained using the second column of R_1 matrix as below:

$$\text{Using second column of } R_1 \rightarrow c = \begin{Bmatrix} 0 \\ 4.1404 \Leftarrow Diagonal \\ 1.2778 \end{Bmatrix}, c_{norm} = 4.3331$$

e vector is obtained as the unit vector in size of *c* vector (zeroing non-diagonal terms in *c vector*) as following:

$$\text{Unit vector} : e = \begin{Bmatrix} 0 \\ 1 \Leftarrow sign(Diagonal) > 0 \\ 0 \end{Bmatrix},$$

In this case, using obtained information, v vector and Householder matrix H_2 can be obtained as below:

$$v = c + c_{norm} e \rightarrow v = \begin{Bmatrix} 0 \\ 8.4735 \\ 1.2778 \end{Bmatrix} \rightarrow H_2 = I - \frac{2(vv^T)}{(v^T v)}$$

$$= \begin{bmatrix} 1.0000 & 0 & 0 \\ 0 & -0.9555 & -0.2949 \\ 0 & -0.2949 & 0.9555 \end{bmatrix}$$

Product of H_2 and R_1 matrices leads to a new R_2 matrix ($H_2R_1 = R_2$) where all under diagonal terms in the second column are zero as following:

$$R_2 = H_2R_1 = H_2H_1\underbrace{R_0}_{B} = \begin{bmatrix} -5.7966 & 2.6558 & 0.3299 \\ 0.0000 & -4.3331 & -1.6851 \\ 0.0000 & 0 & 1.2740 \end{bmatrix}$$

For iteration 1, decomposition process is completed after calculating H_1 and H_2 matrices. After decomposition, Q and R matrices are obtained as $Q = H_1H_2$ and $R = R_2$ as below:

$$R = R_2 = \begin{bmatrix} -5.7966 & 2.6558 & 0.3299 \\ 0.0000 & -4.3331 & -1.6851 \\ 0.0000 & 0 & 1.2740 \end{bmatrix}, Q = H_1H_2$$
$$= \begin{bmatrix} -0.9661 & -0.2467 & -0.0761 \\ 0.2582 & -0.9231 & -0.2849 \\ 0.0000 & -0.2949 & 0.9555 \end{bmatrix}$$

Step 2: Rotation of A matrix

Rotated B matrix is obtained using as $B = R^*Q$ equation as below:

$$B = RQ = \begin{bmatrix} 6.2857 & -1.1188 & -0.0000 \\ -1.1188 & 4.4969 & -0.3757 \\ -0.0000 & -0.3757 & 1.2174 \end{bmatrix}$$

Iterations are repeated similarly. After 16 iterations (using convergence tolerance of $\varepsilon \leq 0.001$) upper triangular form of original A matrix (i.e., rotated A matrix) is obtained as B matrix as following:

$$B = RQ = \begin{bmatrix} 6.8284 & -0.0008 & 0.0000 \\ -0.0008 & 4.0000 & -0.0000 \\ -0.0000 & -0.0000 & 1.1716 \end{bmatrix}$$

From B matrix, eigenvalues are obtained as diagonal values as $\lambda_1 = 6.8284$, $\lambda_2 = 4.0000$, $\lambda_3 = 1.1716$

Calculation of Eigenvectors

Since QR method does not give eigenvectors, they should be calculated separately using eigenvalue equation of $AX = \lambda_i X$ $(X = x_1, x_2, \ldots, x_n)$. In the solution of $AX = \lambda_i X$ equation, x_1 can be assigned a predetermined value such as $x_1 = 1$ and other variables ($x_2, x_3, \ldots, x_n$) can be calculated from the equation. *Note*: after substituting λ_i value and assigning an x_i value, $AX = \lambda_i X$ equation transforms to a linear equation system which can be solved methods in Chap. 4. Solution of $AX = \lambda_1 X$, $AX = \lambda_2 X$, and $AX = \lambda_3 X$ equations yield eigenvectors (X) corresponding to eigenvalues as following:

1. *Eigenvector calculation for* $\lambda_1 = 6.8284$:

   Upon substitution of λ_1 in the eigenvalue equation of $AX = \lambda_1 X$, following linear equation system is obtained:

$$\begin{bmatrix} 4 & -2 & 0 \\ -2 & 4 & -2 \\ 0 & -2 & 4 \end{bmatrix} \underbrace{\begin{Bmatrix} x_1 \\ x_2 \\ x_3 \end{Bmatrix}}_{X} = \lambda_1 \underbrace{\begin{Bmatrix} x_1 \\ x_2 \\ x_3 \end{Bmatrix}}_{X} \rightarrow \text{Linear equation system}$$

$$: \begin{aligned} &(4 - 6.8284)x_1 - 2x_2 = 0 \\ &-2x_1 + (4 - 6.8284)x_2 - 2x_3 = 0 \\ &-2x_2 + (4 - 6.8284)x_3 = 0 \end{aligned}$$

   Since right sides are zero in linear system, a unique solution of X cannot be obtained. However, when one of the unknowns is assigned a value, others can be solved easily using linear equation methods in Chap. 4. In this case, when $x_1 = 1$ is assigned, x_2 and x_3 values can be solved using any two equations (e.g., first and second equations) in the linear system as $x_2 = -1.4142$ and $x_3 = 1$. Thus, eigenvector X for the eigenvalue of $\lambda_1 = 6.8284$ is obtained as

$$X = \begin{Bmatrix} 1.0000 \\ -1.4142 \\ 1.0000 \end{Bmatrix}.$$

2. *Eigenvector calculation for* $\lambda_2 = 4.0000$:

   Upon substitution of λ_2 in the eigenvalue equation of $AX = \lambda_2 X$, following linear equation system is obtained:

$$\begin{bmatrix} 4 & -2 & 0 \\ -2 & 4 & -2 \\ 0 & -2 & 4 \end{bmatrix} \underbrace{\begin{Bmatrix} x_1 \\ x_2 \\ x_3 \end{Bmatrix}}_{X} = \lambda_2 \underbrace{\begin{Bmatrix} x_1 \\ x_2 \\ x_3 \end{Bmatrix}}_{X} \rightarrow \text{Linear equation system}$$

$$\begin{aligned} & (4-4.0000)x_1 - 2x_2 = 0 \\ : \quad & -2x_1 + (4-4.0000)x_2 - 2x_3 = 0 \\ & -2x_2 + (4-4.0000)x_3 = 0 \end{aligned}$$

When $x_1 = 1$ is assigned, x_2 and x_3 values can be solved using any two equations (e.g., first and second equations) in the linear system as $x_2 = 0$ and $x_3 = -1$. Thus, eigenvector X for the eigenvalue of $\lambda_2 = 4.0000$ is obtained as $X = \begin{Bmatrix} 1.0000 \\ 0.0000 \\ -1.0000 \end{Bmatrix}$.

3. *Eigenvector calculation for* $\lambda_3 = 1.1716$:

Upon substitution of λ_3 in the eigenvalue equation of $AX = \lambda_3 X$, following linear equation system is obtained:

$$\begin{bmatrix} 4 & -2 & 0 \\ -2 & 4 & -2 \\ 0 & -2 & 4 \end{bmatrix} \underbrace{\begin{Bmatrix} x_1 \\ x_2 \\ x_3 \end{Bmatrix}}_{X} = \lambda_3 \underbrace{\begin{Bmatrix} x_1 \\ x_2 \\ x_3 \end{Bmatrix}}_{X} \rightarrow \text{Linear equation system}$$

$$\begin{aligned} & (4-1.1716)x_1 - 2x_2 = 0 \\ : \quad & -2x_1 + (4-1.1716)x_2 - 2x_3 = 0 \\ & -2x_2 + (4-1.1716)x_3 = 0 \end{aligned}$$

When $x_1 = 1$ is assigned, x_2 and x_3 values can be solved using any two equations (e.g., first and second equations) in the linear system as $x_2 = 1.4142$ and $x_3 = 1$. Thus, eigenvector X for the eigenvalue of $\lambda_3 = 1.1716$ is obtained as following: $X = \begin{Bmatrix} 1.0000 \\ 1.4142 \\ 1.0000 \end{Bmatrix}$.

Note: Eigenvalues and eigenvectors in this problem can also be calculated using *MATLAB Program 10.3: Eigenvalue Calculation with QR Method* in appendix.

Solution with MATLAB command:

Eigenvalues and eigenvectors can be obtained using *eig* command expressed in *[EigenVector,EigenValue] = eig(A)* form in MATLAB as following:

```
>> A = [4 -2 0 ;
 -2 4 -2 ;
  0 -2 4 ] ;
>> [EigenVector,EigenValue] = eig(A);
>> EigenValue
EigenValue =
   1.1716       0       0
        0  4.0000       0
        0       0  6.8284
>> EigenVector
EigenVector =
   0.5000  -0.7071  -0.5000
   0.7071   0.0000   0.7071
   0.5000   0.7071  -0.5000
```

Same solutions of *QR* method are obtained with MATLAB. Only eigenvectors are normalized differently in MATLAB. Each column in *V* vector expresses an eigenvector. For example, first column in *V* vector expresses the eigenvector corresponding to the first eigenvalue ($\lambda = 1.1716$) in *Lam* vector.

10.5 Summary

- In engineering, some characteristics problems such as free vibration and buckling are expressed as eigenvalue problems. Finding natural frequencies and vibration modes in free vibration problems and finding critical buckling loads and buckling modes in buckling problems require expression and solution of eigenvalue problems.
- Eigenvalue problems are expressed in engineering with differential or algebraic equations.
- For numerical solutions, eigenvalue problems are usually expressed in $AX = \lambda X$ linear equation form where *A* matrix represents coefficients, λ indicates eigenvalue and *X* denotes eigenvector.
- Polynomial method is used in classical solution of eigenvalue problems.
- In polynomial method, eigenvalue problem is transformed into a root finding problem using $\det(A - \lambda I) = 0$ equation. Eigenvalues λ are obtained from the solution of root problem $c_1 + c_2\lambda + \ldots + c_{n+1}\lambda^n = 0$.
- Eigenvalue λ is then substituted in $(A - \lambda I)X = 0$ equation to yield eigenvector *X*.
- Polynomial method is used in solving small-size eigenvalue problems.
- For big size eigenvalue problems, iterative methods are used to find eigenvalues and eigenvectors.

- Power, Jacobi, and QR methods are iterative methods that are widely used in solving eigenvalue problems.
- In iterative methods, starting from an initial solution, eigenvalue problem is solved iteratively (step by step).
- Dividing or multiplying an eigenvector with a constant (i.e., normalization of an eigenvector) does not change its character. Therefore, eigenvectors are usually expressed in normalized form.
- In some engineering problems, it is sufficient to calculate a limited number of eigenvalues, such as the biggest or smallest eigenvalue. For example, in buckling problems, it is sufficient to know the smallest eigenvalue (critical buckling load).
- Power method was developed to find the biggest eigenvalue (or smallest eigenvalue after rearrangement of the eigenvalue equation) and its corresponding eigenvector.
- In Power method, AX product in eigenvalue equation is iteratively decomposed into the product of eigenvalue and eigenvector as following:

$$AX^{(k-1)} \Rightarrow Decomposition \Rightarrow \lambda^{(k)}X^{(k)} \ (k-1 : \text{old value}, k : \text{new value})$$

- Jacobi method finds all eigenvalues and eigenvectors simultaneously where A matrix is symmetric.
- In Jacobi method, to calculate eigenvalues easily, A matrix is iteratively rotated into B matrix using rotation matrix Q as $B = Q^{-1}AQ$. Diagonal terms of B matrix yield eigenvalues and Q matrix yields associated eigenvectors.
- QR method like Jacobi method is used in the solution of eigenvalue problem $AX = \lambda X$ where A matrix is symmetric.
- In QR method, A matrix is iteratively decomposed into the product of an orthogonal Q matrix and an upper triangular R (all elements below diagonal terms are zero) matrix as $A = QR$. Diagonal terms of A matrix yield eigenvalues.
- QR method calculates only eigenvalues. Therefore, eigenvectors need to be calculated separately.

Appendix: MATLAB Programs

MATLAB Program 10.1: Eigenvalue Calculation with Power Method

```
function Eigenvalue_Calculation_Power_Method
% THIS PROGRAM SOLVES A*X=Lam*X EIGENVALUE PROBLEM USING POWER METHOD
A=[4  -2   0 ;
  -2   4  -2 ;
   0  -2   4 ] ;

itmax=30 ;  %  Maximum iteration number
tol=0.005 ;  %  Convergence tolerance
% -----------------------------------------------------------------------
--
n=size(A,1) ;

it=1 ;
Conv=1e10 ; % Initial value for convergence tolerance

X=ones(n,1) ; % Initial value for Eigenvector X

while (Conv>tol) && (it<=itmax)
    Xn=A*X ;

    Lam=abs(Xn(1)) ;
    index=1 ;

    for i=2:n
        if abs(Xn(i))>Lam
            Lam=abs(Xn(i)) ;
            index=i ;
        end
    end

    Xn=(1/Lam)*Xn ;

    if it>=2
         Conv=abs((Lam-Lamp)/Lamp) ;
    end

    Lamp=Lam ;
    X=Xn ;

    it=it+1 ;
end

if it>=itmax
   error('Not converged after %i iterations !!!',it) ;
end

fprintf('Eigenvalue: Lam') ;
Lam
fprintf('Eigenvector: X') ;
X
```

MATLAB Program 10.2: Eigenvalue and Eigenvector Calculation with Jacobi Method

```
function Eigenvalue_and_Eigenvector_Calculation_Jacobi

% THIS PROGRAM SOLVES A*X=Lam*X EIGENVALUE PROBLEM USING JACOBI METHOD

A=[4  -2   0 ;
  -2   4  -2 ;
   0  -2   4 ] ;

itmax=100 ; % Maximum iteration number
tol=0.001 ; % Convergence tolerance

% ----------------------------------------------------------------------
--
n=size(A,1) ;

it=0 ;
maxi=1e10 ;

% Initial values
Q0=eye(n,n) ;
B=A ;

while (maxi>tol) && (it<=itmax)
    it=it+1 ;

    maxi=abs(B(1,2)) ;
    p=1 ;
    q=2 ;

    % Searching maximum in above diagonal terms
    for i=1:n-1
        for j=i+1:n
            if abs(B(i,j))>maxi
                 maxi=abs(B(i,j)) ;
                 p=i ;
                 q=j ;
            end
        end
    end

    % Searching maximum in below diagonal terms
    for j=1:n-1
        for i=j+1:n
            if abs(B(i,j))>maxi
                maxi=abs(B(i,j)) ;
                p=i ;
                q=j ;
            end
        end
    end


    Q=eye(n,n) ;
```

```
    tan2teta=2*B(p,q)/(B(p,p)-B(q,q)) ;
    teta=(atan(tan2teta))/2 ;
    Q(p,p)=cos(teta) ;
    Q(p,q)=-sin(teta) ;
    Q(q,p)=sin(teta) ;
    Q(q,q)=cos(teta) ;

    B=Q'*B*Q ;
    Q=Q0*Q ;
    Q0=Q ;
end

it

fprintf('------  RESULTS  ------ \n ') ;
fprintf('Eigenvalues: \n') ;
for i=1:n
    fprintf(' %f   ', B(i,i)) ;
end

fprintf(' \n\n') ;
fprintf('Eigenvectors: Q \n') ;
Q

if maxi>tol
    error('Jacobi iterations Not Converged with %i repeat !!! ',itmax) ;
end
```

MATLAB Program 10.3: Eigenvalue Calculation with QR Method

```
function Eigenvalue_Calculation_QR_Method

% THIS PROGRAM SOLVES A*X=Lam*X EIGENVALUE PROBLEM USING QR METHOD

A=[4  -2   0 ;
  -2   4  -2 ;
   0  -2   4 ] ;

itmax=100 ;  %  Maximum iteration number
tol=0.001 ;  %  Convergence tolerance
% -----------------------------------------------------------------------
--

n=size(A,1) ;

it=0 ;
maxi=1e10 ;

B=A ;
while (maxi>tol) && (it<=itmax)
    it=it+1 ;

    % Decomposition of B
    [Q,R]=QR_Decomposition(B) ;

    % Rotation of B
    B=R*Q ;

    % Calculation of absolute maximum for convergence
    % Searching maximum in above diagonal terms
    maxi=abs(B(1,2)) ;
    for i=1:n-1
        for j=i+1:n
            if abs(B(i,j))>maxi
                maxi=abs(B(i,j)) ;
            end
        end
    end

    % Searching maximum in below diagonal terms
    for j=1:n-1
        for i=j+1:n
            if abs(B(i,j))>maxi
                maxi=abs(B(i,j)) ;
            end
        end
    end
end

it

fprintf('RESULTS \n') ;
fprintf('Q Matrix: \n') ;
Q
```

```
fprintf('R Matrix: \n') ;
R

fprintf('Rotated A Matrix: B \n') ;
B

fprintf('Eigenvalues: \n') ;
for i=1:n
     Lam(i)=B(i,i) ;
     fprintf('Eigenvalue %i = %f \n',i, Lam(i)) ;
 end

 fprintf(' \n') ;

 fprintf('------  RESULTS  ------ \n ') ;
 fprintf('Eigenvalues: \n') ;
 for i=1:n
     fprintf(' %f     ', Lam(i)) ;
 end

 fprintf(' \n ') ;
 % Eigenvectors column by column for all eigenvalues
 XeigVec=Eigenvectors(A,Lam) ;
 fprintf('Eigenvectors: \n') ;
 XeigVec


 %=======================================================================
 %=======================================================================
 function XeigVec=Eigenvectors(A,Lam)
 %CALCULATION OF EIGENVECTOR X FROM SOLUTION OF A*X=Lam*X EQUATION

 n=size(A,1) ;

 for m=1:n
       I=eye(n,n) ;
       Atemp=A-Lam(m)*I ;

       Ared=zeros(n-1,n-1) ;
       Bred=zeros(n-1,1) ;

       for i=1:n-1
            for j=1:n-1
                % Eigenvector Calculation: Assume x1=1 and calculate
 x2,...,xn
                Ared(i,j)=Atemp(i+1,j+1) ;
            end

            Bred(i)=-Atemp(i+1,1) ;
       end

       if size(Ared,1)==1
           Xred=Bred(1)/Ared(1,1) ;
       else
```

```
        %Xred=inv(Ared)*Bred ;
        Xred=Ared\Bred ;
    end

    XeigVec(1,m)=1;

    for i=2:n
        XeigVec(i,m)=Xred(i-1);
    end
end

%========================================================================
%========================================================================
function [Q,R]=QR_Decomposition(A)
% A=QR decomposition using Householder method

n=size(A,1) ;

I=eye(n,n) ;
Q=I ;
R=A ;

for i=1:n-1
    c=zeros(n,1) ;
    e=zeros(n,1) ;

    for j=i:n
        c(j)=R(j,i) ;

        if R(i,i)>0
            e(i)=1 ;
        else
            e(i)=-1 ;
        end
    end

    sum=0 ;
    for k=1:n
        sum=sum+c(k)^2  ;
    end

    cnorm=sqrt(sum) ;

    v=zeros(n,1) ;
    v=c+cnorm*e ;

    % Householder Matrix
    H=I-2*(v*v')/(v'*v) ;

    Q=Q*H ;
    R=H*R ;
end
```

Problems

Problem 10.1: In the eigenvalue problem of $AX = \lambda X$, A matrix is given as $A = \begin{bmatrix} 2 & 3 \\ 1 & 4 \end{bmatrix}$. Solve the eigenvalue problem with polynomial method.

Problem 10.2: In the eigenvalue problem of $AX = \lambda X$, A matrix is given as $A = \begin{bmatrix} 8 & 5 & 1 \\ 4 & 7 & 3 \\ 2 & 6 & 9 \end{bmatrix}$. Solve the eigenvalue problem with polynomial method.

Problem 10.3: In the eigenvalue problem of $AX = \lambda X$, A matrix is symmetric and is given as $A = \begin{bmatrix} 5 & -1 & 0 \\ -1 & 7 & -3 \\ 0 & -3 & 6 \end{bmatrix}$. Find the biggest eigenvalue and it corresponding eigenvector using Power method.

Problem 10.4: In the eigenvalue problem of $AX = \lambda X$, A matrix is symmetric and is given as $A = \begin{bmatrix} 9 & 2 & 3 & 2 \\ 2 & 7 & 1 & 5 \\ 3 & 1 & 8 & 4 \\ 2 & 5 & 4 & 6 \end{bmatrix}$. Find the biggest eigenvalue and it corresponding eigenvector using Power method.

Problem 10.5: In the eigenvalue problem of $AX = \lambda X$, A matrix is symmetric and is given as $A = \begin{bmatrix} 9 & -2 & 0 \\ -2 & 7 & -1 \\ 0 & -1 & 8 \end{bmatrix}$. Find the eigenvalues and eigenvectors using Jacobi method.

Problem 10.6: In the eigenvalue problem of $AX = \lambda X$, A matrix is symmetric and is given as $A = \begin{bmatrix} 9 & 2 & 3 & 2 \\ 2 & 7 & 1 & 5 \\ 3 & 1 & 8 & 4 \\ 2 & 5 & 4 & 6 \end{bmatrix}$. Find the eigenvalues and eigenvectors using Jacobi method.

Problem 10.7: In the eigenvalue problem of $AX = \lambda X$, A matrix is symmetric and is given as $A = \begin{bmatrix} 8 & 2 & 0 \\ 2 & 6 & 3 \\ 0 & 3 & 7 \end{bmatrix}$. Obtain Householder matrices H_i.

Problem 10.8: In the eigenvalue problem of $AX = \lambda X$, A matrix is symmetric and is given as $A = \begin{bmatrix} 9 & 2 & 3 & 2 \\ 2 & 7 & 1 & 5 \\ 3 & 1 & 8 & 4 \\ 2 & 5 & 4 & 6 \end{bmatrix}$. Obtain Householder matrices H_i.

Problem 10.9: In the eigenvalue problem of $AX = \lambda X$, A matrix is symmetric and is given as $A = \begin{bmatrix} 4 & -2 & 0 \\ -2 & 4 & -2 \\ 0 & -2 & 4 \end{bmatrix}$. Obtain the eigenvalues using QR method.

Problem 10.10: In the eigenvalue problem of $AX = \lambda X$, A matrix is symmetric and is given as $A = \begin{bmatrix} 9 & 2 & 3 & 2 \\ 2 & 7 & 1 & 5 \\ 3 & 1 & 8 & 4 \\ 2 & 5 & 4 & 6 \end{bmatrix}$. Obtain the eigenvalues using QR method.

References

1. Meirovitch, L. (2001). *Fundamentals of vibrations*. McGraw-Hill.
2. Rao, S. S. (2006). *Vibration of continuous systems*. Wiley.
3. Chapra, S. C., & Canale, R. P. (2020). *Numerical methods for engineers*. McGraw-Hill.
4. Bernatz, R. A. (2010). *Fourier series and numerical methods for partial differential equations*. Wiley.
5. Everstine G. C. (2010). *Numerical solution of partial differential equations* [Lecture notes, PDF]. George Washington University. Retrieved from https://www.academia.edu/31518755/Numerical_Solution_of_Partial_Differential_Equations.
6. Epperson, J. F. (2013). *An Introduction to numerical methods and analysis*. John Wiley & Sons.
7. Burden, R. L., & Faires, J. D. (2011). *Numerical analysis*. Cengage Learning.
8. Sert, C. (2006). *ME310 numerical methods* [Lecture notes, PDF]. Middle East Technical University. Retrieved from https://users.metu.edu.tr/csert/me310/me310_1_introduction.pdf.
9. Grewal, B. S. (2019). *Numerical methods in engineering and science*. Mercury Learning and Information.
10. Sastry, S. S. (2012). *Introductory methods of numerical analysis*. PHI Learning.

MIX
Papier aus verantwortungsvollen Quellen
Paper from responsible sources
FSC® C105338

If you have any concerns about our products,
you can contact us on
ProductSafety@springernature.com

In case Publisher is established outside the EU,
the EU authorized representative is:
Springer Nature Customer Service Center GmbH
Europaplatz 3, 69115 Heidelberg, Germany

Printed by Libri Plureos GmbH
in Hamburg, Germany